ANNUAL REVIEW OF ASTRONOMY AND ASTROPHYSICS

EDITORIAL COMMITTEE (1987)

JOHN N. BAHCALL
GEOFFREY BURBIDGE
PETER S. CONTI
CARL HEILES
HUGH S. HUDSON
DAVID LAYZER
JOHN G. PHILLIPS
ALLAN R. SANDAGE

Responsible for the organization of Volume 25
(Editorial Committee, 1985)

JOHN N. BAHCALL
GEOFFREY BURBIDGE
SANDRA M. FABER
JOHN W. HARVEY
CARL HEILES
DAVID LAYZER
JOHN G. PHILLIPS
HUGH S. HUDSON (Guest)
ALLAN R. SANDAGE (Guest)
WAYNE A. STEIN (Guest)
VIRGINIA TRIMBLE (Guest)

Production Editor KEITH DODSON
Subject Indexer   IRENE H. OSTERBROCK

ANNUAL REVIEW OF ASTRONOMY AND ASTROPHYSICS

VOLUME 25, 1987

GEOFFREY BURBIDGE, *Editor*
University of California, San Diego

DAVID LAYZER, *Associate Editor*
Harvard College Observatory

JOHN G. PHILLIPS, *Associate Editor*
University of California, Berkeley

ANNUAL REVIEWS INC 4139 EL CAMINO WAY P.O. BOX 10139 PALO ALTO, CALIFORNIA 94303-0897

ANNUAL REVIEWS INC.
Palo Alto, California, USA

COPYRIGHT © 1987 BY ANNUAL REVIEWS INC., PALO ALTO, CALIFORNIA, USA. ALL RIGHTS RESERVED. The appearance of the code at the bottom of the first page of an article in this serial indicates the copyright owner's consent that copies of the article may be made for personal or internal use, or for the personal or internal use of specific clients. This consent is given on the condition, however, that the copier pay the stated per-copy fee of $2.00 per article through the Copyright Clearance Center, Inc. (21 Congress Street, Salem, MA 01970) for copying beyond that permitted by Sections 107 or 108 of the US Copyright Law. The per-copy fee of $2.00 per article also applies to the copying, under the stated conditions, of articles published in any *Annual Review* series before January 1, 1978. Individual readers, and nonprofit libraries acting for them, are permitted to make a single copy of an article without charge for use in research or teaching. This consent does not extend to other kinds of copying, such as copying for general distribution, for advertising or promotional purposes, for creating new collective works, or for resale. For such uses, written permission is required. Write to Permissions Dept., Annual Reviews Inc., 4139 El Camino Way, P.O. Box 10139, Palo Alto, CA 94303-0897.

International Standard Serial Number: 0066-4146
International Standard Book Number: 0-8243-0925-1
Library of Congress Catalog Card Number: 63-8846

Annual Review and publication titles are registered trademarks of Annual Reviews Inc.

Annual Reviews Inc. and the Editors of its publications assume no responsibility for the statements expressed by the contributors to this *Review*.

TYPESET BY AUP TYPESETTERS (GLASGOW) LTD., SCOTLAND
PRINTED AND BOUND IN THE UNITED STATES OF AMERICA

PREFACE

This volume of the *Annual Review of Astronomy and Astrophysics* was planned at a meeting of the Editorial Committee held on May 4, 1985, in La Jolla, California. Those present were editor Geoffrey Burbridge; Associate Editors David Layzer and John Phillips; Production Editor Keith Dodson; Committee Members John Bahcall, Sandra Faber, and Carl Heiles; and guests Hugh Hudson, Allan Sandage, Wayne Stein, and Virginia Trimble.

In the preface to Volume 24 I pointed out that 29 articles were scheduled for this volume. In fact 17 articles are contained here, so that the default rate is again slightly more than 40%. For volume 26, 27 articles are currently scheduled. I am currently considering possible ways of reducing the default rate in future volumes.

Once again my thanks go to Keith Dodson, who does an excellent job in Palo Alto; to the authors, who continue to produce first-class reviews; and to the members of the Editorial Committee and my guests for their annual day of hard work.

THE EDITOR

SOME RELATED ARTICLES IN OTHER *ANNUAL REVIEWS*

From the *Annual Review of Earth and Planetary Sciences*, Volume 15 (1987):

  The Atmospheres of Venus, Earth, and Mars: A Critical Comparison, Ronald G. Prinn and Bruce Fegley, Jr.

  Origin of the Moon—The Collision Hypothesis, D. J. Stevenson

From the *Annual Review of Fluid Mechanics*, Volume 19 (1987):

  Von Kármán Swirling Flows, P. J. Zandbergen and D. Dijkstra

From the *Annual Review of Nuclear and Particle Science*, Volume 37 (1987):

  A Cosmic-Ray Experiment on Very High Energy Nuclear Collisions, W. Vernon Jones, Yoshiyuki Takahashi, Barbara Wosiek, and Osama Miyamura

Annual Review of Astronomy and Astrophysics
Volume 25, 1987

CONTENTS

CLUSTERING OF ASTRONOMERS, *W. H. McCrea*	1
STAR FORMATION IN MOLECULAR CLOUDS: OBSERVATION AND THEORY, *Frank H. Shu, Fred C. Adams, and Susana Lizano*	23
ELEMENTS AND PATTERNS IN THE SOLAR MAGNETIC FIELD, *Cornelis Zwaan*	83
WOLF-RAYET STARS, *David C. Abbott and Peter S. Conti*	113
THE ART OF N-BODY BUILDING, *J. A. Sellwood*	151
THE IRAS VIEW OF THE EXTRAGALACTIC SKY, *B. T. Soifer, J. R. Houck, and G. Neugebauer*	187
COMETS AND THEIR COMPOSITION, *Hyron Spinrad*	231
ROTATION AND MAGNETIC ACTIVITY IN MAIN-SEQUENCE STARS, *Lee W. Hartmann and Robert W. Noyes*	271
THE LOCAL INTERSTELLAR MEDIUM, *Donald P. Cox and Ronald J. Reynolds*	303
CEPHEIDS AS DISTANCE INDICATORS, *M. W. Feast and A. R. Walker*	345
PHYSICAL CONDITIONS, DYNAMICS, AND MASS DISTRIBUTION IN THE CENTER OF THE GALAXY, *R. Genzel and C. H. Townes*	377
EXISTENCE AND NATURE OF DARK MATTER IN THE UNIVERSE, *Virginia Trimble*	425
VERY LOW MASS STARS, *James Liebert and Ronald G. Probst*	473
THE IRAS VIEW OF THE GALAXY AND THE SOLAR SYSTEM, *C. A. Beichman*	521
DYNAMICAL EVOLUTION OF GLOBULAR CLUSTERS, *Rebecca Elson, Piet Hut, and Shogo Inagaki*	565
THE GALACTIC SPHEROID AND OLD DISK, *K. C. Freeman*	603
VARIATIONS OF SOLAR IRRADIANCE DUE TO MAGNETIC ACTIVITY, *G. A. Chapman*	633
INDEXES	
Subject Index	669
Cumulative Index of Contributing Authors, Volumes 15–25	677
Cumulative Index of Chapter Titles, Volumes 15–25	679

ANNUAL REVIEWS INC. is a nonprofit scientific publisher established to promote the advancement of the sciences. Beginning in 1932 with the *Annual Review of Biochemistry*, the Company has pursued as its principal function the publication of high quality, reasonably priced *Annual Review* volumes. The volumes are organized by Editors and Editorial Committees who invite qualified authors to contribute critical articles reviewing significant developments within each major discipline. The Editor-in-Chief invites those interested in serving as future Editorial Committee members to communicate directly with him. Annual Reviews Inc. is administered by a Board of Directors, whose members serve without compensation.

1987 Board of Directors, Annual Reviews Inc.

Dr. J. Murray Luck, Founder and Director Emeritus of Annual Reviews Inc.
  Professor Emeritus of Chemistry, Stanford University
Dr. Joshua Lederberg, President of Annual Reviews Inc.
  President, The Rockefeller University
Dr. James E. Howell, Vice President of Annual Reviews Inc.
  Professor of Economics, Stanford University
Dr. William O. Baker, *Retired Chairman of the Board, Bell Laboratories*
Dr. Winslow R. Briggs, *Director, Carnegie Institution of Washington, Stanford*
Dr. Sidney D. Drell, *Deputy Director, Stanford Linear Accelerator Center*
Dr. Eugene Garfield, *President, Institute for Scientific Information*
Dr. Conyers Herring, *Professor of Applied Physics, Stanford University*
Mr. William Kaufmann, *President, William Kaufmann, Inc.*
Dr. D. E. Koshland, Jr., *Professor of Biochemistry, University of California, Berkeley*
Dr. Gardner Lindzey, *Director, Center for Advanced Study in the Behavioral Sciences, Stanford*
Dr. William D. McElroy, *Professor of Biology, University of California, San Diego*
Dr. William F. Miller, *President, SRI International*
Dr. Esmond E. Snell, *Professor of Microbiology and Chemistry, University of Texas, Austin*
Dr. Harriet A. Zuckerman, *Professor of Sociology, Columbia University*

Management of Annual Reviews Inc.

John S. McNeil, Publisher and Secretary-Treasurer
William Kaufmann, Editor-in-Chief
Mickey G. Hamilton, Promotion Manager
Donald S. Svedeman, Business Manager

ANNUAL REVIEWS OF
Anthropology
Astronomy and Astrophysics
Biochemistry
Biophysics and Biophysical Chemistry
Cell Biology
Computer Science
Earth and Planetary Sciences
Ecology and Systematics
Energy
Entomology
Fluid Mechanics
Genetics
Immunology
Materials Science
Medicine
Microbiology
Neuroscience
Nuclear and Particle Science
Nutrition
Pharmacology and Toxicology
Physical Chemistry
Physiology
Phytopathology
Plant Physiology
Psychology
Public Health
Sociology

SPECIAL PUBLICATIONS

Annual Reviews Reprints:
  Cell Membranes, 1975–1977
  Immunology, 1977–1979

Excitement and Fascination
  of Science, Vols. 1 and 2

Intelligence and Affectivity,
  by Jean Piaget

Telescopes for the 1980s

A detachable order form/envelope is bound into the back of this volume.

CLUSTERING OF ASTRONOMERS

W. H. McCrea

Astronomy Centre, University of Sussex, Brighton BN1 9QH, England

In the astronomical universe, nature produces clusters of objects of almost every sort. Astronomers spend much effort in the study of such clusters. Then they are sometimes impelled to seek and to study clusters of these clusters.

In every department of human endeavor there are times and places at which notable achievers are markedly more numerous than at other times and places. In basic natural philosophy, for instance, about the time of Newton there were I. Newton (1642–1727) himself and, among others, R. Boyle (1627–91), J. Flamsteed (1646–1719), E. Halley (1656–1742), R. Hooke (1635–1703), C. Huygens (1629–95), G. W. Leibnitz (1646–1716), J. Wallis (1616–1703), and C. Wren (1632–1723). It seems natural to describe these individuals as being members of a "cluster."

This might indeed be the most outstanding example, but readers will readily think of clusters of writers, dramatists, musicians, artists, soldiers, and others, as well as scientists.

This paper is the attempt of an astronomer to describe and discuss some clusters of astronomers that he has been able to observe in his own experience. I seek in a quite naive way to examine the clustering phenomenon as it affects astronomers at levels from undergraduate students to established astronomers of note. The examples are selected for their aptness in illustrating features of the phenomenon, not because of the reputations of the persons involved. I describe the observations first then try to discuss what may be inferred from them.

Here a caveat must be entered: One astronomer might find it more profitable or agreeable to be a member of what I am calling a cluster than not be one; another might find the reverse to be the case. Neither, on that account, must be deemed to excel in virtue. Any comparisons between astronomers or their institutions regarding the presence or absence of a cluster is not to be taken to imply any assignment of relative merit.

A. *Cambridge: Late 1920s*

When in 1923 I arrived in Cambridge as an undergraduate intending to read mathematics, the one and only word of advice I was given was this: Astronomy is in the syllabus—ignore it; there is plenty more in the syllabus, and astronomy won't help you with anything else.

This was true, but doing astronomy might have helped with astronomy; on the other hand, it might have put me off the subject for life. Anyhow, I took the advice.

As an undergraduate I did, however, once see astronomers. It was in the summer of 1925, when I was spending part of the long vacation in Cambridge. At one stage I began to notice around Trinity College the oddest-looking bunch of characters upon whom I had ever set eyes. In those days foreigners looked foreign, and when I asked who they were, I was told "foreign astronomers." In fact I was having an outsider's view of the II General Assembly of the International Astronomical Union— all 100–200 or so of them.

Only a year later I became a "research student," with R. H. Fowler (1889–1944) as my supervisor. Very slowly I became aware of the existence at that time of a cluster of astronomers that included some of the most famous there had ever been.

In Cambridge the then-successor of Newton as Lucasian Professor of Mathematics was Sir Joseph Larmor (1857–1942). He took an uncovenanted interest in what was going on in astronomy; from time to time, he sent speculative comments to the *Observatory* magazine. The Lowndean Professorship of Astronomy and Geometry was called that using "geometry" in its literal sense of Earth measuring. It had in fact been held by the astronomer Sir Robert Ball (1840–1913), but in 1914 H. F. Baker (1866–1956) had succeeded him. Baker was known primarily as an eminent geometer in the ordinary sense. But he took the "astronomy" in the title of his Professorship with seriousness. He conscientiously made it his habit to give a lecture course on celestial mechanics, and he wrote a number of papers on the subject. His course included a treatment of rotating gravitating liquids; this significantly influenced the development of that topic by others, notably R. A. Lyttleton.

The Director of the Solar Physics Observatory and Professor of Astrophysics was H. F. Newall (1857–1944), the first occupant of the chair, who made astrophysics a recognized discipline—to use a term not then current in this sense—in Cambridge. He had been good at encouraging a succession of astronomers younger than himself, including R. A. Sampson (1866–1939), F. J. M. Stratton (1881–1960), E. A. Milne (1896–1950), and J. A. Carroll (1899–1974). The Director of the University Observatory and Plumian Professor of Astronomy and Experimental Philosophy was A. S. Eddington (1882–1944); his *Internal Constitution of the Stars* first appeared in 1926. Eddington's energetic lieutenant at the Observatory was W. M. Smart (1889–1975).

Fowler was the most influential mathematical physicist in Cambridge; he had a strong interest in astrophysics, and in 1926 he published his solution of the white-dwarf problem in terms of the quantum degenerate electron gas. Milne had recently moved to Manchester but was often seen in Cambridge; he was doing work on the outermost part of the solar atmosphere, which happened to be what first brought me into the field. J. H. Jeans (1877–1946) had long before given up a Cambridge appointment; he was very active in the Royal Society and the Royal Astronomical Society (RAS), and he was developing his own ideas about stellar structure. The Astronomer Royal at Greenwich, Sir Frank Dyson (1868–1939), had a vast range of interest and influence. The Astronomer Royal for Scotland was R. A. Sampson, who had contributed pioneering ideas about the importance of radiative transfer in stars. Both of these were Cambridge men; they had worked closely with others of the astronomers already mentioned.

All these astronomers—I wish to suggest—belonged to a recognizable "cluster." The fact all were from Cambridge was no doubt a causal factor. More importantly, they had an implicitly acknowledged community of interest. They may not have been conscious of it at the time, but, as we can now see, this was no less than the laying of the foundations of almost the whole of modern astrophysics. I am suggesting that in one way or another they all contributed significantly to this end, not, of course, that even unconsciously their whole effort was dedicated to it. In the years concerned, around the late 1920s, many of them inevitably were prominent in the activities of the RAS, and it was there perhaps that the essential community of their endeavors became most apparent.

The membership of a cluster is notoriously difficult to specify and determine. There were numerous other astronomers in the United Kingdom, and many had other interests. However, some besides those named should be counted in this "cluster"—H. Jeffreys, A. Schuster, and W. M. H. Greaves are some who come to mind.

Looking at the astronomical world more broadly, it might seem natural to extend the specification of the cluster to include, for instance, R. Emden, E. Hertzsprung, S. Rosseland, H. N. Russell, and K. Schwarzschild—readers may have other favored candidates. One has only to open any astronomical periodical of that time to see names of authors who have now become famous, including some in the context of the cluster.

I have dealt with this example at more length than I shall devote to the rest because it was a great time made great by a considerable body of remarkable astronomers having a discernible cohesion of aim. We may be apt to picture, say, Jeans and Eddington as popping up from nowhere in particular. It was not like that.

B. *Cambridge 1952–53*

The only sabbatical leave I ever requested was for the academic year 1952–53, when I had the good fortune to be elected into a newly founded Comyns Berkeley Bye-fellowship of Gonville and Caius College, Cambridge. I mention it here—out of chronological sequence—in order to compare the Cambridge I then experienced with the Cambridge of a quarter of a century earlier, which was the basis for the previous section.

The earlier cluster as such had of course suffered dissolution. The circumstance that its absence was so noticeable seemed indeed to confirm its former reality as an observable phenomenon.

The clustering that was becoming most evident was that of radio astronomers. It was remarkable how leaders emerged contemporaneously in widespread centers: M. Ryle (1918–84) in Cambridge, A. C. B. Lovell in Manchester, J. S. Hey in Malvern, J. L. Pawsey (1908–62) and others in Australia, and H. C. van de Hulst and others in the Netherlands. The whole development was a splendid affair and certainly more spectacular than the earlier cluster. I should wish to call it a "cluster," but its constitution was obviously different from that of the earlier example. So far as the Cambridge component was concerned, Ryle impressed me as a born leader. He was collecting a group of outstandingly able and enthusiastic young supporters: The ones I came to know at that time were F. Graham Smith and A. Hewish.

Cambridge astronomy as a whole was then much more fragmented than earlier, H. Bondi, T. Gold, F. Hoyle, and R. A. Lyttleton were the chief astronomical theorists. In observational optical astronomy, R. O. Redman (1905–75) had succeeded Stratton (who had succeeded Newall) as Professor of Astrophysics, and he directed the by-then combined observatories. H. Jeffreys had succeeded Eddington as Plumian Professor, but he was involved more in geophysics than astronomy. With also men like R. Stoneley, S. K. Runcorn and R. Hide, geophysics in Cambridge was on the upgrade.

The former underlying community of interest was no longer evident. Indeed, I speculated as to whether some of these groups were made of matter and others of antimatter. But as in other comparisons, I make no suggestion that one state of affairs was "better" than the other.

C. *Göttingen 1928–29*

Returning to the chronological sequence, while still officially a Cambridge research student, I spent the academic year 1928–29 in Göttingen. I have gratefully valued that experience ever since. But as regards my present theme, here I can say little more than that I observed the absence of the celebrated cluster that had been in Göttingen up to only a few months before I went there. M. Born (1882–1970), W. Heisenberg (1901–76), F. Hund (1896–), P. Jordan (1902–81), and W. Pauli (1900–58) had been the best-known members in Göttingen itself. All of them except Born had been called to professorships elsewhere. One could say, of course, that the cluster—in which should be included N. Bohr (1885–1962), L. de Broglie (1892–), P. A. M. Dirac (1902–84), and E. Schrödinger (1887–1961)—still existed in time, but Göttingen was no longer a good place to observe it. In any case it was not in astronomy, and it would have been outside the scope of this paper. If it had still been flourishing in Göttingen, however, I should presumably have reverted to my original interest in mathematical physics—and I should not now have been writing for this *Review*.

D. *Edinburgh 1930–32*

From 1912 to 1946 the Professor of Mathematics in the University of Edinburgh was that man of great heart and mind, (Sir) Edmund Whittaker (1873–1956). For most of the three years 1930–32 I was a lecturer in his department. There I observed a double cluster. One component was the staff (faculty), in which I had colleagues of outstanding talent and fascination. The other was a concentration of gifted students, some of whom later attained much distinction. I thought there could never have been a time quite like it.

It was this experience, in fact, that first made me ponder the clustering phenomenon—and I have pondered it ever since. Alas, this Edinburgh example itself had little astronomical flavor, so I must not pause over it now—except to note one intriguing aspect.

As a young man, Whittaker had been a mathematical don in Cambridge; Jeans, Eddington, and (I believe) Stratton had been among his pupils. Whittaker was a Secretary of the RAS from 1901 to 1907. He was Royal Astronomer of Ireland from 1906 to 1912. Therefore, the Edinburgh outcome was by no means devoid of any astronomical antecedents. Indeed, Whittaker may be cited as a shining example of the benefits of the traditional British coming and going between practitioners of astronomy and of mathematics.

E. *Imperial College 1932–36*

That last comment leads naturally to the account of my next experience in 1932–36 as a Reader in Mathematics in the University of London and Assistant Professor in the Department of Mathematics in the Imperial College of Science and Technology. For the "Chief Professor" of mathematics there was Sydney Chapman (1888–1970), and he too, like Whittaker, had been an astronomical practitioner. He had been a Chief Assistant (to Dyson) at the Royal Observatory, Greenwich 1910–14, 1916–18; later he was President of the RAS 1941–43, Gold Medal 1949. The mathematics faculty was large for those times; other members having some astronomical interest included T. G. Cowling, Gold Medal RAS 1956, President 1965–67, V. C. A. Ferraro (1907–74), J. C. P. Miller (1906–81), who had been a research student of Eddington, A. T. Price (1903–78), RAS Gold Medal 1969, and Bertha Swirles (later Lady Jeffreys). All of these (except, I think, Swirles) collaborated with Chapman in one or another of his numerous research interests, mainly in geophysics. Cowling has told something about this in his own Prefatory Chapter in this series in 1985.

Chapman was a truly great man. Along with his near-contemporary, Harold Jeffreys, he ranks as a principal founder of geophysical science. One does not, however, instinctively think of him as belonging to a cluster; he was remarkably successful in collaborating with other individuals. At the same time, at any rate when I was a colleague, he did little to bring these individuals together with each other. There was not much corporate life in his department.

As a "Reader" I had somewhat more independence of action than other colleagues. So I tried to foster some collective activity by organizing a weekly colloquium in mathematical physics. It was the only thing like it in the University of London at that time. It was moderately well attended, and it seemed to be appreciated by those who came; incidentally, these included Edward Teller, recently arrived from Hungary, and he was very supportive in the group. All this had little immediately to do with astronomy.

Astronomy in London continued to be much centered upon the RAS. I would say that, in contrast to the 1920s, in the 1930s there was little sense of a "cluster" environment in astronomy. The RAS was as vigorous as ever, but astronomers' interests had been diversified, and no one group was predominant. As an example, on the observational side the intensive monitoring of Nova Herculis 1934 became almost an obsession. On the theoretical side, S. Chandrasekhar arrived upon the scene, and he proceeded to enlarge greatly the scope of the study of stellar structure; his controversy with Eddington concerning relativistic degeneracy drew much attention. Nuclear physics was advancing, and I think that already in the years being considered, there was a growing conviction that it must hold the key to the process of stellar energy generation. Cosmology was also receiving much attention from observers and theorists, and the RAS organized several discussions on the subject.

F. *Queen's University, Belfast 1936–44*

I was Professor of Mathematics in the Queen's University of Belfast (QUB) from 1936 to 1944, but as reported in the following section I was absent on war work for part of the time. "Mathematics" here meant pure mathematics. Applied mathematics (mathematical physics) was associated with the department of physics. Its independent head, within that department, was from 1933 to 1938 H. S. W. Massey [later Sir Harrie Massey FRS (1908–83)]; from 1939 to 1949 it was P. P. Ewald (1888–1985). The Director of the Armagh Observatory from 1937 was E. M. Lindsay (1907–74); there was friendly, but largely informal, interaction with QUB.

The University had (and has) a notable tradition in mathematics and mathematical physics. J. Larmor, who was mentioned earlier, was a graduate. Another graduate was my predecessor, J. G. Semple (1904–85), who had moved to King's College, London. It had gone on producing good people, but so far as I was aware, after Semple there had been few of quite comparable distinction. Then, rather suddenly, in the next few years we had a set of students of commanding ability—a "cluster" if ever there was one.

The accompanying list (Table 1) shows some of them. These were the more academically inclined members, mainly because I know more about their careers. There were others of high quality who went into various sorts of work of national importance in World War II. Again there were some who for a year or two showed comparable promise in mathematics, although their intention was to specialize in some science subject—mostly, as it happened, in chemistry—and this they proceeded to do with distinction.

One would expect almost nowhere to find a cluster of undergraduates composed of none but "astronomers." If one is interested in astronomers having been in some cluster as undergraduates, one has to look at a mixed bag like the present example. Here the "astronomical" element consisted of Bates and Ramsey. In his undergraduate career, Bates followed essentially the entire course in mathematics as well as that in physics, but he chose to graduate in physics.

What makes this cluster remarkable, besides the sheer quality of the

Table 1 Some mathematics undergraduates at QUB 1936–44

Individual	Notes	Some appointments
Bates, David R. (Sir David Bates FRS)	RAS Gold Medal 1977 (Graduated in Physics)	Prof. of Applied Math., QUB, Head of Physics Dept.; Pres. Section A, British Assoc. 1986–87
Best, Ernest	Research in math. set theory; author of many books	Prof. of Divinity, Glasgow Univ.
Foster, Frederick G.	Fellow of Trinity Coll. Dublin	Prof. of Statistics, London School of Econ.; Prof. of Statistics, Trinity Coll. Dublin
Hamilton, James	Fellow of Christ's Coll. Cambridge	Prof. of Physics, Univ. Coll. London; Prof. of Physics, Nordic Inst. Copenhagen; Director, Nordita Copenhagen 1984–86
Herivel, John W.	Transferred to Cambridge before graduating in QUB; subsequently returned	Reader and Dept. Head History and Phil. Sci., QUB
Macbeath, A. Murray	Research Fellow of Clare Coll. Cambridge	Mason Prof. of Pure Math., Birmingham Univ.; Prof. of Math. and Statistics, Univ. of Pittsburgh
Ramsey, W. Henderson (1922–65)	Well known for theory of Earth interior	Researth positions at Bristol, Manchester, and Johannesburg
Simpson, Edward H. (CB)	Commonwealth Fund Fellow 1956–57	Deputy Secretary, Dept. of Education and Science
Unwin, J. J. (c.1916–42)	Aircraft research; killed in test flight	Research student, Christ's Coll. Cambridge

individuals, is the fact that they were drawn from such a small total number of students. Nearly all the students who in those years completed the course in mathematics were very good. Because the best were so very good, on looking back I think it became exceptionally hard for a student to gain first-class honours in mathematics.

G. *Admiralty 1943–45*

With wartime leave of absence, first from QUB and later from Royal Holloway College, I was in the Operational Research group in the Admiralty in London from early 1943 to the summer of 1945. This was a cluster with a vengeance! Or some might say not at all—it was simply a picked team (or "squad," perhaps, in modern jargon). We come back to this point in the general discussion.

There were about 30 members all told, with occasional visitors from Canada and the USA. Listed here (Table 2) are those scientists with whom I had dealings at the time. As indicated, some happen to figure elsewhere in this paper.

The object was to provide a reasoned basis for further naval operations and to propose such operations, or to comment upon proposals coming from other quarters. This was to be done by applying scientific common sense to interpret all the evidence provided by any recent actual operations. The whole business was a glorified "learning from experience."

It might be supposed that nothing could be further from astronomy than U-boat warfare and the like. But I have marked on the list those individuals having some association with astronomy. I was surprised when I first noted that these form a quite considerable proportion. Not many of us could be called professional astronomers, but it does appear that scientists with some touch of astronomy were apt to be qualified for work of this kind.

Incidentally, the inclusion of Hamilton, Lindsay, McCrea, and Ramsey seems to show that being Irish was also a help. More seriously—to apply operational research to operational research—it seems that it was a mainly British contribution in the war. If so, this may be more a tribute to the commanders of the British forces, who were ready to listen to intruding British scientists, than a tribute to those intruding scientists themselves.

H. Royal Holloway College 1944–66

In the middle of the time at the Admiralty, the University of London appointed me Professor of Mathematics in Royal Holloway College (RHC). I was head of the department in the College from 1944 to 1966, the longest period of time I have been in one post. Much happened in all those years. Here I confine my attention to graduate students on the "applied" side. I would say that these formed a cluster in the sense in which I am using the term.

I keep to students who worked on topics in astrophysics or in relativity and cosmology—who happened mostly to work with me—and those who worked in parts of mathematical physics having astrophysical significance—who were among a good number of others who worked with my colleague (and present successor) M. R. C. McDowell. The students selected on these criteria are listed here with some indication of their subsequent careers (Table 3).

Table 2 Some wartime members of the Operational Research group (established 1942) of the Admiralty in London

Individual[a]	R. Astron. Soc.	Notes
Director 1942–45		
(Lord) Blackett (OM) FRS (1897–1974)	FRAS	Pres., R. Soc. 1965–70; Nobel Prize Physics 1948
Assistant Director 1944–45		
(Sir) Edward Bullard FRS (1907–80)	FRAS, Gold Medal 1965	Director, Natl. Physical Laboratory 1950–55
G. L. Clark (1914–82)	FRAS	Pupil and collaborator of A. S. Eddington
Sir Ralph Fowler OBE, FRS (1889–1944)	FRAS	Plummer Prof. of Applied Math., Univ. Cambridge 1932–44
Attached SE Asia Command		
J. Hamilton		Prof. of Physics, Nordic Inst. for Theor. Atomic Phys. 1964–86; Director, Nordita Copenhagen 1984–86
H. R. Hulme	Sec. RAS 1943–46	Chief Assistant, R. Obs. Greenwich 1938–45; Chief of Nuclear Research, AWRE 1959–73
(Sir) Andrew Huxley (OM, FRS)		Pres., R. Soc. 1980–85; Nobel Prize Physiology 1963
E. M. Lindsay (OBE) (1907–74)	FRAS	Director, Armagh Obs. 1937–74
(Sir) William McCrea (FRS)	Pres., RAS 1961–63, Gold Medal 1976	
W. H. Ramsey (1922–65)	FRAS	Noted for theory of Earth interior in terms of quantum phase changes in olivine, etc.
K. V. Thimann (For. Member RS)		Prof. of Biology, Univ. Calif., Santa Cruz 1965–
J. H. C. Whitehead FRS (1904–60)		Waynflete Prof. of Math., Oxford Univ. 1947–60
Assistant Director 1944–45		
E. J. Williams FRS (1903–45)		Prof. of Physics, Univ. Coll. Aberystwyth 1938–45
Visitors from Canada		
A. McKellar (MBE) (1910–60)	FRAS	From Dominion Astrophys. Obs., Victoria
P. M. Millman	Professional Astronomer	Now of Herzberg Inst. of Astrophys.
Visitors from USA		
S. Goudsmit (Hon. OBE) (1901–78)		Co-discoverer of electron spin
J. von Neumann (1903–57)		Prof., Inst. for Advanced Study, Princeton

[a] Awards made after the war are in ( ).

Table 3 Some mathematics graduate students with astronomical interests at RHC 1944–46

Individual	Appointment subsequently held[a]	Notes
Agacy, Rex L.	Lecturer in Math., James Cook Univ., North Queensland, Australia	
Crampin, D. Joan	Lecturer in Math., Univ. Reading	Doctorate completed in Cambridge
Davidson, William	Prof. of Applied Math., Univ. Otago, New Zealand	Organized IAU SE Asia Reg. Meeting 1978
Dodd, Kenneth	Advanced computing, RAE Farnborough	Numerous books on computing
Dolan, Patrick	Lecturer in Math., Imp. Coll. London	Sec. Org. Comm. Schrödinger Centenary Meeting 1987
Florides, Petros S.	Assoc. Prof. Mathematics, Trinity Coll. Dublin	Fellow of Trinity Coll. Dublin
Gardiner, Joyce	Lecturer in Math., Univ. Western Australia, Perth	
Goodstein, Peter	Actuarial Adviser	
Hilton, Elizabeth	Lecturer in Applied Math., Univ. Coll. Swansea	
Hogarth, Jacke	Prof. of Math., Queen's Univ., Kingston, Canada	Served as Dean, etc.
Liddell, Heather	Senior Lect. Computer Sci. Q.M.C.	
McNally, Derek	Senior Lecturer in Astron., UCL; Asst. Dir., Univ. London Obs.	Sec. RAS 1966–72; Asst. Gen. Sec. IAU 1986–
Mikhail, Fahmy I.	Prof. of Applied Math., Ain Shams Univ. Cairo	
Myerscough, Valerie P. (1942–80)	Lecturer in Math., QMC	Miller Fellow in Astron. Univ. Calif. Berkeley 1965–67; Vis. Prof. (part time) Univ. Md. 1973–78, Univ. Tex. Austin 1977, 1979; Sec. RAS 1979–80
Pavelle, Richard	Dir., Symbolics, Inc., Cambridge, Mass.	Pioneer of computer algebra
Peach, Gillian	Reader in Physics, UCL	
Rowan-Robinson, Michael	Reader in Astron., QMC	Books: *Cosmology* (1977), *The Cosmological Distance Ladder* (1985)
Williams, Iwan P.	Reader in Math. QMC	Books: *Matrices for Scientists* (1972), *The Origin of the Planets* (1975)

[a] Abbreviations: QMC, Queen Mary College (London); UCL, University College London.

All of these have had success as scientists. The outstanding trait in all of them besides their strictly academic attainments—or so it has always seemed to me—was that every one was very much a "character" in his or her own right. Any of them who read this are by now at a stage when they ought not to deem it a defamation of character for me to describe them so. Their characters showed up in the full parts they played in the general life of the College, and as well as in their continuing readiness to shoulder responsibilities in their communities. It will be seen, too, how enterprising they have been in the appointments they have taken in various parts of the world.

The times these students spent at RHC were spread through about 20 years. But there were no gaps, and so I think they passed ideas and traditions on from one to another and acquired some sense of belonging together. This is borne out by quite a number continuing to keep in touch with each other. This is why I think it justified to call them a cluster. The cluster in section A was associated with one epoch rather than one place; the cluster in this section is associated with one center but extended in time.

Here I am writing for astronomers about astronomers. So I have had to select appropriate individuals for mention. In identifying the cluster concerned, it would, however, be much better to include all the graduate students in both pure and applied mathematics in the department during these years. The department has been growing ever since, and "our" cluster has been succeeded by others.

The primary obligation of the department was the teaching of undergraduates taking degree courses wholly or partly in mathematics. Recently the University of London as a whole has undergone far-reaching restructuring. The College has become "Royal Holloway and Bedford New College" on the RHC site. This is one of the five colleges of the University designated as special "Science Sites" for mathematics and science. The reputation earned by these clusters as a whole may have made some contribution toward this outcome.

I now describe three of the longer visits to various places that I was able to make during the years at RHC. At the time, a department like mine was less geared than it might be now to having members away on leave. So it behoves me gratefully to record my indebtedness to all the colleagues who helped in making these visits possible. And none of the rest who happen to read this will be surprised if I name particularly Dr. Mary Bradburn and Dr. Barbara Yates on whom, by reason of their status and experience, special responsibility inevitably fell. If such visits served useful purposes, the credit must go to all these colleagues.

I. *Berkeley 1956*

Otto Struve (1897–1963) invited me to spend the first semester of 1956 in his department in Berkeley; Royal Holloway College was so good as to arrange for me to have leave of absence so that I could accept.

The Astronomy Department then was still housed in the old Leuschner Observatory buildings. These were one-story, mainly wooden structures enclosing a small square and thereby forming a secluded enclave in the splendid Berkeley campus of the University of California.

Struve was one of the great astronomers of the century. I think that his time in Berkeley, 1950–60, was a relatively untroubled interlude in a career that as a whole had an abnormal share of stresses and strains. And he did get through an immense amount of work in his well-ordered life in that setting.

Here we are thinking of him as the leader of a remarkably outstanding group of astronomers. How did he do it? Very simply—by being the best astronomer around. He was just a very, very good scientist with an aura that made the people about him want to be very, very good scientists too. And he was willing to believe that they could be.

In his department, before being allowed to start work on a thesis and being assigned to a thesis supervisor, a graduate student had to pass a "qualifying examination." I was astonished to observe how exacting was the standard expected of any such candidate. So I felt obliged to report to Struve that one particular young man with whom I had to deal seemed as though he could not imaginably make that grade. His reply was, in effect, "You never can tell." He went on to tell me how a couple of the most renowned observational astronomers of the time had seemed hopelessly slipshod when he first had anything to do with them. Struve would not lower the hurdle, but he would give the man the fullest possible opportunity to try to surmount it.

Mention of this recollection prompts one to say something about all the students and colleagues whom one has helped so much less than one ought to have done. It must be the dilemma of all who are called to serve as "teachers": Unless they do other things as well, they have nothing to teach; if they do not curtail other things, they have no time to teach. One's failures on both sides are a perpetual distress. In recalling what appear to have been happy circumstances for so many, whether one was personally involved or simply in the role of observer, one would wish to let it be known that others for whom the circumstances proved less happy are not forgotten. Without excusing failures in cases where one was in fact involved, it can fairly be said that certain of the individuals concerned have nevertheless discovered other circumstances that have proved more congenial.

The cluster around Struve during my time in Berkeley included the members on the list I have compiled here (Table 4), basically from memory (checked with Professor Bruce Stephenson, to whom I am most grateful). These were faculty and others with a community of interest in some aspect of stellar evolution, which was then Struve's own chief interest. But, whatever they were doing individually, the group as a whole was inspired by a very evident esprit de corps. The list includes graduate students, research assistants, and some long-term visitors. I have noted some information about subsequent appointments, but most readers will instantly recognize a majority of the names. It must be emphasized that this is intended simply as an example of the clustering phenomenon; it is not an attempt to tell about everybody who was in the department at that time.

One of my roles as a visiting faculty member was to give a course of lectures directed primarily to the graduate students who were coming

Table 4 Some research workers, graduate students, and faculty members at UC Berkeley 1956 (January–June)

Individual	Appointment subsequently held/notes
Research workers and graduate students	
Abhyankar, K. D.	Osmania Univ., Hyderabad
Crawford, John	
Cudaback, David D.	Radio Astron. Lab., Berkeley
Huang, Su-Shu (c.1915–77)	
Klemola, Arnold R.	Lick Obs.
Lynds, Beverly T.	Kitt Peak Natl. Obs. (sometime Deputy Director)
Lynds, C. Roger	Kitt Peak Natl. Obs.
Poveda, Arcadio	Inst. Astron., Univ. Nac. Mexico (sometime Director)
Preston, George W.	AAS Helen B. Warner Prize 1965; (sometime Director Mt. Wilson Obs.)
Roberts, Morton S.	Natl. Radio Astron. Obs. (sometime Director)
Roemer, Elizabeth	Lunar & Space Sci. Lab., Univ. Arizona
Sahade, Jorge	President, IAU 1985–
Shane, W. Whitney	Astron. Inst. Nijmegen
Stephenson, C. Bruce	Case Western Reserve Univ., Prof. Astron.
Zebergs, Velta	
Faculty members concerned in the group	
Otto Struve (1897–1963)	
Louis Henyey (1910–70)	
John G. Phillips	
Helen Pillans	
Harold F. Weaver	
Leland E. Cunningham	

toward their "qualifying examination." I was expected to give each such student a project arising from the lectures that required him to do some original investigation. Among these, Morton Roberts produced a minor classic in showing that all Population I stars could almost certainly have been formed in galactic clusters, subsequently dispersed. On his way to that, he and I came to the conclusion that interstellar matter could compose only something like 2% of the mass of the Galaxy, not the 50% or so that had then come to be widely accepted. But we never got around to writing that up before others reached the same conclusion by other routes. Likewise, Bruce Stephenson made a very good job of a study of the effect of radiative pressure upon dust particles in the solar neighborhood.

While at Berkeley, I also organized a weekly seminar on topics of current interest in astrophysics and cosmology. Most of the young astronomers in the group attended both the lectures and the seminars; somewhat more senior members, including Struve himself, came to the seminar when they could and sometimes gave the talk. According to my recollection, both audiences averaged about 20. The object of the weekly seminar normally was to have one of the younger members work up a suitable topic, talk about it, and then open a discussion. My job was mainly to check the suitability of the topic and to help a little with the literature—and to enlist the speaker. This last part of the operation went through three phases. For a couple of weeks I picked upon what appeared to be fairly self-confident young men and applied a certain amount of persuasion. Things went well for all concerned. Then for a few weeks I found myself being accosted in quiet corners by a few young astronomers who insinuated that they might not be averse to being persuaded to speak. After that, there was something like naked competition.

Each volunteer took his commitment with grim seriousness. The young wife of one of them told me that for the week before her husband's talk, she scarcely dared to breathe in the room where he was preparing it.

J. Berkeley 1967

Berkeley did me the honor of inviting me back—for a "quarter"—in 1967. The composition of the astronomy department then was of a very different character from that in 1956. It was larger and contained a greater number of established researchers, all doing very distinguished work. But one had no urge to describe it as a "cluster." There was no less esprit than before; there just seemed to be less corps.

Again the difference made me more certain that the "cluster" phenomenon was something significant. But again nothing made me think of the place being "better" one time than the other.

K. *Warner and Swasey Observatory: Case Institute of Technology (now Case Western Reserve University) 1964*

The neatest example of what I am calling a cluster was what I observed in 1964 during a semester at the Warner and Swasey Observatory. This was part of the Case Institute of Technology in Cleveland, Ohio, which was the doctorate-granting body concerned. As an example of the phenomenon, it is so similar to that of Berkeley in 1956, even if on a somewhat reduced scale, that it might be considered redundant to include both. However, the observation that two astronomy departments in such generally dissimilar environments could experience such similar phases—which were evidently rather special for both—reinforces the conviction that there is here a significant phenomenon.

Sidney McCuskey (1907–79) was Professor of Astronomy in the Institute—where he also gave a lecture-course in applied mathematics—and Director of the Observatory. Throughout my stay in 1964, and in subsequent visits, my respect for his personality and his scientific powers grew ever deeper. He was the most unassuming of men, but he had impressive natural dignity in everything he did.

The research topics in his group (Table 5) had mostly to do with stellar systems and galactic structure. Its only two significant research telescopes were the famous Burrell 24-inch Schmidt telescope with its objective prism—on a somewhat indifferent site a few miles outside Cleveland—and a 36-inch Cassegrain telescope—on the still less favorable site of the Observatory itself. (Both telescopes have been resited in recent years.) But McCuskey had good working relationships with observatories in the UK and Europe, and his group had an excellent reputation nationally and internationally. The group as I knew it attained its peak about 1970, I believe, after which external circumstances imposed considerable restructuring.

McCuskey and his colleagues had wide interests in astronomy. Their students got a first-rate education in it, and in cognate subjects dealt with in Case. As in Berkeley, students had to pass a series of stiff qualifying tests before starting upon their thesis work.

Table 5 shows some of the graduate students I got to know. All of them have continued to contribute effectively to the well-being of astronomy; in the table I show where they are working at the present time. What had attracted this group of talented young scientists from various parts of the USA to work at that time in what some would regard as an unspectacular school in the American Midwest? And they remain devoted. They are the only such group I know who make a habit of holding a gathering of their own during General Assemblies of the IAU. I return to this question later.

Table 5 Some graduate students and faculty members at Case Institute, Warner and Swasey Observatory 1964 (January–July)

Individual	Appointment subsequently held
Graduate students	
Dahn, Conard C.	Flagstaff Station, US Naval Obs.
Fitzgerald, M. Pim	Astron. Programme, Univ. Waterloo, Canada, Faculty
Hidayat, Bambang	Bosscha Obs., Indonesia, Director
Houk, Nancy	Astron. Dept., Univ. Michigan, Senior Research Associate
MacConnell, Darrell Jack	Space Telescope Sci. Inst., Baltimore
Philip, A. G. Davis	Research Professor, Union Coll., Schenectady, N.Y.
Sanduleak, Nicholas	Astron. Dept., Case Western Reserve Univ., Senior Research Associate
Wehinger, Peter H.	Astron. Group, Arizona State Univ., Tempe
Wyckoff, Susan	Astron. Group, Arizona State Univ., Tempe
Faculty members concerned in the group 1964; subsequent appointments	
Sidney W. McCuskey (1907–79)	Director 1959–70; Prof. of Astron. 1959–75
Victor M. Blanco	Prof. of Astron. to 1965; since 1967 at CTIO (sometime Director)
C. Bruce Stephenson	Director (acting) 1984–86; now Prof. of Astron.
Peter Pesch	Director 1975–84, 1986–; now Prof. of Astron.
Jason J. Nassau (1893–1965)	Director 1924–59; still astronomically active in 1964

L. *Sussex From 1965*

For the past 20-odd years, the body of astronomers that I have been able to observe more closely than any other has been at the Astronomy Centre in the University of Sussex, which has always been intimately linked with the Royal Greenwich Observatory. The reader could well have expected this to afford the greatest amount of material for my chapter—I expected so myself when I began. But I have been driven to the conclusion that the record is too complex to allow the isolation of a clear-cut case of the clustering phenomenon.

It was a great experience to be actively associated with the Centre from its inception in 1965 up to my "retirement" in 1972. I am deeply grateful to the University, and to R. J. Tayler (Professor since 1967), M. J. Rees (Professor 1972–73), L. Mestel (Professor since 1973), and their colleagues in the Centre for allowing me the privilege of an emeritus association ever since.

Even if I could not identify a good case of what I should want to call a cluster, I thought I might cite some interesting "subclusters" of overseas visitors, of research fellows, or of graduate students. But this would entail the invidious singling out of individuals for special mention.

M. *Other Places*

I have chosen cases for discussion because they conveniently illustrate certain features or simply because of relative ease of description. There are numerous other places in various parts of the world that I hold in just as high esteem. It is no derogation that I have not used them for the particular purpose of this article.

Discussion

The "observations" here reported show fairly definitely that the phenomenon of clustering of astronomers does occur. Apparently the astronomers concerned may be at any stage—undergraduates, graduates, or established scientists. We ask, what are the causes?

It seems not possible to go through the cases and to identify a main cause in each. The best we can do seems to be to recognize some causes that may be operating, to try to classify these, and then to go back to try to discern which are significant in the various cases.

Simple Effects

Asking whether any case might be ascribed to essentially a single cause leads to a consideration of the following possibilities:

1. RANDOM FLUCTUATION   Since in some cases we have to do with quite small numbers of astronomers, random fluctuations could be appreciable. But the effects we have described are too great to be explained in this way. In any case, random fluctuations are the outcome of many causes, so it may be inappropriate to list them under the heading "simple."

2. DELIBERATE SELECTION   A cluster could be simply a case of picking a team. At least to a first approximation, the operational research group in case G may be such a case.

It might be said that anyone can make himself that sort of cluster; that we are not asking here about "instant" clusters, but about clusters that arise spontaneously without deliberate intent. In answer we can say that deliberate selection plays a part in most of our cases (e.g. where students are concerned, somebody has to select them). So under the present heading we are merely considering the possibility that it may be the determining factor.

Once the relevance of selection is admitted, we find ourselves facing the question, who selects the selector(s)? How far back in the sequence ought we to go? So once again we get into trouble about what is "simple."

3. PERSONALITY EFFECT   A cluster may come into existence because of the presence of one individual. He may be a successful leader, or he may be an individual of such distinction that merely having him present is a powerful attraction. The Berkeley and Cleveland clusters (cases I, K) may be accounted for by the presence of Struve and McCuskey.

4. BOAT-RACE EFFECT   Since the year 1829, Oxford and Cambridge have competed in 132 boat races; Cambridge has won 69, Oxford has won 62, and there has been one dead heat. Thus we might infer that the chances of winning are equal. (It would be interesting to know how many times each side has won the toss, and to see if these numbers are as nearly equal as are the numbers of wins.) Astonishingly, however, Cambridge won 13 successive races from 1924 to 1936; Oxford won 10 successive races from 1976 to 1985. So in the short term, the chances are far from equal. There must be some factor X that when present on one side gives it an overwhelming advantage, but in the long term X is equally likely to be on either side. In the example of the boat race itself, I do not know what X is; it might be the possession of a winning coach.

The preceding "personality effect" may be regarded as a particular instance of the boat-race effect, and in the cited cases Struve and McCuskey might be winning coaches.

Numerous other cases of clustering may be basically of this nature. A number of institutions may be such that in the long term there is nothing to choose between them, but there may be some factor X going about and equally likely to affect any one; any institution affected will produce a cluster as a relatively short-term consequence. Stated this way, the explanation seems rather trivial. But if it applies, it would offer also some explanation of the absence of an earlier type of cluster in Cambridge in 1952–53 and in Berkeley in 1967 (cases B, J).

There is, however, one subtlety about the effect. Where this effect supplies an explanation, in general we shall not know for certain what the factor X is. Were it known, it should in general be controllable, and so its occurrence would not be random.

5. TIME-RIPE EFFECT   Astrophysics is obviously a subject in which a major advance may be possible only if several branches of physics and astronomy have all reached a certain stage of development and thereby, as we say, made the time "ripe" for this advance. Thus 60–70 years ago atomic theory, radiation theory, statistical mechanics, and stellar spectroscopy had together developed to a point at which an understanding of stellar

constitution appeared feasible. Many of the astrophysicists mentioned in my case A appreciated this and found themselves working on some aspect of this subject. Hence, basically, this determined the cluster there described.

6. BANDWAGON EFFECT   Members of a "time-ripe" cluster may acquire followers, and so a "bandwagon" cluster may ensue. My cases, however, seem not to include an example.

Multiple Effects

1. HALLEY EFFECT   A certain I. Newton was an able scientist and a certain E. Halley was an able scientist. Had there been no interaction between the two they would be remembered for such discoveries or inventions as they had been disposed to leave on record. Historians might in passing remark upon the random fluctuation denoted by the two having lived at about the same time.

As things actually turned out, however, Newton's work that revolutionized all thinking about the physical world not only was published, but apparently it was mostly carried out, as a result of Halley's famous visit to Newton in 1684. And among many other consequences, Halley's comet is celebrated only because Halley was interested in a particular application of Newton's theory.

Thus, the minor fluctuation of two careers overlapping in time was transformed into the greatest ever peak in the history of scientific achievement.

This is an extreme example of the sort of interaction that could be taking place between members of any cluster. When it leads to results that significantly would not otherwise have emerged, we can call it the "Halley effect." It is to be noted that it is definitely nonlinear.

This effect alone does not account for the prototype cluster mentioned in the introductory section. But it could be claimed to have played a major part in producing a time-ripeness effect as well.

2. EDDINGTON EFFECT   When, in the fashion mentioned in item 5 above, the time was ripe for the study of stellar constitution, Eddington laid down the main lines of the study as they have been followed ever since. He seemed uniquely to possess the intuition needed to put together all the advances in atomic physics and everything else so as to reveal what a star as a whole is really like. Jeans and Milne, for instance, had some very good ideas, but they had also some very wild ideas, about what goes on inside a star. The evidence is that as the general validity of Eddington's models became accepted, then the valid parts of other people's contributions fell into place, and the tremendous early achievements of astrophysics quickly followed. Without Eddington's guiding concepts, it seems that there would

have been merry chaos, and the cluster in case A would have been far less honored.

"Might-have-been" speculations are perhaps futile, but they seem to be unavoidable. One thing we cannot help asking is whether, in the 70 years or so since Eddington put the study of stellar constitution on what seem undoubtedly to be the right lines, great opportunities for progress in other fields of astrophysics have been missed for want of an Eddington effect. One thinks, for instance, of star formation.

It is perhaps even more futile to speculate the other way round and to wonder what Eddington would have done had he been born some 20 years earlier. Then he would have been in his prime 20 years before the time was ripe for work on the constitution of the stars. Has the world had any number of village Eddington's for whom there happened to be no ripeness of time?

3. GROWTH EFFECTS   If any of the preceding effects are tending to produce a cluster, there are various consequent effects that may help it to grow, e.g. a good reputation of a group may attract more recruits or more funds. These were operative in producing the cluster in all the particular cases here described.

4. ADVERSE EFFECTS   If any of the foregoing effects produce an incipient cluster, clearly there are other effects that may quench it. There is then very little to observe and perhaps no one is any the wiser about what has happened. Or there may be controls of funds, studentships, etc., that simply inhibit clustering.

OTHER FIELDS   Something like the foregoing discussion would presumably apply to any other science besides astronomy. But there seem to be clusters in almost all walks of life—as remarked at the beginning—and explanations of most sorts offered here would not apply to those. This makes one wonder whether we have been missing something essential in our own cases.

Comments

Were this to be a "refereed" contribution it might well be rejected on the ground of not being quantitative, but I cannot see any way of doing much about that.

Cosmogonic studies suggest that all stars may be formed in stellar clusters. It could be that all astronomers have emerged from clusters of astronomers. In that case this paper is about a routine process. At any rate it would be something for grant-giving bodies to recognize—that clusters come and go and that this is no bad thing.

This paper set out, however, to be about what seemed to be a far from routine phenomenon. Then when we came to seek explanations, we found that so far as they go they seem to be commonplace. But there are non-understood aspects, and this is emphasized by the occurrence of corresponding phenomena in nonscientific fields.

Astronomy is an activity pursued by man for the sake of man in order to enlarge his acquaintance with the Universe around him. It is one of the ways for him to approach with awe the great problems and mysteries of his own existence as part of that Universe. Here I have written about some of the less awesome aspects of this human endeavor. I have done so in happy and grateful recollection of all the good humans whom I have been able to mention, and of innumerable others whom I wish I could have mentioned as well.

STAR FORMATION IN MOLECULAR CLOUDS: OBSERVATION AND THEORY

Frank H. Shu, Fred C. Adams, and Susana Lizano

Astronomy Department, University of California, Berkeley, California 94720

1. OVERVIEW

Stars are the fundamental objects of astronomy; thus, the formation of stars constitutes one of the basic problems of astrophysics. Star formation in galaxies is a complex process, which spans roughly 12 orders of magnitude in both mass and linear scale (10^{11} to 10^{-1} $M_\odot$, 10^{23} to 10^{11} cm) and involves many diverse physical phenomena. This review, which was written in October and November of 1986, concentrates on those aspects of the problem that occur on the scale of a giant molecular cloud (roughly, 10^6 $M_\odot$ and 10^{20} cm) and smaller. Moreover, we concern ourselves with the problems of present-day star formation, as posed by observations of star-forming regions in our own Galaxy and in other spiral galaxies. We do not discuss the implications of star formation in dwarf galaxies, the possibility of star formation in the cooling flows of hot gas in galaxy clusters and elliptical galaxies, the nature of primordial star formation, the origin of starburst galaxies, and many similarly intriguing topics. Even with such truncation, our subject is a large one, and we are forced to give cursory treatments of many interesting and controversial ideas.

1.1 Origin of Stellar Masses

The first issue confronting any fundamental discussion of star formation is the question of the origin of stellar masses. Normal stars are self-gravitating balls of gas whose central temperatures are high enough to sustain fusion reactions. Most of the stars that form in the cosmos have masses of a few tenths of a solar mass and, hence, are only marginally capable of burning hydrogen (see also the review chapter by Liebert & Probst in this volume). *Why should the Universe ubiquitously produce objects that have just the right mass range to stably transform the primary product of the Big Bang—hydrogen—into heavier elements?*

There are at least two different approaches to answering this question. The first is to suppose that the result—objects with masses sufficient for thermonuclear fusion—occurs somewhat by chance. Here, the interstellar medium, through some process of hierarchical fragmentation spanning many orders of magnitude in mass scale, happens to preferentially produce objects with masses between 10^{-1} and 10^2 $M_\odot$. The second approach is to suppose that star formation is intrinsically an accretion process, whereby protostellar masses are built up from initially small values, and that this accumulation is eventually halted by some process, possibly a stellar wind, which is itself somehow triggered by the onset of thermonuclear fusion. In what follows, we review the various lines of evidence that bear on the problem of star formation in molecular clouds before returning to evaluate the merits and disadvantages of these two schools of thought.

1.2 Bimodal Star Formation

Bimodal star formation, the notion that the birth of low- and high-mass stars may involve separate mechanisms, originated in its modern guise with Herbig (1962) and Mezger & Smith (1977; see also Elmegreen & Lada 1977). As an empirical idea, it has been invoked in an increasing number of contexts—from helping to resolve difficulties with models of the chemical evolution of the Galaxy (Gusten & Mezger 1982) to solving the problem of Oort's (1960) limit through the placement of the missing disk mass in the form of stellar remnants (Larson 1986). In his original work, Herbig noted the existence of star-forming regions like Taurus that contain no stars more massive than ~ 2 $M_\odot$, and he contrasted this situation with Orion, which contains young stellar objects (YSOs) of both high and low mass. Citing also the discrepancy in the turnoff and contraction time scales of the Pleiades star cluster, Herbig speculated that star births in a given cloud occur over a period of (at least) a few tens of millions of years, and that stars of low mass are born before those of high mass. Mezger & Smith gave various lines of argument that indicated that the sites of high-mass and low-mass stars may be spatially distinct on a global scale, with the formation of OB stars taking place primarily in large cloud complexes situated in the spiral arms of the Galaxy. The formation of T Tauri stars was envisaged to occur in both small and large clouds distributed throughout the disk, but little direct evidence was available on the issue. Mezger & Smith also left open the question whether one type of star-forming cloud could evolve into the other.

More recent investigations have reinforced these basic findings (see, e.g., Elias 1978a, Cohen & Kuhi 1979, Adams et al. 1983, Stauffer 1984, Dame et al. 1986, Solomon & Sanders 1985, Montmerle 1987, Scoville 1987). In particular, Solomon et al. (1985) divide molecular clouds into a cold population (< 10 K) and a warm one (including many clouds with cores > 20 K). The cold clouds, which do not contain stars earlier in spectral type than late B, are found to be smoothly distributed throughout the galactic disk. The warm clouds, which tend to be among the largest and most massive, are associated with radio H II regions and appear to be a spiral arm population (see also Stark 1985, Waller et al. 1987, Lo et al. 1987).

Thus, although the idea of a *temporal* sequence in the formation of stars of different masses remains controversial (see, e.g., Stahler 1985), the *spatial* distinction between regions of high-mass and low-mass star formation is well founded. In particular, stars are now known to form only from the densest portions of a molecular cloud, and the properties of such molecular cloud "cores" differ for regions of high- and low-mass star formation (Evans 1978, Rowan-Robinson 1980, Myers 1986; see Section 3.3 below).

1.3 *Initial Mass Function*

The first empirical derivation of the initial-mass function from the observed stellar luminosity function in the solar neighborhood was made by Salpeter (1955), who found that the number of stars born with masses between m and $m+dm$ satisfied

$$f(m)\, dm \propto \xi\, (\log m)\, d \log m \propto m^{-2.35}\, dm, \qquad 1.$$

for stellar masses m between 0.4 and 10 $M_\odot$, with considerable uncertainties in the shape of the mass spectrum outside of this range. The problem was reanalyzed by Miller & Scalo (1979) and by Scalo (1986). Figures 16 and 17 in the latter paper suggest that the function ξ may have *two* local maxima, one at $m \approx 0.3\ M_\odot$ and another at 1.2 $M_\odot$, with the existence and location of the 1.2 $M_\odot$ maximum dependent on the exact assumptions made for the history of stellar birthrates. The importance of a primary maximum at 0.3 $M_\odot$ is the apparent existence, unlike the Salpeter law (Equation 1), of a preferred mass scale around a few tenths of a solar mass; the importance of a second maximum at 1–2 $M_\odot$ is that it is a possible indicator of the process of bimodal star formation. Given the uncertainties present in transforming luminosity functions into initial-mass functions, however, the subtle presence of a second maximum should be regarded with some caution.

1.4 Spontaneous and Induced Star Formation

The idea that OB star formation is somehow *triggered* by external events has gained wide currency, and the evidence that environmental effects on a galactic scale do play a role seems fairly compelling (see, e.g., de Jong & Maeder 1977, Elmegreen & Elmegreen 1983, Persson 1987). What remains controversial is whether phenomena such as spiral structure, merging galaxies, etc., merely yield conditions that are conducive for the agglomeration of giant complexes of molecular clouds, which subsequently undergo spontaneous star formation (see, e.g., Shu 1985, 1987, Scoville et al. 1986), or whether the larger scale structure is itself (sometimes) produced by induced star formation (e.g. Seiden 1983, Elmegreen 1985a).

It has been proposed that star formation might spread like an infectious disease through an individual cloud (and, perhaps, even through an entire galaxy) as a wave of gravitational collapse is propagated by the compression associated with supernova shocks (e.g. Mueller & Arnett 1976, Cameron & Truran 1977, Gerola & Seiden 1978, Herbst & Assousa 1978) or ionization fronts (e.g. Elmegreen & Lada 1977, Klein et al. 1983, Ho et al. 1986) or stellar winds (e.g. Norman & Silk 1980, Cameron 1985). These mechanisms have various degrees of a priori plausibility; however, they also share a few common shortcomings. First, except for that of Elmegreen & Lada, they predict the collapse of protostars to begin from "outside-in," whereas the currently available evidence suggests that the process occurs from "inside-out" (Ho & Haschick 1986, Walker et al. 1986, Keto et al. 1987, Reid et al. 1987). Second, they have limited ability to induce gravitational collapse if magnetic fields play a major role in the support of the parent clouds (see Section 3.1). And third, they require the prior presence of mature stars; thus, they implicitly invoke the existence of some *spontaneous* mechanism, in any case, to provide a primary generation of stars.

1.5 Binary Stars, Bound Clusters, and Hierarchical Fragmentation

The majority of stars in the Galaxy are found in binary or multiple systems (Batten 1973). A particularly intriguing aspect of binary stars is the apparent dichotomy of primary and secondary masses when the periods P are longer than or shorter than ~ 100 yr (Abt & Levy 1976; see also Abt 1983). For $P > 100$ yr, the mass of the secondary is unrelated to that of the primary; hence, there are many more companions of low mass rather than of high mass (comparable to that of the primary), as if the two stars were formed by independent processes. Thus, the probability distribution for the mass of the secondary is the same as that for a field star. On the other hand, for binaries with $P < 100$ yr, there are disproportionate numbers of high mass ratios (≈ 1) compared to low mass ratios ($\ll 1$), as if the formation of the companion was influenced by the formation of the primary, so that the outcome is weighted toward obtaining stars of more nearly equal mass.

Abt & Levy (1976) interpreted their finding to imply that systems with $P > 100$ yr formed by independent condensation (or capture), whereas systems with $P < 100$ yr formed by fission. Given the observational selections that enter at intervals of 10^2–10^3 yr, however, the nominal dividing point of 100 yr probably should not be taken too literally.

Theorists generally distinguish between fission and fragmentation on the basis of whether the splitting of the two (or more) components takes place in a state of (unstable) equilibrium or against a dynamically collapsing background. The stability of rapidly rotating and self-gravitating equilibria against fission has been studied for both incompressible and compressible fluids (for reviews, see Chandrasekhar 1969, Tassoul 1978). Evolutionary simulations of fairly coarse numerical resolution even seemed to suggest that the outcome is two bodies of nearly equal masses (Lucy & Ricco 1979). However, more recent studies (Durisen et al. 1986) demonstrate that rotationally unstable polytropes tend to shed rings and disks (via the gravitational torques of bars and spiral structure) that absorb the excess angular momenta of the systems and prevent the central bodies from fissioning in the manner envisaged by earlier workers. Equally devastating for the classical fission hypothesis are the observational findings that T Tauri stars start their pre-main-sequence contractions with fairly small radii (Cohen & Kuhi 1979) and with fairly low rotation speeds (Vogel & Kuhi 1981, Bouvier et al. 1986, Hartmann et al. 1986), so that fission from quasi-static configurations into two bodies with orbital separations of AU scales would seem to be impossible. A more attractive scenario for the origin of binaries with periods $< 10^2$–10^3 yr would now seem to be formation from the dust and gas disks of $\sim 10^2$ AU that are being discovered prolifically around young stars (see Section 6.2).

The origin of binaries with periods between a few hundred and a few million years, which constitute about 25–30% of the total population, may require the (single-stage) fragmentation of a rapidly rotating and collapsing cloud core (e.g. Boss 1986) on a scale of 10^{17} cm or smaller. Tidal capture in a dense stellar cluster is another possibility (e.g. Press & Teukolsky 1977), but the small fraction (~ 10%) of all stars likely to be formed in bound clusters makes this an unlikely origin for all except the widest binaries [which are then subject to disruption within the same cluster (Heggie 1975)].

Hoyle's (1953) concept of hierarchical fragmentation—the isothermal collapse of a single Jeans mass of (nonrotating and nonmagnetic) gas followed by successive rounds of dynamical fragmentation—has not been realized in detailed numerical simulations (Tohline 1982, and references therein), which tend to produce a single accreting, centrally condensed object. Including the effects of rotation alleviates some of the difficulties (see the reviews by Bodenheimer 1980, Hayashi 1986, Boss 1987). Also, rapid decoupling of the principal source of mechanical support against self-gravity by shock compression (Woodward 1976, Klein et al. 1985) or catastrophic cooling (D. N. C. Lin, private communication) would produce, effectively, many Jeans masses in the pre-gravitational-collapse configuration and thereby lead to fragmentation. The value of the minimum fragment mass, variously estimated at 0.003 to 0.01 $M_\odot$ (Low & Lynden-Bell 1976, Rees 1976, Silk 1977, Boss 1986), is somewhat embarrassing. Appeal could be made to accretion or agglomeration to increase the final mass, but then there are no obvious reasons why stellar values should result from this scenario.

When applied to the problem of present-day star formation, the concept of hierarchical fragmentation suffers from the observational criticism that large-scale free-fall motions in molecular clouds have never been found (see, however, Section 3.2), and from the theoretical criticism that magnetic fields can strongly inhibit fragmentation for subcritical clouds (see Section 3.1). On the other hand, flattening of a supercritical cloud (along field lines) may allow it to fragment into quasi-spherical pieces with sizes given approximately by the vertical thickness of the cloud (Mestel 1965). This latter process may explain the formation of open clusters on a time scale given by magnetically diluted collapse (see Sections 3.2). The field-free hierarchical fragmentation scenario may apply to gaseous configurations (e.g. proto-globular clusters) where magnetic fields are not a factor.

1.6 *Efficiency of Star Formation*

In order to form a bound system, the minimum efficiency of star formation has been estimated to be between 20–50%, depending on the rate of gas dispersal and the degree of central concentration of the system (Lada et al. 1984, Elmegreen & Clemens 1985). The transformation of a large fraction of the gas of a region into stars in a time comparable to the intrinsic dynamical scale must be a relatively rare occurrence, as only $\sim 10\%$ of all stars today are believed to have been born in bound clusters (Roberts 1957, Wilking & Lada 1985). From an analysis of the total rate of star formation in the first Galactic quadrant, Myers et al. (1986) estimate that the overall efficiency of star formation is $\sim 2\%$ (see also Blitz 1978). The total production rate of stars in the Galaxy is often cited as ~ 3–5 $M_\odot$ yr^{-1}. This value is to be compared with a rate of gas return back to the interstellar medium (primarily in the form of mass loss from evolved stars) of ~ 1–$2\ M_\odot$ yr^{-1} (Knapp & Morris 1985, Jura 1987). Therefore, despite the low overall efficiency of star formation in molecular clouds, an irreversible conversion of interstellar gas into stars seems to be occurring in the Galaxy with a characteristic depletion time of $\sim 10^9$ yr. Whether such a relatively short time scale should be of concern and whether appeal should be made to mechanisms such as galactic infall of fresh gas or different histories for the production of high- and low-mass stars (bimodal star formation) is a matter of some debate (see, e.g., the discussion of Wyse & Silk 1987).

2. PROPERTIES OF MOLECULAR CLOUDS

2.1 Observed Physical Characteristics

It is well established that molecular clouds are the principal sites of active star formation in our Galaxy (Zuckerman & Palmer 1974, Burton 1976). A substantial part of the CO emission, and most of the total molecular mass, is contained in giant clouds with masses between 10^5 and 3×10^6 $M_\odot$ and with sizes of several tens of parsecs (Solomon et al. 1979). The precise values are controversial (see, e.g., Blitz & Thaddeus 1980) and depend on the conversion factor for transforming integrated CO intensity into H_2 column density (see the review of van Dishoeck & Black 1987).

Observations of CO emission with moderately high spatial resolution (see Figure 1) indicate that giant molecular clouds (GMCs) are actually cloud complexes, composed of smaller units ("clumps") with masses $M_{cl} \sim 10^3$–$10^4\ M_\odot$, sizes $R \sim 2$–5 pc, densities $n_{H_2} \sim 10^{2.5}$ cm^{-3}, and temperatures $T \sim 10$ K (Sargent 1977, Evans 1978, Blitz 1978, Rowan-Robinson 1979). It is unclear to what extent the observed clump sizes, masses, and densities are the result of observational selection, since the excitation of ^{12}CO, even with radiative trapping, requires H_2 densities of a few hundred per cm^3 (Kutner & Leung 1985). Subregions within the clumps that contain gas of much higher density—cloud cores—can be mapped in H_2CO (e.g. Evans & Kutner 1976), in CS (e.g. Linke & Goldsmith 1980), in HC_3N (e.g. Avery 1980), in NH_3 (reviewed by Ho & Townes 1983), and even in CO if they are appreciably hotter than their surroundings.

The clumps resemble the nearby dark clouds surveyed by Lynds (1962). A well-known complex of such dark clouds exists in Taurus and contains $\sim 10^4\ M_\odot$ of material actively forming an unbound association of low-mass stars (Cohen & Kuhi 1979, Kleiner & Dickman 1984, Cernicharo et al. 1985). Such clumps have small cloud cores (e.g. Myers & Benson 1983) whose observed sizes and densities depend on the molecular line used for

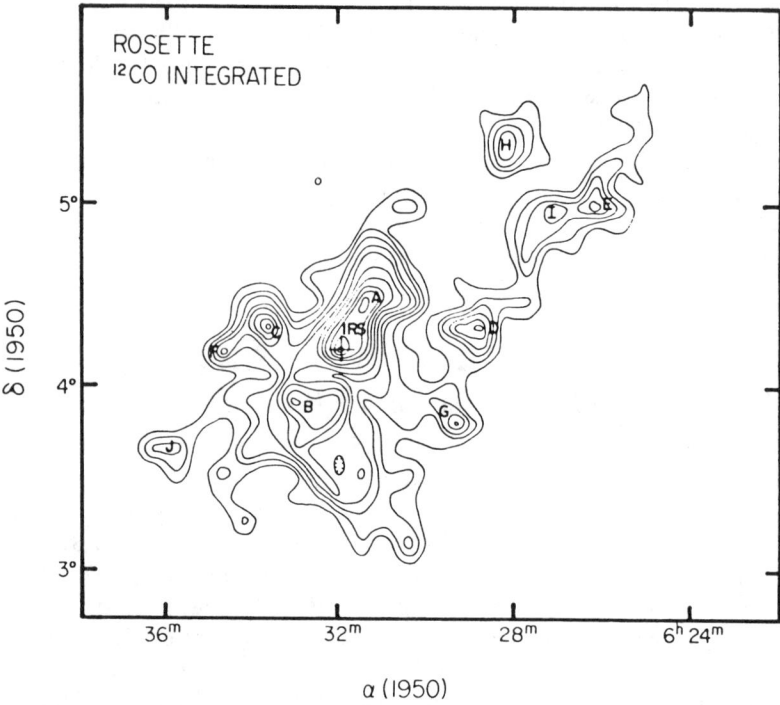

Figure 1 Clumps observed in ^{12}CO within the molecular cloud complex associated with the Rosette Nebula. A hot core has been illuminated by a luminous infrared source at the position of the cross (from Blitz & Thaddeus 1980).

the mapping (Myers 1985, Walmsley 1987). The masses enclosed within the density contours to which ammonia is sensitive (roughly $n_{H_2} > 10^4$ cm^{-3}) are $M_{core} \sim 10^0 \, M_\odot$. The cores defined in this manner have sizes $R \sim 10^{-1}$ pc, temperatures $T \sim 10$ K, and NH$_3$ linewidths that are almost thermal. Perault et al. (1985) find from ^{13}CO observations that the dense cores are embedded in massive envelopes of several hundred $M_\odot$ (see also Falgarone 1987). The contention (Myers & Benson 1983) that the dense, quiet NH$_3$ cores are the sites of low-mass star formation is supported by their close association with known T Tauri stars (see Figure 2), by the correlation of bipolar outflow sources with them (Fuller & Myers 1987), and by the fact that IRAS detected infrared sources in approximately half of the NH$_3$ cores (Beichman et al. 1986). The deeply embedded sources can be identified as protostars by the characteristics of their infrared spectral energy distributions (Adams & Shu 1986, Adams et al. 1987, Myers et al. 1987a).

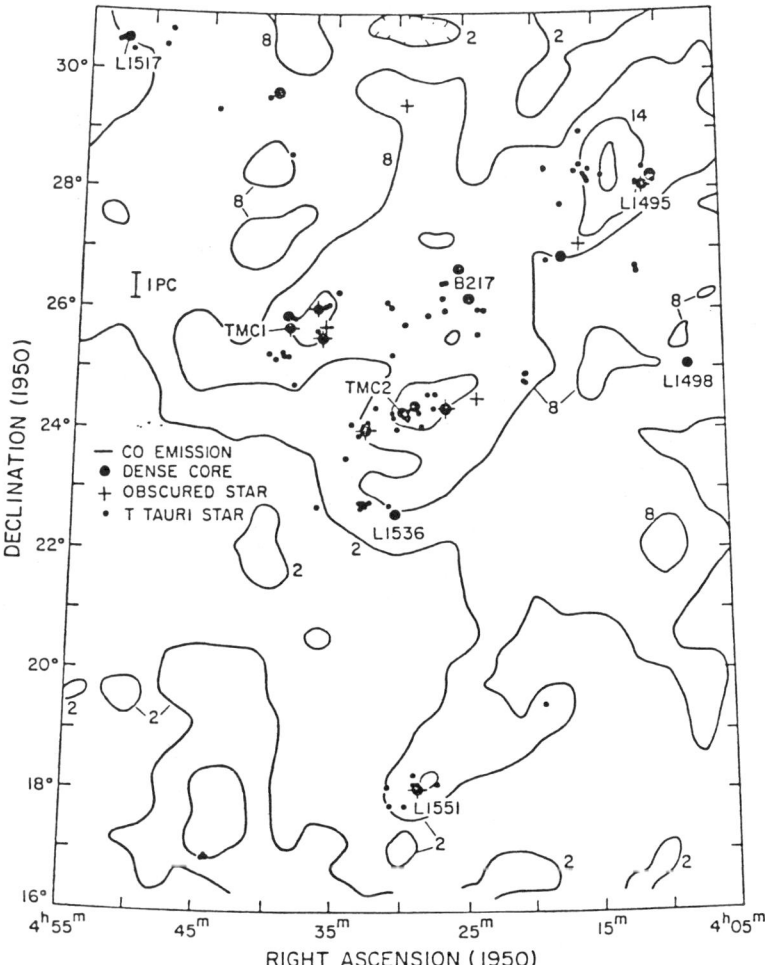

Figure 2 CO contour map of the Taurus molecular cloud with positions of dense NH_3 cores, embedded infrared sources, and visible T Tauri stars (from Myers 1986).

There are also denser clumps, like the R Coronae Australis and ρ Ophiuchi regions, which have a rather high efficiency of star formation (between 25–50%; see Wilking & Lada 1983, Lada & Wilking 1984, Wilking et al. 1986) and seem to be forming bound clusters. These clumps have radii ~ 0.3–0.6 pc, masses ~ 19–$110\ M_\odot$, average densities $\sim 10^3$ cm^{-3} (Loren et al. 1983), and may be closely packed collections of cores with internal densities rising above 10^6 cm^{-3} (Snell et al. 1984).

2.2 Lifetimes of Giant and Dwarf Molecular Clouds

The issue of the lifetimes of molecular clouds attracted considerable controversy at the turn of the decade (Solomon et al. 1979, Blitz & Shu 1980, Cohen et al. 1980), but further analysis of the observational data has led to a greater consensus of opinion. A substantial fraction of the total molecular material is now thought to be contained in dwarf molecular clouds (DMCs), which are not restricted to the spiral arms. These clouds probably survive one or more rotation periods of the Galaxy and thus have lifetimes of 10^8 yr or more. In contrast, molecular clouds with hot cores (presumably GMCs containing OB stars) are probably assembled and dispersed as they cross the spiral arms on a time scale of $\sim 10^7$ yr (e.g. Bash & Peters 1976). In principle, the determination of cloud ages (at least, those parts that are relatively young) could be obtained more directly through the use of molecular clocks (see the review by Glassgold 1985). Unfortunately, large uncertainties dominate present discussions, and definitive results await future developments.

In any case, the existing evidence is consistent with the interpretation that giant cloud complexes are not much older than the crossing times of their individual clumps. Their heterogeneous internal structures pose no severe dynamical problems if they are transient objects (Blitz & Shu 1980); the main theoretical issue for GMCs is then how the individual clumps of gas are gathered. In high-angular-resolution observations of M83 and M51 (Allen et al. 1986, Lo et al. 1987), the ridge of H I emission is observed to be displaced *downstream* from the dust lane and peak molecular and nonthermal emission (presumably the region of maximum density-wave compression; see, e.g., Roberts & Yuan 1970). This finding indicates that GMCs in the regions being observed are assembled from molecular rather than atomic material (e.g. Kwan & Valdes 1983, 1986, Roberts & Stewart 1987), and that the H I gas comes from the partial dissociation of the H_2 gas after the formation of OB stars. The possible role of various instabilities in helping to gather the complex is left unsettled (Mouschovias et al. 1974, Cowie 1981, Elmegreen 1982a, 1987, Balbus & Cowie 1985, Lubow et al. 1986), but there is little doubt that theoretical models must grapple with the basic clumpy nature of the interstellar gas (see, e.g., Bash & Kaufman 1986).

2.3 Mechanical Balance of Molecular Clumps

In contrast to giant molecular clouds, DMCs and individual molecular clumps must be near mechanical equilibrium. The Jeans mass M_J associated with the average conditions of a clump is only a few $M_\odot$, much smaller than the mass of the clump M_{cl}, yet molecular clouds cannot be collapsing on a free-fall time scale or else the rate of star formation in the Galaxy would be far too high (Zuckerman & Palmer 1974). Since $M_J \ll M_{cl}$, thermal pressure could be important for molecular cloud cores, but not for their envelopes. Various alternative mechanisms of support for self-gravitating interstellar clouds have been invoked at one time or another: magnetic fields (Chandrasekhar & Fermi 1953, Mestel 1965, Spitzer 1968, Mouschovias 1976), rotation (Field 1978), and turbulence (Norman & Silk 1980, Larson 1981). Of these possibilities, magnetic fields are probably the crucial ingredient, because without them the observed levels of rotation and turbulence in molecular clouds would be very difficult to understand.

2.3.1 FLUID TURBULENCE Since the discovery of CO lines in molecular clouds, it has been known that the CO linewidths correspond to very supersonic fluid motions. When interpreted as turbulence, the velocities approach virial values $\Delta v \sim (2GM/R)^{1/2}$, where R is the radius of the cloud and Δv is the FWHM of the molecular line. The CO linewidths have also been empirically determined to follow a power-law correlation with the size of the region being observed ($\Delta v \propto R^\alpha$). From a compilation of data from different sources, Larson (1981) derives $\alpha \approx 0.3$; Solomon & Sanders (1985) and Dame et al. (1986) prefer a value of $\alpha \sim 0.5$–0.6. Myers (1987) attributes the difference to Larson's not having subtracted out the thermal contribution to the linewidth before drawing his correlation diagrams.

Theoretically, the difficulty with supersonic turbulence is that in the absence of magnetic fields, it should be highly dissipative (see, e.g., the discussion of Goldreich & Kwan 1974). Scalo & Pumphrey (1982) have performed numerical simulations, in which the turbulence was modeled with moving blobs; such a model might apply to the dissipation of the relative motions of clumps within a cloud complex. In any case, Scalo & Pumphrey claim that the turbulent dissipation rate is usually overestimated because the relevant interaction cross section is (in effect) πR^2 instead of the gas kinetic formula $4\pi R^2$. In addition, the collisions are generally oblique, and coalescence reduces the number of fragments with time. However, most estimates in the literature do not include the extraneous factor of 4 [see, e.g., the discussion of Spitzer (1968) or Blitz & Shu (1980), which discuss what happens to a given *small* parcel of gas], and the other effects discussed by Scalo & Pumphrey are highly model dependent (e.g. gravitational focusing will enhance the collision rate).

For individual clumps the situation is worse, since the turbulent dissipation of energy per unit mass in a nonmagnetic continuous medium (compressible or not) will always occur at a rate $\sim v^3/l$, where v is the typical turbulent velocity and l is the characteristic interaction scale. (Basically, the specific energy $\sim v^2$ in macroscopic motions is ultimately converted to heat in a time $\sim l/v$ for both cascading eddies and radiative shocks. Any acoustic radiation transported away in the process only hastens the drain of energy.) In order for turbulence to provide mechanical support for a cloud, the turbulent velocity v must be of the order of the virial speed, and l should be much smaller than the cloud size R. Consequently, the turbulent dissipation time l/v must be much smaller than the crossing time R/v. Thus, turbulent support of a cloud is not possible unless there is a mechanism to replenish the fluctuations on a sufficiently short time scale. Fleck (1981) has proposed the tapping of shear energy in galactic differential rotation as a possible mechanism, but this has long been considered an unpromising source of interstellar turbulence (Spitzer 1968). A more likely possibility— winds (in a snowplow phase) from young stars (Norman & Silk 1980)— may take too long to pump up the general velocity field (Bally & Lada 1983; P. Goldreich, private communication). Moreover, direct conversion of wind energy into gas motions would most likely deposit the largest velocities at the smallest scales, contrary to the observed correlations (Fleck 1981). Finally, turbulent "pressure" leads to forces only if it has a gradient, yet empirically the combination ρv^2 appears to be nearly a constant.

Observationally, the nature of turbulent velocity fields in dark clouds is further constrained by interstellar polarization maps (Vrba et al. 1976), which show that the directions of the embedded magnetic fields are well ordered over the dimensions of the clouds (see also Monetti et al. 1984). This demonstrates that turbulence cannot be totally dominant over the magnetic fields. If the energy of a turbulent velocity field, e.g. in the form of a cascade of eddies (see Landau & Lifshitz 1959), were much greater than the energy of the magnetic field, a more tangled magnetic field configuration would result. Since this is not the case, the observed "turbulent" motions are probably more wavelike than eddylike; thus, although fluctuations of the field are present (say, in the form of Alfvén waves), the field still has a well-defined mean direction.

2.3.2 MAGNETIC SUPPORT AND ALFVÉNIC TURBULENCE Since, unlike turbulence, magnetic fields have the virtue that they are not easily dissipated, they deserve consideration as the major agent for molecular cloud support. Their longevity makes them a natural candidate as a resilient obstacle to rapid star formation. The strength of the magnetic fields needed to provide appreciable mechanical support of molecular clouds against their self-gravity can be ascertained by virial theorem arguments (Strittmatter 1966) or, more accurately, by detailed model calculations (Mouschovias 1976).

A magnetic flux Φ can support a cloud provided its mass does not exceed a critical value given by the following relation [see the extrapolation in Figure 1 of Mouschovias & Spitzer (1976) to zero $(p_T/p_m)^{1/3}$]:

$$M_{cr} = 0.13 \frac{\Phi}{G^{1/2}} \approx 10^3 \, M_\odot \left(\frac{B}{30 \, \mu G}\right)\left(\frac{R}{2 \, \text{pc}}\right)^2. \qquad 2.$$

Zeeman splitting of thermal OH (the excitation of which requires densities $> 10^3$ cm^{-3}) allows the measurement of the component of **B** along the line of sight for Orion A (125 μG), Orion B (38 μG), W22A (18 μG), W22B (32 μG), W40 (14 μG), W49B (21 μG), S88B (69 μG), and S106 (130 μG) (Troland et al. 1986, Crutcher & Kazes 1983, Kazes & Crutcher 1986, Heiles & Stevens 1986, Troland & Heiles 1986, Crutcher et al. 1987, Kazes et al. 1987; see also the review of Heiles 1987). It should be noted that there are many other cases with only upper limits; on the other hand, for random orientations, the total field strengths should be larger, on average, by a factor of 2 than the values deduced from a Zeeman analysis for differential circular polarization in the line wings. Except for W22 and W49, all the measured sources have nearby H II regions; the last two sources have bipolar flows. In OH maser sources associated with collapsed regions near young stars (with gas densities of the order of 10^5–10^8 cm^{-3}), the derived *total* field strengths (from direct observation of the separate Zeeman components) are typically several milligauss (see the review of Reid & Moran 1981).

If magnetic fields do provide the dominant mechanism for the support of self-gravitating clouds, then the properties of the observed turbulence become explicable (Myers & Goodman 1987, Shu 1987). The energy for the turbulence may originate with many sources—stellar winds, cloud collisions, expanding H II regions, supernovae, etc.—but ultimately the disturbances will excite a spectrum of MHD waves (Arons & Max 1975, Zweibel & Josafatsson 1983, Elmegreen 1985b, Falgarone & Puget 1986). Waves with super-Alfvénic fluid motions will generate compressive shocks and will dissipate rapidly. Hence, the fluctuating part of the fluid velocities will generally become sub-Alfvénic (but still supersonic), i.e. $\delta v \lesssim v_A$. In the ^{12}CO line, there is an observational selection bias toward seeing the highest velocities, $\delta v \sim v_A$, because photon trapping tends to shield regions of common (low) velocities (P. Goldreich, private communication; Kutner & Leung 1985). However, from Equation 2, it is easy to show that for clouds near the critical state $M_{cl} \approx M_{cr}$, the mean Alfvén speed is automatically of the magnitude needed for virial equilibrium, i.e.

$$v_A^2 \equiv \frac{B^2}{4\pi\rho} \sim v_{V.T.}^2 \equiv \frac{GM_{cl}}{R}. \qquad 3.$$

Thus, $\delta v \sim v_A$ implies that $\delta v \sim v_{V.T.}$ (i.e. cloud "turbulence" automatically has a tendency to appear sufficient for virial equilibrium). Moreover, if B does not vary strongly from region to region (in CO envelopes), Equation 2 implies that clouds near the critical state, $M_{cl} \approx M_{cr}$ (i.e. clouds for which magnetic fields provide the dominant means of support), should have nearly the same mean column densities, $M_{cl}/\pi R^2 \propto \rho R \approx$ constant, implying $v_A \propto \rho^{-1/2} \propto R^{1/2}$. Thus, this line of argument provides a simple mechanistic explanation of the observed correlation $\Delta v \propto R^\alpha$, with $\alpha \approx 0.5$. The assumption that B does not vary from region to region may be relaxed if the observed relation, $\rho R \sim$ constant, arises (partially) from selection effects, because the observations may be sensitive to only a relatively narrow range of column densities (Larson 1981).

The presence of nonlinear Alfvén waves circumvents a primary difficulty associated with magnetic support—magnetic fields cannot provide support along the direction of the field, but the cloud clumps are not observed to be highly flattened along field lines. If thermal pressure and static magnetic fields were the only means of support for a self-gravitating clump, the cloud would have a characteristic scale height in the direction parallel to the field equal to (Spitzer 1942)

$$z_0 = \frac{kT/m}{\pi G \sigma}, \quad\quad 4.$$

where σ is the surface mass density projected along **B**. For $T \sim 10$ K, a mean molecular weight $m = 2.3\, m_H$, and $\sigma = 0.01$ g cm^{-2} (corresponding to 2.5 mag of visual extinction), Equation 4 yields $z_0 \sim 0.06$ pc. Since the lateral dimensions of molecular cloud clumps are typically a parsec or more, some theorists (e.g. Pudritz 1986) have suggested that the cloud clumps should essentially be sheets. Indeed, the L204 cloud (McCutcheon et al. 1985) and the denser condensations in the Taurus complex (Monetti et al. 1984) appear quite flattened and have often been interpreted in terms of either sheets or filaments. However, the flattening is probably not as severe as indicated by Equation 4 because bipolar flows in the Taurus region tend to be aligned with the ambient field (Strom et al. 1985), and some of these flows have apparently swept up ambient molecular gas out to a parsec and are still within the cloud (see also Figure 18 of Bally & Lada 1983). Moreover, in Ophiuchus and R Coronae Australis, dust filaments are found that are aligned *parallel* to the mean field (Vrba et al. 1976, Heyer et al. 1986).

Alfvén waves may provide the requisite support along **B** (Dewar 1970). Consider a field **B** made up of a static (mean) part $\mathbf{B}_0$ pointing in the z direction, and a fluctuating (Alfvén-wave) part $\delta \mathbf{B}$ with magnitude

≈ $(\delta v/v_A)B_0$ and with direction perpendicular to B_0. For simplicity, assume δB has the form of a spatially damped traveling wave,

$$\delta B = b \sin(k_R z - \omega t)\exp(-k_I z), \qquad 5.$$

with **b** lying in the x-y plane. This wave provides a secular transfer of momentum to the ambient medium, which is given by the time average of the quadratic term in the Lorentz force per unit volume:

$$\frac{1}{4\pi}(\nabla \times \delta\mathbf{B}) \times \delta\mathbf{B}. \qquad 6.$$

The substitution of Equation 5 into Equation 6, averaged over one cycle, gives

$$\frac{1}{8\pi}|\mathbf{b}\exp(-k_I z)|^2 k_I \hat{\mathbf{z}}, \qquad 7.$$

which equals the average value of $-\nabla(|\delta\mathbf{B}|^2/8\pi)$. The force (Equation 7) is directed along the direction of wave propagation (parallel to $\mathbf{B}_0$) and has magnitude comparable to the Lorentz force associated with the mean field,

$$\frac{1}{4\pi}(\nabla \times \mathbf{B}_0) \times \mathbf{B}_0 \qquad 8.$$

(which supports the cloud in the lateral directions), provided that $\delta\mathbf{B} \sim \mathbf{B}_0$ (i.e. if $\delta v \sim v_A$) and that the characteristic damping length k_I^{-1} is comparable to the size R of the cloud clump. We have already commented upon the evidence from the observed CO linewidths that suggests that δv is comparable to v_A; standard formulae (see the references cited by Zweibel & Josafatsson 1983) for the damping of Alfvén waves by ion-neutral collisions or by nonlinear steepening suggest that the condition $k_I^{-1} \sim R$ is also likely to occur in molecular clouds. (Here, the real part of the wave number k_R may be associated with wave generation by molecular outflows.) Presumably, an observed cloud clump with a given distribution of sources has contracted to a size Z in the direction parallel to the mean field, so that the available set of waves can provide support against gravity for that configuration.

Notice that the "Reynolds stress" in the transverse fluid motions associated with an Alfvén wave provides no momentum transfer along the direction of the field, $\mathbf{B}_0 \cdot \nabla \cdot (\rho \delta\mathbf{v}\delta\mathbf{v}) = 0$, and hence does not support the cloud in the direction of the mean field. In addition, only outwardly propagating waves support the cloud against self-gravity; inwardly propagating waves would tend to *compress* the cloud. Thus, cloud support by this mechanism requires that there be a negative gradient in wave sources from inside to outside the cloud.[1]

2.3.3 MAGNETIC BRAKING The generation of (torsional) Alfvén waves may also explain the small values of the angular velocity Ω commonly deduced for molecular clouds (see below). There are exceptional cases with relatively large rotation rates (see, e.g., Vogel & Welch 1983), but these clouds probably represent collapsed objects in later stages of evolution (Vogel 1984). Usually, when the envelopes of molecular clumps have measurable rotation rates, the rates are of order $\Omega \sim 1$ km s^{-1} pc$^{-1} = 3 \times 10^{-14}$ rad s^{-1}. An angular velocity of the same order (1 km s^{-1} pc^{-1}) is often observed for the ammonia cores in the Taurus complex (Myers 1987), which suggests that the cores of clumps may be rotationally coupled to their envelopes by magnetic braking (Gillis et al. 1974, 1979, Mouschovias & Paleologou 1979, 1980). Since the rotation rates of NH$_3$ cores in Taurus are always less than 5 km s^{-1} pc^{-1}, rotation cannot be important for the support of these objects (Myers et al. 1987c). In addition, there appears to be no clear correlation between the projected clouds' rotation axes and the axis of Galactic rotation (Goldsmith & Sernyak 1984).

The process of magnetic braking of cloud cores by their envelopes is important because it produces a potential reservoir of low-angular-momentum material for the formation of stars, planets, and binary systems (Mouschovias 1978). In general, Alfvén waves generated by the spin of the core relative to its envelope will tend to bring the two into corotation; significant braking occurs when the amount of envelope material affected by the outwardly propagating Alfvén wave has a moment of inertia equal to that of the core. In an approximation that models the core as a cylinder of half-height Z and radius R with uniform density ρ_{core} embedded within an envelope of uniform density ρ_{env}, Mouschovias & Paleologou (1980) show that the characteristic time scale for braking the component of Ω parallel to $\mathbf{B}_0$ is given by

$$\tau_\parallel = \frac{\rho_{\text{core}}}{\rho_{\text{env}}} \frac{Z}{v_{\text{A}}}. \qquad 9.$$

On the other hand, the corresponding time scale for braking the component of Ω perpendicular to $\mathbf{B}_0$ is given by

$$\tau_\perp = \frac{1}{2}\left[\left(1 + \frac{\rho_{\text{core}}}{\rho_{\text{env}}}\right)^{1/2} - 1\right]\frac{R}{v_{\text{A}}}. \qquad 10.$$

[1] The same comment applies to proposals that invoke stellar outflows as a direct mechanism for cloud support (P. Goldreich, private communication).

In Equations 9 and 10, the Alfvén speed v_A in the envelope is to be evaluated just outside the surface of the core under the assumption that farther out it is independent of z but declines as $\tilde{\omega}^{-1}$.

For the case $\rho_{core} \gg \rho_{env}$, the time scale $\tau_\perp$ for braking the perpendicular component will be much less than that for the parallel component $\tau_\parallel$; this situation occurs because the greater lever arm in the perpendicular directions offsets the assumed decrease of propagation speed with increasing $\tilde{\omega}$. The lower efficiency of parallel magnetic braking will tend to leave the spin axes of molecular cloud cores (and the objects to which they collapse) aligned parallel to $\mathbf{B}_0$; this effect may explain why the bipolar flow axes of several sources in Taurus line up with the mean direction of the ambient magnetic field (Strom 1985, Strom et al. 1985, Heyer 1986).

2.3.4 MEASURED ROTATION RATES Goldsmith & Arquilla (1985; see also Fuller & Myers 1987) have compiled a list of rotation rates in 16 dark cloud regions, with masses in the range 6×10^3 to 0.7 $M_\odot$, sizes from 17 to 0.1 pc, and velocity gradients from 0.2 to 6 km s^{-1} pc^{-1}; they find a tendency for the smallest and densest regions to have the shortest rotation periods (see their Figure 6).[2] Unfortunately, the various correlations derived on the basis of these data may not be extremely meaningful, since they may represent limits below which rotation in dark clouds cannot be detected. What is significant about the rotation rates for ordinary dark clouds is that they are uniformly low or undetectable (see also Table 3 of Arquilla & Goldsmith 1986), which suggests that the clouds are magnetically controlled, i.e. they are subcritical regions (see Section 3.3) and are not contracting rapidly as a whole.

Observations of rotation in molecular cloud cores in regions other than Taurus are fairly sparse. Two cases are known in ρ Ophiuchi (Wadiak et al. 1985), where H_2CO measurements show two fragments with diameters ~ 0.01–0.02 pc and masses ~ 1–2 $M_\odot$ enclosed within density contours $\sim 5 \times 10^6$ cm^{-3} and having rotation rates ~ 40 km s^{-1} pc^{-1}. In Orion, five cases have been found and measured (Harris et al. 1983; L. Mundy, private communication); NH_3 observations show cores with diameters ~ 0.05 pc and masses ~ 10 $M_\odot$ enclosed within density contours $\sim 3 \times 10^4$ cm^{-3} and having rotation rates 25–50 km s^{-1} pc^{-1}. The faster rotation rates and

---

[2] It is tempting to interpret the apparent spin-up as being due to contraction and to conclude that it has proceeded with considerable braking because the specific angular momentum J/M of the various regions decreases with decreasing size (see their Figure 5). This conclusion may well be correct; however, it is not warranted on the basis of the evidence presented. The specific angular momentum in any stably stratified medium will generally increase outward, so that averaging over progressively larger volumes will naturally produce larger values of J/M, independent of whether there is any evolutionary connection between the material on the inside and that on the outside.

larger masses within a given radius are all consistent with a picture in which these cores have formed from a supercritical background that is itself contracting (see Section 3.2), thereby lowering the available time for magnetic braking.

Vogel & Welch (1983), Zheng et al. (1985), and Ho & Haschick (1986) have studied the kinematics of the molecular gas in the neighborhood of the H II regions W58, ON1, and G10.6−0.4; these regions presumably represent the core environments in which high-mass stars form. The gas masses within ~ 0.5–1 pc contours and mean densities $\sim 10^5$ cm^{-3} are large (of order 10^2–10^3 $M_\odot$). The measured angular velocities are all of order 10 km s^{-1} pc^{-1} and may again represent the spin-up associated with the overall contraction of a supercritical region.

The central regions of G10.6−0.4 studied by Ho & Haschick (1986; see also Keto et al. 1987) seem to be collapsing toward a compact group of OB stars, with the inflow velocities increasing inward. The infalling gas appears to spin faster than the radially stationary material farther out; these are characteristics naturally produced from an "inside-out" collapse such as that discussed in Section 4.1.2.

2.3.5 IONIZATION FRACTION The ionization fraction in typical regions of molecular clouds is believed to be very low, although modern attempts to measure it precisely (see the review of Langer 1985) have been foiled by the revision of the dissociative recombination coefficient for H_3^+ (Smith & Adams 1984). The lower limit on the ionization number fraction, as determined from the abundance of $HC^{18}O^+$, is still 10^{-8}. An upper limit of $\sim 10^{-6}$ can be set from deuterium fractionization reactions (Dalgarno & Lepp 1984), and a value of $\sim 10^{-7}$ can be derived from the abundance of HCO^+ if the cosmic-ray ionization rate is assumed to be $\zeta \sim 10^{-17}$ s^{-1} (Smith & Adams 1984). It may be possible to determine the cosmic-ray flux in dense cores by measuring (in self-absorption) the amount of atomic hydrogen formed by the dissociation of H_2 (M. Jura, private communication).

Until better empirical results are available, the most accessible determinations at present are theoretical. When cosmic-ray ionization at a rate $\zeta = 1 \times 10^{-17}$ s^{-1} is balanced by two-body recombinations of charged particles and on charged grains, the ion mass density ρ_i depends on the neutral mass density ρ_n according to the law (Elmegreen 1979)

$$\rho_i = C\rho_n^{1/2}, \qquad 11.$$

where C is a weak function of gas temperature and is proportional to the square root of the metal depletion. For average metal depletions of 0.1 and gas temperatures of 10–30 K, Elmegreen (1979) obtains results that correspond to $C = 3 \times 10^{-16}$ cm$^{-3/2}$ g$^{1/2}$. Thus, with a typical ion mass $m_i \approx 30\, m_H$ and a typical neutral mass of $m_n \approx 2.3\, m_H$ (de Jong et al. 1980), the ionization number fraction is about 10^{-7} at a density of 10^4 cm^{-3}. At densities much higher than 10^8 cm^{-3}, natural radioactivity becomes dominant over cosmic rays as the primary ionizing agent (Nakano & Tademaru 1972), grains begin to carry a major fraction of the total charge, and ρ_i approaches a constant value (Umebayashi & Nakano 1980).

2.3.6 AMBIPOLAR DIFFUSION AND MAGNETIC FLUX LOSS When the ionization fraction is very low, magnetic support of molecular clouds becomes synonymous with ambipolar diffusion—the process by which magnetic fields (and the charged plasma to which they are coupled) drift relative to a background of neutrals (Mestel & Spitzer 1956). The magnetic forces are felt directly only by the charged particles; the neutrals can be supported against their self-gravity only through the frictional drag that they experience as they slip relative to the ions. To see this quantitatively, note that with the low ionized mass fractions implied by Equation 11, the pressure and gravitational forces acting on the fluid of charged species are always very small in comparison with the Lorentz force,

$$\frac{\mathbf{j}}{c} \times \mathbf{B} = \frac{1}{4\pi}(\nabla \times \mathbf{B}) \times \mathbf{B}, \qquad 12.$$

where we have used Ampere's law, $\mathbf{j} = (c/4\pi)\nabla \times \mathbf{B}$. The mean Lorentz force will drive the ions through a sea of neutrals at a (mean) relative terminal velocity given by a balance with the drag force (per unit volume):

$$\mathbf{f}_d = \rho_n \rho_i \gamma (\mathbf{u}_i - \mathbf{u}_n), \qquad 13.$$

where $\mathbf{u}_i$ and $\mathbf{u}_n$ are the fluid velocities of the ions and neutrals, respectively, and γ is the drag coefficient associated with momentum exchange in ion-neutral collisions. When the collision process is dominated by the dipole moment induced in the neutral by the passing ion, we have $\gamma \approx 3.5 \times 10^{13}$ cm^3 g^{-1} s^{-1} (Draine et al. 1983). For drift speeds in excess of ~ 10 km s^{-1}, the Langevin approximation breaks down, and γ should approach the geometric value $\sim \pi(r_i + r_n)^2 |\mathbf{v}_d|/(m_i + m_n)$ (Mouschovias & Paleologou 1981). A balance between Equations 12 and 13 leads to a drift velocity of the ions (which in turn nearly comove with the electrons) relative to the neutrals given by (Spitzer 1958)

$$\mathbf{v}_d \equiv \mathbf{u}_i - \mathbf{u}_n = \frac{1}{4\pi\gamma\rho_i\rho_n}(\nabla \times \mathbf{B}) \times \mathbf{B}. \qquad 14.$$

Because the cyclotron frequency $eB/m_i c$ of the typical ion is much greater than the mean collision frequency of ions with neutrals, $\rho_n \gamma$, ions gyrate about a given field line many times before they are knocked off by collisions; thus, the ions are effectively tied to **B**. The time evolution of the magnetic field itself may then be obtained from the approximation that it is frozen in the plasma of ions and electrons:[3]

$$\frac{\partial \mathbf{B}}{\partial t} + \nabla \times (\mathbf{B} \times \mathbf{u}_i) = 0. \qquad 15.$$

The substitution of Equation 14 into Equation 15 yields the nonlinear diffusion equation

$$\frac{\partial \mathbf{B}}{\partial t} + \nabla \times (\mathbf{B} \times \mathbf{u}_n) = \nabla \times \left\{ \frac{\mathbf{B}}{4\pi \gamma \rho_i \rho_n} \times \left[\mathbf{B} \times (\nabla \times \mathbf{B}) \right] \right\}. \qquad 16.$$

If the right-hand side were zero, the field would be well coupled to the motion of the neutrals. As it is, Equation 16 states that the effective "diffusion coefficient" $\mathscr{D}$ associated with the drift of the field relative to the neutrals has order of magnitude $v_A^2 t_{ni}$, where $v_A \equiv (B^2/4\pi\rho_n)^{1/2}$ is the Alfvén speed in the (combined) medium, and $t_{ni} \equiv (\rho_i \gamma)^{-1}$ is the mean collision time of a neutral molecule in a sea of ions. If the magnetic field has characteristic length R, the time scale for ambipolar diffusion is $t_{AD} \sim R^2/\mathscr{D} \sim R/v_d$. For a magnetically supported self-gravitating cloud that satisfies Equations 3 and 11, the ratio of t_{AD} to the dynamic time scale $t_{dyn} \sim R/v_A \sim (G\rho_n)^{-1/2}$ is approximately given by the dimensionless combination

$$\frac{t_{AD}}{t_{dyn}} \sim \frac{\gamma C}{2(2\pi G)^{1/2}}, \qquad 17.$$

where the numerical coefficient has been adjusted to agree with a detailed analysis in a slab geometry (Shu 1983, Lizano & Shu 1987). Notice that the ratio (Equation 17) is independent of M_{cl} or R (and therefore the mean ρ_{cl}). For values of γ and C quoted above, the coupling constant on the right-hand side of Equation 17 has the value 8; thus, ambipolar diffusion can be generally expected to take place at a relatively slow rate compared with dynamical events until the ionization fraction begins to depart appreciably from the law (Equation 11).

[3] Because grains are negatively charged in opaque clouds, Elmegreen (1979, 1986) and Nakano & Umebayashi (1980) considered the extent to which they are also tied to field lines and can impede the slip of the neutrals. Large grains have small gyrofrequencies and are not well coupled to **B**, while small grains present cross-sectional areas for collisions with neutrals that are not much larger than the Langevin cross sections of ions with neutrals. Consequently, the effects of charged grains on the rate of ambipolar diffusion are important only at high densities, when the charges are mostly carried by grains.

The equation of motion for the neutrals reads as

$$\rho_n \left[\frac{\partial \mathbf{u}_n}{\partial t} + (\mathbf{u}_n \cdot \nabla)\mathbf{u}_n \right] = -\nabla P_n - \rho_n \nabla \phi + \mathbf{f}_d, \qquad 18.$$

where P_n is the neutral gas pressure, and ϕ is the gravitational potential satisfying

$$\nabla^2 \phi = 4\pi G \rho_n \qquad 19.$$

in the approximation $\rho_i \ll \rho_n$. In a heuristic treatment that does not follow fluctuating motions (Alfvén waves), e.g. the adoption of Equation 14, one might include in P_n the "turbulent" contributions discussed in Section 2.3.2. In any case, if we now substitute Equations 13 and 14 into Equation 18, we see that the magnetic forces provide indirect support to the neutrals against their self-gravity via the intermediary of frictional drag. Thus, in a lightly ionized gas like a molecular cloud, the processes of magnetic support and ambipolar diffusion are synonymous; one cannot have one without having the other. As is shown in Section 3.4, this automatically leads to a mechanism for the formation of substructures (cores and envelopes) in molecular clouds.

Ambipolar diffusion is believed to be the process by which the classic flux problem is resolved for forming stars (see the review of Mestel 1985). Moreover, the issue of when the magnetic field decouples from the matter may determine the range of internal densities present in a molecular cloud core at the point when it becomes unstable to dynamical collapse (see Section 4.1.1). The value of the cloud density required in order for dynamical decoupling of the field to occur is controversial. Nakano (1984, 1985) asserts that it happens when the density becomes greater than 10^{11} cm^{-3}; Mouschovias et al. (1985, and references therein) claim that decoupling will occur for 10^5 cm^{-3}. The difference in opinion arises because of an unorthodox definition that leads the latter authors (see their Equation C1a) to label relatively slow motions as being "dynamical" even when the evolution proceeds quasi-hydrostatically in the core by ambipolar diffusion. (The large infall velocities in the outer layers are artifacts of the slab geometry, in which the gravitational field approaches a constant at infinity instead of falling off as $1/r^2$.) In any case, Equation 17 favors a high density (that for which Equation 11 begins to break down) for decoupling; however, the argument depends sensitively on the precise values adopted for γ, C, and the numerical coefficient in the denominator. Observations of systematic velocity differences between neutral and ionized species with similar critical densities for excitation can help to clarify this issue (M. Walmsley, private communication).

2.4 Thermal Balance

Temperatures of about 10 K are generally inferred from CO observations of molecular cloud envelopes and can be understood in terms of a balance between cosmic-ray heating and CO cooling (e.g. de Jong et al. 1980). In the outer parts, heating by photoelectrons from dust grains is an important additional contribution (for a review of several possible mechanisms for heating molecular clouds, see Goldsmith & Langer 1978). The cloud cores associated with regions of low-mass star formation also have low temperatures (Myers & Benson 1983, Menten & Walmsley 1985), but those near OB stars are considerably hotter (for a review, see Wynn-Williams 1982). The conventional explanation for this latter observation has been that luminous embedded stars heat the surrounding dust grains, which in turn heat the gas (Goldreich & Kwan 1974). This interpretation has been questioned in a few well-studied cases where the gas appears to be warmer than the dust (e.g. Wilking & Lada 1983) or where the densities are insufficient to provide good thermal coupling between gas and dust (e.g. Evans et al. 1981); also, there now appears to be some reason to believe that a general elevation of the gas temperature may have taken place in such regions before the appearance of massive young stars (see Sections 3.2 and 3.4).

3. ORIGIN OF SUBSTRUCTURE IN MOLECULAR CLUMPS

3.1 Two Modes of Cloud Contraction

If we accept the importance of magnetic fields for molecular cloud support, there are logically two regimes of interest in the problem of star formation (see, e.g., the discussion of Mestel 1985):

1. In the *supercritical* regime, $M_{cl} > M_{cr}$, the clump's self-gravity can overwhelm the magnetic support *even if the fields remain frozen in the fluid*. Cloud evolution in this state is characterized by magnetically diluted collapse (e.g. Scott & Black 1980). The flow of gas along field lines (cloud flattening) would eventually provide subregions, of size comparable to the vertical dimension, with the same mass-to-flux ratio (on average) as the entire cloud; thus, these regions would themselves be supercritical and may be able to fragment from the contracting background (Mestel 1965).

2. In the *subcritical* regime, $M_{cl} < M_{cr}$, indefinite gravitational collapse (star formation) cannot be induced by *any* amount of increased external load (external pressure) if Φ is conserved (field freezing) because the mass-to-flux ratio M_{cl}/Φ remains fixed and subcritical. Neither can *dynamical*

fragmentation of the cloud take place. Cloud evolution in this state is most likely driven by ambipolar diffusion (e.g. Nakano 1979).

3.2 Supercritical Case: High-Mass Star Formation or High Star Formation Efficiency

The two cases discussed in Section 3.1 may provide a natural basis for the phenomenon of bimodal star formation (Shu et al. 1986). In particular, supercritical clouds may arise naturally when clumps are agglomerated in large cloud complexes (Shu 1987). The condition $M_{cl} > M_{cr}$ is equivalent to the existence of a critical surface density (see, e.g., Field 1970):

$$\frac{M_{cl}}{\pi R^2} > 80 \frac{M_\odot}{\text{pc}^2}\left(\frac{B}{30\ \mu\text{G}}\right). \qquad 20.$$

In the absence of other means of support, a cloud with a supercritical mass will suffer relatively rapid contraction and compress the embedded magnetic fields well above the starting values. Cloud contraction under these circumstances may be expected to efficiently form stars (see Section 1.5). If significant core heating also takes place (see Section 3.4), the stars that form could have relatively high masses (see Section 4.5). Although the contraction process can be expected to increase the mean surface density well beyond the starting value, we note that if one were to put even a healthy fraction of 80 $M_\odot$ pc^{-2} into O and B stars, the resulting areal luminosity densities would be $\sim 10^4$–10^5 $L_\odot$ pc^{-2}. This is close to the value seen in the region of the Trapezium stars in Orion.

For a canonical gas-to-dust ratio, Equation 20 is equivalent to a critical mean visual extinction:

$$A_V > 4\ \text{mag}\left(\frac{B}{30\ \mu\text{G}}\right). \qquad 21.$$

The value 30 μG may typify the average conditions only in the envelopes of small dark clouds; after gravitational contraction, dense cores and protostellar regions may have considerably larger values. For example, Simonetti & Cordes (1986) observe significant variations in the rotation measure toward L1551 that they interpret as being due to a magnetic field of strength 300 μG. In any case, it is interesting to note the following observed progression:

1. The visual extinction through the envelope of the Taurus molecular cloud is ~ 1–2 mag (e.g. Dickman 1978, Cernicharo et al. 1985). Its cores have $A_V \sim 10^1$ mag (Myers & Benson 1983, Cernicharo et al. 1984) and probably condensed from clumps that are subcritical (see below). The gas

temperatures in the cores are generally 10–11 K. Taurus is, of course, a region with low star formation efficiency and seems to be forming an unbound association of low-mass stars (see the review of T Tauri stars by Cohen 1984).

2. The visual extinction through the envelope of the ρ Ophiuchi molecular cloud is ~ 6 mag (e.g. Encrenaz et al. 1975, Elias 1978b, Frerking et al. 1982); the cores in its densest portion have $A_V \sim 10^2$ mag (Wilking & Lada 1983). The gas temperatures in the cores may be higher than in Taurus (Zeng et al. 1984), and ^{12}CO and ^{13}CO measurements indicate temperatures of 30–35 K for the general ρ Ophiuchi region (Loren et al. 1983, Wilking & Lada 1983). This portion of the cloud has a high star formation efficiency and may be forming a bound cluster containing mostly low-mass stars but also a few B stars (Grasdalen et al. 1973, Elias 1978b, Lada & Wilking 1984). Wilking & Lada (1983) observed CO line profiles that were asymmetrically self-reversed over the entire core region of the ρ Ophiuchi cloud (C. Lada, private communication; see also Encrenaz et al. 1975). The sense of this asymmetry corresponds to overall contraction at 1–2 km s^{-1}, consistent with the entire dense region being supercritical.

3. The average visual extinction through GMCs is controversial; estimates range from 4 mag (Blitz & Shu 1980) to 12 mag (Sanders et al. 1984, Solomon et al. 1987). High-angular-resolution studies near the water masers in W3(OH) suggest a young stellar object embedded in a dense core with $A_V \sim 10^3$ mag (Turner & Welch 1984). GMCs are very inhomogeneous, and portions of them are very likely to be supercritical; hence, it is informative that an inverse P Cygni profile has been found (after the submission of this review) in HCO$^+$ toward the brightest compact H II region in W49, which indicates that about 10^5 $M_\odot$ is collapsing on itself at 10–20 km s^{-1} to form the most luminous association of OB stars in the Galaxy (W. J. Welch, private communication). Since HCO$^+$ is a molecular ion, embedded magnetic fields must be dragged in by the infall. The gas kinetic temperature in the region is in excess of 50 K, which is typical of hot-core sites containing OB stars.

3.3 *Subcritical Case: Low-Mass Star Formation and Low Star Formation Efficiency*

When a cloud has less mass than M_{cr}, it can attain a stable quasi-equilibrium state by adjusting its size and shape in accordance with an external medium of finite pressure and/or finite magnetic field (Mouschovias 1976). However, such a cloud cannot remain forever in the same state because the neutral particles slip relative to the ions (their means of magnetic support), resulting in the quasi-static condensation of a dense pocket of gas and dust (Nakano 1979). If $M_{cl} > M_{cr}$, ambipolar diffusion will also take place, but against the backdrop of a dynamically shrinking envelope (Black & Scott 1982).

The physical principle of ambipolar diffusion is illustrated by the idealized problem of the quasi-static evolution of a plane-parallel, self-gravitating slab of isothermal gas that is lightly ionized (Shu 1983). As time increases, the field tends to decay to its background value, and the gas evolves toward the configuration that applies in the absence of a magnetic field, i.e. support against self-gravity by thermal pressure alone. With a large initial ratio of magnetic to thermal pressure, the outcome is the production of a small dense core in the background of a more extended envelope.

The one-dimensional problem is unrealistic because the gravitational field of the slab saturates at a finite value proportional to the total surface density. Thus, one-dimensional slabs with finite thermal pressures can never undergo true gravitational collapse. In a more realistic three-dimensional problem, where the original cloud clump contains many Jeans masses but M_{cl} is still less than M_{cr}, we may expect the clump to fragment *quasi-statically* into many "cores," each asymptotically evolving toward a configuration in which the field lines become almost uniform and straight (the background condition) and where the total velocity dispersions approach purely thermal values as the Alfvén velocity drops because of the increase in density. This picture provides an attractive evolutionary explanation for the small quiet cores that are observed in the Taurus dark cloud.

However, *stable* asymptotic support by thermal pressure alone in a quasi-spherical geometry is impossible if the potential reservoir of matter is very large in comparison with the Jeans mass in the initial state. Thus, a three-dimensional isothermal core must ultimately undergo gravitational collapse when the core becomes sufficiently centrally condensed. This has apparently happened to roughly half of the NH_3 cores in Taurus (Beichman et al. 1986).

The relatively long time scale for core formation by ambipolar diffusion (see Equation 17) probably explains why star formation is generally an inefficient process in subcritical molecular clouds. On the time scale of the dynamical collapse of an unstable molecular cloud core, star formation is badly synchronized. Newly formed T Tauri stars have ample time to turn on their winds and disrupt the incipient condensation of neighboring cores, which makes star formation in T associations loosely aggregated. There is no guarantee, however, that the coupling coefficient (Equation 17) is large compared with unity in all circumstances. Severe metal depletions or high initial mass-to-flux ratios (which, in a three-dimensional geometry, can lead to cloud contraction even in the absence of ambipolar diffusion) may well lead to relatively high star formation efficiencies.

3.4 Heating of Cloud Cores by Ambipolar Diffusion

The friction between ions and neutrals due to the ambipolar diffusion of the field is a source of heating for the cloud (cf. Scalo 1977, Mouschovias 1978, Elmegreen 1982b). Quantitative estimates of this heating (Lizano & Shu 1987) indicate that significant elevation of gas temperatures can occur in cores with high column density (see Figure 3). In particular, this heating mechanism provides a natural means by which the gas might become hotter than the surrounding dust, a characteristic of the ρ Ophiuchi region (Loren et al. 1983, Wilking & Lada 1983). In contrast, ambipolar diffusion yields little extra heating above that provided by cosmic rays ($T \sim 10$–11 K) for cases that simulate the ammonia cores in Taurus.

In Ophiuchus, the higher gas temperature cannot be due to heating of gas by warm dust grains (which are in turn heated by luminous stars) because the entire region only has three embedded B stars (Elias 1978b). In addition, the presence of large quantities of warm dust would produce far-infrared emission greatly exceeding that measured by balloon-borne observations (Fazio et al. 1976; see also Wilking 1985) and by IRAS (Young et al. 1986). The situation is more ambiguous for regions of true

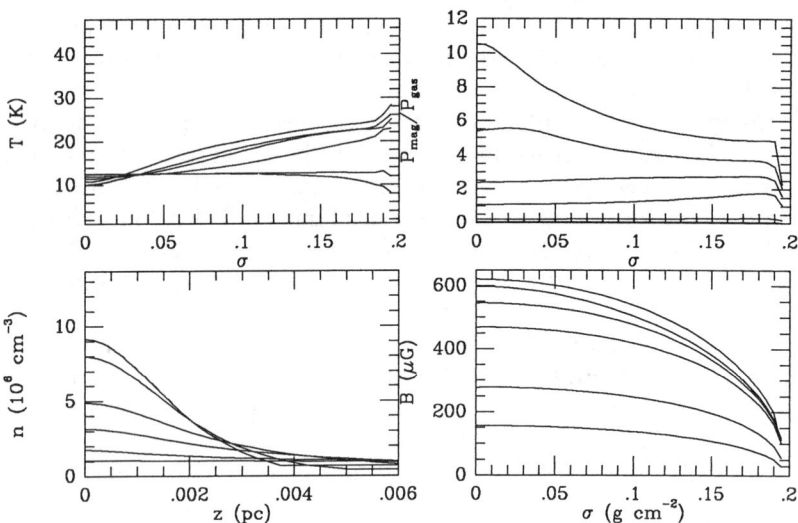

Figure 3 Time evolution of the temperature, magnetic to gas pressure, density, and magnetic field for a condensing cloud core with a column density corresponding to a (double-sided) visual extinction of 100 mag. The profiles shown occur at times 0, 0.06, 0.2, 0.3, 0.6, and 1.2 Myr (from Lizano & Shu 1987).

high-mass star formation. Additional observations made with high angular resolution are needed to help to sort out this problem.

4. GRAVITATIONAL COLLAPSE AND PROTOSTARS

4.1 *Gravitational Collapse of Molecular Cloud Cores*

At some point, as a cloud core gradually loses magnetic and "turbulent" support, the growing central concentration will cause it to become unstable to gravitational collapse. For stable (nonmagnetic and nonrotating) isothermal spheres bounded by an external medium of negligible inertia and constant pressure (i.e. a very hot gas), the maximum density contrast from center to edge is approximately 14 (Ebert 1955, Bonnor 1956). This result is, however, probably irrelevant to the present problem, where the core has access to a very large reservoir of cold matter and where the residual magnetic fields may greatly affect the stability criterion without appreciably modifying the equilibrium configuration (e.g. consider the role of quasi-uniform fields). A maximum contrast of 14 also contradicts the observational evidence that molecular clouds typically have H_2 densities ranging from 10^2 cm^{-3} to well in excess of 10^6 cm^{-3}.

4.1.1 INITIAL CONDITIONS AT THE ONSET OF COLLAPSE Consider a cloud of gas that is nearly isothermal and rotating very slowly (due to efficient braking), and suppose that the residual magnetic fields become increasingly unimportant for mechanical support of a core. Under these conditions, the continued *subsonic* evolution of the gas (with any mass) will tend to produce a $1/r^2$ density distribution (Bodenheimer & Sweigart 1968), provided that the thermal pressure gradients remain nearly in balance with the gravitational field.[4] In particular, a slowly forming molecular cloud core that is not artificially separated from its surroundings by a boundary will tend to acquire the density distribution of a singular isothermal sphere (see, e.g., Chandrasekhar 1939):

$$\rho = \frac{a^2}{2\pi G r^2}, \qquad\qquad 22.$$

where $a = (kT/m)^{1/2}$ is the isothermal sound speed, and T and m are the (constant) temperature and mean molecular weight of the gas, respectively.

---

[4] Numerical simulations of protostellar collapse ubiquitously establish $1/r^2$ density profiles, even when the initial state is taken to be homogeneous (e.g. Larson 1969a). The analytic explanation of this phenomenon given by Larson (1969b) and by Penston (1969a,b) is probably incorrect (Shu 1977; see, however, Hunter 1977) because it relies on the notion of (constant) *supersonic* inflow velocities at large r. Such an explanation is in conflict with the quasi-equilibrium condition that different parts of the cloud remain in good acoustic contact.

The configuration (Equation 22) is subject to the criticism (e.g. Whitworth & Summers 1985) that it is singular both at the origin, where the density is infinite, and at infinity, where the enclosed mass $M_{core}(r) = 2a^2 r/G$ increases without limit. Nevertheless, it has some desirable properties that make it a useful initial state in modeling the collapse of real molecular cloud cores. Like the many observational claims of $1/r^2$ density distributions surrounding forming stars (e.g. Bally & Lada 1983, Myers et al. 1987a, Keto et al. 1987), Equation 22 is a power law. As a consequence, although the system has a definite temperature, it has no characteristic density and, therefore, no characteristic Jeans mass. (It has one Jeans mass at each radius r, which is the only possible equilibrium accessible to a system with a virtually unlimited supply of matter.) Regarded in this sense, the infinite total mass merely simulates a situation in which the gas potentially available for star formation greatly exceeds the mass of the object finally formed. Similarly, the infinite density at the origin may be taken to represent an idealization of the tendency for magnetic decoupling to take place only at very high central densities (see, however, the controversy cited at the end of Section 2.3.6).

4.1.2 INSIDE-OUT COLLAPSE As long as the densities in a condensing molecular cloud core span several orders of magnitude before a stage of true dynamic instability is reached, its subsequent collapse properties should closely resemble those of the singular isothermal sphere (Equation 22). The latter can be found *analytically* (Shu 1977), thereby avoiding the problems of dynamical range associated with numerical simulations of protostar formation (cf. the central "sink cell" in the calculations by Boss & Black 1982). The solution at any instant in time looks like a stretched version of the solution at previous times. The self-similarity of the wave of infall, whose head propagates outward at the speed of sound, arises because of the lack of characteristic time and length scales; hence, except for scaling factors, all quantities must be functions of the similarity variable $x \equiv r/at$.

An accreting protostar of mass $M \equiv \dot{M}t$ is formed at the center, where $\dot{M}$ is the mass infall rate and t is the elapsed time since core collapse. Notice that this solution defines no characteristic mass scale that can be associated with the forming star. Instead of a mass scale, it is the (constant) *rate* $\dot{M}$ at which the star accumulates matter that is well defined (provided that T is well defined):

$$\dot{M} = m_0 a^3/G, \qquad 23.$$

where $m_0 = 0.975$. This result (Equation 23) is more robust than its formal derivation. The mass-infall rate of any critical cloud of gas of radius R

and mass M_{cl} must be given, in order of magnitude, by the estimate $\dot{M} \sim M_{cl}/t_{dyn}$, where $t_{dyn} = R/a$ and a is now the virial speed. If this cloud of gas was originally (marginally) supported against its self-gravity, then $a^2 \sim GM_{cl}/R$, which yields (upon elimination of M_{cl}) $\dot{M} \sim a^3/G$, independent of R [i.e. independent of any assumption about a (nonexistent) characteristic density]. In particular, the strong central concentration implied by Equation 22 is a basic consequence of self-gravitation and should almost always cause a spontaneous collapse to occur from inside-out. However, $\dot{M}$ will not necessarily be a constant in all cases; its variation with time will depend on the adopted boundary conditions and other assumptions (see, e.g., Zinnecker & Tscharnuter 1984).

To summarize, in any general configuration that is initially near critical equilibrium, the mass-infall rate can be expected to be given by Equation 23, within factors of order unity, as long as the calculation of the characteristic or signal speed a appropriately accounts for *all* mechanisms of mechanical support. Thus, if we use α and β to denote the ratios of the (nonfluctuating) magnetic pressure $B_0^2/8\pi$ and the (characteristic) turbulent pressure $\rho\langle\delta v^2\rangle$ to the thermal pressure $\rho kT/m$, respectively, we are tempted to identify a with the generalized magnetosonic speed (Stahler et al. 1980a):

$$a = [(1+2\alpha+\beta)kT/m]^{1/2}. \qquad 24.$$

Given the likely anisotropic nature of the magnetic and "turbulent" support mechanisms, however (see Section 2.3.2), Equation 24 must represent an overestimate for the actual "effective speed of sound."

In this context, it is interesting to note that the one-dimensional simulations of core formation by ambipolar diffusion (see Section 3.3) tend to produce almost *spatially* constant values of α (of order unity when magnetic fields are still important for core heating). On the other hand, multidimensional models of clouds that flatten as they contract under a constraint of field freezing show a tendency for α (of a given fluid element) to stay *temporally* constant (Mouschovias 1976, Scott & Black 1980). If the turbulence is further constrained to be sub-Alfvénic ($\beta < 2\alpha$), the maximum correction for a in cloud cores from its purely thermal values cannot exceed a factor of ~ 2 (at least, for subcritical clouds), a conclusion that seems empirically validated by an examination of molecular linewidths (Goodman 1987). Unfortunately, certain key formulae (e.g. Equations 23 and 25) are very sensitive to the parameter a. Realistic MHD calculations of cloud core collapse are needed if we are to replace heuristic devices like Equation 24 with honest a priori estimates. Until such estimates are available, the only pragmatic approach is to regard a as a free parameter that is to be determined by best empirical fits to the observational data (see, e.g., Adams et al. 1987).

4.1.3 THE EFFECTS OF ROTATION

The effects of (slow) rotation at an initial uniform rate Ω can be incorporated in a rigorous treatment using singular perturbation theory (Terebey et al. 1984). The axisymmetric time-dependent collapse, with variations in two spatial dimensions, can be followed analytically if the initial state is taken to be a slowly rotating, singular isothermal sphere. Ahead of the expanding wave of infall but where the equilibrium rotation is still subthermal, $at < r < a/\Omega$, the density distribution very nearly satisfies Equation 22. For r much less than at and much greater than a centrifugal radius,

$$R_C \equiv G^3 M^3 \Omega^2 / 16 a^8 \qquad 25.$$

(where $M \equiv \dot{M}t$ is now the mass of the central star plus disk), the density law has an approximate free-fall form:

$$\rho = \frac{\dot{M}}{4\pi (2GM)^{1/2}} r^{-3/2}, \qquad 26.$$

appropriate for quasi-steady infall on nearly radial streamlines. Inside of R_C (the position where infalling matter in the equatorial plane encounters a centrifugal barrier if it conserves its initial specific angular momentum), the isodensity contours are highly flattened and the rotationally modified similarity solution asymptotically joins onto free-fall ballistic trajectories (see, e.g., Ulrich 1976), which are parabolas in the limit of a very concentrated central mass distribution. When R_C is greater than the stellar radius R_*, a nebular disk forms around the protostar, whose outer dimension extends at least to R_C (and beyond if the disk has internal mechanisms for appreciable transport of mass and angular momentum in an infall time scale $M/\dot{M}$; see Cassen & Moosman 1981). Averaged over all angles, the density outside the disk between R_C and R_* behaves approximately as $\rho \propto r^{-1/2}$ (Chevalier 1983, Adams & Shu 1986).

To fix ideas, consider a typical numerical example: a molecular cloud core having negligible magnetic field and "turbulence" with $T = 10$ K, $m = 2.3\, m_H$, and $\Omega = 1$ km s^{-1} pc^{-1}. With these values of T and m, Equation 24 implies that $a = 0.2$ km s^{-1}; Equation 22 then predicts that densities $\sim 3 \times 10^4$ cm^{-3} should be reached at $r \sim 10^{17}$ cm (in reasonably good agreement with the NH$_3$ maps of Taurus cores by Myers & Benson 1983); and Equation 23 yields $\dot{M} = 2 \times 10^{-6}\, M_\odot$ yr^{-1}, which would build up a central mass $M = 0.5\, M_\odot$ in $t = 8 \times 10^{12}$ s, at a time when a stellar wind may already have started to blow (see Section 4.4). Infall speeds on the order of 1 km s^{-1} would be acquired only for $r \sim 10^{16}$ cm (see Figure 3b of Shu 1977) and would be detectable by the current generation of single-dish millimeter-wave telescopes only through special spectroscopic techniques (see, e.g., Walker et al. 1986). With the adopted value of Ω, Equation 25 gives $R_C = 45$ AU, which would be in rough agreement with the sizes of dust disks inferred to surround T Tauri stars (see Section 6.2).

The above description gives a representative accounting of the (isothermal) collapse dynamics for situations where centrifugal support of the collapsed object arises at scales that are small in comparison with the original dimensions of the interstellar gas mass from which it formed. Thus, the results can probably be applied to the generic problems of the formation of single stars with planetary systems and (ultimately) binary stars with relatively short orbital periods (e.g. less than 10^3 yr). The fragmentation of cloud cores with large original rotation rates, or of clumps with supercritical initial mass-to-flux ratios, requires other approaches (see Section 1.5).

4.2 Evolution of High-Mass Protostars

As long as the relevant regions remain reasonably optically thin, the core collapse proceeds nearly isothermally (see, e.g., the discussion of Hayashi 1966). Indeed, the rapid loss of any compressional heat generated guarantees the existence of a dynamical phase in protostellar evolution (Cameron 1962, Gaustad 1963, Gould 1964). The runaway increase of the density at the center is halted only after the formation of an optically thick protostar (of mass $\sim 10^{-3}$–10^{-2} $M_\odot$) whose thermal time scale exceeds the sound crossing time. The protostar then grows hydrostatically (after a transient phase associated with the dissociation of molecular hydrogen; see Larson 1969a) by accreting matter from infalling gas and dust.

Certain general points concerning the structure of protostars were recognized early—for example, protostars of high mass should evolve differently from those of low mass. The reason is that the protostar changes on a Kelvin-Helmholtz time scale GM_*^2/R_*L_*—the time required to reach internal thermal equilibrium—whereas the infall takes place at a rate that is not highly variable. For a massive protostar the Kelvin-Helmholtz time scale will be shorter than the infall time scale, and the star will reach the main sequence while still gaining mass from the infalling envelope (Kahn 1974; see also Yorke & Krugel 1977). This conclusion seems in reasonable accord with the observations (see the review by Wynn-Williams 1982).

A complication arises because the radiation pressure acting on dust grains may become large enough to halt the infall. Indeed, Larson & Starrfield (1971) and Kahn (1974) proposed this as a possible mechanism (in addition to expanding H II regions) to set an upper limit on stellar masses; however, the observed upper limits (see, e.g., the review by Humphreys & Davidson 1984) disagree with the best theoretical value using modern dust opacities. Within factors of order unity, radiation pressure acting on grains can reverse spherical infall if the ratio of luminosity to mass, L/M, of the central source exceeds the critical value,

$$L/M = 6\pi cG/\kappa_P(T_d) \approx 700 \ L_\odot/M_\odot, \qquad 27.$$

where $\kappa_P(T_d)$ is the Planck-mean of the opacity at the dust destruction temperature and has been taken to be 30 cm^2 g^{-1} for standard dust abundances. Main sequence stars heavier than 7 $M_\odot$ have L/M that exceed this critical value. A proposed escape from this dilemma involves the postulate that massive stars form only in regions relatively depleted of dust, especially graphite grains (Wolfire & Cassinelli 1987). Another possibility is that the infall is highly nonspherical because of rotation, which thereby allows matter to fall into a disk (which is shielded from the direct radiation of the star) at radii substantially larger than the dust destruction front. Accumulation of a high-mass star may then proceed via disk accretion. The newly born OB star may later reveal itself optically, like its low-mass counterparts, by producing a stellar wind that removes the surrounding placenta of gas and dust. Another argument against terminating the infall by radiation pressure (as contrasted with a wind) is that the radial momentum carried in a typical cold outflow (probably swept-up molecular gas) exceeds the momentum contained in all the photons that have ever left the star by ~ 2–3 orders of magnitude (see the review of Lada 1985).[5]

4.3 *Evolution of Low-Mass Protostars*

The longer Kelvin-Helmholtz contraction times associated with lower stellar masses imply that they may end their accretion phase without having reached the main sequence. To determine where in the Hertzsprung-Russell diagram such objects might first become optically visible requires accurate and detailed computations of the ensuing radiative hydrodynamics. However, the collapse of an interstellar cloud is highly non-homologous (McNally 1964); in order to follow the thermal and mechanical structure of the central protostar, the calculation must include a range of densities spanning more than 20 orders of magnitude. This enormous dynamical range strained the capabilities of most numerical attacks on the problem, even in its simplest spherical formulation, and it led to considerable early controversy (Larson 1969a, 1972, Narita et al. 1970, Hayashi 1970, Appenzeller & Tscharnuter 1974, 1975, Westbrook & Tarter 1975, Kondo 1978, Tscharnuter & Winkler 1979).

Improvements in numerical techniques (Winkler & Newman 1980a,b) and reformulation of the computational problem to take advantage of

---

[5] Multiple scatterings in a dust envelope cannot help because the photons would be degraded (by true absorption and reradiation) to the far-infrared; the photons can then easily escape before experiencing the required number of scatterings.

certain analytical insights (Stahler et al. 1980a,b, 1981) have resolved most of the controversy in the spherical case. In particular, the radius R_* of a protostar that grows in mass at a given rate $\dot{M}$ is now known with good accuracy. When the freely falling material in the envelope encounters the hydrostatic surface of the star (at R_*), it dissipates its energy in an accretion shock, producing the luminosity

$$L = GM\dot{M}/R_*. \qquad 28.$$

Notice that Equation 28 assumes that essentially all of the kinetic energy of the infall is converted into radiation; this holds to a high degree of approximation in the low-mass protostar problem. If, instead, a substantial fraction of the infall energy were to be converted into internal energy of the gas, the resulting stellar radius would be much larger than the computed value of a few solar radii.

The formula (Equation 28) yields the system luminosity only for low-mass protostars with spherical symmetry. For high-mass protostars, the interior luminosity will dominate the shock luminosity. Also, for a low-mass protostar with rotation, the (infall) luminosity will be lower (for given $\dot{M}$) if angular momentum conservation prevents material from falling all the way to the stellar surface and if part of the gravitational energy released can be stored as rotational energy in the star or in an accompanying circumstellar disk (Adams & Shu 1986).

4.4 *Termination of the Infall*

In the spherical calculations for a mass infall rate $\dot{M} = 1 \times 10^{-5} M_\odot \text{ yr}^{-1}$ (Stahler et al. 1980b), the protostar accumulated matter (processed through an accretion shock) of ever increasing specific entropy. Hence, the star remained radiative until the ignition of deuterium, which occurred when the stellar mass was $\sim 0.3 M_\odot$. A convection zone then spread outward through the star until the protostar became almost entirely convective at a mass of about $0.5 M_\odot$. Except for this event, nothing happened to distinguish any particular mass scale for the accreting protostar; in the actual calculations, the infall was terminated artificially after $1 M_\odot$ was accumulated.[6] After a short period of readjustment, the newly formed star then descended a convective pre-main-sequence track (Hayashi et al. 1962), developed a radiative interior, and followed a radiative track to the main sequence (Henyey et al. 1955).

The above scenario resurrects the fundamental issue of star formation outlined in Section 1.1. In the present context, we may pose the question

---

[6] As commented upon by the authors, however, models for lower final masses can be simply recovered from the published calculations by merely terminating the accretion earlier (see, e.g., Stahler 1983).

as follows: What mechanism actually determines the end of the infall phase? The large reservoir of gas in molecular clouds and the general inefficiency of star formation make it unlikely that the star continues to gain mass until there is no available material left. A more plausible explanation is that stellar winds eventually reverse the inflow and halt the accretion (see, e.g., Shu & Terebey 1984, Lada 1985, Myers et al. 1987b). This viewpoint, however, raises another important issue—what triggers the onset of these winds?

4.5 Deuterium Burning and the Stellar Birthline of Low-Mass Stars

An empirical clue for low-mass stars exists in the Hertzsprung-Russell (H-R) diagrams of T Tauri stars (Cohen & Kuhi 1979). Stahler (1983) was the first to note that H-R diagrams for very different stellar groupings—Taurus-Auriga, Orion, NGC 7000/IC 5070, and Ophiuchus—all tended to show the same upper envelope (see Figure 4). It was as if pre-main-sequence stars first appeared optically visible on a universal "birthline." Stahler's explanation for this amazing fact in terms of a special accretion rate in spherical geometry was criticized by Mercer-Smith et al. (1984). Shu (1985) pointed out that the observed "birthline" could be understood

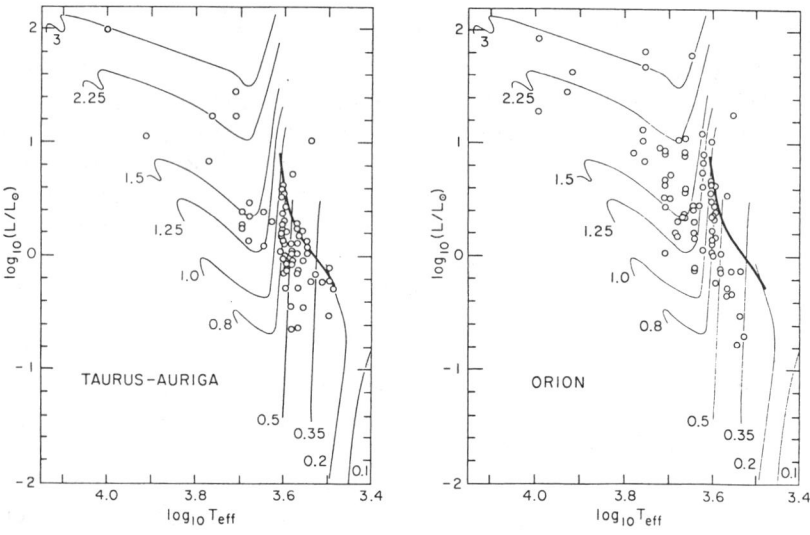

Figure 4 Hertzsprung-Russell diagrams from Cohen & Kuhi (1979) showing theoretical pre-main-sequence contraction tracks and T Tauri stars in the Taurus-Auriga and Orion cloud complexes. The heavy solid curve is the theoretical "birthline" of Stahler (1983).

as a condition that the star must be able to burn deuterium, thereby giving a connection with the fusion capabilities of bona fide stars. In a low-mass star, deuterium burning can drive convection, which, when coupled with the presence of differential rotation, can produce dynamo action and generate magnetic activity (Parker 1979). The released energy (which had been stored in rotation) may then power the intense stellar surface activity observed in YSOs. The resulting stellar wind eventually reverses the infall and thereby defines the mass of the formed star at the point when it first becomes optically visible.

For any given mass, there is a unique radius for which a completely convective star will have a central temperature (Chandrasekhar 1939),

$$T_c = 0.54 G M_* m / k R_*,\qquad 29.$$

high enough for deuterium to burn. In Equation 29 we have assumed that the convective star is an ideal gas of constant mean molecular weight m (i.e. a polytrope of index 1.5). For a completely ionized gas of cosmic abundances, $m = 0.62\, m_H$, and for $0.01\, M_\odot < M_* < 2\, M_\odot$, we can obtain an approximate criterion for deuterium burning by setting $T_c = 1 \times 10^6$ K in Equation 29:

$$R_*/R_\odot \approx 0.15 + 7.6(M_*/M_\odot),\qquad 30.$$

where the small correction term 0.15 (and the slight readjustment of the coefficient 7.6) on the right-hand side enters to take into account the effects of partial degeneracy at low stellar masses (cf. Nelson et al. 1986). Equation 30 produces a birthline in the H-R diagram for pre-main-sequence stars that depends only on the physics of stars and not on the details of their formation. Given the observational uncertainties and the possibility that the stellar luminosity can be contaminated by radiation from an active disk (especially at low masses; see Sections 6.4 and 6.5), the predicted birthline agrees reasonably well with the observed upper envelope for T Tauri stars.

The rate of stellar mass accumulation, $\dot{M}_*$, affects only the timing of deuterium ignition (i.e. the eventual mass of the star, M_*, on the birthline). Higher rates of stellar mass accumulation allow less time for the radiation of the stellar binding energy, which leads to larger values of R_* at every M_* and thus produces higher masses before the onset of deuterium burning. The final stellar masses will also depend on whether the infall is sufficiently intense to suppress a breakout of the stellar wind despite the convection induced by deuterium burning, and on the amount of stellar accretion that occurs from a disk. In any case, for spherical stars that accumulate $\sim 2\, M_\odot$ before deuterium ignites, the extra interior luminosity released by deuterium burning can be carried out by radiative diffusion without the inducement of convection. [See the discussions of Cassen et al. (1985) and of Stahler et al. (1986).] Therefore, the termination of stellar accumulation for high-mass stars must occur by a different process than that outlined above (possibly deuterium burning in a pseudodisk).

4.6 Emergent Spectral Energy Distributions of Protostars

A test of any theory of star formation is to compare the theoretical predictions of the emergent spectral energy distributions of protostars with actual observations. Historically, calculations of emergent protostellar spectra have fallen into two basic categories. The parametric approach ignores the hydrodynamic portion of the problem, assumes a known central source of luminosity, constructs a series of model dust shells, and conducts a survey of the resulting parameter space. Such a strategy has been used to study emission from dust shells around late-type stars (e.g. Jones & Merrill 1976), from stars embedded in extended dust envelopes (Rowan-Robinson 1980), and from protostellar dust clouds forming high-mass stars (Yorke & Shustov 1981). The self-consistent approach begins with a hydrodynamic model of protostellar collapse (which includes an internally determined central source of luminosity) and either solves the radiative transfer problem simultaneously or postprocesses the results of a rigorous collapse calculation. Such an approach typically has fewer adjustable parameters and has been used to study the dynamical evolution and spectral appearance of protostars of low mass (Larson 1969b, Bertout 1976), intermediate mass (Yorke 1979, 1980), and high mass (Yorke & Krugel 1977, Yorke 1977, Wolfire & Cassinelli 1986). All of these basic models have been restricted to spherical symmetry. Recently, a fast approximate technique has been developed, which is applicable to multi-dimensional situations as well (Adams & Shu 1985, 1986).

From the parametric studies, given a known central source, the emergent (infrared) spectra are found to be sensitive primarily to the total optical depth of the dust envelope and to the optical properties of the dust grains (usually taken to be a mixture of graphite and silicate grains), which provide the dominant contribution to the opacity κ_ν. If the dust envelope is sufficiently optically thick so that most of the original radiation from the central source is absorbed and reradiated by the grains, the resulting emergent spectra consist of a single broad hump of emission with a peak in the far-infrared (~ 60–$100\ \mu$m), and absorption features at $10\ \mu$m (from silicate grains) and $3.1\ \mu$m (from water ice). If the column density through the dust envelope is not large, a substantial fraction of the central source radiation can escape, and a double-peaked spectrum is produced. In all cases, the emergent spectra are broader than that of a single-temperature blackbody. This is a general feature of "extended atmospheres" (Mihalas

1978) and implies limited utility to the concept of a dust "photospheric temperature."

Good infrared opacities for dust grains have become available in recent years through better observational constraints (see Hildebrand 1983, and references therein) and through better self-consistent calculations (see Draine & Lee 1984). On the other hand, our theoretical knowledge of the properties of the central source and of the total optical depth through the cloud core to the dust destruction front (along any given line of sight in rotating models) has also greatly improved (see previous section). Thus, detailed comparison of theoretical predictions with specific protostellar candidates can go substantially beyond the purely parametric studies because the adjustable parameters can be related to measurable mechanical properties of the source that are independent of the observed spectrum. In addition, far-infrared observations by the IRAS satellite (e.g. Beichman et al. 1986) and follow-up ground-based observations in the mid- and near-infrared (such as those of Myers et al. 1987a) have considerably expanded the relevant data base. Thus, it is encouraging that theoretical calculations based on the collapse models of slowly rotating cloud cores discussed in Section 4.1.3 and the protostars discussed in Section 4.4 yield spectral energy distributions that are in close agreement with observations of a class of low-luminosity infrared sources—those that have negatively steep spectra in the near- and mid-infrared and are found near the centers of dense molecular cloud cores (Adams et al. 1987; see Figure 5 for an

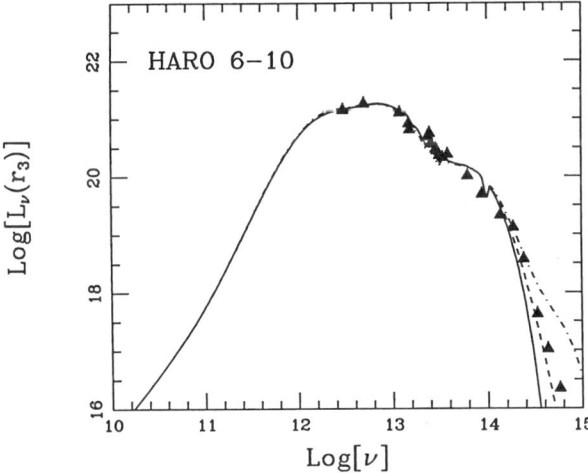

Figure 5 Model fit to spectral energy distribution (cgs units) of Haro 6–10. The solid curve assumes purely absorbing dust grains; the dashed-dotted and dashed curves make different heuristic corrections for scattering (adapted from Adams et al. 1987).

example). Major theoretical uncertainties remain concerning the appropriate values to adopt for the efficiencies η_* and η_D with which the energy of rotation in dissipated in the star and disk (see Adams & Shu 1986); however, within reasonable bounds, the inferred values of the parameters of the rotating cloud core, a and Ω, and the resulting mass M_* and radius R_* of the central star, seem consistent with current observational knowledge of these quantities (see also the discussion of Myers et al. 1987a).

5. BIPOLAR OUTFLOWS FROM YOUNG STELLAR OBJECTS

5.1 *Basic Energetics*

The growing base of observational data concerning outflows from YSOs suggests that all stars pass through an outflow stage as a fundamental part of the star formation process (see the reviews by Lada 1985, Welch et al. 1985, and Snell 1986). The outflows are highly energetic (observed kinetic energies are $\sim 10^{43}$–10^{47} erg) and exhibit a wide variety of observable phenomena. Related manifestations include Herbig-Haro (H-H) objects (see the review by Schwartz 1983), high-velocity water maser sources (see the review by Reid & Moran 1981), and "cometary" reflection nebulae (e.g. Strom et al. 1985, Goodrich 1987). Some level of collimation usually exists for the outflows; the most notable examples are massive bipolar flows of cold molecular gas (e.g. Snell et al. 1980), and long and narrow optical jets (e.g. Mundt & Fried 1983). The optical jets (often linear chains of H-H objects) can often be traced very close to the exciting (low-mass) stars and almost certainly originate from them. The molecular outflows (which occur in YSOs of all masses) may also be driven by strong stellar winds, but this suggestion is more controversial. The case for stellar winds is supported by the claim that the CO gas in the bipolar lobes of L1551 exists in spatially thin shells (Snell & Schloerb 1985, Kaifu 1986).

The best-known alternative possibility—centrifugally driven flows (Pudritz & Norman 1983) or magnetohydrodynamically driven flows (Uchida & Shibata 1985, and references therein) from large ($\sim 10^{17}$ cm) and massive ($\sim 10^2$ $M_\odot$) circumstellar disks—has severe mechanistic drawbacks. In the Pudritz & Norman model, in order to get molecular outflows with a speed $\sim 10^1$ km s^{-1} from the parts of disks where the typical rotational velocities are $\sim 10^0$ km s^{-1}, Alfvénic enforcement of corotation (constant angular velocity) must occur over a range of radii that typically spans a factor of 10. Since the moment of inertia per gram of the outflow gas is then larger (by a factor of 10^2) than that of the material in the disk that ejects it, the back reaction (i.e. the conservation of total angular momentum) produces a mass flux through the disk (leading to protostellar accretion) that is typically 10^2 times larger than the mass outflow. Since the observed *molecular* outflows (which in a more conventional interpretation would be swept-up cloud gas rather than material from a circumstellar disk) involve masses and time scales of the order of solar masses and 10^4 yr, respectively, the deduced disk accretion rates are $\sim 10^{-2}\ M_\odot\ \mathrm{yr}^{-1}$ or larger. This severe requirement leads to some unusual conclusions (see, e.g., Table 1 of Pudritz & Norman 1986): The disk masses must be $\sim 10^2\ M_\odot$ or larger (to be able to last 10^4 yr), the central protostars must be built up very quickly (at a rate that would make them very transient objects), and they must have very small mass-to-radius ratios ($\sim 10^{-3}$ in solar units) in order to avoid an overly large accretion luminosity from feeding material into a deep gravitational potential well. These requirements are not substantiated, to our knowledge, by any known properties of molecular cloud structures or the stars that they form.

The problem is, of course, that very big disks (size $\sim 10^{17}$ cm) have very little specific energy; thus, to use them to drive the observed molecular outflows requires making them very massive. Although spatially large disks have often been claimed in the observational literature (e.g. Kaifu et al. 1984, but see Batlra & Menten 1985), the only well-substantiated case, S106 (which is both highly flattened and rapidly rotating), has a probable mass of 2 $M_\odot$ (Bieging 1984), and the star at its center has a (large) mass-to-radius ratio (judging from its luminosity and ionizing flux) appropriate to a main-sequence star of $\sim 20\ M_\odot$. The paucity of specific energy contained in spatially big disks makes them unlikely candidates for the power behind bipolar flows.

In all self-gravitating systems, most of the specific binding energy is contained at the smallest radii. Rapidly rotating protostars, which are $\sim 10^6$ times smaller than 10^{17} cm, potentially have 10^6 times more energy per gram that can be dissipated to power winds. On energetic grounds alone, therefore, the model of Draine (1983), which makes magnetohydrodynamic use of a rotating protostar, is preferable to that of Uchida & Shibata, which is an otherwise beautiful numerical simulation. A problem with Draine's model is that it requires a tightly wrapped magnetic field configuration for the initial state, which may be difficult to produce from a smoothly evolving scenario for the formation of the star.

In any case, supplying enough mechanical energy is not a problem for young stars, but supplying enough (radial) momentum may be. The momenta contained in the measured ionized winds from YSOs are insufficient to account for the cold molecular flows by one to two orders of magnitude (see, e.g., Levreault 1985); a heretofore undetected neutral component of the stellar wind may be necessary (Rodriguez & Canto 1983, Snell et al. 1985). In the famous bipolar flow source L1551, indirect evidence for hidden mechanical luminosity in excess of the known stellar wind (and jet) may have been found in the presence of far-infrared emission that is too strong and too extended for direct heating by the central protostar, IRS 5 (Clark & Laureijs 1986, Edwards et al. 1986, Clark et al. 1986).

5.2 Influence of Outflows on the Surroundings

It has been proposed that outflows from young stars represent an important input that maintains the "turbulence" in molecular clouds (see the discussion in Section 2.3.1). In further developments along such lines, feedback from such effects has been proposed as an important step in the self-regulation of star formation (e.g. Norman & Silk 1980, Franco 1984). In such scenarios, star formation will affect the subsequent evolution of the parent molecular clouds, which in turn will affect further star formation: cloud → protostar → outflow → cloud → .... However, the detailed manner in which these schematic arrows work requires more elucidation.

Stellar outflows are also useful as diagnostics of their surroundings. For example, warm H_2 molecules moving at high velocities have been amply detected in regions containing molecular outflows [e.g. Lane & Bally (1986); see Shull & Beckwith (1982) for a review]. These observations are most simply understood in terms of shock-accelerated gas produced by the interaction of powerful stellar winds with the ambient medium. However, ordinary gas-dynamic shocks of high speed should dissociate the hydrogen molecules (see the review of McKee & Hollenbach 1980). Since the molecules are left intact, warm, and radiating, they must be accelerated smoothly in magnetized "C shocks" (Draine 1980, Chernoff et al. 1982), an ungentle version of ambipolar diffusion. The required magnetic fields typically correspond to strengths of ~ 1 mG in regions where the ambient density is $\sim 10^6$ cm^{-3}.

The properties of outflows have also been used to deduce the density of the ambient gas before it was swept up into a thin shell. Dividing the total mass of swept-up gas by the volume of the lobes (as computed from the geometric mean of the major and minor diameters) in a number of flow sources, Bally & Lada (1983) show that the statistics are consistent with the radial distribution of gas density falling off as $1/r^2$ in star-forming regions.

5.3 Collimation Mechanisms

Proposed mechanisms for collimating stellar jets and bipolar molecular flows fall into two generic categories. One school of thought holds that outflows are produced in a well-collimated form near the star (e.g. Hartmann & MacGregor 1982, Jones & Herbig 1982). The other viewpoint holds that an initially isotropic stellar wind is channeled into a bipolar form by an anisotropic distribution of matter in the surrounding circumstellar environment (e.g. Canto et al. 1981, Konigl 1982, Torrelles et al. 1983; but see Heyer et al. 1986). Both possibilities in a very general sense may arise naturally (as a progression in time) in the scenario for star formation posed in Section 4 (Shu & Terebey 1984, Shu et al. 1986).

The basic idea is that during the infall phase, the swirling inflowing matter acts like the lid of a pressure cooker and suppresses *any* incipient stellar outflow. As the lid weakens (because the gas falls increasingly onto the disk rather than directly onto the star), breakout will occur through the channel of least resistance (the safety valve) at the rotational pole(s) of the accreting protostar, where the ram pressure of the infalling material is least during the period when R_C is not much larger than R_*. Since the system is smoothly transforming from an inflow state to an outflow state, breakout in the absence of perfect spherical symmetry should first occur at one point (on each hemisphere). Thus, the existence of collimated, dual-exhaust jets in the youngest outflow objects becomes a natural part of protostellar evolution. Later, the stellar outflow may widen and become nearly isotropic, but the swept-up shell of gas may still make less outward progress in the equatorial directions because of the larger amounts of obstructing (inflowing) gas (see Figure 6*d* of Terebey et al. 1984). Detailed modeling is needed to provide quantitative comparisons with observations.

In the interim, it is significant that the rotating infall models described in Section 4.5 correctly predict the emergent spectra for some infrared sources that are known bipolar outflow sources [see Adams et al (1987) for examples]. This finding, coupled with the observation that well-collimated sources tend to be heavily embedded objects, supports the idea that they represent systems in which inflow and outflow are taking place *simultaneously*. In this view, a bipolar flow source is the transitional phase of evolution between a purely accreting protostar and a fully revealed pre-main-sequence star. A configuration of combined inflow and outflow has been claimed on completely different grounds for the Becklin-Neugebauer object in Orion (see the review of Scoville 1985), which is probably a case of massive star formation.

If magnetic activity plays an intrinsic role in the generation of the stellar winds from low-mass protostars (e.g. Lago & Penston 1982; see also Section 4.4), the outflows will carry away material with a higher specific angular momentum than the average for the star. Intense mass loss will naturally produce intense magnetic braking. Thus, the slow rotation rates observed for (revealed) T Tauri stars (Vogel & Kuhi 1981, Bouvier et al. 1986, Hartmann et al. 1986), which were quite surprising when first discovered, can now be attributed to an earlier epoch of heavy mass loss.

6. VISIBLE YOUNG STELLAR OBJECTS

Since high-mass stars are believed to join the main sequence while still in the accretion phase (see Section 4.2), they do not convey much information concerning their origins when they become visible as optical objects. The study of pre-main-sequence stars is thus operationally confined primarily to low-mass stars, although YSOs of intermediate mass, the Herbig Ae and Be stars (Herbig 1960, Strom et al. 1975, Finkenzeller & Mundt 1984), deserve more attention.

6.1 *T Tauri Stars*

Soon after their discovery (Joy 1942, 1945, 1949), T Tauri stars were recognized as being a very young population of (low-mass) stars, newly born within the dark clouds with which they are invariably associated both spatially and kinematically (e.g. Herbig 1977). The optical spectra of T Tauri stars are characterized by high variability and prominent emission lines, most notably the Balmer series of hydrogen and Fe II lines (see Herbig 1962). The underlying photospheres are relatively cool, usually less than 6000 K and typically ~ 4000 K. The strength of the ultraviolet emission lines (and continuum excesses) have traditionally been interpreted to indicate heavy chromospheric activity (see, e.g., Calvet et al. 1984, Herbig & Goodrich 1986), although the boundary layer associated with an accretion disk is another possibility (Bertout 1987). T Tauri stars have strong mass loss (Kuhi 1964), and periodic components to the light variations suggest that these stars have very inhomogeneous surfaces, with an extensive network of magnetically active and quiescent zones (e.g. Herbig & Soderblom 1980, Bertout et al. 1987). Thus, T Tauri stars seem to be bubbling with surface activity, which together with appreciable surface lithium abundances (Herbig 1960) is yet another indication of their extreme youth. Since a comprehensive review of the properties of T Tauri stars is available elsewhere (Cohen 1984), we concentrate here on the characteristics of T Tauri stars that bear most directly on the theory of low-mass star formation.

T Tauri stars presumably represent the postinfall phase of young star evolution. When the object becomes optically revealed, has its bolometric luminosity measured, and can be given a spectral type, it can then be placed in the H-R diagram. The task has been completed for ~ 450 T Tauri stars in an extensive survey by Cohen & Kuhi (1979). Their results indicate that T Tauri stars are mostly descending convective tracks, in apparent accord with the classical pre-main-sequence theory of Hayashi et al. (1962; see also Iben 1965, Ezer & Cameron 1965). The implied mass range of these objects is 0.2–3 $M_\odot$; the formal contraction ages range from $\sim 2 \times 10^5$ to 2×10^7 yr.

The activity of T Tauri stars may result simply from the adjustment needed to accommodate material recently acquired from the interstellar medium to stellar conditions. A finding that supports this general point of view is the observed decay, as T Tauri stars age, in the properties that earmark these objects as a distinct class: surface activity, infrared excesses, and emission lines (Cohen et al. 1987, Walter 1987).

The overall spread of ages ($\sim 10^7$ yr) in a cloud like Taurus is consistent with the theory of molecular cloud core formation by ambipolar diffusion outlined in Section 3.3. The most recently born stars require corrections to take into account the dynamical stages of protostellar evolution (e.g. Stahler 1983). The derived "ages" are further uncertain to the extent that a substantial contribution to the bolometric luminosity may not originate with the star but with an active accretion disk (see Section 6.5). Nevertheless, within a factor of 2 or so, the "contraction ages" (or, more precisely, the locations in the H-R diagram) of the youngest T Tauri stars are consistent with the mass accumulation rates implied by the core collapse scenario discussed in Section 4.1.2.

Luminosity overestimates would introduce mass overestimates if the stars are on the radiative portions of their tracks. Mass determinations for T Tauri stars may, therefore, be somewhat uncertain at the highest values. Since convective tracks are nearly vertical in the H-R diagram, variations in the stellar luminosity estimates produce little error in the mass determinations at low values. More serious for the latter are the assignment of accurate effective temperatures for objects that have strong emission lines, anomalous continuum fluxes, and (sometimes) composite spectral types. It would be important to check whether, within the uncertainties, the masses and radii of T Tauri stars really are consistent with their protostellar progenitors having had their accretion phases halted by the onset of deuterium burning (see especially the discussion of the "birthline" in Section 4.4).

6.2 *Dusty Nebular Disks*

In recent years there has been growing evidence for dusty disks of size $\sim 10^2$ AU around nearby stars. The high degree of optical linear polarization from a few sources (in excess of 10%) has been interpreted (Elsasser & Staude 1978) as scattering from an elongated or flattened dust distribution [one that partially extinguishes the starlight (i.e. is not completely spatially thin; otherwise, the direct unpolarized component coming from the star would reduce the degree of polarization)]. More recently, IRAS observations and follow-up ground-based work have produced resolved images of disklike structures around somewhat older stars (Aumann et al. 1984, Smith & Terrile 1984). Lunar occultation studies at infrared wavelengths also reveal similar elongated configurations around stars newly born from molecular clouds (Simon et al. 1985).

The star HL Tau has proven to be especially intriguing. One unique aspect of this T Tauri star is the presence of a deep water-ice feature at 3.1 μm in its spectrum; this feature has been interpreted (Cohen 1975) as evidence indicating that the object is especially young. Further observations of the unusual infrared emission and absorption features of HL Tau led Cohen (1983) to suggest that in this system we are looking through an edge-on disk, which had a radius $\sim 10^2$ AU, an (estimated) axial ratio of 100:1, and $\sim 10^{-9}$ $M_\odot$ in silicate and ice grains. Direct near-infrared imaging (Grasdalen et al. 1984) and speckle interferometry (Beckwith et al. 1984) seemed to confirm this suggestion, with the scattering optical depths yielding a revised (but still small) estimate of $\sim 10^{-7}$ $M_\odot$ for the total mass of the grains (assuming a standard interstellar size distribution). Follow-up radio-interferometric observations (Beckwith et al. 1986) in the $J = 1$-0 line of ^{12}CO and in the 2.7-mm continuum found the source to be essentially unresolved with a beam size of 6" (corresponding to a radius of ~ 400 AU at 140 pc, the distance of the Taurus cloud). The continuum flux at 2.7 mm, interpreted as thermal emission from warm grains, implied a dust mass between 1×10^{-4} and 2×10^{-3} $M_\odot$, depending on the emissivity law adopted for the grains. Preliminary ^{13}CO measurements by Sargent & Beckwith (1987) confirm the likelihood of up to 0.2–0.3 $M_\odot$ of gas and dust lying in a rotating disk centered on the system. Thus, although Cohen's original suggestion of a dusty disk in HL Tau of roughly solar-system dimensions has proven exceptionally fertile, there is an unsettling 6 orders of magnitude difference between the smallest and largest estimates for the mass of solid material present. In Section 6.5 we offer a resolution of this discrepancy.

The above discussion summarizes the direct observational evidence concerning small disks ($\sim 10^2$ AU, as compared to the $\sim 10^4$-AU objects occasionally claimed by molecular radio astronomers) around a few YSOs. However, there is *indirect* evidence that suggests the almost universal existence of such structures around T Tauri stars; this evidence comes from the examination of the infrared excesses in T Tauri spectra.

6.3 *Infrared Excesses of T Tauri Stars*

As early as two decades ago, the spectral energy distributions of T Tauri stars were observed to exhibit excess infrared radiation (Mendoza 1966, 1968). Since then, a multitude of data have been accumulated [cf. Rydgren et al. (1984) for a catalog of T Tauri stars in Taurus-Auriga]. Using several

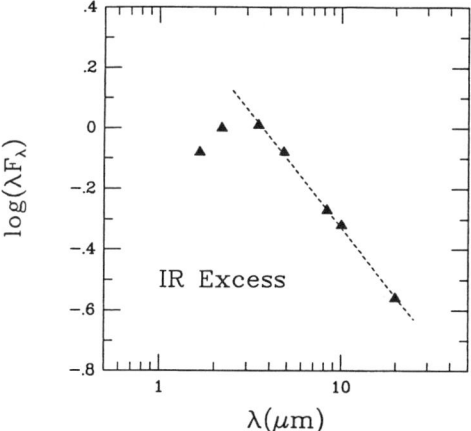

Figure 6 Composite spectral energy distribution of nine T Tauri stars. The dashed line shows that the near- and mid-infrared excess is well represented as a power-law spectrum with index $n = 0.75$. A stellar blackbody would give $n = 3$ (from Rydgren & Zak 1987).

representative sources, Rydgren & Zak (1987) have produced a composite spectrum, as shown in Figure 6, that clearly displays the nature of the infrared excess. If we define the spectral index n by

$$n \equiv \frac{d \log (\nu F_\nu)}{d \log \nu} = -\frac{d \log (\lambda F_\lambda)}{d \log \lambda},\qquad 31.$$

then $n \approx 0.75$ for the composite spectrum at near- and mid-infrared wavelengths. Individual sources in the catalog of Rydgren et al. (1984) have a wide range of spectral indices n, varying from somewhat greater than 1 (steep spectrum sources) to $n \approx 0$ (flat spectrum sources).[7]

Mendoza (1966) suggested that the infrared excesses of T Tauri stars were caused by circumstellar dust grains, which absorb stellar photons and reradiate them at longer wavelengths. The alternative hypothesis—that the infrared excesses are given by free-free emission—cannot explain the most extreme cases (Cohen & Kuhi 1979). A wide variety of geometries for the distribution of dust have been proposed: shells, disks, and grain formation in outflows from the young stars.

Until recently, the most popular approach has been to assume that all infrared excesses are created by more or less spherical circumstellar dust shells that surround the stars. These models generally treat the properties of the star (the effective temperature T_* and the radius R_*) and those of

---

[7] It has not escaped our attention that an analogous situation holds for the spectral energy distributions of active galactic nuclei. Indeed, comparisons of jets and disks in YSOs with those in quasi-stellar objects have been frequently made in the literature.

the dust shell (the inner and outer radii and the density distribution, which combine to give the total optical depth) as adjustable quantities. The resulting spectra are reasonably successful in reproducing the spectral energy distributions of many observed sources (see, e.g., Rowan-Robinson et al. 1986, Wolfire & Churchwell 1987).

For evolved sources near the end point of stellar evolution, where dusty outflows may be present, the adoption of spherical symmetry is a natural first approximation. However, dust distributed in a spherical manner *close* to an optically revealed pre-main-sequence star (so that the dust has temperatures high enough to account for the near-infrared emission) is much less plausible. Since the infall has been reversed, not much dust can be infalling; but the dust cannot be outflowing either, as the stellar temperatures are generally not low enough to allow solid grains to condense. Thus, the closer and warmer dust grains must actually be in *orbit* around the star, i.e. distributed in a flat disk. Such a model is also consistent with the ideas developed in Section 5, in which stellar outflows break out at the poles and widen to the equator, so that an equatorial disk should be the last remnant circumstellar material to be cleared away.

6.4 Passive Disks

Given that T Tauri stars have dusty disks around them, it can be easily shown that unless the masses of the disks (gas plus dust) are much less than $\sim 10^{-1}\ M_\odot$, the disks will be optically thick to their own thermal radiation, from the near- to far-infrared, out to $\sim 10^2$ AU. Even a minimum-mass solar nebula ($\sim 10^{-2}\ M_\odot$) should be optically thick for radii within the orbit of Neptune (Lin & Papaloizou 1985), and modest gas opacities are likely to keep the disk optically thick in the dust-free environment adjacent to the star. Thus, essentially all disks will be optically thick and reprocess a significant portion of the luminosity from the central star. Such disks can be divided into two conceptual categories: those that also have appreciable radiation arising from intrinsic luminosity in the disk (an *active* disk) and those that do not (a *passive* disk).

The theory of passive disks is especially simple: An optically thick disk, radially extended and vertically thin, will intercept and reradiate 25% of the stellar luminosity and produce a spectral energy distribution that is approximately a power law in the near- and mid-infrared with a spectral index $n = 4/3$ (Adams & Shu 1986). The simple theoretical models, which contain *no* free parameters (apart from viewing geometry) when the properties of the central star are known, are in good agreement with the observed spectra of (positively) steep spectrum T Tauri stars (see Adams et al. 1987).

In a passive disk, the equilibrium temperature distribution in the disk is given approximately by $T \propto r^{-3/4}$, which is also the classical result for a Keplerian accretion disk (Lynden-Bell & Pringle 1974); hence, the corresponding spectral energy distributions for the two types of disk have the same shape. Thus, simple reprocessing of stellar photons will mask the effects of Keplerian disk accretion if the accretion rate is low enough that the intrinsic disk luminosity is less than $\sim 25\%$ of the stellar luminosity (see also Friedjung 1985). In particular, the typical accretion rate $\dot{M}_a$ is $\sim 5 \times 10^{-8} M_\odot$ yr^{-1} in a minimum-mass solar nebula driven by thermal convection (Ruden & Lin 1986), and it produces a total disk luminosity $\sim G\dot{M}_a M_*/2R_*$ that is too small to compete with disk reprocessing. Such small accretion rates will be observable only if a viscous boundary layer between the rapidly rotating disk and the slowly spinning star radiates $\sim G\dot{M}_a M_*/2R_*$ in the ultraviolet, where there is competition only from the chromosphere. Ultraviolet excesses (cf. Bertout 1987) therefore may be especially valuable diagnostics in deciding whether systems like FU Orionis (for a review, see Herbig 1977) are sources in which an outburst has taken place in the star or in the disk (Hartmann & Kenyon 1985, Adams et al. 1987).

For a given $\dot{M}_a$, the accretion luminosity scales as M_*/R_*, which is constant for a star on the deuterium "birthline" and is proportional to $M_*^{0.4}$ on the upper main sequence. But the intrinsic stellar luminosity L_* varies as a relatively high power of M_*. Thus, the disks around stars of high-enough mass must be dominated by reprocessing of starlight. This fact, coupled with the fierce tendency of high-mass stars to ablate any nearby material, may explain why near-infrared excesses associated with optically revealed YSOs of high mass are not prominent.

6.5 Active Disks

There is another class of T Tauri stars where n is significantly less than 4/3; in particular, there are some T Tauri stars with flat spectra ($n = 0$). Adams et al. (1987) proposed that such sources could be understood if the disks were *active* and possessed intrinsic luminosity comparable to or larger than their central stars. It is easy to show that in order to produce a spectrum with $n < 4/3$, it is necessary to have a temperature distribution that falls off less steeply than $T \propto r^{-3/4}$. In particular, a spectrum with $n = 0$ will result if $T \propto r^{-1/2}$ (Elmegreen 1982b). If such a temperature distribution were produced in a viscously evolving axisymmetric system, the rotation curve would have to be flat (like in a spiral galaxy) rather than Keplerian, which would imply that the mass distributions are extended $[M(r) \propto r]$.

However, if the relation $M(r) \propto r$ extends from the inner disk ($\sim$ a few stellar radii) out to the outer regions of the disk (~ 100 AU), the resulting disk would be very massive (R. Thompson, private communication). Using this method to explain extreme cases such as T Tau or HL Tau would require disk masses in excess of 10^3 $M_\odot$. Such models can be ruled out on the basis of observations at millimeter wavelengths, where the thermal emission from dust grains and rare isotopes of CO is optically thin (e.g. Sargent & Beckwith 1987), and on the basis of theoretical considerations of disk stability (e.g. Lin & Bertin 1985). Indeed, the latter consideration suggests that if circumstellar disks become roughly as massive as their central stars, further growth of the disk would be limited by the onset of nonaxisymmetric gravitational instabilities (bars or spiral waves) that would redistribute the material in the disk into a more centrally condensed configuration. In other words, nonaxisymmetric processes in such systems may drive disk accretion and generate waves in the inner regions of the disk that propagate and dissipate their energies at outer locations, thereby producing a less steeply declining equilibrium temperature gradient. The *nonlocal* transport implied by such mechanisms (whose forced analogues have been found in Saturn's rings; see Borderies et al. 1985, Shu et al. 1985) would distinguish them from true viscous transport. This should be contrasted with axisymmetric instabilities that do not propagate and may therefore be amenable to a pseudoviscous formalism (Lin & Pringle 1987). A more fundamental approach to the process is needed to obtain theoretical progress on this issue. In the future, spatial maps of very high resolution may show whether the disks around flat-spectrum T Tauri stars have interesting nonaxisymmetric structures.

In the meantime, the case for moderately massive disks can be stated more robustly. The objects in question must often have intrinsic disk luminosities of several $L_\odot$ in order to explain the observed infrared excesses. Accretion onto a star (either the central source or a companion embedded in the disk) is the most likely energy source and hence must occur at a rate of 10^{-5} to 10^{-6} $M_\odot$ yr^{-1}. Since many T Tauri stars have such infrared excesses, and their lifetimes are on the order to 10^5 to 10^6 yr, the reservoirs of material for accretion in the disks must often be 0.1–1.0 $M_\odot$, a value that compares well with Sargent & Beckwith's (1987) estimate for the disk of HL Tau.

What about the lower estimates for the mass of HL Tau's disk? In our opinion, the absorption and scattering measurements (Cohen 1983, Beckwith et al. 1984, Grasdalen et al. 1984) refer to nonplanar material that is still undergoing *residual infall* onto this very young T Tauri star. The amount of gas and dust involved in such residual infall is always very small, and it produces a characteristic signature—a hump of far-infrared radiation—that is, in fact, seen in the spectral energy distribution of HL Tau. In contrast, the 2.7-mm continuum emission measurement of

Beckwith et al. (1986) detects emission from *all* warm dust. Most of this warm dust lies in a spatially flat disk, and it can contain a large fraction of a solar mass without coming into any conflict with the determinations of absorption or scattering optical depths, provided we are *not* viewing the disk edge-on.

6.6 *Implications for Binary Star and Planetary System Formation*

Given the above discussion, it is interesting to speculate further that active disks are active precisely because they are massive, and that passive disks are passive precisely because they lack the gravitational instability mechanisms that would give them appreciable accretion rates (Shu et al. 1987). If this speculation has any foundation, then there may be a natural basis for two modes of companion formation in T Tauri disks. If a nebular disk turns out to have enough mass to make another star, it may be an excellent candidate for forming a relatively close binary system (with a period shorter than $\sim 10^2$–10^3 yr; see Abt 1983). A passive or nearly passive disk will contain a smaller amount of material, and it may be a likely candidate for forming a planetary system [see Wetherill (1980) and Pollack (1984) for reviews].

7. CONCLUSION

7.1 *Summary: Four Stages of Star Formation*

We summarize this review with a proposed outline of the various stages involved in the birth of stars from molecular clouds; this includes most of what we know theoretically and observationally about the process. We begin with a molecular cloud, after its basic constituent material has been assembled somehow and somewhere in a galaxy. The first stage of star formation is then the formation of slowly rotating cloud cores (see Figure 7a). In subcritical clumps this occurs through the slow leakage of magnetic (and turbulent) support by ambipolar diffusion; in supercritical clumps, the clump may also fragment as it contracts and flattens as a whole. As long as the core formation process occurs relatively slowly (i.e. with a time scale that is longer than the characteristic signal crossing time), the cores probably asymptotically approach centrally concentrated states that resemble singular isothermal spheres. However, such end states cannot actually be reached because they are unstable.

The second phase begins when a condensing cloud core passes the brink of instability and collapses dynamically from "inside-out" (see Figure 7b). This evolutionary phase is characterized by a central protostar and disk, deeply embedded within an infalling envelope of dust and gas. The infalling

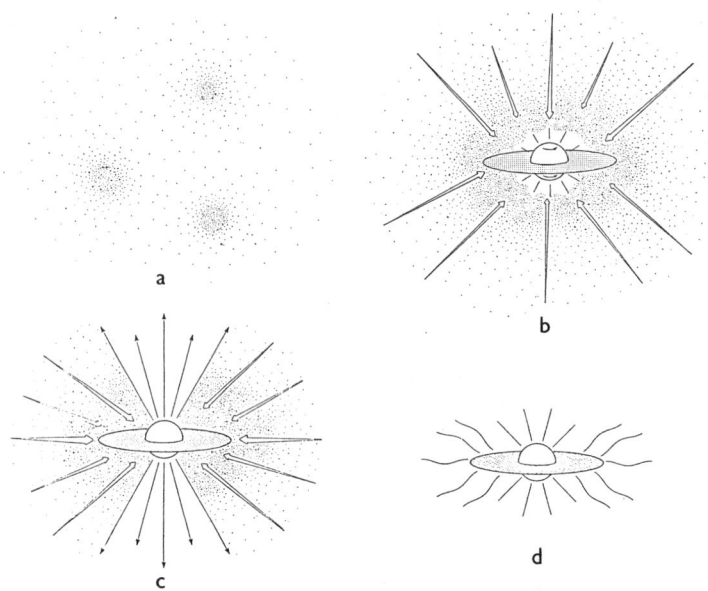

Figure 7 The four stages of star formation. (*a*) Cores form within molecular clouds as magnetic and turbulent support is lost through ambipolar diffusion. (*b*) A protostar with a surrounding nebular disk forms at the center of a cloud core collapsing from inside-out. (*c*) A stellar wind breaks out along the rotational axis of the system, creating a bipolar flow. (*d*) The infall terminates, revealing a newly formed star with a circumstellar disk.

material passes through an accretion shock as it falls onto the central star and disk, which, along with accretion within the disk, produces the main contribution to the luminosity for low-mass protostars. The emergent spectral energy distributions of theoretical models in the infall stage are in close agreement with those of recently found infrared sources with negatively steep spectra in the near- and mid-infrared. Protostars of high mass, in a pure accretion phase, have yet to be found, although the source near the water masers in W3(OH) is probably close to being such an object.

As a protostar accretes matter, deuterium will eventually ignite in the central regions and drive the star nearly completely convective if its mass is less than about 2 $M_\odot$. If the convection and the differential rotation of the star combine to produce a dynamo, the star can naturally evolve toward a state with a stellar wind. However, at first the ram pressure from material falling directly onto the stellar surface suppresses breakout. Gradually, the "lid" of direct infall will weaken as the incoming material falls preferentially onto the disk rather than onto the star. The stellar wind then rushes through the channels of weakest resistance (the rotational poles), which leads to collimated jets and bipolar outflows. Thus, the protostar enters the next step of evolution—the bipolar outflow phase (see Figure 7c). Examples of massive YSOs in this phase of stellar evolution (combined inflow and outflow) include the Becklin-Neugebauer object in Orion, but the trigger for the stellar wind in such cases is still theoretically obscure.

As time proceeds, more and more of the rotating inflowing matter will fall preferentially onto the disk rather than onto the star. In any reasonable picture of stellar outflows, the opening angle of the wind will widen with time, eventually sweeping outward over all 4π steradians and revealing the fourth evolutionary stage—in the case of a low-mass YSO—a T Tauri star with a surrounding remnant nebular disk (see Figure 7d). Radiation from the disk adds an infrared excess to the expected spectral energy distribution of the revealed source. The detailed shape of this infrared excess depends on whether the disk is largely passive and merely reprocessing stellar photons, or whether it is relatively massive and actively accreting. Both extremes of spectral shapes are observed in T Tauri stars; the amount of circumstellar material in the form of disks around nearly formed stars may be related to the dual issues of the origins of binary star and planetary systems. Disks around high-mass YSOs may also be common; in at least one well-documented case, S106, the disk is very large spatially but not extremely massive (i.e. not more massive than the central star).

A fifth phase of evolution can also be considered—namely, the final disappearance of the nebular disk as matter becomes incorporated into planets or stellar companions or is dispersed by the energetic outflow. For low-mass YSOs, such objects are probably the "naked T Tauri stars" (Mundt et al. 1983, Walter 1986, 1987). Their story more properly belongs to the subject of pre-main-sequence evolution than to star formation, although it can also be viewed as logically flowing from the developments of the four prior stages. High-mass stars apparently join the main sequence while still in the accretion phase; thus, they have no "pre-main-sequence story" to tell.

7.2 *Outstanding Problems*

There has been considerable theoretical and observational progress in recent years in our understanding of how individual stars form and evolve. In particular, the discoveries of the low general efficiency of star formation in molecular clouds and of the ubiquity of outflows from YSOs have challenged older conceptions of the ultimate origin for the masses of forming stars. Still, many gaps remain in our knowledge of how stars are born. We have only relatively rough ideas of the properties of the final system that is produced from the collapse of a cloud core with given initial conditions. Much of the uncertainty arises from ignorance about basic disk transport processes; new ideas may not be needed so much as rigorous calculations. Observations carried out with high angular resolution at millimeter, submillimeter, and infrared wavelengths, in the continuum and in spectral lines, could be of crucial help in unraveling the puzzle.

In contrast, there are few viable ideas concerning the generation of stellar jets and bipolar flows that are amenable to rigorous theoretical treatment. Quantitative work has focused, for good reason, on the easier problem of jet propagation and wind interaction with the surrounding medium. Steady progress on the latter front can be expected as the models being developed become more quantitative and can be compared fruitfully with the wealth of existing observational data, but fundamental advances on the former issue may lie in the distant future (especially if the generation of the winds requires a deep understanding of the basic processes underlying stellar activity).

Computations can be done (and are in the works) to follow in detail how a molecular cloud core condenses from a parent clump of given properties and evolves to the point of gravitational collapse. Indeed, if magnetic fields and ambipolar diffusion control the process, then in some sense theorists may be ahead of observers on this problem. The current empirical data base contains very little secure information concerning magnetic field strengths and the ionization fractions in molecular cloud cores. Resolved velocity maps at scales less than 10^{16} cm, in both neutral and ionized molecular species, would also be very useful to kinematically confirm or reject the basic concepts of the first three phases of star formation discussed in Section 7.1.

Many of the detailed comparisons between theory and observation in recent years have focused on the problem of low-mass star formation. The reasons frequently cited are that the developed theory for low-mass stars is more complete, and that high-mass stars, once formed, do so much damage to their surroundings that it becomes difficult to sift through the remaining clues to reconstruct prior events. Yet in order to progress to the point where the initial-mass functions can be predicted (e.g. for use in models of galactic evolution), we must have an a priori theory for how high-mass stars are formed. A viable, if conservative, approach may be a greater concerted effort, both observational and theoretical, on the somewhat neglected problem of the formation of intermediate-mass stars.

Another fruitful area for future research will be the problem of the formation of star clusters. Largely as a result of the reviewers' own prejudices and expertise, this article has concentrated on the problem of star formation in clouds that are dominated by the presence of magnetic fields.

However, as discussed in Sections 3.1 and 3.2, there is a second regime logically and probably empirically as well—the supercritical regime—in which cloud evolution is gravitationally dominated. In the extreme case of a very supercritical cloud, the behavior should resemble those calculations in which the magnetic field is neglected altogether. A rich picture of fragmentation and collapse is characteristic of such calculations, but to date little success has been achieved when the theoretical results are compared with observations. This lack of agreement, we believe, arises because of a misplaced sense of where the results should apply. Ignoring magnetic fields altogether is probably not valid for most present-day molecular clouds; however, it may be an acceptable first approximation for proto-clusters (especially proto–globular clusters). In a similar vein, the concept of induced star formation gains credibility in environments that are essentially free of magnetic fields or are decoupled from them.

ACKNOWLEDGMENTS

This work was supported in part by NSF grant AST83-14682 and in part under the auspices of a special NASA astrophysics theory program that supports a joint Center for Star Formation Studies at NASA–Ames Research Center, UC Berkeley, and UC Santa Cruz. In addition, SL is grateful for fellowship support from the National University of Mexico and the Amelia Earhart Foundation.

Literature Cited

Abt, H. A. 1983. *Ann. Rev. Astron. Astrophys.* 21: 343
Abt, H. A., Levy, S. G. 1976. *Ap. J. Suppl.* 30: 273
Adams, F. C., Shu, F. H. 1985. *Ap. J.* 296: 655
Adams, F. C., Shu, F. H. 1986. *Ap. J.* 308: 836
Adams, F. C., Lada, C. J., Shu, F. H. 1987. *Ap. J.* 312: 788
Adams, M. T., Strom, K. M., Strom, S. E. 1983. *Ap. J. Suppl.* 53: 893
Allen, R. J., Atherton, P. D., Tilanus, R. P. J. 1986. *Nature* 319: 296
Appenzeller, I., Tscharnuter, W. 1974. *Astron. Astrophys.* 30: 423
Appenzeller, I., Tscharnuter, W. 1975. *Astron. Astrophys.* 40: 397
Arons, J., Max, C. E. 1975. *Ap. J. Lett.* 196: L77
Arquilla, R., Goldsmith, P. F. 1986. *Ap. J.* 303: 356
Aumann, H. H., Gillett, F. C., Beichman, C. A., de Jong, T., Houck, J. R., et al. 1984. *Ap. J. Lett.* 278: L23

Avery, L. W. 1980. In *Interstellar Molecules*, ed. B. Andrew, p. 47. Dordrecht: Reidel
Balbus, S. A., Cowie, L. L. 1985. *Ap. J.* 297: 61
Bally, J., Lada, C. J. 1983. *Ap. J.* 265: 824
Bash, F. M., Kaufman, M. 1986. *Ap. J.* 310: 621
Bash, F. M., Peters, W. L. 1976. *Ap. J.* 205: 786
Batlra, W., Menten, K. M. 1985. *Ap. J. Lett.* 298: L19
Batten, A. H. 1973. *Binary and Multiple Systems of Stars.* London: Pergamon
Beckwith, S., Zuckerman, B., Skrutskie, M. F., Dyck, H. M. 1984. *Ap. J.* 287: 793
Beckwith, S., Sargent, A. I., Scoville, N. Z., Masson, C. R., Zuckerman, B., Phillips, T. G. 1986. *Ap. J.* 309: 755
Beichman, C. A., Myers, P. C., Emerson, J. P., Harris, S., Mathieu, R. D., et al. 1986. *Ap. J.* 307: 337
Bertout, C. 1976. *Astron. Astrophys.* 51: 101
Bertout, C. 1987. In *Circumstellar Matter*, *IAU Symp. No. 122*, ed. K.-H. Bohm. Dordrecht: Reidel. In press

Bertout, C., Bouvier, J., Bouchet, P. 1987. In preparation
Bieging, J. H. 1984. *Ap. J.* 206: 591
Black, D. C., Scott, E. H. 1983. *Ap. J.* 263: 696
Blitz, L. 1978. PhD thesis. Columbia Univ., New York, N.Y.
Blitz, L., Shu, F. H. 1980. *Ap. J.* 238: 148
Blitz, L., Thaddeus, P. 1980. *Ap. J.* 241: 676
Bodenheimer, P. 1980. In *Fundamental Problems in the Theory of Stellar Evolution, IAU Symp. No. 93*, ed. D. Sugimoto, D. Q. Lamb, D. N. Schramm, p. 5. Dordrecht: Reidel
Bodenheimer, P., Sweigart, A. 1968. *Ap. J.* 152: 515
Bonnor, W. B. 1956. *MNRAS* 116: 351
Borderies, N., Goldreich, P., Tremaine, S. 1985. *Icarus* 63: 406
Boss, A. P. 1986. *Ap. J. Suppl.* 62: 519
Boss, A. P. 1987. In *Interstellar Processes*, ed. D. Hollenbach, H. Thronson. Dordrecht: Reidel. In press
Boss, A. P., Black, D. C. 1982. *Ap. J.* 258: 270
Bouvier, J., Bertout, C., Benz, W., Mayor, M. 1986. *Astron. Astrophys.* 165: 110
Burton, W. B. 1976. *Ann. Rev. Astron. Astrophys.* 14: 275
Calvet, N., Basri, G., Kuhi, L. V. 1984. *Ap. J.* 277: 725
Cameron, A. G. W. 1962. *Icarus* 1: 13
Cameron, A. G. W. 1985. *Ap. J. Lett.* 299: L83
Cameron, A. G. W., Truran, J. W. 1977. *Icarus* 30: 447
Canto, J., Rodriguez, L. F., Barral, J. F., Carral, P. 1981. *Ap. J.* 244: 102
Cassen, P., Moosman, A. 1981. *Icarus* 48: 353
Cassen, P., Shu, F. H., Terebey, S. 1985. In *Protostars and Planets II*, ed. D. C. Black, M. S. Matthews, p. 448. Tucson: Univ. Ariz. Press
Cernicharo, J., Guelin, M., Askne, J. 1984. *Astron. Astrophys.* 138: 371
Cernicharo, J., Bachiller, R., Duvert, G. 1985. *Astron. Astrophys.* 149: 273
Chandrasekhar, S. 1939. *An Introduction to Stellar Structure.* Chicago: Univ. Chicago Press
Chandrasekhar, S. 1969. *Ellipsoidal Figures of Equilibrium.* New Haven, Conn: Yale Univ. Press
Chandrasekhar, S., Fermi, E. 1953. *Ap. J.* 118: 116
Chernoff, D. F., Hollenbach, D. J., McKee, C. F. 1982. *Ap. J. Lett.* 259: L97
Chevalier, R. A. 1983. *Ap. J.* 268: 753
Clark, F. O., Laureijs, R. J. 1986. *Astron. Astrophys.* 154: L26
Clark, F. O., Laureijs, R. J., Chiewicki, G., Zhang, C. Y., van Oesterom, W., Kester, D. 1986. *Astron. Astrophys.* In press
Cohen, M. 1975. *MNRAS* 175: 279
Cohen, M. 1983. *Ap. J. Lett.* 270: L69
Cohen, M. 1984. *Phys. Rep.* 116(4): 173
Cohen, M., Emerson, J. P., Beichman, C. A. 1987. Preprint
Cohen, M., Kuhi, L. V. 1979. *Ap. J. Suppl.* 41: 743
Cohen, M., Cong, H., Dame, T. M., Thaddeus, P. 1980. *Ap. J. Lett.* 239: L53
Cowie, L. L. 1981. *Ap. J.* 245: 66
Crutcher, R. M., Kazes, I. 1983. *Astron. Astrophys.* 125: L23
Crutcher, R. M., Kazes, I., Troland, T. H. 1987. Preprint
Dalgarno, A., Lepp, S. 1984. *Ap. J. Lett.* 287: L47
Dame, T. M., Elmegreen, B. G., Cohen, R. S., Thaddeus, P. 1986. *Ap. J.* 305: 892
de Jong, T., Maeder, A., eds. 1977. *Star Formation, IAU Symp. No. 75.* Dordrecht: Reidel
de Jong, T., Dalgarno, A., Boland, W. 1980. *Astron. Astrophys.* 91: 68
Dewar, R. L. 1970. *Phys. Fluids* 13: 2710
Dickman, R. L. 1978. *Ap. J. Suppl.* 37: 407
Draine, B. T. 1980. *Ap. J.* 241: 1021
Draine, B. T. 1983. *Ap. J.* 270: 519
Draine, B. T., Lee, H. M. 1984. *Ap. J.* 285: 89
Draine, B. T., Roberge, W. G., Dalgarno, A. 1983. *Ap. J.* 264: 485
Durisen, R. H., Gingold, R. A., Tohline, J. E., Boss, A. P. 1986. *Ap. J.* 305: 281
Ebert, R. 1955. *Z. Ap.* 37: 217
Edwards, S., Strom, S. E., Snell, R. L., Jarrett, T. H., Beichman, C. A., Strom, K. M. 1986. *Ap. J. Lett.* 307: L65
Elias, J. H. 1978a. *Ap. J.* 224: 857
Elias, J. H. 1978b. *Ap. J.* 224: 453
Elmegreen, B. G. 1979. *Ap. J.* 232: 729
Elmegreen, B. G. 1982a. *Ap. J.* 253: 655
Elmegreen, B. G. 1982b. In *Formation of Planetary Systems*, ed. A. Brahic, p. 63. CNRS: Cepadues Ed.
Elmegreen, B. G. 1985a. In *Star Formation, Lectures at Les Houches Summer School*, ed. R. Lucas, A. Aumont, R. Stora, p. 215. Amsterdam: North-Holland
Elmegreen, B. G. 1985b. *Ap. J.* 299: 196
Elmegreen, B. G. 1986. In *Light on Dark Matter*, ed. F. Israel. Dordrecht: Reidel. In press
Elmegreen, B. G. 1987. *Ap. J.* 312: 626
Elmegreen, B. G., Clemens, C. 1985. *Ap. J.* 294: 523
Elmegreen, B. G., Elmegreen, D. M. 1983. *Ap. J.* 267: 31
Elmegreen, B. G., Lada, C. J. 1977. *Ap. J.* 214: 725
Elsasser, H., Staude, H. J. 1978. *Astron. Astrophys.* 70: L3
Encrenaz, P. J., Falgarone, E., Lucas, R. 1975. *Astron. Astrophys.* 44: 73

Evans, N. J. 1978. In *Protostars and Planets*, ed. T. Gehrels, p. 152. Tucson: Univ. Ariz. Press
Evans, N. J., Kutner, M. L. 1976. *Ap. J. Lett.* 204: L131
Evans, N. J., Blair, G. N., Harvey, P., Israel, F., Peters, W. L., et al. 1981. *Ap. J.* 250: 200
Ezer, D., Cameron, A. G. W. 1965. *Can. J. Phys.* 43: 1497
Falgarone, E. 1987. In *NATO/ASI Physical Processes in Interstellar Clouds*, ed. M. Scholer. Dordrecht: Reidel. In press
Falgarone, E., Puget, J. L. 1986. *Astron. Astrophys.* 162: 235
Fazio, G. G., Wright, E. L., Zeilik, M., Low, F. J. 1976. *Ap. J. Lett.* 206: L165
Field, G. B. 1970. *Mem. Soc. R. Sci. Liège* 19: 29
Field, G. B. 1978. In *Protostars and Planets*, ed. T. Gehrels, p. 243. Tucson: Univ. Ariz. Press
Finkenzeller, U., Mundt, R. 1984. *Astron. Astrophys. Suppl.* 55: 109
Fleck, R. C. Jr. 1981. *Ap. J. Lett.* 246: L151
Franco, J. 1984. *Astron. Astrophys.* 137: 85
Frerking, M. A., Langer, W. D., Wilson, R. W. 1982. *Ap. J. Suppl.* 262: 590
Friedjung, M. 1985. *Astron. Astrophys.* 146: 336
Fuller, G. A., Myers, P. C. 1987. In *NATO/ASI Physical Processes in Interstellar Clouds*, ed. M. Scholer. Dordrecht: Reidel. In press
Gaustad, J. E. 1963. *Ap. J.* 138: 1050
Gerola, H., Seiden, P. E. 1978. *Ap. J.* 223: 129
Gillis, J., Mestel, L., Paris, R. B. 1974. *Astrophys. Space Sci.* 27: 167
Gillis, J., Mestel, L., Paris, R. B. 1979. *MNRAS* 187: 311
Glassgold, A. E. 1985. In *Protostars and Planets II*, ed. D. C. Black, M. S. Matthews, p. 641. Tucson: Univ. Ariz. Press
Goldreich, P., Kwan, J. 1974. *Ap. J.* 189: 441
Goldsmith, P. F., Arquilla, R. 1985. In *Protostars and Planets II*, ed. D. C. Black, M. S. Matthews, p. 137. Tuscon: Univ. Ariz. Press
Goldsmith, P. F., Langer, W. D. 1978. *Ap. J.* 222: 881
Goldsmith, P. F., Sernyak, M. Jr. 1984. *Ap. J.* 283: 140
Goodman, A. 1987. In *Interstellar Processes*, ed. D. Hollenbach, H. Thronson. Dordrecht: Reidel. In press
Goodrich, R. W. 1987. Preprint
Gould, R. J. 1964. *Ap. J.* 140: 638
Grasdalen, G. L., Strom, K. M., Strom, S. E. 1973. *Ap. J. Lett.* 184: L53
Grasdalen, G. L., Strom, S. E., Strom, K. M., Capps, R. W., Thompson, D., Castelaz, M. 1984. *Ap. J. Lett.* 283: L57
Gusten, R., Mezger, P. G. 1982. *Vistas Astron.* 26: 159
Harris, A., Townes, C. H., Matsakis, D. N., Palmer, P. 1983. *Ap. J. Lett.* 265: L63
Hartmann, L., Kenyon, S. J. 1985. *Ap. J.* 299: 462
Hartmann, L., MacGregor, K. B. 1982. *Ap. J.* 257: 264
Hartmann, L., Hewett, R., Stahler, S., Mathieu, R. D. 1986. *Ap. J.* 309: 275
Hayashi, C. 1966. *Ann. Rev. Astron. Astrophys.* 4: 171
Hayashi, C. 1970. *Mem. Soc. R. Sci. Liège* 19: 127
Hayashi, C. 1986. In *Star-Forming Regions*, *IAU Symp. No. 115*, ed. M. Peimbert, J. Jugaku. Dordrecht: Reidel. In press
Hayashi, C., Hoshi, R., Sugimoto, D. 1962. *Prog. Theor. Phys. Suppl.*, Vol. 22
Heggie, D. C. 1975. *MNRAS* 173: 729
Heiles, C. H. 1987. In *NATO/ASI Physical Processes in Interstellar Clouds*, ed. M. Scholer. Dordrecht: Reidel. In press
Heiles, C. H., Stevens, M. 1986. *Ap. J.* 301: 331
Henyey, L. G., LeLevier, R., Levee, R. D. 1955. *Publ. Astron. Soc. Pac.* 67: 154
Herbig, G. H. 1960. *Ap. J. Suppl.* 4: 337
Herbig, G. H. 1962. *Adv. Astron. Astrophys.* 1: 47
Herbig, G. H. 1977. *Ap. J.* 214: 747
Herbig, G. H., Goodrich, R. W. 1986. *Ap. J.* 309: 294
Herbig, G. H., Soderblom, D. R. 1980. *Ap. J.* 242: 628
Herbst, E., Assousa, G. E. 1978. In *Protostars and Planets*, ed. T. Gehrels, p. 368. Tucson: Univ. Ariz. Press
Heyer, M. H. 1986. PhD thesis. Univ. Mass., Amherst
Heyer, M. H., Snell, R. L., Goldsmith, P. F., Strom, S. E., Strom, K. M. 1986. *Ap. J.* 308: 134
Hildebrand, R. H. 1983. *Q. J. R. Astron. Soc.* 24: 267
Ho, P. T. P., Haschick, A. D. 1986. *Ap. J.* 304: 501
Ho, P. T. P., Townes, C. H. 1983. *Ann. Rev. Astron. Astrophys.* 21: 239
Ho, P. T. P., Klein, R. I., Haschick, A. D. 1986. *Ap. J.* 305: 714
Hoyle, F. 1953. *Ap. J.* 118: 513
Hunter, C. 1977. *Ap. J.* 218: 834
Humphreys, R. M., Davidson, K. 1984. *Science* 223: 243
Iben, I. 1965. *Ap. J.* 141: 993
Jones, B. F., Herbig, G. H. 1982. *Astron. J.* 87: 1223
Jones, T. W., Merrill, K. M. 1976. *Ap. J.* 209: 509
Joy, A. H. 1942. *Publ. Astron. Soc. Pac.* 54: 15

Joy, A. H. 1945. *Ap. J.* 102: 168
Joy, A. H. 1949. *Ap. J.* 110: 424
Jura, M. 1987. In *Interstellar Dust and Related Topics, Fermi School Lectures*, ed. S. Aiello. New York: Academic. In press
Kahn, F. D. 1974. *Astron. Astrophys.* 37: 149
Kaifu, N. 1986. In *Star-Forming Regions, IAU Symp. No. 115*, ed. M. Peimbert, J. Jugaku. Dordrecht: Reidel. In press
Kaifu, N., Suzuki, S., Hasegawa, T., Morimoto, M., Inatani, J., et al. 1984. *Astron. Astrophys.* 134: 7
Kazes, I., Crutcher, R. M. 1986. *Astron. Astrophys.* 164: 328
Kazes, I., Troland, T. H., Heiles, C. H., Crutcher, R. M. 1987. In preparation
Keto, E. R., Ho, P. T. P., Haschick, A. D. 1987. Preprint
Klein, R. I., Sandford, M. T., Whitaker, R. W. 1983. *Ap. J. Lett.* 271: L69
Klein, R. I., Whitaker, R. W., Sandford, M. T. 1985. In *Protostars and Planets II*, ed. D. C. Black, M. S. Matthews, p. 340. Tucson: Univ. Ariz. Press
Kleiner, S. C., Dickman, R. L. 1984. *Ap. J.* 286: 255
Knapp, G. R., Morris, M. 1985. *Ap. J.* 292: 640
Kondo, M. 1978. *Moon and Planets* 19: 245
Konigl, A. 1982. *Ap. J.* 261: 115
Kuhi, L. V. 1964. *Ap. J.* 140: 1409
Kutner, M. L., Leung, C. M. 1985. *Ap. J.* 291: 188
Kwan, J., Valdes, F. 1983. *Ap. J.* 271: 604
Kwan, J., Valdes, F. 1986. *Ap. J.* In press
Lada, C. J. 1985. *Ann. Rev. Astron. Astrophys.* 23: 267
Lada, C. J., Wilking, B. A. 1984. *Ap. J.* 287: 610
Lada, C. J., Margulis, M., Dearborn, D. 1984. *Ap. J.* 285: 141
Lago, M. T. V. T., Penston, M. V. 1982. *MNRAS* 198: 429
Landau, L. D., Lifshitz, E. M. 1959. *Fluid Mechanics*. Oxford: Pergamon
Lane, A. P., Bally, J. 1986. *Ap. J.* 310: 820
Langer, W. D. 1985. In *Protostars and Planets II*, ed. D. C. Black, M. S. Matthews, p. 650. Tucson: Univ. Ariz. Press
Larson, R. B. 1969a. *MNRAS* 145: 271
Larson, R. B. 1969b. *MNRAS* 145: 297
Larson, R. B. 1972. *MNRAS* 157: 121
Larson, R. B. 1981. *MNRAS* 194: 809
Larson, R. B. 1986. *MNRAS* 218: 409
Larson, R. B., Starrfield, S. 1971. *Astron. Astrophys.* 13: 190
Levreault, R. M. 1985. PhD thesis. Univ. Texas, Austin
Lin, C. C., Bertin, G. 1985. In *The Milky Way, IAU Symp. No. 106*, ed. H. van Woerden, R. J. Allen, W. B. Burton, p. 513. Dordrecht: Reidel
Lin, D. N. C., Papaloizou, J. 1985. In *Protostars and Planets II*, ed. D. C. Black, M. S. Matthews, p. 981. Tucson: Univ. Ariz. Press
Lin, D. N. C., Pringle, J. E. 1987. Preprint
Linke, R. A., Goldsmith, P. F. 1980. *Ap. J.* 235: 437
Lizano, S., Shu, F. H. 1987. In *NATO/ASI Physical Processes in Interstellar Clouds*, ed. M. Scholer. Dordrecht: Reidel. In press
Lo, K. Y., Ball, R., Masson, C. R., Phillips, T. G., Scott, S. L., Woody, D. P. 1987. Submitted for publication
Loren, R. B., Sandqvist, A., Wootten, A. 1983. *Ap. J.* 270: 620
Low, C., Lynden-Bell, D. 1976. *MNRAS* 176: 367
Lubow, S. H., Balbus, S. A., Cowie, L. L. 1986. *Ap. J.* 309: 496
Lucy, L. B., Ricco, E. 1979. *Astron. J.* 84: 401
Lynden-Bell, D., Pringle, J. E. 1974. *MNRAS* 168: 603
Lynds, B. T. 1962. *Ap. J. Suppl.* 7: 1
McCutcheon, W. H., Vrba, F. J., Dickman, R. L., Clemens, D. P. 1985. *Ap. J.* 309: 619
McKee, C. F., Hollenbach, D. J. 1980. *Ann. Rev. Astron. Astrophys.* 18: 219
McNally, D. 1964. *Ap. J.* 140: 1088
Mendoza, E. E. 1966. *Ap. J.* 143: 1010
Mendoza, E. E. 1968. *Ap. J.* 151: 977
Menten, K. M., Walmsley, C. M. 1985. *Astron. Astrophys.* 146: 369
Mercer-Smith, J. A., Cameron, A. G. W., Epstein, R. I. 1984. *Ap. J.* 287: 445
Mestel, L. 1965. *Q. J. R. Astron. Soc.* 6: 161
Mestel, L. 1985. In *Protostars and Planets II*, ed. D. C. Black, M. S. Matthews, p. 320. Tucson: Univ. Ariz. Press
Mestel, L., Spitzer, L. 1956. *MNRAS* 116: 503
Mezger, P. G., Smith, L. F. 1977. In *Star Formation, IAU Symp. No. 75*, ed. T. de Jong, A. Maeder, p. 133. Dordrecht: Reidel
Mihalas, D. 1978. *Stellar Atmospheres*. San Francisco: Freeman
Miller, G. E., Scalo, J. M. 1979. *Ap. J. Suppl.* 41: 513
Monetti, A., Pipher, J. L., Helfer, H. L., McMillan, R. S., Perry, M. L. 1984. *Ap. J.* 282: 508
Montmerle, T. 1987. In *Star Formation in Galaxies*, ed. C. Persson. In press
Mouschovias, T. Ch. 1976. *Ap. J.* 207: 141
Mouschovias, T. Ch. 1978. In *Protostars and Planets*, ed. T. Gehrels, p. 209. Tucson: Univ. Ariz. Press
Mouschovias, T. Ch., Paleologou, E. V. 1979. *Ap. J.* 230: 204
Mouschovias, T. Ch., Paleologou, E. V. 1980. *Ap. J.* 237: 877

Mouschovias, T. Ch., Paleologou, E. V. 1981. *Ap. J.* 246: 48
Mouschovias, T. Ch., Spitzer, L. 1976. *Ap. J.* 210: 326
Mouschovias, T. Ch., Shu, F. H., Woodward, P. R. 1974. *Astron. Astrophys.* 33: 73
Mouschovias, T. Ch., Paleologou, E. V., Fiedler, R. A. 1985. *Ap. J.* 291: 772
Mueller, M. W., Arnett, W. D. 1976. *Ap. J.* 210: 670
Mundt, R., Fried, J. W. 1983. *Ap. J. Lett.* 274: L83
Mundt, R., Walter, F. M., Feigelson, E. D., Finkenzeller, U., Herbig, G. H., Odell, A. P. 1983. *Ap. J.* 269: 229
Myers, P. C. 1985. In *Protostars and Planets II*, ed. D. C. Black, M. S. Matthews, p. 81. Tucson: Univ. Ariz. Press
Myers, P. C. 1986. In *Star-Forming Regions*, ed. M. Peimbert, J. Jugaku. Dordrecht: Reidel. In press
Myers, P. C. 1987. In *Interstellar Processes*, ed. D. Hollenbach, H. Thronson. Dordrecht: Reidel. In press
Myers, P. C., Benson, P. J. 1983. *Ap. J.* 266: 309
Myers, P. C., Goodman, A. 1987. *Ap. J.* In press
Myers, P. C., Dame, T. M., Thaddeus, P., Cohen, R. S., Silverberg, R. F., et al. 1986. *Ap. J.* 301: 398
Myers, P. C., Fuller, G. A., Mathieu, R. D., Beichman, C. A., Benson, P. J., et al. 1987a. *Ap. J.* In press
Myers, P. C., Heyer, M., Snell, R., Goldsmith, P. 1987b. *Ap. J.* In press
Myers, P. C., Goodman, A., Benson, P. J. 1987c. Submitted for publication
Nakano. T. 1979. *Publ. Astron. Soc. Jpn.* 31: 697
Nakano T. 1984. *Fundam. Cosmic Phys.* 9: 139
Nakano. T. 1985. *Publ. Astron. Soc. Jpn.* 37: 69
Nakano. T., Tademaru, E. 1972. *Ap. J.* 173: 87
Nakano. T., Umebayashi, T. 1980. *Publ. Astron. Soc. Jpn.* 32: 613
Narita, S., Nakano. T., Hayashi, C. 1970. *Progr. Theor. Phys.* 43: 942
Nelson, L. A., Rappaport, S. A., Joss, P. C. 1986. *Ap. J.* In press
Norman, C., Silk, J. 1980. *Ap. J.* 238: 158
Oort, J. H. 1960. *Bull. Astron. Inst. Neth.* 15: 45
Parker, E. N. 1979. *Cosmical Magnetic Fields*. Oxford: Oxford Univ. Press
Penston, M. V. 1969a. *MNRAS* 144: 425
Penston, M. V. 1969b. *MNRAS* 145: 457
Perault, M., Falgarone, E., Puget, J. L. 1985. *Astron. Astrophys.* 152: 371
Persson, C., ed. 1987. *Star Formation in Galaxies*. In press
Pollack, J. B. 1984. *Ann. Rev. Astron. Astrophys.* 22: 389
Press, W. H., Teukolsky, S. A. 1977. *Ap. J.* 213: 183
Pudritz, R. E. 1986. *Publ. Astron. Soc. Pac.* 98: 709
Pudritz, R. E., Norman, C. A. 1983. *Ap. J.* 274: 677
Pudritz, R. E., Norman, C. A. 1986. *Ap. J.* 301: 571
Rees, M. 1976. *MNRAS* 176: 483
Reid, M. J., Moran, J. M. 1981. *Ann. Rev. Astron. Astrophys.* 19: 231
Reid, M., Myers, P. C., Bieging, J. 1987. Preprint
Roberts, M. S. 1957. *Publ. Astron. Soc. Pac.* 69: 59
Roberts, W. W., Stewart, G. L. 1987. *Ap. J.* In press
Roberts, W. W., Yuan, C. 1970. *Ap. J.* 161: 887
Rodriguez, L. F., Canto, J. 1983. *Rev. Mex. Astron. Astrofiz.* 8: 163
Rowan-Robinson, M. 1979. *Ap. J.* 234: 111
Rowan-Robinson, M. 1980. *Ap. J. Suppl.* 44: 403
Rowan-Robinson, M., Lock, T. D., Walker, D. W., Harris, S. 1986. *MNRAS* 222: 273
Ruden, S. P., Lin, D. N. C. 1986. *Ap. J.* 308: 883
Rydgren, A. E., Zak, D. S. 1987. *Publ. Astron. Soc. Pac.* In press
Rydgren, A. E., Schmelz, J. T., Zak, D. S., Vrba, F. J. 1984. *Publ. US Nav. Obs.*, Vol. 25, Part 1
Salpeter, E. E. 1955. *Ap. J.* 121: 161
Sanders, D. B., Solomon, P. M., Scoville, N. Z. 1984. *Ap. J.* 276: 182
Sargent, A. I. 1977. *Ap. J.* 218: 736
Sargent, A., Beckwith, S. 1987. In preparation
Scalo, J. M. 1977. *Ap. J.* 213: 705
Scalo, J. M. 1986. *Fundam. Cosmic Phys.* 11: 1
Scalo, J. M., Pumphrey, W. A. 1982. *Ap. J. Lett.* 258: L29
Schwartz, R. D. 1983. *Ann. Rev. Astron. Astrophys.* 21: 209
Scott, E. H., Black, D. C. 1980. *Ap. J.* 239: 166
Scoville, N. Z. 1985. In *Protostars and Planets II*, ed. D. C. Black, M. S. Matthews, p. 188. Tucson: Univ. Ariz. Press
Scoville, N. Z. 1987. In *Star Formation in Galaxies*, ed. C. Persson. In press
Scoville, N. Z., Sanders, D. B., Clemens, D. P. 1986. *Ap. J. Lett.* 310: L77
Seiden, P. E. 1983. *Ap. J.* 266: 555
Shu, F. H. 1977. *Ap. J.* 214: 488
Shu, F. H. 1983. *Ap. J.* 273: 202
Shu, F. H. 1985. In *The Milky Way, IAU*

Symp. No. 106, ed. H. van Woerden, W. B. Burton, R. J. Allen, p. 561. Dordrecht: Reidel
Shu, F. H. 1987. In *Star Formation in Galaxies*, ed. C. Persson. In press
Shu, F. H., Terebey, S. 1984. In *Cool Stars, Stellar Systems, and the Sun*, ed. S. Baliunas, L. Hartmann, p. 78. Berlin: Springer-Verlag
Shu, F. H., Dones, L., Lissauer, J. J., Yuan, C., Cuzzi, J. N. 1985. *Ap. J.* 299: 542
Shu, F. H., Lizano, S., Adams, F. C. 1986. In *Star-Forming Regions, IAU Symp. No. 115*, ed. M. Peimbert, J. Jugaku. Dordrecht: Reidel. In press
Shu, F. H., Adams, F. C., Lizano, S. 1987. In *Interstellar Dust and Related Topics, Fermi School Lectures*, ed. S. Aiello. New York: Academic. In press
Shull, J. M., Beckwith, S. 1982. *Ann. Rev. Astron. Astrophys.* 20: 163
Silk, J. 1977. *Ap. J.* 214: 152
Simon, M., Peterson, D. M., Longmore, A. J., Storey, J. W. V., Tokunaga, A. T. 1985. *Ap. J.* 298: 328
Simonetti, J. H., Cordes, J. M. 1986. *Ap. J.* 303: 130
Smith, B. A., Terrile, R. J. 1984. *Science* 226: 1421
Smith, D., Adams, N. G. 1984. *Ap. J. Lett.* 284: L13
Snell, R. L. 1986. In *Star-Forming Regions, IAU Symp. No. 115*, ed. M. Peimbert, J. Jugaku. Dordrecht: Reidel. In press
Snell, R. L., Schloerb, F. B. 1985. *Ap. J.* 295: 490
Snell, R. L., Loren, R. B., Plambeck, R. L. 1980. *Ap. J. Lett.* 239: L17
Snell, R. L., Mundy, L. G., Goldsmith, P. F., Evans, N. J., Erickson, N. R. 1984. *Ap. J.* 276: 625
Snell, R. L., Bally, J., Strom, S. E., Strom, K. M. 1985. *Ap. J.* 290: 587
Solomon, P. M., Sanders, D. B. 1985. In *Protostars and Planets II*, ed. D. C. Black, M. S. Matthews, p. 81. Tucson: Univ. Ariz. Press
Solomon, P. M., Sanders, D. B., Scoville, N. Z. 1979. In *The Large Scale Characteristics of Galaxies, IAU Symp. No. 84*, ed. W. B. Burton, p. 35. Dordrecht: Reidel
Solomon, P. M., Sanders, D. B., Rivolo, A. R. 1985. *Ap. J. Lett.* 292: L19
Solomon, P. M., Rivolo, A. R., Mooney, T. J., Barrett, J. W., Sage, L. J. 1987. In *Star Formation in Galaxies*, ed. C. Persson. In press
Spitzer, L. 1942. *Ap. J.* 95: 329
Spitzer, L. 1958. In *Electromagnetic Phenomena in Cosmical Physics, IAU Symp. No. 6*, p. 169. Cambridge: Cambridge Univ. Press
Spitzer, L. 1968. In *Nebulae and Interstellar Matter, Stars and Stellar Systems*, ed. B. Middlehurst, L. H. Aller, 7: 1. Chicago: Univ. Chicago Press
Stahler, S. W. 1983. *Ap. J.* 274: 822
Stahler, S. W. 1985. *Ap. J.* 293: 207
Stahler, S. W., Shu, F. H., Taam, R. E. 1980a. *Ap. J.* 241: 637
Stahler, S. W., Shu, F. H., Taam, R. E. 1980b. *Ap. J.* 242: 226
Stahler, S. W., Shu, F. H., Taam, R. E. 1981. *Ap. J.* 248: 727
Stahler, S. W., Palla, F., Salpeter, E. 1986. *Ap. J.* 308: 697
Stark, A. A. 1985. In *The Milky Way, IAU Symp. No. 106*, ed. H. van Woerden, R. J. Allen, W. B. Burton, p. 445. Dordrecht: Reidel
Stauffer, J. R. 1984. *Ap. J.* 280: 189
Strittmatter, P. A. 1966. *MNRAS* 132: 359
Strom, S. 1985. In *Protostars and Planets II*, ed. D. C. Black, M. S. Matthews, p. 17. Tucson: Univ. Ariz. Press
Strom, S. E., Strom, K. M., Grasdalen, G. L. 1975. *Ann. Rev. Astron. Astrophys.* 13: 187
Strom, S. E., Strom, K. M., Grasdalen, G. L., Sellgren, K., Wolff, S., et al. 1985. *Astron. J.* 90: 2281
Tassoul, J.-L. 1978. In *Theory of Rotating Stars*. Princeton, NJ: Princeton Univ. Press
Terebey, S., Shu, F. H., Cassen, P. 1984. *Ap. J.* 286: 529
Tohline, J. E. 1982. *Fundam. Cosmic Phys.* 8: 1
Torrelles, J. M., Rodriguez, L. F., Canto, J., Carral, P., Marcaide, J., et al. 1983. *Ap. J.* 274: 214
Troland, T. H., Heiles, C. H. 1986. *Ap. J.* 301: 339
Troland, T. H., Crutcher, R. M., Kazes, I. 1986. *Ap. J. Lett.* 304: L57
Tscharnuter, W., Winkler, K.-H. A. 1979. *Comput. Phys. Commun.* 18: 171
Turner, J. L., Welch, W. J. 1984. *Ap. J. Lett.* 287: L81
Uchida, Y., Shibata, K. 1985. *Publ. Astron. Soc. Jpn.* 37: 515
Ulrich, R. K. 1976. *Ap. J.* 210: 377
Umebayashi, T., Nakano, T. 1980. *Publ. Astron. Soc. Jpn.* 32: 405
van Dishoeck, E. F., Black, J. H. 1987. In *NATO/ASI Physical Processes in Interstellar Clouds*, ed. M. Scholer. Dordrecht: Reidel. In press
Vogel, S. 1984. *Bull. Am. Astron. Soc.* 16(4): 921
Vogel, S., Kuhi, L. 1981. *Ap. J.* 245: 960
Vogel, S., Welch, W. J. 1983. *Ap. J.* 269: 568
Vrba, F. J., Strom, K. M., Strom, S. E. 1976. *Astron. J.* 81: 958
Wadiak, E. J., Wilson, T. L., Rood, R. T., Johnston, K. J. 1985. *Ap. J. Lett.* 295: L43

Walker, C. K., Lada, C. J., Young, E. T., Maloney, P. R., Wilking, B. A. 1986. *Ap. J. Lett.* 309: L47
Waller, W. H., Clemens, D. P., Sanders, D. B., Scoville, N. Z. 1987. *Ap. J.* In press
Walmsley, M. 1987. In *Interstellar Dust and Related Topics, Fermi School Lectures*, ed. S. Aiello. New York: Academic. In press
Walter, F. W. 1986. *Ap. J.* 306: 573
Walter, F. W. 1987. *Publ. Astron. Soc. Pac.* In press
Welch, W. J., Vogel, S. N., Plambeck, R. L., Wright, M. C. H., Bieging, J. H. 1985. *Science* 228: 1329
Westbrook, C. K., Tarter, C. B. 1975. *Ap. J.* 200: 48
Wetherill, G. W. 1980. *Ann. Rev. Astron. Astrophys.* 18: 77
Whitworth, A., Summers, D. 1985. *MNRAS* 214: 1
Wilking, B. A. 1985. In *Nearby Molecular Clouds*, ed. G. Serra, p. 104. Berlin: Springer-Verlag
Wilking, B. A., Lada, C. J. 1983. *Ap. J.* 274: 698
Wilking, B. A., Lada, C. J. 1985. In *Protostars and Planets II*, ed. D. C. Black, M. S. Matthews, p. 297. Tucson: Univ. Ariz. Press
Wilking, B. A., Taylor, K. N. R., Storey, J. W. V. 1986. *Astron. J.* 92: 103
Winkler, K. H., Newman, M. J. 1980a. *Ap. J.* 236: 201
Winkler, K. H., Newman, M. J. 1980b. *Ap. J.* 238: 311
Wolfire, M. G., Cassinelli, J. P. 1986. *Ap. J.* 310: 207
Wolfire, M. G., Cassinelli, J. P. 1987. *Ap. J.* In press
Wolfire, M. G., Churchwell, E. 1987. *Ap. J.* In press
Woodward, P. R. 1976. *Ap. J.* 207: 484
Wynn-Williams, C. G. 1982. *Ann. Rev. Astron. Astrophys.* 20: 587
Wyse, R., Silk, J. I. 1987. *Ap. J.* 313: In press
Yorke, H. W. 1977. *Astron. Astrophys.* 58: 423
Yorke, H. W. 1979. *Astron. Astrophys.* 80: 308
Yorke, H. W. 1980. *Astron. Astrophys.* 85: 215
Yorke, H. W., Krugel, E. 1977. *Astron. Astrophys.* 54: 183
Yorke, H. W., Shustov, B. M. 1981. *Astron. Astrophys.* 98: 125
Young, E. T., Lada, C. J., Wilking, B. A. 1986. *Ap. J. Lett.* 304: L45
Zeng, Q., Batrla, W., Wilson, T. L. 1984. *Astron. Astrophys.* 141: 127
Zheng, X. W., Ho, P. T. P., Reid, M. J., Schneps, M. H. 1985. *Ap. J.* 293: 522
Zinnecker, H., Tscharnuter, W. M. 1984. In *Proceedings of the Workshop on Star Formation*, ed. R. D. Wolstencroft, p. 83. Edinburgh: Royal Obs.
Zuckerman, B., Palmer, P. 1974. *Ann. Rev. Astron. Astrophys.* 12: 279
Zweibel, E. G., Josafatsson, K. 1983. *Ap. J.* 270: 511

ELEMENTS AND PATTERNS IN THE SOLAR MAGNETIC FIELD

Cornelis Zwaan

Sterrewacht "Sonnenborgh," Zonnenburg 2, 3512 NL Utrecht, The Netherlands

1. Introduction

The intricate time-dependent structure and the nonthermal processes in the solar atmosphere indicate the importance of magnetic fields in turbulent cosmical bodies. We are now aware that magnetic processes also occur in other stars, in protostars, in supernova remnants, in gaseous disks of galaxies, in accretion disks around compact stars, and in active nuclei of galaxies. The solar atmosphere, however, is the prime focus for the observational study of astrophysical magnetohydrodynamics (MHD) and plasma physics.

This selective review concentrates on observational studies that may reveal the MHD processes in the solar interior and photosphere that generate and shape the magnetic field.

The intrinsically strong magnetic field is contained in seemingly isolated elements, ranging from the thick bundles of flux tubes in sunspots to the hypothetical thin flux fibers (Section 3). These elements are arranged in the typical patterns observed in active regions and in the magnetic network (Sections 4 and 5). The processes of emergence of magnetic flux into the atmosphere (Section 6) and removal of flux from the photosphere (Section 8) are directly related to the magnetic structure and dynamics in the solar convective envelope.

Particular attention is paid here to the ends of the size spectrum of magnetic structures. The smallest observable scales are involved in the processes of interaction between magnetic field and convective turbulence, emergence of magnetic flux, and removal of flux from the photosphere. There are new results, and the still-open questions need to be formulated for planning subarcsecond studies from space. In Section 5 the observational data on the so-called intranetwork fields are summarized. It is argued that these fields probably are intrinsically weak, unlike the fields in the magnetic network and the active regions.

Large-scale patterns in solar magnetism and processes extending over years (Section 7) reflect the global dynamo action in the interior. Some of the recent developments are prompted by coronal studies, and some result from concerted solar-stellar studies. In this framework, chromospheric and coronal indicators of magnetic structure are discussed in Section 7.

2. Magnetic-Field Measurements: Possibilities and Limitations

The scope of observational studies of solar magnetism depends on the possibilities and limitations of magnetic-field measurement. Here we briefly introduce the technique using the Zeeman splitting of spectral lines; for a more comprehensive review, see Semel (1985). We consider a two-component model: Magnetic flux tubes of strength B and an inclination angle γ with respect to the line of sight fill a fraction f of the atmosphere, with the rest of the atmosphere being nonmagnetic.

Our discussion is restricted to measurements using the two spectra $I_L(\lambda)$ and $I_R(\lambda)$ in the opposite directions of circular polarization. From these spectra the *total intensity* $I(\lambda)$ and the *V parameter* of Stokes, $V(\lambda)$, may be obtained by

$$I(\lambda) = I_L(\lambda) + I_R(\lambda) \qquad \text{1a.}$$

and

$$V(\lambda) = I_L(\lambda) - I_R(\lambda). \qquad \text{1b.}$$

Consider a spectral line of the simplest magnetic splitting, yielding one undisplaced π component and two σ components equidistant from the π component located in the line center. The displacement of the σ components is proportional to $Bg\lambda^2$, where g is the Landé factor (which is dependent on the transition producing the spectral line).

If the line splitting is so large that the line components do not significantly overlap, then the wavelength separation between the extrema in the V profile is a linear measure of $Bg\lambda^2$, and hence of the field strength B. The amplitude A_V of the V profile depends linearly on the factor $f \cos \gamma$ and on a factor depending on the formation of the spectral line. Note that in this case the amplitude A_V does not depend on B.

In cases where the σ components overlap, the separation $\Delta\lambda_V$ is no longer a linear measure for $Bg\lambda^2$. In the extreme case where the magnetic splitting is much smaller than the width of the spectral line (component), it follows from Equation 1b that

$$V(\lambda) \propto Bgf \cos \gamma \cdot \lambda^2 \frac{\partial I(\lambda)}{\partial \lambda}, \qquad 2.$$

where $I(\lambda)$ is the line profile for $B = 0$. The wavelength separation $\Delta\lambda_V$ between the extrema λ in the V profile contains no information on the field strength B; $\Delta\lambda_V$ measures the distance between the inflection points in the $I(\lambda)$ profile. All that may be derived from a V profile in the case of small Zeeman splitting is a (model-dependent) measure for the magnetic flux density in the line of sight: $Bf \cos \gamma$.

For strongly inclined fields, the amplitude of the V profile is lost in the noise. However, if the average magnetic field strength Bf is sufficiently large, the presence of such a field is apparent in the total $I(\lambda)$ profile by the "ears" that are produced by the σ components. Brants (1985a,b,c) demonstrated the use of the width of the $I(\lambda)$ profile near the line base as a diagnostic for strong magnetic fields, particularly in the case of a very small amplitude of the $V(\lambda)$ profile. Another indication for strongly inclined fields is the magnetic enhancement of spectral line strength, which increases with increasing inclination $\gamma \to \pi/2$. This effect occurs in saturated spectral lines.

If the polarities are strongly mixed, the net V profile may be lost in the noise. In the case of strong fields, with sufficiently large filling factors, the fields are detectable by nearly unpolarized σ components in the I profile. This fact is used in the improved Unno (1956) method by Robinson et al. (1980) and Marcy (1983) to measure fields in cool stars.

The radial velocities of the plasma within the magnetic flux tubes cause Doppler shifts of the V profiles. If the radial velocity within the resolution element correlates with the field strength, then the net V profile departs from strict antisymmetry. Such departures have been detected (e.g. Stenflo et al. 1984, Solanki & Stenflo 1984, 1985). Landi Degl'Innocenti (1985) pointed out that such departures might be caused by departures from statistical equilibrium in the magnetic-substate populations. Wiehr (1985) found that the departures from antisymmetry and the mean Doppler shifts of the V profiles are quite different for different magnetic elements.

A longitudinal magnetograph measures the degree of circular polarization in the two flanks of a spectral line. Usually the spectral line and the measurement windows are chosen so that the signal is proportional to the mean flux density $Bf \cos \gamma$ over the resolution element. For sufficiently intense fields the intrinsic strength B may be determined separately from the factor $f \cos \gamma$ by using two spectral lines, or two window pairs in one line. One such line, or one window pair, is chosen so that the magnetograph signal does *not* depend linearly on B. The ratio of the two signals is a measure of the intrinsic field strength B.

Videomagnetographs, equipped with narrow birefringent filters that function as the spectroscopic analyzer and with two-dimensional detectors, use the available photon flux much more efficiently than do the scanning spectrographic magnetographs. Hence, videomagnetographs are suitable for the detection of fields of low flux density and for following the rapid evolution of the field.

Most of the photospheric magnetic flux measured with modern scanning magnetographs is present in elements with field strengths larger than about 1000 G. Outside sunspots, these strong magnetic elements are grouped in plages and in the magnetic network (see Section 4); unfortunately, the magnetic elements cannot be resolved with presently available instruments. The field strengths within these elements range from about 1000 to 2000 G. This range has been confirmed by at least three different techniques:

1. The shape of the V profile of the sensitive infrared spectral line Fe I $\lambda 15,648.5$ Å maps the field strength B directly. This approach has been used by D. N. B. Hall & J. Harvey (see Harvey 1977); with the Fourier Transform Spectrometer at the National Solar Observatory (Kitt Peak), the measurements of B have been confirmed with precision (Harvey et al. 1980).
2. Howard & Stenflo (1972) found that more than 90% of the magnetic flux in active regions (but outside sunspots) is contained in strong fields with strengths between about 1 and 2 kG using the line-ratio technique with scanning magnetographs. Stenflo (1973) deduced similar values for the quiet network. These results have since been confirmed by perfections of the line-ratio method (see Wiehr 1978, Stenflo 1985).
3. The detailed study of Stokes' I and V profiles across an active region by Tarbell & Title (1977) indicates field strengths exceeding 1000 G in points where the magnetograph signal exceeded 125 G.

Less certain are the flux densities, the filling factors, and the velocity field derived from V profiles. The uncertainties in the determinations from magnetograph signals are even larger. The precise interpretation of the data would require detailed three-dimensional models of the magnetic elements that incorporate detailed temperature and density structure, the magnetic field, the velocity field, and departures from local thermodynamic equilibrium (LTE). A setup for a simple two-dimensional flux-tube model demonstrates the complexity of the interpretation problem (see van Ballegooijen 1985a). Van Ballegooijen (1985b) points out that the magnetic flux contained in very thin flux tubes (diameters $\lesssim 50$ km, fluxes

$\phi \lesssim 3 \times 10^{16}$ Mx) is underestimated (if not altogether missed) if the flux tube is slightly inclined with respect to the line of sight.

Field configurations that may escape present techniques of measurements in $I(\lambda)$, $V(\lambda)$ profiles are (a) strong fields of a very low filling factor, and (b) weak fields, particularly if the polarities are mixed. The detection limit depends on the angular resolution and on the signal-to-noise ratio in the spectral windows.

The determination of the magnetic field *vector* is difficult because of the extremely high demands on the quality of the measurements of the four Stokes parameters I, Q, U and V, and because of the complexity of the schemes to interpret the measurements. There is progress, however: Vector magnetographs are operational in several countries, and the interpretation problems are being vigorously studied in various institutes. The recent developments are surveyed in workshop proceedings edited by Hagyard (1985). Most of the published vector-magnetograph studies deal with complex active regions, with particular attention paid to the occurrence of major solar flares. This topic, however, is beyond the scope of this review.

3. *The Magnetic Elements and the Hierarchy*

Observations discussed in Section 2 have indicated that most of the magnetic flux passing through the photosphere is concentrated in "elements," i.e. patches of high-field strength that are embedded in relatively field-free plasma (for the intranetwork field, see Section 5). The properties of an element mainly depend on the total magnetic flux passing through it, with the elements forming a hierarchy (see Zwaan 1978, 1981). Here we consider the quasi-static elements, postponing the discussion of transient phenomena occurring during rapid growth or decay to Sections 6 and 8. Table 1 presents an updated summary of the properties of the elements; in this section we discuss some of those properties that seem important for understanding the elements and their hierarchic order. For the semiquantitative interpretation in terms of models of magnetostatic flux tubes and flux-tube bundles, see summaries by Zwaan (1978) and C. Zwaan & L. E. Cram (in preparation).

Note that the magnetic axes of the elements are nearly vertical in the photosphere, an observation that is readily explained by the buoyancy of the tops of magnetostatic tubes. Strongly inclined fields are found in the penumbrae of spots, and hence as parts of large-scale structures, or in transient configurations, such as those that occur during flux emergence (Section 6).

In Table 1 the umbral field strengths have been bracketed between 2900 and 3300 G. Occasionally field strengths larger than 3300 G have been

Table 1 The hierarchy of magnetic elements

Property[a]	Sunspot with penumbra		Pore	Magnetic knot (micropore)	Faculae, network clusters	Flux fiber
	Large	Small				
Φ (10^{18} Mx = 10^{10} Wb)	3×10^4	500	250–50	≈ 10	$\lesssim 20$	$\lesssim 0.5$?
R (Mm)	28	4	—	—	—	—
R_u (Mm)	11.5	2.0	1.8–0.7	≈ 0.5	—	$\lesssim 0.01$
B (in G = 10^{-4} T)	2900 ± 400	2400 ± 200	2200 ± 200	≈ 1500–2000	—	≈ 1500

Overall contrast:	dark	bright
Cohesion:	single compact structure	cluster of
Behavior in time:	remain sharp during decay, shrinking	modulated by granulation
Occurrence:	exclusively in active regions	both inside and outside active regions

[a] Φ is magnetic flux, R is the average radius of a sunspot, R_u is the radius of sunspot umbrae and of smaller elements, and B is the magnetic field strength (at the axis of the element).

measured, sometimes even exceeding 4000 G (e.g. von Klüber 1947). In well-documented cases, however, these high values appear to refer to only a small fraction of a sunspot umbra or penumbra, the strong fields are nearly horizontal, and in cases where the evolution of the particular sunspot has been followed, these strong fields have been observed to be transient phenomena in the evolution of the complex sunspot (Livingston 1974).

There is a marked intrinsic scatter in the umbral field strengths; in the largest umbrae, the field tends to be somewhat stronger than in small umbrae. In small umbrae and in pores, the field strengths between 2000 and 2500 G listed in Table 1 are free from effects of stray light because a "purely umbral" spectral line has been used for the measurements (Brants & Zwaan 1982).

Probably the structure of a sunspot is best described as a bundle of flux tubes held tightly together somewhere well below the photosphere, with the top parts floating together by buoyancy (e.g. Parker 1979). Because of the overall order in sunspot structure, the term "composite model" seems more appropriate than "spaghetti model." In this description, the bright umbral dots are explained as the tops of essentially field-free columns between the intense flux tubes. The intrinsic spreads in the umbral field strength and in the mean umbral brightness translate into differences in the tightness of the flux-tube bundle. Over periods of several days to some weeks, sunspots may decay slowly by shrinking without much change of shape. Suddenly, the spot dissolves rapidly, and this deterioration is announced by the appearance of a system of very bright umbral dots (Zwaan 1968). This pattern suggests that the not-yet-determined constriction of the flux-tube bundle holds for a considerable time and then lets go.

In the hierarchy the magnetic knots take positions between the dark and the bright magnetic structure. At maximum resolutions of about 1″, they are not conspicuous in continuum or line wings. Hence, they show up as mere concentrations of magnetic flux—this is why Beckers & Schröter (1968) called these elements magnetic knots (see also Spruit & Zwaan 1981). At better resolutions, these features presumably correspond to the micropores observed by Dunn & Zirker (1973)—a slightly darker core surrounded by a bright ring. In the cores of the Ca II H and K lines, these knots or micropores appear as very bright faculae.

Magnetic structure that is bright both in the continuum and in most spectral lines consists of small elements with a strong tendency to cluster (see Section 4). These bright clusters are called *faculae* in active regions and *network clusters* in areas outside active regions. When observed in the photosphere or low chromosphere at high resolution ($<0.″5$), these clusters resolve into tiny bright points that are called *facular points* or *network bright points*. These bright points are found at the edges of the convective cells: In particular, they are found in intergranular lanes and at the boundaries of supergranules [see Muller (1985a) for a review and references]. Locally the points are aligned, and thus they form the crinkles of the "filigree" structure (Dunn & Zirker 1973).

The present magnetographs do not resolve the faculae and network clusters, but from arguments explained in Section 2 we know that intrinsically strong magnetic structure fills only a fraction of the resolution elements [>1 (arcsec)^2]. We describe this fine magnetic structure as collections of small elements, which we term magnetic *flux fibers*.

Because of the lack of resolution, the spatial and temporal relations between the bright points and the flux fibers are not yet clear. Down to scales of about $1.″5$, the spatial coincidence between the magnetic structures and the photospheric or chromospheric bright features is excellent (see Chapman & Sheeley 1968). At scales smaller than 1″, the relation is not known. The network bright points are short lived: Their mean lifetime is 18 min, with the individual lifetimes varying from 5 to 50 min (Muller 1983). It is difficult to accept such short lifetimes for corresponding magnetic fibers. A possible explanation is that a flux fiber is visible as a bright point only part of the time, possibly because the photometric visibility is modulated by the granulation. A fundamental question is: do clusters of flux fibers, each fiber having a specific (long) lifetime, properly describe the magnetic fine structure? Or should we rather adopt the picture of continuous sheets of vertical field "creeping" between the granules, as is suggested by Nordlund's (1983) numerical simulations?

Note that the typical flux per fiber ($\Phi = 5 \times 10^{17}$ Mx in Table 1) is set equal to the average flux per bright point derived by Mehltretter (1974); hence, this estimate is based on the assumption of a one-to-one relation between bright points and fibers.

Quite recently, the interpretation of the small-scale magnetic structure was relieved from the problem of having to explain a steady downdraft of about 2 km s^{-1} within the flux tubes—recent measurements using V profiles of high spectral resolution indicate no systematic downdraft in excess of 0.1 km s^{-1} (e.g. Stenflo & Harvey 1985).

4. *Patterns of Magnetic Elements*

In a young, large active region the complete hierarchy of solar magnetic elements is present—from large sunspots down to small faculae. For a description of the structure and evolution of an active region, see Sheeley (1981) and Zwaan (1981, and references therein). The emergence of an active region and the formation of sunspots are summarized in Section 6. Here we note that the typical active-region features—the compact plages of relatively high filling factor, the sunspots, the pores, and the micropores—are all formed during the emergence phase, within several days. Apparently these features are prefabricated in the convection zone, in a toroidal flux strand somewhere deep in or just below the convection zone, from where the magnetic loops forming the active region are assumed to originate. In a young active region, up to about 60% of the magnetic flux in one polarity may be present in one or more sunspots (N. R. Sheeley, private communication, 1980; Schrijver 1986, Chap. 11).

Once all the available magnetic flux has emerged, the decay of the active region begins: Sunspots diminish in size and in magnetic flux, the area of the region expands, and the average filling factor decreases (see Sheeley 1981, Figure 2.2). The spectrum of the magnetic elements becomes simpler. The dark elements are the first to disappear; one spot may last much longer than the other spots, but eventually only bright facular elements are left.

In a young active region, the part of the magnetic flux outside dark spots is largely in *plages*, areas that are more or less evenly covered with magnetic elements. Plages are made of faculae visible in photospheric lines, in chromospheric lines (such as Ca II H and K, Mg II *h* and *k*, Lα), and in ultraviolet emission lines originating in the transition region between the chromosphere and corona (such as the C IV and Si IV resonance lines).

Schrijver's (1986, Chap. 11) investigation of the relation between the chromospheric, transition-region, and coronal emission from active regions (as observed by Skylab) and the magnetic field (as recorded at Kitt Peak) has provided new insight concerning the plage phenomenon. Schrijver found that on Kitt Peak magnetograms of $2''\!.5$ resolution, a flux density level of 50 G defines a suitable outline for an active region. Within that contour, the flux density varies by an appreciable factor, but the *mean* flux density turns out to be nearly the same: $\langle B \rangle = 100 \pm 20$ for all active regions in Schrijver's sample, regardless of whether the region is large or small, young or old. This mean flux density for plages agrees with the value found by Tarbell et al. (1979). From the intrinsic field strength $B = 1500 \pm 500$ G, one finds typical filling factors between 5 and 10% for plages. Schrijver's comparison confirms that the plages as recorded in C II $\lambda 133.5$ nm have similar shapes to those of the magnetic plages in the photosphere; significant differences in area may be explained by the geometrical effects of flux loops of different temperatures.

The cohesiveness shown by plages is symptomized by the absence of "empty cells" of supergranular size. This suggests that at mean flux densities in excess of about 100 G, supergranular cells cannot develop. Within plages there is granulation, and locally there may be roughly circular open areas with sizes larger than granules but smaller than supergranules (Ramsey et al. 1977, Zwaan 1978, 1979). With the gradual dispersion of the active region, open areas of supergranular size appear, and the plages are gradually replaced by *enhanced network*. Note that Schrijver's (1986) value for the mean flux density (100 G) also applies to enhanced network within the 50-G contour. Gradually the enhanced network expands, becomes fainter, and eventually is lost in the "quiet" network. By then, the bipolar signature of the original active region is completely lost. At somewhat higher heliographic latitudes, some network remnants of following polarity (i.e. the polarity in the part trailing in the solar rotation) from a succession of decayed active regions may merge to form large areas of unipolar network (see Sections 7 and 8).

The roughly circular "holes" in the magnetic structure of active regions are attributed to convective cells. The sequence of phenomena in active regions indicates that the interaction between convection and magnetic field is *reciprocal: Field and convection expel each other*. Strong magnetic field is found outside convective cells. Convection adjusts to the relative freedom left by the field; if the convective heat flow finds its way blocked, then it bypasses the obstacle (e.g. sunspot). This reciprocal interaction is probably responsible for the three levels of magnetic flux concentration:

1. In umbrae of sunspots, the filling factor of the composite flux-tube bundle is nearly 100%, i.e. all normal convection is excluded.
2. In plages, the filling factor of between 5 and 10% apparently is large enough to suppress supergranular convection. Convection on smaller scales does occur, however: There is granulation, and occasionally there are local convective cells larger than granules but smaller than supergranules.
3. In enhanced network, there are cells of the size of supergranules. These cells live for several days, however, which is a period much longer than that of normal supergranules (Sheeley 1969, Livingston & Orrall 1974). This suggests that the magnetic structure is not dense enough to stop supergranulation but still strong enough to keep the supergranules arrested.

Another example of a long-lived convective cell of supergranular size that is kept in place by magnetic field is the *moat cell*, which may surround (part of) the periphery of a mature sunspot (Sheeley 1969, 1972, Harvey & Harvey 1973, Vrabec 1974).

The sizes of active regions, measured in surface area or in magnetic flux, cover several orders of magnitude (see Table 2). Only during the peak of their development do the large active regions show full-fledged sunspots. The lifetimes estimated in Table 2 refer to the visibility of the active region as a bipolar system. The lifetime is proportional to the magnetic flux at maximum development, with the constant of proportionality being about 2×10^{20} Mx day^{-1} (Golub 1980).

The number of active regions present on the Sun as a function of area increases rapidly with decreasing area. The tiny short-lived *ephemeral active regions* (a term often shortened to *ephemeral regions*) are very numerous, and they also appear well outside the activity belts defined by the large active regions (see Section 7).

The active regions stand out in chromospheric, transition-region, and coronal emission. Any of the emission lines from the chromosphere

Table 2 Active regions

	Magnetic flux[a] at maximum development	Lifetime
Large active region (with sunspots)	5×10^{21}–4×10^{22}	months
Small active region (no complete spots, maybe pores)	1×10^{20}–5×10^{21}	days–weeks
Ephemeral active regions	3×10^{18}–1×10^{20}	hours–one day

[a] Magnetic flux in one polarity (in Mx).

($T \lesssim 50{,}000$ K) and the transition region between the chromosphere and the corona ($100{,}000 \lesssim T \lesssim 500{,}000$ K) render magnetic plage, enhanced network, and "quiet" network with a remarkable similarity in shape, but there are systematic differences in areal extent. These relations may be explained qualitatively by attributing the origins of the emissions to low-lying parts of magnetic flux loops (Schrijver 1986, Chap. 11). The projection of the coronal region ($T \gtrsim 1$ MK) onto the solar disk observed in some EUV emission lines or soft X rays is not similar to the shape of a magnetic plage; this is readily explained by the coronal radiation originating in a system of high magnetic loops.

The relations between areas and total fluxes measured in the magnetic field and in various chromospheric, transition-region, and coronal emissions, integrated over entire active regions (see Schrijver 1986), are remarkable in their tightness. The departures from linearity are small but significant in some of these relationships. These relations may be used as constraints on theories of chromospheric and coronal heating and as building material in simulating solar-type stars as a function of the level of activity (Schrijver 1986, Chap. 12). Here we emphasize the conclusion that the radiative losses from the chromospheric, transition-region, and coronal temperature regimes in active regions depend tightly and nearly linearly on just the total magnetic flux in the plage, i.e. on the plage area. There is some spread in the average specific intensities in all of the emissions, but again these specific intensities correlate very well—even the specific intensities in the "quiet" network seem to fit the relations established for active regions. Moreover, Schrijver (1986, Chap. 11) demonstrates that the average specific intensity of active regions in one emission line (Mg X), and hence in all emissions, depends slightly on the total magnetic flux in the photosphere, with little scatter. This leaves little room for effects of the overall topology on chromospheric and coronal heating—intuitive ideas, such as the one that young regions or regions with a complex magnetic structure would be brighter than other active regions, apparently do not hold. (Young complex regions, however, do show more flares and flarelike activity.)

The ephemeral regions are frequently overlain by coronal bright points, which are small, short-lived emission features that are visible in soft X rays and in EUV spectral lines (Golub et al. 1977, Golub 1980). The converse—that coronal bright points always overlie ephemeral regions (Golub et al. 1977)—is a too hasty generalization that has created confusion as to the occurrence of ephemeral regions during the activity cycle (see Section 7).

Harvey (1985) analyzed the relation between coronal bright points and magnetic features using transient "dark points" in He I $\lambda 10{,}830$ Å spectroheliograms as proxies for coronal bright points. (For the correspondence between He I dark points and X-ray bright points, see the paper by Harvey et al. 1975a). Indeed, nearly 90% of the He I dark points appear to correspond to small magnetic bipoles, but no more than one third of these are associated with ephemeral regions. The other two thirds correspond to close "chance encounters" of small magnetic elements of opposite polarity (see Section 8). Working the other way around, Harvey (1985) found that only 40% of the ephemeral regions stand out as distinct dark points in He I $\lambda 10,830$ Å.

One consequence is that a convincing identification of an ephemeral region requires more than one high-resolution magnetogram to show the characteristic evolution: An ephemeral active region starts out as a very close bipole, with steadily increasing distance between the poles. Martin et al. (1985a) report that Hα filtergrams show that the poles of some of the ephemeral regions are *connected* by dark fibrils or arch filaments, such as occur in other emerging flux regions (Section 6). In contrast, the other temporary bipoles, called "pseudo-ephemeral regions" by Martin and coworkers, sometimes show a dark fibril *dividing* the polarities, which is readily interpreted as a miniature filament indicating the polarity inversion line. While these Hα features are interesting indications for the different origins of the two types of magnetic bipoles, they too do not present a reliable criterion for the discovery of ephemeral regions from single frames.

5. *Ubiquitous Fields: The Network Field and the Intranetwork Field*

Large active regions occupy only a small fraction of the photosphere, and they are not observed outside the activity belts. In this section I summarize the magnetic fields that cover the entire photosphere, although with only small filling factors. The practice of relating these fields to the "quiet" Sun ignores the fact that some components in these fields depend on the solar activity cycle (Section 7), and this is why the term "ubiquitous fields" is used here.

The *network clusters* stand out as relatively steady in both location and flux. The clusters are constantly changing in shape, but usually they retain a recognizable coherence throughout an observing day [see illustrations in Martin (1984) and in Livi et al. (1985)]. Over periods of hours, some clusters change markedly, either by splitting, by merging with a field fragment of the same polarity, or by a complete or partial cancellation with a field fragment of opposite polarity (see Section 8). Magnetic fluxes range from about 2×10^{18} Mx up to about 3×10^{19} Mx. The network clusters consist of intrinsically strong flux fibers, with field strengths between 1000 and 2000 G (Sections 2, 3). The crude network indicated by the magnetic clusters has been found to correspond to the boundaries and, in particular, to the vertices of supergranular velocity cells with diameters of about 30,000 km (Simon & Leighton 1964).

On single magnetograms the poles of ephemeral active regions cannot be distinguished from network clusters. A few magnetograms recorded at a time interval of a few hours reveal their nature as poles moving apart. Although ephemeral regions may occur far outside the sunspot belts (see Section 7), I do not classify them with ubiquitous fields but rather consider them as the smallest active regions (see Sections 4 and 6).

The second category of the ubiquitous fields, the *intranetwork field* (INF), can only be discovered by sensitive magnetographs during fair seeing conditions. With the 512-channel magnetograph at Kitt Peak, Livingston & Harvey (1975; see also Harvey 1977) found magnetic flux within the magnetic network cells. The polarities are mixed on a scale of a few arcseconds, and they are independent of the dominant polarity of the surrounding network field. The existence of the intranetwork field is confirmed with the videomagnetograph at Big Bear Solar Observatory (see Martin 1984). The magnetic fluxes in the fragments in the INF are typically from a few times 10^{16} Mx (the detection limit; see Zirin 1985) to a few times 10^{17} Mx—hence, they are one to two orders of magnitude smaller than the fluxes in network clusters. The fragments of the INF are in motion, with typical velocities of 0.3 km s^{-1}. In some network cells INF fragments are seen to appear near the center and then to be swept to the boundaries, but frequently nonradial displacements occur. INF fragments of the same polarity may coalesce; in this way, fluxes of up to a few times 10^{18} Mx may be built up.

Because of internal motions, the pattern of the intranetwork field markedly changes during one hour. If seeing permits, however, individual fragments can be traced for several hours or more. Eventually the fragments disappear by merging or by cancellation with network clusters, with poles of ephemeral regions, or with other INF fragments (Martin 1984, Livi et al. 1985).

The description above suggests that the intranetwork field is qualitatively different from the network field and the ephemeral regions. Tarbell et al. (1979) found that the intranetwork field must be rather diffuse and *intrinsically* weak, with an upper limit of 50 G on the rms vertical component of the field. They derived this figure by observing the absence of any intranetwork field on their photodigital magnetograms of limited sensitivity (50 G) but high angular resolution (0″.5). Note that this upper limit and the Big Bear videomagnetograph estimates of flux densities, ranging from about 5 G to about 40 G, are compatible by a narrow margin.

While the measured flux densities in the network field decrease rapidly toward the solar limb, the magnetic signal in INF fragments decreases only slightly toward the limb (S. F. Martin, private communication, 1986). This is compatible with an intrinsically weak INF because, unlike the strong flux tubes, a weak field is not set vertical by buoyancy (H. C. Spruit, private communication, 1986).

One important consequence of the weakness and the mixed polarities of the INF fragments is that the INF does not penetrate into the outer atmosphere. The magnetic field in the transition region and the corona is determined by the photospheric fields in active regions, in ephemeral regions, and in the magnetic network.

The indications for intrinsic weakness of the INF make one wonder about the long lifetimes of the fragments. The fragments are expected to respond to the granulation by fluctuations in the field strength—this response would be qualitatively different from the mutual expulsion of granulation and strong network elements. Subarcsecond angular resolution is also required for the study of the relation between INF fragments and network elements, because this interaction is expected to occur in the intergranular lanes.

6. *Emergence of Magnetic Flux*

Some recent developments, such as the case study in Brants's thesis of an emerging flux region (Brants 1985a,b,c, Brants & Steenbeek 1985), have been reviewed in the context of the previous literature by Zwaan (1985b); hence, I restrict the discussion here to a brief summary.

The emergence of the top of a loop bundle of flux tubes, as pictured in Figure 1, qualitatively describes the events in an emerging flux region (EFR): the appearance of an initially small, compact, bright bipolar plage; the growing separation between the poles; the arch filament system, which is seen in the Hα line core to connect the plages of opposite polarity; and the coalescence of small magnetic elements into sunspots. Active regions grow from the inside outward, at rates of up to about 1×10^{20} Mx hr^{-1}. Usually several bipolar emerging flux regions take part in the formation of one large active region.

The well-known downward flow in the chromospheric legs of the arch filament system continues in a photospheric downdraft of about 2 km s^{-1}. [For illustrations, see Zwaan et al. (1985, Figure 4) and Brants (1985b, Figure 5).] A well-observed photospheric downdraft area was seen to develop into a dark sunspot pore (Brants & Steenbeek 1985, Figure 10). Zwaan (1985b) interprets this process as the *convective collapse* that transforms the emergent field of intermediate strength ($B \lesssim 500$ G) into a strong field; Spruit (1979) calculated that thin flux tubes in the final collapsed

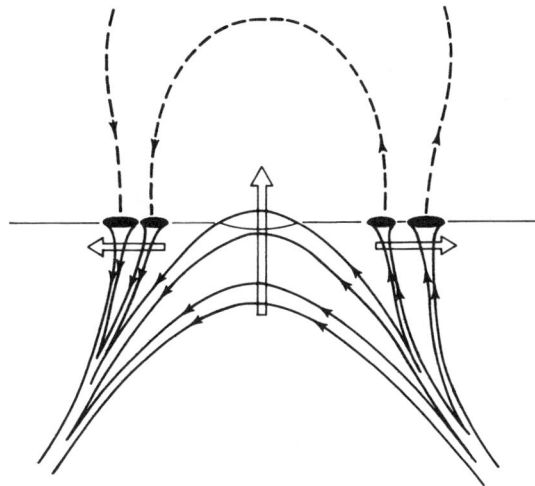

Figure 1 Model for the emergence of magnetic flux, separation of polarities, and coalescence of sunspots. Broad open arrows indicate local displacements of flux tubes. (Real emerging flux regions consist of many more separate flux loops.)

state have field strengths in excess of 1500 G. Once a dark pore is formed, with a field strength larger than 2000 G, the downward flow stops.

In the rising-loop model (Figure 1), the coalescence of magnetic elements of the same polarity into a sunspot may be explained by the buoyancy of the tops of the collapsed flux tubes, provided that somewhere in the convection zone a constriction keeps the flux bundle together.

The majority of the pores in Brants's data grew, but some were seen to decay; one small pore dissolved again within two hours after its birth. Note that the dissolution of pores may be caused by the same convective instability as the collapse: A flux tube collapses if the initial flow is downward, but it expands if there is an upward flow (see Spruit 1979). For a comprehensive description of growth and decay of pores, see Brants & Steenbeek (1985).

In the parts of the active regions where magnetic flux is still emerging, there are alignments of abnormally dark intergranular lanes in the granulation (Brants & Steenbeek 1985, Figure 9). These lanes probably correspond to tops of magnetic loops surfacing through the photosphere. Such an alignment lives for about 10 minutes. At the emergence site, magnetic field and granular convection are in a transient state: There are no pores and no faculae with the filigree structure. The magnetic field is of intermediate strength and strongly inclined, as is indicated by the polarity reversals. Brants (1985a,b) estimates that the local field strength is $B = 500 \pm 300$ G, which is close to the equipartition field strength in the top of the convection zone. The scanty data combine into the following tentative measures for a single emerging flux loop: diameter 1500 km, flux 1×10^{19} Mx [which agrees with Born's (1974) estimate for the flux per filament in an arch filament system], rising through the photosphere at an Alfvén speed of a few kilometers per second.

Ephemeral regions and emerging flux regions that eventually contribute to the formation of large active regions cannot be distinguished during the early phase of their development. The larger ephemeral regions develop arch filament systems that are easily visible in Hα. These data suggest that ephemeral regions are also formed by the emergence of Ω-shaped flux loops (Figure 1). The flux tubes in ephemeral regions probably experience the convective collapse, which suggests that the magnetic field becomes intrinsically strong. Another indication of the strength of the field is that the visibility of ephemeral regions on magnetograms decreases steeply toward the solar limb (Harvey et al. 1975b). This indicates that the field is very nearly vertical, an observation that is compatible with the buoyancy of strong flux tubes.

The appearance of the intranetwork fragments is quite different: There are no distinct bipoles in which the poles separate. At present there is no more definite description than "INF fragments spontaneously appear in the intranetwork space" (Martin 1984). Current speculations on the nature of the INF are summarized in Sections 8 and 9.

7. *The Activity Cycle: Large-Scale Patterns*

The solar activity cycle is variable in several respects: Neither the period nor the amplitude is constant. There are some rules, however, that apply to the vast majority of the active regions, and these rules may reveal fundamental properties of the magnetic structure below the atmosphere. The majority of the bipolar active regions, with the exception of the smallest regions, are oriented nearly east-west, at a small inclination such that the polarity preceding in the solar rotation is closer to the equator. Moreover, most active regions follow Hale's polarity law. These rules suggest that active regions arise as Ω-shaped loops from a toroidal flux strand, and that this toroidal magnetic flux system is intrinsically strong. Only if the field strength exceeds the equipartition strength B_{eq} may the field withstand the disordering effect of turbulent convection. Even for a large active region the magnetic flux may be stored in a toroidal bundle with a diameter conveniently smaller than the local pressure scale height, provided that the location is close to the bottom of the convection zone and the field strength exceeds B_{eq} (Zwaan 1978, 1985a).

Many observers have noted that active regions (or sunspot groups) tend to appear in the immediate vicinity of an existing active region or at the site of a previous active region. Such a sequence of active regions is often called an *activity complex*. If there is more than one bipolar active region at one time, then such a cluster is called a *complex region* or a *complex sunspot group*. The properties of the clustering tendency stand out most clearly if one focuses on the *emergences* of active regions, or on sequences in sunspot groups (Becker 1955). Here we call complexes defined as the loci of flux emergences *active nests*. Gaizauskas et al. (1983) studied strong bipolar magnetic features in the Kitt Peak synoptic magnetic maps, and Castenmiller et al. (1986) investigated sunspot groups.

Nests form within 1 month, and then they are maintained for typically 3 to 6 months by fresh injections of new active regions (Gaizauskas et al. 1983). Typical lifetimes of sunspot nests range from 6 to 15 months, however, if gaps of up to 2 months are allowed in the presence of sunspots (Castenmiller et al. 1986). Adopting 4 months as the arbitrary lower limit to the lifetime of a sunspot nest, we find that at least 30% of the sunspot groups are members of nests.

Many nests form double structures, in which the components may remain equidistant or may diverge. Each single nest, as well as each component of a double nest, rotates at its own constant and well-defined rate about the Sun. The intrinsic scatter in the rotation rates appears to be much larger than the latitude-dependent trend in the differential rotation. The displacements of nests in latitude are very small (less than a few meters per second). Nests do not show a meridional circulation in excess of 3 m s^{-1}.

Single nests and components of double nests are quite compact: Centroids of active regions fall within an area comparable to that of a medium-large sunspot group (Becker 1955, Castenmiller et al. 1986). Note that the term "active longitude" is misleading: Active nests are also bounded to a small range in latitude.

During the active lifetime of a large nest, its total magnetic flux remains steady to within a factor of two; the magnetic polarities are closely balanced (Gaizauskas et al. 1983). Since most of the magnetic flux appears to remain within the nest, this flux must emerge and be removed at equally large rates, and in situ. In a large nest some tens of large active regions are seen to emerge, and hence a large amount of magnetic flux is processed (see Section 8). Only after the emergence of new active regions has stopped does magnetic flux disperse. Usually the remnant flux does not rotate at the rate of the previous nest, and thus it is sheared by differential rotation.

There are some indications for a large-scale pattern in active regions and nests. Gaizauskas et al. (1983) suggested a wavelike pattern in longitude, with a wavelength of 45° (i.e. 5.5×10^5 km).

Stenflo & Vogel (1986) analyzed 25 years of magnetograms obtained at Mount Wilson and at Kitt Peak in terms of spherical harmonics. Their analysis is restricted to zonal modes (i.e. $m = 0$) that are symmetric about the rotation axis. Power spectra of the harmonic coefficients bring out just one dominant frequency v for all odd harmonic degrees in the range $1 \leq l \leq 13$; this resonance frequency corresponds to the magnetic-cycle period of 22 years. For the even modes, the pattern in the v-l diagram is entirely different: The resonance frequency increases with degree l, at least up to $l = 14$. Stenflo & Vogel do not offer an explanation, but it seems clear that the v-l diagrams provide important constraints on theories concerning the origin of the magnetic fields.

The magnetic network varies in polarity characteristics and in mean flux density with the phase in the activity cycle and with latitude. Hence, the magnetic Sun is variable everywhere—there are no really quiet regions. Giovanelli (1982) distinguished two types of network: *mixed-polarity* network, in which the ratio between the mean flux densities in the opposite polarities $|\bar{B}_{major}/\bar{B}_{minor}|$ is between 1 and 2.5; and the *unipolar* network, in which the dominant polarity contributes more than 80% of the absolute flux, i.e. $|\bar{B}_{major}| > 0.80\{|\bar{B}_{major}|+|\bar{B}_{minor}|\}$. For latitudes of up to 60°, mixed polarities dominate during the 5 years around sunspot minimum. About 2 years after the minimum in 1976, the mixed polarities were rapidly replaced by large areas of unipolar network over a large part of the sunspot belts (up to 40° latitude). These unipolar network fields originate in large active regions of the new cycle.

The dominance of mixed-polarity network around sunspot minimum favors "chance encounters" between network clusters of opposite polarity. The bipoles thus created are mainly responsible for the X-ray bright points (Harvey 1985), which explains why these bright points reach a maximum around sunspot minimum. Therefore, X-ray bright points may not be considered as tracers for ephemeral regions. This eliminates the basis for the suggestion by Golub et al. (1979) that the solar activity cycle would be an oscillation in the spectrum of the emerging magnetic flux rather than a variation in the amount of flux.

In fact, the number of ephemeral regions varies nearly in phase with the number of large active regions, with the minimum of ephemeral regions preceding sunspot minimum by about one year (Martin & Harvey 1979, Harvey 1984). Ephemeral regions are more broadly distributed in heliographic latitude than are sunspot groups, but small active regions indicate a gradual transition between these latitude distributions (see Figure 2 in

Martin & Harvey 1979). Hence, there is no sharp demarcation between ephemeral regions and active regions—ephemeral regions may constitute the extreme end of the size spectrum of active regions.

The normalized distribution function for sizes (areas, or magnetic fluxes) of active regions is surprisingly constant throughout the solar cycle, with the exception of the smallest regions (Schrijver 1986, Chap. 12; see also Tang et al. 1984). The relative number of the smallest regions (i.e. regions with Ca II plage area smaller than about 2×10^{-4} of the visible hemisphere) is somewhat larger during sunspot minimum than during the rest of the cycle. This departure is probably caused by the new cycle starting earliest in the smallest active regions and ephemeral regions, as noticed by Martin & Harvey (1979) and Harvey (1984).

The brightness variation of large sunspot umbrae with the phase of the cycle (see Maltby et al. 1986) probably is an important indication of the time-dependent condition of the subphotospheric flux bundles; it is not yet explained.

Coronal holes stand out as dark in all coronal radiations, and large coronal holes are visible in He I $\lambda 10,830$ Å as a lack of absorption (see Kahler et al. 1983). The magnetic field in coronal holes is predominantly unipolar, and it extends away from the Sun in diverging open lines of force. High-speed solar-wind streams originate in coronal holes; these streams cause geomagnetic storms, which may recur several times because many coronal holes live for more than five solar rotations, and some endure for more than a year.

Although coronal holes were discovered in optical data by Waldmeier (1957), it was the Skylab mission (May 1973–February 1974) that boosted the research of these phenomena. An excellent monograph on coronal holes has been edited by Zirker (1977). Note, however, that several conclusions and ideas drawn from the Skylab data, obtained during declining activity, have not been confirmed by observations since 1975. For instance, the nearly rigid rotation derived from a few large coronal holes during the Skylab mission is not found in 1978–79 data (Sheeley & Harvey 1981, Levine 1982) and in 1982–83 data (Shelke & Pande 1985).

Only parts of the unipolar network areas contain coronal holes, and a hole covers just a part of the corresponding unipolar area. Individual holes cover between 0.1 and 3% of the visible hemisphere; polar holes are often larger, up to about 10% of a solar hemisphere. During Skylab, up to 20% of the Sun was covered with coronal holes, and three quarters of the hole area was over the polar caps.

In holes at low and middle latitudes, the mean magnetic flux density varies during the cycle, albeit with a large intrinsic scatter. Values range between 1 and 7 G around minimum and between 3 and 36 G near maximum (Harvey et al. 1982). The total-flux ranges are $2-5 \times 10^{21}$ Mx (around minimum) and $10-15 \times 10^{21}$ Mx (near maximum); these ranges correspond to the flux in one polarity in one medium-large active region or one large active region, respectively. The coronal holes at low latitudes largely determine the measured flux density averaged over the solar disk (which rarely exceeds the 1-G level), but a fraction of the net (i.e. locally not balanced) flux is rooted in active regions (Levine 1977, 1982).

Polar coronal holes are present throughout most of the solar cycle. They are absent or intermittent only during a few years near solar maximum, when the dominant polarity of the polar caps changes sign. Even in well-developed polar holes the mean flux density is low, "no more than a few gauss" (Howard & LaBonte 1981), which corresponds to a net flux of no more than 10×10^{21} Mx (Harvey et al. 1982). This figure agrees with Sheeley's (1976) estimate from an analysis of polar faculae. Somewhat larger values follow from Giovanelli's (1982) study. From data obtained with the Stanford magnetograph in 1977, Svalgaard et al. (1978) derived a net flux of about 30×10^{21} Mx for either polar cap. The discrepancies remind us that the interpretation of magnetograph signals from elements close to the limb is uncertain.

The processes of formation and maintenance of coronal holes are not yet well understood. The holes at low and middle latitudes particularly pose problems; for example, how are they maintained against differential rotation? One prerequisite for a hole is the existence of a sufficiently large unipolar network area; these areas are formed by one-polarity remnants of large active regions. Moreover, the magnetic field of the network area should have been opened to space. Bohlin (1977) pointed out that a lower latitude hole is formed as a result of the emergence of a new active region nearby. Thus these holes, although located above quiet areas, lie directly adjacent to active regions.

Levine (1977) studied the formation of coronal holes, comparing Skylab data with models of open magnetic structures derived with the potential-field extrapolation technique from magnetograms. The growth of coronal holes lags behind the formation of the open magnetic field pattern by one-half to one solar rotation. More specifically, only some of the calculated open structures exist as coronal holes—most of the rest of the open structures are associated with parts of active regions. A large open magnetic structure, which may show a coronal hole during part of its existence, is usually associated with a succession of different active regions. This suggests that long-lived coronal holes are caused by a *specific pattern* in the *emergence* of *active regions*, such that flux of the appropriate polarity is added to the neighboring open structure.

We conclude that there must be long-lived patterns in the emergence of active regions such that two types of configurations are created: the nests (or complexes), and the coronal holes (or more generally, the open-field structures). Note that nests and coronal holes share several properties: long lifetimes, individual and steady rotation rates (for coronal holes, see figures in Shelke & Pande 1985), and the involvement of many active regions. Apparently there are no reports on the relationship between coronal holes and active nests, but probably both patterns reveal fundamental processes in solar magnetism.

8. *Transport of Magnetic Flux*

Whereas the emergence of magnetic flux is conspicuous enough, flux leaves the photosphere with little show. Yet magnetic flux must be removed from the photosphere on a variety of time scales. In many ephemeral regions the flux in at least one of the polarities is lost within a few hours after emergence. Wallenhorst & Howard (1982) and Wallenhorst & Topka (1982) found that magnetic flux disappears from decaying active regions without a visible transport of flux. Howard & LaBonte (1981) deduced that the average rate of flux removal from active regions is about 10% day^{-1}. Large rates of flux emergence and flux removal occur in active nests, on time scales of a few weeks (Section 7). The transformation of unipolar network areas into mixed-polarity network, and vice versa (see Section 7), involves years. The polar-field reversals are produced by episodic streams of monopolar flux transported from the activity belts to the poles (Howard & LaBonte 1981). It is clear that the amount of magnetic flux that emerges during one activity cycle must be removed again from the photosphere during one cycle period.

Before we turn to direct observational indications of flux removal from the photosphere, let us first consider possible modes. Because of the low electrical resistivity of the solar gas, the magnetic field cannot free itself from the gas and escape. In order to remove magnetic flux from the photosphere, it must be transported together with the gas: Flux removal requires a proper velocity field. Note that diffusion spreads magnetic flux over a larger area but does not remove flux from the atmosphere. But the spreading of flux requires an appropriate velocity field, if this spreading should happen at an observable rate. For instance, the dissolution of sunspot pores (Section 6) may be caused by a convectively enhanced upward flow in the flux tube.

Figure 2 illustrates modes for removal of magnetic flux from the photosphere, starting with two patches of magnetic flux of opposite polarity. In all cases a velocity is required that converges on the polarity dividing line in order to bring patches of opposite flux together and to start the convective downdraft that is needed to pull the (lowest) flux loop downward. If the

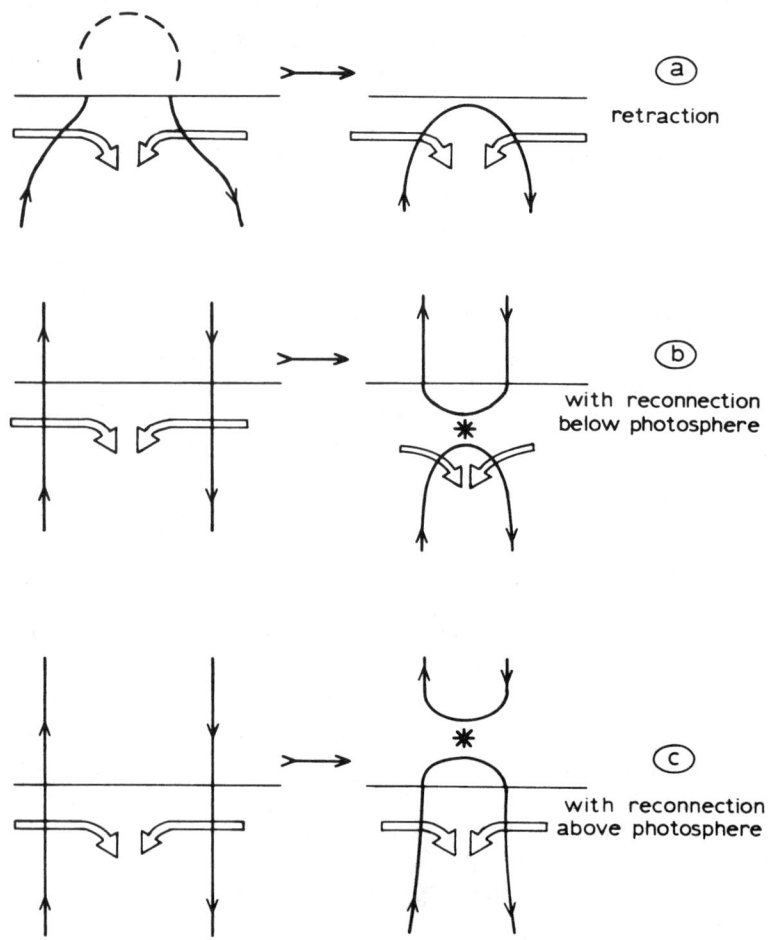

Figure 2 Modes for removal of magnetic flux from the photosphere (horizontal line). Broad open arrows symbolize the required converging fluid flows, and asterisks indicate sites of field reconnection. In order to complete the flux removal, in panel (*b*) the upper loop must be lifted out of the photosphere, and in panel (*c*) the lower loop must be pulled down.

two poles are still connected by a coronal field loop, then the flux removal is just a retraction of the flux loop (Figure 2*a*). In other cases there is another prerequisite: The fields need to be reconnected, either below or above (or in) the photosphere. In the case of flux reconnection below the photosphere (Figure 2*b*), the removal of the upper U-shaped flux loop is hampered by the load of gas trapped in the loop. Hence, flux removal by retraction (Figure 2a), if necessary after reconnection above the photosphere (Figure 2c), seems a more efficient process to remove flux from the photosphere. Yet there are reasons to consider the partial emergence of U-loops in the Sun as a viable process and to search for observational evidence in the atmosphere.

U-loops may be formed by two adjacent Ω-loop emergences of active regions from the same toroidal flux strand in the convection zone (see Parker 1984, his Figure 2a). When the horizontal tube is carried upward over several density scale heights, it expands enormously, and thus the field strength drops: The field becomes intrinsically weak. The buoyant instability suggests that the originally horizontal tube becomes undulatory, so that it surfaces in the atmosphere like a sea serpent. Parker (1984) points out that by reconnection in the upper part of the sea-serpent structure, a small fraction of the flux in the original U-loop may escape into space.

A. A. van Ballegooijen, A. M. Title & H. C. Spruit (preprint) invoke U-loops as the origin for the intranetwork field, and they indicate how through the sea-serpent process two distant magnetic-flux regions of opposite sign (the original vertical legs of the U-loop) may lose flux, seemingly in situ if the sea-serpent process is not noticed owing to a lack of resolution or sensitivity.

We now turn to observational indications of transport of magnetic flux. Pertinent observations have been obtained by means of the Caltech video magnetograph, both during its early stages (Smithson 1973) and particularly during its recent level of sophistication (Martin 1984, Livi et al. 1985, Martin et al. 1985a,b, Zirin 1985).

Flux transport involving magnetic elements of one polarity is seen in the splitting of network clusters (Smithson 1973) and in the merging of magnetic elements (in any combination of network clusters, poles of ephemeral regions, or INF fragments). Splitting and merging of network clusters do contribute to changes in the magnetic network over a period of days, but these processes do not drastically alter the field structure. The largest magnetic elements that occasionally are seen to form by merging of elements in enhanced network are sunspot pores (S. H. B. Livi & S. F. Martin, in preparation), but the great majority of pores are formed in emerging flux regions.

Magnetic structure is effectively changed by *cancellation* of magnetic flux in the encounter of two magnetic elements of opposite polarity. As in recent papers by S. F. Martin and coworkers, the word "cancellation" is used here as a purely descriptive term: Two elements of opposite polarity are seen to approach each other, whereupon there is a steady loss of magnetic flux in both of the encountering features. This loss usually continues until the smallest flux element has completely disappeared.

Livi et al. (1985) describe in detail the cancellation processes occurring between opposite-polarity fragments present in the network, ephemeral regions, and intranetwork field. The estimated flux-loss rate per cancelling pair is found to range from 10^{17} Mx hr^{-1} to 4×10^{18} Mx hr^{-1}. The present data do not allow a distinction between the modes of flux removal. Most of the cancelling flux elements were previously unrelated, which indicates that reconnection of field lines—either above or below the photosphere— is a prerequisite (Figure 2b,c).

In the first part of this section, we mentioned evidence that in many active regions large parts of the magnetic flux do somehow disappear at a large rate. Zirin (1984) shows a striking example of an active region in which, after all flux had emerged, the faculae of opposite polarity come back together and vanish completely. Martin et al. (1985b) describe in detail the flux disappearance observed during the decay phase in an active region of modest size. The magnetic plages of opposite polarity are seen to fragment; small fragments of opposite polarity move toward the polarity dividing line, where they cancel. The flux cancellations occur mainly in a narrow strip of a few arcseconds around the polarity dividing line; several bipoles are observed to be cancelling simultaneously or in rapid succession. In this decaying active region, flux cancels at a typical rate of 10^{19} Mx hr^{-1}.

Flux removal near the polarity dividing line within an active region suggests submergence of the field, but it need not be a simple retraction: Some previous reconnection of the field may be necessary. The case of flux disappearance from the active region described by Rabin et al. (1984) does suggest simple retraction. But certainly retraction of original bipolar pairs is not the only mode of flux removal from active regions; for example, Martin (1984) has never seen an ephemeral region contract and then submerge again. Usually the separating poles of an ephemeral region remain sharp until they merge or cancel with a network cluster or with a pole of some other ephemeral region. As a result of such collisions, 90% of the ephemeral regions disappear within about 1 day, but 10% of these regions live longer, up to about 6 days (K. L. Harvey, private communication).

The fast removal of flux that must occur in some large active regions, particularly in active nests, has not yet been observed in sufficient detail.

The rapid growth and decay of isolated magnetic features, with no indication of where the magnetic flux came from or went to, have been observed by Wilson & Simon (1983), Simon & Wilson (1985), and Topka et al. (1986). Topka et al. followed for one hour the magnetic fluxes in 28 isolated unipolar features in an old active region. Seventeen features did not change significantly, but the fluxes in the other 11 features changed markedly. For the majority of the changing features, no plausible origin or destiny of the flux changes could be found. Yet Maxwell's law $\mathbf{V} \cdot \mathbf{B} = 0$ requires flux transport. Apparently there are covert ways of flux transport (perhaps by the sea-serpent process mentioned above) that have not yet been observed.

The urgent problems of magnetic-flux changes require new measurement techniques, with both increased sensitivity and subarcsecond resolution.

The large-scale but slow changes in the magnetic field, such as the creation of large unipolar network areas, mixed-polarity areas, and polarity changes in the polar caps, require large-scale velocity fields of long duration. Since these changes involve magnetic flux that has emerged at different times, reconnection of fields is a prerequisite. One attractive idea is that the long quiescent filaments mark the sites of field reconnection in the corona, and that the photospheric strips underneath are the sites of flux removal.

9. *Concluding Comments*

In this section several observational aspects are combined in an attempt to elucidate how various zones in the Sun participate in the creation of the time-dependent magnetic field and to identify the processes that are involved.

The intrinsically strong magnetic fields emerge in *pulses*. The smallest "building block" is the emerging flux region (Section 6). A large active region is formed by a succession of several of these bipolar emerging flux regions. A sequence of active regions appearing within a small area forms an active nest or complex (Section 7). The east-west orientation of active regions indicates that the fields are rooted in strong toroidal fields; Schüssler (1983) lists several reasons why the toroidal field system must be rooted near the bottom of the convection zone. Hence, the large-scale magnetic structure observed in the atmosphere is prefabricated in the deepest part of the convection zone. We do not know the processes that constrict the flux-tube bundle in the composite sunspot or the processes that maintain the repetition of active regions in an active nest for many months. Usually some large-scale convective velocity field ("giant cells") is held responsible for the appearance of sunspots, active regions, and active nests. The distinct effects suggest, however, that the processes themselves involve some kind of interaction between large-scale convection and the magnetic field.

Processes in the top of the convection zone and in the photosphere help shape the magnetic field in the atmosphere. Radiative cooling in the photosphere and the strong superadiabaticity of the layers underneath cause the convective collapse that concentrates emerging flux tubes to field strengths over 1 kG (Section 6). Convection presumably erodes large-scale magnetic elements, such as sunspots. In parts of active regions where the average magnetic flux density is large enough, the interaction between convection and the magnetic field leads to plages with mean flux densities of 100 G that are without supergranular flow (Section 4). The flux distribution of network clusters is determined by the splitting of clusters, by the merging of elements of the same polarity, and by the cancellation of elements of opposite polarity (Section 8). Occasionally, sunspot pores are formed by the merging of clusters in enhanced network; apparently pores are the largest elements that may be formed long after flux emergence.

The corona probably plays an important role in solar magnetism if indeed the reconnection required for flux removal between large areas of opposite unipolar flux occurs in the corona. In particular, the reconnection near the polar caps is extremely important if the field in the caps is an indication of the poloidal magnetic field in the Sun.

At present the relation between ephemeral regions and (large) active regions is not quite clear. There is no doubt, however, that the number of ephemeral regions varies with the solar cycle (Section 7). In fact, ephemeral regions at high latitudes herald the new solar cycle a few years before sunspot minimum. Their appearance on magnetograms and the indication for Ω-loop emergence suggest that ephemeral regions contain intrinsically strong fields.

Howard & LaBonte (1981) confirmed that the magnetic field in large-scale structures originates exclusively in the sunspot belts. Many of the large-scale processes in the solar magnetic field also require large-scale velocity fields; the diffusion supposedly produced by supergranular flow cannot explain all of the features of the solar activity cycle. The inclusion of a meridional circulation (e.g. in Sheeley et al. 1985) is but a first step. Specific flows are required in order to create and to maintain activity nests and coronal holes, to cause the episodic streams of monopolar flux toward the poles, and to bring the opposite polarities together in the removal of flux from the photosphere.

Although new observations are becoming available, our knowledge of the intranetwork field is still quite limited. The data at hand suggest that the INF is quite different from the rest of the solar magnetic field: The INF probably is diffuse and intrinsically weak (Section 6). One interesting suggestion is that the INF is produced by the partial emergence of U-loops; such a process is complementary to the emergence of Ω-loops, which produce the intrinsically strong fields.

Progress in the study of solar magnetic structure and dynamics depends particularly on developments in two fields of research: observations of *very high spatial resolution*, and dedicated *synoptic observations*.

Sophisticated instruments in spacecraft are required to obtain the time series of magnetograms, monochromatic images, and spectrograms at a resolution $\lesssim 0''.3$ that are needed to tackle the physical problems still hidden in fine structure. These include the following: What is the fundamental small-scale structure in the intrinsically strong magnetic field? Does it consist of long-lived flux fibers, or is it a flexible system of vertical sheets? How do the strong field and the granular convection interact? What is the nature of the intranetwork field, and how does it relate to the network field? The emergence and removal of magnetic flux, which are important processes that must reveal essential features of solar magnetism, involve extremely small scales. The chromospheric and coronal heating and the driving of the solar wind are most probably powered by some specific convective actions on the feet of the thin flux tubes in the top of the convection zone—again, very high spatial resolution is required for clues about the precise mechanisms. Note that the above-mentioned problems of magnetohydrodynamics and plasma physics are of broad relevance, with applications to stars and other astrophysical plasmas. It is the solar atmosphere, however, that provides the observational opportunity to study the processes at their appropriate scales.

Progress in handling large amounts of data has stimulated the study of synoptic solar data that have been and are being collected in full-disk white-light images, monochromatic images, and magnetograms. These data are indispensable for the study of the solar activity cycle. Solar and stellar studies complement each other: For example, the active nests discussed in Section 7 are probably the solar counterparts of the long-lived features in stellar atmospheres that cause the rotational modulation observed in specific radiative fluxes from cool stars.

Techniques in synoptic observations should be updated regularly and new techniques added, as, for instance, is done in solar and stellar seismology. Synoptic observations should cover many years, since studies in solar and stellar activity cycles require several decades of uninterrupted data. Such time spans are much longer than the typical periods of grants, and thus long-term planning is essential. The recent history of Mount Wilson Observatory—with its records of long uninterrupted series of fundamental solar data—demonstrates that synoptic programs are vulnerable. The scientific community should convince the funding authorities of the cost-effectiveness of synoptic observations aiming at a deeper understanding of the fundamental processes in solar and stellar interiors.

Acknowledgments

I wish to thank the many colleagues who have sent reprints, preprints, and suggestions that proved very useful in the preparation of this paper. Both the content and the presentation of this paper have been greatly improved by comments on drafts of parts of the paper given by V. Gaizauskas, J. W. Harvey, P. Hoyng, S. F. Martin, H. C. Spruit, and A. M. Title.

Literature Cited

Becker, U. 1955. *Z. Astrophys.* 37: 47
Beckers, J. M., Schröter, E. H. 1968. *Sol. Phys.* 4: 142
Bohlin, J. B. 1977. See Zirker 1977, p. 27
Born, R. 1974. *Sol. Phys.* 38: 127
Brants, J. J. 1985a. Thesis. Univ. Utrecht, Neth.
Brants, J. J. 1985b. *Sol. Phys.* 95: 15
Brants, J. J. 1985c. *Sol. Phys.* 98: 197
Brants, J. J., Steenbeek, J. C. M. 1985. *Sol. Phys.* 96: 229
Brants, J. J., Zwaan, C. 1982. *Sol. Phys.* 80: 251
Castenmiller, M. J. M., Zwaan, C., van der Zalm, E. B. J. 1986. *Sol. Phys.* 105: 237
Chapman, G. A., Sheeley, N. R. 1968. *Sol. Phys.* 5: 442
Dunn, R. B., Zirker, J. B. 1973. *Sol. Phys.* 33: 281
Gaizauskas, V., Harvey, K. L., Harvey, J. W., Zwaan, C. 1983. *Ap. J.* 265: 1056
Giovanelli, R. G. 1982. *Sol. Phys.* 77: 27
Golub, L. 1980. *Philos. Trans. R. Soc. London Ser. A* 207: 595
Golub, L., Davis, J. M., Krieger, A. S. 1979. *Ap. J. Lett.* 299: L145
Golub, L., Krieger, A. S., Harvey, J. W., Vaiana, G. S. 1977. *Sol. Phys.* 53: 111
Hagyard, M. J., ed. 1985. *Measurements of Solar Vector Magnetic Fields, NASA Conf. Publ. 2374.*
Harvey, J. 1977. In *Highlights of Astronomy*, ed. E. A. Müller, 4(2): 223
Harvey, J., Brault, J., Stenflo, J., Zwaan, C. 1980. *Bull. Am. Astron. Soc.* 12: 676
Harvey, J. W., Krieger, A. S., Timothy, A. F., Viana, G. 1975a. *Oss. Mem. Oss. Arcetri* 104: 50
Harvey, K. L. 1984. *Proc. Eur. Meet. Sol. Phys., 4th, ESA SP-220*, p. 235
Harvey, K. L. 1985. *Aust. J. Phys.* 38: 875
Harvey, K. L., Harvey, J. W. 1973. *Sol. Phys.* 28: 61
Harvey, K. L., Harvey, J. W., Martin, S. F. 1975b. *Sol. Phys.* 40: 87
Harvey, K. L., Sheeley, N. R., Harvey, J. W. 1982. *Sol. Phys.* 79: 149
Howard, R., LaBonte, B. J. 1981. *Sol. Phys.* 74: 131
Howard, R., Stenflo, J. O. 1972. *Sol. Phys.* 22: 402
Kahler, S. W., Davis, J. M., Harvey, J. W. 1983. *Sol. Phys.* 87: 47
Landi Degl'Innocenti, E. 1985. See Hagyard 1985, p. 279
Levine, R. H. 1977. See Zirker 1977, p. 103
Levine, R. H. 1982. *Sol. Phys.* 79: 203
Livi, S. H. B., Wang, J., Martin, S. F. 1985. *Aust. J. Phys.* 38: 855
Livingston, W. 1974. In *Flare-Related Field Dynamics, Conf. Rep.*, p. 269 Boulder, Colo: NCAR
Livingston, W. C., Harvey, J. 1975. *Bull. Am. Astron. Soc.* 7: 346
Livingston, W. C., Orrall, F. Q. 1974. *Sol. Phys.* 39: 301
Maltby, P., Avrett, E. H., Carlsson, M., Kjeldseth-Moe, O., Kurucz, R. L., Loeser, R. 1986. *Ap. J.* 306: 284
Marcy, G. W., 1983. In Stenflo 1983, p. 3
Martin, S. F. 1984. In *Small-Scale Dynamical Processes in Quiet Stellar Atmospheres* ed. S. L. Keil, p. 30. Sunspot, NM: Natl. Sol. Obs.
Martin, S. F., Harvey, K. L. 1979. *Sol. Phys.* 64: 93
Martin, S. F., Livi, S. H. B., Wang, J. 1985b. *Aust. J. Phys.* 38: 929
Martin, S. F., Livi, S. H. B., Wang, J., Shi, Z. 1985a. In Hagyard 1985, p. 403
Mehltretter, J. P. 1974. *Sol. Phys.* 38: 43
Muller, R. 1983. *Sol. Phys.* 85: 113
Muller, R. 1985a. *Sol. Phys.* 100: 237
Muller, R., ed. 1985b. *High Resolution in Solar Physics.* Berlin: Springer-Verlag
Nordlund, A. 1983. See Stenflo 1983, p. 79
Parker, E. N. 1979. *Ap. J.* 230: 905
Parker, E. N. 1984. *Ap. J.* 281: 839
Rabin, D., Moore, R., Hagyard, M. J. 1984. *Ap. J.* 287: 404
Ramsey, H. E., Schoolman, S. A., Title, A. M. 1977. *Ap. J. Lett.* 215: L41
Robinson, R. D., Worden, S. P., Harvey, J. W. 1980. *Ap. J. Lett.* 236: L155
Schrijver, C. J. 1986. Thesis. Univ. Utrecht, Neth.

Schüssler, M. 1983. See Stenflo 1983, p. 213
Semel, M. 1985. See Muller 1985b, p. 178
Sheeley, N. R. 1969. *Sol. Phys.* 9: 347
Sheeley, N. R. 1972. *Sol. Phys.* 25: 98
Sheeley, N. R. 1976. *J. Geophys. Res.* 81: 3462
Sheeley, N. R. 1981. In *Solar Active Regions*, ed. F. Q. Orrall, p. 17. Boulder: Colo. Assoc. Univ. Press
Sheeley, N. R., Harvey, J. W. 1981. *Sol. Phys.* 70: 237
Sheeley, N. R., DeVore, C. R., Boris, J. P. 1985. *Sol. Phys.* 98: 219
Shelke, R. N., Pande, M. C. 1985. *Sol. Phys.* 95: 193
Simon, G. W., Leighton, R. B. 1964. *Ap. J.* 140: 1120
Simon, G. W., Wilson, P. R. 1985. *Ap. J.* 295: 241
Smithson, R. S. 1973. *Sol. Phys.* 29: 365
Solanki, S. K., Stenflo, J. O. 1984. *Astron. Astrophys.* 140: 185
Solanki, S. K., Stenflo, J. O. 1985. *Astron. Astrophys.* 148: 123
Spruit, H. C. 1979. *Sol. Phys.* 61: 363
Spruit, H. C., Zwaan, C. 1981. *Sol. Phys.* 70: 207
Stenflo, J. O. 1973. *Sol. Phys.* 32: 41
Stenflo, J. O., ed. 1983. *Solar and Stellar Magnetic Fields: Origins and Coronal Effects, IAU Symp. No. 102.* Dordrecht: Reidel
Stenflo, J. O. 1985. *Sol. Phys.* 100: 189
Stenflo, J. O., Harvey, J. W. 1985. *Sol. Phys.* 95: 99
Stenflo, J. O., Harvey, J. W., Brault, J. W., Solanki, S. 1984. *Astron. Astrophys.* 131: 333
Stenflo, J. O., Vogel, M. 1986. *Nature* 319: 285
Svalgaard, L., Duvall, T. L., Scherrer, P. H. 1978. *Sol. Phys.* 58: 225
Tang, F., Howard, R., Adkins, J. M. 1984. *Sol. Phys.* 91: 75
Tarbell, T. D., Title, A. M. 1977. *Sol. Phys.* 52: 13
Tarbell, T. D., Title, A. M., Schoolman, S. A. 1979. *Ap. J.* 229: 387
Topka, K. P., Tarbell, T. D., Title, A. M. 1986. *Ap. J.* 306: 304
Unno, W. 1956. *Publ. Astron. Soc. Jpn.* 8: 108
van Ballegooijen, A. A. 1985a. See Hagyard 1985, p. 322
van Ballegooijen, A. A. 1985b. In *Theoretical Problems in High Resolution Solar Physics, Max Planck-Inst. No. 212*, p. 167
von Klüber, H. 1947. *Z. Astrophys.* 24: 121
Vrabec, D. 1974. In *Chromospheric Fine Structure, IAU Symp. No. 56*, ed. R. G. Athay, p. 201. Dordrecht: Reidel
Waldmeier, M. 1957. *Die Sonnenkorona*, Vol. 2. Basel: Verlag Birkhauser
Wallenhorst, S. G., Howard, R. 1982. *Sol. Phys.* 76: 203
Wallenhorst, S. G., Topka, K. P. 1982. *Sol. Phys.* 81: 33
Wiehr, E. 1978. *Astron. Astrophys.* 69: 279
Wiehr, E. 1985. *Astron. Astrophys.* 149: 217
Wilson, P. R., Simon, G. W. 1983. *Ap. J.* 273: 805
Zirin, H. 1984. *Ap. J.* 291: 858
Zirin, H. 1985. *Aust. J. Phys.* 38: 961
Zirker, J. B., ed. 1977. *Coronal Holes and High Speed Wind Streams.* Boulder: Colo. Assoc. Univ. Press
Zwaan, C. 1968. *Ann. Rev. Astron. Astrophys.* 6: 135
Zwaan, C. 1978. *Sol. Phys.* 60: 213
Zwaan, C. 1979. In *Small Scale Motions on the Sun, Mitt. Kiepenheuer Inst. No. 179*, ed. E. H. Schröter, p. 71
Zwaan, C. 1981. In *The Sun as a Star, NASA SP-450*, ed. S. Jordan, p. 163
Zwaan, C. 1985a. See Muller 1985b, p. 263
Zwaan, C. 1985b. *Sol. Phys.* 100: 397
Zwaan, C., Brants, J. J., Cram, L. E. 1985. *Sol. Phys.* 95: 3

WOLF-RAYET STARS

David C. Abbott and Peter S. Conti

Joint Institute for Laboratory Astrophysics, University of Colorado and National Bureau of Standards, Boulder, Colorado 80309-0440

1. INTRODUCTION

Wolf-Rayet (W-R) stars are hot, luminous objects with strong, broad ($\sim 10^3$ km s^{-1}) emission lines in the optical region (189) due to a substantial stellar wind. The W-R phase is a normal stage of evolution for all stars initially more massive than $\sim 40\ M_\odot$, and these extreme Population I stars are the topic of this review. A W-R phase is also present in some central stars of planetary nebulae, which we do not discuss here. Some 160 individual stars of this class are known in our own Galaxy, and many others have been identified among galaxies of the Local Group.

W-R stars come in two major subtypes (22): the WN, in which lines of helium and nitrogen ions are seen; and the WC, which contain lines of carbon and oxygen along with the helium ions. An additional minor subtype (WO) with strong O VI lines has also recently been identified (19). The main characteristic of the optical spectrum of W-R stars is the dominance of emission lines. In a few WN stars, hydrogen appears to be present (52, 159), but generally the spectra of W-R stars are notable for the absence of this element. Absorption lines are generally not found in W-R stars, except in some instances as a result of a binary or a more distant companion or as P Cygni profiles associated with certain emission features. An upper Balmer-line absorption spectrum appears to be an intrinsic feature in a very few WN stars (54).

A previous review of the *optical* spectra of W-R stars has been given in these volumes (165). A brief historical perspective and bibliography on W-R stars up to the middle of 1981 is contained in the "Sixth Catalogue" (171). Recent IAU Symposia concerned with W-R stars include No. 99 ["Wolf-Rayet Stars: Observations, Physics, Evolution" (61)], No. 105 ["Observational Tests of Stellar Evolution Theory" (110)], and No. 116 ["Luminous Stars and Associations in Galaxies" (62)]. A recent monograph containing more detailed information and controversies about many aspects of W-R stars is that of Conti & Underhill (55).

In this review we consider the properties and evolutionary status of W-R stars both from the observable data and from the theoretical models used to describe the stellar structure, atmospheres, and winds. Ideally, we should like to describe the characteristics of *luminosity, effective temperature, chemical composition, radius, mass,* and (for stars such as these) the *mass-loss rates and other stellar wind parameters*. In practice, however, not all of these are well known given the complex physical phenomena present in W-R stars. We thus proceed from the better-known quantities, such as the absolute visual magnitudes and line strengths, through those that can be reasonably inferred from the existing data with very simplified models, such as the composition and mass-loss rates, to more controversial issues, such as the luminosities and effective temperature scale.

2. OBSERVABLE PROPERTIES

In this section we discuss the properties of individual W-R stars that can be deduced from the data with little need of theoretical interpretation or model construction.

2.1 *Spectral Types and Line Strengths*

W-R star classification is a one-dimensional system. The WN stars are classified into an excitation/ionization sequence depending on the line ratios among the nitrogen ions ($\lambda\lambda$4634, 4640 N III; λ4057 N IV; $\lambda\lambda$4603, 4610 N V). These range from the highest to the lowest excitation (WN3, WN4, ..., WN9) and are similarly called "early" and "late" types by analogy with the MK classification. The WC stars are also assigned to excitation/ionization subtypes, which are labeled WC4, WC5, ..., WC9. These depend on the ratios of the carbon ions (λ5696 C III; $\lambda\lambda$5801, 5812 C IV) and an oxygen ion (λ5592 O III). The basis of the subtypes (157) has a modern updating (171). The ionization sequence for the WO stars ranges from WO1 to WO4, depending on the strengths and line ratios of λ3400 O IV and $\lambda\lambda$3811, 3834 O VI lines (19). The WN stars show the anomalous abundances representative of "equilibrium" CNO-cycle hydrogen-burned material with enhanced helium and nitrogen at the expense of hydrogen, carbon, and oxygen; the WC stars show the composition characteristic of helium-burned material, in which carbon and oxygen are produced from the previously formed helium (e.g. 160). The WO stars have proportionately more oxygen and less helium and carbon. We are observing stars that are highly evolved and in which the products of nuclear reactions in the stellar cores are seen at the surface.

In the far-UV regions, emission lines dominate the appearance of W-R stars, with P Cygni profiles of species such as He II, C IV, N IV, and N V being found in the WN subtypes and C III, C IV, O VI, and Si IV being very strong in WC stars. Lines of nitrogen ions are not identified in WC stars, nor is C III in WN stars (187). The subordinate lines are usually seen with pure emission profiles. In the near IR regions, W-R stars show a predominant emission-line spectrum with ions of helium, carbon, nitrogen, and oxygen being found, again depending on the WN or WC nature (177).

Figure 1 shows spectrophotometry from 1200–7000 Å of a typical WN star, HD 192163, corrected for interstellar extinction [following Savage & Mathis (151)]. The dominance of the emission-line spectrum may readily be seen. The solid curve is an attempt to fit a plane-parallel model to the continuum data, a completely unsatisfactory state of affairs to which we return later.

Measurements of the strengths of numerous optical emission lines have been carried out for many WN (52) and WC stars (164) using image-tube spectrograms. This work quantified the spectral-type classifications of the WN and WC sequences. The strongest line in WN stars is typically $\lambda4686$ He II, with measured line strengths (equivalent widths) up to 400 Å. In

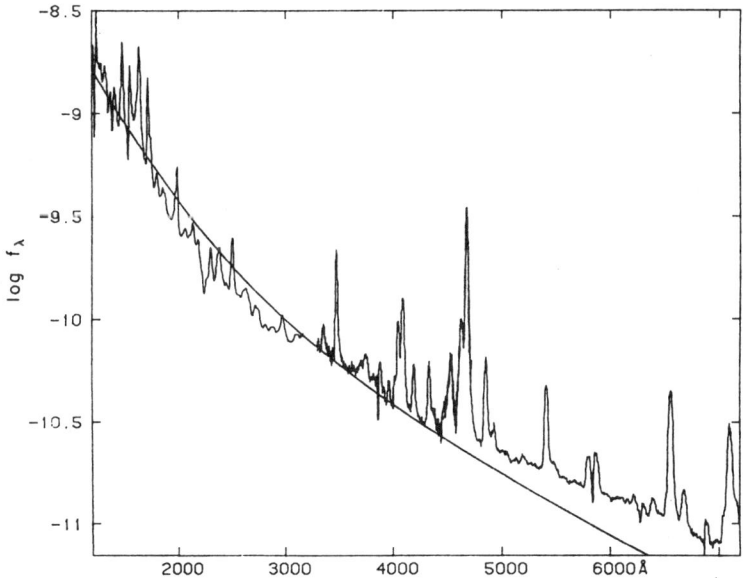

Figure 1 Spectrophotometry of a typical WN star, HD 192163 [adapted from (75)], corrected for interstellar extinction. The solid curve is a not very successful attempted fit with a plane-parallel model (see text).

WC stars the strongest optical feature is the C III/He II blend at 4650/4686 Å, with line strengths up to 2000 Å being found. Many other lines with strengths of a few angstroms are seen, and weaker lines (although present) cannot readily be measured. The emission-line strengths of ionized helium typically have a dispersion of a factor of 10 among the various stars, even amongst stars of the same subtype, thus suggesting the heterogeneity of the WN stars. In WC stars, the line strengths are more homogeneous, although differences among stars of the same subtype are still found. W-R stars with absorption lines invariably have the weakest emission-line features because of dilution by the brighter continuum of the companion O-type star. A few objects with no evidence of a companion also have relatively weak emission lines.

2.2 Magnitudes and Colors

A photometric system was devised and used by Smith (158) to derive magnitudes and colors of W-R stars that would be as free as possible from contamination by the numerous strong emission lines. This system succeeds reasonably well, but for the strong WC stars the indices are still affected by the emission lines (112). Absolute spectrophotometry of northern W-R stars (112) and of faint southern objects (105) has been published. Massey (112) has derived line-free reddening-independent indices that can be used to find the intrinsic continuum colors and magnitudes of W-R stars. The considerable dispersion in intrinsic colors within a spectral subtype often exceeds the differences in colors between subtypes.

Absolute magnitudes M_v can be found for Galactic W-R stars that are members of clusters and associations with well-established distances. A detailed compilation of these values for a number of Galactic objects has recently been made (106). Similarly, the W-R stars known in the Large Magellanic Cloud (LMC) can be used as an independent calibrator (148). From the combined data (29, 49), the M_v of a W-R subtype can be estimated from its spectrum to an accuracy of ± 0.5, not unlike what can be done for OB supergiants.

Figure 2 shows the individual stars used for the M_v calibration. A substantial dependence of M_v on subtype is found, particularly for the WN classes in which the later types are certainly brighter. The M_v dependence on spectral subtype is similar in both the Galaxy and the LMC for the WN and WC stars, although the difference at WN6 is a little larger than for the other subclasses.

2.3 Intrinsic Variability

A recent workshop on stellar instabilities (101) contains the first review (175) on this topic for W-R stars [see also (15)]. W-R stars show variability

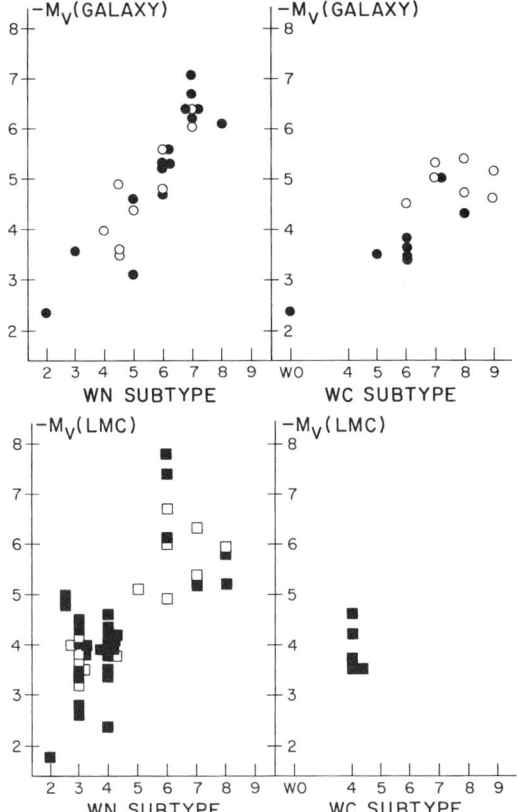

Figure 2 W-R stars with M_v calculated from their (*top*) association/cluster or (*bottom*) LMC membership [from 49)]. Note the decrease in magnitude with increasing spectral subtypes, as well as the dispersion in M_v. The filled symbols have better-determined photometry.

in both the continuum and the emission-line shapes and strengths, although in many cases the effects are very subtle and at the limit of observational techniques.

Many W-R stars appear to show variability in the continuum magnitudes and in the colors at the few-percent level (126). The time scales are typically a few days, but strict periodicity has not been firmly established except for the known binary systems. Some W-R stars show no variability, and the WN8 objects appear to have the largest amplitudes.

The detection of subtle effects in line profiles in W-R stars is just now gaining currency as the signal-to-noise characteristics of spectroscopic detectors improve. Many W-R stars are observed to have line variability at the few-percent level, even though the overall characteristics of W-R spectra remain constant over many tens of years. Precise line-profile data are scanty, and complications in the interpretation can arise if the W-R star has an O-type companion or a collapsed companion, such as a neutron star or a black hole (see, e.g., 118 and references therein).

We return to the issue of collapsed companions below. Vreux (174) has drawn attention to the possible existence of periods of a few hours in the emission-line profiles in some W-R stars that were otherwise masked by aliasing with other periods of a few days duration. Specifically, a period of a few hours is apparently present in HD 192163 from the changing line profiles of a N III/N IV blend near $\lambda 4075$ (176); the few-days period previously reported may be the rotation period. Vreux suggests that the short period found is indicative of nonradial pulsations (e.g. 136), although this is by no means certain theoretically (109). Much more extensive data bases are needed before the existence of nonradial pulsations in W-R stars can be demonstrated without ambiguity.

2.4 Infrared and Radio Observations

The vast size of the circumstellar envelope formed by the wind, coupled with the λ^3 dependence of free-free opacity, makes W-R stars among the brightest of all stellar infrared and radio sources. For example, the radius of optical depth unity to free-free absorption at 6 cm is typically 10^3 R_*, or 10–20 AU. At current sensitivities, the VLA can detect free-free radio emission from typical W-R stars to a distance of roughly 3 kpc. Free-free emission at 10 and 25 μm from the same stars has been detected by the IRAS satellite. Because the wind is flowing with constant velocity and relatively uniform temperature very far from the star, the infrared and radio observations are ideal diagnostics of the wind mass-loss rates and composition. Even more exciting, two entirely different physical processes have been discovered in the outer wind by infrared and radio observations. One is the formation of dust, which is observationally confirmed in roughly 15% of all W-R stars. The second is a nonthermal emission component, which is observationally confirmed in roughly 12% of all W-R stars.

2.4.1 RADIO OBSERVATIONS   The most detailed study at radio wavelengths of a W-R star is the VLA interferometry of γ^2 Vel, which spatially resolves the emission at 2, 6, and 20 cm (89). Figure 3 shows the fit of the 5-GHz data to an isothermal, spherically symmetric, constant-velocity outflow model. The agreement is remarkable considering the simplicity of the model. Analyzing the emissivity as a function of radius confirms that the radio flux from γ^2 Vel is free-free emission from a stellar wind that is symmetric on the plane of the sky, with a mass-loss rate of 8.6×10^{-5} $M_\odot$

yr^{-1} and a roughly isothermal temperature of 5600 K over the radius range of 10^{+15} to 10^{+17} cm. While the emission from other stars is generally too weak to be spatially resolved, continuum fluxes for a distance-limited sample of 40 W-R stars have now been observed with the VLA (6, and references within). Because knowledge of the structure of the emissivity was lacking, the radio emission was categorized as "free-free" or "nonthermal" based on the empirical criteria of spectral index, variability, and disagreement with optical diagnostics of the mass loss rate.

In this sample, 5 stars (12%) were probable or definite nonthermal sources, 33 stars (83%) were probable or definite free-free sources, while 2 stars (5%) were undetermined. Both mechanisms may operate in all stars, of course, and the description only identifies the dominant process. The incidence of measurable nonthermal radio radiation is roughly a factor of two smaller in the W-R stars than in a comparable OB sample (5). Also, in contrast to the OB nonthermal sources, the W-R sources tend to be peculiar at other wavelengths.

The prototype nonthermal radio source is HD 193793, which has been observed over a period of nearly 10 years (23, 71). On two occasions (1976 and 1983) it exhibited a high state of emission [S(5 GHz) $\cong$ 25 mJy, spectral index $\cong$ -0.2], which was unresolved at 5 GHz by the longest baselines of the VLA. This cannot be free-free emission, and presumably it is a tracer of a nonthermal, very energetic population of particles that exists in the outer reaches of the stellar wind, beyond the free-free photosphere. On both occasions, the presence of strong nonthermal radio emis-

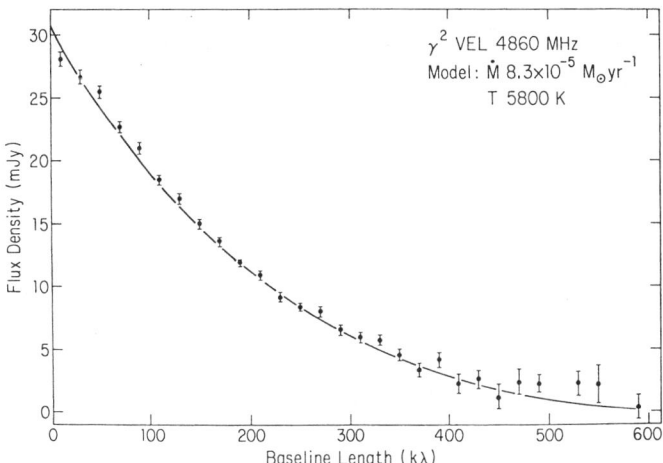

Figure 3 Comparison of the observed visibility flux measured with the VLA to a model wind (solid line) having the parameters shown [from (89)].

sion was correlated with the appearance of strong infrared emission from circumstellar dust (78, 183, 185). Preliminary data (122, 185) suggest that HD 193793 is a W-R+O binary with an eccentric orbit and a period of roughly 7.9 yr, and that the orbital motion provides the timing for the outbursts.

The cause of the nonthermal radiation in W-R stars is not established. White (180) proposed that the strong shocks thought to permeate all hot-star winds accelerate a small population of relativistic electrons. The synchrotron emission from these particles is visible once the particles are advected by the wind beyond the free-free photosphere. In this model, the rarity of nonthermal emission in W-R stars may be related to the much higher density of the wind, which increases the energy loss of the relativistic particles and also increases the radius of the free-free photosphere. The variable nonthermal emission from HD 193793 could then be related to the presence of shocks from colliding winds as the O star orbits the W-R star.

2.4.2 INFRARED OBSERVATIONS The presence of dust in some W-R winds is detectable from observations of the shape of the infrared continuum (10, 46, 172). In Figure 4, a large infrared excess is present, well above any amount that can be explained by free-free emission. Strong dust emission is prevalent in the WC8 and WC9 spectral types and recent IRAS observations found evidence for dust in two peculiar, late-type WN stars (172). From the shape of the dust feature, the dust has a grain temperature on the order of 1000 K. No stellar silicate absorption or emission features are identifiable in the spectrum of the WC stars, which together with the known overabundance of carbon suggests a graphite grain composition. The dust shell has been resolved by infrared speckle interferometery in one WC9 star, MR 80 (9, 67). The size of the shell is a few $\times$ 10^{15} cm, which is consistent with the radii deduced from the grain temperatures and emission energies.

The infrared spectrum of the dust shells is generally unvarying, which implies a constant rate of dust formation. In three W-R stars, however, the formation of dust occurs in episodes. Two of the stars, HD 192641 and HD 193793, are of spectral type WC7+abs, and the time scale for the formation and decay of the dust shell is on the order of a few years (78, 183, 184). The third star is a recently discovered WC9 star (57), and little is known about the variations from this object.

An emerging area of study is infrared spectroscopy (8, 85, 182). Observations of emission lines at infrared wavelengths offer several distinct advantages. The analysis of infrared recombination lines is simplified because the lines are usually optically thin and because the upper levels of

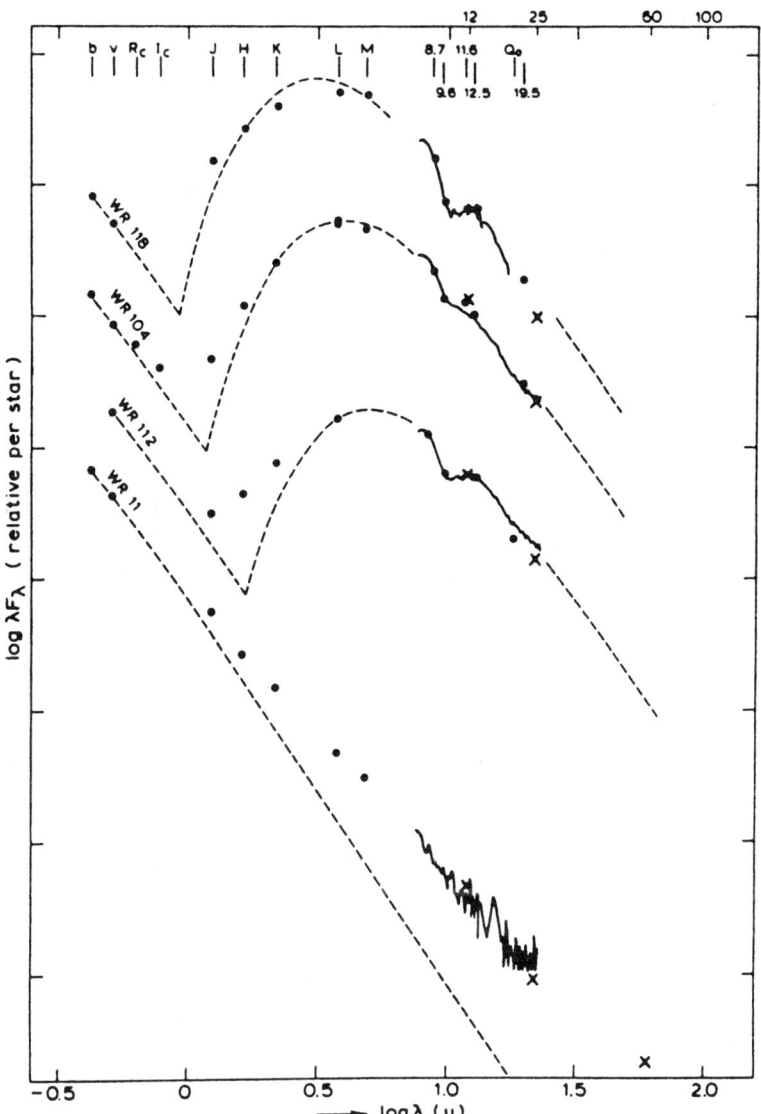

Figure 4 Optical-infrared energy distributions of three late-type-WC stars that exhibit dust emission (top three curves) [from (172)]. The dashed curve is a model combining a plane-parallel atmosphere and a Planck function with $T \cong 900$ K. The silicate absorption feature at 9.7 μm is of interstellar origin. The bottom curve is the energy distribution of γ Vel (WC8+O9I), which has no evidence for dust; note, however, the strong [Ne III] emission line at 15.5 μm.

the transitions are usually close to local thermodynamic equilibrium (LTE). This makes the infrared lines more useful for deriving abundances (90, 99). Since the infrared lines are formed against a background continuum opacity that strongly increases with wavelength, line measurements at different wavelengths probe wind conditions at different radii (88). While this effect complicates the determination of abundances from infrared lines, it provides the possibility of mapping velocity and excitation conditions as a function of radius in the wind. The infrared spectroscopy also shows that broadband continuum filters may be severely contaminated by line emission, as is also the case in the optical.

Despite this uncertainty, several conclusions are evident from broadband studies of the infrared-radio continuum energy distributions (46, 79, 83, 88). The analyses confirm that free-free emission is the dominant emission mechanism for the infrared continuum in the majority of W-R stars, and that the winds of W-R stars are generally opaque to continuum radiation at all frequencies. The models are unclear as to the exact mapping between infrared wavelength and the emitting region in the wind. Several lines of evidence suggest that all infrared continuum emission beyond a few microns in wavelength originates in the wind far from the star, where the velocity has attained its asymptotic or terminal value. On the other hand, the spectral index between the 10-μm flux and the radio flux is significantly larger than the 0.6 value predicted for free-free emission in a constant-velocity outflow (20, 26). In the models of HD 50896, this has been explained by adding a second zone of acceleration far from the star (83, 88) or an ionization front between the infrared and radio emission regions (87). Preliminary results (D. C. Abbott & R. Russell, in preparation) suggest that shock compression in the inner wind could also create sufficient excess infrared emission to explain the discrepancy.

2.5 X-Ray Observations

The X-ray emission from W-R stars presents a complex pattern. Pollock (146a) summarizes all published W-R X-ray observations plus new data from the Einstein archives. The X-ray spectrum of the W-R stars, as measured by the IPC instrument of Einstein, resembles that observed in the O stars (155, 181). This suggests that the X-ray-emitting gas has a temperature of a few million degrees, that the emission measure of this hot gas is a small fraction of the total of the wind ($\sim 10^{-4}$), and that the observed X rays must be emitted far out in the wind, where the absorption of the soft X-ray flux is greatly reduced. Variable X-ray fluxes were observed in three W-R stars—HD 50896, V444 Cyg (120), and HD 193793 (185). The variations appear to phase with the binary orbits. Rapid variability (shorter than 1 hr) was observed on two occasions in HD 50896

(181). In two stars with multiple observations, γ Vel and HD 93162, the X-ray flux did not vary.

In O stars, the X-ray luminosity is tightly correlated with the bolometric luminosity according to the ratio $L_x/L_{bol} \sim 10^{-7}$. Most of the W-R data are also consistent with this scaling; however, the comparison is hampered by the large uncertainty in L_{bol} for W-R stars. One star, HD 93162, has an unusually large ratio of L_x/L_{bol}. Present data suggest that the WC stars are intrinsically weaker X-ray sources than the WN stars.

The W-R stars detected in X rays have a higher than usual incidence of composite systems, which suggests that the presence of a companion may enhance the X-ray emission. Of 14 detected stars, 7 are spectroscopic binaries and 5 are W-R + abs systems. It is possible that the X-ray emission in binary systems is enhanced by hot gas formed at the shock interface between colliding winds (149).

Also unusual is the incidence of nonthermal radio sources in the X-ray sample. All three of the definitely identified nonthermal sources were detected in X rays (145). Pollock (146) has suggested that the X-ray emission in these objects is enhanced by Compton scattering of the stellar radiation field by the same energetic particles that are responsible for the nonthermal radio flux. In addition, relativistic bremsstrahlung by the particles could produce a detectable gamma-ray flux. Pollock (146) finds a correlation between W-R nonthermal radio emitters and COS-B gamma-ray sources. However, the large positional error boxes of the gamma-ray sources make any identification somewhat speculative.

3. THE W-R POPULATION

3.1 *Galactic Distribution*

Garmany (72) has plotted the longitude distribution of OB stars more massive than 40 $M_\odot$ projected onto the plane of the Galaxy within 3 kpc. In Figure 5 the analogous distribution of W-R stars is indicated. We see considerable numbers in the inner Carina-Sagittarius arm and in the Cygnus arm but very few outward from the Sun in the Perseus arm. This anomalous galactocentric distribution is similar to that of the massive O stars, which suggests that the W-R stars are the descendants of these objects (51). There are 245 progenitor massive OB stars in the 3-kpc volume; this is to be compared with 63 W-R descendants in Figure 5. Thus, if all stars more massive than 40 $M_\odot$ become W-R stars, the number ratio O/W-R is about 4:1.

The vertical z distribution of massive OB stars within 3 kpc (72) and of the W-R stars (48) shows that they are closely confined to the Galactic

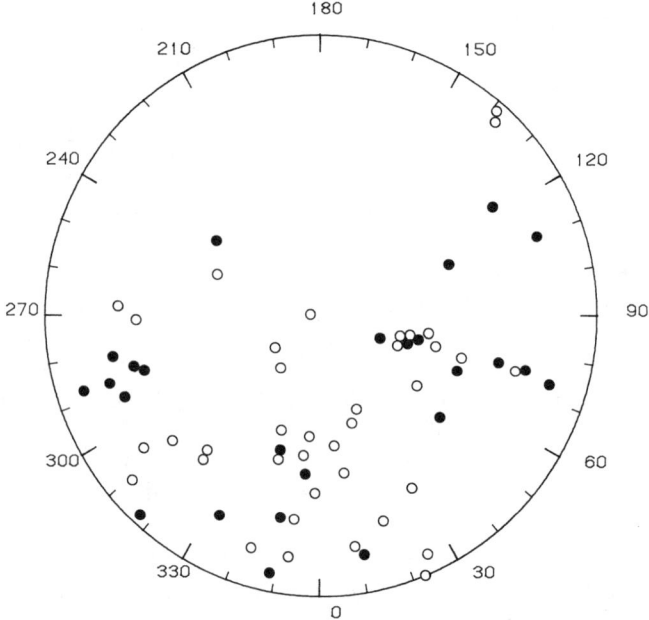

Figure 5 Longitude distribution of Galactic W-R stars within 3 kpc [from (49)]. Open circles are WC stars; filled circles are WN stars.

plane, with a vertical extent of some ±100 pc. A very few OB and W-R stars are found substantially farther from this range and could be fairly described as "runaway" stars. Gies (77) has recently considered the nature of such objects and concludes that many have escaped from close star-star interactions in the cores of young clusters and associations.

Lundstrom & Stenholm (106) and Humphreys et al. (92) discuss the membership of W-R stars in clusters and associations. The most luminous stars in these groups are invariably the blue supergiants; red supergiants are not found associated with W-R stars. This anticorrelation is an important clue to the evolutionary status of W-R objects.

3.2 W-R Stars in Other Galaxies

Breysacher (28) lists 101 known W-R stars in the LMC; a few more have since been fortuitously discovered (14, 50, 56, 127). Most of the LMC W-R population are WN types, and late WC stars are not found (28, 157, 158, 167). Many W-R stars are found in the 30 Dor region, and several objects of the central core also show W-R–like spectra (44, 124). Overall in the LMC the WN/WC ratio is 4.5, which is rather different from the solar vicinity, where the value is near unity (cf. Fig. 5). Azzopardi &

Breysacher (13) list 8 W-R stars in the Small Magellanic Cloud (SMC), but probably one or two of them are really better classified as Of (74). One other is a WC (WO) star, so the WN/WC ratio is large. The statistics of W-R stars in the Magellanic Clouds is believed to be nearly complete (14).

There are nearly 100 W-R stars identified in M33 (115), which is believed to be a substantial fraction of the total. The WN/WC ratio changes appreciably with galactocentric distance and is smaller in the nuclear regions. All the WC stars appear to be of early type, as are most of the WN types. W-R spectra are seen in the stellar components in some of the giant H II regions of M33 (53, 63, 64, 150). In M31, 17 W-R stars have been identified (125) and an additional 20 or so candidates have been suggested (114). The statistics of W-R stars in M31 is very incomplete at the present time. A WN star has been found in NGC 6822 (178) and a WC type in IC 1613 (58, 64). Armandroff & Massey (12) have identified some additional W-R candidates in each of these small irregular Local Group galaxies. In NCG 6822, the candidates all seem to be WN types; in IC 1613, several each of the WN and WC types are suggested.

The numbers and types of W-R stars are not the same in the various galactic environments. The WN/WC ratio in particular is different, as is the apparent absence of late-type WC stars anywhere but in the solar vicinity. Various initial composition differences (107) or the numbers of massive star progenitors (113) might affect the evolution and the production of W-R stars, but the present data are insufficient to disentangle these effects (or others).

W-R type spectra are found in the central exciting objects of giant H II regions in our Galaxy, e.g. NGC 3603 (HD 97950) (123, 124), and in other galaxies (53, 63, 64, 97, 124). Emission-like features (*notably* $\lambda 4686$ He II) have been found in integrated light in a handful of galaxies beyond the Local Group, e.g. He 2–10 (11), IZw 18 (25), Tol 3 (97), NGC 5430 (93), NGC 6764 and Mkn 309 (141), and Tol 89 (6). The presence of W-R emission lines seen against the Galaxy continuum implies that extraordinary numbers of these stars are present.

3.3 *Binaries*

Vanbeveren & Conti (167) have considered the statistics of W-R binaries among a large spectroscopic sample of Galactic and LMC W-R stars. They noted that less than 40% of the W-R stars in their extensive sample showed evidence of absorption lines, which are normally indicative of an O-type companion [a few exceptions can be found; see (54)]. In a few absorption-line systems the companion O-type star is nearby, *but not close*. The O-type companions are usually sufficiently bright that the absorption lines are visible. Thus, the binary frequency of W-R stars with O-type companions was suggested to be near the observed 40% limit. Massey (111) has also addressed the question of absorption-line W-R systems and finds a binary frequency of this same order. Moffat et al. (121), in an updated study that allows for more distant companions, find a binary frequency of 43%. These numbers are similar to those for the progenitor O stars (73).

A list of the known W-R binary systems is given in Table 1 [adapted from (111), with additional references as indicated]. Are all W-R stars binaries? At one time, this was a favored scenario [see, e.g., IAU Symposium No. 49 (16)]. However, careful statistical arguments strongly suggest that single W-R stars exist. Furthermore, morphological similarities in the spectra led Conti (47) to suggest that evolution by mass loss could lead directly from an Of phase to the late WN phase. It seems hard to imagine that the anomalous spectra of the W-R systems are related solely to their binary nature, since the known single stars have similar characteristics.

Do any W-R stars exist in systems with collapsed companions? These are expected theoretically, following the van den Heuvel (170) binary scenario. One would expect that some known X-ray binary systems, containing an O star and a collapsed companion, should evolve further into

Table 1 Wolf-Rayet binaries: double-lined

Star	Spectral type	Period (days)	$M_{W-R} \sin i$ ($M_\odot$)	Mass ratio W-R/O	M_{W-R} ($M_\odot$)	Basis	Reference
AB6 (SMC)	WN3+O7	6.861	6.4	0.17	6.4	$M_\odot \sim 37$	117
HD 5980 (SMC)	WN4+O7I:	19.266	2.0?	0.3?	8?	See ref.	30
SK 188 (SMC)	WO4+O4V	16.644	1.7	0.27	13.5	$M_\odot \sim 50$	119
γ Vel	WC8+O9I	78.5002	17	0.54	25	$M_\odot \sim 35$	111
HD 90657	WN4+O4–6	8.255	8.4	0.52	11	See ref.	132
HD 94305	WC6+O6–8	18.82	15	0.47	16	$M_\odot \sim 35$	131
HD 94546	WN4+O8V	4.831	2.7	0.43	7	See ref.	130
HD 97152	WC7+O7V	7.886	3.6	0.59	20	$M_\odot \sim 35$	111
HDE 311884	WN6+O5V	6.34	40	0.84	50	$M_\odot \sim 50$	111
HD 152270	WC7+O5	8.893	1.8	0.36	20	$M_\odot \sim 60$	111
HDE 320102	WN3+O7	8.83	1.8	0.33	11	$M_\odot \sim 35$	129
CV Sen	WC8+O8V	29.707	11	0.48	13	Eclipse	111
HD 186943	WN4+O9V	9.5548	9–11	0.52	13	$M_\odot \sim 25$	111
HD 190918	WN4.5+O9I	112.8	0.7	0.26	9	$M_\odot \sim 35$	111
V 444 Cyg	WN5+O6	4.21238	9.3	0.40	11	Eclipse	111
GP Cep	WN6+O	6.6884	—	>0.22	10–25	Eclipse	111
CQ Cep	WN7+O?	1.6	23	1.19	≥23	Eclipse	111

a W-R object and the same collapsed companion as time proceeds. Simple arguments (e.g. 167) suggest that about as many W-R stars with these companions as with O-type stars should exist. However, no known strong X-ray binary system contains W-R stars. Could there be systems containing an X-ray "quiet" W-R plus a collapsed companion?

Moffat and associates have studied photometrically and spectroscopically a sample of W-R stars (generally without absorption lines) and find apparent periodic variability in a number of them (117, 118). The periods seem to be a few days in many cases, the same period is often found in the photometric and the spectroscopic data, and the mass functions are small; these findings suggest that the masses of the inferred companion objects are also low. A few systems have more compelling data than others. For example, HD 197406 has a well-determined radial velocity curve which is unquestionably due to a companion object (65). Polarimetric observations suggest a nonspherically symmetric W-R star, since periodic variability with the same period as the orbit is found. The authors contend that a black hole companion to HD 197406 is the most likely explanation for the data. In many other systems, the periodicities are not well established or may represent only rotational periods.

The question of collapsed companions accompanying many W-R stars is confused by the issue of intrinsic variations (174), and the answer will require substantially more data than we have available at the present time. The binary scenario would suggest that W-R stars with collapsed companions should exist: If they are in fact not observed, an explanation will have to be found to explain this. If those W-R objects identified by Moffat and associates are indeed such systems, one will alternatively have to explain why they are not strong X-ray sources (102, 168).

4. INFERABLE PROPERTIES OF W-R STARS

4.1 *Masses*

The spectroscopic analysis of these systems is often complicated by the fact that one is measuring emission lines in the W-R component that are formed in the stellar wind, along with absorption lines in the O-type companion that are formed in a more static atmosphere. The velocity amplitudes for various W-R emission lines often differ appreciably: In most cases those of highest excitation give the smallest values, which are presumed representative of the deepest layers. The systemic velocities of the emission lines often differ from those of the absorption lines; the latter are thought to best represent the binary motion. Some W-R binary systems contain components in close contact with one another, and the interactions between the individual stellar winds lead to very complicated and phase-dependent variability. This prevents easy interpretation of the radial velocity curves and the inferred masses and mass ratios.

In nearly all W-R systems the W-R component is the less massive, consistent with its expected more advanced evolutionary state compared with its companion. A thorough discussion of the values of masses and mass ratios in W-R binary systems is given by Massey (111), and the updated values are listed in Table 1. The mass values range from some 6.4 to 50 $M_\odot$; there is no strong dependence on spectral subtype. Although one might expect that WC stars are more evolved than WN stars, given their compositions, the former are on average *more* massive than the latter. Thus, with the present data, it is not possible to conclude that all WN stars become WC stars, nor that all WC stars were once WN stars.

4.2 Composition

Composition studies of W-R stars are still relatively primitive, but the anomalies are so striking that order-of-magnitude estimates of the abundances may be made (52, 138, 139, 160, 164). These results suggest that the WN stars have greatly enhanced helium and nitrogen and little or no hydrogen present. Carbon and oxygen are present in WN stars. In WC stars, helium, carbon, and oxygen are all found in great abundance, and there is no evidence for hydrogen or nitrogen. These data are qualitatively in agreement (to a factor of a few) with the predictions of massive-star evolution with mass loss and mixing, in which the underlying nuclear reaction core is revealed (108, 135, 147).

4.3 Stellar Winds

Because the winds of W-R stars are so opaque, reliable measurements of the stellar mass-loss rate $\dot{M}$ and the asymptotic wind velocity, or terminal velocity V_∞, are straightforward. As a class, W-R stars have the highest known mass fluxes of any stellar type.

4.3.1 TERMINAL VELOCITIES The maximum outflow velocity of the wind is directly measured from the maximum Doppler shift observed in a spectral line. The only requirement is that the line must be sufficiently opaque so that the measured feature extends all the way to the terminal velocity. That this condition is readily met for W-R stars is demonstrated by the C III intercombination line at 1909 Å in γ Vel, which exhibits P Cygni absorption to V_∞ despite an oscillator strength of only 1.6×10^{-7} (137).

Ultraviolet lines The violet edge of the absorption component of a P

Cygni profile is the usual diagnostic of V_∞ (186). As shown in Figure 6, the method has two problems. The first is line blending, which is illustrated by the unidentified emission feature that peaks at 1527 Å and blends with the violet edge of C IV. The second, more fundamental problem is the ambiguity of the definition of terminal velocity. If terminal velocity refers to the asymptotic velocity of the bulk of the outflow, which seems reasonable, then the proper definition is the blueward extent of the saturated absorption, which corresponds to roughly 2500 km s^{-1} for the resonance lines of HD 165763. If terminal velocity refers to the maximum velocity attained by any outflowing gas, then the proper definition is the point where the optically thin part of the violet edge joins the continuum, which Willis (186) gives as 3700 km s^{-1} in this example. The difference between the two values presumably results from gas in very strong shocks that attains velocities higher than that of the mean flow, but that has either a very small column density or a very small covering factor. Unfortunately, the majority of the W-R stars are too reddened for high-resolution UV observations with the presently available IUE satellite.

Infrared lines The velocity width of emission lines in the infrared is nearly as reliable an indicator of the terminal velocity, because these lines are formed against the free-free continuum, which forms at large radii. An infrared transition of particular note is [Ne II] at 12.8 μm, which has been resolved in the spectrum of γ^2 Vel with a full width of 3000 km s^{-1}, implying a terminal velocity of 1500 km s^{-1} (18). Since the critical density for the formation of this fine-structure transition is of the order 10^5 cm^{-3}, this line

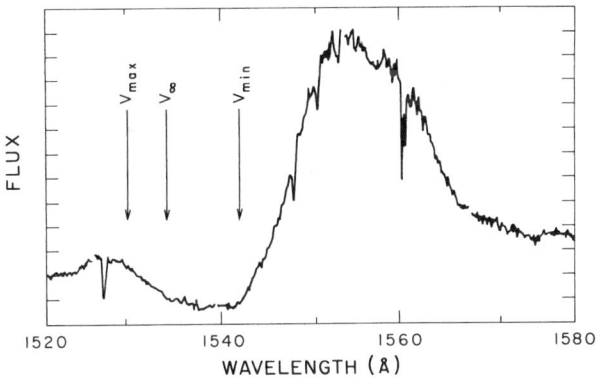

Figure 6 A high-resolution IUE spectrum of the C IV 1550 Å resonance doublet in the WC5 star HD 165763 [from (187)]. The arrows show the Doppler-shifted wavelengths corresponding to the velocities of maximum wind absorption (V_{max}), the maximum saturated wind absorption (V_∞), and the minimum saturated wind absorption (V_{min}).

is formed at very large radii in a similar region as the free-free radio emission (17).

Optical lines The width of the optical emission lines can be a reliable indicator of V_∞ except for two circumstances. In the WN sequence, the optical lines may not be opaque enough for their emission to reach the highest velocities, especially for the WN7 and WN8 spectral types. In the WC sequence, the width of the optical lines depends on the excitation potential of the line, and an extrapolation of the optical data to zero excitation potential is necessary to estimate V_∞ (164).

A final tabulation of known terminal velocities of W-R winds is shown in Figure 7. For the WC sequence there is a clear correlation of terminal velocity with spectral class, in the sense that higher velocities are found in stars with the highest level of excitation. Line breadth is, in fact, a spectral classification criterion in the WC sequence. For the WN sequence there is no correlation with spectral type, except perhaps for the very earliest and very latest types. Further, there is a large dispersion in terminal velocities within a WN spectral type. In the OB stars, the terminal velocity scales with the effective escape velocity (1). One possible interpretation of the correlation in the WC sequence is that the early spectral types have smaller radii and hence higher effective temperatures and higher gravities. By this

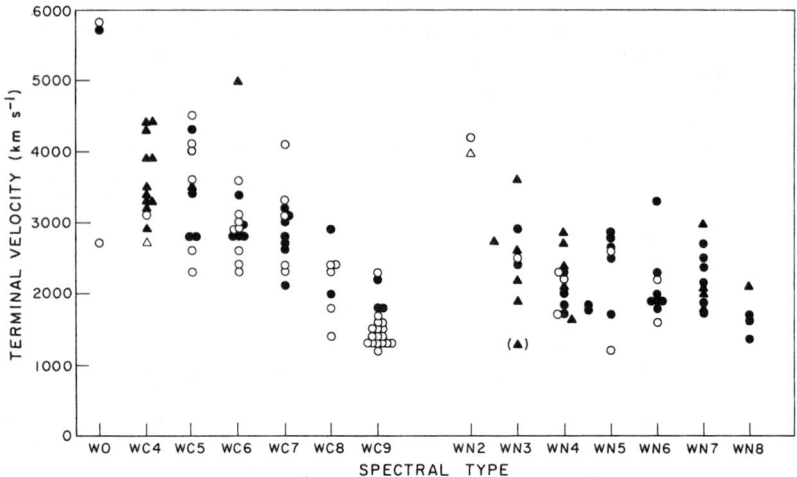

Figure 7 The terminal wind velocity versus W-R spectral type. Triangles represent LMC stars, and circles represent Galactic stars. Filled symbols are V_∞ measurements from UV data, while open symbols are from optical measurements. For the WC stars, data are from Torres et al. (164), while for the WN stars, data are from Willis (186), Perry (144), Abbott et al. (6), plus 20 newly derived values from spectra in the IUE archives.

explanation, a large range in luminosity and/or mass must exist within a WN spectral type, so that high- and low-gravity stars can have similar effective temperatures.

4.3.2 MASS-LOSS RATES There are a variety of diagnostics of mass loss in W-R stars (17); however, $\dot{M}$ is most readily measured from the free-free flux at radio wavelengths. In a steady, isotropic wind the stellar mass-loss rate can be derived from the measured radio flux if V_∞, the distance to the star, and the mean mass and charge per ion are given (190). Using this method, Abbott et al. (6) derived mass-loss rates for all W-R stars thought to be free-free radio sources. All 24 detected stars and all 9 upper limits were confined to the narrow range $0.8 \times 10^{-5} \lesssim \dot{M} \leq 8.0 \times 10^{-5} M_\odot \, \mathrm{yr}^{-1}$, with a mean value of $\dot{M}(\mathrm{W\text{-}R}) \sim 2 \times 10^{-5} M_\odot \, \mathrm{yr}^{-1}$. The distribution of mass-loss rates versus spectral type is shown in Figure 8. There is no measurable correlation between the mass-loss rate and the chemical composition, state of excitation, absolute visual magnitude, terminal velocity, or binarity of the W-R star. There is a tentative correlation between $\dot{M}$ and stellar mass, based on data for five double-lined spectroscopic binaries in the radio sample, as shown by Abbott et al. (6). They speculate that the narrow range in observed $\dot{M}$ results from an intrinsically narrow range in stellar mass for the W-R class.

An unresolved uncertainty in these values of $\dot{M}$ is the ionization balance in the outer wind. Abbott et al. (6) assumed that the ionization in the region of radio emission was the same as that observed near the star, where the optical recombination lines are formed. Recent theoretical studies (87,

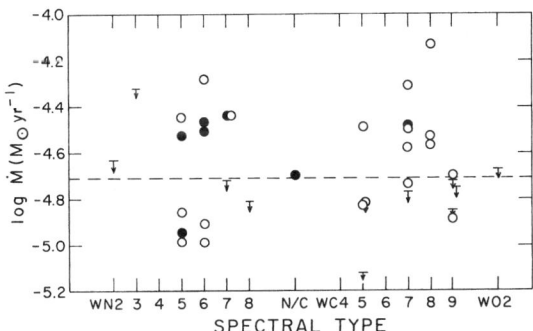

Figure 8 Mass-loss rate derived from radio fluxes that are definite free-free emission (filled symbols) and probable free-free emission (open symbols). There is no measurable correlation with spectral type, which implies no dependence of the mass-loss rate on chemical composition (WN vs. WC sequence) or on the excitation equilibrium of the wind (early vs. late spectral types). The dashed line is the mean rate for all W-R stars, $\dot{M} = 2 \times 10^{-5} \, M_\odot \, \mathrm{yr}^{-1}$ [from (6)].

139, 153) show that the wind may recombine between the optical and radio regions. The maximum error is a factor of roughly 2.7 difference between the cases of a wind of pure He^{++} and a wind of pure He^+. Thus, the $\dot{M}$ in Figure 8 may be systematically too low, especially for the early spectral types, where Abbott et al. assumed that He^{++} was dominant.

4.3.3 MOMENTUM FLUX   Abbott et al. (6) tabulate the momentum flux $\dot{M}V_\infty$ for the 40 W-R stars within 3 kpc. This is usually compared with the photon momentum flux L_*/c to form a measure of the efficiency of the transfer of momentum from the radiation to the gas. For W-R stars, the comparison is hampered by the uncertainty in L_*. Abbott et al. (6) estimated the ratio $4 \lesssim \dot{M} V_\infty/(L_*/c) \lesssim 14$ for five spectroscopic binaries using the mass/luminosity relation of helium stars to estimate L_*. Barlow et al. (20) found ratios about a factor of three larger using luminosities estimated from M_v and bolometric corrections. Whatever the correct ratio, it is clear that W-R winds have the highest momentum flux of any stellar type.

4.3.4 WIND INSTABILITIES   The radiation-driven winds of OB stars are highly unstable, and the development of strong shocks is inevitable (142). The presence of strong shocks in the wind provides a natural explanation for almost all of the observed X-ray emission in OB stars (e.g. 35, 103). Many strong shocks will also create a very nonmonotonic velocity law, which explains the extended region of zero residual flux, or blackness, that is observed in the absorption component of the P Cygni profiles of OB stars (104). Since radiation pressure probably dominates the acceleration in the majority of the W-R wind, since the X-ray emission from W-R stars is qualitatively similar to that seen in OB stars, and since the P Cygni profiles of W-R stars also have very extended blackness (e.g. the difference between V_∞ and V_{min} in Figure 6), it is likely that W-R winds are also permeated by strong shocks. In this interpretation, the extent of the blackness implies velocity amplitudes of 500–1000 km s^{-1} for the shocks of W-R winds.

5. MODEL ATMOSPHERES

Classic model atmospheres are clearly inappropriate for W-R stars because of their flagrant violation of hydrostatic equilibrium and plane-parallel geometry. There is no analog to the "photospheric analysis" of OB stars, which yields such fundamental parameters as T_{eff}, $\log g$, chemical composition, and the continuum radiation field (e.g. 95). In addition, the treatment of line formation in the winds of OB stars, which gives the radiation force (e.g. 38) and the empirical analyses of P Cygni profiles (e.g. 39), is not valid in W-R stars because of the following:

1. The continuum radiation field in the wind is not known a priori but rather depends sensitively on the run of density and temperature with radius, i.e. the "core-halo" approximation breaks down. This means that the ionization-excitation equilibrium also depends on the density/temperature structure in the wind.
2. Pure scattering is no longer a good approximation to describe resonance-like formation, because thermal creation of photons is now a major source of line photons, as evidenced by the large ratio of emission to absorption observed in the P Cygni profiles of W-R stars. Thus, the line profile shapes also are sensitive to the run of density and temperature in the wind.
3. Line blanketing becomes very severe in the highly opaque winds of W-R stars, so that the Doppler-shifted frequencies of distinct lines overlap in the wind. This couples the statistical equilibrium of different elements through the line radiation field. In addition, continuum radiative transfer is now complicated by the fact that there are large bandwidths in the UV and EUV, where there is no such thing as a line-free continuum (e.g. 7).

Because of these difficulties, it is presently impossible to solve the expanding-atmosphere problem in W-R stars, which requires the self-consistent solution of the equation of transfer, the equation of motion, the statistical equilibrium of the major wind species, and the constraint of radiative equilibrium. All models discussed in this section solve a substantially simplified version of the general problem, which is usually the solution of the radiative transfer and statistical equilibrium of helium, while adopting ad hoc distributions for the density and temperature structure.

5.1 *Line Formation*

Such models are sufficient, however, to demonstrate that expansion is the dominant mechanism of line formation in W-R atmospheres. This was first proposed by Beals (21) as the only satisfactory explanation for the P Cygni profiles that are evident in the optical spectra of most W-R stars. Castor (36) developed radiative transfer techniques for line formation in an expanding atmosphere and showed that all of the observed types of W-R line profiles—pure emission, P Cygni, and flat-topped emission—are natural consequences of expanding atmospheres. Using these techniques, Castor & Van Blerkom (40) solved the statistical equilibrium of helium for a "representative point" in the wind and demonstrated that the observed line intensities are also consistent with expansion. This model has also been used to provide crude estimates of the abundances of He and CNO in W-R stars (186). Recently, models have been developed that solve the statistical equilibrium of helium at the full range of depths in the expanding atmosphere (80, 87). These more elaborate calculations also confirm that it is the large emitting volume of the extended atmosphere that is producing the broad emission lines in W-R stars.

Direct observational evidence for geometric extension as the line formation mechanism comes from the interferometry of γ Vel (82). The angular diameter of the C III 4650 Å emission line was found to be roughly a factor of four greater than the angular diameter of the neighboring continuum. Additional evidence comes from eclipse observations of V444 Cyg by Eaton et al. (68), who found that the strong UV lines are formed beyond the orbit of the O star, a region that is at least a factor of three greater than the radius of continuum formation. A crude check on the line formation mechanism is also provided by recent measurements of W-R mass-loss rates and terminal velocities. Abbott et al. (6) showed that the observed strength of the optical emission lines is broadly consistent with the emission measure of the wind, calculated with the observed $\dot{M}$ and V_∞, and the assumption that the lines are formed by recombination.

From all of these studies, it is clear that the emission-line spectra of W-R stars are mainly determined by conditions in the stellar wind and contain little or no direct information about the fundamental properties of the underlying star. As a result, the basic quantities such as effective temperature, luminosity, gravity, and composition are poorly determined and quite controversial.

An example of a comparison between observation and current best theory is shown in Figure 9. Note the extended emission at velocities

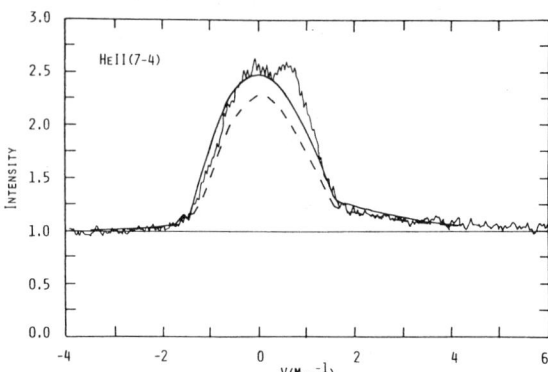

Figure 9 Observed and theoretical profiles for the 5411.5 Å line of He II in the star HD 50896. The dashed line is a model with $\dot{M} = 2.5 \times 10^{-5}\ M_\odot\ \text{yr}^{-1}$, while the solid line is a model with $\dot{M} = 5.0 \times 10^{-5}\ M_\odot/\text{yr}$. The radial velocity of HD 50896 is not known, so the redshift of the observed emission peak relative to the model calculations may not be real [from (86)].

redward of $V_\infty \cong 2000$ km s^{-1}, which result from the high-electron-scattering optical depth of the wind. The model curves in Figure 9 demonstrate the dependence of the profile on the adopted mass-loss rate. However, similar changes in the profile shape are obtained by appropriate changes in the adopted stellar radius, temperature, luminosity, and presumably the velocity law, so that the interpretation of line profiles remains ambiguous at present (81, 87).

5.2 Continuum Formation

The continuum-forming regions of W-R atmospheres are also geometrically extended. This is most evident from the total absence of any absorption lines in most W-R stars. Additional evidence that the ionizing continua are very extended comes from the fact that in the observable wavelength range between 1300 and 10,000 Å, WC stars emit roughly 50–60% of their energy in recombination lines, while WN stars emit roughly 30% in recombination lines (e.g. 20). The most direct evidence that the continuum is extended in W-R winds comes from the eclipse observations of V444 Cyg, which demonstrate that the size of the continuum photosphere increases with wavelength as the continuum opacity increases, and that electron-scattering optical depth unity occurs at a velocity of roughly 500 km s^{-1} (41).

The atmospheric extension can be visualized in terms of the relationship between the depth of continuum formation and the sonic point in the wind, since the atmosphere will resemble hydrostatic equilibrium interior to the sonic point (e.g. 34). If the continuum photosphere (as represented, for example, by the radius of Rosseland mean optical depth unity) lies deep in the subsonic region, the star will appear as a normal O star, well described by hydrostatic equilibrium and plane-parallel geometry, except for the most opaque of all UV resonance lines. As the photosphere approaches the sonic point, the stronger recombination lines go into emission, and the star develops an "Of" spectrum, in which the weaker lines are still well described by classical model atmospheres (e.g. 96) but the stronger lines require formation in the wind (e.g. 94). For example, the supersonic flow of the prototype Of star, ζ Puppis, has an optical depth to electron scattering of roughly a few tenths. Only when the continuum photosphere moves well beyond the sonic point can the spectrum become purely emission lines (or P Cygni profiles), regardless of the optical depth of the line or the nature of the transition (e.g. resonance versus subordinate line).

The velocity width of an emission line can never be smaller than the velocity at which the neighboring continuum is formed. Thus, one estimate of the depth of continuum formation in a W-R star comes from measuring the width of the narrowest emission line observed in a W-R spectrum. Typical values are on the order of a few hundred kilometers per second. On the other hand, the continuum photosphere cannot be too extended without violating two other diagnostics. The first is the absence of any continuum emission jumps in W-R spectra (e.g. 40). The second is the velocity extent of the absorption feature in P Cygni profiles. Absorption cannot occur at velocities smaller than the depth of continuum formation. The velocity of minimum absorption therefore provides an upper bound on the location of the photosphere (see Figure 6). This velocity is usually in the range 500–1000 km s^{-1}. These general considerations suggest that the continuum optical depths of W-R winds are confined to a fairly narrow range of values, corresponding to a velocity of the continuum photosphere between a few hundred and 1000 km s^{-1}.

The effects of this sphericity on the continuous spectrum emergent from the star are well known (e.g. 32, 37, 116, 169). The main result is to flatten the shape of the observable continuum, so that stars with the same effective temperature, but with increasing degree of extension, will have a continuum characterized by a strongly decreasing color temperature. The effect is quite strong, so that an atmosphere whose scale height is only 3% of the stellar radius may still have a color temperature that is 20% less than a comparable plane-parallel atmosphere. In the limit of extreme extension, where the continuum is formed in the outer wind flowing at the terminal velocity, the breakdown between the spectrum and the temperature is complete, and the spectrum becomes independent of effective temperature and luminosity, depending only on $\dot{M}$ and V_∞ (e.g. 190).

5.3 Effective Temperatures

When the radius of continuum formation varies with wavelength, as it does in spherical atmospheres, there is no generally accepted definition of the effective temperature T_{eff}. We adopt the definition $T_{\text{eff}} = [L_*/4\pi\sigma(R_{\text{ph}})^2]^{1/4}$, where R_{ph} is the radius of Rosseland mean optical depth unity (37). As discussed above, R_{ph} will generally occur above the sonic point, so the T_{eff} from this atmosphere definition will be smaller than the T_{eff} of stellar evolution models, which is defined by the radius of a wind-free, hydrostatic, plane-parallel atmosphere. The ratio of T_{eff}(atmosphere)/T_{eff}(evolution) can easily be a factor of ~ 0.5 from this effect alone (e.g. 60).

With the advent of ultraviolet satellite observations, several studies have derived $T_{\text{eff}} \cong 30,000$ K for W-R stars by using the color temperature of the ultraviolet-optical continuum or by comparing the observed continuum with plane-parallel, hydrostatic model atmospheres (140, 166, 188). Even if the large intrinsic uncertainties in the corrections for interstellar extinction are overcome (75; see also Figure 1), these estimates are totally unreliable because of the effects of sphericity discussed above. This is demonstrated most convincingly by the fact that self-consistent spherical model atmospheres with high T_{eff} can also reproduce the observed UV-optical continuum of W-R stars (e.g. 81, 87). Figure 10 shows a fit to the observations of HD 50896 by a model with $T_{eff} \cong 75,000$ K. The shape of the continuum longward of ~ 1000 Å is insensitive to T_{eff} because the bulk of the energy is emitted in the EUV (228–1000 Å).

One way to minimize the effects of spherical extension is to measure the maximum color temperature of the continuum, which in W-R stars always occurs at the shortest wavelength observed (see Figure 10). In atmospheres in which the extension is fairly mild, there is still some relationship between the UV continuum and T_{eff}, although the color temperature will still underestimate T_{eff} (81). The T_{eff} values derived in this manner range from

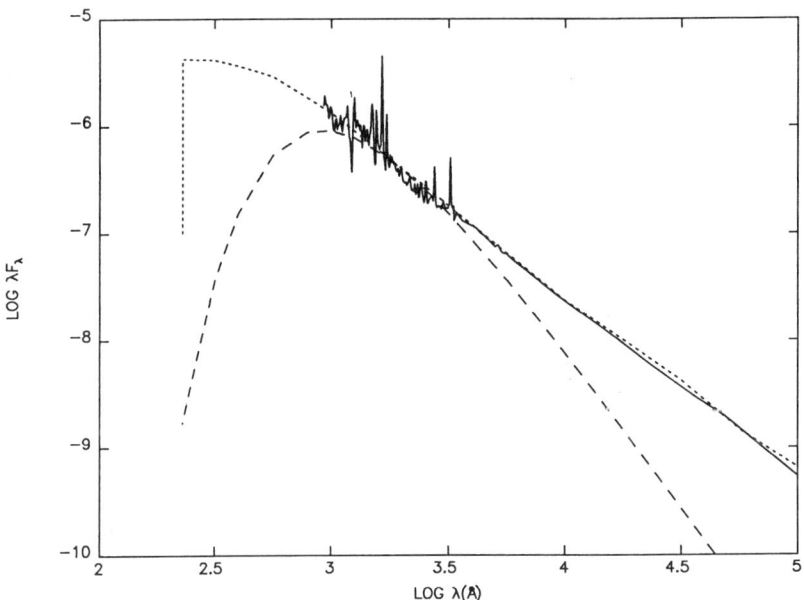

Figure 10 Comparison of the observed energy distribution of the WN5 star HD 50896 (solid line) to the model calculations of Hillier (87) characterized by $L_* = 2.5 \times 10^5 \ L_\odot$, $R_c = 2.5 \ R_\odot$, $T_{eff} = 75,000$ K, and $\dot{M} = 5 \times 10^{-5} \ M_\odot \ yr^{-1}$ (short dashed line). The observations are from Voyager data between 912 and 1200 Å (R. Polidan, private communication), IUE data between 1200 and 3000 Å (140), and optical-infrared photometry beyond 3000 Å (46). Also shown for comparison is a Planck curve with $T = 40,000$ K (long dashed line). Note how the color temperature of the observed continuum increases with decreasing wavelength. The observations were normalized by assuming a distance of 1 kpc for HD 50896 and were dereddened assuming $E(B-V) = 0.04$.

38,000 K for the WC8 and WN8 types to 60,000 K for the WC5 and WN6 stars (20).

One color temperature that should avoid the uncertainties of the unobserved Lyman continuum is the Zanstra temperature, which is based on comparison of the integrated ionizing photon flux to the observed flux at $\lambda 5556$ Å. Radio observations of nine WN stars associated with ring nebulae have yielded T_{eff} values that range from 25,000 K for a WN8 star to 55,000 K for a WN5 star (128). Two uncertainties of this method are (a) the question of whether or not the nebulae are optically thick, so that every ionizing photon is counted; and (b) the calibration between the Zanstra temperature and T_{eff}, which has so far been done with plane-parallel, hydrostatic atmospheres. Efforts have also been made to derive Zanstra temperatures based on the He II recombination lines (140, 188). However, the results are inconclusive because they are sensitive to the statistical equilibrium of He II (152).

It is a fact of stellar evolution that once the nuclear-processed material appears at the stellar surface, the T_{eff} must increase because the mean molecular weight has become both larger and more uniform (162, 163). In the limit that a W-R star has no hydrogen and a uniform composition, which is likely given the observed $\dot{M}$ (108), it will lie on the helium zero-age main sequence (ZAMS) in the H-R diagram, with a T_{eff} in excess of 100,000 K. Support for a high T_{eff} also comes from a study of the eclipsing binary V444 Cyg, which found that the brightness temperature of the radiation at the electron-scattering photosphere was of the order 90,000 K (41).

One would naively expect the observed excitation-ionization state of a W-R wind to correlate with T_{eff}, implying that the earlier spectral types are indeed hotter. This is especially true in the late-type W-R stars, where the spectral lines of low-excitation species like He I, C II, N III, etc., definitely give the impression of a cool T_{eff}. Such a correlation has yet to emerge from theoretical models, however, and the meaning of spectral type remains unsolved. Preliminary calculations suggest that no W-R star with He I recombination lines in the spectrum can be hotter than $T_{\text{eff}} = 60,000$ K (154). This value may depend, however, on an assumed distribution of density and temperature of the model.

5.4 Luminosity

The controversy over T_{eff} carries over directly into uncertainty about the intrinsic luminosity L_* of W-R stars. The most common method of estimating L_* is to calculate a bolometric correction from T_{eff} and then apply it to the observed absolute visual magnitudes (20, 47). These give bolometric luminosities in the range $-8 \gtrsim M_{\text{bol}} \gtrsim -10$, with the exception of the late WN spectral types, which are brighter (see Figure 11). In an empirical approach, Humphreys et al. (92) selected W-R and O stars in the same associations. They then estimated an empirical bolometric correction by finding the difference between the M_v of the W-R star and the M_{bol} of the most luminous O star. Under the reasonable assumption that the W-R star evolved from a star at least as massive as the most luminous O star, they find a minimum bolometric correction that is larger than usually assumed.

A check on these estimates is also provided by the rather tight mass-luminosity relation found for the helium stars by stellar evolution calculations with mass loss (108). We have applied this relation to the 17 W-R+O binaries with mass estimates in Table 1 and compared the resultant L_* with a spectral-type luminosity calibration (20). The luminosities pre-

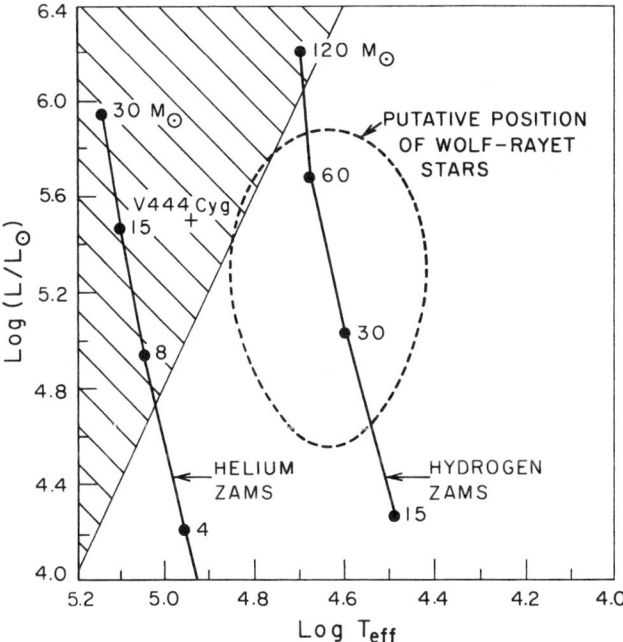

Figure 11 Schematic H-R diagram showing the hydrogen and helium zero-age main sequences (ZAMS) and the location of W-R stars following the empirical luminosity calibration of Barlow et al. (20). The "+" symbol shows the location in the H-R diagram of V444 Cygni from the eclipse analysis (41). The shaded region shows where radiation-driven winds become optically thick in the continuum, based on estimates we have made with the multiple scattering model of Abbott & Lucy (7). If W-R stars have the high T_{eff} and L_* suggested by the evolution calculations, then their winds can probably be explained by radiation pressure.

dicted by stellar evolution are roughly a factor of three larger at all spectral types except WC5 (for which there is only one star). By any standard, W-R stars are overluminous for their mass when compared with OB stars.

5.5 Mass-Loss Mechanisms

The major challenge to any theory for the mass loss from W-R stars is to explain the incredibly high efficiency of the winds. Using the estimates of Abbott et al. (6), which make only the most conservative assumptions concerning $\dot{M}$ and L_*, the wind kinetic energy is from 1% to 8% of the total stellar luminosity. Comparable numbers for OB stars are a few tenths of a percent, while in the Sun the kinetic energy of the wind is of the order $10^{-7} L_*$. Since W-R stars are hot and very overluminous for their mass, a natural explanation for the cause of the mass loss is radiation pressure. This idea is supported by the fact that W-R winds resemble OB star winds in terms of their velocities, the ions present in the wind, the basic character of the UV P Cygni profiles, and their X-ray emission. If W-R winds are high-density extensions of OB star winds, it is likely that the basic driving mechanism is also the same. Other mechanisms that have been proposed to explain the mass loss, which we do not discuss further, are pulsations (109, 156) and magnetic-centrifugal forces (33).

A rather complete theory of radiation-driven mass loss exists for OB star winds (e.g. 4), but so far all models assume a core-halo approximation for the continuum radiation field that is invalid for W-R stars; thus, a direct confrontation between theory and observation is impossible. As an exploratory calculation, Pauldrach et al. (143) showed that the mass loss from V444 Cyg could be explained by radiation pressure. However, the model was subject to the following assumptions: (a) Radiation pressure was described by the core-halo formulas appropriate for OB stars, and (b) the luminosity was near the Eddington limit, so that continuum radiation pressure was dominant in the inner wind.

The central question is, why are W-R winds so efficient at converting radiative momentum to the gas? A typical photon must transfer roughly 10 times its momentum to explain the mass loss, so extensive multiple scattering is essential to any explanation. Abbott (2) showed that the change of composition from OB stars to W-R stars produced no gain in radiative driving in the WN sequence, and only a modest (factor of three) increase in the mass-loss rate in the WC sequence.

The efficiency of radiation pressure is greatly increased, however, in certain regions of the H-R diagram. We demonstrate this using the multiple scattering model of Abbott & Lucy (7). While this is a core-halo model, and hence it is unable to represent a W-R star, it can be used to calculate where in the H-R diagram a radiation-driven wind becomes so dense that its continuum optical depth exceeds unity. The results of applying the Abbott & Lucy model to various (L_*, T_{eff}) combinations are sketched in Figure 11. These results are preliminary because they were only calculated for one combination of velocity law, composition, and ionization-excitation equilibrium. However, the conclusion is clear: Radiation pressure can drive the winds of W-R stars only if the stars are very hot and/or very luminous.

The gain in efficiency with increasing T_{eff} occurs mainly because of the decreasing radius. The efficiency of the radiative acceleration is determined primarily by the number of optically thick lines that result from a given mass-loss rate. The more lines that are found, the greater is the force. With all other factors being held constant, a decreasing radius causes an increasing density, which gives an increasing optical depth. The amplification in the radiation pressure is nonlinear because the higher mass-loss rate resulting from the increased force also gives a higher density. When the number of optically thick lines is high enough that the frequency separation between lines is small compared with the Doppler shift of the wind, multiple scattering becomes very efficient, so that the average photon can experience a redshift on the order of 10%.

Thus, in this scenario, the large $\dot{M}$ in W-R stars is in fact caused by chemical composition, but this effect is indirect. The removal of the hydrogen envelope causes the hydrostatic core of the star to evolve to high T_{eff} and small R_*, and this movement in the H-R diagram causes the increased efficiency of the radiation driving. To confirm this hypothesis, it remains to be shown that stars within the W-R domain of Figure 11 can actually achieve the high observed mass loss. Some preliminary results from a multiple scattering model that self-consistently accounts for continuum sphericity are encouraging. D. C. Abbott & L. B. Lucy (in preparation) find that a model based on the stellar parameters of HD 50896 can achieve $\dot{M} V_\infty/(L_*/c) \sim 5$ with an emergent energy distribution that matches observations.

6. W-R STARS AND THE GALACTIC ENVIRONMENT

The return of mass and energy to the interstellar medium by W-R winds can be directly calculated from their measured $\dot{M}$, V_∞, and Galactic distribution (e.g. 3, 6). These calculations show that W-R stars play a very significant role in the "Galactic ecology," despite their rarity. In the solar vicinity W-R stars input wind kinetic energy at a rate of 1.1×10^{38} erg s^{-1}

kpc^{-2}, which is roughly 50% of the total wind energy from all stellar types. Integrating over the lifetime of the W-R phase, the stellar wind energy input is comparable to that from the supernova stage that presumably follows.

W-R winds input mass to the interstellar medium (ISM) at a rate of $5 \times 10^{-5} M_\odot$ yr^{-1} kpc^{-2}, which is comparable to the net mass input from all OB stars but is roughly a factor of four smaller than the contribution of late-type stars and planetary nebulae. What makes the W-R mass input interesting is the fact that nearly all of it represents material processed by either hydrogen or helium burning in the interior of a massive star. Thus, according to stellar evolution models (108), W-R winds strongly contribute to the Galactic enrichment of ^{4}He, ^{12}C, ^{17}O, and ^{22}Ne, while contributing moderately to the enrichment of ^{14}N, ^{26}Mg, ^{25}Mg, and ^{16}O. W-R winds are also a significant source of interstellar dust. Assuming that dust condenses in the winds of all WC8 and later spectral types, roughly 15% of the mass loss involves the formation of dust, presumably of the graphite variety.

The predicted overabundance of ^{22}Ne in W-R winds has been used to explain the high ^{22}Ne/^{20}Ne ratio observed in Galactic cosmic rays (e.g. 31). If ^{22}Ne is a factor of 120 overabundant in WC stars, then 1 in 60 cosmic rays must be accelerated in W-R winds to explain the observations. The cosmic rays are presumed to be accelerated in the terminal shock of the W-R wind. One possible difficulty with this idea is the recent observation of two forbidden Ne lines in the infrared spectrum of γ Vel (8, 173; see spectrum in Figure 4). In the recombination analysis of Barlow (18), these lines indicate a solar ratio of Ne/He.

One observed interaction between W-R stars and the interstellar medium is the formation of "ring nebulae" (43, 84), which are arcs of nebulosity centered on and ionized by the W-R star. Chu et al. (45) classify the 15 known Galactic ring nebulae as being stellar ejecta, wind-blown interstellar bubbles, or quiescent H II regions. The nebulae presumably can evolve from one type to another in time. Since W-R stars are born in a stellar wind cavity created by the wind of the progenitor O star, the existence of ring nebulae is a puzzle. They may indicate that the transition between an O star and a W-R star sometimes involves a very rapid phase of mass loss, i.e. an ejection episode. Wind-blown nebulae are generally associated with WN stars, while quiescent H II regions are generally associated with WC stars, which suggests that WN stars are younger than WC stars. Abundance analyses of ring nebulae have found that the material comes from a mixture of processed stellar material and unprocessed interstellar gas (e.g. 98). With one exception (a stellar ejecta nebulae), the stellar contribution is of the order of a few percent.

7. W-R STARS AS A PHASE OF STELLAR EVOLUTION

7.1 *Production Scenarios*

A recent review of massive-star evolution has been given in these volumes (42). A review of some empirical aspects of W-R star evolution has also recently appeared (113). Let us put together a logical path by which W-R stars may appear and then review the data that suggest they are, indeed, the descendants of massive OB stars. These progenitor stars in the hydrogen-burning state have substantial winds and mass loss (e.g. 95) that must be taken into account in evolution. Mixing processes, which are not yet fully understood, also play a role in massive-star evolution (107). The hydrogen envelope is thus removed both by mixing with the convective core of the star and by a stellar wind. As the star evolves toward lower effective temperatures at a more or less constant luminosity, some combination of these effects will lead to the initial stellar core of the massive star being "uncovered" during its normal evolution (108, 134, 147). As the surface hydrogen content drops from the normal values, the star will turn back toward higher effective temperatures. This we tentatively identify with a W-R phase, specifically with the WN objects containing hydrogen (47). Possibly other intermediate stages, represented by the S Doradus types (161) or P Cygni stars (100), Hubble-Sandage variables, or Eta Car or like objects (59) [all generically referred to as "luminous blue variables" (LBV)], may play a role in separating the OB phase from the W-R phase.

What evidence supports this general scenario? We see that the massive OB stars and the W-R stars share a similar Galactic longitude and latitude distribution. In particular, there are fewer of both objects toward the anticenter region. In associations in our Galaxy and in M33, the W-R stars and the red supergiants are not found together. Humphreys & Davidson (91) had previously noted that in the galaxies of the Local Group, red supergiants are never found with M_v brighter than -8.0 magnitude, corresponding to an initial mass between 30 and 50 $M_\odot$ depending on whose evolution tracks are used. It seems clear that above some initial mass of the order of 40 $M_\odot$ (51), stars do not become red supergiants when they begin to burn helium in the core. But helium burning must go on someplace, which one can readily identify with the W-R stars. We have already noted the evidence that these objects are, in fact, helium rich. They must also be helium burning for the most part, since the mass/luminosity ratio of those in binaries indicates that they are overluminous for their mass, and that their companions are typically more massive (111).

7.2 Evolution of W-R Stars

Massive-star evolution normally ends in a W-R phase, which is followed by the most rapid and advanced stages. An interesting issue is the competition between the time scale for further significant mass loss during the W-R phase and the evolution time of the nuclear core. We have seen that the mass loss may decrease as the star evolves to lower masses and luminosities, and the final evolution may then depend on the core physics alone. Theoretical models (147) have not settled this issue owing to the present uncertainties in the physics.

Observationally, there appear to be certain types of supernovae that have properties suggestive of an immediate past history as a W-R star. These supernovae make up the newly identified type "Ib"; they have similar light curves to the "classical" Ia supernovae yet are less luminous. In the Ib supernovae, lines of hydrogen are not found, nor is the characteristic feature at $\lambda 6150$ Å, which is attributed to Si II and is so very prominent in the Ia supernovae. Examples of the Ib supernovae are 1985f (24, 76), 1983n (27, 70), and 1984l (179). These objects also appear to be found in H II regions and in Galaxy locations with other massive-star populations, rather than the lower masses inferred for the Ia types. There are not yet any spectra of the type Ib stellar progenitors, so this scenario is not conclusive; however, the circumstantial evidence is very suggestive.

A spectrum of the type II supernova 1983k before maximum light (133) was very much like that of a late WN star. However, this period was one after which the explosion has already begun, and it is possible that the shock conditions at that time mimic conditions in a W-R star; hydrogen is present in this spectrum and others of type II, so the progenitor star is unlikely to be a W-R. There is no spectrum of the object that exploded as Cas A, only information about its remnants. Fesen et al. (69) suggest that the progenitor object was a late-type WN star based on the helium- and nitrogen-rich composition of the fast-moving knots ejected in the initial phases of the explosion.

These data taken together suggest that W-R stars may end their lives as supernovae, either as type Ib or in the peculiar phase indicated by the Cas A remnant. Perhaps, given the Cas A composition, not all WN stars become WC objects before ending their evolution (69). This area of research is a very active one in which additional data and calculations are rapidly becoming available, and the complete details of W-R star death scenarios are still open.

7.3 Major Unsolved Issues

While it is clear that W-R stars are the evolved remnants of initially more massive O stars, it is not clear how the evolution proceeds. Is it mass loss during the O-star phase, mass loss during a red giant phase, envelope ejection during the LBV phase, binary mass transfer, or mixing between the convective core and the hydrogen envelope? Each of these ideas has some observational or theoretical support, and all of these mechanisms may operate individually or in combination to give the observed population of W-R stars. Does each of these "channels" for the formation of a W-R star lead to a particular spectral type? For example, are the transitional WN7 objects the only W-R stars formed solely by mass loss in the O-star phase? We do not know the answers to these questions, but current observations give tantalizing hints.

Most striking are observed differences in the WN/WC number ratio between different galaxies and between different locations within a galaxy. In the LMC, both early and late WN stars are found, but only early WC objects are present. In the SMC and M33, a preponderance of early WN types is found. Late WC stars are only found near the Sun and toward the central regions of our Galaxy. These differences may suggest that early WN stars evolve from the less-massive O stars, whereas the late WN and early WC stars evolve from the most-massive O stars. Differences in metallicity will alter the mean mass-loss rates in earlier stages of evolution, which can also affect which channel forms the W-R stars.

Do WN stars evolve into WC stars? The theory of stellar evolution would say yes, since mass loss peels away the products of hydrogen burning to reveal the products of helium burning. Observations, however, are not so clear. The connection in location and numbers between the luminous late WN stars and some of the early WC stars supports this scenario. On the other hand, mass determinations from spectroscopic binaries show no difference between the mass of a typical WN star and that of a typical WC star.

What causes W-R stars to have stronger winds than any other spectral type, winds so strong that the underlying star is totally obscured from view? Presumably W-R stars are on the brink of instability. Is this an instability to radiation pressure, vibrations, or an as yet unidentified cause? Although it seems that radiation pressure models will work for a star like HD 50896, which is very hot and has only a moderate mass-loss rate for a W-R star, can these same models explain a star like γ^2 Vel, which has a significantly cooler atmosphere but a larger mass-loss rate?

What does the spectral sequence mean? Presumably, the early types with high excitation-ionization have a higher atmospheric T_{eff}, although this conclusion is by no means certain. The relationship between the atmospheric T_{eff} and the T_{eff} of the wind-free models of stellar evolution is unknown. A low atmospheric T_{eff} could mean that the underlying stellar core has a low T_{eff}, or else it could mean that the atmosphere has greater extension. Stellar evolution models give limited guidance. If W-R stars evolve homologously, they would move to hotter T_{eff} in the H-R diagram with time. If W-R stars evolve with stratification in their interiors, they would move to cooler T_{eff} with time. The fact that the WN subtypes contain stars that span a factor of 5 in mass and at least a factor of 10 in hydrogen abundance strongly suggests that stars with very different locations in the theoretical H-R diagram can be of the same W-R spectral type.

What of the future? Current progress in further understanding of W-R stars is stymied mainly by inadequate atmosphere models. Literally hundreds of high-resolution optical and UV spectra exist, yet no model is presently adequate to fully analyze these data. Once this roadblock is overcome, and the recent impressive progress in atmospheric models suggests it will be soon, further breakthroughs in our ability to understand W-R stars will be forthcoming. Our progress in the last 20 years since the last *Annual Reviews* article (165) on this subject has been impressive: We now broadly understand their central role in massive-star evolution and have some relatively good estimtes of the nominal stellar parameters. As the evolution details improve, we may be able to make sense out of the varied distributions and numbers of the W-R subtypes observed in the different environments of nearby galaxies

ACKNOWLEDGMENTS

We greatly appreciate discussions with and preprints from Drs. Hillier, Polidan, Pollock, and Wheeler. Our colleagues here at JILA have, as usual, provided stimulating discussions concerning W-R stars, and our publications office is thanked for its exceptional service. We are indebted to our funding agencies for continued support under NSF grants AST-8520728 (PSC) and AST-8505919 (DA) and NASA contract NAGW-766 (DA).

Literature Cited

1. Abbott, D. C. 1978. *Ap. J.* 225: 893
2. Abbott, D. C. 1982. See Ref. 61, p. 185
3. Abbott, D. C. 1982. *Ap. J.* 263: 723
4. Abbott, D. C. 1985. In *Relations Between Chromospheric-Coronal Heating and Mass Loss in Stars*, ed. R. Stalio, J. B. Zirker, p. 265. Italy: Tabographics
5. Abbott, D. C., Bieging, J. H., Churchwell, E. 1985. In *Radio Stars*, ed. R. M. Hjellming, D. M. Gibson, p. 219. Dordrecht: Reidel
6. Abbott, D. C., Bieging, J. H., Churchwell, E., Torres, A. V. 1986. *Ap. J.* 303: 239
7. Abbott, D. C., Lucy, L. B. 1985. *Ap. J.* 288: 679
8. Aitken, D. K., Roche, P. F., Allen, D. A. 1982. *MNRAS* 200: 69P
9. Allen, D. A., Barton, J. R., Wallace, P. T. 1981. *MNRAS* 196: 797
10. Allen, D. A., Swings, J. P., Harvey, P. M. 1972. *Astron. Astrophys.* 20: 333
11. Allen, D. A., Wright, A. E., Gross, W. N. 1976. *MNRAS* 177: 91
12. Armandroff, T., Massey, P. 1985. *Ap. J.* 291: 685
13. Azzopardi, M., Breysacher, J. 1979. *Astron. Astrophys.* 75: 120

14. Azzopardi, M. 1986. *Astron. Astrophys.* In press
15. Baade, D. 1987. See Ref. 55. In press
16. Bappu, M. K. V., Sahade, J., eds. 1973. *Wolf-Rayet and High Temperature Stars. IAU Symp. No. 49*, Dordrecht: Reidel
17. Barlow, M. J. 1982. See Ref. 61, p. 149
18. Barlow, M. J. 1986. *Proc. Br.-Span. Sch. Astrophys., 4th*, ed. M. Moles. In press
19. Barlow, M. J., Hummer, D. G. 1982. See Ref. 61, p. 387
20. Barlow, M. J., Smith, L. J., Willis, A. J. 1981. *MNRAS* 196: 101
21. Beals, C. S. 1929. *MNRAS* 90: 202
22. Beals, C. S., Plaskett, H. 1935. *Trans. Int. Astron, Union* 5: 184
23. Becker, R. H., White, R. L. 1985. *Ap. J.* 297: 649
24. Begelman, M. C., Sarazin, C. L. 1986. *Ap. J. Lett.* 302: L59
25. Bergeron, J. 1977. *Ap. J.* 211: 62
26. Bieging, J. H., Abbott, D. C., Churchwell, E. 1982. *Ap. J.* 263: 207
27. Blair, W. P., Panagia, N. 1986. In *Scientific Accomplishments of the IUE*, ed. Y. Kondo. Dordrecht: Reidel. In press
28. Breysacher, J. 1981. *Astron. Astrophys.* 43: 203
29. Breysacher, J. 1986. *Astron. Astrophys.* 160: 185
30. Breysacher, J., Moffat, A. F. J., Niemela, V. 1982. *Ap. J.* 257: 116
31. Casse, M., Paul, J. A. 1982. *Ap. J.* 258: 860
32. Cassinelli, J. P. 1971. *Astrophys. Lett.* 8: 105
33. Cassinelli, J. P. 1982. See Ref. 61, p. 173
34. Cassinelli, J. P., Castor, J. I. 1973. *Ap. J.* 179: 189
35. Cassinelli, J. P., Swank, J. H. 1983. *Ap. J.* 271: 681
36. Castor, J. I. 1970. *MNRAS* 149: 111
37. Castor, J. I. 1974. *Ap. J.* 189: 273
38. Castor, J. I., Abbott, D. C., Klein, R. I. 1975. *Ap. J.* 195: 157
39. Castor, J. I., Lamers, H. J. G. L. M. 1979. *Ap. J. Suppl.* 39: 481
40. Castor, J. I., Van Blerkom, D. 1970. *Ap. J.* 161: 485
41. Cherepashchuk, A. M., Eaton, J. A., Khaliullin, K. F. 1984. *Ap. J.* 281: 774
42. Chiosi, C., Maeder, A. 1986. *Ann. Rev. Astron. Astrophys.* 24: 329
43. Chu, Y.-H. 1981. *Ap. J.* 249: 195
44. Chu, Y.-H., Daod, N. A. 1984. *Publ. Astron. Soc. Pac.* 96: 999
45. Chu, Y.-H., Treffers, R. R., Kwitter, K. B. 1983. *Ap. J. Suppl.* 53: 937
46. Cohen, M., Barlow, M. J., Kuhi, L. V. 1975. *Astron. Astrophys.* 40: 291
47. Conti, P. S. 1976. *Mem. Soc. R. Sci. Liège* 9: 193
48. Conti, P. S. 1986. See Ref. 62, p. 199
49. Conti, P. S. 1987. See Ref. 55. In press
50. Conti, P. S., Garmany, C. D. 1983. *Publ. Astron. Soc. Pac.* 95: 411
51. Conti, P. S., Garmany, C. D., de Loore, C., Vanbeveren, D. 1983. *Ap. J.* 274: 302
52. Conti, P. S., Leep, E. M., Perry, D. N. 1983. *Ap. J.* 268: 228
53. Conti, P. S., Massey, P. 1981. *Ap. J.* 249: 471
54. Conti, P. S., Niemela, V. S., Walborn, N. R. 1979. *Ap. J.* 228: 206
55. Conti, P. S., Underhill, A. B., eds. 1987. *O, Of, and Wolf-Rayet Stars. NASA Spec. Publ.* In press
56. Cowley, A. P., Crampton, D., Hutchings, J. B., Thompson, I. B. 1984. *Publ. Astron. Soc. Pac.* 99: 968
57. Danks, A. C., Dannefeld, M., Wamsteker, W., Shaver, P. A. 1983. *Astron. Astrophys.* 118: 301
58. Davidson, K., Kinman, T. D. 1982. *Publ. Astron. Soc. Pac.* 94: 634
59. Davidson, K., Walborn, N. R., Gull, T. R. 1982. *Ap. J. Lett.* 254: L47
60. de Loore, C. W. H., Hellings, P., Lamers, H. J. G. L. M. 1982. See Ref. 61, p. 53
61. de Loore, C. W. H., Willis, A. J., eds. 1982. *Wolf-Rayet Stars: Observations, Physics, and Evolution, IAU Symp. No. 99*. Dordrecht: Reidel
62. de Loore, C. W. H., Willis, A. J., Laskarides, D., eds. 1986. *Luminous Stars and Associations in Galaxies, IAU Symp. No. 116*. Dordrecht: Reidel
63. D'Odorico, S., Benvenuti, P. 1983. *MNRAS* 203: 157
64. D'Odorico, S., Rosa, M. 1981. *Ap. J.* 258: 1015
65. Drissen, L., Lamontagne, R., Moffat, A. F. J., Bastien, P., Seguin, M. 1986. *Ap. J.* 304: 188
66. Durret, F., Bergeron, J., Bokesenberg, A. 1985. *Astron. Astrophys.* 143: 347
67. Dyck, H. M., Simon, T., Wolstencroft, R. D. 1984. *Ap. J.* 277: 675
68. Eaton, J. A., Cherepashchuk, A. M., Khaliullin, Kh. F. 1985. *Ap. J.* 297: 266
69. Fesen, R. A., Becker, R. H., Blair, W. P. 1987. *Ap. J.* 313: 378
70. Filippenko, A. V., Sargent, W. L. W. 1986. *Astron. J.* 91: 691
71. Florkowski, D. R. 1982. See Ref. 61, p. 63
72. Garmany, C. D. 1986. See Ref. 62, p. 19

73. Garmany, C. D., Conti, P. S., Massey, P. 1980. *Ap. J.* 242: 1063
74. Garmany, C. D., Massey, P. 1984. *Bull. Am. Astron. Soc.* 16: 508
75. Garmany, C. D., Massey, P., Conti, P. S. 1984. *Ap. J.* 278: 233
76. Gaskell, C. M., Cappellaro, E., Dinerstein, H. L., Garnett, D., Harkness, R. D., Wheeler, J. C. 1986. Preprint
77. Gies, D. 1985. PhD thesis. Univ. Toronto, Can.
78. Hackwell, J. A., Gehrz, R. D., Grasdalen, G. L. 1979. *Ap. J.* 234: 133
79. Hackwell, J. A., Gehrz, R. D., Smith, J. R. 1974. *Ap. J.* 192: 383
80. Hamann, W. R. 1985. *Astron. Astrophys.* 145: 443
81. Hamann, W. R., Schmutz, W. 1987. *Astron. Astrophys.* In press
82. Hanbury-Brown, R., Davis, J., Herbison-Evans, D., Allen, L. R. 1970. *MNRAS* 148: 103
83. Hartmann, L., Cassinelli, J. P. 1977. *Ap. J.* 215. 155
84. Heckathorn, J. N., Bruhweiler, F. C., Gull, T. R. 1982. *Ap. J.* 252: 230
85. Hillier, D. J. 1982. See Ref. 61, p. 225
86. Hillier, D. J. 1984. *Ap. J.* 280: 744
87. Hillier, D. J. 1987. *Ap. J. Suppl.* In press
88. Hillier, D. J., Jones, T. J., Hyland, A. R. 1983. *Ap. J.* 271: 221
89. Hogg, D. E. 1985. In *Radio Stars*, ed. R. M. Hjellming, D. M. Gibson, p. 117. Dordrecht: Reidel
90. Hummer, D. G., Barlow, M. J., Storey, P. J. 1982. See Ref. 61, p. 79
91. Humphreys, R. M., Davidson, K. 1979. *Ap. J.* 232: 409
92. Humphreys, R. M., Nichols, M., Massey, P. 1984. *Astron. J.* 90: 101
93. Keel, W. C. 1982. *Publ. Astron. Soc. Pac.* 94: 765
94. Klein, R. I., Castor, J. I. 1978. *Ap. J.* 220: 902
95. Kudritzki, R. P., Hummer, D. G. 1986. See Ref. 62, p. 3
96. Kudritzki, R. P., Simon, K. P., Hamann, W. R. 1983. *Astron. Astrophys.* 118: 245
97. Kunth, D., Sargent, W. L. W. 1981. *Astron. Astrophys.* 101: L5
98. Kwitter, K. B. 1984. *Ap. J.* 287: 840
99. Lambert, D. L., Hinkle, K. H. 1984. *Publ. Astron. Soc. Pac.* 96: 222
100. Lamers, H. J. G. L. M., de Groot, M., Cassatella, A. 1983. *Astron. Astrophys.* 128: 299
101. Lamers, H. J. G. L. M., de Loore, C. W. H., eds. 1986. *Stabilities in Luminous Early-Type Stars*. Dordrecht: Reidel. In press
102. Lipunov, V. M. 1982. *Sov. Astron. Lett.* 8(3): 194
103. Lucy, L. B. 1982. *Ap. J.* 255: 286
104. Lucy, L. B. 1983. *Ap. J.* 274: 372
105. Lundstrom, I., Stenholm, B. 1984. *Astron. Astrophys. Suppl.* 56: 43
106. Lundstrom, I., Stenholm, B. 1984. *Astron. Astrophys. Suppl.* 58: 163
107. Maeder, A. 1982. *Astron. Astrophys.* 105: 149
108. Maeder, A. 1983. *Astron. Astrophys.* 120: 113
109. Maeder, A. 1985. *Astron. Astrophys.* 147: 300
110. Maeder, A., Renzini, A., eds. 1984. *Observational Tests of Stellar Evolution Theory, IAU Symp. No. 105*. Dordrecht: Reidel
111. Massey, P. 1981. *Ap. J.* 246: 153
112. Massey, P. 1984. *Ap. J.* 281: 789
113. Massey, P. 1985. *Publ. Astron. Soc. Pac.* 97: 5
114. Massey, P., Armandroff, T., Conti, P. S. 1986. *Astron. J.* 92: 1303
115. Massey, P., Conti, P. S. 1983. *Publ. Astron. Soc. Pac.* 95: 440
116. Mihalas, D. M., Hummer, D. G. 1974. *Ap. J. Suppl.* 28: 343
117. Moffat, A. F. J. 1982. *Ap. J.* 257: 110
118. Moffat, A. F. J. 1982. See Ref. 61, p. 263
119. Moffat, A. F. J., Breysacher, J., Seggewiss, W. 1985. *Ap. J.* 292: 511
120. Moffat, A. F. J., Firmani, C., McLean, I. S., Seggewiss, W. 1982. See Ref. 61, p. 577
121. Moffat, A. F. J., Lamontagne, R., Shara, M. M., McAlister, H. A. 1986. *Astron. J.* 91: 1392
122. Moffat, A. F. J., Lamontagne, R., Williams, P. M., Horn, J., Seggewiss, W. 1987. *Ap. J.* 312: 807
123. Moffat, A. F. J., Niemela, V. S. 1984. *Ap. J.* 284: 631
124. Moffat, A. F. J., Seggewiss, W., Shara, M. M. 1985. *Ap. J.* 295: 109
125. Moffat, A. F. J., Shara, M. M. 1983. *Ap. J.* 273: 544
126. Moffat, A. F. J. 1986. *Astron. J.* 92: 952
127. Morgan, D. H., Good, A. R. 1985. *MNRAS* 216: 459
128. Morton, D. C. 1973. See Ref. 16, p. 54
129. Niemela, V. S. 1982. See Ref. 61, p. 299
130. Niemela, V. S., Mandrini, C. H., Mendez, R. H. 1985. *Rev. Mex. Astron. Astrofis.* 11: 143
131. Niemela, V. S., Mendez, R. H., Moffat, A. F. J. 1983. *Ap. J.* 272: 190
132. Niemela, V. S., Moffat, A. F. J. 1982. *Ap. J.* 259: 213
133. Niemela, V. S., Ruiz, M. T., Phillips, M. M. 1985. *Ap. J.* 289: 52

134. Noels, A., Conti, P. S., Gabriel, M., Vreux, J.-M. 1980. *Astron. Astrophys.* 92: 242
135. Noels, A., Gabriel, M. 1981. *Astron. Astrophys.* 101: 215
136. Noels, A., Scuflaire, R. 1986. *Astron. Astrophys.* 161: 125
137. Noerdlinger, P. D. 1979. In *Mass Loss and Evolution of O-Type Stars, IAU Symp. No. 83*, ed. P. S. Conti, C. W. H. de Loore, p. 253. Dordrecht: Reidel
138. Nugis, T. 1975. In *Variable Stars and Stellar Evolution, IAU Symp. No. 67*, ed. V. E. Sherwood, L. Plaut, p. 291. Dordrecht: Reidel
139. Nugis, T. 1982. See Ref. 61, p. 131
140. Nussbaumer, H., Schmutz, W., Smith, L. J., Willis, A. J. 1982. *Astron. Astrophys. Suppl.* 47: 257
141. Osterbrock, D. E., Cohen, R. D. 1982. *Ap. J.* 261: 64
142. Owocki, S. P., Rybicki, G. B. 1985. *Ap. J.* 299: 265
143. Pauldrach, A., Puls, J., Hummer, D. G., Kudritzki, R. P. 1985. *Astron. Astrophys.* 148: L1
144. Perry, D. N. 1983. PhD thesis. Univ. Colo., Boulder
145. Pollock, A. M. T. 1985. *Space Sci. Rev.* 40: 63
146. Pollock, A. M. T. 1987. *Astron. Astrophys.* In press
146a. Pollock, A. M. T. 1987. *Ap. J.* In press
147. Prantzos, N., Doom, C., Arnould, M., de Loore, C. W. H. 1986. *Ap. J.* 304: 695
148. Prévot-Burnichon, M.-L., Prévot, L., Rebeirot, E., Rousseau, J., Martin, N. 1981. *Astron. Astrophys.* 103: 83
149. Prilutskii, O. F., Usov, V. V. 1976. *Sov. Astron. AJ* 20: 2
150. Rosa, M., D'Odorico, S. 1982. *Astron. Astrophys.* 108: 339
151. Savage, B. D., Mathis, J. S. 1979. *Ann. Rev. Astron. Astrophys.* 17: 73
152. Schmutz, W. 1982. See Ref. 61, p. 23
153. Schmutz, W., Hamann, W. R. 1986. *Astron. Astrophys.* 166: L11
154. Schmutz, W., Hamann, W. R., Wessolowski, U. 1986. In *Circumstellar Matter, IAU Symp. No. 122*, ed. I. Appenzeller, C. Jordan. Dordrecht: Reidel. In press
155. Seward, F. D., Chlebowski, T. 1982. *Ap. J.* 256: 530
156. Simon, N. R., Stothers, R. 1969. *Ap. J.* 155: 247
157. Smith, L. F. 1968. *MNRAS* 140: 409
158. Smith, L. F. 1968. *MNRAS* 141: 317
159. Smith, L. F. 1973. See Ref. 16, p. 15
160. Smith, L. J., Willis, A. J. 1982. *MNRAS* 201: 451
161. Stahl, O., Wolf, B., Klare, G., Cassatell, A., Krauter, J., et al. 1983. *Astron. Astrophys.* 127: 49
162. Stothers, R., Chin, C.-W. 1979. *Ap. J.* 233: 267
163. Tanaka, Y. 1966. *Prog. Theoret. Phys. (Kyoto)* 36: 844
164. Torres, A. V., Conti, P. S., Massey, P. 1986. *Ap. J.* 300: 379
165. Underhill, A. B. 1968. *Ann. Rev. Astron. Astrophys.* 6: 39
166. Underhill, A. B. 1983. *Ap. J.* 266: 718
167. Vanbeveren, D., Conti, P. S. 1980. *Astron. Astrophys.* 88: 239
168. Vanbeveren, D., van Rensbergen, W., de Loore, C. W. H. 1982. In *The Most Massive Stars*, ed. S. D'Odorico, D. Baade, K. Kjär, p. 199. Garching: ESO Publ.
169. Van Blerkom, D. 1973. See Ref. 16, p. 165
170. van den Heuvel, E. P. J. 1978. In *Physics and Astrophysics of Neutron Stars and Black Holes*, ed. R. Giacconi, R. Ruffini, p. 828. Amsterdam: North-Holland
171. van der Hucht, K. A., Conti, P. S., Lundstrom, I., Stenholm, B. 1981. *Space Sci. Rev.* 28: 227
172. van der Hucht, K. A., Jurriens, T. A., Olnon, F. M., Thé, P. S., Wesselius, P. R., Williams, P. M. 1985. In *Birth and Evolution of Massive Stars and Stellar Groups*, ed. W. Boland, H. van Woerden, p. 167. Dordrecht: Reidel
173. van der Hucht, K. A., Olnon, F. M. 1985. *Astron. Astrophys.* 149: L17
174. Vreux, J.-M. 1985. *Publ. Astron. Soc. Pac.* 97: 274
175. Vreux, J.-M. 1986. See Ref. 101. In press
176. Vreux, J.-M., Andrillat, Y., Gosset, E. 1985. *Astron. Astrophys.* 149: 337
177. Vreux, J.-M., Dennefeld, M., Andrillat, Y. 1983. *Astron. Astrophys. Suppl.* 54: 37
178. Westerlund, B. E., Azzopardi, M., Breysacher, J., Lequeux, J. 1983. *Astron. Astrophys.* 123: 159
179. Wheeler, J. C., Harkness, R. P., Barkat, Z., Swartz, D. 1986. *Publ. Astron. Soc. Pac.* 98: 1018
180. White, R. L. 1985. *Ap. J.* 289: 698
181. White, R. L., Long, K. S. 1986. *Ap. J.* 310: 832
182. Williams, P. M. 1982. See Ref. 61, p. 73
183. Williams, P. M., Beattie, D. H., Lee, T. J., Stewart, J. M., Antonopoulou, E. 1978. *MNRAS* 185: 467
184. Williams, P. M., Longmore, A. J., van der Hucht, K. A., Talvera, A., Wamsteker, W. M., et al. 1985. *MNRAS* 215: 23P

185. Williams, P. M., van der Hucht, K. A., van der Woerd, H., Wamsteker, W. M., Geballe, T. R., et al. 1986. See Ref. 101. In press
186. Willis, A. J. 1982. *MNRAS* 198: 897
187. Willis, A. J., van der Hucht, K. A., Conti, P. S., Garmany, C. D. 1986. *Astron. Astrophys. Suppl.* 63: 417
188. Willis, A. J., Wilson, R. 1978. *MNRAS* 182: 559
189. Wolf, C. J. E., Rayet, G. 1867. *Comptes Rendus* 65: 292
190. Wright, A. E., Barlow, M. J. 1975. *MNRAS* 170: 41

THE ART OF N-BODY BUILDING

J. A. Sellwood

Department of Astronomy, The University, Manchester M13 9PL, England

1. INTRODUCTION

Ever since the early work of Jeans (1915), astronomers have struggled with the dynamical problems posed by large aggregates of stars. As more observational data have poured in, the problems have become still harder. For example, elliptical galaxies now appear to be triaxial objects supported by anisotropic velocity dispersions, presenting great difficulties in constructing even an equilibrium model. Galaxies may also be embedded in unseen halos of material whose distribution and extent can only be probed indirectly. This discovery does nothing to simplify the old problems of spiral structure and bar stability, and makes it very difficult to estimate total masses for galaxies and consequently the extent to which their interactions might be significant. We are also currently faced with trying to understand how groups and clusters of galaxies form in a universe that appears so uniform on yet larger scales.

Analytical treatments of these problems are characterized by simplifying approximations, linearizations, etc., and are still far from clear answers to many of the major questions that arise. Holmberg (1941) recognized that simulations could break free from this type of restriction and offer new insight. With the advent of modern computers, this approach has opened up a whole new industry; scarcely an issue of a current major journal appears without a paper reporting new results from simulations. However, simulations can only complement, not replace, analytical studies; they usually possess a different set of weaknesses. Understanding is most rapidly advanced when the numerical and analytical approaches are used in concert, each guiding the other in small steps.

It is neither necessary nor desirable to use a full-blooded N-body simulation for every problem. In fact, it is best to avoid this extreme if at all possible—more limited methods, if used intelligently, can be much more informative. A classic example is provided by Toomre & Toomre (1972), who used a restricted three-body technique to study interacting galaxies; because of the relatively low cost of their method, they were able to explore a large set of parameters and consequently reach a degree of understanding that would have been impossible with a fully self-consistent treatment.

Nevertheless, I concentrate here on techniques for full N-body simulations. I do not aim to provide a detailed description of the algorithms, but I do hope to give the reader some idea of both the power and especially the limitations of the various methods, as well as some guidance as to which applications each is most suited. I omit all discussion of techniques to introduce dissipation, confining myself to pure stellar dynamical problems; I also do not discuss methods suitable for globular cluster problems (see Elson et al.'s article in this volume) or for smaller clusters of objects where close encounters between point masses dominate the dynamics, as these have recently been described in some detail by Aarseth (1985).

The techniques discussed here might apply to the following problems:

1. Internal dynamics of galaxies, including both elliptical and disk systems with or without bulges or halos.
2. Interactions of galaxies with small companions, such as the decay of satellite orbits, formation of shells around elliptical galaxies, and rings in disk galaxies.
3. Interacting galaxies of more nearly equal mass, principally mergers of disk and elliptical systems.
4. Groups of galaxies.
5. Clustering in an expanding universe.

I conclude, however, that several of these problems cannot yet be tackled properly by any of the methods currently available. The characteristic feature of the problems listed above is that the dynamics is controlled by mean field forces, or equivalently, that the systems are collisionless.

I first discuss relaxation in Section 2 and the circumstances under which a system made up of massive particles can be considered collisionless. This section also describes the techniques that try to suppress the effects of encounters in simulations with a practicable number of particles. Simulations, of course, can only approximate the collisionless behavior they are designed to mimic, and I discuss in Section 3 the basic limitations of any model. In Section 4 I summarize the different types of code currently in use and their relative strengths and weaknesses. Most are good for some particular application, but none is best for all. Sections 5–7 describe a few of the major methods for efficient estimation of the gravitational field.

Subsequent sections deal with time integration, initial conditions, and code verification.

2. COLLISIONS

Particle trajectories in a collisionless system are not substantially deflected either by the gravitational forces from nearby particles or by the cumulative effects of distant encounters. The dynamics of such a system is governed by the collisionless Boltzmann equation. The stellar components of both elliptical and spiral galaxies, outside their nuclei, are widely believed to approximate this state, although grouping of stars into clusters and the presence of massive objects such as giant molecular clouds considerably enhance the relaxation rate in many situations of interest.

2.1 *Classical Theory*

Spitzer & Hart (1971) find the ratio of the relaxation time τ_r (due to distant encounters) to the crossing time $\tau_c (= R_h/v)$ for N points of mass m moving with isotropic rms velocity v in a cluster with half-mass radius R_h to be

$$\tau_r/\tau_c \simeq N/(26 \log_{10} \Lambda), \qquad 1.$$

where $\Lambda = R_h/p_0$ and $p_0 (\simeq 2Gm/v^2)$ is the smallest impact parameter for which the deflection during a single encounter can be considered small. In the problem considered by Spitzer & Hart, the value of Λ was given by $\Lambda \simeq 0.4N$. Provided that $\log_{10} \Lambda \gg 1$, relaxation due to close encounters can be neglected.

Rybicki (1972) drew attention to an important difference between the relaxation rates in two- and three-dimensional stellar systems. Relaxation in strictly planar systems is dominated by close encounters, whereas in three-dimensional systems the cumulative effect of distant encounters is most important, when N is large. Rybicki showed that the relaxation time for a two-dimensional system is always short, *independent* of N. However, strictly planar stellar systems are constructed only by theorists and have no counterparts in nature. The stellar disks of real galaxies are finitely thick; the two-dimensional formula applies only when the disk thickness is not much greater than p_0, which is never true (for star-star encounters) in even the thinnest of galactic disks.

The mutual gravitational attraction between particles changes the probability distribution of near neighbors from what would be expected if the particles were uniformly distributed—the medium is said to be "polarized." Polarization has long been known to give rise to dynamical friction (Chandrasekhar 1943): The "wake" of a massive particle moving through a sea of background particles exerts a retarding force on the source particle.

There are, therefore, two processes that affect the trajectory of a star in a cluster: It is braked by its polarization wake, and accelerated (on average) by encounters (e.g. Hénon 1973a).

2.2 Collective Effects

The classical theory discussed so far has treated encounters as independent. However, the density enhancement of a star's wake increases its effective mass, which further reinforces the wake. This collective response is discussed in a rather formal manner by Gilbert (1968), who finds (to first order in $1/N$) that each star is subject to forces from its own (now enhanced) wake and to fluctuations arising from other "dressed" particles.

Kalnajs (1972) discusses how the collective response might be estimated but notes that a proper calculation is impossible in most circumstances, since it requires a complete solution for all collective modes of the stellar system under consideration. If the stellar system is stable, the collective response converges to a value proportional to the mass of the source particle (to first order in $1/N$). Wakes must grow indefinitely if the system is unstable, but where global instabilities grow rapidly the concept of a wake loses much of its usefulness. It is simpler in this situation to regard the overall particle distribution as determining the initial amplitude of the unstable modes.

Julian & Toomre (1966) use a local approximation to calculate the collective response of a stellar disk to a point mass orbiting within it, which is, in effect, a calculation of the polarization cloud. Under this approximation, the system is stable and the response is therefore bounded. An example, taken from their paper, is shown in Figure 1, where it is clear that the integrated mass of the "wake" must be several times the mass of the original perturber. The mass of the polarization cloud decreases as the velocity dispersion of the stellar system rises; Julian & Toomre find that it drops by a factor of more than 20 as the velocity dispersion rises from 1.2 to 2 times the minimum required for local stability.

Julian & Toomre (1966) also show physically that such strong responses are due to the low relative velocities of particles in a rotationally supported disk. This property is important for two reasons: Firstly, the long encounter times produce large deflections, and secondly, the highly ordered motions permit very vigorous collective enhancements. It is only to be expected, therefore, that the mass of the wake would drop off so steeply with rising velocity dispersion. In stellar systems supported by (roughly) isotropic random motion and not rotation (e.g. elliptical galaxies), these effects must be very weak indeed.

Thus, it is much more doubtful that a cool stellar disk can be considered collisionless; it has been known since the classical work of Spitzer &

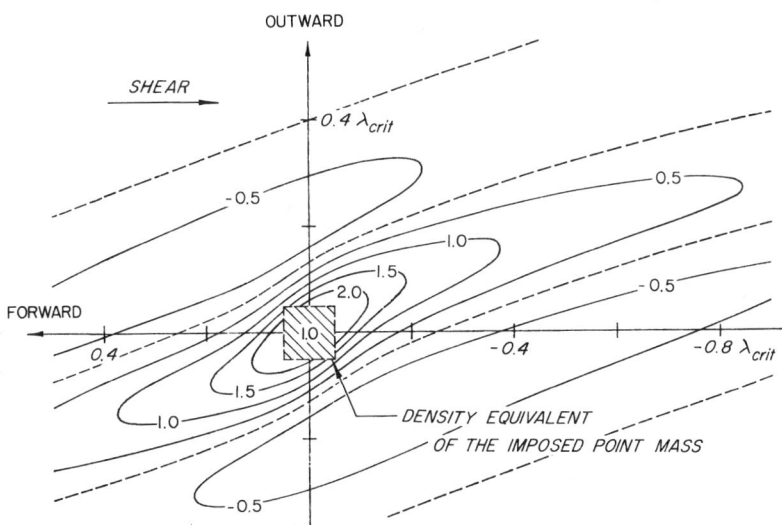

Figure 1 Contours of surface density showing the steady response of a uniformly shearing stellar sheet to a unit point mass placed in the flow. The shear rate corresponds to a flat rotation curve, and the velocity dispersion of the stars in the sheet is 1.4 times the critical value for stability. The slewed region of high density clearly has a total mass exceeding that of the perturber by a substantial factor [reproduced from Julian & Toomre (1966), with permission].

Schwarzschild (1951, 1953) that massive gas clouds or star clusters scatter disk stars, but the calculations of Spitzer & Schwarzschild neglected the enhancement due to the wake of each cloud—a substantial correction that has been considered only by Julian (1967). Collective responses are responsible not just for enhanced relaxation, as A. Toomre's recent work (so far disseminated only through private communications) has demonstrated; substantial wakes also produce quite authentic-looking spiral structure.

What is the relevance of all this interesting theory to N-body experiments? Firstly, one must regard the relaxation time predicted by Equation 1 merely as an upper limit; the more organized the particle motions in an N-body system, the more its relaxation rate will be enhanced and the more particles will be required to approximate a collisionless system. (This is a quite separate effect from the distinction between planar and three-dimensional systems noted by Rybicki.) Secondly, the behavior of a smooth stellar fluid can be studied using N-body models only if the lumpiness of the particle distribution can be artificially masked (Section 9). Thirdly, to the extent that real galaxies contain discrete lumps of material,

N-body simulations embody an important aspect of galaxy dynamics that is omitted whenever a stellar disk is approximated as a smooth Vlasov fluid.

2.3 Softening

Because *N*-body simulations have to use many orders of magnitude fewer particles than exist in the systems of interest in this review, the mass of each particle is correspondingly larger. The minimum impact parameter for and frequency of strong deflections are both increased by the same factor.

Scattering by these close encounters may be substantially reduced by softening. One common form of this is to replace the Newtonian potential of a point mass m by

$$-Gm/(r^2+\varepsilon^2)^{1/2}, \qquad 2.$$

where ε is called the softening length. The greatly weakened forces at short range that result from this substitution reduce that part of relaxation caused by large-angle scattering (e.g. White 1976).

In direct *N*-body codes, softening is easily introduced by using interparticle forces derived from Equation 2. In the case of grid codes, particles within the same grid cell attract each other very weakly, if at all, and the full Newtonian attraction is not reached until they are several mesh spaces apart (see Figure 2*a*). Thus, even where softening is not introduced directly, a grid has a similar effect.

Softened forces imply that a system of particles in equilibrium no longer satisfies the conventional virial theorem $2T+W = 0$, where T and W are the total kinetic and potential energies of the system, respectively. However, a quantity known as the "virial of Clausius" (e.g. Chandrasekhar 1957, p. 50) can be used in place of W, which restores the virial theorem for any arbitrary force law. The virial of Clausius is simply $\Sigma\, m\mathbf{x} \cdot \ddot{\mathbf{x}}$, where the summation is over all particles, and $\mathbf{x}$ and $\ddot{\mathbf{x}}$ are, respectively, the position and acceleration vectors for each.

It is useful to regard softening (whether explicit or implicit) as replacing each point mass by a finite-size mass cloud, and the codes are sometimes dubbed "finite-size particle" codes. The mass cloud associated with the potential of Equation 2 is a Plummer sphere, for which 85% of the mass is contained within a radius of 3ε. In grid codes, particles have an effective size of rather more than one grid cell.

2.3.1 THREE DIMENSIONS Relaxation is dominated by distant encounters in large-*N* three-dimensional systems. Softening reduces this part of the relaxation rate only to the extent that the $\log_{10} \Lambda$ factor in the denominator of Equation 1 is reduced. Farouki & Salpeter (1982) find empirically that $\Lambda \approx R_h/3\varepsilon$ (when $\varepsilon \ll R_h$), which should be used for a softened particle simulation in place of Spitzer & Hart's (1971) value for point masses of $0.4N$.

2.3.2 TWO DIMENSIONS The rules are different in two dimensions. Assuming deflections at impact parameters less than ε to be negligible, Rybicki (1972) found that

$$\tau_r/\tau_c \approx \lambda^3 N\varepsilon/2R$$

for a strictly planar N-body system of softened particles, where λ is the ratio of a typical peculiar velocity to the circular streaming velocity, and R is the radius of the disk. At face value, this formula suggests that quite acceptably long relaxation times should be possible for warm disks, provided that a large enough ε/R is chosen. This appeared to be confirmed empirically by Hohl (1973), who measured the rate of energy equipartition between groups of particles with differing masses and found excellent agreement with theory—disposing of doubts about the validity of strictly planar simulations still raised from time to time (e.g. Mishurov 1984). Hohl had to use a stable model for this test; he chose the only type known at the time—a uniformly rotating hot disk. Unfortunately (though not coincidentally), this is the very type of model in which collective enhancements are weakest, and Hohl's empirical relaxation rate is likely to be a severe underestimate of the rate in more typical circumstances.

There is an alternative physical interpretation of softening in the case of disk systems, namely that it introduces a quite reasonable thickness correction for a two-dimensional model. Equation 2 may be viewed as the potential in a plane offset a distance ε from a point mass. Particles confined to a plane but interacting through mutual forces derived from Equation 2 will behave much as though their mass were distributed throughout a layer of half-thickness ε, which therefore ideally should be chosen to correspond to this physical quantity.

2.4 Avoiding Softening

Softening suppresses the small-scale, steep fluctuations that characterize the field of a set of point masses and that are largely responsible for relaxation. A similar result can be achieved by approximating the global density distribution as a truncated expansion in a suitable set of basis functions and solving for the gravitational field term by term (Section 4.4). Individual particles will pursue orbits in the global field, and relaxation will occur only to the extent that each contributes to the field.

This approach, therefore, gives the Newtonian field of a smoothed density distribution, as distinct from the smoothed field of a collection of mass points that results from the softening approach. Because short-range fluctuations are already suppressed, there should be no need for further softening of the potential. [Villumsen (1982) and White (1983), however, add it to their spherical harmonics codes for other reasons (see Section 6)].

3. FUNDAMENTAL LIMITATIONS

3.1 Orbit Quality

A particle will conserve quantities such as energy along its orbit in a stable collisionless system, but it cannot do so when encounters are not negligible. Computer simulations can never be totally collisionless, and other unwanted deflections may also be caused by the rough nature of the force determination used in many methods. These problems are closely related; the density distribution that would give a smooth potential can only be approximated by the finite number of particles, but the more faithfully the forces represent the precise field of those particles, the more severe the relaxation problem becomes. It is therefore inherent in the N-body approach to modeling collisionless systems that the integrals of individual particles cannot be conserved, although the larger the number of particles and the higher the quality of the force determination, the smaller the fluctuations. Van Albada (1986) reported an extensive investigation (with A. May) that confirmed this to be the case in simulations of quite high quality. However, they found that global properties of the models changed on a much longer time scale than that for energy fluctuations of an individual particle. These slower variations set a hard limit for the time scale over which secular changes in the model can be trusted.

As we cannot treat orbits of individual particles as reliable, changes in the gravitational field of the system that would result from a "small" displacement of one particle must be considered unimportant. This uncertainty is equally well expressed by giving the particles a finite size or by considering the particles as representative, in a Monte-Carlo sense, of those actually present in the system being modeled.

3.2 Experiments with Small N

Many useful results have been obtained from experiments with extremely few particles, although they have to be interpreted with caution. White (1978, 1979), for example, studied mergers between "galaxies" of 250 particles each using a softening length $\varepsilon \approx R_h/6$. These choices give $\tau_r \approx 30\tau_c$, and some relaxation must be expected as the systems merge, as he demonstrated in his empirical tests (White 1979). He also found, however, that relaxation affected the density profile on a longer time scale;

secular changes in the energies of particles were an order of magnitude slower (White 1982, Farouki & Salpeter 1982), and the global structure evolved on the time scale over which the distribution function for the system changed. Thus he could justifiably claim the merger rates he observed were not too seriously affected by relaxation, but he could not be so confident about the structure of the resulting merged system.

Calculations with still smaller N per "galaxy," such as those of Roos & Norman (1979), are more difficult to assess. Each system contains so few particles (typically 40) and uses such a large softening length (as much as $\frac{1}{2}R_h$ in some models) that the relaxation time formula ceases to apply. The statistical fluctuations within such low-N systems are also very substantial.

Worries about small-N experiments are not solely confined to relaxation rates and statistical fluctuations: Many aspects of the behavior of galaxies are dependent upon resonant interactions between stars and the temporally changing global potential (e.g. Tremaine 1981). Even with quite substantial numbers of particles, phase space cannot be densely populated in any sense and in general there will be no particle close to an important resonance. The situation is saved for calculations such as mergers because resonances are very broad owing to the rapid changes in the orbits of the two galaxies on a time scale of about an orbital period. Broad resonances enable particles very far from exact resonance to contribute to the resonant response, and it does not matter too much that phase space is sparsely populated. From this it follows that the slower the changes one wishes to study, the larger the number of particles that are required for an accurate representation of the behavior.

Collective response to particle noise is the dominant problem in rotationally supported disks and demands much higher N than appears to be adequate for pressure-supported systems (White 1982). Nevertheless, some qualitatively correct behavior was seen by Ostriker & Peebles (1973) with $N = 500$, although later work by Carnevali (1983) showed that relaxation totally dominated all but the initial behavior. Increasing N helps, of course, but Sellwood (1986) found that $N \geq 20,000$ is required to prevent a cool stable disk from heating rapidly as a result of collectively enhanced particle noise.

3.3 *Spatial Resolution*

Spatial resolution in an N-body simulation is limited in two ways. Firstly, features described by just a few particles are highly uncertain, be they small scale in dense regions or large scale in low-density regions. Secondly, in simulations using an expansion for the potential, resolution is obviously limited to variations on the scale of the highest order component retained. For all other methods the finite size of each particle imposes a lower resolution limit below which no meaningful structure can be resolved. As it is customary, and desirable, to employ many particles per softening volume, the second restriction is usually the more severe.

However, softening can also affect the development of even quite large-scale features. Efstathiou et al. (1985) and Bouchet & Kandrup (1985) point out that a Cartesian grid method returns forces from a sinusoidal density perturbation that are significantly in error even when the wavelength is as large as eight grid spaces.

The problem is again particularly troublesome in disks. As softening is equivalent to sampling the potential in a plane offset a distance ε from the plane of the disk (Section 2.3.2), perturbing forces from disturbances in the plane of wave number k are lessened by the factor $e^{-k\varepsilon}$. This substantially weakens even global instabilities: Sellwood (1983) attempted to reproduce the bar mode predicted by linear theory for a uniformly rotating disk and found that the growth rate dropped rapidly as the softening length increased (i.e. it was halved when ε was as little as 5% of the disk radius). This phenomenon misled Berman & Mark (1979) into believing their disk galaxy model to be stable, whereas the instability was merely suppressed by the large effective softening length implied by their coarse grid (Sellwood 1981).

4. SUMMARY OF TECHNIQUES

In this section I introduce the principal techniques available for N-body simulations and discuss their relative merits and disadvantages for particular applications. More technical details of a few major methods are given in Sections 5–7.

4.1 *Restricted Methods*

Although these techniques determine the trajectories of large numbers of particles, they are not usually classed as N-body methods. I discuss them briefly here, however, since they can display the essential behavior, especially where such behavior is largely kinematic, much more economically than do fully self-consistent simulations. An additional very substantial advantage is that a system of test particles cannot relax. They have been used most extensively for studies of galaxy interactions.

Toomre & Toomre (1972) used a restricted three-body technique to study the response of a rotating stellar disk to the passage of a companion. The stellar disk was represented by one or two hundred test particles on circular orbits in the central attraction of a point mass. As a perturbing "galaxy" passed, the test particles moved under the combined attraction of the two mass points alone; all interactions between the test particles themselves were simply ignored, which led to an extremely fast integration. The bridges and tails, which were the principal results of interest, may not be too seriously compromised by the neglect of self-gravity, since they are rather tenuous chains of material largely stripped from the outer regions of the disks.

Borne's (1982, 1984) "multiple three-body" and Lin & Tremaine's (1983) "semi-restricted" methods are an extension of this basic technique, already suggested by Toomre & Toomre in their (1972) paper. These codes attempt to calculate the drag on a satellite orbiting a spherical galaxy by computing the back reaction of the particles on the orbiting satellite. The other restricted aspects are retained, i.e. the stars of the primary galaxy do not interact with each other and move only in response to the rigid central potential of the primary and the (also rigid) satellite. Borne's method appears to be superior because it avoids two serious defects, pointed out by White (1983), in Lin & Tremaine's scheme. Firstly, the satellite was attracted toward the primary both by its central potential and by the particles, whereas the particles felt only the former and would have moved on orbits with different angular frequencies. Secondly, Lin & Tremaine did not allow the central potential of the primary galaxy to move, which caused the tidal response of the particles to be predominantly dipole rather than quadrupole. White (1983) avoided these problems in a manner rather different from Borne's and went on to examine whether the inclusion of self-gravity seriously altered the decay rates. White appeared to find that it did, but Bontekoe & van Albada (1987) were unable to confirm this and suggested that White was misled because his self-consistent experiments did not start from the identical initial state as his restricted experiments. (He created an equilibrium primary galaxy for each experiment by allowing the particle distribution to relax, a process that must have followed a substantially different course with each type of code.) Thus, it appears after all that a careful semi-restricted three-body code is remarkably accurate for this problem.

Quinn & Goodman (1986) have recently applied a technique similar to Borne's to study satellites of disk galaxies, and again they find some evidence to suggest that the restricted approach is inadequate. It is unlikely that their paper will reopen the above debate for spheroidal systems, however, since collective responses in disks are far more likely to be important.

Quinn (1984) adopted an even more drastic simplification in an attempt to account for shells around elliptical galaxies. He modeled the gravitational field as a single potential well representing that of a massive elliptical galaxy and computed the motion of test particles, imagined to be the debris resulting from the complete disruption of a small nearby companion. Such a calculation can obviously be refined by considering the self-gravity of the companion during its disruption phase (Wilkinson & James 1986), but the gravitational forces from the tenuous shells themselves, once formed, must be negligible.

4.2 Direct N-Body

This is the least suitable, but most widely used, technique for self-gravitating simulations of collisionless systems. It employs a collection of directly interacting softened particles that, in its most naive form (all particles moving with the same time step with an acceleration computed from the direct sum of the contributions from all the other particles), has a computational cost proportional to $N(N-1)$. Refinements, such as individual time steps (Aarseth 1972) and Ahmad & Cohen's (1973) scheme to update the slowly varying contribution from distant particles less frequently than that from neighbors, lead to very significant economies, but calculation time still rises very steeply with N. There is no need for special treatment of binary and multiple interactions because softening prevents these from forming. Techniques for all current applications of direct N-body simulations are described in detail by Aarseth (1985).

The most desirable aspect of such codes is that they make no assumptions about the geometry of the system; they are easy to adapt to a variety of applications and, unlike grid codes for example, can follow gross changes in shape and radial profile (e.g. Aarseth & Binney 1978). Their principal drawback is that simulations become prohibitively expensive at quite modest N. Farouki & Shapiro (1980), using an array processor, were limited to about 1000 particles; and even with a supercomputer, Duncan (1986) was only able to employ a maximum of 2500 particles. The substantial savings realizable by computing in comoving coordinates (Aarseth 1979, 1985) enables several thousand particles (e.g. Aarseth et al. 1979, Dekel & Aarseth 1984) to be used in cosmological experiments. Valuable as these early experiments were, the technique has been largely superseded by grid and particle-particle-particle-mesh methods.

4.3 Tree Schemes

These recent methods may prove more efficient than a direct N-body calculation and enable substantially more particles to be used. The basic principle is to arrange that the forces from a group of distant particles are approximated by a single contribution from their center of mass. The more distant the clump, the less its internal structure matters and the larger the grouping that can be considered as a unit.

The technique shares the advantage of direct methods in that it can treat arbitrary density distributions but has computer-time requirements that increase only as fast as $O(N \log N)$. It is more efficient (for very large N), but the price for this is that the algorithms are more difficult to program and very ill suited to vectorization.

Appel (1985) was able to employ 10,000 particles in his tree-code simulations of gravitational clustering in an expanding universe; this was an ideal application, since many separate clusters developed. Barnes & Hut (1986) use their method for a galaxy merger simulation employing just over 4000 particles—a great improvement over direct N-body techniques—but two-center expansion methods (Section 4.4.3) may be still faster for this problem and have the added advantage of suppressing relaxation to a considerable extent.

The principal power of tree codes is their ability to compute forces over many different length scales and to retain a fairly accurate representation of short-range interactions, a feature that suggests that they may be better suited to collisional applications. However, the technique favors the use of a fixed time-step, which is inefficient in star cluster problems. [Regularization, introduced by Jernigan (1985) and Porter (1985) in their tree code, overcomes this handicap to a considerable extent.] If the codes are to be used for collisionless problems, they must employ softened forces, effectively eliminating the need to compute accurate forces between nearby particles. It is also hard to see how the tree codes so far proposed can compete with the efficiency of grid methods for the internal structure of galaxies. They may, however, prove useful for simulations of groups of galaxies for which direct N-body techniques are all that is currently available.

4.4 *Potential Expansions*

4.4.1 PURE EXPANSIONS   The discussion in Section 3.1 strongly suggests that it would be quite reasonable to approximate the density distribution of the particles in a collisionless N-body simulation by a truncated expansion in some set of basis functions, provided that the gross features can be described by the first few members of the set. The gravitational field of this drastically truncated expansion is much smoother than that of the original collection of point masses, but hopefully has the large-scale structure of the system one is trying to mimic. Particles in the simulation are therefore treated as a representative sample, in a Monte-Carlo sense, of the stars in the galaxy, or whatever.

The success of techniques based on this approach is crucially dependent upon the basis set chosen. Many members from an inappropriate set will be required to give a reasonable representation, but a large number of functions will also begin to reproduce the lumpiness of a set of mass points as well as being more time consuming to evaluate. The technique, therefore, lacks the flexibility to follow gross changes in the distribution of stars—a defect that limits its usefulness, in its purest form, to systems close to equilibrium.

An example of this type of scheme was proposed by Clutton-Brock (1972a), who devised Hankel-Laguerre functions for simulations of flat but extensive disk galaxies. He used the method (Clutton-Brock 1972b) to study interactions between a disk galaxy and a satellite, which frequently produced long filamentary bridges and tails. As he felt he had to carry the expansion to very high order to resolve the gravitational field of the filaments, Clutton-Brock concluded that the method was too expensive. This conclusion, however, may be overly pessimistic. The method is very powerful for problems not requiring high spatial resolution, and it is hard to see the need to resolve the field of the filaments in his application. The technique is probably more expensive than grid methods when even moderate spatial resolution is required, but it has the advantage of not requiring softening, which although providing a natural correction for disk thickness (Section 2.3.2), can be a nuisance in some applications (Section 3.3). Thielheim & Wolff (1984) employed it for a study of bar-driven spiral patterns.

4.4.2 SPHERICAL HARMONICS Expansion methods have recently become popular for spheroidal systems, although the technique is quite old (e.g. Aarseth 1967). It overcomes the inflexibility problem of the pure basis-set approach by using a hybrid particle/expansion method: Particles are regarded as a system of concentric spherical shells, with a nonuniform density distribution over the surface of each. Smoothing is confined to the angular variables with very little degradation of the detailed radial distribution of the particles, although some radial smoothing is required on practical grounds (see Section 6). The scheme is able to follow very substantial radial variations in the density profile, such as occur during collapse (Hénon 1964, van Albada 1982, McGlynn 1984), which would not be possible if the functions used in the expansion had a predetermined radial structure. For this application the scheme appears to be superior to Cartesian grid methods, which suffer from inflexible spatial resolution, even though few calculations have employed more than 5000 particles.

Expansions in spherical harmonics have also been used to study equilibrium models of spheroidal galaxies (e.g. Norman et al. 1985, Merritt & Aguilar 1985, Barnes et al. 1986). This application still requires very high resolution near the center but no adjustment as the models evolve.

4.4.3 TWO-CENTER EXPANSIONS This technique, well suited to binary galaxy simulations, was used by van Albada & van Gorkom (1977) for head-on encounters with axial symmetry and by Villumsen (1982, 1983) for fully three-dimensional cases. These authors expand the potential about two centers and are able to employ 1000–2000 particles with quite modest computing resources. Difficulties arise, however, as the two galaxies begin to merge, since each particle must still be associated with one or the other center. Villumsen (1986) has recently suggested that bispherical harmonics be used to avoid this difficulty, and it will be interesting to see how such a code performs.

4.5 Grid Methods

These techniques (also known as particle-mesh) are the most computationally efficient for large N—they are the only methods to have utilized several hundred thousand particles. The gravitational field is calculated on a raster of points within a fixed volume, and the acceleration of each particle is obtained by interpolation. (In some respects it is another, particularly efficient, expansion to quite high order in some general set of basis functions.) It makes good sense to use considerably fewer grid points than particles and a grid geometry that allows the field to be evaluated efficiently. Under these circumstances the force calculation part of a time-step cycle takes less time than that required to advance the coordinates of the particles. The overall time is then proportional to N with a fixed overhead for the force determination, an efficiency that can scarcely be improved upon. Moreover, these codes are very easy to vectorize.

The price for such efficiency, however, is quite high. In the same way as the pure expansion methods, the grid possesses an inflexibility that can be very restrictive. It limits spatial resolution to roughly the distance between mesh points, maintains (and, if badly used, can impose) a preferred geometry that cannot adjust to the changing morphology of the system under study, and is unable to determine accelerations for, or from, particles that leak outside the region enclosed by the grid. Resolution limitations are particularly severe in the case of Cartesian meshes, since strong density concentrations, characteristic of many star systems, cannot be modeled adequately (unless such a region is singled out for special treatment). The unnatural "boxiness" of a Cartesian grid, on the other hand, can be very effectively masked using the techniques developed by plasma physicists for this purpose (see Section 5.2.4). Particles that have leaked outside the grid may be moved in an approximate force field (e.g. Hohl 1970) and can again contribute to the field if they reenter the grid.

Grid codes have been used very successfully for studies of the internal dynamics of galaxies, particularly for disks where polarization problems require very large numbers of particles to prevent rapid relaxation. Two-dimensional grids appear to be adequate for many aspects of disk dynamics: Both Cartesian grids, first used by Miller & Prendergast (1968) and Hohl & Hockney (1969), and two-dimensional polar grids, proposed by Miller (1976) and later refined by Sellwood (1981), have been used to great effect. The polar grid has proved to give superior results for little extra cost.

The straightforward generalization of two-dimensional Cartesian codes to three dimensions (e.g. Brownrigg 1975) has proved less successful because memory and computer-time limitations prevent the use of fine grids. The largest such calculations (e.g. Miller & Smith 1979) employ a 64^3 grid and most use 32^3, whereas 128^2 or 256^2 grids were required (e.g. Hohl 1970) in two dimensions. This limitation casts a shadow over all results from these codes, although much useful work has been done with them.

Miller & Smith (1979), Hohl & Zang (1979), and Wilkinson & James (1982) used three-dimensional Cartesian grid codes to study the dissipationless formation, structure, and stability of elliptical galaxies, which provided much useful information. Unfortunately, the resolution limit of the grid prevents realistic central concentrations from building up, and spherical harmonic codes have been found to be superior for these problems. Spherical harmonic codes are, however, unsuited to models containing disks, and a three-dimensional Cartesian grid, used by Hohl (1978), Sellwood (1980), and May & James (1984), is the best available technique. A superior algorithm is really required in order to study warps or angular momentum exchange between a well-resolved disk and a realistically dense bulge/halo population.

In addition to problems of resolution, grids are not well suited to simulations of interacting galaxies, which should begin with the galaxies well separated and which may splash debris over a large volume of space. James & Weeks (1986) are developing a two-grid scheme for this problem, however, which should prove quite powerful as long as most of the material stays within the grids.

Cartesian grid codes have also proved extremely useful for simulations of clustering in an expanding universe. Some models (e.g. Doroshkevich et al. 1980, Melott 1983, Bouchet et al. 1985) are confined to two dimensions in order to obtain high spatial resolution; others (e.g. Klypin & Shandarin 1983, White et al. 1983, Centrella & Melott 1983, Miller 1983, Shapiro et al. 1983) are fully three dimensional. The latter codes, however, do not offer the dynamic range in resolution required by the problem, a limitation that led Quinn et al. (1986) to change horses midstream. They switched to a direct N-body code after perturbations had grown for some time in order to follow a fraction of their original particles, in a smaller region of space, at much higher resolution.

The particle-particle–particle-mesh (or P^3M) method, first applied to an astronomical problem by Efstathiou & Eastwood (1981), substantially improves the spatial resolution by using the grid to compute the distant field and direct N-body methods to calculate the nearby contribution. This has proved the most successful strategy for cosmological experiments so far, although it has yet to be tested against Appel's (1985) tree scheme. The problem of relaxation again appears (for both these schemes) as dense objects containing rather few particles begin to form.

4.6 *Collisionless Boltzmann Codes*

Although not N-body techniques, these recently developed codes offer another alternative for computer simulations of collisionless stellar systems. They are basically multidimensional fluid dynamical codes, although they are still somewhat in their infancy. Only two such codes are currently in use: Nishida et al. (1981) use the Cheng-Knorr splitting scheme, whereas P. R. Woodward & R. L. White (in preparation; see also White 1986) have developed a piecewise parabolic scheme.

The greatest strength of such codes is that they avoid many of the problems that arise from using particles! Numerical diffusion will, however, mimic a form of relaxation, and these codes have other very substantial drawbacks that limit their usefulness considerably. The principal snag is their computational cost; they are limited to problems possessing very high degrees of symmetry so that the number of dimensions can be drastically reduced. The use of grids in these models also imposes the usual restriction on ability to follow wholesale rearrangements of the density distribution. I do not describe these codes further here.

5. PARTICLE-MESH TECHNIQUES

Tabulating the gravitational field on a regular lattice is efficient for two principal reasons. Firstly, as the locations of the grid points are fixed at the outset, the calculation of the field at these points can be very highly optimized. Secondly, the number of points where the gravitational field has to be determined is usually much less than the number of particles in the model—the acceleration for each particle is obtained by reference to the table, requiring only a few operations per particle. This strategy also has the physically important virtue that the grid smoothes unwanted small-scale variations in the gravitational field. (The disadvantages of this inflexibility were noted in Section 4.5.)

5.1 Solution for the Field on Cartesian Grids

The full efficiencies of field determination on a Cartesian grid are best illustrated by tracing the historical development of the techniques.

5.1.1 CONVOLUTION METHODS The gravitational potential Φ at grid points (i,j,k) due to masses M at grid points (p,q,r) is given by the convolution

$$\Phi(i,j,k) = \sum_{p=1}^{N_x} \sum_{q=1}^{N_y} \sum_{r=1}^{N_z} M(p,q,r)g(p-i,q-j,r-k),$$

where g is the Green's function

$$g(i,j,k) \begin{cases} = g_{000}, & i,j,k \text{ all zero} \\ = -G[(i\Delta_x)^2+(j\Delta_y)^2+(k\Delta_z)^2]^{-1/2}, & \text{otherwise.} \end{cases} \qquad 3.$$

The Δ's are the spacings of the grid points in each coordinate direction, and g_{000} is a constant, frequently taken as equal to $g(1,0,0)$. This expression for g is the most obvious, but not necessarily the best, choice (see Section 5.2.4).

Hohl (1970) experimented with a direct evaluation of this summation in two dimensions, avoiding the need for repeated determination of distances between grid points by precalculating and storing the Green's function. Despite experiments with grouping distant points (a forerunner of the tree schemes), Hohl found the method to be less efficient than the Fourier transformation technique described in Hohl & Hockney (1969). Fourier transformation of the mass array and of the Green's function reduces the convolution to a simple multiplication, and the potential is recovered by Fourier synthesis. This procedure is more economical only because of the development of fast Fourier transforms.

The technique is not quite straightforward, however, since one has to allow for the periodicity of the Fourier representation, which implies that images of the mass array are repeated ad infinitum in all directions, giving what Hockney & Brownrigg (1974) describe as "a squadron of galaxies." Contributions to the field from mass in the adjacent images, which should be avoided for isolated systems, can be eliminated altogether if the grid dimensions are doubled in each direction and the additional grid points left empty of mass. The Fourier technique still produces contributions from images, but the parts of the images that contribute to the field in the active region contain no mass, and nonzero contributions arise only in the blank areas where the field is not required. It is sometimes claimed that this strategy merely reduces the tidal forces from the images, which are now more distant, but this is incorrect: The field obtained is an accurate solution, over the region of interest, for an isolated mass distribution.

The problem does not arise in cosmological applications, where periodic boundary conditions are used and the field of neighboring images is needed. Boundary-value techniques, described in great detail in Hockney & Eastwood (1981), are faster than convolution algorithms in three dimensions when periodic boundary conditions are imposed, but they are rarely used in practice (e.g. Efstathiou et al. 1985).

The required doubling of the grid in each dimension adds a considerable overhead to the Fourier convolution method for isolated systems. There is no avoiding the additional work required for the longer Fourier transforms, but sensible organization can save on storage (e.g. Hohl 1972), which is helpful in two dimensions but essential in three dimensions (Brownrigg 1975).

5.1.2 JAMES'S METHOD The convolution method for three-dimensional isolated systems requires Fourier transforms on a grid eight times the size of the active mesh. James's (1977) technique avoids this additional work, replacing it with a much more economical strategy.

Poisson's equation can be solved directly once boundary conditions are specified, but all that is known for an isolated mass distribution is that the potential tends to zero at infinity. James gets around this difficulty with a two-stage solution that is more readily comprehended when couched in electrostatic terms. The first stage is to imagine the positive charge distribution to be surrounded on all sides by the walls of a perfectly conducting grounded box; Poisson's equation may be solved by standard techniques for boundary-value problems (e.g. Hockney & Eastwood 1981), since the potential is known to be zero over the box surface. In order to fulfill this boundary condition, a system of negative charges must have been induced on the surface of the box. It is straightforward to deduce their magnitude at every point on the box by applying Poisson's equation, assuming the potential outside the box also to be zero.

The field resulting from the first stage is that of the isolated positive charges superposed on the field of the unwanted induced boundary charges. James's trick is to recognize that the unwanted part of the field can be canceled out relatively easily. He does this by calculating the field of a system of positive charges spread around the boundary, having everywhere a magnitude equal to the known induced negative charges. If this second field is added to that determined at stage one, it will exactly cancel the field of the induced negative charges everywhere, leaving the desired field of the isolated charges. Note that only positive charges contribute to the final field, so the method can be used for gravitational problems. Note also that any real charges actually on the boundary can be added to the boundary correction charges at the beginning of the second stage.

The second stage of this scheme—calculating the field of the boundary charges—is not straightforward and occupies much of James's paper. In particular, it is possible to always work with transformed masses and potentials, so avoiding the need to synthesize the field at stage one. This and other refinements achieve savings over the convolution technique amounting to a factor of about three for a 33^3 mesh and perhaps four for very large meshes.

5.1.3 VILLUMSEN'S METHOD None of the methods so far described requires equal spacing of the grid points in the different coordinate directions, but all require regular spacing in any one direction. J. V. Villumsen (in preparation) has devised a technique whereby the planes of the grid can be spaced irregularly, but in only one coordinate direction (z say). The standard Fourier convolution method is used, plane by plane, to determine the contribution to the potential from the mass in each plane in isolation. The potential of each Fourier component of wave number k decays away from the plane as $\exp(-k|z|)$; thus, suitably weakened contributions from all other planes may simply be added to the potential transform of each before Fourier synthesis to determine the total field.

Villumsen devised this scheme in order to study the vertical structure of a galactic disk, which can scarcely be resolved if grid cells are cubic (Hockney & Brownrigg 1974, James & Sellwood 1978). A finer grid in the z direction, however, has the disadvantage of introducing substantial anisotropies in the interparticle forces—the effective softening length depends upon the orientation of the line of centers with respect to the grid. It is not clear to what extent this might influence the behavior.

5.2 *Interpolation Techniques and Suppression of Grid Effects*

The particles in a simulation are usually free to move continuously over the grid [though Miller & Prendergast's (1968) scheme is an exception]. It is therefore necessary to assign their masses to the grid points before the field is calculated, and afterwards to interpolate between the grid points to determine the forces acting on each. While these two stages appear entirely independent, and thus we may feel free to adopt arbitrary strategies for each, the best schemes consider the two procedures together. In this way desirable features, such as conservation of momentum or energy, can be built in. I briefly summarize these schemes here; a comprehensive account has recently been given in Hockney & Eastwood (1981).

5.2.1 POTENTIAL DIFFERENCING The force components at each grid point are usually computed by differencing the grid values of the gravitational potential. Central differences are required to avoid self-forces, and the simplest two-point scheme is usually thought to be most cost effective (e.g. Efstathiou et al. 1985). A more elaborate difference procedure was proposed by Hockney & Brownrigg (1974), but their scheme introduces self-forces. May & James (1984) adopted a better multidimensional difference method that included values at additional grid points off the principal coordinate axes; they found it substantially reduced angular anisotropies in the forces.

Differencing the potential has the effect of smoothing the field over two grid spaces and further degrades the spatial resolution of the grid. It may therefore be advantageous to convolve with each component of the force separately, rather than with the potential; the mass array need be transformed just once, but each force component would need to be synthesized and stored separately. The result would be a force field with twice the spatial resolution obtained with about half the additional work required simply to double the grid for a potential convolution. I have used this procedure for a polar grid (Sellwood 1981), but no one (as far as I am aware) has done so for Cartesian grids.

5.2.2 MOMENTUM-CONSERVING SCHEMES The most popular class of schemes constructs the interpolation procedures so as to conserve momentum in each coordinate direction. This is achieved by avoiding self-forces and arranging that the mutual attraction between a pair of particles is equal and oppositely directed. However, forces do not act along the line of centers, in general, which gives rise to a small couple that destroys angular momentum conservation. Energy is not conserved either, as forces cannot be derived from a Hamiltonian.

There is a series of such schemes that give progressively smoother results as the computational cost rises. The higher order schemes also reduce spatial resolution because they increase the effective size of each particle. The cheapest is the nearest-grid-point (NGP) method (Hockney 1966), in which the mass of each particle is assigned to, and forces are computed from, the nearest grid point, with no interpolation. This is clearly very fast but also very rough; forces change discontinuously whenever particles cross cell boundaries.

A better scheme, known as either the cloud-in-cell (CIC; Birdsall & Fuss 1969) or the particle-in-cell (PIC; Morse & Nielson 1970) method, is essentially a multilinear interpolation. Each particle is considered as a uniformly dense cubic (in three dimensions) cloud centered on the particle and having a side equal to one grid space. Its mass is shared between the eight nearest grid points in proportion to the volume of the cloud closest to each. The force acting on the particle is taken as the weighted average of the forces at the same grid points, with weights equal to the fractions of the mass assigned to each point (thereby ensuring momentum conservation). Forces now vary continuously as particles move over the grid, with discontinuities only in the gradients, but the price for this smoother behavior is that many more grid values must be accessed for each particle.

The third level of sophistication of these schemes is a multiquadratic interpolation, known as the triangular-shaped-cloud (TSC; Buneman 1973) method. The name arises because the density of the cloud of mass associated with each particle is now considered to be nonuniform, varying linearly from a maximum at the position of the particle to zero at the edges. Because the cubic cloud in this scheme has a side of two mesh spaces, the mass is distributed over 27 grid points (in three dimensions); the force acting on the particle is the weighted average of values at the same 27 grid points. Discontinuities now arise in only the second derivative of the force, but the computational cost has again risen sharply. It has never proved worthwhile to extend this sequence of schemes further.

5.2.3 ENERGY-CONSERVING SCHEMES   Lewis (1970) proposes a class of interpolation schemes derived from a Lagrangian so as to conserve energy (in the limit of zero time step). As the simpler versions of these schemes do not conserve linear momentum and generally introduce self-forces (Langdon 1973), they are less widely used. Lewis et al. (1972) succeeded in constructing a scheme that conserves both energy and momentum, but the computational costs are so hefty that it has never been used in astronomical problems.

5.2.4 GRID NOISE   As a particle moves across the grid, the force acting on it varies, in part, with a periodicity of the grid spacing. Of course, the more elaborate the scheme, the weaker these variations are, but no scheme is able to eliminate them entirely. These effects, in whatever form they are expressed, are referred to as grid noise.

Langdon (1970) has analyzed this behavior and has demonstrated that these unphysical variations, which have a wavelength of less than one grid space, are coupled through aliases to the larger scale forces. He shows that the extent of the coupling depends upon the strength of the "sidelobes" in the Fourier transform of the effective particle shape; these sidelobes are stronger for NGP than for CIC, which in turn are stronger than for TSC.

Grid noise was recognized by Hockney et al. (1974) as arising principally from fluctuations in the forces between two particles as their position within the mesh changed; the force varies both with the position relative to a cell boundary and with angle to the grid axes. They proposed that much of this unphysical behavior could be removed by tailoring the Green's function. Anisotropies can be substantially reduced by *adjusting* the values of g (Equation 3) for small values of (i, j, k), an adjustment that therefore has no effect on the long-range part of the forces. This strategy, which they call quiet-particle-mesh (QPM), adds nothing to the running time required; the optimal Green's function can be determined and stored (in transformed form of course) at the start. However, it cannot be employed when the field is obtained by boundary-value methods (such as James's).

The effectiveness of this technique was very nicely demonstrated by Efstathiou et al. (1985) in a diagram reproduced here (Figure 2). This shows the magnitude of the mean interparticle force for a number of different schemes: the standard uncorrected NGP, CIC, and TSC schemes; one energy-conserving (EC) scheme; and the QPM scheme, in which the Green's function was tailored for the TSC scheme. Different values of the Green's function would be needed if the CIC scheme were used instead. (The theoretical interparticle force is not precisely r^{-2} in a periodic system.)

5.2.5 PARTICLE-PARTICLE–PARTICLE-MESH An additional point made very clearly by Figure 2a is that the price of a higher quality scheme is a softer interparticle force and lower spatial resolution. This factor discourages the use of high-quality schemes on the already quite coarse meshes, which are the most that can be afforded in three dimensions.

Hockney et al. (1974) proposed a hybrid scheme (abbreviated P^3M) to overcome this limitation. The scheme has two distinct parts: Forces on each particle are first computed using the QPM method, but the part arising from nearby particles is then subtracted and replaced with a more accurate direct calculation, as in the direct N-body methods. Efstathiou et al. (1985), however, found the QPM method resulted in such small residual force variations that it was not worth subtracting its local contribution; these authors proposed simply adding a short-range correction chosen so as to match onto the mesh-calculated force at about two grid spaces, although this gives a small discontinuity at most angles. The short-range correction obviously adds substantially to the cost of the calculation, depending upon the number of near neighbors. In the cosmological clustering experiments of Efstathiou et al., this overhead rises as the clustering develops.

It should also be borne in mind that the much harder forces provided by this scheme increase the rate of relaxation (primarily through close encounters) and render P^3M unsuitable for most collisionless problems. In order to prevent short-range relaxation, many neighboring particles should lie within one softening length, a condition that would make P^3M very

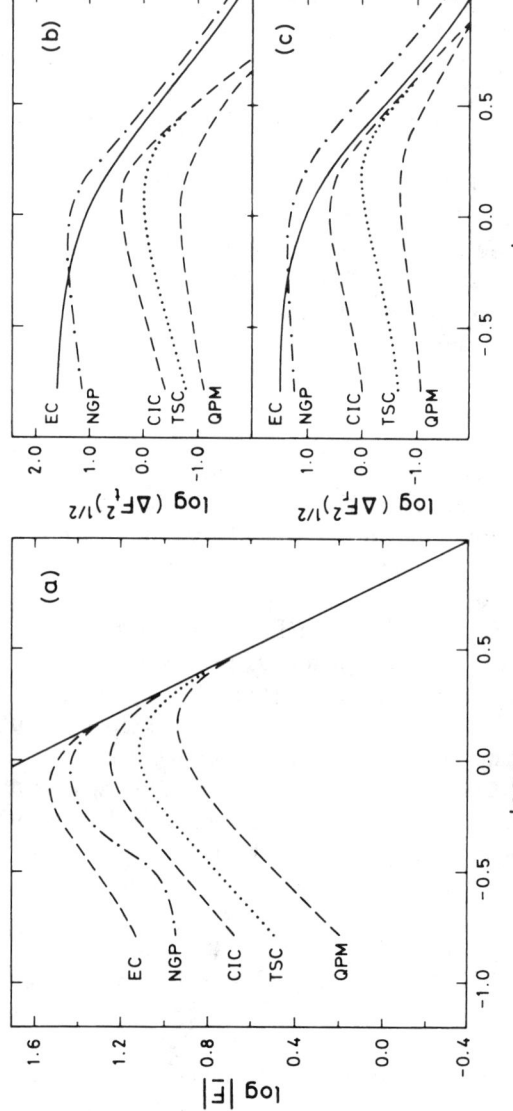

Figure 2 (*a*) The mean interparticle force as a function of separation in mesh spaces for several particle-mesh schemes. The solid line in (*a*) shows the unsoftened force. The other panels show the rms fluctuations in the tangential (*b*) and radial (*c*) directions [reproduced from Efstathiou et al. (1985), with permission].

expensive; the desired improvements to short-range behavior could be achieved more economically with a finer mesh. The peculiar properties of cosmological clustering, however, with many centers of dense concentrations of particles combined with the simultaneous need for accurate long-range forces, make P^3M a reasonable algorithm for the problem.

5.3 Other Grid Geometries

With the exception of cosmological applications, where periodic boundary conditions are an advantage, Cartesian grids can be justified only on the grounds of the efficiency offered by fast Fourier transforms (FFTs). The coordinate system is naturally suited neither to centrally condensed objects, such as elliptical galaxies, nor to nearly axisymmetric objects, such as disk galaxies. This consideration has led to the development of alternative grid geometries: polar grids in two dimensions and spherical polar grids in three dimensions. These, too, offer very substantial advantages in efficiency over gridless codes.

5.3.1 TWO-DIMENSIONAL POLAR GRID   Miller (1976) introduced a polar grid for disk galaxy simulations as a test to ensure that the "squareness" of the Cartesian grids was not promoting the formation of bars. However, he discovered that the code could compete with the efficiency of a Cartesian grid because very many fewer grid points were required for equivalent resolution. Miller spaced the grid points equally around rings, which in turn were spaced logarithmically, giving uniform grid-cell shapes of a size that increased linearly with radius.

The solution for the gravitational field is again performed by convolution. As Poisson's equation is rotationally invariant, each azimuthal harmonic may be calculated separately. After Fourier analysis in angle, however, the radial part of the convolution proceeds directly, requiring the storage of a large table of values for the Green's function. Whereas Miller used a single convolution only for the potential, I found it advantageous to convolve separately for the radial and tangential forces (Sellwood 1981). This avoids the need to difference the potential on a nonuniform radial grid, and it also improves resolution more cheaply than by refining the grid.

Whichever method is used to find the forces at grid points, any form of interpolation in azimuth leads to self-forces on the particles—essentially because the grid lines are not parallel. The self-forces are readily calculated and can be subtracted, particle by particle, before each is accelerated.

These codes are remarkably cheap to run: Very high quality results can be obtained with a grid of 63 by 96 points (Sellwood 1983), whereas the cells of a 128^2 grid are much larger near the center. The convolution part of the calculation is typically only 10–20% of the cpu time required for a complete time-step cycle for $N \sim 50{,}000$. The remainder of the time is spent assigning masses to the mesh, looking up acceleration components, and advancing the coordinates by one step, all of which is directly proportional to N.

5.3.2 SPHERICAL POLAR GRID   A grid scheme for axisymmetric problems is described in some detail by van Albada & van Gorkom (1977); the generalization to three dimensions has yet to be published (T. S. van Albada & Tj. R. Bontekoe, in preparation) but is quite straightforward. The basis of this method is very similar to the gridless techniques described in the next section.

6. SPHERICAL HARMONIC CODES

The codes described in this section use a hybrid potential expansion/particle method that has recently become very popular. It appears to be the best available technique for nearly spherical systems with a steep (and/or varying) radial density gradient. The technique was first reported by Aarseth (1967) but has been considerably refined since. Probably the clearest description is given by McGlynn (1982, 1984), who also presents a historical survey.

The codes are based on the expansion for the potential $\Phi(r, \theta, \phi)$, of a mass m at the point (r', θ', ϕ') as

$$\Phi(r,\theta,\phi) = -4\pi Gm \sum_{l=0}^{\infty} \sum_{m=-l}^{l} \frac{1}{2l+1} \frac{r_<^l}{r_>^{l+1}} Y_{lm}^*(\theta', \phi') Y_{lm}(\theta, \phi). \qquad 4.$$

Here (r, θ, ϕ) are spherical polar coordinates, $r_>$ and $r_<$ are, respectively, the greater and lesser of r and r', and the Y_{lm}'s are the usual surface spherical harmonics (e.g. Jackson 1962, Chapter 3). The same expansion was employed by Villumsen (1982), although he used the alternative name of tesseral harmonics, and by White (1983), who couched it in terms of multipoles.

The expansion converges everywhere except on the sphere $r = r'$, but convergence is slow when $r \approx r'$. The truncated expansion gives a larger and larger spurious result, but over a narrower and narrower range around $r = r'$, as the expansion is taken to higher order, in a manner closely analogous to Gibbs's phenomenon.

In codes that do not use a grid, particles move in response to the local gravitational field, which is the sum of the truncated expansion for all other particles. An essential strategy is to sort particles according to radius and to compute contributions from interior particles in a separate summation from those exterior; forces can then be evaluated in a single sweep through the particles. In practice, the acceleration of each particle is determined from a term-by-term derivative of Equation 4.

A truncated expansion, in effect, replaces the mass of each particle by a certain distribution over the surface of a sphere. Using the $l = 0$ term alone (Hénon 1964, 1973b) makes each particle a uniform shell of material and thus imposes spherical symmetry. Whenever two shells cross, there is a discontinuity in the radial attraction acting on each—making it very difficult to integrate the equations of motion accurately. Moreover, the discontinuity worsens as the expansion is taken to higher (but finite) order—particles at similar radii exert large and unequal forces on each other that should be avoided if possible.

A curious, and possibly even dangerous, aspect of schemes based around Equation 4 is that the radial coordinate of each star is, in effect, deemed to be highly significant, whereas the angular coordinates are given little weight. The inadequacies of orbit integration (Section 3.1) suggest that no significant information would be discarded by introducing an appropriate radial smoothing. Moreover, smoothing also avoids the discontinuities associated with "shell crossings."

Villumsen (1982) and White (1983) smooth the potential of Equation 4 by replacing the factor $r_>^{l+1}$ in the denominator with $(r_>^2 + \varepsilon^2)^{(l+1)/2}$. White noticed that this led to an instability that drove the center of mass away from the coordinate origin, but that could be cured by using half the value of ε^2 for the $l = 0$ term used for all other terms. McGlynn (1984), however, argues that softening is undesirable on the grounds that it immediately limits the resolution of the code by an equal amount at all radii; he advocates a radial grid instead.

Van Albada & van Gorkom (1977) describe a grid method for an axisymmetric solution for the gravitational field; the generalization to nonaxisymmetric systems is quite straightforward (T. S. van Albada & Tj. R. Bontekoe, in preparation). The scheme employs a three-dimensional grid on which masses are assigned and forces tabulated in a similar manner as for other grid codes. McGlynn (1982, 1984), on the other hand, uses a grid in radius only, tabulating the coefficients for the angular expansion at each grid point, which avoids any discretization in azimuth.

Radial gridding removes the discontinuities because the force on each particle is obtained by interpolation. The radial grid points can have any convenient spacing, which can even be varied as the simulation proceeds. Smoothing in this manner appears to be superior to softening because the averaging length decreases wherever the grid rings are closely spaced. This is usually where the particles are most densely packed and the mean orbital dimensions much smaller, which implies that the density distribution is known to higher precision and justifies a shorter smoothing length. Moreover, the logarithmic grid spacing adopted by McGlynn and by van Albada's group is the natural choice, since the width of the region of poor convergence varies with the radius of the shell.

The expansion of Equation 4 is valid for any arbitrary origin of coordinates, but it is important to keep the coordinate system centered on the density peak. Failure to do so will degrade spatial resolution near the peak, where it is most required; more seriously, force errors will be magnified, causing poor energy conservation. As the distribution of particles outside the core is likely to be asymmetrical, the density spike will move relative to the origin, and thus the coordinates should be recentered regularly (McGlynn 1984).

Finally, McGlynn (1982, 1984) worries about "sidelobes" in the truncated expansion—which refers to a weak image of each particle induced on the opposite side of its shell. As this may favor the formation of bisymmetric structures in simulations, he suggests a modification to the standard expansion to weaken the image.

7. TREE SCHEMES

A tree (in these codes) is an organizational structure used to arrange particles in a hierarchy of groups and subgroups. It is introduced in order to reduce the number of contributions to the force acting on any particle that need be evaluated to estimate the attraction by distant particles. The particles are regarded as the leaves of the tree; close neighbors are grouped into pairs; pairs in turn to neighboring pairs or more isolated particles on the same branch, and so on, until the whole system is finally connected to the stem of the tree.

When estimating the force acting on a particular particle, the contributions from nearby masses should be summed exactly, but the substructure of more distant groups can usually be ignored and their attraction approximated by a monopole term. If the linear dimensions of a group are "large" in comparison with the distance of its center of mass to the particle being considered, then its substructure must be resolved. This is achieved by working up the tree, successively breaking down the group, until every subgroup is either a single particle or is far enough away, relative to its spatial extent, to be not worth further subdivision. Clearly, the maximum permitted solid angle for the substructure of a group to be ignored determines the accuracy of the resulting force. The smaller this tolerance parameter is set, the more precise the final force, but the more subdivision will be required and the more time consuming the calculation will become.

As the particles move during the simulation, the original groupings will quickly cease to be optimal and a certain amount of reorganization becomes essential. Appel (1985) achieves this by local rearrangements within the tree structure after every time step. Barnes & Hut (1986), on the other hand, recommend rebuilding the tree ab initio for every time step. They achieve this quite efficiently by subdivision of the space into ever smaller cubic boxes until each particle is located in a separate box.

Although simpler in concept than Appel's scheme, the rigid spatial divisions of the Barnes & Hut scheme will frequently interpose boundaries between adjacent particles while grouping others that are farther apart. Appel's scheme, though not perfect in this respect, should achieve a better tree structure: He reports a factor of two speed-up in the overall force determination after an imperfect tree structure (such as the Barnes & Hut algorithm might devise) has been improved by his algorithm. He also saves time by not having to rebuild the entire tree at each step. In order to achieve a given degree of approximation to exactly calculated forces, however, Appel's scheme requires the evaluation of many more force components than does the Barnes & Hut method. Barnes & Hut justify their approach on the grounds that it is simpler (it may be vectorizable to some extent) and that force errors are amenable to a rigorous analysis. Finally, forces between pairs of particles are equal and opposite in Appel's scheme, which therefore conserves momentum, unlike the Barnes & Hut method. It would appear that both algorithms have substantial drawbacks; tree schemes clearly require further development before they can be recommended by this reviewer for collisionless problems.

8. TIME INTEGRATION

The cheapest, and therefore most popular, time integration scheme is the time-centered leap-frog scheme. Particle velocities are stored at half-integral, and positions at integral, multiples of the time step. It is a second-order scheme but requires only one force determination per time-step cycle. Accuracy and stability of the scheme are discussed at some length by Hockney & Eastwood (1981).

Forces in collisionless N-body simulations are rarely determined with sufficient accuracy to merit higher order schemes (Section 3.1). However, higher order schemes are used in direct N-body codes, generally based on Aarseth's (1972, 1985) method, and in recent spherical harmonics experiments (White 1983, McGlynn 1984, Meritt & Aguilar 1985, Barnes et al. 1986, and others). A fourth-order predictor-corrector scheme (e.g. Acton 1970) requires only two force determinations per time step and is therefore faster than a fourth-order Runge-Kutta method (which needs four). Williamson (1980) gives techniques for efficient storage of data when using Runge-Kutta methods.

Leap-frog is time-reversible; Runge-Kutta and predictor-corrector schemes are not. Buneman (1967) advocates time reversibility for simulation codes on the philosophical grounds that it embodies an important aspect of the physical equations governing the system. If a time-reversible scheme is used, the evolution of any dissipationless simulation will be intrinsically reversible (except for round-off error). However, if the method is not intrinsically reversible, reversibility of the behavior becomes a stringent test of the quality of the integration. For example, Dawson (1972) shows it is possible to retrace a two-stream instability in a plasma even from a quite advanced stage of mixing.

Since most efficient techniques evaluate forces for all particles at once, it is natural to advance the coordinates of all particles with equal time steps. This is not too restrictive in many applications, since close binaries, etc., which demand short time steps in collisional codes, never occur. However, as most galaxies are highly centrally condensed objects, much larger accelerations occur near the center than near the edge, and it would be advantageous to use more flexible schemes. Sellwood (1985), using leap-frog, and Aguilar (1985), using predictor-corrector, have both divided their simulations into inner and outer zones, using a time step for particles in the outer zone several times longer than that in the inner. There is no limit, in principle, to the number of such zones that may be employed, except that bookkeeping becomes tedious. Forces from particles in the outer zone must be estimated when advancing particles in the inner zone. Both authors do this by extrapolation from two previous force determinations, which destroys the time reversibility of the leap-frog scheme. I am indebted to Prof. A. Toomre for pointing out that leap-frog allows the particles in the outer zone to be advanced before those in the inner, so that forces on the latter can be estimated by *interpolation*. This apparently minor change suppressed a spurious instability, which had proved particularly obdurate in some delicate experiments.

9. INITIAL CONDITIONS

The importance of a careful initial setup for an equilibrium model has become increasingly apparent in recent years. There are two separate aspects to this: suppression of particle noise and choosing coordinates from the desired distribution function. The first is relatively simple; the second, though closely related, is somewhat more subtle.

9.1 Quiet Starts

Randomly generated initial positions automatically introduce shot-noise density fluctuations in the initial model on all scales. These are larger in the simulation, by a factor of $(N_s/N)^{1/2}$, than would be expected in the system of N_s bodies it is designed to mimic. In models close to equilibrium at the start, the initial behavior is dominated by the collective response to the initial noise, which may totally mask the "dominant" modes of the continuous system (Sellwood 1983). Noise of this kind is easy to suppress by arranging the particles regularly at the outset (Byers & Grewel 1970). A regular spacing gives a discrete noise spectrum instead of a continuous one, with very large amplitudes at the wavelength of the particle spacing (and its overtones) that must be suppressed during the force determination. If this can be achieved, the particle distribution behaves as a smooth fluid.

A regular spacing in each coordinate direction (Melott 1983, Efstathiou et al. 1985) gives a quiet start to cosmological simulations on a Cartesian grid, provided there are at least as many planes of particles as grid spaces in each direction. The desired initial spectrum of fluctuations has then to be added with some care (e.g. Efstathiou et al. 1985). (The turn-up in the initial power spectrum at high k in their Figure 4a is the alias of their imposed small-k spectrum at wave numbers $k_g - k$, where $k_g = \pi/N_x$, but the force calculation scheme strongly inhibits subsequent growth of these short-wavelength fluctuations.)

A quiet start for a cold disk galaxy simulation is also readily achieved on a polar grid. Particles are placed on rings, with the rings spaced in radius as the desired mass-radius relation. As for the Cartesian case, the number of particles on each ring must be related to the number of azimuthal Fourier harmonics m_{max} that contribute to the force determination. However, in order to prevent coupling of modes through aliases, $2(m_{max}+1)$ particles are required on each ring and NOT $m_{max}+1$, as I recommended in Sellwood (1983) and Sellwood & Athanassoula (1986). The error introduced by mode coupling is insignificant when the behavior is dominated by a few low-order modes, but it substantially affects the disruption rate of the quiet start in stable, or nearly stable, models.

Quiet starts are also possible for warm stellar disks, but it is not practicable to suppress both radial and azimuthal density variations at the same time (Sellwood 1983). Noise in azimuthal forces must be suppressed if the simulation is intended for a study of nonaxisymmetric instabilities. The trick is simply to give each particle on the ring identical velocity components so that the initial orbits remain congruent and regularity can be preserved. The radial density profile will fluctuate at a quite high amplitude, but this should not affect the linear growth of nonaxisymmetric modes, especially if the axisymmetric part of the radial attraction is held fixed at its theoretical value.

Unfortunately, quiet-start models do not remain quiet for very long, even in stable systems. Small density fluctuations, which can either be imposed or arise from round-off error, seed an exponential breakup of the regular arrangement, and polarization sets in as the true graininess is unmasked. Dawson (1983) discusses the break-up of quiet starts in plasma simulations, but strong collective enhancements substantially accelerate this process in cool stellar disks (J. A. Sellwood, in preparation).

Quiet starts are a device to permit the continuum behavior to be exhibited by a collection of comparatively massive particles. However, it should be recognized that continuum disks are a theorist's idealization; many galactic disks, containing massive clumps of gas and star clusters, are much more lumpy than a uniformly distributed disk of 10^{10} stars. Noisy-start simulations may, in this sense, be a closer approximation to reality.

9.2 *Choosing From a Distribution Function*

Randomly generated velocity components will lead to statistical fluctuations about the prescribed mean and dispersion. A deterministic procedure to eliminate these, extending Byers & Grewel's (1970) idea, is possible for plasma simulations (e.g. Dawson 1972, 1983), but it demands too many particles in multidimensional problems. Sternlieb (1977) advocates adjusting randomly generated velocities (averaged over some volume) to start plasma simulations with precisely the intended lowest two moments. The distribution function in a stellar system, however, rarely leads to means and dispersions uniform over large volumes, but initial coordinates can be generated more elegantly and "quietly" directly from a distribution function (if one is known).

Polyachenko (1981) and Lin & Tremaine (1983) advocate choosing integrals for each particle at random from the distribution function, but this still results in statistical fluctuations about the intended function. These can be eliminated by choosing integrals for each particle in a deterministic manner such that their distribution is as close as possible to the required form. The technique, first used by Hénon (1968, 1973b) for spherical polytropes, is described for disks by Sellwood & Athanassoula (1986), who use energy and angular momentum as the independent variables; however, any other set of isolating integrals in which the distribution function can be expressed would work equally well.

10. VERIFICATION OF RESULTS

There are many ways in which simulation codes can, and should, be checked. At the lowest level are verifications of energy, momentum, and angular momentum conservation. Minor fluctuations in those quantities not intrinsically conserved by the integration scheme give a measure of the quality of the simulation. Somewhat stronger checks for more subtle errors are tests for force-free rectilinear motion of a single particle (which would reveal self-forces) and the ability to compute an easily checked orbit for just a pair of particles.

However, checks of the global behavior are also required. The strongest of these is to check against standard results wherever possible: Hohl (1972), Zang & Hohl (1978), Sellwood (1983), and Sellwood & Athanassoula (1986) have demonstrated, with increasing stringency, that simulations of stellar disks can reproduce the normal modes predicted from linear-stability analyses of the equivalent smooth system. Efstathiou et al. (1985) compare results from simulations with theoretical predictions to determine empirically the length scales over which results can be trusted.

Of course, such tests do not verify that the code performs correctly under all circumstances, nor that the behavior can be trusted at large amplitude, which is precisely the regime that simulations are usually employed to investigate. Other checks must be sought before major results can be thought secure. These include tests to determine the extent to which the behavior depends upon numerical parameters (e.g. the number of particles, the size of grid cells, and the length of the time step). Time reversibility can be another very strong test; van Albada & van Gorkom (1977) were able to retrace a well-advanced merger using an integration scheme that is not intrinsically time reversible.

Still more impressive are comparisons between codes (e.g. Miller 1976). Probably the most stringent check to date of global nonlinear behavior has been made by Inagaki et al. (1984), who compared an N-body simulation against the behavior of a collisionless Boltzmann simulation using the same bar-unstable model and initial perturbation. The mode saturated at the same epoch and amplitude in both simulations, although the evolution diverged thereafter, probably because gradients in the distribution function became too steep for the fluid code to handle on a coarse grid.

11. CONCLUDING REMARKS

N-body simulations are used to tackle questions that cannot be answered in any other way. It should, however, be clear from this review that they are something of a blunt instrument; they are extremely demanding of computer time and should be a weapon of last resort. A series of experiments to answer a specific question requires careful planning, and the results demand detailed appraisal in order to sift out solid conclusions. Where possible, results should be related to existing work, both theoretical and numerical. It is vital to save the fullest possible information from the calculations for detailed postrun analysis. Though time consuming for the practitioner, this process is essential if the full potential of the simulation technique is to be exploited.

Comparison of recent simulations with those from only a few years ago shows that their quality is rising extremely rapidly. This improvement, of course, owes much to advances in the power and availability of computers, but probably still more to development of superior algorithms. There seems little reason to doubt that these trends will continue.

The new-generation vector machines, while impressively fast when they get going, are extremely demanding and inflexible; optimization strategies for one machine are totally inappropriate for another (e.g. Duncan 1986, James 1986). This inflexibility may present a significant disincentive to experimentation with new algorithms. The routine repetitive computations of the time-step cycle are ideal for the supercomputer, once the code has been optimized, but the analysis stage and sometimes also the setup stage require much more flexibility. A typical large simulation employing many hundreds of thousands of particles will produce vast amounts of data for analysis that must be readily transferable elsewhere for the vector machine to be useful.

The ultimate sacrifice of flexibility for speed was made by Applegate et al. (1986), who built a small but fast parallel computer specifically designed for simulations of the outer solar system. The machine has answered questions that could not be addressed on conventional mainframes, but it is not at all adaptable to other problems. It is greatly to be hoped that we will not all be required to sacrifice flexibility for speed to quite this extent!

ACKNOWLEDGMENTS

I would like to thank all those who sent reprints and preprints, as well as S. J. Aarseth, R. G. Carlberg, S. D. M. White, and my colleagues in the Manchester Astronomy Department for their comments on the manuscript. The financial support of an SERC Advanced Fellowship is acknowledged.

Literature Cited

Aarseth, S. J. 1967. *Bull. Astron.* 2: 47
Aarseth, S. J. 1972. In *Gravitational N-Body Problem, IAU Colloq. No. 10*, ed. M. Lecar, p. 29. Dordrecht: Reidel
Aarseth, S. J. 1979. In *Instabilities in Dynamical Systems*, ed. V. G. Szebehely, p. 69. Dordrecht: Reidel
Aarseth, S. J. 1985. In *Multiple Time Scales*, ed. J. W. Brackbill, B. J. Cohen, p. 377. New York: Academic

Aarseth, S. J., Binney, J. 1978. *MNRAS* 185: 227
Aarseth, S. J., Gott, J. R., Turner, E. L. 1979. *Ap. J.* 228: 664
Acton, F. S. 1970. *Numerical Methods That Work*. New York: Harper & Row. 540 pp.
Aguilar, L. A. 1985. PhD thesis. Univ. Calif., Berkeley
Ahmad, A., Cohen, L. 1973. *J. Comput. Phys.* 12: 389

Appel, A. W. 1985. *SIAM J. Sci. Stat. Comput.* 6: 85
Applegate, J. H., Douglas, M. R., Gürsel, Y., Sussman, G. J., Wisdom, J. 1986. *Astron. J.* 92: 176
Barnes, J., Goodman, J., Hut, P. 1986. *Ap. J.* 300: 112
Barnes, J., Hut, P. 1986. *Nature* 324: 446
Berman, R. H., Mark, J. W.-K. 1979. *Astron. Astrophys.* 77: 31
Birdsall, C. K., Fuss, D. 1969. *J. Comput. Phys.* 3: 494
Bontekoe, Tj. R., van Albada, T. S. 1987. *MNRAS* 224: 349
Borne, K. D. 1982. PhD thesis. Calif. Inst. Technol., Pasadena
Borne, K. D. 1984. *Ap. J.* 287: 503
Bouchet, F. R., Adam, J.-C., Pellat, R. 1985. *Astron. Astrophys.* 144: 413
Bouchet, F. R., Kandrup, H. E. 1985. *Ap. J.* 299: 1
Brownrigg, D. R. K. 1975. PhD thesis, Univ. Reading, Engl.
Buneman, O. 1967. *J. Comput. Phys.* 1: 517
Buneman, O. 1973. *J. Comput. Phys.* 11: 250
Byers, J. A., Grewel, M. 1970. *Phys. Fluids* 13: 1819
Carnevali, P. 1983. *Ap. J.* 265: 701
Centrella, J., Melott, A. L. 1983. *Nature* 305: 196
Chandrasekhar, S. 1943. *Ap. J.* 97: 255
Chandrasekhar, S. 1957. *An Introduction to the Theory of Stellar Structure.* New York: Dover. 509 pp.
Clutton-Brock, M. 1972a. *Astrophys. Space Sci.* 16: 101
Clutton-Brock, M. 1972b. *Astrophys. Space Sci.* 17: 292
Dawson, J. M. 1972. In *Gravitational N-Body Problem, IAU Colloq. No. 10*, ed. M. Lecar, p. 315. Dordrecht: Reidel
Dawson, J. M. 1983. *Rev. Mod. Phys.* 55: 403
Dekel, A., Aarseth, S. J. 1984. *Ap. J.* 283: 1
Doroshkevich, A. G., Kotok, E. V., Novikov, I. D., Polyudov, A. N., Shandarin, S. F., Sigov, Yu. S. 1980. *MNRAS* 192: 321
Duncan, M. J. 1986. In *Use of Supercomputers in Stellar Dynamics*, ed. S. McMillan, P. Hut. Berlin: Springer-Verlag. In press
Efstathiou, G., Davis, M., Frenk, C. S., White, S. D. M. 1985. *Ap. J. Suppl.* 57: 241
Efstathiou, G., Eastwood, J. W. 1981. *MNRAS* 194: 503
Farouki, R. T., Salpeter, E. E. 1982. *Ap. J.* 253: 512
Farouki, R. T., Shapiro, S. L. 1980. *Ap. J.* 241: 928
Gilbert, I. H. 1968. *Ap. J.* 152: 1043
Hénon, M. 1964. *Ann. Astrophys.* 27: 83
Hénon, M. 1968. *Bull. Astron. Paris* 3: 241
Hénon, M. 1973a. In *Dynamical Structure and Evolution of Stellar Systems*, ed. L. Martinet, M. Mayor, p. 182. Sauverny: Geneva Obs.
Hénon, M. 1973b. *Astron. Astrophys.* 24: 229
Hockney, R. W. 1966. *Phys. Fluids* 9: 1826
Hockney, R. W., Brownrigg, D. R. K. 1974. *MNRAS* 167: 351
Hockney, R. W., Eastwood, J. W. 1981. *Computer Simulation Using Particles.* New York: McGraw-Hill. 540 pp.
Hockney, R. W., Goel, S. P., Eastwood, J. W. 1974. *J. Comput. Phys.* 14: 148
Hohl, F. 1970. *NASA TR R-343*
Hohl, F. 1972. *J. Comput. Phys.* 9: 10
Hohl, F. 1973. *Ap. J.* 184: 353
Hohl, F. 1978. *Astron. J.* 83: 768
Hohl, F., Hockney, R. W. 1969. *J. Comput. Phys.* 4: 306
Hohl, F., Zang, T. A. 1979. *Astron. J.* 84: 585
Holmberg, E. 1941. *Ap. J.* 94: 385
Inagaki, S., Nishida, M. T., Sellwood, J. A. 1984. *MNRAS* 210: 589
Jackson, J. D. 1962. *Classical Electrodynamics.* New York: Wiley. 621 pp.
James, R. A. 1977. *J. Comput. Phys.* 25: 71
James, R. A. 1986. In *Use of Supercomputers in Stellar Dynamics*, ed. S. McMillan, P. Hut. Berlin: Springer-Verlag. In press
James, R. A., Sellwood, J. A. 1978. *MNRAS* 182: 331
James, R. A., Weeks, A. 1986. In *Use of Supercomputers in Stellar Dynamics*, ed. S. McMillan, P. Hut. Berlin: Springer-Verlag. In press
Jeans, J. H. 1915. *MNRAS* 76: 70
Jernigan, J. G. 1985. In *Dynamics of Star Clusters, IAU Symp. No. 113*, ed. J. Goodman, P. Hut. Dordrecht: Reidel
Julian, W. H. 1967. *Ap. J.* 148: 175
Julian, W. H., Toomre, A. 1966. *Ap. J.* 146: 810
Kalnajs, A. J. 1972. In *Gravitational N-Body Problem, IAU Colloq. No. 10*, ed. M. Lecar, p. 13. Dordrecht: Reidel
Klypin, A. A., Shandarin, S. F. 1983. *MNRAS* 204: 891
Langdon, A. B. 1970. *J. Comput. Phys.* 6: 247
Langdon, A. B. 1973. *J. Comput. Phys.* 12: 247
Lewis, H. R. 1970. In *Methods in Computational Physics*, 9: 307. New York: Academic
Lewis, H. R., Sykes, A., Wesson, J. A. 1972. *J. Comput. Phys.* 10: 85
Lin, D. N. C., Tremaine, S. 1983. *Ap. J.* 264: 364
McGlynn, T. A. 1982. PhD thesis. Princeton Univ., Princeton, N.J.
McGlynn, T. A. 1984. *Ap. J.* 281: 13

May, A., James, R. A. 1984. *MNRAS* 206: 691
Melott, A. L. 1983. *MNRAS* 202: 595
Merritt, D., Aguilar, L. A. 1985. *MNRAS* 217: 787
Miller, R. H. 1976. *J. Comput. Phys.* 21: 400
Miller, R. H. 1983. *Ap. J.* 270: 390
Miller, R. H., Prendergast, K. H. 1968. *Ap. J.* 151: 699
Miller, R. H., Smith, B. F. 1979. *Ap. J.* 227: 407
Mishurov, Yu. N. 1984. *Sov. Astron. AJ* 28: 628
Morse, R. L., Nielson, C. W. 1970. *Phys. Fluids* 12: 2418
Nishida, M. T., Yoshizawa, M., Watanabe, Y., Inagaki, S., Kato, S. 1981. *Publ. Astron. Soc. Jpn.* 33: 567
Norman, C. A., May, A., van Albada, T. S. 1985. *Ap. J.* 296: 20
Ostriker, J. P., Peebles, P. J. E. 1973. *Ap. J.* 186: 467
Polyachenko, V. L. 1981. *Sov. Astron. Lett.* 7: 79
Porter, D. 1985. PhD thesis. Univ. Calif., Berkeley
Quinn, P. J. 1984. *Ap. J.* 279: 596
Quinn, P. J., Goodman, J. 1986. *Ap. J.* 309: 472
Quinn, P. J., Salmon, J. K., Zurek, W. H. 1986. *Nature* 322: 329
Roos, N., Norman, C. A. 1979. *Astron. Astrophys.* 76: 75
Rybicki, G. B. 1972. In *Gravitational N-Body Problem, IAU Colloq. No. 10*, ed. M. Lecar, p. 22. Dordrecht: Reidel
Sellwood, J. A. 1980. *Astron. Astrophys.* 89: 296
Sellwood, J. A. 1981. *Astron. Astrophys.* 99: 362
Sellwood, J. A. 1983. *J. Comput. Phys.* 50: 337
Sellwood, J. A. 1985. *MNRAS* 217: 127
Sellwood, J. A. 1986. In *Use of Supercomputers in Stellar Dynamics*, ed. S. McMillan, P. Hut. Berlin: Springer-Verlag. In press
Sellwood, J. A., Athanassoula, E. 1986. *MNRAS* 221: 195
Shapiro, P. R., Struck-Marcell, C., Melott, A. L. 1983. *Ap. J.* 275: 413
Spitzer, L., Hart, M. H. 1971. *Ap. J.* 164: 399
Spitzer, L., Schwarzschild, M. 1951. *Ap. J.* 114: 385
Spitzer, L., Schwarzschild, M. 1953. *Ap. J.* 118: 106
Sternlieb, A. 1977. *J. Comput. Phys.* 25: 118
Thielheim, K. O., Wolff, H. 1984. *Ap. J.* 276: 135
Toomre, A., Toomre, J. 1972. *Ap. J.* 178: 623
Tremaine, S. 1981. In *The Structure and Evolution of Normal Galaxies*, ed. S. M. Fall, D. Lynden-Bell, p. 67. Cambridge: Cambridge Univ. Press
van Albada, T. S. 1982. *MNRAS* 201: 939
van Albada, T. S. 1986. In *Use of Supercomputers in Stellar Dynamics*, ed. S. McMillan, P. Hut. Berlin: Springer-Verlag. In press
van Albada, T. S., van Gorkom, J. H. 1977. *Astron. Astrophys.* 54: 121
Villumsen, J. V. 1982. *MNRAS* 199: 493
Villumsen, J. V. 1983. *MNRAS* 204: 219
Villumsen, J. V. 1986. In *Structure and Dynamics of Elliptical Galaxies, IAU Colloq. No. 127*, ed. P. T. de Zeeuw. Dordrecht: Reidel. In press
White, R. L. 1986. In *Use of Supercomputers in Stellar Dynamics*, ed. S. McMillan, P. Hut. Berlin: Springer-Verlag. In press
White, S. D. M. 1976. *MNRAS* 174: 467
White, S. D. M. 1978. *MNRAS* 184: 185
White, S. D. M. 1979. *MNRAS* 189: 831
White, S. D. M. 1982. In *Morphology and Dynamics of Galaxies*, ed. L. Martinet, M. Mayor, p. 182. Sauverny: Geneva Obs.
White, S. D. M. 1983. *Ap. J.* 274: 53
White, S. D. M., Frenk, C. S., Davis, M. 1983. *Ap. J. Lett.* 274: L1
Wilkinson, A., James, R. A. 1982. *MNRAS* 199: 171
Wilkinson, A., James, R. A. 1986. In *Structure and Dynamics of Elliptical Galaxies, IAU Symp. No. 127*, ed. P. T. de Zeeuw. Dordrecht: Reidel. In press
Williamson, J. H. 1980. *J. Comput. Phys.* 35: 48
Zang, T. A., Hohl, F. 1978. *Ap. J.* 226: 521

THE IRAS VIEW OF THE EXTRAGALACTIC SKY

B. T. Soifer

Division of Physics, Mathematics, and Astronomy, California Institute of Technology, Pasadena, California 91125

J. R. Houck

Department of Astronomy, Cornell University, Ithaca, New York 14853

G. Neugebauer

Division of Physics, Mathematics, and Astronomy, California Institute of Technology, Pasadena, California 91125

> 'God lives at 100 μm'
> Allan R. Sandage, private communication

1. INTRODUCTION

The lure of finding a new class of "infrared galaxies" was a strong motivation for an all-sky survey in the infrared. The survey by the Infrared Astronomical Satellite (IRAS) has realized this objective with the discovery of many hundreds of galaxies emitting well over 95% of their total luminosity in the infrared. Galaxies were detected ranging from the closest galaxies to those with redshifts $z > 0.4$. Far-infrared luminosities were found spanning seven orders of magnitude, from less than $10^6 L_\odot$ to $\sim 10^{13} L_\odot$. The "starburst" phenomenon was found to be ubiquitous. Finally, the IRAS observations have led to a more complete picture for "normal" galaxies and have raised new questions about "active" galaxies.

In this review, we describe what has been learned to date about the extragalactic sky from the IRAS data. This review complements that of Beichman (1987), which summarizes the advances made by IRAS in studying the Galaxy and its contents. We concentrate on the IRAS observations, rather than discussing in depth the observations at other wavelengths. Since this review is a snapshot of an exciting, rapidly expanding field of research, we cannot hope to be complete.

In Sections 2 and 3 we present an overview of the numerical and morphological properties of the IRAS extragalactic observations and a brief discussion of the mechanisms that give rise to radiation at the survey wavelengths. The observations of normal galaxies are discussed in Sections 4, 5, and 6, while the observations of active galactic nuclei are described in Section 7. Many galaxies have been found to have large infrared luminosities; these are described in Section 8. Finally, cosmology and deeper surveys are addressed in Section 9.

2. OVERVIEW OF THE IRAS EXTRAGALACTIC SKY

The IRAS all-sky survey covered over 96% of the sky with a completeness limit of roughly 0.5 Jy at 12 μm, 25 μm, and 60 μm, and 1.5 Jy at 100 μm. Major advantages of the IRAS survey include a uniform calibration for point sources of better than 10% over nearly the entire sky, and virtually negligible Galactic extinction at the survey wavelengths; at Galactic latitudes $|b| > 30°$, extinction in the Galaxy is $\lesssim 1\%$ at 25, 60, and 100 μm, and at most a few percent at 12 μm.

The 60 μm band is by far the most sensitive for the detection of most extragalactic objects. Roughly 25,000 galaxies have been detected, about half of which had been previously listed in optical catalogs. Well over 75% of the 60 μm sources in the *IRAS Point Source Catalog* (1985; hereafter PSC) at $|b| > 30°$ are extragalactic (*IRAS Explanatory Supplement* 1985, Soifer et al. 1986a, Lawrence et al. 1986), while of the previously identified extragalactic objects cataloged in the PSC, $\sim 75\%$ are detected at 60 μm, with all other bands having significantly lower detection rates (*Cataloged Galaxies and Quasars Detected in the IRAS Survey* 1985). The positional accuracy of the IRAS catalogs is crucial in making such associations; typical uncertainties are $4'' \times 15''$ (1 σ).

The vast majority of the extragalactic objects detected in the IRAS survey are late-type spiral galaxies; elliptical and S0 galaxies are rarely detected. In a survey of the brightest 324 extragalactic objects in the IRAS survey (Soifer et al. 1987a,b), only one was found with a stellar appearance on the Palomar Sky Survey. Lawrence et al. (1986) found no quasarlike objects out of 496 sources studied in a survey of PSC 60 μm sources to the survey completeness limit. Searches at 100 μm show no higher detection rate of quasars, but searches at 25 μm (Carter 1984, De Grijp et al. 1985) have detected Seyfert galaxies with higher frequency. Systematic identifications of IRAS sources with optical counterparts (Savage et al. 1987) have shown that nearly all high-latitude far-infrared PSC sources that are not artifacts of the Galactic "cirrus" emission (Low et al. 1984, Beichman 1987) have reasonably bright optical counterparts.

At the sensitivity limit of the PSC, the median redshift of the extragalactic objects selected at 60 μm is $z = 0.03$, corresponding to a 60 μm luminosity of $\sim 2 \times 10^{10}$ $L_\odot$.[1] Luminosities of the galaxies range from $< 10^6$ $L_\odot$ for the closest dwarf ellipticals to $\sim 10^{13}$ $L_\odot$ for the most luminous galaxies detected (Kleinman & Keel 1987). Figure 1a shows the luminosity functions derived for different samples of IRAS galaxies (Soifer et al. 1986a, 1987a,b, Lawrence et al. 1986, Rieke & Lebofsky 1986, Smith et al. 1987) adjusted to the same Hubble constant and making small adjustments for the slightly different definitions of luminosities used by Rieke & Lebofsky and Smith et al. The agreement between the different determinations of the luminosity functions, both in shape and absolute calibration, is excellent.

A comparison of the luminosity functions of infrared bright galaxies with other classes of extragalactic objects is shown in Figure 1b. For $L_{bol} \lesssim 10^{11}$ $L_\odot$, the number of infrared selected galaxies is ~ 20–25% of the number of optically selected galaxies at a given bolometric luminosity, while the integrated luminosity emitted at far-infrared wavelengths is roughly 25% of that emitted by stars in normal galaxies (Soifer et al. 1987a). For $L_{bol} > 2 \times 10^{11}$ $L_\odot$ the infrared-bright galaxies become the dominant population in the local Universe, having space densities greater than normal and optically selected starburst galaxies; their space density approximately equals that of optically selected Seyfert galaxies. For $L_{bol} > 10^{12}$ $L_\odot$ the infrared selected galaxies exceed in space density the quasars, the only other known objects with such luminosities (Soifer et al. 1986a).

The tightest correlation found between far-infrared emission and other observables is that between radio nonthermal emission and far-infrared

---

[1] In this review we adopt $H_0 = 75$ km s^{-1} Mpc^{-1} and $q_0 = 0$ whenever we quote luminosities. In addition, when a luminosity is quoted at a wavelength (e.g. 60 μm luminosity), we mean νL_ν (60 μm) and give the luminosity in units of solar bolometric luminosities. When a far-infrared luminosity L_{fir} is quoted, it means effectively the luminosity from 40–400 μm, and it is taken as the integrated flux of a Planck function multiplied by a ν^1 emissivity function fitted to the 60 and 100 μm flux densities and then converted to luminosity. The details of the far-infrared luminosity calculation are given in *Cataloged Galaxies and Quasars Detected in the IRAS Survey* (1985). In some of the early work with the IRAS data, the far-infrared luminosity of objects was sometimes approximated as νL_ν (80 μm), where L_ν was calculated from the flux density $f_\nu(80\ \mu m) = [f_\nu(60\ \mu m) + f_\nu(100\ \mu m)]/2$. Most authors have now adopted the definition L_{fir} described above because it is a more accurate approximation of the total far-infrared luminosity of a source.

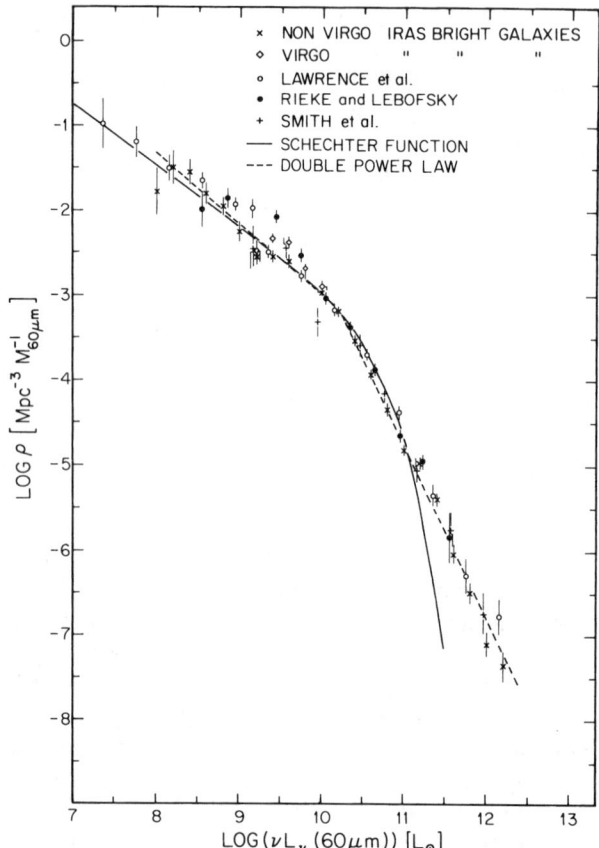

Figure 1a The 60 μm luminosity functions derived from different samples of galaxies from the IRAS catalogs are compared. The luminosity functions have all been adjusted to $H_0 = 75$ km s^{-1} Mpc^{-1} and common units. The data are from Soifer et al. (1986a, 1987b) (x, ◇), Lawrence et al. (1986) (○), Rieke & Lebofsky (1986) (●), and Smith et al. (1987) (+). The figure is from Soifer et al. (1987b), who visually fit the Schechter function (Schechter 1976) to the luminosity function for galaxies in the IRAS bright-galaxy sample. The "double power law" is a best fit of two power laws to this same bright-galaxy luminosity function.

emission (de Jong et al. 1985, Helou et al. 1985, Sanders & Mirabel 1985, Dickey & Salpeter 1984, Eales et al. 1987, Dickey et al. 1987). The radio flux densities are proportional to the far-infrared flux within a factor of ~2 over more than three orders of magnitude in observed infrared flux, and over a larger range in luminosity. Figure 2 illustrates these correlations as discussed by Helou et al. Dressel (1987) has found that this relation also holds for a sample of non-Seyfert Markarian galaxies. It is hard to

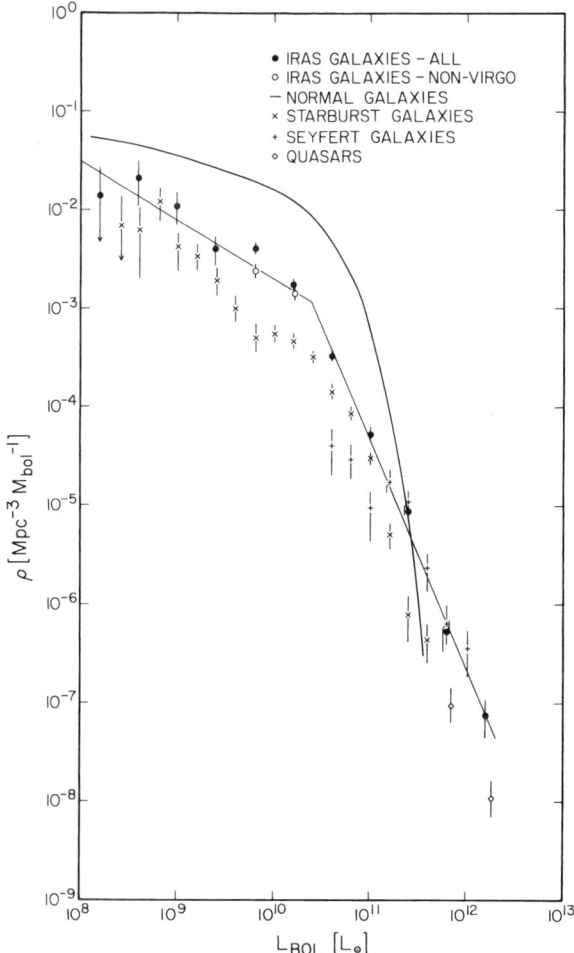

Figure 1b The luminosity functions of a variety of classes of extragalactic sources taken from Soifer et al. (1987b), normalized to the same Hubble constant ($H_0 = 75$ km s^{-1} Mpc^{-1}), and plotted in units of bolometric luminosity. The (●) and (○) represent the far-infrared luminosity function derived for the IRAS bright-galaxy sample taken from Soifer et al., including and excluding the Virgo cluster, respectively. The solid curve represents the analytical fit to the normal-galaxy luminosity function taken from Schechter (1976); the (x) symbols represent the optically selected starburst galaxies and (+) the optically selected Seyfert galaxies, both taken from Huchra (1977). The (◇) symbols represent optically selected quasars taken from Schmidt & Green (1983). The corrections applied to convert from blue luminosity to bolometric luminosity are described in Soifer et al. (1987b). The straight lines represent the best fit of two power laws to the bright-galaxy luminosity function excluding the Virgo galaxies.

understand the tightness of this correlation, since much larger scatter in this relation would be expected with the variety of known emission processes and time scales associated with the infrared and radio emission.

It should be noted that Eales et al. (1987), using subsets of the same infrared data, find much greater scatter in the ratio of radio nonthermal emission to far-infrared emission than do Dickey et al. (1987). Eales et al. used the Very Large Array (VLA), which is insensitive to low-surface-brightness extended emission, whereas Dickey et al. used the Arecibo telescope, which provides a good measure of the total continuum emission in the ~3′ beam. When the different sensitivities to low-surface-brightness emission are considered, the results are consistent and do indeed appear to confirm the narrow relation.

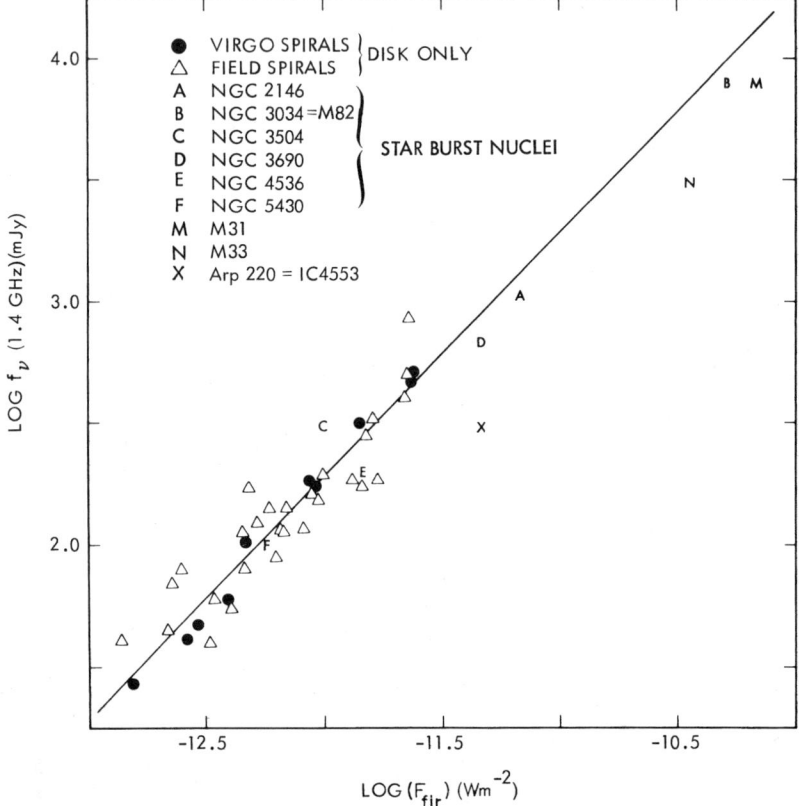

Figure 2 The flux density at 1.4 GHz is plotted vs. the far-infrared flux (F_{fir}) for a wide variety of galaxies (from Helou et al. 1985). The range in luminosities represented by the galaxies in this plot is almost four orders of magnitude.

The strongest known radio sources and the strongest IRAS sources form two almost disjoint samples, with only ~1% of the discrete radio sources in the Green Bank 1400 MHz survey (Condon & Broderick 1985, 1986a) having IRAS identifications (Condon & Broderick 1986b). For the small fraction of the radio sources that are also detected at infrared wavelengths, a correlation exists between far-infrared and radio continuum that is similar to that discussed above, but the ratio of far-infrared to radio continuum is higher by a factor of 1.4 for the infrared flux-limited sample. The radio sources from the Green Bank survey not identified with infrared sources are radio galaxies and quasars. The results of this work are consistent with all of the far-infrared detected galaxies having a similar far-infrared to radio flux ratio; in particular, there is no evidence for a populous class of infrared-bright, radio-quiet extragalactic galaxies.

The ratio F_{fir}/F_{blue} (or equivalently L_{fir}/L_{blue})[2] has been adopted by many authors as a convenient way of measuring the infrared activity of galaxies, normalized to the total starlight or (presumably) mass of the galaxy. A correlation has also been found between this measure of activity and the 60 to 100 μm color temperature of most galaxies (de Jong et al. 1984, Iyengar et al. 1985, Soifer et al. 1987b). This correlation is illustrated in Figure 3, which shows schematically the ratio L_{fir}/L_{blue} increasing as the 60 to 100 μm flux density ratio increases for galaxies in the IRAS survey. The correlation exists whether the sample is an optically selected sample of galaxies, as in de Jong et al., or an infrared selected sample, as in Soifer et al. The optical magnitude–limited samples represent the lower bound to the range of F_{fir}/F_{blue} found for the infrared samples.

The relation between the far-infrared emission and the emission in the 2.6 mm line of CO has been studied by many workers. Prior to the IRAS mission, Rickard & Harvey (1984) noted a general correlation between the far-infrared and CO luminosities in galaxies. Young et al. (1984), Sanders & Mirabel (1985), Kenney & Young (1987), Solomon (1987), Sanders et al. (1986), Stark et al. (1986), Rengarajan & Verma (1986), and Young et al. (1986a) have extended this work by comparing the CO and far-infrared luminosities of galaxies detected in the IRAS survey. The ratio L_{fir}/L_{co} shows significant scatter (by a factor of almost 100), much more than the total scatter (a factor of 4) in the far-infrared vs. nonthermal radio relation shown in Figure 2.

There are significant trends in the L_{fir}/L_{co} relation that could be the cause

---

[2] F_{fir} refers to the far-infrared flux and was described above. F_{blue} has been adopted by many authors as the quantity νf_ν (0.44 μm), and this is the definition used here. Most workers have taken this definition, as did all the papers describing the initial IRAS results (1984, *Ap. J. Lett.* 278: L1–85) even though several of those papers did not explicitly state their definition of F_{blue}. Note that this F_{blue} is almost a factor of 5 larger than the flux in the B bandpass.

of the observed large dispersion. Young et al. (1986a) and Young (1987) have noted that there seems to be a systematic change in L_{fir}/L_{co} depending on the observed color temperature of the far-infrared emission. The suggestion from these observations is that L_{fir}/L_{co} (i.e. the ratio of the luminosity of young massive stars, presumed to power the far-infrared luminosity, to the CO luminosity, presumed to be measuring the mass in H_2) is a measure of the conversion rate of molecular gas into stars (i.e. the inverse lifetime for the process). The correlation of this ratio with far-infrared color, i.e. dust temperature, suggests that the dust temperature, which measures the mean radiation field in a galaxy, is a measure of the star formation "activity" of the galaxy, whereas the total far-infrared (and CO) luminosity of a galaxy measures both its activity and its total dust and gas content. Sanders et al. (1986) suggest that L_{fir}/L_{co} correlates directly with the star

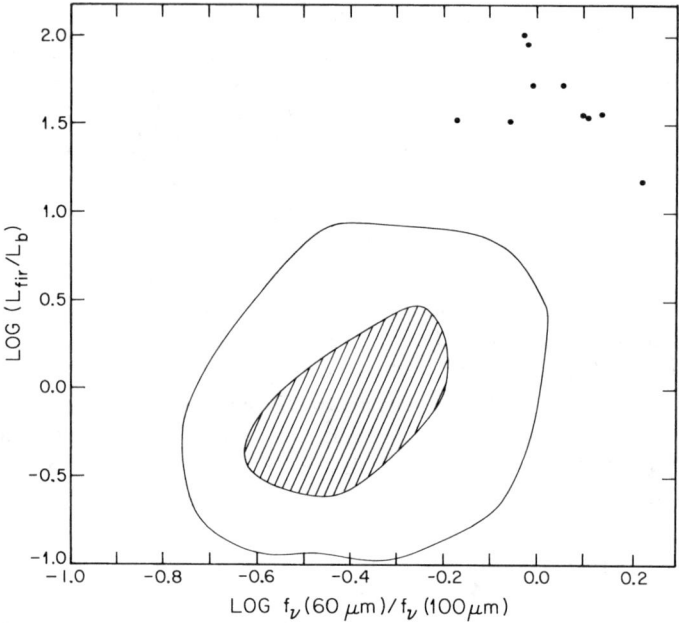

Figure 3 The ratios of the far-infrared to blue luminosity vs. the ratios of 60 μm to 100 μm flux density for galaxies in the UGC catalog with optical diameters $<3.5'$ and $B < 14.5$ mag are shown schematically, with the hatched area having the highest concentration; the data have been provided by W. L. Rice. The spiral galaxies in the sample generally show increasing L_{fir}/L_{blue} with increasing 60 μm/100 μm flux density ratio. The Sd/Irr and E/S0 galaxies dominate the population for log (L_{fir}/L_{blue}) < -0.3 and log [f_ν(60 μm/f_ν(100 μm)] > -0.4. The points in the upper right-hand portion of the plot correspond to the 10 luminous galaxies listed in Table 1.

formation activity of a galaxy: "normal galaxies" have $L_{\text{fir}}/M(H_2)^3 \sim 4$–10, "classic starburst" galaxies (as typified by M82 and NGC 253) have $L_{\text{fir}}/M(H_2) \sim 20$–40, and the most active infrared galaxies (typified by Arp 220 and NGC 6240) have $L_{\text{fir}}/M(H_2) \sim 100$–200.

3. OVERVIEW OF PROCESSES PROMINENT AT IRAS WAVELENGTHS

Emission Mechanisms

The three mechanisms thought to produce the infrared emission in bright extragalactic sources are photospheric emission from stars, synchrotron emission from relativistic electrons spiraling in the sources' magnetic fields, and emission by dust heated by another luminosity source. In many early-type galaxies, photospheric emission makes a substantial contribution to the 12 μm emission, although it is never significant at longer wavelengths. Synchrotron emission dominates the emission observed by IRAS in luminous BL Lac objects and optically violent variable (OVV) quasars, and it may be important in some other quasars and Seyfert galaxies. Dust heated by the various radiation fields is probably responsible for the infrared emission in the vast majority of galaxies detected in the IRAS survey. The observational signatures of dust emission in a given source are threefold: (*a*) extended emission in those sources where sufficient linear resolution can be achieved, (*b*) large infrared excesses over the expected photospheric emission or a reasonable extrapolation from the radio spectrum, and (*c*) the shape of the continuum energy distribution (sometimes) providing an excellent fit to a single-temperature Planck function modified by a wavelength-dependent emissivity.

Because small dust grains have high ratios of optical/ultraviolet opacity to mass, they are extremely efficient at absorbing incident higher energy photons and converting them into infrared photons. The resulting very high infrared luminosity to mass ratio makes far-infrared emission an extremely efficient tracer of cold material in galaxies.[4] As an example,

---

[3] Here $M(H_2)$ is the mass of molecular hydrogen and is related to L_{co} by a constant of proportionality, typically $\sigma[H_2/I_{co}] = 4 \times 10^{20}$ cm^{-2}/(K km s^{-1}) (Sanders et al. 1984, Young & Scoville 1982). This constant is comparatively insensitive to many details of the cloud environment (N. Z. Scoville & D. B. Sanders, private communication).

[4] The dust mass M_d necessary to produce a luminosity L in optically thin far-infrared emission is approximately

$$M_d = 10^4 \left(\frac{L}{10^8 L_\odot}\right)\left(\frac{T_d}{40 \text{ K}}\right)^{-5} [M_\odot], \qquad 1.$$

where T_d is the temperature of the dust. Normal parameters for interstellar dust have been adopted (Draine & Lee 1984, Hildebrand 1983).

Virgo galaxies at the limit of the IRAS survey with dust temperatures of 40 K and masses of dust as small as $\sim 5 \times 10^4$ $M_\odot$ are detectable. For Galactic gas-to-dust ratios, this corresponds to $\sim 10^7$ $M_\odot$ of gas. Reprocessing survey data achieves ~ 3 times fainter flux levels, while the IRAS pointed observations are generally ~ 5–10 times more sensitive than the PSC (Neugebauer et al. 1984). The IRAS survey can therefore detect extremely small amounts of cold interstellar material in distant galaxies; for a normal gas-to-dust ratio, IRAS matches or surpasses the sensitivity of 21 cm H I line radio astronomy in detecting cold matter in galaxies.

In most situations, the individual dust grains reach a nearly constant temperature in a steady-state balance between the absorbed and emitted radiation. At 12 and 25 μm, however, some of the dust emission has been attributed to transient heating of small dust grains (Sellgren 1984, Leger & Puget 1984). For these small grains the temperature rises substantially following the absorption of a single optical/UV photon. At the longer wavelengths, steady-state heating dominates, although transient heating can be important even at 60 μm in radiation fields similar to those found in the solar neighborhood (Draine & Anderson 1985).

Origin of the Luminosity

An important problem is the identification of the mechanism responsible for generating the luminosity in galaxies. In the early model of de Jong et al. (1984), the far-infrared emission was ascribed to emission, analogous to the cirrus emission from the Galaxy (Low et al. 1984), from a cold disk of interstellar dust that is heated by the ambient radiation field and to warm emission in and surrounding regions of active star formation. The coldest galaxies predominantly show the disk component, while the addition of a (warm) star formation component increases the amount of infrared luminosity compared with the blue-light output of the galaxy. More recent models (Rowan-Robinson & Crawford 1987) have included components associated with strong nuclear sources, i.e. active galactic nuclei. In a more speculative vein, Harwit et al. (1986) have pointed out that the kinetic energy of gas clouds colliding within interacting galaxies can provide very high luminosities for times on the order of 10^5–10^6 yr.

Many galaxies are thought to be undergoing episodes of extremely rapid star formation—"starbursts"—and these objects are especially prominent in the infrared. Although composed of individual regions thought to be qualitatively similar to the star formation regions in normal galaxies, these bursts often have such high star formation rates that they cannot persist for more than 10^8 or 10^9 yr or else they would consume more mass than is available in the interstellar medium of the galaxy. The most luminous of the starburst galaxies have infrared luminosities much greater than

10^{11} $L_\odot$. For example, in a galaxy with a luminosity from young stars of $\sim 10^{11}$ $L_\odot$ and a "normal" amount of interstellar matter, the entire supply of interstellar gas ($\sim 4 \times 10^9$ $M_\odot$) would be consumed in $\sim 5 \times 10^8$ yr (Scoville & Young 1983).[5]

Evidence for starbursts comes from observations in virtually every wavelength region. X rays are produced by compact stellar remnants and medium- to low-mass binary systems. Photospheric absorption lines of Si II and C IV from O and B stars are present in the ultraviolet, and a strong blue continuum is observed, a further indication of the presence of hot young stars. The visual spectrum often resembles a low-excitation H II region with narrow (< 500 km s^{-1}) forbidden and Balmer lines. In the near-infrared, strong photospheric CO lines indicate the presence of giants and supergiants resulting from the starburst. The Brackett and Paschen recombination lines of atomic hydrogen and strong shock-excited vibration/rotation lines of H_2 are usually quite strong as well. At the infrared wavelengths probed by IRAS, the strong thermal continuum characteristic of the huge number of H II regions often contains more luminosity than the rest of the energy distribution combined. The radio continuum is nonthermal and is generally assumed to arise from the synchrotron emission from the relativistic electrons generated by supernova explosions of the most massive stars. The nonthermal radio emission is usually stronger than that associated with the individual remnants and probably results from the electrons that radiate throughout a combined volume that is much larger than that occupied by the remnants. The dense interstellar medium responsible for fueling the burst is also a strong source of molecular lines at radio frequencies.

Several key questions about starbursts remain to be answered. Does every spiral galaxy have one or more bursts during its evolution? Why do some bursts seem to be confined to the nuclear region while others involve the entire disk? Are the stars formed with initial-mass functions that are like the initial-mass function observed locally? In particular, what is the low-mass cutoff to the initial-mass function? Finally, what triggers the burst, and does it proceed until all of the interstellar gas is consumed?

---

[5] Scoville & Young (1983) show that the mass consumed by O, B, and A stars is given, in convenient units, by the expression

$dM/dt = 7 \times 10^{-11} (L/L_\odot) [M_\odot \text{ yr}^{-1}]$.

The relation is based on considerations of the energy released in the CNO cycle and of the fraction of the mass in such stars that process hydrogen. The calculation neglects both the mass returned to the interstellar medium from these stars and the mass locked up in white dwarfs and in lower mass stars as an extension of an initial-mass function to lower mass stars, and it further assumes that new stars are forming at the rate that hydrogen is consumed in the present generation of stars.

Unfortunately, there are no simple answers to any of these questions at this time. Part of the problem is that the situation may differ for galaxies of differing mass, metallicity, and morphological type. In addition, both external and internal factors such as interactions and mergers or the presence of a bar appear to influence the onset and magnitude of a starburst.

Active galactic nuclei (AGN) are generally thought to be powered by a massive compact object, a black hole, located at or near their centers. The luminosity is ultimately derived from the release of gravitational potential energy of the material being accreted onto the compact object, while the nature and spectrum of the emission depend on the density and optical depth of the material surrounding the central object. In some sources (luminous BL Lac objects, for example) we are observing synchrotron emission from relativistic electrons produced by the accretion process. In other situations, we are seeing line and continuum emission from the dense plasma surrounding the central object. The infrared appearance of the object depends on the degree of reprocessing of the energy in the material surrounding the core. A simple correlation between the observed spectrum and the emission process has not yet been established and in fact may not exist. Rees (1984) has presented an excellent review of the accretion and emission processes.

In many cases it now appears that star formation and an AGN are present at the same time. In particular, the most luminous infrared galaxies show the characteristics of both a strong starburst and an active nucleus. There is no real understanding of why these two processes should contribute roughly equally to the luminosity for so many sources.

4. SPIRAL AND DISK GALAXIES

Spiral galaxies are by far the dominant population of far-infrared bright galaxies. De Jong et al. (1984) showed that a large fraction of the spiral galaxies in the Shapley-Ames catalog (Sandage & Tammann 1981; hereafter RSA) were detected by IRAS, and that the late-type galaxies were more often detected than the early-type galaxies. The two largest spirals in the sky, M31 and M33, are comparatively inactive galaxies in terms of their infrared luminosities, with $L_{fir}/L_{blue} \sim 0.07$ and 0.2, respectively (Habing et al. 1984, Rice et al. 1987a, Walterbos & Schwering 1986). Figure 5 (see p. 206) illustrates the energy distribution for the integrated light of M33 taken from Rice et al. The ratio of the integrated 60 and 100 μm flux densities is significantly different in the two galaxies, 0.17 for M31 and 0.34 for M33 (Rice et al. 1987b). Qualitatively, the giant spiral galaxy M101 does not differ in its global properties from those of M31 and M33, even though it is a more active galaxy in the infrared, with $L_{\text{fir}}/L_{\text{blue}} \sim 0.4$ (Beichman et al. 1987).

Spatial Distribution of Infrared Emission

The infrared emission extends as far as the optical extent of both M31 and M33. The spatial distribution of the infrared emission in M33 follows that of the blue light reasonably well. In M31, the presence of the dominant bulge causes the correlation to be poorer in the inner portion of the galaxy, since the photospheric emission significantly contributes to the total emission of the bulge at 12 μm (Soifer et al. 1986b).

Little radial variation of the infrared colors is observed in either galaxy. In M33 (Rice et al. 1987a), the ratios of the 25, 60, and 100 μm flux densities vary by no more than $\sim 25\%$ over 20–30' radii, with a factor of ~ 2.5 decrease seen in the 12 to 100 μm flux ratio over this same range of distances. In M31, the 12 to 60 μm flux density ratio is nearly constant with radius, while outside the central few kiloparsecs the ratio of the 100 μm flux density to the 60 μm flux density shows little variation with radius (Walterbos & Schwering 1986). In both M31 and M33, the radial variation of the ratio of far-infrared to H I emission is substantial, decreasing by more than an order of magnitude from the center to the optical edge. Walterbos & Schwering (1986) interpret this gradient in M31 as reflecting a metallicity gradient. In M33, the ratio of far-infrared flux to thermal radio continuum is found to be nearly constant as a function of galactic radius. Rice et al. (1987a) find a correlation in the point-by-point comparison of the far-infrared to radio continuum ratio that has a much larger scatter than that shown in Figure 2, although the integrated ratio for M33 is consistent with the figure.

In both galaxies, the 12 and 25 μm emission is well above that expected from stellar photospheres, circumstellar dust shells, or equilibrium heating of large grains by the interstellar radiation field, but it appears to track the distribution of starlight reasonably well. The dominant source of the emission at these wavelengths appears to be single-photon transient heating of small grains (see Section 3). This process contributes significantly to the 60 μm diffuse emission as well.

Burstein & Lebofsky (1986) have found a puzzling result that the apparent rate of detection of Sc galaxies in the IRAS catalog depends on the inclination of the galaxies, with face-on galaxies being detected at twice the rate of edge-on galaxies. If confirmed, this result would require that the far-infrared emission from such galaxies be optically thick at 100 μm, which suggests that the emission comes mainly from the nuclear regions. Another analysis (W. L. Rice & C. J. Persson, private communication) proposes that this result is largely a selection effect, predominantly resulting from the difficulties of establishing magnitudes and classifications for nearly edge-on galaxies. The Burstein & Lebofsky result is also inconsistent with that of Devereux et al. (1987), who show that Sc galaxies do not have excess nuclear emission.

The study of the environmental effects on the infrared properties of galaxies is only beginning. Bicay & Giovanelli (1987) find that the properties of the infrared bright galaxies in seven Abell clusters are similar to those of field galaxies, with no significant correlation between the infrared properties and either the projected location in the cluster or the H I deficiency. There is also no evidence of star formation being stimulated by the interaction of the cluster galaxies with an intercluster medium.

Origin of the Emission in the Normal Galaxies

In the lower luminosity galaxies ($L < 10^{10} L_\odot$) there is not a one-to-one identification of far-infrared emission with current star formation. For both M31 and M33, the observations are explained by a model where the infrared emission is comprised of two components—that from interstellar dust heated by the diffuse interstellar radiation field and that from star formation activity in H II regions/molecular cloud complexes. In M31, Walterbos & Schwering (1986) separate the observations into two components at each position to derive a star formation luminosity of 2×10^8 $L_\odot$ and a diffuse luminosity of $1.4 \times 10^9 L_\odot$; this result is sensitive to the assumed temperatures of the two components. In M33 Rice et al. (1987a) consider the overall infrared excess in M33 as compared with that found for the individual H II region complexes in galaxies to estimate that half of the far-infrared luminosity of $6.4 \times 10^8 L_\odot$ derives from star formation complexes and half from dust heated by the diffuse interstellar radiation field.

The infrared colors of galaxies detected in the IRAS survey show systematic trends that do not appear to be artifacts of the selection criteria. Figure 4 shows the 60 μm/100 μm flux density ratio plotted vs. the 12 μm/25 μm flux density ratio for galaxies selected from the PSC using the requirement that the 25 μm/60 μm flux density ratio R is less than 0.18. For those galaxies with $R > 0.18$ (mostly Seyfert galaxies) the ratios show much more scatter. The sense of this plot is that galaxies with warmer 60 to 100 μm color temperatures show colder color temperatures at 12 and 25 μm. To explain this plot, Helou (1986a) has invoked a two-component model, with variable amounts of emission from a region of active star formation and from the cirrus in the galaxy disk. The active component has a warm 60 to 100 μm color temperature and a cold 12 to 25 μm color temperature, whereas the cirrus component has a warm 12 to 25 μm color temperature and a cold 60 to 100 μm color temperature.

Persson & Helou (1987) have studied the correlation of the far-infrared luminosity vs. the observed Hα flux and vs. the blue light in a sample of disk galaxies carefully chosen so as not to be "starburst" galaxies. They find that the correlations between either the Hα or the blue light and the far-infrared emission are equally good. Since the correlation between Hα and the far-infrared emission is not better than that between the blue light and the far-infrared emission, Persson & Helou conclude that the star formation component is significantly weaker than was suggested by de

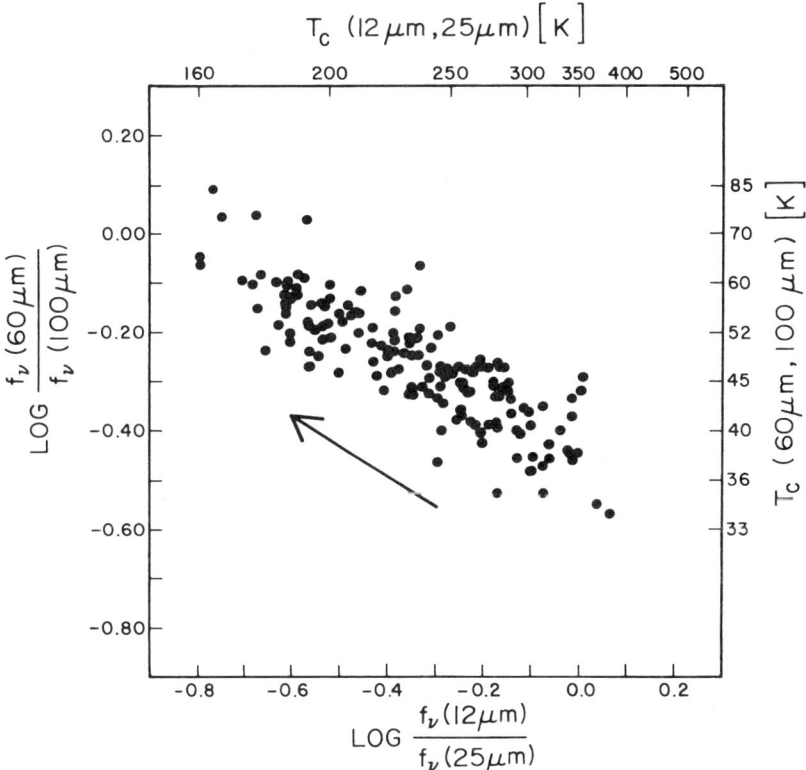

Figure 4 The ratios of 60 to 100 μm flux densities are plotted vs. the ratios of 12 to 25 μm flux densities for galaxies in the PSC; the figure is adapted from Helou (1986a). Note the clear trend of increasing 60–100 μm color temperature as the color temperature measured between 12 and 25 μm decreases. Helou has interpreted the colors as resulting from an admixture of a disk population (*lower right*) and starburst activity (*upper left*); the arrow points in the direction of increasing star formation.

Jong et al. (1984) (see Section 3). Rather, they argue that most of the emission in many cases can be attributed to the interstellar radiation field being recycled through thermal dust emission. A similar conclusion has been reached for galaxies in the RSA by de Jong & Brink (1987) from a consideration of the F_{fir}/F_{blue} vs. 60 μm/100 μm color diagram and the internal extinction in the galaxies.

Nearly all investigators who have attempted to infer the distribution between interstellar emission and star formation emission have done so by beginning with an assumed dust temperature for star formation regions and for interstellar emission, as was done by Walterbos & Schwering (1986) for M31. As mentioned above, the allocation of luminosity depends critically on the assumed temperatures for the two components. The technique applied by Rice et al. (1987a) to determine the contribution of star formation to the luminosity of M33 (i.e. by comparing the infrared to thermal radio flux ratio of the H II regions with that of the galaxy as a whole) should be utilized in as many galaxies as possible in order to assess the adequacy of the cruder techniques that must be applied to the vast majority of the more distant galaxies observed by IRAS.

Barred and Ring Galaxies

Enhanced infrared emission occurs in early-type spiral galaxies having bars. Harwarden et al. (1986) discovered that the 12 μm/25 μm flux density ratio in galaxies is a sensitive indicator of the presence of bars. Devereux (1987) finds that for galaxies of type Sb or earlier with $L_{fir} > 4 \times 10^9 L_\odot$, 40% of the galaxies exhibit conspicuous nuclei. The barred galaxies in the sample as a whole have significantly higher 10 μm luminosities and lower 12 μm/25 μm flux density ratios than do the nonbarred galaxies. This distinction does not occur in later-type galaxies. Puxley et al. (1987) suggest that the infrared activity originates within the central 2 kpc in the barred galaxies. If the 12 μm/25 μm color is a measure of star formation activity (see above), then perhaps the barred galaxies have more star formation activity, triggered by the efficient funneling of material into the nuclear regions of the galaxies by the bars, than do the nonbarred galaxies. The question then arises as to why the same effect is not present in the later-type galaxies.

A sample of ring galaxies has been found by Appleton & Struck-Marcell (1987) to have higher far-infrared luminosities, 60–100 μm color temperatures, and F_{fir}/F_{blue} ratios than do normal galaxies. Since optical observations place a large fraction of the young stars in these galaxies in the rings, these authors suggest that the ring galaxies contain a unique form of nonnuclear coherent starbursts that extend over many kiloparsecs.

5. IRREGULAR, DWARF, AND BLUE COMPACT GALAXIES

The irregular galaxies show approximately the same range of $L_{\rm fir}/L_{\rm blue}$ as found in normal galaxies (Gallagher & Hunter 1987, Hunter et al. 1986, 1987), while the observed 60 μm/100 μm flux density ratios appear significantly greater for the irregulars, as compared with "normal" spirals, at a given value of $L_{\rm fir}/L_{\rm blue}$. While no explanation for these observations has been developed in detail, they can be interpreted as indicating a lack of the cold "disk" component of interstellar dust found in spiral galaxies, which suggests that a greater fraction of the far-infrared emission in these galaxies is directly associated with star formation than in spiral galaxies. This idea is supported by the finding of Hunter et al. (1987) that the average 12 μm/25 μm color temperatures in irregulars is lower than for spirals, consistent with the spectral synthesis model of Helou (1986a) in which galaxies with colder 12 μm/25 μm ratios are dominated by star formation. Further evidence for a significant (perhaps dominant) contribution to the far-infrared luminosity from star formation comes from the observations of a correlation between the far-infrared and Hα luminosities in these galaxies (Gallagher & Hunter 1987, Hunter et al. 1987).

No correlation between the 60 μm/100 μm flux density ratio and oxygen abundance is found in the irregular galaxies. Hunter et al. (1987) also find that the mass of dust, relative to that of interstellar gas, is lower in the irregular galaxies than in the spirals. Thus, while the total dust emission is consistent with the observations of lower metallicity in irregular galaxies, no correlation is found between metallicity and the dust properties. Although there is less dust in these objects, the optical properties of the dust appear to be the same as those of dust in normal spiral galaxies.

Detailed observations of the nearest irregulars, the Magellanic Clouds, have been summarized by Frogel (1987). Because of this proximity, the Clouds represent unique galaxies for study at all wavelengths. The IRAS angular resolution of $\sim 1'$ corresponds to a linear resolution of ~ 16 pc at the Large Magellanic Cloud (LMC) and ~ 20 pc at the Small Magellanic Cloud (SMC). The Clouds are not particularly active in the infrared: The LMC has a total far-infrared luminosity $L_{\rm fir} \sim 6 \times 10^8\ L_\odot$ and $L_{\rm fir}/L_{\rm blue} \sim 0.18$, while the SMC has $L_{\rm fir} \sim 7 \times 10^7\ L_\odot$ and $L_{\rm fir}/L_{\rm blue} \sim 0.09$ (Rice et al. 1987b). Several workers (Jones et al. 1987, Elias & Frogel 1987) have found compact sources associated with Magellanic Cloud H II regions to be quite rare; most infrared H II regions in the Clouds are extended on a scale of a few parsecs. Elias & Frogel note a significantly lower level of star formation activity in the SMC than in the LMC. This is consistent with the relative far-infrared luminosities of the two systems as noted above. Jones et al. note a lack of compact "protostellar" sources associated with H II regions in the Clouds, although a few such objects have been identified (Gatley et al. 1981, 1982). There is also a lower-than-expected infrared luminosity from the H II region complexes, relative to the radio continuum, as compared with Galactic H II regions, and Jones et al. suggest that this indicates a lack of intermediate- (nonionizing) mass stars being formed in the Clouds.

The Magellanic Clouds are the only galaxies, apart from the Milky Way, where individual stars can be detected by IRAS. The Clouds present a unique opportunity to locate the late-type supergiants with significant circumstellar emission that are still very young stars. Two of these stars have been identified by Elias et al. (1986), and one of these is the only known OH/IR star in the Magellanic Clouds (Wood et al. 1986). The IRAS data have, in principle, the sensitivity to locate most OH/IR stars in the Clouds, but confusion limits such surveys to the outer parts of the LMC.

IRAS observations of blue compact dwarf galaxies support the suggestion that these are undergoing periods of active star formation. Most of the dwarf galaxies studied with IRAS data are selected from lists, such as that of Thuan & Martin (1981), that were selected on the basis of blue colors and large Hα equivalent widths. Analyses of samples of these galaxies have found a range of F_{fir}/F_{blue} from 0.5 to > 10; this range appears to be greater than that found in normal spirals (Helou 1986b, Kunth & Sevre 1986, Thronson & Telesco 1986). No correlation has been found between optical color indices (or apparent optical reddening) and far-infrared emission in these galaxies. The low metallicity of the galaxies also appears to have no significant correlation with their infrared emission (Kunth & Sevre). In comparison, low-surface-brightness dwarf galaxies appear to be relatively weak infrared emitters, having an average $F_{fir}/F_{blue} \sim 0.4$ (Helou 1986b).

The correlation between far-infrared and radio continuum emission in blue compact galaxies is virtually identical, both in proportionality factor and dispersion, to that found in spiral galaxy samples (Kunth & Sevre 1986, Klein & Wunderlich 1987). This result is surprising if the luminosity sources for the far-infrared emission arise from stellar populations having significantly different evolutionary time scales and if the radio emission is related to the magnetic field structures, which differ significantly in the different classes of galaxies (Shu 1987).

A large, heterogeneous sample of "small" galaxies has been found by Thronson & Bally (1987) to have higher ratios of L_{fir}/L_{co} than do large spirals (Young et al. 1986a, Sanders et al. 1986). Thronson & Bally attribute the high L_{fir}/L_{co} ratios in these galaxies to very efficient star formation, but such high ratios may be the result of a relative lack of CO emission.

6. ELLIPTICAL AND S0 GALAXIES

From the earliest IRAS results, it was clear that very few early-type galaxies would be included in the PSC. De Jong et al. (1984) showed that 0 of 26 ellipticals and 6 of 21 S0 galaxies were detected in IRAS observations of optically bright RSA galaxies. More recent work analyzing the entire PSC has improved the detection statistics significantly. For a sample of ellipticals, Jura (1986) found the IRAS detection rate to be strongly correlated with the blue magnitude of the galaxies, with 50% of the ellipticals brighter than $m_B = 11.0$ mag being detected and over 20% detected for $11 < m_B < 12$ mag. In contrast, spiral galaxies are often detected to $m_B \sim 15$ mag.

From a search of the PSC for all E, S01, and S03 galaxies in the RSA, Tytler (1986) found detection rates of 6, 4, and 93%, respectively. More recently, Jura et al. (1987) have further expanded the detection statistics by using co-added IRAS data that achieve ~ 3 times greater sensitivity than the PSC. They found that more than 50% of the RSA galaxies with total blue magnitude $B_T^0 < 11.0$ mag were detected by IRAS at 100 μm, with a slightly higher fraction being detected at 12 μm. Ellipticals are detected as efficiently at 100 μm as at 60 μm, which reflects the lower effective color temperature of the far-infrared radiation from these galaxies as compared with spirals. The energy distribution of the bulge of M31 resembles that of elliptical galaxies and is plotted in Figure 5.

The mechanism for producing the infrared radiation in ellipticals is almost certainly thermal emission by dust. With only the survey detections, there is a strong correlation between detectable flux at 100 μm and the presence of a strong central radio source (Tytler 1986), but the most recent work (Jura et al. 1987) has expanded the detections to the point where it is no longer possible to argue that the far-infrared emission is an extrapolation of the radio emission. The 12 μm emission from ellipticals is in excess of that expected from the stellar photospheres (Soifer et al. 1986b), and the excess could arise either in the circumstellar shells of late-type stars undergoing mass loss or as emission from transiently heated grains. In principle, spectroscopy at 3 or 10 μm could identify the latter type of emission, but the galaxies are too extended and too faint for such observations using current ground-based instrumentation. Soifer et al. argued that the transient grain heating can be neglected in ellipticals, since the ultraviolet component of the interstellar radiation field is much weaker than that found in spiral galaxies (Coleman et al. 1980). Tytler (1986) has suggested that heating by electrons in a hot gas (Dwek 1986) could provide a significant amount of heating for the dust in these galaxies.

To the extent that the transient heating of small grains can be neglected in producing the observed 12 μm emission of galaxies and that radiation pressure drives the mass loss, the 12 μm excess over photospheric emission provides a good estimate of the amount of material being injected into the interstellar medium of a galaxy (Soifer et al. 1986b). In the case of the nuclear bulge of M31, this estimate (0.015 $M_\odot$ yr^{-1}; Soifer et al. 1986b) is in excellent agreement with that predicted by Faber & Gallagher (1976) on the basis of stellar evolution considerations in an old stellar population.

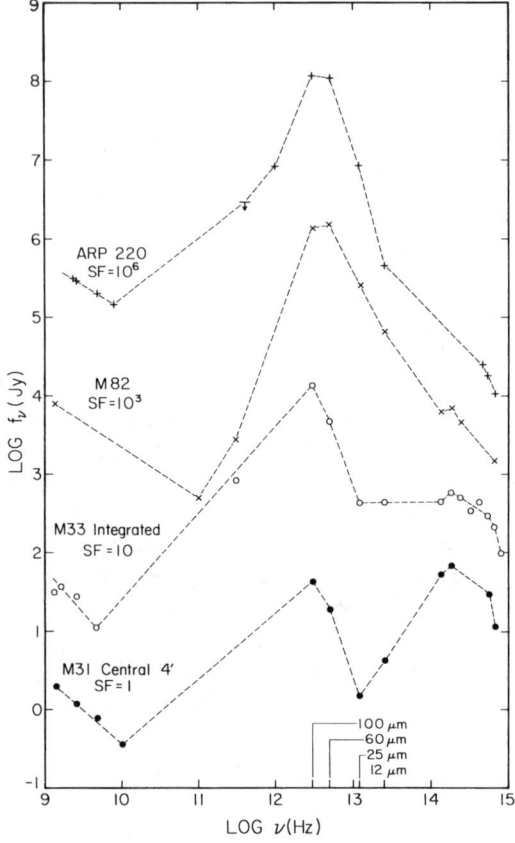

Figure 5 The continuum energy distributions from 10^9 to 10^{15} Hz are shown for four galaxies. The ordinates should be divided by the scale factors to get the flux densities of the objects. The dashed lines are not meant to trace out the continua; they are drawn only to guide the eye. The nucleus of M31 represents an old stellar population (Soifer et al. 1986b, Walterbos 1986), M33 represents a typical spiral galaxy (Rice et al. 1987a), and M82 and Arp 220 are archetypical of starburst galaxies (Rieke et al. 1980) and of extremely luminous galaxies with large infrared-to-blue ratios, respectively (Soifer et al. 1984). The observations are from a large number of published references, and those cited here are only the primary sources.

The far-infrared emission can measure the mass of the cold-dust component in the interstellar medium in elliptical galaxies, but the determination of dust mass is extremely sensitive to the temperature adopted for the dust (see Equation 1). Jura et al. (1987) find 60–100 µm color temperatures in the range of 32–42 K, which corresponds to grain temperatures of 28–33 K. These temperatures lead to dust masses in the range 10^4–10^5 $M_\odot$ (Tytler 1986). This range must be considered a lower limit on the actual masses, since effects such as distributions of dust temperatures or transient heating effects (Draine & Anderson 1985) will enhance the observed 60 µm flux and result in an overestimate of the mean dust temperature and hence an underestimate of the inferred mass. Theoretical considerations of equilibrium grain temperatures suggest that most dust is significantly colder (Jura 1982), which if true will increase the estimated amount of dust contained in ellipticals by as much as a factor of ten.

The measurement of the cold material in elliptical galaxies, combined with determinations of the input rate to the interstellar medium, allows a direct estimate of the time necessary to accumulate the observed mass. Soifer et al. (1986b) estimated $M/(dM/dt)$ of $\sim 10^7$ yr for the bulge of M31, while estimates for the accumulation times based on the observations of the total mass from the far-infrared observations fall in the range 10^7–10^8 yr. These estimates require a sink for the interstellar medium, a problem recognized years ago on the basis of optical studies and discussed comprehensively by Faber & Gallagher (1976). Of the mechanisms discussed there, removal of dust and gas from the galaxy by galactic winds might be precluded on the basis of X-ray observations of halos surrounding many early-type galaxies (Forman et al. 1985). Sputtering of dust could account for the observed dust mass (Tytler 1986), but the question of the sink for the corresponding gas remains. Low-mass star formation could provide an appropriate sink for such interstellar matter (Jura 1977, 1986).

7. IRAS OBSERVATIONS OF ACTIVE GALACTIC NUCLEI

The quasar 3C 273 is near the threshold of detectability in the PSC. Thus the expectation was that only those active galaxies with large thermal components (e.g. some Seyfert galaxies) or those with strong nonthermal sources [e.g. the optically violent variable (OVV) quasars] would be detected. So far, the survey has had few surprises with respect to active galaxies, and many of the measurements of AGN have been limited to preselected observations using the pointed observation mode of the IRAS satellite (Neugebauer et al. 1984). In this section we summarize the infrared properties of the well-established active galaxies as seen by IRAS. The discussion is presented roughly in the order of how well we suspect the infrared emission mechanism is understood.

Radio Galaxies

Among the early results of the IRAS survey was the discovery that the broad-line radio galaxy 3C 390.3 emits most of its energy ($7 \times 10^{10} L_\odot$) in the infrared (Miley et al. 1984). As seen in Figure 6, the continuum has

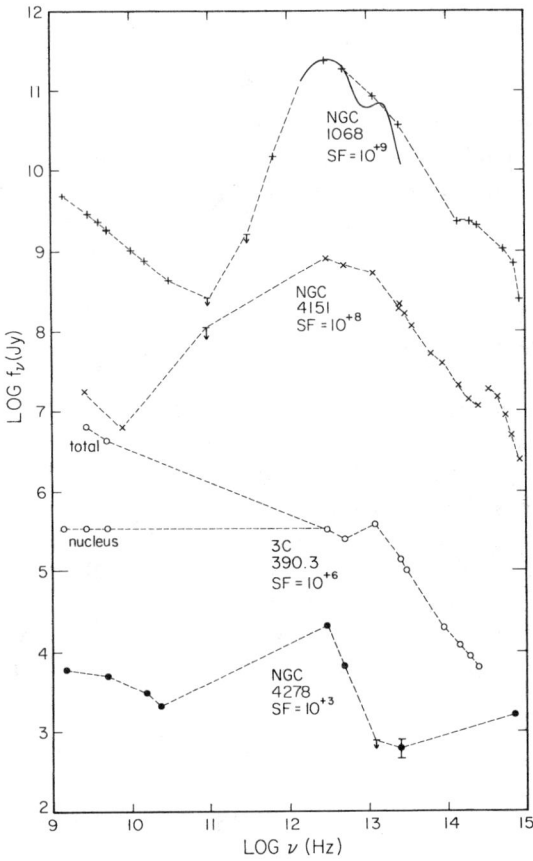

Figure 6 The continuum energy distributions from 10^9 to 10^{15} Hz are shown for four active galaxies. The conventions are as in Figure 5. Two radio galaxies showing apparent thermal emission are represented by 3C 390.3 (Miley et al. 1984) and NGC 4278, an E/S0 galaxy with a compact radio core (Wrobel et al. 1986). The galaxy NGC 1068 is representative of Seyfert 2 galaxies; the solid line represents a series of closely packed photometric observations (Telesco et al. 1984), while NGC 4151 typifies the Seyfert 1 galaxies; the break in the continuum corresponds to two different beam sizes (Beall et al. 1981, Rieke & Lebofsky 1981).

an excess near 25 μm, which is most easily interpreted as thermal radiation from dust at $T \sim 180$ K. Likewise, 4 of 10 E/S0 galaxies with compact, nonthermal radio cores studied by Wrobel et al. (1986) were detected in at least two IRAS bands and show large far-infrared excesses; the energy distribution of one of these, NGC 4278, is included in Figure 6. The infrared excesses are probably thermal radiation from heated dust, but again this interpretation cannot be proven.

More than half of some 106 radio galaxies studied by G. K. Miley & P. Golombek (private communication) have $L(60 \text{ μm}) > 10^{10} L_\odot$. The most luminous sources in the radio and optical are also very luminous in the infrared; the luminosity distribution function apparently does not depend on the optical properties. More than half have steep 60 μm to 100 μm slopes [$\alpha(100/60) \sim -2$], statistically more consistent with thermal than nonthermal slopes (see Figure 8).[6]

BL Lacs and OVV Quasars

It was known from a large number of pre-IRAS measurements that BL Lac objects and OVV quasars (or blazars) were bright in the infrared [see, e.g., the reviews of Rieke & Lebofsky (1979) and by Angel & Stockman (1980)]. Thus, although several were easily detected in the IRAS survey, many were also selected as targets in the program of pointed observations. Since BL Lac objects may present a direct view of the central engine of AGNs, a critical issue was whether or not BL Lacs showed signs of unusual excesses, presumably the result of thermal emission, in the far infrared.

The IRAS measurements of nine BL Lac objects or OVVs were included in a series of nearly simultaneous measurements from 0.14 μm to 20 cm of some 27 active extragalactic sources by Landau et al. (1986). The energy distributions look qualitatively like the smooth continuum of 3C 345 shown in Figure 7. The continua of all the OVVs observed could be fit, within $\sim 15\%$ accuracy in f_ν, by parabolas in the log (f_ν) vs. log (ν) plane for $9 < \log [\nu(\text{Hz})] < 15$. As seen for the case of 3C 345 in Figure 7, the IRAS observations fit smoothly into the overall energy distribution of that object as determined from the multifrequency observations (see also Bregman et al. 1986). Similarly, the IRAS observations of Mrk 501 smoothly fill the large gap in frequency between the radio and near-infrared/ultraviolet measurements (Sembay et al. 1985). Thus, again the IRAS measurements are consistent with the general picture for BL Lacs and OVVs of a smooth synchrotron spectrum powered by a central non-thermal source. The range of spectral indices $\alpha(100/60)$ and $\alpha(60/25)$ is

---

[6] Spectral indices α are defined such that $f_\nu \propto \nu^{+\alpha}$; $\alpha(60/25) = 2.63 \log [f_\nu(25 \text{ μm})/f_\nu(60 \text{ μm})]$ and $\alpha(100/60) = 4.51 \log [f_\nu(60 \text{ μm})/f_\nu(100 \text{ μm})]$.

illustrated in Figure 8; the two indices are often nearly equal, but there is a larger spread in $\alpha(100/60)$ than in $\alpha(60/25)$.

About 30% of the BL Lacs compiled from the lists of Hewitt & Burbidge (1987), Veron-Cetty & Veron (1985), Angel & Stockman (1980), and Urry (1985) have been detected in at least one IRAS band, generally at 60 μm. Impey et al. (1987; personal communication) conclude that the infrared emission of the BL Lacs with $L > 3 \times 10^{10} L_{\odot}$ is nonthermal synchrotron emission. Although the sample is nonhomogeneous, the parameters of the

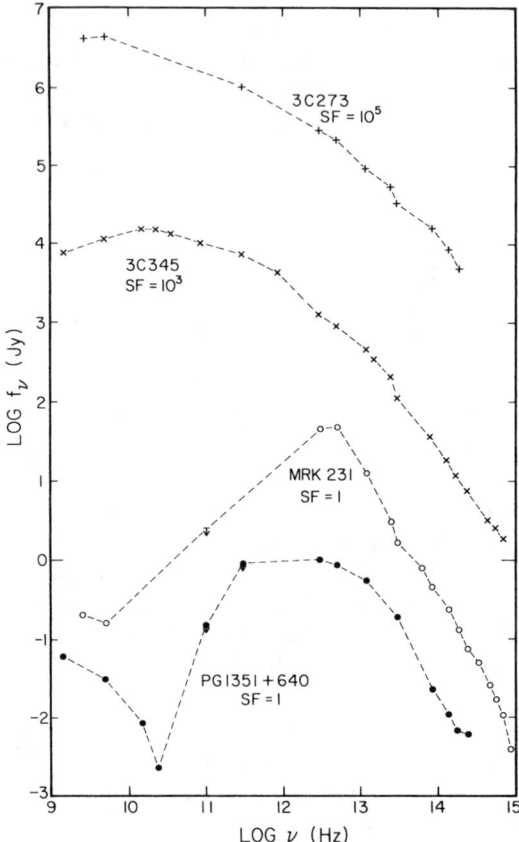

Figure 7 The continuum energy distributions from 10^9 to 10^{15} Hz are shown for four active galaxies. The conventions are as in Figure 5. The quasar 3C 273 is radio loud (Clegg et al. 1983), while the quasar PG 1351+640 is classified as radio quiet (Neugebauer et al. 1985). 3C 345 is an OVV quasar whose energy distribution is representative of OVVs and BL Lacs; an effort was made to obtain nearly simultaneous observations (Bregman et al. 1986). Mrk 231 is variously classified as a luminous Seyfert 1 galaxy or quasar and is the most luminous object within the local Universe (Neugebauer et al. 1985).

synchrotron model, such as the turnover frequencies, are strikingly similar. Below $3 \times 10^{10} L_\odot$, Impey et al. find that thermal emission by dust heated by the nuclear source contributes significantly to the infrared emission in BL Lacs.

Only a few repeated observations of OVVs were made in search of variability because of the short IRAS lifetime (10 months), the scanning geometry, and the emphasis on the all-sky survey in the IRAS mission. So far, significant variability has been found in 3C 345 (Bregman et al. 1986), 3C 446 (Neugebauer et al. 1986b), OJ 287, and BL Lac (C. Impey, personal communication). Near the start of 1983, 3C 345 showed an increase in flux by almost a factor of three at visible and near-infrared wavelengths, which gradually died out during the year. The amplitude of the outburst decreased with increasing wavelength, but the effects of the outburst were still observed at 12 μm at the start of the IRAS mission (Bregman et al. 1986). No sign of the outburst was observed at 100 μm or at radio wavelengths. Although this behavior has not been analyzed in detail, the appearance of the pulse with different amplitudes at different wavelengths is consistent with radiation emitted from a relativistic plasma moving at high velocities with respect to the observer. The OVV quasar 3C 446

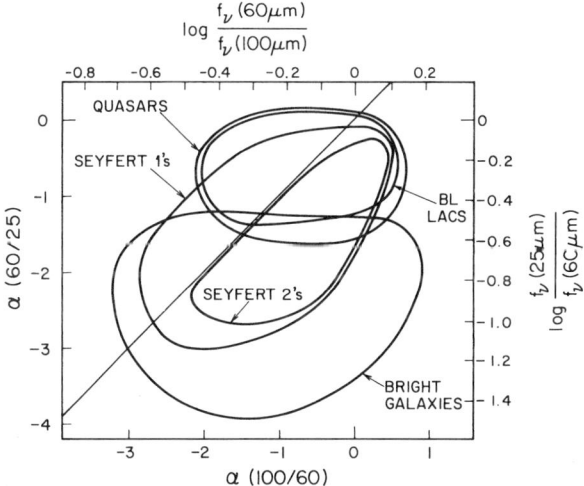

Figure 8 A schematic 25 μm/60 μm vs. 60 μm/100 μm color-color diagram based on IRAS observations is given for several classes of objects. The straight line represents constant spectral index. For reference, the indices of M82, the archetypical starburst galaxy, are $\alpha(100/60) = 0.0$ and $\alpha(60/25) = -1.7$. The quasar indices are based on the work of Neugebauer et al. (1986b), the Seyfert galaxy indices on that of Miley et al. (1985), the IRAS bright galaxies on work by Neugebauer et al., and the BL Lacs on studies by C. Impey & G. Neugebauer (private communication).

showed an outburst in late 1983 at near-infrared wavelengths (Neugebauer et al. 1986b). Two IRAS observations, one taken before and one after the outburst, show significant changes at 100 μm and smaller changes at 25 and 12 μm. This trend is different from standard outburst models, which generally show greater variability at shorter wavelengths, but it is similar to behavior seen in the infrared/submillimeter emission of 3C 273 (Robson et al. 1986). One explanation is that a thermal peak near 25 μm remains unchanged while the nonthermal radiation decreases from a more dominant contribution to a level where the 25 μm thermal peak is revealed.

Seyfert Galaxies

The IRAS survey has proved to be a rich field for finding new Seyfert galaxies. In fact, the surface density of Seyfert galaxies discovered by examination of IRAS galaxies is comparable with that of previously known Seyferts (De Grijp et al. 1985). Ten of a sample of 13 IRAS galaxies whose far-infrared continua are representative of "warm" galaxies were shown by Carter (1984) to be Seyfert galaxies. De Grijp et al., basing their search on the excess observed near 25 μm in the radio galaxy 3C 390.3 (Miley et al. 1984), obtained spectra of 40 PSC sources with "warm" 25 μm to 60 μm colors and found approximately half to be Seyfert galaxies. When a sample of 563 sources with "warm" $\alpha(100/60)$ and Galactic latitudes $>20°$ was restricted to those galaxies with flat infrared spectra, a 70% success rate for detecting Seyfert galaxies was achieved (Miley & De Grijp 1985). Osterbrock & De Robertis (1985) confirm that a large fraction of a sample of some 30 "warm" galaxies are indeed previously unknown Seyfert galaxies, which form the low-luminosity tail of the Seyfert luminosity function. As might be expected, many of the galaxies are significantly more heavily reddened than previously-known Seyferts. The Seyfert galaxies found by DeGrijp et al. are similar in their near-infrared properties to those known from optical work (Glass 1985).

Seyfert galaxies show evidence of both thermal and nonthermal emission, depending on the wavelength region and the spatial scale observed. The IRAS observations by themselves cover too limited a wavelength region to distinguish between the various types of emission, but a comparison of the flux in the IRAS bands to that at the radio and visual wavelengths can potentially identify the dominant emission mechanism. In some cases there is a definite sign of an excess or a "bump" in the infrared, which is most easily explained as thermal reradiation from heated dust. In many cases, however, the continuum over the IRAS bands is smooth with slopes that lie in the range of overlap of nonthermal sources and of thermal sources (Figure 8); thus the measurements provide ambiguous evidence, which can be used to bolster preconceived prejudices for either view. There has been no case where variability of a Seyfert galaxy has been established at IRAS wavelengths, but as mentioned above, the mission was not optimized to search for variability in specific objects.

One hundred and sixteen Seyferts, culled from a search of the PSC at the positions of all Markarian and NGC galaxies classified as Seyfert 1 (104) or Seyfert 2 (84) galaxies, are detected at 60 μm (Miley et al. 1985); about 50% of these have 60 μm luminosities in excess of $\sim 10^{10}$ $L_\odot$. Luminosity distributions at 60 μm of the Seyfert 1 and Seyfert 2 galaxies appear similar, although it is possible to statistically separate the two types in the color-color diagram shown in Figure 8. A striking feature of Figure 8 is that there is a broad overlap in the infrared colors of the various objects. The Seyfert galaxies bridge the colors representing the bright galaxies, which in general have steeper 60 μm to 25 μm slopes, and the colors representing the quasars and BL Lac objects, which have flatter 60 μm to 25 μm slopes. The Seyfert 2 galaxies lie preferentially closer to the colors of the bright IRAS galaxies, whereas the Seyfert 1 galaxies show a broader dispersion. One interpretation of these trends is that the nonthermal sources have flatter $\alpha(60/25)$, while the steeper indices [especially $\alpha(60/25)$ but also $\alpha(100/60)$] correspond to thermal components. This has been emphasized by Wilson (1987), who points out that the shape of the IRAS continuum serves as one observational diagnostic for the recognition of circumstellar star formation in Seyfert galaxies, and that the region of steep $\alpha(60/25)$ and flatter $\alpha(100/60)$ corresponds to the continuum being dominated by emission from dust heated by stars rather than by the emission from the Seyfert nucleus itself. Thus, Figure 8 "reflects primarily the relative contributions of the active nucleus and dust heated by stars (circumnuclear or disk) to the infrared fluxes." The Seyfert 1 galaxies have statistically less curvature in their spectra near 60 μm than do the Seyfert 2 galaxies; these latter galaxies show curvatures approaching those of galaxies with emission-line spectra characteristic of H II regions.

Finally, there is also an apparent correlation between the 60 μm luminosity and $\alpha(100/60)$. As is the case with the infrared selected normal galaxies, the Seyfert galaxies with higher luminosities have flatter continua. This correlation is expected in thermal models, where the volume of the emitting region does not increase as fast as the luminosity.

The observations are consistent with a simplified picture in which Seyferts consist of various amounts of the following three components: (*a*) a nonthermal nuclear power-law component extending to X-ray wavelengths, (*b*) a cool component, presumably the result of a starburst, and (*c*) a mid-infrared nuclear component that peaks between 20 and 50 μm (Miley et al. 1985). The details to fill in this general picture are often ambiguous and occasionally contradictory. The nonthermal component

(a) is generally prominent in Seyfert 1 galaxies, especially at infrared wavelengths shorter than 25 μm; the cool component is conspicuous in Seyfert 2 galaxies. Edelson & Malkan (1986) have fit models to energy distributions between ~ 0.1 and 100 μm that select these and similar components. They find reasonable power-law fits to the infrared continua of Seyfert 1 galaxies, while the continua of Seyfert 2 galaxies cannot be accurately fit with power laws. A different emphasis has been placed on the interpretation of the continua by Ward et al. (1987) and Carleton et al. (1987), who have studied the infrared continua of a sample of Seyfert galaxies selected on the basis of their hard X-ray emission. They contend that the effects of dust are dominant in creating the wide variety of observed infrared-visible continuum shapes, and that differences in the intrinsic central source continuum may make only minor contributions. In addition, Rodriguez-Espinosa et al. (1986, 1987) find that the infrared emission at wavelengths longer than 25 μm is apparently not correlated with the emission of Seyferts at wavelengths shortward of 10 μm and has a significant extended component. They emphasize that the nonthermal component cannot supply all the luminosity of the Seyferts, and thus they require a prominent cool, extended component (perhaps dust heated by hot stars). It is clear from Figures 6 and 7 that many of the gross differences in the energy distributions of the active galaxies occur at wavelengths longer than 100 μm, and that one must go well beyond the IRAS data in wavelength coverage to get a comprehensive picture of the energy distributions.

Many IRAS measurements have been integrated into multifrequency studies of individual Seyfert galaxies, and these studies are often extremely informative and generally support variations on the simplified composite picture of Seyfert galaxies. Often, the IRAS data of the Seyfert 1 galaxies lie near the basic synchrotron spectrum [e.g. Mrk 507, 5C 3.100, 1 Zw 1 (Halpern & Oke 1987), Mrk 79, Mrk 279, and MCG 8-11-11 (Spinoglio et al. 1985)], although a thermal emission bump at 25 μm is present in MCG 8-11-11. In NGC 1275 (3C 84), a Seyfert 2 galaxy that is also identified as a flat-spectrum radio source, the IRAS observations indicate that the infrared emission is dominated by thermal emission superposed on a synchrotron spectrum (Gear et al. 1985). The bright radio source PKS 1345+125, which Gilmore & Shaw (1986) model as a merger of a spiral galaxy and an elliptical galaxy containing a Seyfert nucleus, is associated with strong far-infrared flux; the role or origin of the infrared is not understood. Finally, it should be noted that the galaxy 0421+040P06, which was selected because of its infrared brightness (Beichman et al. 1985), may represent a transition between the weak radio sources in active spiral galaxies and the larger radio sources associated with radio galaxies.

At visual wavelengths it shows a Seyfert 2 spectrum, while radio measurements show a large extended structure with nonthermal radio emission.

Quasars

Some dozen quasars were included in the PSC, although the exact number identified depends on the positional agreement required (Kidger & Beckman 1986, Miley & De Grijp 1985, Neugebauer et al. 1986a). The PSC plus the pointed observations provide observations of 197 previously known quasars selected from the Veron-Cetty & Veron (1985) quasar catalog (Neugebauer et al. 1986b). Generally, the IRAS observations fit smoothly into the continua determined from measurements at other wavelengths; the IRAS data define local maxima in 3C 48, PG 1351+640, Mrk 1014, and 3C 459. At 60 μm, the luminosities extend up to $10^{14}\,L_\odot$ and are comparable to the visible luminosities. It is seen from Figure 8 that the quasars occupy much the same area of the 25/60/100 μm color-color plane as do the Seyfert galaxies, but that they tend to have flatter $\alpha(60/25)$. A subclassification of the quasars according to the radio emission, or lack thereof, yields no strong correlation between the radio properties and the location on the color-color plot, which implies that the turnover in radio-quiet quasars must be well into the millimeter region. This lack of correlation is an example of the common result that the differences between the infrared properties of the types of quasars are small and not well correlated with the radio or optical properties of the quasar.

For those quasars whose radio emission has a flat spectrum, the spectrum continues smoothly from the radio through the infrared into the visible wavelength regions, circumstantial evidence that the infrared emission is predominantly nonthermal in origin. The luminosities of the quasars with flat radio spectra are systematically higher by an order of magnitude than those with steep radio spectra and the radio-quiet quasars. Thus, if these quasars have thermal components, they are presumably masked by the larger nonthermal emission. All the quasars detected by IRAS at high redshifts have flat radio spectra.

The arguments for the origin of the infrared emission from the quasars with steep radio spectra and the radio-quiet quasars are less clear, and there is evidence that the emission must be a mixture of thermal and nonthermal components. For many quasars with steep radio spectra, and certainly for the radio-quiet quasars, their infrared luminosity is considerably greater than their radio luminosity. Most likely, the thermal emission rises from heated dust in narrow-line regions (Carter 1984, Miley & De Grijp 1985) and/or from a larger dust-rich galaxy such as associated with 3C 48 (Neugebauer et al. 1985) and, presumably, Mrk 1014.

The studies discussed above were all measurements of either optically or radio selected quasars. The first quasar discovered from its infrared emission is IRAS 13349+2438, which Beichman et al. (1986) identified as a quasar with redshift $z = 0.107$ while studying sources found at 12 µm in the PSC. The object is a weak radio source with a bolometric luminosity of $\sim 3 \times 10^{12} L_\odot$, emitted mostly between 5 and 12 µm. It may serve as a prototype of radio-quiet, infrared-loud quasars. Vader & Simon (1987) have recently discovered another infrared-loud quasar at a redshift $z \sim 0.3$. The discovery of these quasars suggests that infrared-loud quasars may constitute a significant, previously unsuspected population.

8. OBSERVATIONS OF HIGHLY LUMINOUS INFRARED GALAXIES

High-luminosity, infrared-bright galaxies are a major constituent of the overall population of galaxies. Although such objects as NGC 253 and Mrk 231 were known in the early 1970s to emit the bulk of their total luminosity in the infrared, IRAS has demonstrated that infrared-bright galaxies comprise a substantial fraction of all galaxies with bolometric luminosities exceeding $\sim 3 \times 10^{10} L_\odot$.

In a seminal paper, Rieke et al. (1980) analyzed the broad range of observations of M82 and NGC 253 and successfully interpreted them in the context of models of starbursts. Subsequent observations, e.g. the high-resolution observations of variable radio sources in M82 (Kronberg & Sramek 1985), have served to support the basic ideas in the model proposed by Rieke et al. Analysis suggests that infrared-bright galaxies in the luminosity range $3 \times 10^{10} L_\odot$ to $3 \times 10^{11} L_\odot$ are predominantly undergoing outbursts of star formation activity. Roughly half the galaxies in infrared selected samples of galaxies have luminosities $L_{\text{fir}} > 3 \times 10^{10} L_\odot$ (Elston et al. 1985, Carico et al. 1986, Moorwood et al. 1986, Soifer et al. 1987a,b). More than half the objects in the infrared selected sample discussed by Elston et al. have starburst optical spectra.

In the luminosity range from $3 \times 10^{10} L_\odot$ to $3 \times 10^{11} L_\odot$, the space densities of infrared selected galaxies and (non-Seyfert) Markarian galaxies are virtually identical (Soifer et al. 1986a). The far-infrared luminosities of the Markarian galaxies are on average substantially greater than those of normal galaxies (Sekiguchi 1987, Deutsch & Willner 1986, 1987), while the far-infrared color temperatures of these galaxies are higher than those of "normal" galactic disks. These properties suggest that the predominant luminosity source in these galaxies is star formation, and that the Markarian and infrared selected galaxies represent basically the same population over the luminosity range 3×10^{10}–$3 \times 10^{11} L_\odot$.

Effects of Interactions

Interactions have long been thought to play an important role in triggering starbursts (e.g. Larson & Tinsley 1978, Lonsdale et al. 1984, Cutri & McAlary 1985). Several groups have tried to establish this link using IRAS observations, with mixed success. Haynes & Herter (1987) found that a sample of optically selected binary galaxies showed little or no enhanced far-infrared emission compared with a sample of isolated galaxies. Kennicutt et al. (1987) have studied samples of close pairs of galaxies and strongly perturbed systems, and they find substantially enhanced star formation activity in many of the highly disturbed galaxies but little enhanced star formation in the other systems. While interaction-enhanced star formation seems to be at best a marginal effect in the optically selected samples, interaction does seem to be significant in the infrared selected samples. Sanders et al. (1987b) noted that the fraction of interacting galaxies increases monotonically with increasing luminosity in the IRAS bright-galaxy sample for $L_{\rm fir} > 3 \times 10^{10} \, L_\odot$, with interacting and close pairs of galaxies accounting for one fourth of the galaxies with $L_{\rm fir} \sim 3 \times 10^{10} \, L_\odot$, two thirds of the galaxies at $L_{\rm fir} \sim 10^{11} \, L_\odot$, and all of the galaxies with $L_{\rm fir} > 5 \times 10^{11} L_\odot$.

There is also a significant difference in both the mean far-infrared color temperature and the $L_{\rm fir}/M({\rm H}_2)$ ratio between isolated galaxies and strongly interacting galaxy pairs (Young et al. 1986b). While the average molecular gas content and $L_{\rm blue}/M({\rm H}_2)$ ratios are similar for the two samples, the far-infrared color temperature and $L_{\rm fir}/M({\rm H}_2)$ values are significantly greater in the interacting galaxies than in the isolated galaxies. How much this result depends on selection effects is presently unclear.

Houck et al. (1984) searched the early IRAS data for bright 60 μm sources without obvious optical identifications. Of an original sample of ~ 200 candidates, a total of nine objects seemed to lack obvious optical identifications on the Palomar Schmidt Survey. Follow-up studies by Antonucci & Olszewski (1985, 1986), Aaronson & Olszewski (1984), and Houck et al. (1985) have shown that all but one of these nine objects can be identified with an unusual galaxy; the one remaining source has been shown to be a wisp of Galactic infrared cirrus. Optical studies of the candidate galaxies show them to be mostly emission-line objects with redshifts ranging from 0.05 to 0.2. The failure of the initial identification search implies, of course, that these objects have very high infrared/blue luminosity ratios, ranging up to several hundred. The resultant infrared luminosities range up to $\sim 3 \times 10^{12} \, L_\odot$, making these objects among the most luminous infrared galaxies. Six of the eight objects are members of small (two to four member) groups of galaxies. Two galaxies show short linear features indicating the possibility of strong tidal interactions.

For luminosities $L > 3 \times 10^{11} L_\odot$, infrared-bright galaxies have a higher space density than do normal galaxies (Figure 1b, Soifer et al. 1986a). Of the 324 objects in the bright-galaxy sample discussed above, 25 galaxies have $L_{fir} > 3 \times 10^{11} L_\odot$. Table 1, adapted from Sanders et al. (1987b), lists the 10 most luminous galaxies in the bright-galaxy sample; all have $L(8-1000 \ \mu m) \geq 10^{12} L_\odot$. These luminosities are equivalent to the bolometric luminosities of optically selected quasars (Schmidt & Green 1983, Veron-Cetty & Veron 1985), but the wavelengths over which this luminosity emerges are grossly different. Besides their extreme luminosity, Table 1 shows that these galaxies all have ratios of $L_{fir}/L_{blue} > 10$ and show evidence for merging or interacting galaxy structures.

Origins of the Luminosity

Both a nonthermal nucleus and extreme star formation activity can result in prominent luminosity in the far infrared. Indeed, as noted below, there is growing belief that both mechanisms may be at work in the most luminous sources. In addition, the conversion of kinetic energy into infrared luminosity can release such luminosity for short periods (Harwit et al. 1986).

The two best-studied extreme-infrared galaxies are Arp 220 ($=$ IC 4553; Soifer et al. 1984), whose energy distribution is included in Figure 5, and NGC 6240 (Wright et al. 1984). Arp 220 and NGC 6240 both show complicated morphologies, presumably evidence for either tidal interactions or mergers (see, e.g., Arp 1966, Fried & Schulz 1983, Sanders et al. 1987b). Thus, they are specific examples of the fact that those "galaxy-galaxy interactions which have resulted in a merger are, as a class, ultraluminous" (Joseph & Wright 1985). Both galaxies also show evidence for thermal and nonthermal luminosity sources. The evidence for an active nucleus is strong in both Arp 220 and NGC 6240. The Brackett α line has been detected in Arp 220 with a width of 1200 km s^{-1} (DePoy 1987), which indicates the presence of an active nucleus. Beck et al. (1986) find the strength of the Brackett α and γ lines in Arp 220 to be inconsistent with O stars contributing significantly to the luminosity in these galaxies. Likewise, DePoy et al. (1986) conclude that their measurements of the Paschen α emission line indicate that there is insufficient ionizing flux from O and B stars with any reasonable initial-mass function in NGC 6240 to provide the measured bolometric luminosity, and thus that their measurements are more consistent with an active galactic nucleus such as seen in Seyfert galaxies. Norris (1985, Norris et al. 1985) has concluded from visible

Table 1 Ultraluminous IRAS galaxies

	RA (1950)	Decl. (1950)	cz^a (km s^{-1})	L_{fir} ($10^{12} L_\odot$)	$L(8-1000\ \mu m)$ ($10^{12} L_\odot$)	$\dfrac{\nu f_\nu(80)}{\nu f_\nu(B)}$	Morphology^b	Optical^c spectra
IRAS	05 18 58.6	−25 24 40	12,706	0.8	1.23	24	S	Sey 1.5
IRAS	08 57 13.0	+39 15 40	17,480	1.0	1.25	10	M	H II
UGC 5101	09 32 04.6	+61 34 37	12,000	0.8	1.02	22	M	Sey 2
IRAS	12 11 12.2	+03 05 20	21,703	1.5	1.94	68	M	LINER
Mrk 231	12 54 04.8	+57 08 38	12,623	2.1	3.45	22	M	Sey 1
Mrk 273	13 42 51.6	+56 08 13	11,400	1.3	1.47	33	T	Sey 2
IRAS	14 34 52.3	−14 47 24	24,332	1.5	1.88	33	M	LINER/H II
IRAS	15 25 03.1	+36 09 00	16,009	0.8	1.00	25	M	LINER
Arp 220	15 32 46.3	+23 40 08	5,450	1.3	1.58	59	M	Sey 2
IRAS	22 49 09.6	−18 08 20	22,807	1.2	1.33	25	M	LINER

^{a} Mean optical LSR redshift.
^{b} M—advanced merger; tidal tails observed; T—evidence of a single long tail for Mrk 273 due to a recent strong interaction with disturbed companion 2' north; S—appears starlike on POSS print; obvious nebulosity cn short CCD exposure.
^{c} Based on linewidth of Hα+[N II] and/or O[III]/Hβ line ratio from long-slit spectrum.

spectra, near-infrared observations, and radio maps that Arp 220 is "an edge-on Seyfert galaxy with a stellar halo, perhaps similar to those of starburst galaxies, around the nucleus."

A starburst energy source in the ultra-high-luminosity infrared galaxies has been inferred from the presence of substantial amounts of molecular gas (Sanders & Mirabel 1985, Young et al. 1984, Sanders et al. 1986); many of these galaxies have $L_{\text{fir}}/M(H_2)$ ratios comparable to those found in the archetypical starburst galaxies M82 and NGC 253. Molecular hydrogen has been detected in both Arp 220 and NGC 6240 (Joseph et al. 1984), and Joseph et al. (1987) have detected extended emission in the 2.1 μm line of H_2 from the most luminous IRAS galaxies. They argue that this implies a starburst origin for the luminosity. Rieke et al. (1985) have also found the 3.3 μm emission feature usually associated with star formation rather than active nuclei (see, e.g., Aitken & Roche 1985) in the spectrum of Arp 220 and the apparent dominance of the 2.2 μm light by cool giants in both Arp 220 and NGC 6240. Bottinelli et al. (1987) have suggested that the OH megamasers associated with many ultraluminous galaxies are associated with star formation processes.

The interpretation of some of the observations of these objects are ambiguous and can fit into either picture of the luminosity sources. In Arp 220 and Mrk 171, Scoville et al. (1986) and Sargent et al. (1987) find that the CO emission is concentrated into diameters of < 1.5 kpc and 1 kpc, respectively. This does not distinguish starburst from AGN models, but it does show that there is an abundance of molecular gas available to fuel either energy generation mechanism. Likewise, the presence of a strong, extended 2.1 μm H_2 line demonstrates the presence of shocked molecular gas; the relation to star formation activity is indirect at best. In the most extreme galaxies, starburst models have difficulty in accounting for the total luminosity (Rieke et al. 1985). Furthermore, in Arp 220 the ratio of CO emission to total luminosity is low compared with its value in the starburst galaxies like M82 and NGC 253, which again suggests that star formation does not account for all of the luminosity (Sanders et al. 1986).

In principle, high-spatial-resolution infrared measurements of these galaxies would distinguish between the models. Unfortunately, while numerous measurements of the size of the infrared source in Arp 220 have been published, several of these are contradictory. Rieke et al. (1985) have concluded that the nuclear source is extended on the order of 5–10" (2–4 kpc) at 10 μm, which would argue strongly for a distributed luminosity (i.e. a starburst model). On the other hand, Becklin & Wynn-Williams (1987), with more extensive observations, find that at 10 μm and 20 μm the source is less than 2" in diameter. Joy et al. (1986), using the Kuiper

Airborne Observatory, give limits to the diameter of 7.5" at 50 μm and 8.5" at 100 μm. Norris (1985) has shown that at 2.2 μm the nucleus is concentrated into an area <2" in diameter; slit scans at 2.2 μm show a source ~1" in diameter containing a significant fraction of the light (Neugebauer et al. 1987). The 2.2 μm observations do not preclude starburst models, but they do force the conditions in the center of the galaxy to be extreme.

Thus, there seems to be growing evidence in individual galaxies that the high-infrared-luminosity galaxies are powered by a combination of Seyfert-type activity and star formation in the nucleus (Becklin 1987, Rieke 1987). It should be noted that this duality is clearly the rule rather than the exception, even in galaxies of lesser luminosity; for example, Telesco et al. (1984) have pointed out that in the well-studied galaxy NGC 1068, about half the luminosity of $3 \times 10^{11} L_\odot$ is associated with a Seyfert nucleus and half is associated with a 3 kpc diameter disk surrounding the nucleus. In this connection, it is perhaps relevant that the IRAS observations of the quasar 3C 48 give evidence for an accompanying galaxy with $L \sim 3 \times 10^{12} L_\odot$ (Neugebauer et al. 1985). This apparent combination of an active nucleus and star formation activity seems not to be accidental.

Mrk 231 represents perhaps the closest example of an ultra-high-luminosity infrared galaxy that shares many properties of "classical" quasars. While this galaxy has been known as a high-luminosity infrared galaxy for over a decade (Young et al. 1972, Rieke & Low 1972), the IRAS survey has shown it to be the most luminous object known within 300 Mpc. The energy distribution of this galaxy (Figure 7) is intermediate between those of the extreme IRAS galaxies typified by Arp 220 (Figure 5) and the "classical" radio quiet quasars typified by PG 1351+640 (Figure 7). Spectroscopically, Mrk 231 has been classified as a Seyfert 1 galaxy (Boksenberg et al. 1977, Adams & Weedman 1972) having an Hα emission-line width of ~ 6000 km s^{-1}; in addition, it shows absorption lines indicative of a young, hot stellar population, and optical and infrared absorption indicating large reddening. The recent detection of CO emission in this galaxy (Sanders et al. 1987a) shows that it contains a mass of molecular gas of $1.4 \times 10^{10} M_\odot$, a value ~ 4 times that in the Milky Way, while new optical images of the system show the tidal tails characteristic of a recent merger (Sanders et al. 1987a).

A tentative picture of the role of the ultraluminous infrared galaxies is evolving. Sanders et al. (1987b) have speculated that as two galaxies containing large amounts of gas merge together, a large burst of luminosity occurs as a result of a combination of star formation and a black hole fed by the interstellar medium; the black hole may be either preexisting or formed in the interaction. Arp 220 could possibly be a galaxy at this point in the evolutionary sequence. As the luminosity of the system increases, the gas is gradually consumed or swept away until the material around the nucleus is thin or patchy enough that the nucleus is visible. Mrk 231 may be an example of a galaxy at this stage. Eventually, the dust content becomes negligible, the dominant power is the result of accretion onto the black hole, and a "classical" quasar becomes visible.

9. COSMOLOGICAL RESULTS

The characteristics of the IRAS survey that make it useful for cosmological studies are its uniformity across the sky in terms of completeness, reliability, calibration, and lack of extinction. The results are thus much less susceptible to observational bias than are ground-based optical studies, which necessarily require the use of more than one telescope and observing location. The most serious complication for deep surveys with the IRAS data, even at the limit of the PSC, is confusion caused by the cirrus emission from dust in the Galaxy (Low et al. 1984). Since cirrus in other galaxies is at least partly responsible for their infrared emission, it is not surprising that it is difficult to separate Galactic cirrus clumps from external galaxies. Several different techniques have been developed by different authors to aid in this discrimination. Luckily, these different techniques do not seem to affect the outcome of the various studies unduly.

The differential number counts for 60 μm sources in the two Galactic pole caps ($|b| > 60°$, away from Virgo) are consistent with a slope in the log (number) vs. log (f_v) of -1.5, but they show a 20% excess of sources in the northern cap over the southern cap at all flux levels down to about 1.5 times the survey limit (Rowan-Robinson et al. 1986). This anisotropy has been explored in detail by Meiksin & Davis (1986) and by Yahil et al. (1986). Although these groups used very different criteria for source selection and analysis procedures, their results for the amplitude of the anisotropy and for the direction of the anisotropy relative to the direction of the anisotropy of the microwave background are in remarkable agreement (see Table 2).

From these results the authors suggest they have identified the material responsible for the acceleration of the Local Group with respect to the microwave background. From indirect estimates of the distances to the galaxies, the authors derive an estimate for the cosmological density parameter of $0.5 < \Omega_0 < 1$. The lack of significant systematic calibration errors over very large angular scales is crucial to the validity of this result. Although there is no evidence for such errors, the data have not been tested at the level needed to validate the assumptions needed for the

Table 2 Dipole moment directions

	Amplitude (%)	l (°)	b (°)	θ^a (°)
Meiksin & Davis (1986):				
full sample	4.1	235	43.5	29
$f_\nu(60) > 1.94$ Jy	7.0	247	44.8	22
Yahil et al. (1986)	—	248 ± 9	40 ± 8	26 ± 10
Microwave background[b]	—	268	30	—

[a] θ is the angle between infrared and microwave anisotropies.
[b] Lubin et al. (1983), Fixsen et al. (1983).

anisotropy analysis. Whether the results are to be interpreted in the straightforward manner suggested by the authors in light of the apparent streaming motion (e.g. Burstein et al. 1987) of the Local Supercluster is not clear at this time.

An attempt has been made to determine the magnitude of a uniform 100 μm background (Hauser et al. 1984, Rowan-Robinson 1986). Although there appears to be such a component in the IRAS data, it is very difficult to assess its validity, since the IRAS telescope did not have an absolute-zero reference level, and instrumental offsets may be responsible for the observed "background." If the observed background is real and due to starburst galaxies at large redshifts, it implies substantial evolution in the luminosity function for these objects (see Rowan-Robinson 1986, Hacking & Houck 1987b). The final resolution of this problem must await the launch of an instrument capable of the required absolute measurements, such as the Cosmic Background Explorer (COBE).

To the limit of the PSC, there is no indication of evolution in the number counts of galaxies. Hacking & Houck (1987a) have co-added data from the 6 square degrees surrounding the IRAS secondary point source calibrator, NGC 6543, at the ecliptic pole. The resultant map contains 60 μm sources down to a 5σ detection level of 50 mJy, 10 times fainter than the PSC. Although there are many areas in the sky where there is less cirrus emission, the hundreds of scans of this field over a wide range of position angles provide additional filtering against cirrus clouds, making this the deepest survey achieved by IRAS to date. It is likely to remain the deepest view of the infrared sky until the launch of the next generation of space infrared missions.

The number count analysis of the data in the NGC 6543 field show an excess of sources at the lowest flux levels (Hacking & Houck 1987b,

Hacking et al. 1987; see Figure 9). The excess counts can be fit by a model with pure luminosity evolution similar to that used to explain the turnup in the radio counts observed by Condon (1984) and Windhorst et al. (1985). The increased counts also scale roughly as the increased interaction frequency expected as a result of the higher physical density of galaxies at large look-back times. The existence of evolution and the type of evolution, if present, can be determined from the redshift distribution of the galaxies.

Weedman (1987) has used the measured luminosity function for bright starburst galaxies to predict several cosmologically interesting quantities. Given the ratio $f_v(60\ \mu\mathrm{m})/f_v(20\ \mathrm{cm}) = 150$ (Helou et al. 1985), he predicts that more than 10% of the radio counts in the 100 mJy range are due to

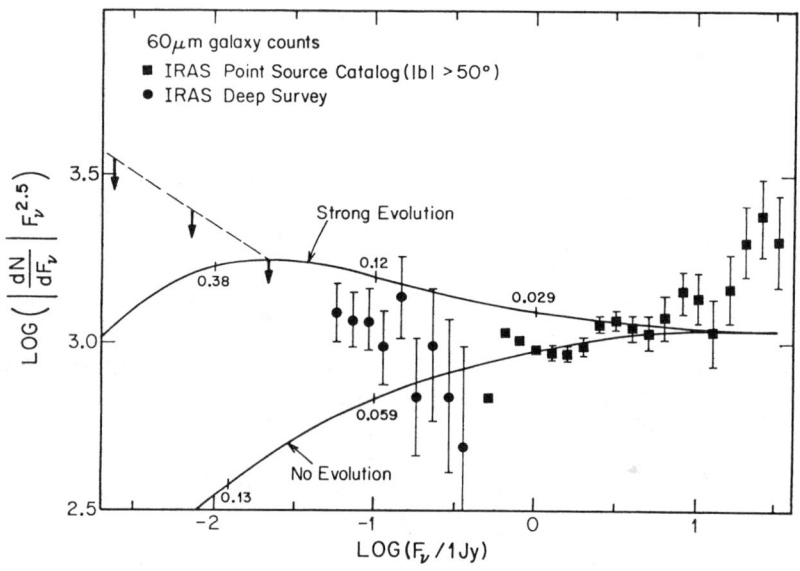

Figure 9 Differential source counts at 60 μm taken from Hacking et al. (1987) are presented. The ordinate is the differential number of sources normalized by the number expected in a static Euclidian universe. For flux densities greater than 0.6 Jy, the counts are taken from the PSC, while for flux densities less than 0.6 Jy, the counts are taken from the field around NGC 6543. The two curves plotted show models fitted to the observed counts at $f_v(60\ \mu\mathrm{m}) > 0.6$ Jy. The curve marked "no evolution" assumes constant space densities of infrared bright galaxies as a function of redshift, while the curve marked "strong evolution" assumes a model of luminosity evolution consistent with the model of Condon (1984). The values marked along each curve show the median redshift in that flux bin predicted from the appropriate model. The upper limits below 50 mJy are derived from an analysis of the fluctuations in the baseline "between" detected sources.

starburst galaxies (see also Biermann et al. 1985). Weedman also points out that X rays from starburst galaxies may give rise to a significant fraction of the isotropic X-ray background. On a more speculative note, he suggests that the extinction produced by the dust in the galaxies may be responsible for the falloff in the optical source counts at $z > 3$. These speculations will gain added strength if the turnup in the infrared source counts reported by Hacking et al. (1987) prove to be caused by an excess of objects at high redshifts.

10. SUMMARY

The availability of the IRAS data has changed the numbers of extragalactic sources observed between 10 μm and 100 μm from a few dozen to 20,000. With this enormous change, the study of the extragalactic sky in the infrared has gone from a study of a few unique and peculiar objects selected at noninfrared wavelengths to the systematic study of the infrared properties of virtually every known class of extragalactic source. Understanding the view of the extragalactic sky presented by IRAS will, if not revolutionize, at least lead to major advances in our understanding of normal galaxies, starburst galaxies, active galactic nuclei, and perhaps even the structure of the Universe.

ACKNOWLEDGMENTS

It is a pleasure to thank Rosanne Hernandez, Helen Knudsen, and the IPAC library staff for assistance in gathering and checking the references for this review. We also thank all the authors who provided us with preprints of their work. Walter Rice, George Helou, and Perry Hacking kindly provided figures for this article, while Dave Sanders allowed us to use Table 1 prior to publication. Discussions with Dave Sanders and Jay Elias were most illuminating. We thank Eric Becklin, Chas Beichman, Roger Blandford, Jay Elias, Chris Impey, Barry Madore, Sterl Phinney, Nick Scoville, Charlie Telesco, and Rene Walterbos for reading and commenting on the manuscript. Final manuscript preparation was expertly done by Julie Serpa.

Finally, we extend our deepest appreciation to all those individuals and their families whose diligence, dedication, and sacrifice made the IRAS mission a success. Without their efforts, we would not know what a rich sky the infrared presents.

This research was funded by NASA under the IRAS Extended Mission program.

Literature Cited

Aaronson, M., Olszewski, E. W. 1984. *Nature* 309: 414–17
Adams, T. F., Weedman, D. W. 1972. *Ap. J. Lett.* 173: L109–11
Aitken, D. K., Roche, P. F. 1985. *MNRAS* 213: 777–88
Angel, J. R. P., Stockman, H. S. 1980. *Ann. Rev. Astron. Astrophys.* 18: 321–61
Antonucci, R. R. J., Olszewski, E. W. 1985. *Astron. J.* 90: 2203–6
Antonucci, R. R. J., Olszewski, E. W. 1986. *Astron. J.* 91: 56–57
Appleton, P. N., Struck-Marcell, C. 1987. *Ap. J.* 312: 566–73
Arp, H. C. 1966. *Ap. J. Suppl.* 14: 1–20
Beall, J. H., Rose, W. K., Dennis, B. R., Crannell, C. J., Dolan, J. F., et al. 1981. *Ap. J.* 247: 458–63
Beck, S. C., Turner, J. L., Ho, P. T. P. 1986. *Ap. J.* 309: 70–75
Becklin, E. E. 1987. In *Star Formation in Galaxies*, ed. C. Persson, pp. 753–56. Washington, DC: US Govt. Print. Off.
Becklin, E. E., Wynn-Williams, G. C. 1987. In *Star Formation in Galaxies*, ed. C. Persson, pp. 643–50. Washington, DC: US Govt. Print. Off.
Beichman, C. A. 1987. *Ann. Rev. Astron. Astrophys.* 25: 521
Beichman, C. A., Boulanger, F., Rice, W., Persson, C., Viallefond, F. 1987. In *Star Formation in Galaxies*, ed. C. Persson, pp. 643–50. Washington, DC: US Govt. Print. Off.
Beichman, C. A., Soifer, B. T., Helou, G., Chester, T. J., Neugebauer, G., et al. 1986. *Ap. J. Lett.* 308: L1–5
Beichman, C. A., Wynn-Williams, C. G., Persson, C. J., Persson, S. E., Heasley, J. N., et al. 1985. *Ap. J.* 293: 148–53
Bicay, M. D., Giovanelli, R. 1987. In *Star Formation in Galaxies*, ed. C. Persson, pp. 277–81. Washington, DC: US Govt. Print. Off.
Biermann, P., Eckart, A., Witzel, A. 1985. *Astron. Astrophys.* 142: L23–24
Boksenberg, A., Carswell, R. F., Allen, D. A., Fosbury, R. A. E., Penston, M. V., et al. 1977. *MNRAS* 178: 451–56
Bottinelli, L., Dennefeld, M., Gouguenheim, L., Martin, J. M., Paturel, G., et al. 1987. In *Star Formation in Galaxies*, ed. C. Persson, pp. 597–600. Washington, DC: US Govt. Print. Off.
Bregman, J. N., Glassgold, A. E., Huggins, P. J., Neugebauer, G., Soifer, B. T., et al. 1986. *Ap. J.* 301: 708–26
Burstein, D., Davies, R. L., Dressler, A., Faber, S. M., Lynden-Bell, D., et al. 1987. In *Galaxy Distance and Deviations From Universal Expansion*, ed. B. F. Madore, R. B. Tully, pp. 123–30. Dordrecht: Reidel
Burstein, D., Lebofsky, M. J. 1986. *Ap. J.* 301: 683–88
Carico, D. P., Soifer, B. T., Beichman, C., Elias, J. H., Matthews, K., et al. 1986. *Astron. J.* 92: 1254–61
Carleton, N. P., Elvis, M., Fabbiano, G., Willner, S. P., Lawrence, A., et al. 1987. *Ap. J.* In press
Carter, D. 1984. *Astron. Express* 1: 61–74
Cataloged Galaxies and Quasars Detected in the IRAS Survey. 1985. Prepared by C. J. Persson, G. Helou, J. C. Good, W. L. Rice. *JPL D1932*, Jet Propul. Lab., Pasadena, Calif. (XCAT)
Clegg, P. E., Gear, W. K., Ade, P. A. R., Robson, E. I., Smith, M. G., et al. 1983. *Ap. J.* 273: 58–63
Coleman, G. D., Wu, C. C., Weedman, D. W. 1980. *Ap. J. Suppl.* 43: 393–416
Condon, J. J. 1984. *Ap. J.* 287: 461–74
Condon, J. J., Broderick, J. J. 1985. *Astron. J.* 90: 2540–49
Condon, J. J., Broderick, J. J. 1986a. *Astron. J.* 91: 1051–57
Condon, J. J., Broderick, J. J. 1986b. *Astron. J.* 92: 94–102
Cutri, R. M., McAlary, C. W. 1985. *Ap. J.* 296: 90–105
De Grijp, M. H. K., Miley, G. K., Lub, J., de Jong, T. 1985. *Nature* 314: 240–42
de Jong, T., Brink, K. 1987. In *Star Formation in Galaxies*, ed. C. Persson, pp. 323–28. Washington, DC: US Govt. Print. Off.
de Jong, T., Clegg, P. E., Soifer, B. T., Rowan-Robinson, M., Habing, H. J., et al. 1984. *Ap. J. Lett.* 278: L67–70
de Jong, T., Klein, U., Wielebinski, R., Wunderlich, E. 1985. *Astron. Astrophys.* 147: L6–9
DePoy, D. L. 1987. In *Star Formation in Galaxies*, ed. C. Persson, pp. 701–5. Washington, DC: US Govt. Print. Off.
DePoy, D. L., Becklin, E. E., Wynn-Williams, C. G. 1986. *Ap. J.* 307: 116–22
Deutsch, L. K., Willner, S. P. 1986. *Ap. J. Lett.* 306: L11–15
Deutsch, L. K., Willner, S. P. 1987. *Ap. J. Suppl.* In press
Devereux, N. 1987. In *Star Formation in Galaxies*, ed. C. Persson, pp. 219–26. Washington, DC: US Govt. Print. Off.
Devereux, N. A., Becklin, E. E., Scoville, N. Z. 1987. *Ap. J.* 312: 529–41
Dickey, J. M., Garwood, R. W., Helou, G. 1987. In *Star Formation in Galaxies*, ed. C. Persson, pp. 575–78. Washington, DC: US Govt. Print. Off.
Dickey, J. M., Salpeter, E. E. 1984. *Ap. J.* 284: 461–70

Draine, B. T., Anderson, N. 1985. *Ap. J.* 292: 494–99
Draine, B. T., Lee, H. M. 1984. *Ap. J.* 285: 89–108
Dressel, L. L. 1987. In *Star Formation in Galaxies*, ed. C. Persson, pp. 579–82. Washington, DC: US Govt. Print. Off.
Dwek, E. 1986. *Ap. J.* 302: 363–70
Eales, S., Wynn-Williams, C. G., Beichman, C. A. 1987. In *Star Formation in Galaxies*, ed. C. Persson, pp. 531–36. Washington, DC: US Govt. Print. Off.
Edelson, R. A., Malkan, M. A. 1986. *Ap. J.* 308: 59–77
Elias, J. H., Frogel, J. A. 1987. In *Star Formation in Galaxies*, ed. C. Persson, pp. 241–44. Washington, DC: US Govt. Print. Off.
Elias, J. H., Frogel, J. A., Schwering, P. B. W. 1986. *Ap. J.* 302: 675–79
Elston, R., Cornell, M. E., Lebofsky, M. 1985. *Ap. J.* 296: 106–14
Faber, S., Gallagher, J. S. 1976. *Ap. J.* 204: 365–78
Fixsen, D. J., Cheng, E. S., Wilkinson, D. T. 1983. *Phys. Rev. Lett.* 50: 620–23
Forman, W., Jones, C., Tucker, W. 1985. *Ap. J.* 293: 102–19
Fried, J. W., Schulz, H. 1983. *Astron. Astrophys.* 118: 166–70
Frogel, J. A. 1987. In *Star Formation in Galaxies*, ed. C. Persson, pp. 161–66. Washington, DC: US Govt. Print. Off.
Gallagher, J., Hunter, D. 1987. In *Star Formation in Galaxies*, ed. C. Persson, pp. 167–77. Washington, DC: US Govt. Print. Off.
Gatley, I., Becklin, E. E., Hyland, A. R., Jones, T. J. 1981. *MNRAS* 194: 17P–21P
Gatley, I., Hyland, A. R., Jones, T. J. 1982. *MNRAS* 200: 521–26
Gear, W. K., Gee, G., Robson, E. I., Nolt, I. G. 1985. *MNRAS* 217: 281–90
Gilmore, G., Shaw, M. A. 1986. *Nature* 321: 750–53
Glass, I. S. 1985. In *Light on Dark Matter*, ed. F. P. Israel, pp. 487–88. Dordrecht: Reidel
Habing, H. J., Miley, G., Young, E., Baud, B., Boggess, N., et al. 1984. *Ap. J. Lett.* 278: L59–62
Hacking, P., Condon, J. J., Houck, J. R. 1987. Submitted for publication
Hacking, P., Houck, J. R. 1987a. *Ap. J. Suppl.* 63: 311–33
Hacking, P., Houck, J. R. 1987b. In *Star Formation in Galaxies*, ed. C. Persson, pp. 547–52. Washington, DC: US Govt. Print. Off.
Halpern, J. P., Oke, J. B. 1987. *Ap. J.* 312: 91–100
Harwarden, T. G., Fairclough, J. H., Joseph, R. D., Leggett, S. K. 1986. In *Light on Dark Matter*, ed. F. P. Israel, pp. 455–62. Dordrecht: Reidel
Harwit, M., Houck, J. R., Soifer, B. T., Palumbo, C. G. 1986. *Ap. J.* In press
Hauser, M. G., Gillett, F. C., Low, F. J., Gautier, T. N., Beichman, C. A., et al. 1984. *Ap. J. Lett.* 278: L15–18
Haynes, M., Herter, T. L. 1987. *Ap. J.* In press
Helou, G. 1986a. *Proc. Workshop Star-Forming Dwarf Galaxies and Related Objects*, ed. D. Kunth, T. X. Thuan, pp. 319–30. Gif sur Yvette: Ed. Front. Transl. by J. Thanh Van
Helou, G. 1986b. *Ap. J. Lett.* 311: L33–36
Helou, G., Soifer, B. T., Rowan-Robinson, M. 1985. *Ap. J. Lett.* 298: L7–11
Hewitt, A., Burbidge, G. 1987. *Ap. J. Suppl.* 63: 1–246
Hildebrand, R. 1983. *Q. J. R. Astron. Soc.* 24: 267–82
Houck, J. R., Schneider, D. P., Danielson, G. E., Beichman, C. A., Persson, C. J., et al. 1985. *Ap. J. Lett.* 290: L5–8
Houck, J. R., Soifer, B. T., Neugebauer, G., Beichman, C. A., Aumann, H. H., et al. 1984. *Ap. J. Lett.* 278: L63–66
Huchra, J. 1977. *Ap. J. Suppl.* 35L: 171–95
Hunter, D. A., Gillett, F. C., Gallagher, J. S., Rice, W. L., Low, F. J. 1986. *Ap. J.* 303: 171–85
Hunter, D., Rice, W., Gallagher, J., Gillett, F. 1987. In *Star Formation in Galaxies*, ed. C. Persson, pp. 253–56. Washington, DC: US Govt. Print. Off.
Impey, C., Neugebauer, G., Miley, G. K. 1987. In *Star Formation in Galaxies*, ed. C. Persson, pp. 731–35. Washington, DC: US Govt. Print. Off.
IRAS Explanatory Supplement. 1985. Ed. C. A. Beichman, G. Neugebauer, H. J. Habing, P. E. Clegg, T. J. Chester. Washington, DC: US Govt. Print. Off.
IRAS Point Source Catalog. 1985. Washington, DC: US Govt. Print. Off.
Iyengar, K. V. K., Rengarajan, T. N., Verma, R. P. 1985. *Astron. Astrophys.* 148: 43–51
Jones, T. J., Hyland, A. R., Harvey, P. M. 1987. In *Star Formation in Galaxies*, ed. C. Persson, pp. 245–46. Washington, DC: US Govt. Print. Off.
Joseph, R. D., Wright, G. S. 1985. *MNRAS* 214: 87–95
Joseph, R. D., Wright, G. S., Wade, R. 1984. *Nature* 311: 132–33
Joseph, R. D., Wright, G. S., Wade, R., Graham, J. R., Gatley, I., et al. 1987. In *Star Formation in Galaxies*, ed. C. Persson, pp. 421–33. Washington, DC: US Govt. Print. Off.
Joy, M., Lester, D. F., Harvey, P. M., Frueh, M. 1986. *Ap. J.* 307: 110–15

Jura, M. 1977. *Ap. J.* 212: 634–36
Jura, M. 1982. *Ap. J.* 254: 70–74
Jura, M. 1986. *Ap. J.* 306: 483–89
Jura, M., Kim, D. W., Knapp, G. R., Guhathakurta, P. 1987. *Ap. J. Lett.* 312: L11–15
Kenney, J. D., Young, J. S. 1987. In *Star Formation in Galaxies*, ed. C. Persson, pp. 287–92. Washington, DC: US Govt. Print. Off.
Kennicutt, R. C., Roettiger, K. A., Keel, W. C., van der Hulst, J. M., Hummel, E. 1987. In *Star Formation in Galaxies*, ed. C. Persson, pp. 401–8. Washington, DC: US Govt. Print. Off.
Kidger, M. R., Beckman, J. E. 1986. *Astron. Astrophys.* 164: L25–28
Klein, U., Wunderlich, E. 1987. In *Star Formation in Galaxies*, ed. C. Persson, pp. 583–88. Washington, DC: US Govt. Print. Off.
Kleinmann, S. G., Keel, W. 1987. In *Star Formation in Galaxies*, ed. C. Persson, pp. 559–62. Washington, DC: US Govt. Print. Off.
Kronberg, P. P., Sramek, R. A. 1985. *Science* 227: 28–31
Kunth, D., Sevre, F. 1986. *Proc. Workshop on Star-Forming Dwarf Galaxies and Related Objects*, ed. D. Kunth, T. X. Thuan, pp. 331–49. Gif sur Yvette: Ed. Front. Transl. by J. Thanh Van
Landau, R., Golisch, B., Jones, T. J., Jones, T. W., Pedelty, J., et al. 1986. *Ap. J.* 308: 78–92
Larson, R. B., Tinsley, B. M. 1978. *Ap. J.* 219: 46–59
Lawrence, A., Walker, D., Rowan-Robinson, M., Leech, K. J., Penston, M. V. 1986. *MNRAS* 219: 687–701
Leger, A., Puget, J. L. 1984. *Astron. Astrophys.* 137: L5–8
Lonsdale, C., Persson, E., Matthews, K. 1984. *Ap. J.* 287: 95–107
Low, F. J., Beintema, D. A., Gautier, T. N., Gillett, F. C., Beichman, C. A., et al. 1984. *Ap. J. Lett.* 278: L19–L22
Lubin, P. M., Epstein, G. L., Smoot, G. F. 1983. *Phys. Rev. Lett.* 50: 616–19
Meiksin, A., Davis, M. 1986. *Astron. J.* 91: 191–98
Miley, G. K., De Grijp, M. K. 1985. In *Light on Dark Matter*, ed. F. P. Israel, pp. 471–86. Dordrecht: Reidel
Miley, G. K., Neugebauer, G., Clegg, P. E., Harris, S., Rowan-Robinson, M., et al. 1984. *Ap. J. Lett.* 278: L79–81
Miley, G. K., Neugebauer, G., Soifer, B. T. 1985. *Ap. J. Lett.* 293: L11–14
Moorwood, A. F. M., Veron-Cetty, M.-P., Glass, I. S. 1986. *Astron. Astrophys.* 160: 39–50
Neugebauer, G., Elias, J., Matthews, K., McGill, J., Scoville, N., Soifer, B. T. 1987. *Astron. J.* In press
Neugebauer, G., Habing, H. J., van Duinen, R., Aumann, H. H., Baud, B., et al. 1984. *Ap. J. Lett.* 278: L1–6
Neugebauer, G., Miley, G. K., Soifer, B. T., Clegg, P. E. 1986b. *Ap. J.* 308: 815–28
Neugebauer, G., Soifer, B. T., Miley, G. K. 1985. *Ap. J. Lett.* 295: L27–31
Neugebauer, G., Soifer, B. T., Rowan-Robinson, M. 1986a. In *Structure and Evolution of Active Galactic Nuclei*, ed. G. Giuricin, F. Mardirssion, M. Mezzetti, M. Ramella, pp. 11–19. Dordrecht: Reidel
Norris, R. P. 1985. *MNRAS* 216: 701–11
Norris, R. P., Baan, W. A., Haschick, A. D., Diamond, P. J., Booth, R. S. 1985. *MNRAS* 213: 821–31
Osterbrock, D. E., De Robertis, M. M. 1985. *Publ. Astron. Soc. Pac.* 97: 1129–41
Persson, C. J., Helou, G. 1987. *Ap. J.* 314: In press
Puxley, P. J., Hawarden, T. G., Mountain, C. M., Leggett, S. K. 1987. In *Star Formation in Galaxies*, ed. C. Persson, pp. 619–22. Washington, DC: US Govt. Print. Off.
Rees, M. J. 1984. *Ann. Rev. Astron. Astrophys.* 22: 471–506
Rengarajan, T. N., Verma, R. P. 1986. *Astron. Astrophys.* 165: 300–5
Rice, W. L., Boulanger, F., Soifer, B. T., Freedman, W., Viallefond, F. 1987a. Submitted for publication
Rice, W. L., Persson, C. J., Soifer, B. T., Neugebauer, G., Kopan, E. L. 1987b. Submitted for publication
Rickard, L. J, Harvey, P. M. 1984. *Astron. J.* 89: 1520–30
Rieke, G. H. 1987. In *Star Formation in Galaxies*, ed. C. Persson, pp. 633–41. Washington, DC: US Govt. Print. Off.
Rieke, G. H., Cutri, R. M., Black, J. H., Kailey, W. K., McAlary, C. W., et al. 1985. *Ap. J.* 290: 116–24
Rieke, G. H., Lebofsky, M. J. 1979. *Ann. Rev. Astron. Astrophys.* 17: 477–511
Rieke, G. H., Lebofsky, M. J. 1981. *Ap. J.* 250: 87–97
Rieke, G. H., Lebofsky, M. J. 1986. *Ap. J.* 304: 326–33
Rieke, G. H., Lebofsky, M. J., Thompson, R. I., Low, F. J., Tokunaga, A. T. 1980. *Ap. J.* 238: 24–40
Rieke, G. H., Low, F. J. 1972. *Ap.. J. Lett.* 176: L95–100
Robson, E. I., Gear, W. K., Brown, L. M. J., Courvoisier, T. J. L., Smith, M. G., et al. 1986. *Nature* 323: 134–36
Rodriguez-Espinosa, J. M., Rudy, R. J., Jones, B. 1986. *Ap. J.* 309: 76–79
Rodriguez-Espinosa, J. M., Rudy, R. J.,

Jones, B. 1987. In *Star Formation in Galaxies*, ed. C. Persson, pp. 669–74. Washington, DC: US Govt. Print. Off.

Rowan-Robinson, M. 1986. In *Light on Dark Matter*, ed. F. P. Israel, pp. 499–506. Dordrecht: Reidel

Rowan-Robinson, M., Crawford, J. 1987. In *Star Formation in Galaxies*, ed. C. Persson, pp. 135–48. Washington, DC: US Govt. Print. Off.

Rowan-Robinson, M., Walker, D., Chester, T., Soifer, T., Fairclough, J. 1986. *MNRAS* 219: 273–83

Sandage, A., Tammann, G. A. 1981. *A Revised Shapley-Ames Catalog of Bright Galaxies*. Washington, DC: Carnegie Inst. Washington

Sanders, D. B., Mirabel, I. F. 1985. *Ap. J. Lett.* 298: L31–35

Sanders, D. B., Scoville, N. Z., Soifer, B. T., Young, J. S., Danielson, G. E. 1987a. *Ap. J. Lett.* 312: L5–9

Sanders, D. B., Scoville, N. Z., Young, J. S., Soifer, B. T., Schloerb, F. P., et al. 1986. *Ap. J. Lett.* 305: L45–49

Sanders, D. B., Soifer, B. T., Neugebauer, G., Scoville, N. Z., Madore, B., et al. 1987b. In *Star Formation in Galaxies*, ed. C. Persson, pp. 411–20. Washington, DC: US Govt. Print. Off.

Sanders, D. B., Solomon, P. M., Scoville, N. Z. 1984. *Ap. J.* 276: 182–203

Sargent, A. I., Scoville, N. Z., Sanders, D. B., Soifer, B. T. 1987. *Ap. J. Lett.* 312: L35–38

Savage, A., Clowes, R. G., MacGillivray, H. T., Wolstencroft, R. D., Leggett, S. K., et al. 1987. In *Star Formation in Galaxies*, ed. C. Persson, pp. 537–45. Washington, DC: US Govt. Print. Off.

Schechter, P. 1976. *Ap. J.* 203: 297–306

Schmidt, M., Green, R. F. 1983. *Ap. J.* 269: 352–74

Scoville, N. Z., Sanders, D. B., Sargent, A. I., Soifer, B. T., Scott, S. L., Lo, K. Y. 1986. *Ap. J. Lett.* 311: L47–50

Scoville, N. Z., Young, J. 1983. *Ap. J.* 265: 148–65

Sekiguchi, K. 1987. In *Star Formation in Galaxies*, ed. C. Persson, pp. 507–11. Washington, DC: US Govt. Print. Off.

Sellgren, K. 1984. *Ap. J.* 277: 623–33

Sembay, S., Coe, M. J., Clement, R., Dean, A. J., Hanson, C. G., et al. 1985. *MNRAS* 216: 121–25

Shu, F. H. 1987. In *Star Formation in Galaxies*, ed. C. Persson, pp. 743–52. Washington, DC: US Govt. Print. Off.

Smith, B. J., Kleinmann, S. G., Huchra, J. P., Low, F. J. 1987. In *Star Formation in Galaxies*, ed. C. Persson, pp. 565–68. Washington, DC: US Govt. Print. Off.

Soifer, B. T., Helou, G., Persson, C. J., Neugebauer, G., Hacking, P., et al. 1984. *Ap. J. Lett.* 283: L1–4

Soifer, B. T., Rice, W. L., Mould, J. R., Gillett, F. C., Rowan-Robinson, M., et al. 1986b. *Ap. J.* 304: 651–56

Soifer, B. T., Sanders, D. B., Madore, B., Neugebauer, G., Danielson, G. E., et al. 1987b. *Ap. J.* In press

Soifer, B. T., Sanders, D. B., Madore, B. F., Neugebauer, G., Persson, C. J., et al. 1987a. In *Star Formation in Galaxies*, ed. C. Persson, pp. 523–30. Washington, DC: US Govt. Print. Off.

Soifer, B. T., Sanders, D. B., Neugebauer, G., Danielson, G. E., Persson, C. J., et al. 1986a. *Ap. J. Lett.* 303: L41–44

Solomon, P. M. 1987. In *Star Formation in Galaxies*, ed. C. Persson, pp. 37–59. Washington, DC: US Govt. Print. Off.

Spinoglio, L., Persi, P., Ferrari-Toniolo, M., Giovannelli, F., Bassani, L., et al. 1985. *Astron. Astrophys.* 153: 55–63

Stark, A. A., Knapp, G. R., Bally, J., Wilson, R. W., Penzias, A. A., et al. 1986. *Ap. J.* 310: 660–78

Telesco, C. M., Becklin, E. E., Wynn-Williams, C. G., Harper, D. A. 1984. *Ap. J.* 282: 427–35

Thronson, H. A., Bally, J. 1987. In *Star Formation in Galaxies*, ed. C. Persson, pp. 267–70. Washington, DC: US Govt. Print. Off.

Thronson, H. A., Telesco, C. M. 1986. *Ap. J.* 311: 98–112

Thuan, T. X., Martin, G. E. 1981. *Ap. J.* 247: 823–48

Tytler, D. 1986. *Ap. J.* In press

Urry, M. 1985. PhD thesis. Mass. Inst. Technol., Cambridge

Vader, J. P., Simon, M. 1987. *Nature*. In press

Veron-Cetty, M. P., Veron, P. 1985. *ESO Sci. Rep. No. 4*

Walterbos, R. A. M. 1986. PhD thesis. Leiden Univ., Neth.

Walterbos, R. A. M., Schwering, P. B. W. 1986. *Astron. Astrophys.* In press

Ward, M., Elvis, M., Fabbiano, G., Carleton, N. P., Willner, S. P., et al. 1987. *Ap. J.* 315: In press

Weedman, D. W. 1987. In *Star Formation in Galaxies*, ed. C. Persson, pp. 351–61. Washington, DC: US Govt. Print. Off.

Wilson, A. S. 1987. In *Star Formation in Galaxies*, ed. C. Persson, pp. 675–91. Washington, DC: US Govt. Print. Off.

Windhorst, R. A., Miley, G. K., Owen, F. N., Kron, R. G., Koo, D. C. 1985. *Ap. J.* 289: 494–513

Wood, P. R., Bessell, M. S., Whiteoak, J. B. 1986. *Ap. J. Lett.* 306: L81

Wright, G. S., Joseph, R. D., Meikle, W. P. S. 1984. *Nature* 309: 430–31

Wrobel, J. M., Neugebauer, G., Miley, G. K. 1986. *Ap. J. Lett.* 310: L11–14

Yahil, A., Walker, D., Rowan-Robinson, M. 1986. *Ap. J. Lett.* 301: L1–5

Young, E. T., Knacke, R. F., Joyce, R. R. 1972. *Nature* 238: 263

Young, J. S. 1987. In *Star Formation in Galaxies*, ed. C. Persson, pp. 197–215. Washington, DC: US Govt. Print. Off.

Young, J. S., Kenney, J. D., Lord, S., Schloerb, F. P. 1984. *Ap. J. Lett.* 287: L65–68

Young, J. S., Kenney, J. D., Tacconi, L., Claussen, M. J., Huang, Y. L., et al. 1986b. *Ap. J. Lett.* 311: L17–21

Young, J. S., Schloerb, F. P., Kenney, J. D., Lord, S. D. 1986a. *Ap. J.* 304: 443–58

Young, J. S., Scoville, N. Z. 1982. *Ap. J.* 258: 467–89

COMETS AND THEIR COMPOSITION

Hyron Spinrad

Astronomy Department, University of California, Berkeley, California 94720

1. INTRODUCTION AND THE BIG PROBLEM

Cometary science is an old topic, yet one in which significant advances have only been attained in the last few years, especially through the magnificent encounters of the International Cometary Explorer (ICE) spacecraft with comet P/Giacobini-Zinner in September 1985 and of the spacecraft "swarm" with P/Halley in March 1986.

This review attempts to incorporate the space-borne close-up but very brief looks at two periodic comets with the extensive ground-based science of this decade. Naturally the conventional observational and theoretical studies of comets have also been motivated by the Halley opportunity of 1985–86 (usefully coordinated by the International Halley Watch; Newburn & Rahe 1986). Whenever possible, I emphasize available or potentially available physical evidence on controversial or important questions.

To remind the reader that comets are also attractive and fascinating for photographic study, a modern plate of Comet P/Halley is shown in Figure 1. While a few comets are available each year for conventional observations within the Jovian orbit ($r \leq 5$ AU), the vast majority of comets are thought to be quasi-permanent residents of the extreme outer boundaries of our solar system. Or so we are told; the evidence on cometary origins and evolutionary history is indeed very vague and at some level a subject of controversy. Cosmic relic records have always provided astrophysicists with weak clues and difficult challenges, and the presumed history of comets is still more at the deductive level of a plausible theory than a nonspecialist reader might suppose. We first look into this subject and at least suggest some future observational programs that could be useful in differentiating extant cometary origin models. While many workers do

Figure 1 A blue-light photograph of Comet P/Halley, obtained by Drs. A. Dressler and R. Windhorst, with the 2.5-m duPont reflector at the Las Campanas Observatory (Chile) on 16.5 March 1986. Note the wavy ion-tail streamers: The tail and coma structure of Halley changed markedly from night to night in March 1986, presumably because of different rates of gas sublimation at the asymmetric and rotating nucleus.

agree that comets are primitive frozen relics of the original volatile material taking part in an episode of typical star formation (the Sun's birth), this general genesis for comets is not totally accepted. Extra–solar system or "recyclable" inner solar system origins have been proposed too; clearly a long-term scientific goal is to clarify the origin of comets, but this is not an easy task. This review first discusses the origin theories, then moves to empirical studies of cometary nuclei from the ground and from space. We also review gas production rates, the cometary ionosphere, and future expectations.

An interested reader can, of course, find other valuable review material to supplement this contribution. I would recommend the Whipple & Huebner (1976) review in this annual series; the articles by Wyckoff (1982), A'Hearn (1982), and Delsemme (1982) in *Comets* (ed. L. L. Wilkening); and a general perusal of two more modern books, *Asteroids, Comets, Meteors II* [Uppsala University Press, 1986] and *Cometary Exploration*, a three-volume edition produced by the Hungarian Academy of Sciences (1983) after a 1982 conference in Budapest. By the time this *Annual Review* is published, we can expect to also be reading the proceedings of an exciting international conference directed toward the exploration of Halley's comet. The 20th ESLAB Symposium, held in Heidelberg in late October 1986, is publishing proceedings on the "The Exploration of Halley's Comet" (ESA, 1987). It should be a very important contemporary volume.

Finally I note two lengthy monographs on cometary physics that aided me substantially in the preparation of this review. They are by Mendis et al. (1985) and by Fernandez & Jockers (1983).

2. CLUES ON THE ORIGIN AND PERMANENCE OF THE OORT COMETARY CLOUD

This theoretical subject is well disguised in the haze of time. But there are new physical clues worth mentioning. The types of theories can be categorized on the basis of two distinct themes. *First* and most popular is the normal star-formation theory, if I may call it that. Comets formed *somewhere* in the outer solar system contemporarily with the Sun and protoplanets. They have been "stored" at even larger heliocentric distances ($r \geq 10^4$ AU, at least) in an "Oort cloud" (or clouds) for about 4.5×10^9 yr. A typical cloud comet will now have a semimajor axis $a \cong 10^5$ AU.

The popularity of the original theory by Oort (1950) is mainly (and probably correctly) due to his classic observation that the $1/a$ distribution of *new comets* is very asymmetric. Oort noted that the histogram spike of $1/a$ near zero (cometary semimajor axis *exceeding* 10^4 AU) must represent the source of new and long period comets. He then suggested that a fairly massive outer cloud of comets surrounded our solar system, and that the observed ($r \leq 10$ AU) comets were the mere trickle of objects perturbed to enter the Jovian region of the system by the impulses supplied by randomly passing stars. Of course, when this happens, many comets can escape the system and thus deplete the Oort cloud. [Modern modeling by Weissman (1982) gives a breakdown of the loss mechanisms, which are listed as "end states" for perturbed comets as a function of the rms velocity perturbation.] The original estimate of the cloud population comes from an extrapolation of the statistics of actually observed new comets. Weissman's (1982, 1985) Monte Carlo modeling of the Oort cloud follows Oort's general conclusions but requires a larger number of comets—roughly 10^{12} of them with a total mass of $\approx 10^{29}$ g.

The *second* main theme is that our present-day comet cloud was really formed in the Galactic interstellar medium (ISM) and is a relatively recent capture-prize of the Sun. The scenario deemed likely by Clube & Napier (1982, 1984), and slightly revised by Bailey (1983), proposes a series of comet clouds lost and occasionally recaptured as the Sun passed through Galactic giant molecular clouds (GMCs). This is a very controversial stance, and Weissman (1985, 1986) has argued that the perturbations and tides of the GMCs, which could disturb or conceivably strip away the outer Oort cloud, may not be catastrophic to the older theory. The GMCs in the Galactic plane can be fairly important tidal perturbers of the classical Oort cloud at $r \sim 10^5$ AU, and their tides/encounters might make a dense *inner cloud* of comets a necessity.

Another long-lived mechanism to perturb comets in the Oort cloud is the Galactic tidal effect proposed by Heisler & Tremaine (1986) and Delsemme (1986). The tidal source of the Galactic disk slowly changes a distant comet's angular momentum. This will be true no matter how the comets originally filled the Oort cloud. We now examine these two concepts in more detail; where appropriate, I suggest observational tests that could be useful in discriminating between the competing scenarios for comet formation and their dynamical evolution.

2.1 The Star-Formation Origin For Comets— Where Do the Cometesimals Start Out?

Dynamically, and perhaps physically, the favorite "spot" for forming comets in the proto–solar nebula is near the Uranus-Neptune zone (Fernandez 1980), some 20–30 AU from the Sun. Perturbations by the proto-Neptunian group could move the young comets out to the distance of the conventional Oort cloud. On the other hand, Cameron (1973), Hills (1981), and Weissman (1986) have advocated cometary formation far out in the solar system at $r \approx 10^2$–10^3 AU. Unfortunately, it is not possible to discriminate the cometary formation zone from *dynamical observations* today, but perhaps we can do so from physical observations or by a reckless analogy to other possibly forming solar systems within 100 pc of the Sun.

The physical argument depends upon measurement of the chemical composition of the ice composing much of the typical cometary nucleus (and, hopefully, the survival of the early imprinted thermal history in these cometary snows). It is likely that the cometary ice is a mixture of volatile gases condensed out either in the ISM before the formation of the solar system or in the early preplanetary disk as "cometesimals." The depletion of the most volatile gases (H_2, He, Ne) and the next-most volatile diatomic species, such as S_2 and N_2, would presumably be due to collisional heating of the precometary grains and snows by collisions with gas molecules.

All modelers of comets agree that water is the dominant volatile molecule in the nuclei of the comets we can observe; the relatively high evaporation temperature for *water-ice* in space ($T \leq 150$ K, $r \leq 10$ AU in sunlight coupled with a low albedo) makes its permanence in comets a nondiscriminant for cometary origin models. Our main interest here is to differentiate the condensates in the Uranus-Neptune zone ($T \sim 50$ K) from those of a possible origin at a considerably larger heliocentric distance, say $r \geq 100$ AU. To test whether comets might have included gases even more volatile than CO_2 ($T \sim 80$ K), we have listed in Table 1 the potential *parent molecules* and some abundant atomic gases with the lowest sublimation temperatures, following Yamamoto (1985), Greenberg et al. (1986), and Delsemme (1982).

Methane (CH_4) would also be easily lost to a warming cometesimal, but it could be chemically bound up in the water-ice-lattice trap of a clathrate, so its effective vaporization threshold *might* be higher than the nominal 31 K ($r_0 \sim 40$ AU). We ignore this possible chemical complexity in the discussion to follow.

Now, what can we say about the presence of some of these volatile species in comets? Ideally we would like to physically analyze a *new* comet, one that is entering the inner solar system for the first time. That wish has not proven possible to date.

Spectroscopy of long-period comets from the ground has, as previously noted, shown water to be the dominant coma species—presumably also dominating the nuclear ices. However, Feldman (1983) and Woods et al. (1986) have recently argued that CO could be a fairly abundant parent molecule in some comets (including P/Halley). Carbon monoxide can be observed by its fluorescence emission band near $\lambda 1500$ Å in the rocket-ultraviolet; special rocket flights have been required to show the bands clearly. CO is most readily detected in cometary spectra by its ion; the

Table 1 Sublimation temperatures for possible cometary constituents

Gas/molecule	T_s (K)	r_0 (AU)[a]
H_2	5	>1000
Ne	7	700
S_2	20	90
N_2	22	78
Ar	25	63
CO	25	63
H_2O	160	4

[a] Values of r_0 from Delsemme (1982); r_0 is the heliocentric distance beyond which the gas vaporization becomes negligible if 2.5% of the available solar flux is used for vaporization.

blue-violet CO^+ bands are, in fact, the dominant emitters in the blue-region spectra of almost all cometary ion tails (Wyckoff 1982). These CO^+ bands were particularly strong in the spectra of Comet Morehouse (1908) and more recently for Comet Humason (1961) and Comet West (1976). However, no neutral molecular ratios of CO/H_2O have been calculated from the few quantitative measurements; the f values for the ionic lines are in poor shape too (Meyer-Vernet et al. 1987). It is of course conceivable that the observable CO and CO_2/H_2O ratio may be a function of a comet's recent thermal history (Houpis et al. 1985). In this scenario our suggested analysis would not well reflect the bulk (primordial) condition of the comet nucleus. Indeed, Strauss et al. (1986b) suggest that the CO^+/H_2O^+ ion-line ratio in P/Halley may have been higher for their available postperihelion spectra than it was before perihelion.

Nevertheless, *neutral* CO has been detected for three comets now from the UV bands (Feldman 1986, Festou et al. 1986, Woods et al. 1986). Using Feldman's new fluorescence efficiencies, Festou et al. calculate a large relative abundance of CO for P/Halley ($\gtrsim 10\%$ that of H_2O), while the rocket spectra of Woods et al. are modeled to suggest an even higher (10–20%) relative production rate of CO compared with that of water! It may be a major constituent. The critical question is then, is CO a true parent molecule, coming directly from the subsurface of the Halley nucleus, or is it a photodissociative product from CO_2? (The spacecraft observations of P/Halley in March 1986 definitely showed a minor contribution by neutral CO_2.) Also, is the CO distributed throughout the Halley solid nucleus? The latter query is answerable indirectly for P/Halley; this frequent visitor to the inner solar system must have lost its outer mantle to a depth of meters many times over on its previous 76-yr period passes. If CO is *an abundant parent*, as seems to be indicated by the spatial dependence of the CO bands measured by Woods et al., then we have a strong clue that the main body of comet Halley was not heated to over 25 K during its formative years.

Another constraint is provided by the observation of the S_2 emission bands ($\lambda 2900$ Å) in the International Ultraviolet Explorer (IUE) spacecraft spectra of the Earth-grazing small comet IRAS-Araki-Alcock (1983d) (A'Hearn et al. 1983, Feldman et al. 1984). These authors argue that S_2 is a parent molecule in comets, but that it has been detected only in 1983d because of the S_2 molecule's very short lifetime before photodestruction (450 s at $r = 1$ AU). For a typical outflow velocity of 1 km s^{-1} and a typical geocentric distance of 0.7 AU or so, an angle of 1″ subtends 500 km at the comet, so angular resolution limits are critical in the detection of S_2 emission features; this explains the singular measurement for 1983d. (It will be important for cometary observers using the Hubble Space

Telescope to continue the S_2 search with high angular/spectral resolution on comets approaching Earth after 1988 or so.)

Since Comet IRAS-Araki-Alcock is also periodic, its "1983 surface" was not always exposed to the solar ultraviolet radiation or to Galactic cosmic rays. Thus, processes that could recently manufacture S_2 are not likely to be important; following Greenberg et al. (1986), we consider the presence of free S_2 ($T_s = 20$ K) a serious blow to the origins of cometary snows at heliocentric distances under 100 AU. Models for the aggregation of cold interstellar "dust" grains (made up of these snows plus some veneer of refractory atoms) could also satisfy our requirement of a very low-temperature condensation site for cometary bodies. All of these molecular sublimation arguments, of course, rest on simple ideas of thermodynamics; if water-lattice chemistry holds the minority of more volatile species intact until the much higher H_2O sublimation point is attained, then my discussion here would become largely superfluous.

To be really speculative about this matter, we pose the following question: Is there any chance to detect the noble gases, which would leave a forming cometesimal at even lower temperatures than 20 K? The ground-state spectral transitions of neutral neon and argon might show up in fluorescence in deep-UV cometary spectra ($\lambda 736$ Å for Ne, $\lambda 1048$ Å for Ar). Mass spectrometry of a comet coma from a future spacecraft mission could also yield a mass distribution or mass/charge ratio—unfortunately, the most abundant argon isotope, ^{36}Ar, has the same molecular mass as the $^{12}C_3$ radical, which is common in comets.

A quite different and indirect type of information on the possible location of "dusty" cometesimals in the early solar system comes from recent studies of the A5 V star β Pictoris. β Pictoris is one of the few main-sequence stars with a noticeable infrared excess, as found by the IRAS satellite (Aumann 1985; M. Cohen, private communication, 1986). It is thought that a large IR excess (say, at 25–60 μm) indicates a dust envelope around the parent star; in the case of β Pic, Smith & Terrile (1984) found much more. They obtained a spectacular visual-light CCD image, using a coronographic star-blocker, that clearly shows substantial non-axisymmetric emission extending away from the star. In fact, this emission can be traced to a distance of 400 AU from the star! The β Pictoris extension has been interpreted as a disk-shaped protoplanetary system, seen in the optical region now by "dust"-scattered starlight. Presumably both the disk and the A5 V star are relatively young.

The connection to our Sun's family of comets is, of course, indirect speculation. The disk mass of β Pic is *roughly estimated* at 200 $M_\oplus$ (Smith & Terrile 1984, Weissman 1986), perhaps twice the mass estimated now for the hypothetical *inner cloud* of comets suggested by Hills (1981) and

Weissman (1986). This inner comet cloud is not directly observable, but it could lie from 100 to 10^4 AU from the Sun and provide a safe reservoir for comets, one that would be relatively invulnerable to the tides of passing GMCs. Whether a small fraction of the $\geq 10^{12}$ inner comets can be occasionally perturbed to populate the conventional Oort cloud at $r > 0.1$ p.c. ($>2 \times 10^4$ AU) is beyond the scope of this paper, but Hills' (1981) calculations suggest that a combination of planetary and occasional stellar perturbations can do just that. Anyway, the key point in the β Pic analogy is the large extension of the disk, 100–400 AU from the star. At 200 AU from an A5 V star, the radiation balance temperature is about 36 K. Thus, the arguments in this section slightly favor an origin beyond 100 AU for the comets now populating the far outer reaches of our solar system.

The "hybrid" cometary origin theory of O'Dell (1986) is worthy of brief review here. O'Dell supposes that icy comets have formed around small asteroidal cores and dust particle accumulations, which move in permanent, albeit very eccentric orbits with aphelia at $r \geq 10^4$ AU. The virtue of his concept is that it requires about 10^6 (not 10^{12}) comets to live in the outer solar system. O'Dell's eccentric-orbit comets all return. The unusual aspect of the theory is that the dust grains/small asteroids have to continually accumulate a frosting of snow and ice from the present-day interstellar medium (ISM) through which the Sun is plunging at 18 km s^{-1}. If this accretion were to work, the comet's snow mantles would possess isotope ratios typical of the current Galactic ISM rather than those original to our solar system, well imprinted some 4×10^9 yr ago. We discuss in the next subsection future ways to differentiate the isotope ratios found in the "permanent" solar system rocks from those now extant in molecular clouds of the ISM.

2.2 Cataclysmic Theories of Cometary Origin: Replenishment Required

Clube & Napier (1982, 1984, 1986) argue for virtual destruction of the Oort cloud by GMC encounters. The discussion is difficult, since our knowledge of typical molecular cloud parameters is still limited. But it seems likely that "distant" encounters could play a role in disturbing the "conventional" outer cloud boundary, near $r \sim 10^5$ AU.

We all recall the traditional concern and objection to the hypothesis of a direct ISM origin for comets: that there are *no* observed hyperbolic orbits. But the GMC cataclysmic hypothesis implies just that; periodic loss of the Oort comet cloud to the Galactic ISM must then imply that the Sun regains a new comet cloud—either from the GMCs themselves (unlikely) or possibly from a less-understood inner-cloud comet reservoir.

The passage of the Sun through the Galactic plane may also modestly disturb the Oort cloud, but according to the most recent work of Clube & Napier (1986), about 10 *penetrations*, or at least near-encounters (heavy tides), with GMCs have taken place over the last 4.5 Gyr. Clube & Napier suggest that these events have caused a significant cometary depletion beyond 10^4 AU, while Hut & Tremaine (1985) find a smaller effect.

If cometary replenishment were to be from an *external* source, then these comets should also carry the isotope ratios measured contemporaneously in Galactic molecular clouds. The whole idea is naturally very antiestablishment, but it should be considered.

We can measure some cometary isotope ratios today, using quantitative spectroscopy, but not as well as we would like. The $^{12}C/^{13}C$ ratio is available through the strong C_2 and CN bands, but the literature contains only three $^{12}C/^{13}C$ isotope ratios measured in modern times. The two most reliable published ratios for Comet Tago-Sato-Kosaka (Owen 1973) and Comet Kohoutek (Danks et al. 1974) both exceed 100. (Recall the terrestrial/inner solar system ratio of 89 ± 2.) However, the uncertainties are fairly large in the cometary spectroscopy, and the isotope ratios could be plagued with systematic errors (cf. Lambert & Danks 1983). The observational situation will certainly be strengthened when the extensive extant P/Halley data base is further analyzed (Wehinger et al. 1986). Of course, we know next to nothing about what chemical fractionation in the $^{12}C/^{13}C$ ratio might occur during the accumulation of the C_2 and CN parent molecular species.

The current situation regarding Galactic molecular cloud isotope ratios is also a bit "messy," but not because of a dearth of microwave line data. The transitions of $H_2^{12}CO$ and $H_2^{13}CO$ yield a galactocentric gradient of $^{12}C/^{13}C$ that goes from 50 to 4 kpc from the Galactic center to a value of 80 beyond the solar circle. There is scatter, but a good mean ratio for the Sun's galactocentric distance would be near 75 for the formaldehyde data summarized by Churchwell (1983). However, Wannier (1980) made H_2CO, $^{12}C^{34}S/^{13}CS$, and $C^{18}O/^{13}CO$ radio observations that did not show a gradient across the Galaxy, but (important for our present concern) they did show a "mean" carbon isotope ratio of 70 for *today's* ISM. So, if we could be sure that the cometary ratio exceeded 100, the hypothesized Oort cloud replenishment from the contemporary ISM would be physically ruled out.

3. INDIRECT INFERENCES ON THE COMETARY NUCLEUS: OBSERVATIONS FAR FROM THE SUN

We badly need to understand the cometary nucleus—the small, but permanent structure that gives rise to all we see. A worthwhile long-term goal is the study of a *distant new* comet, far enough from the Sun to still maintain its pristine surface. For now, even the discovery of such a visitor would involve plenty of luck! (We discuss such possibilities at the end of this review.)

Right now, we have some sketchy data on just a few "bare-nucleus" comets: P/Halley at $r > 5$ AU, and a few, very inactive "wimpy" comets that have a sufficiently low output of gas and dust so that their comae are faint. In the latter cases the small, solid-body nucleus makes a measurable contribution to the reflected sunlight or reemitted IR thermal radiation. However, I emphasize that *generally*, even at $r \geq 3$ AU, apparently faint and almost "stellar-looking" comets still have *dust comae* that *overwhelm* the radiation coming from the tiny (1–10 km) nucleus.

A key to the recognition of the bare nucleus is a periodic light curve. (I thank Dr. M. Belton for his emphasis of this point.) We know now from ground-based studies of comets P/Arend-Rigaux, P/D'Arrest, and P/Neujmin 1 (three short-period comets with low activity levels) and from the spacecraft images of P/Halley that comets have very *nonspherical* nuclei. They vary in the brightness of their reflected nuclear light, in part simply because of a changing geometric cross section presented to the Sun and Earth. I list in Table 2a some rotational periods for these small comets; I was disappointed to note the discordant periods (0.28, 0.56, 1.13) determined by the three observing teams working on P/Arend-Rigaux. However, note that the two shorter periods are the 1/4 and 1/2 submultiples of the longest one, so perhaps agreement may be reached when the observers combine all the photometric data. In such a case, without a

Table 2a Cometary nuclear rotation

Comet	Observers	Technique	Period (days)
P/Neujmin 1	Wisniewski & Fay 1985	Photometry with photomultipliers	1.05
P/D'Arrest	Fay & Wisniewski 1978	Photometry with photomultipliers	0.215
P/Arend-Rigaux	Jewitt & Meech 1985	CCD photometry	0.28
P/Arend-Rigaux	Wisniewski et al. 1986	Photometry with photomultipliers	0.57 or
P/Arend-Rigaux	Wisniewski et al. 1986	Photometry with photomultipliers	1.13
P/Arend-Rigaux	A'Hearn 1986	Photometry with photomultipliers	0.56

detailed comparison or overlap of the data, the longer period is generally the preferred true rotational period.

The situation for P/Halley also remains confused at this writing; the rotational modulation found by Belton et al. (1986) appears to be responsible for only part of the total cometary light fluctuations at large heliocentric distances (Jewitt & Danielson 1984, West & Pederson 1984, Meech et al. 1986). More puzzling and provocative is the newly recognized multiple periodicities (or at least activity modulations) measured by ground-based coma observers of P/Halley near its 1986 perihelion. There is a dichotomy in the results (see Table 2*b*); *perhaps* the active Halley nucleus is both rotating (2.2 day?) and precessing (7.4 day?), so that some active region is exposed to sunlight and sends out strong jets of dust and gas into the inner coma in a clearly modulated fashion. The nucleus itself is very oblong in shape. Photometric study of P/Halley as it leaves the inner solar system in 1987 and 1988 may resolve this question.

We note parenthetically that there is physical importance to the cometary rotation period determinations; for example, the fact that P/Halley is a *slow* rotator ($P > 1$ day) implies a larger total gas sublimation from the warm sunlit hemisphere than would be the case if it rotated rapidly (Whipple & Huebner 1976, Weissman & Kiefer 1981, Belton 1985).

Compared with asteroids of 5–10 km radius, comets are moderately slow rotators. Burns & Tedesco (1979) show a *mean* rotation period of 1/3 day for their sample of small asteroids; presumably these asteroids are rocky Apollo/Amor objects and are not necessarily representative of the main asteroid belt. No small asteroids rotate as slowly as does P/Halley, but we are not sure whether this is owing to the presumed spin-up of little asteroids by collisions or to the cometary jet action changing their original spin vectors. The connection between cometary and asteroidal rotations has not yet led to a useful physical interrelationship.

We can discuss cometary nuclear sizes in a somewhat model-dependent way now; observations of comets when they are "quiet," far from the Sun, yield area-albedo products (Spinrad et al. 1979). To apply this procedure to the more physically interesting concept of an effective nuclear size R_e, we must measure the albedo. The situation becomes tractable for asteroids when thermal IR measures are obtained, but 10–14 μm photometry of very faint bare-nuclei comets presents a far greater technical challenge to IR observers. These hurdles have been overcome now for several small comets without obvious comae (A'Hearn 1986, Hanner et al. 1985, Tokunaga & Hanner 1985). The albedos of the nuclei of P/Arend-Rigaux and P/Neujmin 1 are well determined and are remarkably low ($p_v = 0.02!$). Such low values make these comets darker than any other solar system object.

Table 2b Rotational dichotomy table for P/Halley

Set	Reference	Technique	P_{mod}[a] (day)	Comment
1	Kaneda et al. 1986	Lα "breathing"	2.2	Conflict with Set 7
2	Sekanina & Larson 1986a,b	Dust envelope shapes	2.2	Assumes $v_{ej} \approx 1$ km s^{-1}
3	Belton et al. 1986	Fourier analysis of photometry	2.2	Also some 7^d.4 amplitude
4	Schlosser et al. 1986	CN shells	2.2	Recurrent shell expansions; imprecise
5	Sagdeev et al. 1986a,b	Vega nuclear orientation	2.2	Comparison, Vegas 1 and 2
6	Keller et al. 1986	Vega plus Giotto orientation	2.2	Comparison of Giotto and Vegas
7	Stewart 1986	Lα photometry	7.4	Conflict with Set 1
8	Festou et al. 1986	Analysis of distant photometry	7.4	May allow 2^d.2 also
9	Schleicher et al. 1986	Photometry at encounters	7.4	Production rates modulated
10	Millis et al. 1986	Photometry after perihelion	7.4	Production rates modulated

[a] P_{mod} is the modulation period; it may or may not be the true axial rotation P of the cometary nucleus.

Equation 1 describes the formula for the cometary area-albedo determination; following Spinrad et al. (1979), we use

$$R_e^2 p_v \varphi(\alpha) = \frac{F_0}{F_\odot} r^2 \Delta^2, \qquad 1.$$

where F_0 is the reflected flux from the comet, $F_\odot$ the flux from the Sun, $\varphi(\alpha)$ is a phase function for the cometary light scattering (usually assumed to be like that of an asteroid), and R_e is the effective cometary radius.

Why is the measured albedo so low? A mantle of very black "dust" may cover a periodic comet's surface, especially before it becomes actively subliming (Brin & Mendis 1979). Of course, many of the Halley flyby data have suggested a consistently low albedo for much of that comet's nonactive surface.

In any case, we list in Tables 3a and 3b the area-albedo products (really just $p_v R_e^2$) and the effective radii for P/Halley and for several distant, less-active comets with photometry that should be safe. The persistent danger that there may be a coma component additional to the nuclear reflection makes the radii of Table 3b reliable upper bounds (for $p_v = 0.02!$). Since the photometric data in Table 3a are not wholly independent, the apparent agreement seen for the Halley radius is slightly misleading.

I advocate a mean $p_v R_e^2$ of 1.0 km^2 and thus a R_e value of 7.0 km as a working average for the distant P/Halley *bare-nucleus* size. Recall that the last step *assumes* that $p_v = 0.02$; an albedo of 4% would decrease R_e to 5.0 km. We compare this value with the spacecraft size measures (at $r = 0.9$ AU) in the next section and find that the agreement is quite good.

Of course, P/Halley is the high-water mark in this tabulation, for we now *know* that the comet's dimensional measures yield effective radii satisfyingly close to our Table 3a value for $p_v = 0.02$. The Giotto volume is consistent with $R_e = 6.2$ km, although the comet is *very* nonspherical!

We note in this conclusion of our discussion of measurements for com-

Table 3a P/Halley $p_v R_e^2$ photometry ($r > 5$ AU)

Set	Reference	$p_v R_e^2$ (km^2)	R_e ($p_v = 0.02$) (km)
1	Belton 1985	1.1	7.2
2	Hughes 1985	1.5	8.5
3	Spinrad et al. 1986	0.8	6.5
4	Wyckoff et al. 1985	1.5±0.3	8.5
5	Jewitt & Danielson 1984	1.5±0.4	8.5
6	Meech et al. 1986	0.97±0.03	7.0

Table 3b Effective radii for selected comets

Comet	R_e (km)	Reference	Notes
P/Encke	3.5	Mendis et al. 1985	Disagrees somewhat with radar (Table 4)
P/Arend-Rigaux	5	A'Hearn 1986	Actually measured to have $p_v = 0.02$
P/Arend-Rigaux	5	Brooke & Knack 1986	
P/Neujmin 1	9.5	A'Hearn 1986	
Meier	13	Fernandez & Jockers 1983	Large gas production earlier
Wirtanen (1957 VI)	~50	Mendis et al. 1985	Giant comet?
Schwassman-Wachmann 1	48	Mendis et al. 1985	Giant comet?

etary nuclei that two comets have passed close enough to Earth to allow coarse radar cross sections to be observed. Table 4 lists these radar data. Again, likely nonspherical nuclei make these radii a bit artificial.

The radii measured for P/Encke by the two quite different techniques we have discussed are discordant, although clearly it is a small comet. One way to "force" size determination equivalence for P/Encke would involve a higher albedo surface for the Encke's snows or some evidence that the nucleus flux (magnitude) used in Equation 1 by Roemer (1966) and tabulated by Mendis et al. (1985) was coma contaminated.

On the other hand, the few *possible* giant comets listed in Table 3b are potentially exciting objects; we never come close to them, but such objects might be the harbingers of the size distribution tail of comets. Conceivably some comets could even have 100-km radii!

Finally, we can use these radii to estimate cometary masses. (There are *no direct* measurements!) Only weak hints on mass can come from the nongravitational forces, which can change the cometary motion (Yeomans 1986). If we compute the mass for an $R_e = 7$ km sphere of mean density ~ 0.5 g cm^{-3}, parallel to the procedure advocated by Hughes (1985), our result is $M_c = 1 \times 10^{18}$ g. More generally, for P/Halley one should use

Table 4 Radar measures of R_e

Comet	R_e (km)	Reference
P/Encke	1:	Kamoun et al. 1982
IRAS-Araki-Alcock	4:	Goldstein et al. 1984

$$M_{\text{Halley}} = 6 \times 10^{15} \, \rho p_v^{-1.5} \text{ g}. \qquad 2.$$

Recall that one Earth mass is given by $m_\oplus = 6.0 \times 10^{27}$ g. It would make cometary astrophysics easier if the cometary brightness close in (e.g. its activity level) were closely tied to the total cometary mass. There may or may not be any relationship between M and "current L," but for exploratory purposes I compare in Table 5 cometary "absolute magnitudes" H_{10}, based upon observations and an activity model, with their indicative *volumes* (M_c).

A simple way to parameterize a comet's brightness, as seen from Earth while it is active, is to use the power law

$$M(r) = M_0 + 2.5n \log r + 5 \log \Delta, \qquad 3.$$

from Fernandez & Jockers (1983) following many previous authors. Here M_0 is the magnitude at $r = 1 = \Delta$. Then, *assuming* $n = 4$ (a typical activity power-law index), we find H_{10} to be the corresponding magnitude for M_0.

In this admittedly small sample, there is no indication that L(active) $\sim M^{2/3}$, a relation that would be expected if the activity were simply proportional to the cometary surface area. The comets apparently do not have fully "usable" surfaces that can be expected to sublime evenly in the sunlight. In P/Halley, jet activity, rather than general sublimation, accounted for much of the gas and dust outflow.

There is a bit more physical information of importance to understanding bare nuclei; I briefly review here the reflection spectrum of P/Halley at the largest heliocentric distances, where it showed at least some rotational modulation (the Belton "nucleus-observed" criterion).

Belton (1985) and Spinrad et al. (1986a) describe the two available low-resolution optical-region spectra of P/Halley at $r = 8.78$ and $r = 7.96$ AU. The spectra are, of course, very noisy, since the bare nucleus was very faint. The spectral data at $r \simeq 5$ AU are somewhat coma contaminated. The earlier pair of spectra were taken when P/Halley's nuclear magnitude was $V = 23.3$ mag, using the Kitt Peak 4-m reflector and a cooled CCD

Table 5 Cometary absolute magnitudes and indicative volumes

Comet	H_{10} (mag)	R_e (km)	R_e^3 (km³)
P/Encke	9	2:	8:
P/Arend-Rigaux	11	5	125
IRAS-Araki-Alcock	10:	4	64
P/Halley	4	7	343

detector behind a low-resolution spectrograph. These observations were a difficult technical task.

These Halley spectra are only continua, but with a well-defined shape of their reflectivity, compared with the solar spectral energy distribution. The shape of this reflectivity curve is quite peculiar; it would look green to the eye, being depressed at both the blue end ($\lambda \sim 4600$ Å) and red end ($\lambda > 7000$ Å) compared with the curve of a G2 V star. This reflection spectrum is quite unlike that of most other small solar system objects, and it has not yet been satisfactorily interpreted. These unique spectra were obtained under difficult conditions, and while they appear satisfactory to me on a second critical analysis, it is impossible to check them independently. As Belton (1985) states, they could be "highly significant or totally worthless!" Probably the slightly coma-contaminated spectra at $r \sim 5$ AU will help us make the decision. It is unclear as to whether a color comparison with the spacecraft data (while the comet surface was active at $r = 0.9$ AU) is relevant. In any case, the images taken by the cameras on the Vegas and Giotto suggest a rather flat reflection spectrum; more analysis on this subject is underway. An asteroidal color comparison with P/Halley seems requisite here, but it may be aphysical. Most asteroids are slightly to substantially redder than the Sun; there is one small, eccentric-orbit asteroid (2201 Oljato) whose reflective albedo does resemble the "convex" P/Halley shape (McFadden et al. 1984). These authors find 2201 Oljato quite different from most of their other small asteroids in spectral reflectance, so perhaps there is a cometary connection.

4. NUCLEAR STUDIES OF P/HALLEY BY SPACECRAFT: A GLANCE AT THE COMET'S "THROBBING HEART"

We are fortunate to live in the era when direct exploration of comets is beginning! The International Cometary Explorer ventured through the tail of P/Giacobini-Zinner in September 1985, and two Japanese probes [Sakigake and Suisei] crept near P/Halley in early March 1986. The exciting and informative close Halley encounters of the two Soviet spacecraft Vega 1 and 2 on 6.3 and 9.3 (UT) March 1986 and the extraordinary close passage of Giotto, the armored European spacecraft, to within 600 km on 14.0 March 1986 were epoch-making events. It is rather too early for the resulting science to be "well digested"; however, the initial colorful Halley issue of *Nature* (Vol. 321), which describes the experiments on board the various spacecraft, is a good preview of the extensive and integrated results that we can probably expect in 1987 and 1988. To temper our immediate enthusiasm slightly, recall that we were studying the partly active nucleus of a well-worn (albeit large) periodic comet. Some day it will be necessary and even more valuable to obtain a close-up view of a *new* comet.

The two Vega spacecraft observed P/Halley's nucleus from a minimum range of about 8500 km; the CCD television systems on board showed that the actual solid surface of the nucleus was difficult to differentiate from the bright overlying clouds of dust and gas. Many of the images presented by Sagdeev et al. (1986a,b) are dominated by active jets of dust that appear very asymmetric. Only on heavily processed early images does the dark nucleus ($p_v \leq 0.04$) show with any contrast. At the Heidelberg meeting on Halley in October 1986, the Vega investigators showed that they had done a quite sophisticated processing job. The Halley nucleus appears to be of a "peanut" shape, 14 ± 1 by 7.5 ± 1 by 7.5 ± 1 km being the dimensions of the principal axes seen by the Vegas. The Vega investigators were also the first to directly observe the low albedo and the large size of the heart of P/Halley. A rotation period of 53 ± 3 hr (2.2 day) was also obtained from comparison of Vega 1 and 2 images (Sagdeev et al. 1986a,b).

One key measurement from Vega 1 was that of the high thermal infrared (6–11 μm) brightness at the P/Halley nucleus (Combes et al. 1986). A nuclear temperature *over* 300 K was observed; it seems to arise at the solid nucleus, and in fact it could come from a small fraction of the total surface area. Since this temperature is well above that expected for an icy surface in sublimation at 0.8 AU from the Sun, it may represent the equilibrium temperature of the dark dust. A surface with a very low thermal conductivity is implied.

The Giotto spacecraft encountered Halley a few days later; its daring close approach did lead to communication loss (at -14 s from its closest approach), so the best imaging data on the nucleus were obtained a fraction of a minute earlier. Excellent images (see Figure 2) were obtained at a range of a few thousand kilometers (Keller et al. 1986).

These images reveal an elongated nucleus having a major-axis length of 16 km, and minor-axis lengths of 8 and 7.5 km. There is some residual size uncertainty because of a bright dust jet that partially obscures the sunward side of the minor axis. A last gasp at spherical geometry yields R_e(Halley) = 6.2 km. The nuclear surface appears irregular and shows cratering; the raggedness of the terminator indicates a moderately rough surface, and a hill was seen as well. The processed Giotto images (Schwarz et al. 1986) show one large, shallow crater on the nucleus that is about 1.5 km in diameter. Again the visual albedo is found to be low (3–4%).

The most striking extranuclear features imaged were two bright jets of dust emanating from the sunlit side of the Halley nucleus. There are about five other, fainter jets also detected (Keller et al. 1986). Most of the

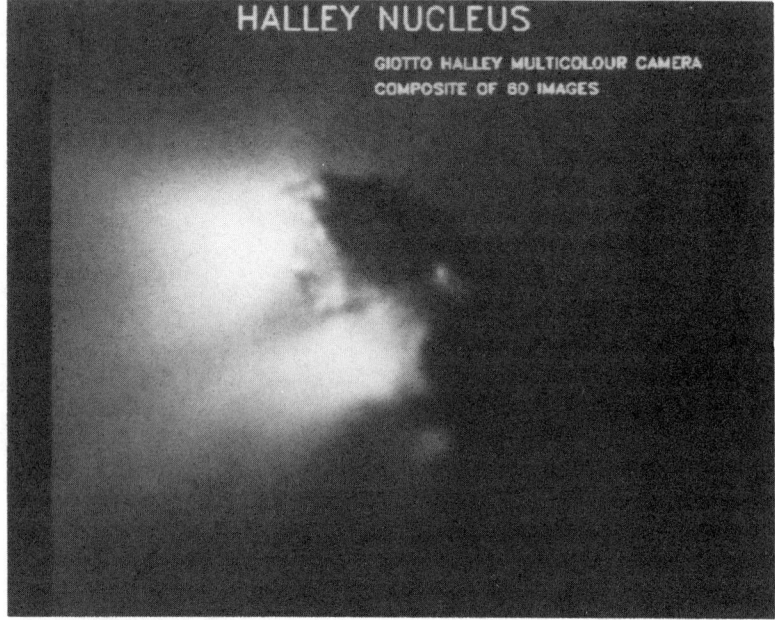

Figure 2 The nucleus of Comet P/Halley, as seen by the fly-through probe Giotto on 14.0 March 1986. This image is a composite of 60 multicolor camera images and thus has an uneven linear resolution over the frame. North is up and the Sun is to the left, 15° below the image plane. This illustration was kindly provided by Dr. H. U. Keller and the Halley Multicolor Camera Team, with headquarters at the MPAE, Germany.

cometary dust activity (and presumably its gas outflow too) must originate in discrete strips (jets, linear surface interface regions?) that occupy only a modest fraction of the nuclear surface area. We cannot yet say whether the *dust* and *gas* jets are coincident.

The brighter Giotto jets can hopefully be matched with the ground-based images of 1986 and 1910 apparitions. This could be a difficult task, however, since we need to extrapolate the preserved rotation period (2.2 day), the survival of the discrete dust jet sources, and the position of the "prime meridian"[1] of the comet at two returns (76 yr apart) past the Sun. Sekanina & Larson (1986b) suggest that the major jet pointing sunward starts from the intersection of their dust sources III, IV, and XI.

A "fatality" of the near-nucleus imaging studies by the Vega and Giotto probes is, of course, the previous generation of cometary models that

---

[1] The prime meridian is defined by Sekanina & Larson (1986b) as the sunward-facing meridian at perihelion.

describe the nucleus and its vaporization as uniform. The next generation of models will be nonhomogeneous (cf. Gombosi & Houpis 1986) and complicated; since the dust/gas emission for Halley takes place in specific (collimated) jets, the general surface area must be protected against sublimation by a solid crustal mantle. Of course, one wonders about the homogeneity of the interior also. From indirect evidence, Delsemme (1985) has argued for internal homogeneity in the nucleus. But clearly direct subsurface exploration is far in the future.

5. COMPOSITIONAL AND PHYSICAL DATA FROM COMAE STUDIES: THE PARENT MOLECULES IN COMET ICES

The onset of activity in a comet approaching the Sun should, in principle, be very dependent on the composition of the volatiles in the outer layers of the icy nucleus, and to a lesser degree on physical parameters, such as its rotation rate and polar inclination. The hope is that the most volatile major species will be recognized by the distance of the onset of gaseous emission in the coma at large heliocentric distance. The reader should examine Figure A19 of Mendis et al. (1985) and also the paper by Rickman & Froeschle (1983). Both of these illustrate graphically the sublimation distances of simple comet nuclei composed of pure H_2O, CO, and CO_2 ices. Water comets should turn on fairly rapidly near 4 or 5 AU; this is in excellent accord with the P/Halley spectroscopy by Wyckoff et al. (1985) and Spinrad et al. (1986a). These authors detected [O I] and CN gas emissions from Halley near $r = 4.8$ AU. Meech et al. (1986) reached the same conclusion from their CCD photometry, which was most sensitive to dust raised by the outgassing flow. Water must dominate the Halley sublimation; models with mostly volatile CO or CO_2 gases show only a *slow increase* in their sublimation over the range $4 \leq r \leq 8$ AU, which contradicts the observations. But surely comets are impure; we are less clear about the activity onset for mixtures of ices, especially if H_2O is the dominant parent. One idea in vogue is that the physical chemistry of the water-ice plays a role; the CO, CO_2, or even CH_4 molecules *may be* bound up in the water-ice-lattice structure (Delsemme & Swings 1952) called a clathrate-hydrate. A maximum of one guest molecule can be trapped in the lattice for each six H_2O molecules. The key point is that the latent heat of sublimation for such a hybrid clathrate-hydrate is almost identical to that of pure H_2O, and it will vaporize rapidly only if $r < 5$ AU. Thus, *if* minority guest molecules are captives of water, they cannot help us much in differentiating the cometary composition by the heliocentric distance dependence of sublimation. Unfortunately, we do not know whether cometary CO (the apparent second-most-abundant species—see the end of this section) does exist in the hydrated form.

As Halley approached Earth in 1983–84, it presented a mystery: There was a conspicuous dust outburst (Djorgovski & Spinrad 1984) at $r = 6.2$ AU. What blew off the dust? Presumably it had *something* to do with a pocket of volatile gas (an "outburst"), with CO or CO_2 (Feldman et al. 1986) being obvious candidates. But it is conceivable that nongaseous processes, such as electrostatic charging of the surface (Mendis et al. 1985), can lift off dust in a sporadic episode, even at $r > 6$ AU. Smoluchowski (1981) has suggested another novel idea: At $r > 5$ AU, a sudden phase change in the crystallization of amorphous water-ice could provide a brief energy source to a dynamically new, warming comet nucleus. Study of a new comet at $r > 10$ would be more valuable even than the heavy early monitoring of P/Halley.

The traditional place to look for parent molecules is in the gas phase near the comet's central coma. However, at the Heidelberg meeting there was some speculation that the organic light-element "dust" grains, coined "CHON," might contain molecules that could form the CO, C_2, or C_3 parents.

The renewed interest in the study of cometary spectra in nontraditional spectral domains has been very helpful in the identification of three or four parent molecular species since 1983! We note with pleasure that (neutral) H_2O emission has finally been observed and even studied physically by Mumma et al. (1986) and by Knacke et al. (1986) in their IR spectra of P/Halley. It is certainly the dominant parent in P/Halley. The other candidate molecules that I discuss in the remainder of this section are HCN (sporadically detected in the microwave region at 3.4 mm; Schloerb et al. 1986), CO (Woods et al. 1986) seen in the UV with rocket spectra, and the S_2 molecule observed in Comet IRAS-Araki-Alcock by the IUE (A'Hearn et al. 1983). By the time this review is published, summary articles on the direct Halley measures by Huebner et al. (1986) and Huntress et al. (1986) should be available.

5.1 H_2O Molecules in Emission—An Infrared Story

The water molecule radiates in the near-IR through its rotation-vibration spectrum. It is an asymmetric rotator, and the v_2 and v_3 vibrations are intensely active! The cometary infrared H_2O bands appear in resonance fluorescence (the fundamental bands), and to a lesser degree their overtones and combinations are excited by sunlight in the $\lambda \leq 2.65$ μm region.

In absorption spectra of laboratory sources or of astronomical objects observed through the Earth's 270-K atmosphere, many rotational levels in each vibrational band are populated. In cometary comae, however, the densities expected are low, and thermalizing collisions will be infrequent and probably too rare to keep the rotational levels of water in equilibrium. This statement is clearly dependent upon the comet model and the actual distance of the water molecules from the cometary nucleus ($\rho_{H_2O} \sim R^{-2}$); Mumma et al. (1986) suggest that only within 1000 km of the nucleus might the equilibrium be appropriate for the kinetic temperature of ~ 200 K. Farther away, the rotational population relaxes by radiation down to the lowest levels. The number of spectral lines in each rotation-vibration band should then decrease, but their *individual* fluorescence efficiencies will be greatly increased, for the same reason. If 5 or 10 lines are favored in, say, the v_3 fundamental, instead of the hundreds of lines seen in the telluric H_2O (absorption) spectrum, the cometary emissions may be detectable with moderate column densities. The theoretical line details have been worked out for the fundamentals and some combination bands by Crovisier & Encrenaz (1983) and Weaver & Mumma (1984).

Observations of cometary water are still difficult; they require a state-of-the-art spectroscopic instrument, such as an infrared Fourier transform spectrometer (FTS). Of course the data should be taken from jet altitudes to minimize the telluric H_2O interference; a large ($v > 20$ km s^{-1}) geocentric Doppler shift also helps. Obviously the job can also be done well from space (cf. Combes et al. 1986).

Strikingly, the FTS observations of P/Halley by Mumma et al. (1986) showed nine sharp emission lines of H_2O in the v_3 band (see Figure 3). The rotational populations did deviate strongly from local thermal equilibrium (LTE), showing an apparent excitation temperature of ≤ 70 K. No Halley water lines with rotational angular momentum quantum states $J > 2$ were detected within the observing aperture of $D = 41''$ (projecting to some 15,000 km at the comet). Using a standard outflow model, Mumma et al. derived a large water production rate $Q(H_2O) \sim 1.7 \times 10^{29}$ mol s^{-1} for their 24 December 1985 spectra. This is a striking confirmation of the dominance of the water molecule in cometary vaporization. Post-perihelion observations by Larson et al. (1986) and Weaver et al. (1986) support this contention.

Now we ask, what can we do to improve or extend this research to other comets, for which high-altitude aircraft observations may not be readily available? One similar method that my colleagues and I have considered is to use CCD technology (CCDs are fine two-dimensional detectors for $\lambda < 9700$ Å) to attempt to detect the much weaker overtone bands in the $\lambda\lambda 6900$–9700 Å region. A practical reason for such an attempt would be to achieve high spatial resolution (10^3 km for a typical Halley distance, but perhaps ≤ 500 km will be resolvable in a future near-Earth comet; naturally, we anticipate parent molecules to be concentrated in the nuclear

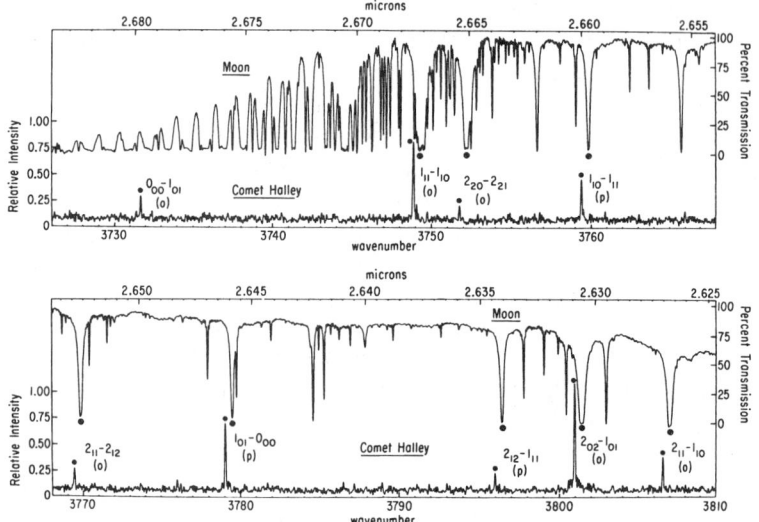

Figure 3 The neutral water IR emission spectrum of Comet P/Halley, seen in "rotational nonequilibrium" by Mumma et al. (1986) with the Kuiper Airborne Observatory at jet altitude. Note that the observed spectrum reveals nine narrow emission lines of H_2O from the comet. The terrestrial absorption spectrum is a hazard, even at jet altitudes. The observations were obtained on 24.1 December 1985. Please see Mumma et al. (1986) for more details.

region). On the other hand, the prospective H_2O lines in this region are weak, and they could be confused with emissions from radicals with short scales, like NH_2. Worst of all, near-IR lines are often superposed on a bright and very concentrated dust continuum of scattered sunlight (as was observed with Halley).

Recently my colleagues P. Wehinger, S. Wyckoff, and I attempted to detect the lowest-lying H_2O lines in the $(301)^2$ band near $\lambda 7220$ Å on long-slit CCD spectra of P/Halley taken at times of favorable velocity. Our initial result appears to be negative; no certain H_2O line pattern was convincingly detected with 1-Å spectral resolution ($\lambda/\Delta\lambda = 7000$). Some higher dispersion spectra remain to be analyzed.

For the convenience of future observers who may wish to improve upon this important search with other bright comets, I list in Tables 6a, 6b, and 6c the wavelengths and identifications of some low-J H_2O lines in the "CCD infrared." A consultation of the solar spectral atlases will determine which cometary geocentric velocity window is best.

[2] This designates the vibrational quantum states (v_1, v_2, v_3).

Table 6a Low-lying water lines in the (201) band

λ (Å) (air)	Designation	λ (Å)	Designation
9377.7	$R(1)$	9417.7	$Q(1)$
9381.2	$R(1)$	9426.9	$Q(1)$
9386.7	$R(1)$	9440.6	$P(1)$
9387.1	$R(1)$		
9399.0	$R(0)$		
9401.6	$R(0)$		

Table 6b The (301) band

λ (Å) (air)	Designation	λ (Å)	Designation
7204.3	$R(1)$	7227.5	$Q(1)$
7206.4	$R(1)$	7232.9	$Q(1)$
7209.5	$R(1)$	7240.6	$P(1)$
7216.5	$R(0)$		

Table 6c The (103) band

λ (Å) (air)	Designation	λ (Å)	Designation
6959.4	$R(1)$	6981.5	$Q(1)$
6961.3	$R(1)$	6986.6	$Q(1)$
		6993.5	$P(1)$

5.2 Other Parent Molecules Worth Consideration

CO seems likely to be an important parent molecule, at least for Comet West (which showed strong CO^+) and P/Halley. It is perhaps surprising that such a volatile species as CO might be a surviving parent in a periodic inner solar system visitor like Comet Halley. Mendis et al. (1985) point out that a *pure* CO nucleus would lose a skin to a depth of about 60 m per perihelion pass. For the putative CO parent ices to survive even when "pocketed" with water-ice chunks, most likely a protective dust-mantle surface layer would be required. The new direct observations of CO by Woods et al. (1986) in the UV found that the spatial profile and column density of the CO bands suggested that the molecule is a fairly abundant parent species. Whether the CO molecules really originate at the cometary nucleus was a point of contention at Heidelberg. Their production rate for P/Halley is surprisingly high, perhaps anywhere from 5–20% that of H_2O. Thus CO might be a constituent *several* times more abundant than CO_2 in the inner coma of the comet (Krankowsky et al. 1986). (CO_2 was well measured in the Giotto flyby.) The Giotto experiments placed coarse upper limits (<10% that of H_2O) for the cosmically common low-temperature compounds NH_3 and CH_4, but the presence of CH_4 derivatives indirectly suggests that this compound is also present. The same may be said for H_2CO.

The S_2 parent story is an unusual one; the molecular emissions were discovered by A'Hearn et al. (1983) on IUE spectrograms of the little Earth-grazing comet, IRAS-Araki-Alcock. As I mentioned earlier in Section 2, S_2 is surprising in that it is easy to sublimate away and has such a short scale length to photodissociation that it can only be glimpsed *near* the nucleus. It is an excellent candidate for a *trace* parent molecule, roughly at the 10^{-3} abundance level compared with water. S_2 is not important to cometary thermodynamics, but it could be an excellent tracer of short-term activity fluctuations because of its very short lifetime (500 s, or 5×10^{-3} that of H_2O).

One other possible parent molecule was observed in P/Halley. The radio spectra of Schloerb et al. (1986) detected the HCN molecule through its microwave $J = 1 \rightarrow 0$ rotational line at 3.4-mm wavelength on 14 observations of Halley. On some dates it was undetectable. They found the average HCN production to be correlated with the general visible brightness trends of the comet; it is also probably a minor constituent parent, with a production rate compared with water's of $Q(HCN)/Q(H_2O) \simeq 10^{-3}$. The evidence for parentage is not overwhelming, but it is thought that HCN may be one of the parents for the CN radical, which is observed to be the strongest band in the blue portion of cometary spectra. This is because $Q(HCN) \simeq Q(CN)$ to about a factor of two; we can expect these rates to be made more precise soon. What is badly needed here is the spatial resolution of the HCN emission in a nearby comet; such information may become observationally possible with the microwave interferometers (resolution $\simeq 1''$) of the next decade.

In concluding this section, I mention that it is still worthwhile to search for unidentified spectral emission lines on two-dimensional spectra of comets. The point is to utilize the information on the spatial emission profile perpendicular to the dispersion. Using the Lick Observatory CCD to obtain long-slit spectrograms of P/Halley, P. McCarthy and I have searched the red spectrum on processed images for emission features as concentrated to the cometary nuclear condensation as is the dust continuum. None were found with any certainty; the dust continuum of

Halley was very concentrated (its FWHM at Δ = 0.73 AU was only 9", or 4100 km), as it is in most comets. The λλ6300, 6363 Å [O I] lines are the most concentrated of the obvious red emission features, and they clearly show a coma distribution that is flatter than that of the dust; a parent molecule might be more concentrated than [O I], and it might or might not be as nucleated as the dust (which, after all, needs no "inner creation zone" to scatter back sunlight) in the inner comet coma. I show in Figure 4 the *spatial ratio* plots of the λ6363 [O I] line and an unidentified spectral line near λ7375 Å, which *may* be due to the NH_2 radical. Both are fairly concentrated, but not as much as the dust. Some of our later observations of P/Halley are a bit asymmetric in the dust/gas spatial distribution, and this makes the above ratio test for scale length less reliable.

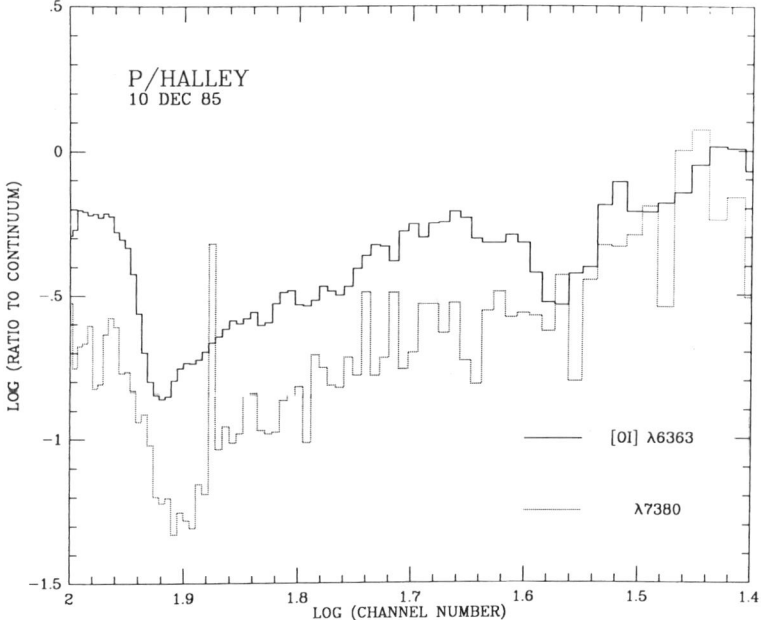

Figure 4 A plot of the *ratio* of the spatial profile along a slit for a weak spectral emission line (λ7380 Å) in P/Halley's red region, compared with the local continuum. With it is shown the relative spatial *ratio* for the weak [O I] water-daughter line at λ6363 Å. See the text discussion. A straight line (Δ log *ratio* = 0.0) would indicate equal concentration of the gas progenitor with the cometary dust continuum light. In this presentation, the nuclear light peak lies at log (channel number) = 1.92; the depth of the depression there indicates that the dust continuum is "peaked" by ≃3 times as steep a spatial distribution as either emission line.

6. QUANTITATIVE DEVELOPMENTS IN COMA PRODUCTION RATES

The quantitative analysis of cometary species, usually the "daughter" product radicals or atoms from parent molecule photodissociation, has been a practical topic of only recent research. Of course the general theme of molecule production, expansion, and destruction has been woven into the theory of comae for some time. In the classic Whipple & Huebner *Annual Review* in 1976 we note that only a few quantitative mass-loss measures were then available, although these have stood the test of time rather well.

Recall that a molecular production rate of $Q_{gas} = dM_g/dt = 10^{30}$ water mol s^{-1}, a value that is reasonable for a bright comet like Halley (near perihelion), is equivalent to a mass-loss rate of 3×10^7 g s^{-1}, or 30 ton s^{-1}! Even a more typical cometary rate of $Q_{H_2O} = 3 \times 10^{29}$ mol s^{-1} = 10^7 g s^{-1} represents a substantial sublimation rate (compared with the nominal cometary mass) if maintained for a half-year.

In the general sense, establishing accurate and self-consistent production rates is important for physically understanding comets as individuals as well as making possible comparisons between them. Production rates have now been established for a wide variety of species that are (mostly) spectroscopically active. We do not discuss all of them, just the most important molecules.

6.1 *Conceptualizing the Production Rate Calculations*

Gas production rates have been reviewed by Wyckoff (1982) and A'Hearn (1982); following the notation and theme of Wyckoff, we recall that the coma molecules we observe with modern quantitative spectrophotometric techniques are generally photodissociative products of parent molecules. Clearly there is a production scale for them. These "daughter" molecules usually have moderate ($\gtrsim 10^5$ s) lifetimes against various destruction processes (normally photoionization in the coma). Thus, the final production rates are somewhat model dependent.

For an observed species (like CN) the emission rate L is related to the number of emitters N in the observing aperture by

$$L = gN, \qquad\qquad 4.$$

where g is the excitation rate. If all of the emission from our molecular species could somehow be spatially integrated (generally almost an impossibility), then the production rate $Q = dM_g/dt$ over the comet would be given by

$$Q = L(g\tau)^{-1} \text{ mol s}^{-1}, \qquad 5.$$

where τ is the species (CN) lifetime. For nominal cometary excitation by resonance fluorescence, the g-factor depends upon the details on the solar flux at the line or band wavelength. The *lifetimes* or effective scale lengths (assuming an outflow velocity v—perhaps near the canonical 1 km s^{-1}—in a uniform outflow model) must be determined empirically for our visual data on the comets. Such lengths are discussed and tabulated by A'Hearn (1982) and Newburn & Spinrad (1984). Of course, the applied photodissociative lifetime will scale as r^2.

In practice we also need to know the scale length of the parent molecule (say, HCN in our sample case here). This is because the parent lifetime implicitly contains a "creation zone size" for the observed daughter radical; it may not really be a true physical length scale from the comet nucleus, since a parent molecule released into the dense inner coma will actually random-walk its way outward from the nucleus. It will not be photodissociated at $l_p = v\tau_p$. The commonly used Haser model (Haser 1957) equation for the daughter particle density in an unmodified isotropic expansion gives

$$n_d(r) = n(r_0) \frac{V_p}{V_d}\left(\frac{r_0}{r}\right)^2 \frac{\tau_d V_d}{\tau_d V_d - \tau_p V_p}\left[\exp\left(\frac{-r}{\tau_p V_p}\right) - \exp\left(\frac{-r}{\tau_d V_d}\right)\right], \qquad 6.$$

where $n(r_0)$ is the scale factor proportional to the total parent production rate at some distance, and the other symbols and subscripts are obvious. A'Hearn & Cowan (1975) and Newburn & Spinrad (1984) calculate "parent scales" that fit the two-process Haser model and go through the somewhat tedious mathematics to derive the *projected* Haser model expression that allows observers to calculate column densities and production rates with spectrophotometric data obtained with small measuring apertures or slits (small compared with either the parent scale length or that of the daughter); these scales are typically $\approx 2 \times 10^4$ km for the parent scale and around 10^5 km for the radicals at $r = 1$ AU (A'Hearn 1982). They are also pretty aphysical but make a conceptually simple model work quite well. The Haser theory has wide applicability because of its simplicity and our ability to calculate self-consistent production rates with it.

To indicate to more sophisticated modelers that their portion of cometary astrophysics has not been completely dormant in the last decade, I note that Combi & Delsemme (1980a) and Festou (1981) have independently suggested inner coma models involving collisions of parent molecules, rather than the purely radial expansion of the Haser paradigm. The linear scales are not of interest in this case, but the techniques do yield lifetimes for prospective parents that can be usefully applied to comparisons with laboratory photodissociation rates. For example, Combi & Delsemme (1980b) suggest that HCN has a lifetime consistent with it being a likely CN parent, in agreement with the recent radio data on P/Halley.

Festou's vectorial model has been widely applied to UV spectroscopy of coma constituents; it also includes the nonradial motion of daughter products after photodissociation of the parent molecule. Festou deals specifically with the case in which the photoproduct carries away kinetic energy transferred during the photodissociation—a good example is the most likely path for H_2O molecule breakup in the UV sunlight ($H_2O \rightarrow H + OH$), where the H atom will attain a considerably larger velocity than that of its water parent. The vectorial model employed tends to yield slightly smaller production rates of water from the IUE observations of OH than would reductions using a Haser model.

6.2 Some Results on the Production Rates for Important Radicals and Inferences on the Parents

The ready availability of an observatory-class satellite, the IUE, has almost revolutionized the cometary measures of H $L\alpha$ and OH ($\lambda 3080$ Å) emissions and yielded acceptable H_2O parent production rates. Feldman (1982, 1983) has argued that the general spectral similarity between the 10 or so comets studied by IUE to that time suggests a common and fairly uniform composition of the cometary ices.

Feldman's early observations of OH yielded maximum water production rates over the large range $1-30 \times 10^{28}$ mol s^{-1}; P/Halley observed within $r = 0.9$ AU will exceed these maxima. The production of radicals of much lower abundance generally behaves in phase with H_2O production; for example, Newburn & Spinrad (1984) showed that all the comets that they studied have a constant mixing ratio of CN to H_2O near 10^{-3}. C_3 and C_2 are also minor constituents (Cochran 1987, Newburn & Spinrad 1984). These famous bright emitting radicals are no more than superficially important to cometary physics. It would be useful to see if any minor abundance trends do eventually correlate with dynamical or other physical properties of comets.

6.3 The Special Case of Oxygen Production

The red auroral oxygen lines are produced in comets by the photodissociation of water; about 10% of the water molecules become excited oxygen atoms. Ninety percent of these O I atoms decay down to the ^1D state (Festou & Feldman 1981, Feldman 1983), which then radiates the red lines at $\lambda\lambda 6300$, 6363 Å. These forbidden oxygen lines are easily detectable in most comets within 2 AU of the Sun if modern sky-subtraction spectroscopic techniques are employed (Spinrad 1982). Of course the cometary geocentric radial velocity can also be employed to clear the cometary [O I] from the ubiquitous airglow line if sufficient spectral resolution is available—as in the case of Fabry-Perot spectroscopy, for example. Measures of the oxygen lines prove to be an important method for determining $Q(\text{O I})$ and hence, by extrapolation, $Q(\text{H}_2\text{O})$, so we discuss this technique in detail.

The procedure is valuable because (*a*) the observations can be continued over long orbital arcs (if ground-based telescope time is available!), and (*b*) it is simple. The linearity of the spectra must be assured; this is because the telluric [O I] lines, the nearby blending NH_2 emissions (Spinrad 1982, Arpigny et al. 1986), and the cometary dust continuum all must be subtracted out before a suitable [O I] $\lambda 6300$-Å flux is available. On spectra of bright comets such as P/Halley (Figure 5) the operational uncertainties in this initial step should be about 20% or less over the comet's projected inner coma. Also useful in this potential long-term project is our increasing capability to compare oxygen/water production rates for different comets.

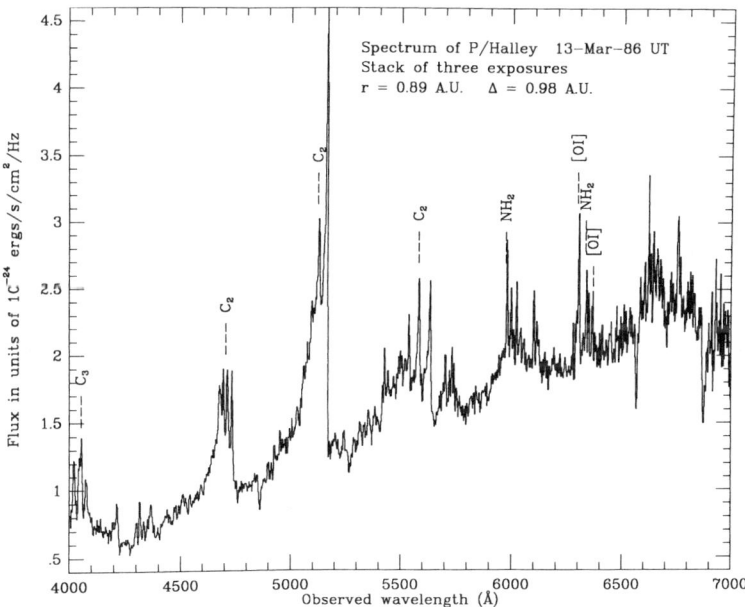

Figure 5 A low-resolution spectrogram of Comet P/Halley obtained at Cerro Tololo Inter-American Observatory by the author and colleagues P. Wehinger, S. Wyckoff, and M. Belton. Reductions by K. Ebneter. Note the C_2, NH_2; and [O I] emission features superposed on a strong scattered-sunlight continuum.

The end-product, $Q(\text{O I})$, determined using the Haser model scheme is only modestly model dependent.

The small-aperture spectrophotometric data on cometary [O I] yield immediately the total number of oxygen atoms $T(s)$ in a cylinder of projected radius s, centered on the comet. In Spinrad's (1982) paper, the symbol s was deemed an aperture size, but it is actually the projected radius of an aperture or an effective (slit-shaped) aperture for long-slit application:

$$T(s) = f_{\text{corr}} \Delta^2 \times 1.72 \times 10^{41} \text{ atom,} \qquad 7.$$

where f_{corr} is the observed $\lambda 6300$-Å line flux (corrected for NH_2 blending), and Δ is the geocentric distance of the comet (in AU).

To place the cylindrical abundance into the Haser model for an oxygen production rate, Spinrad has utilized a modified Haser formula, which is correct for observing apertures that project to a size larger than the Haser "daughter" scale length. For the ^1D oxygen lines, this scale is simply set by the total upper-state lifetime for the excited oxygen atom, $\tau = 150$ s. Thus, with a nominal expansion velocity of 1 km s^{-1}, we find that $l_0 = 150$ km. Most available cometary oxygen observations have been made with effective aperture radii over $1''$, so at $\Delta \geq 0.5$ there is no problem of "undersampling," since $s \gg l_0$. The equation for the oxygen production rate in this normal case is then

$$Q(\text{O I}) = \frac{v}{s} \frac{l_p}{l_0} \frac{2}{\pi} T(s) \text{ atom s}^{-1}. \qquad 8.$$

We use a nominal $v = 1$ km s^{-1}, $l_0 = 150$ km, and $l_p = 7.6 \times 10^4 r^2$ km (the H_2O photodissociation scale with $v = 1$); the $2/\pi$ factor comes from the asymptotic behavior of the integral of a Bessel function in the projected Haser model. Note that Spinrad's original 1982 rates should be increased by a factor of two to properly employ s as the projected radius in kilometers. For the few "Earth-grazing" comets like IRAS-Araki-Alcock, with $\Delta < 0.1$, the simplified Equation 8 above is inadequate; R. Newburn (private communication, 1986) recommends full computation of the Haser Bessel functions by numerical evaluation. The net result will be another approximate doubling of the derived production rates for the few close-approach comets.

Spinrad et al. (1986b) have recently incorporated these revisions and their new observations of P/Halley to explore several aspects of cometary oxygen production—its variation during one long cometary apparition, the rederivation of the ratio $R \equiv Q(H_2O)/Q(\text{O I})$, and the comparison of oxygen production among different comets.

Cometary water production, as we have seen, is best measured by rocket/satellite spectrophotometry of OH in the ultraviolet. It can also be estimated through $Q(H)$ (two H atoms per water molecule, upon eventual breakup), which in turn requires quantitative flux measures of cometary Lα or Hα emission (Scherb et al. 1986). The Hα production rates require a detailed knowledge of the solar Lβ emission-line strength and profile.

Spinrad compared his early (pre-1982) cometary oxygen production rates with the water rates for those comets measured by Weaver et al. (1981). He found that $\bar{R} = 18 \pm 6\sigma$, but this is a value we would now scale down to $\bar{R} = 9$. It turns out that for the near-solar maximum conditions of 1980/81, the strong Lα in the solar chromosphere photodissociates H_2O molecules to excited oxygen rapidly. The theoretical branching ratio of Festou & Feldman (1981) yields $R \cong 10$ at solar maximum. For the "quiet" Sun of 1985/86, this ratio would be expected to rise to $R = Q(H_2O)/Q(O\ I) \cong 15$, and in fact the ratio of the Feldman et al. (1986) water production rate to the Spinrad et al. (1986b) Halley data yields $\bar{R} = 15 \pm 6$ (error estimated). This seems a very satisfactory comparison.

We now scale up the most recent oxygen production rates by the factor 15 to approximately account for the total volatile gas loss by a comet. (This scaling assumes that water *dominates* other volatiles, a proposal safely factual for at least P/Halley.) For Halley at $r = 1.5$ AU (the normalizing distance Spinrad chose for cometary comparisons), we find that $Q^{1.5}(O\ I) = 10^{28}$ atom s^{-1} and $Q^{1.5}(H_2O) = 1.5 \times 10^{29}$ mol s^{-1}. If each molecule has $\langle \mu \rangle = 20$ amu, then the total gas mass-loss rate is $\dot{m}_g = 4 \times 10^6$ g s^{-1}. The total mass loss at this rate for the half-year P/Halley spends in its inner orbit integrates to about 1.2×10^{14} g of gas. Adding in some refractory grain contribution to the loss, we would likely raise our total sublimated mass loss to 1.5–2.0×10^{14} g per orbit revolution. We know Halley's nuclear mass to be near 10^{18} g. Thus P/Halley will be a recognizable survivor in the inner solar system for $< 10^4$ more trips around the Sun if its gas production is constant in future apparitions. Smaller comets, such as P/Tuttle, may not last so long (Spinrad 1982).

We would now like to be able to compare our oxygen production rates in all of the comets with ^1D $\lambda 6300$-Å observations; to do so, it is necessary to place them all at some standard heliocentric distance. I have chosen $r_0 = 1.5$ AU to avoid unnecessary extrapolation; the cometary production rates are usually fit to power laws in r, and the exponents are difficult to determine with precision. Even for the numerous production rates now being published for Comet P/Halley, authors disagree about the power-law exponent appropriate for H_2O production (Scherb et al. 1986, Feldman et al. 1986). One sees representations using $n = -2$ to $n = -5$. I have decided to utilize $n = -3.5$ for all the comet Q comparisons; this "compromise" exponent replaces the bimodal n values advocated by Spinrad (1982). The oxygen/water production rates of both qualitatively "gassy" and "dusty" comets can be fit fairly well with $n = -3 \pm 1$. In any case, the choice of $r_0 = 1.5$ AU means that we will usually be able to interpolate to the reference distance and production rate, rather than needing to extrapolate with an uncertain power. Table 7 lists my updated $Q^{1.5}$(O I) rates for 19 comets; note that the large comets like Meier and Kohoutek with long periods and/or large perihelion distances ($q > 1.5$) have production rates far above average. P/Halley also is fairly productive. The least productive trio—P/Brooks 2, P/Grigg-Skjellerup, and Comet Sugano-Siagusa-Fujikawa (1983e)—lose mass at a rate 100–500 times lower. In conventional astronomical terms, their absolute gas luminosities, defined as $G = -2.5 \log Q^{1.5}$(O I), range over seven magnitudes! Smaller comets with $Q^{1.5}$(O I) $< 0.5 \times 10^{26}$ atom s^{-1} are unlikely to be discovered by conventional means. They can be detected only when they are very near Earth.

We have also noticed the general fact that the comets with the highest oxygen (and thus the highest total volatile) production rates are quite dusty and usually have rather strong NH_2 lines in the red. The direct

Table 7 Absolute cometary oxygen production rates ($r_0 = 1.5$ AU, $n = -3.5$)

Comet	$Q^{1.5}$ (O I) (10^{26} atom s^{-1})
Panther (1980u)	260.0
Meier (1980q)	220.0
Kohoutek (1973f)	130.0
P/Halley	120.0
P/Stephan-Oterma	49.0
P/Borrelly	45.0
P/Wild 2	23.0
Bradfield (1979l)	15.0
P/Swift-Gehrels	14.0
P/Tuttle	12.0
P/Giacobini-Zinner	12.0
P/Wild 3	5.0
P/Encke	4.5
P/Kearns-Kwee	4.0
P/Crommelin	4.0
IRAS-Araki-Alcock (1983d)	3.0
P/Brooks 2	1.2
P/Grigg-Skjellerup	0.5
Sugano-Siagusa-Fujikawa (1983e)	0.4

correlation between Q and the $NH_2/[O\ I]$ flux ratio suggests a decrease in the latter ratio of a factor of two or so for a hundredfold decrease in the absolute gas production. It is unclear now whether this spectroscopic correlation implies a true composition difference or a change in the "weathering" of the surface of "worn-out," short-period comets.

7. SOME HIGHLIGHTS IN THE TAIL OF COMET P/GIACOBINI-ZINNER

While there is much new of interest on the ionosphere and magnetosphere of P/Halley, the analysis of this research is so contemporary that it probably should not be reviewed here. I also recommend an excellent modern review of general cometary ion tails and cometary ionosphere plasma processes that may be found in Mendis et al. (1985).

On 11 September 1985 the International Cometary Explorer (ICE) flew through the ion tail of the periodic Comet Giacobini-Zinner, 7800 km behind its nucleus. A 1986 issue of *Science* (Vol. 232, pp. 297–385) was devoted to the experimental results. Worldwide scientific interest was briefly focused on this small $[Q^{1.5}(O\ I) = 12 \times 10^{26}$ atom s$^{-1}]$ comet, known for years to exhibit a plasma (Type I) tail near perihelion. Chronologically, as the ICE approached the comet, the phenomena associated with the interaction of the solar wind and the cometary atmosphere (cf. von Rosenvinge et al. 1986) set in early; cometary ions picked up and accelerated by the wind were detected as far as 2×10^6 km from P/Giacobini-Zinner (Bame et al. 1986, Ipavich et al. 1986). The magnetometer on ICE did not detect a conventional bow shock, but strong turbulence was measured $\sim 10^5$ km out. The electron temperature at large cometary distance was fairly hot, $T_e \sim 400,000$ K (Bame et al. 1986).

During the fly-through of the central region of the ion tail, which we can envision with the classic barrier overflow picture of field lines draping over an obstacle, the strong section of the *B*-field of the comet was found to have a base diameter of 15,000 km. The central peak field was 60 nT, in both polarities (Smith et al. 1986).

The highest electron density (670 cm^{-3}) was located in the thin neutral sheet between the *B*-field maxima (Meyer-Vernet et al. 1986)—it represents the core or sheath of the ion tail and has a very low electron temperature of $T_e = 13,000$ K (showing the least solar wind influence).

Importantly, the mass/charge data (Coplan et al. 1986) indicate a CO^+/H_2O^+ group ion ratio near 1/10, a value consistent with the ground-based compositional limits of Strauss et al. (1986a). The actual $n(CO^+)/n(H_2O^+)$ ratio is near 0.2, since there is a substantial contribution of H_3O^+ to the mass $\simeq 18$ "water group" (ions near mass 18).

Meyer-Vernet et al. (1987) have combined the ICE plasma data with a calibrated Lick ground-based H_2O^+ profile to deduce a ratio of $n(e^-)/n(H_2O^+) \simeq 4$ (again, of course, ~ 8000 km behind the nucleus). This result implies much H_3O^+ is likely ($\approx 3/4$ of the water ions), as is also suggested by the ionosphere models for water-rich comets (cf. Ip 1980).

Analysis of these H_2O^+ profiles (Strauss et al. 1986a) by Slavin et al. (1986) also suggests that the thin Type I tails of comets imaged from the ground generally correspond to the slab-shaped plasma sheet separating the magnetic inner-tail lobes. This finding agrees well with the conclusion of Meyer-Vernet et al. (1987), who also propose an additional diffuse "ion coma" that is spherical around the nucleus.

One mystery still remains in the ICE mass/charge ratio data: A possible "extra" mass near atomic weight 23 or 24 is still not identified with any confidence (Coplan et al. 1986). The two most likely candidates are C_2^+ and Na^+, but the ionization of their probable parents would seem insufficient to make the peak mass/charge abundance at all noticeable on a scale where the water group gives only twice as large an ion number peak!

8. DESIDERATA AND SPECULATIONS FOR THE FUTURE

8.1 *Some Speculations on the Detection of Distant Giant Comets*

It would be wonderful to actually *detect* a comet situated in the hypothesized *inner* Oort cloud—say, a big one at $r = \Delta = 200$ AU! Of course this cannot be accomplished in 1986, but let us assess our chances to locate *possible giant* comets at such a large distance.

Firstly, we examine the visible region. Recall that from Earth the comet's brightness B will go as $B \sim r^4 d^2$, where d is the comet diameter. If we work in analogy to P/Halley, which was detected at $V = 24$ mag at $r \cong 10$ AU, then at $r = 200$ AU a 10–20 km size comet will be 10^5 times fainter than P/Halley was ($V \cong 31.5$). This is *really* faint, and for a remote comet to be reached by the Space Telescope Faint Object Camera with a few hours of integration and $S/N \geq 2$ ($V \sim 27^m$), the comet would have to have a *diameter* of *120 km*! This would be a giant indeed, but such large objects could easily exist in the outer solar system without our knowledge. Of course, one does not know where to start looking (for a large-parallax object!). What about the infrared thermal radiation from a distant comet? We try this exercise by comparing our hypothesized giant to the outer system object Chiron (Lebofsky et al. 1984), which at $r = 16$ AU had a 22-μm flux of $S_\lambda = 17$ mJy. Chiron has a diameter of 180 ± 50 km, and thus is perhaps a giant asteroid or comet itself. We assume that $T_e = 50$ K, which is a bit warm for $r = 200$ AU but at least gives us a starting point. The one-dimensional detector *survey* type *limit* on the future SIRTF satellite is thought to be about $S_\lambda \cong 3 \times 10^{-4}$ Jy at $\lambda = 100$ μm. Chiron at 200 AU will be near $S = 10^{-5}$ Jy, too faint for a one-dimensional strip survey detector. Again, a two-dimensional detector, "staring" at the correct location, might make a successful detection, but I fear that such a detection would not be easy, even with the technology of the next decade.

8.2 *Instrumental Novelties for Future Coma Observations*

High-resolution spectroscopy has not yet been applied to comets in all possible ways. It should be possible to detect deuterium in the radio regime; one could try OD vs. OH or look at Lα spectroscopically with the Space Telescope high-resolution spectrograph. The deuterium/hydrogen ratio will be difficult to understand chemically, however, because fractionation could dominate such a large molecular mass ratio as for the parent HDO/H$_2$O. Another imminent and useful *instrumental* possibility would be the use of an imaging Fabry-Perot interferometer to study the inner coma kinematics of Hα and [O I] (again water-daughters). The current University of Wisconsin instrument has begun such work (Scherb et al. 1986, Huppler et al. 1975), and a new Rutgers University instrument (Williams et al. 1984) has a field of view $\sim 2'$, with $1''$ pixels and a velocity resolution of 0.6 Å $\cong 30$ km s^{-1} with its high-resolution etalon. So velocities should be measurable to ≤ 5 km s^{-1}, adequate for Hα outflow spatial mapping in a bright comet. Also, one would like to map the H$_2$O$^+$ ion velocity fields from a comet nucleus, over the coma, and out toward the ion tail in a Halley-class comet's future apparition.

8.3 *Post-Halley Spacecraft Missions*

I believe we should aspire for a *very unique* mission plan to these least-explored and most-primitive solar system bodies. Flyby probes such as the ICE are too limited in their observational capabilities, and most importantly they take only a snapshot view of a comet. Even a sophisticated flyby fails to provide any synoptic coverage of cometary activity; moreover, they are "dangerous" missions, with survival probabilities well less than 1 if the comet is active. One mission that in part aims to satisfy our ambitious exploration plans is NASA's *Comet Rendezvous and Asteroid Flyby* (CRAF). The spacecraft will rendezvous with a short-period comet at its aphelion; thus, the comet can be first studied close up while it is relatively inactive. The spacecraft then tags along the comet's orbit, watching and maneuvering near the nucleus for extended periods of time while the comet becomes more active.

This detailed and long-lived CRAF mission will have an even larger payoff if a lander can safely reach the cometary nuclear surface for a direct physical and chemical analysis. Even more sophisticated would be a sample-return portion of such a mission; age dating on a single-grain basis might even be possible. To characterize the nucleus of even a worn-out periodic comet would be an important step forward; right now, the targets for the mid-1990s seem to include the following comets (perihelion year in parentheses): P/Kopff (1996), P/Encke (1997), P/Wild 2 (1997), and/or P/Tempel 2 (1999).

At the conclusion of such an extended cometary exploration voyage, we may gain a real understanding of the early times in the outer solar system and even some clues on the process of star formation. In any case, at that time someone should prepare an up-to-date review paper on comets!

ACKNOWLEDGMENTS

I wish to acknowledge many informative discussions with my colleagues Ivan King, Martin Cohen, Mike Belton, Peter Wehinger, Sue Wyckoff, Mike A'Hearn, Paul Feldman, David Jewitt, and especially Ray L. Newburn. I also thank my hard-working and intellectually curious graduate students, Patrick McCarthy and Michael Strauss. My research on comets is supported by a grant from NASA, which is also acknowledged with gratitude.

Finally, I thank Drs. H. U. Keller and H. Reitsema for the Giotto nuclear composite image, and Drs. A. Dressler and R. Windhorst for the beautiful Las Campanas reflector photograph (Figure 1). Dr. H. Larson kindly supplied the H_2O spectrum of Figure 3.

Literature Cited

A'Hearn, M. F. 1982. In *Comets*, ed. L. L. Wilkening, pp. 433–60. Tucson: Univ. Ariz. Press

A'Hearn, M. F. 1986. In *Astroids, Comets, Meteors II*, ed. C.-I. Lagerkvist, B. A. Lindblad, H. Lundstedt, H. Rickman, pp. 187–90. Uppsala: Uppsala Univ.

A'Hearn, M. F., Cowan, J. J. 1975. *Astron. J.* 80: 852–60

A'Hearn, M. F., Feldman, P. D., Schleicher, D. G. 1983. *Ap. J. Lett.* 274: L99–103

Arpigny, C., Magain, F., Manfroid, J., Dossin, F., Danks, A. C., Lambert, D. L. 1986. Paper presented at the ESA/ESTEC Halley Conf., Heidelberg

Aumann, H. H. 1985. *Publ. Astron. Soc. Pac.* 97: 885–91

Bailey, M. E. 1983. *MNRAS* 208: 575–88

Bame, S. J., Anderson, R. C., Asbridge, J. R., Baker, D. N., Feldman, W. C., et al. 1986. *Science* 232: 356–61

Belton, M. J. S. 1985. *Science* 230: 1229–36

Belton, M. J. S., Wehinger, P., Spinrad, H., Wyckoff, S. 1986. Paper presented at the ESA/ESTEC Halley Conf., Heidelberg

Brin, G. D., Mendis, D. A. 1979. *Ap. J.* 229: 402–8

Brooke, T. Y., Knacke, R. F. 1986. *Icarus* 67: 80–87

Burns, J. A., Tedesco, E. F. 1979. In *Asteroids*, ed. T. Gehrels, pp. 494–527. Tucson: Univ. Ariz. Press

Cameron, A. G. W. 1973. *Icarus* 18: 407–50

Churchwell, E. 1983. *Bull. Astron. Soc. India* 11: 189–213
Clube, S. V. M., Napier, W. A. 1982. *Q. J. R. Astron. Soc.* 23: 45–66
Clube, S. V. M., Napier, W. A. 1984. *MNRAS* 208: 575–88
Clube, S. V. M., Napier, W. A. 1986. In *The Galaxy and the Solar System*. Tucson: Univ. Ariz. Press. In press
Cochran, A. 1987. *Astron. J.* In press
Combes, M., Moroz, V. I., Crifo, J. F., Lamarre, J. M., Charra, J., et al. 1986. *Nature* 321: 266–68
Combi, M. R., Delsemme, A. H. 1980a. *Ap. J.* 237: 633–40
Combi, M. R., Delsemme, A. H. 1980b. *Ap. J.* 237: 641–45
Coplan, M. A., Ogilvie, K. W., A'Hearn, M. F., Bochsler, P., Geiss, J. 1986. Preprint (Univ. Md.)
Crovisier, J., Encrenaz, T. 1983. *Astron. Astrophys.* 126: 170–82
Danks, A. C., Lambert, D. L., Arpigny, C. 1974. *Ap. J.* 194: 745–51
Delsemme, A. H. 1982. In *Comets*, ed. L. L. Wilkening, pp. 85–130. Tucson: Univ. Ariz. Press
Delsemme, A. H. 1985. *Publ. Astron. Soc. Pac.* 97: 861–70
Delsemme, A. H. 1986. Paper presented at the DPS/AAS Meet., Paris
Delsemme, A. H., Swings, P. 1952. *Ann. Astrophys.* 15: 1–6
Djorgovski, S., Spinrad, H. 1984. *IAU Circ.* No. 3996
Fay, T. D., Wisniewski, W. 1978. *Icarus* 34: 1–9
Feldman, P. D. 1982. In *Comets*, ed. L. L. Wilkening, pp. 461–79. Tucson: Univ. Ariz. Press
Feldman, P. D. 1983. *Science* 219: 347–54
Feldman, P. D. 1986. In *Asteroids, Comets, Meteors II*, ed. C.-I. Lagerkvist, B. A. Lindblad, H. Lundstedt, H. Rickman, pp. 263–67. Uppsala: Uppsala Univ.
Feldman, P. D., et al. 1986. *IUE Symp.* In press
Feldman, P. D., A'Hearn, M. F., Millis, R. F. 1984. *Ap. J.* 282: 799–802
Fernandez, J. A. 1980. *MNRAS* 192: 481–91
Fernandez, J. A., Jockers, K. 1983. *Rep. Progr. Phys.* 46: 665–772
Festou, M. C. 1981. *Astron. Astrophys.* 95: 69–79
Festou, M. C., Feldman, P. D. 1981. *Astron. Astrophys.* 103: 154–59
Festou, M. C., Lecacheux, J., Kohl, J. L., Encrenaz, T. 1986. Paper presented at the ESA/ESTEC Halley Conf., Heidelberg
Goldstein, R. M., Jurgens, R. F., Sekanina, Z. 1984. *Astron. J.* 89: 1745–54
Gombosi, T. I., Houpis, H. L. F. 1986. *Nature* 324: 43–44
Greenberg, J. M., Grim, R., van Izendoorn, L. 1986. In *Asteroids, Comets, Meteors II*, ed. C.-I. Lagerkvist, B. A. Lindblad, H. Lundstedt, H. Rickman, pp. 225–27. Uppsala: Uppsala Univ.
Hanner, M. S., Aitken, D. K., Knacke, R., McCorkle, S., Roche, P. F., Tokunaga, A. T. 1985. *Icarus* 62: 97–109
Haser, L. 1957. *Bull. Acad. R. Belg.* 43: 740–50
Heisler, J., Tremaine, S. 1986. *Icarus* 65: 13–26
Hills, J. G. 1981. *Astron. J.* 86: 1730–40
Houpis, H. L., Ip, W.-H., Mendis, D. A. 1985. *Ap. J.* 295: 654–67
Huebner, W. F., et al. 1986. Paper presented at the ESA/ESTEC Halley Conf., Heidelberg
Hughes, D. W. 1985. *MNRAS* 213: 103–9
Huntress, W., et al. 1986. Paper presented at the ESA/ESTEC Halley Conf., Heidelberg
Huppler, D., Reynolds, R. J., Roessler, F. L., Scherb, F., Trauger, J. 1975. *Ap. J.* 202: 276–82
Hut, P., Tremaine, S. 1985. *Astron. J.* 90: 1548–57
Ip, W.-H. 1980. *Astron. Astrophys.* 92: 95–100
Ipavich, F. M., Galvin, A. B., Gloeckler, G., Hovestadt, D., Klecker, B., Scholer, M. 1986. *Science* 232: 366–69
Jewitt, D., Danielson, G. E. 1984. *Icarus* 60: 435–44
Jewitt, D., Meech, K. 1985. *Icarus* 64: 329–35
Kamoun, P. G., Campbell, D. B., Ostro, S. J., Pettengill, G. H., Shapiro, I. I. 1982. *Science* 216: 293–95
Kaneda, E., Hirao, K., Takagi, M., Ashihara, O., Itoh, T., Shimizu, M. 1986. *Nature* 320: 140–41
Keller, H. U., Arpigny, C., Barbieri, C., Bonnet, R. M., Cazes, S., et al. 1986. *Nature* 321: 320–26
Knacke, R. F., Noll, K. S., Geballe, T. R., Tokunaga, A. T., Broske, T. Y. 1986. Paper presented at the ESA/ESTEC Halley Conf., Heidelberg
Krankowsky, D., Lammerzahl, P., Herrwerth, I., Woweries, J., Eberhardt, P., et al. 1986. *Nature* 321: 326–29
Lambert, D. L., Danks, A. C. 1983. *Ap. J.* 168: 428–46
Larson, H. P., Davis, D. S., Mumma, M. J., Weaver, H. J. 1986. *Ap. J.* In press
Lebofsky, L. L., Tholen, D. J., Rieke, G. H., Lebofsky, M. 1984. *Icarus* 60: 532–37
McFadden, L. A., Gaffey, M. J., McCord, T. B. 1984. *Icarus* 59: 25–40
Meech, K. J., Jewitt, D., Ricker, G. R. 1986. *Icarus* 60: 561–74
Mendis, D. A., Houpis, H. L. F., Marconi,

M. L. 1985. *Fundam. Cosmic Phys.* 10: 1–380

Meyer-Vernet, N., Couturier, P., Hoang, S., Perche, C., Steinberg, J. L., et al. 1986. *Science* 232: 370–74

Meyer-Vernet, N., Strauss, M. A., Steinberg, J. L., Spinrad, H., McCarthy, P. J. 1987. *Astron. J.* In press

Millis, R. L., et al. 1986. Paper presented at the ESA/ESTEC Halley Conf., Heidelberg

Mumma, M. J., Weaver, H. A., Larson, H. P., Davis, D. S., Williams, M. 1986. *Science* 232: 1523–28

Newburn, R. L., Rahe, J. 1986. In *Asteroids, Comets, Meteors II*, ed. C.-I. Lagerkvist, B. A. Lindblad, H. Lundstedt, H. Rickman, pp. 473–76. Uppsala: Uppsala Univ.

Newburn, R. L., Spinrad, H. 1984. *Astron. J.* 89: 289–309

O'Dell, C. R. 1986. *Icarus* 67: 71–79

Oort, J. 1950. *Bull. Astron. Inst. Neth.* 11: 91–110

Owen, T. 1973. *Ap. J.* 184: 33–43

Rickman, H., Froeschle, C. 1983. In *Cometary Exploration*, ed. T. Gombosi, pp. 75–84. Budapest: Hung. Acad. Sci.

Roemer, E. 1966. *Mém. Soc. R. Sci. Liège* 12: 15–22

Sagdeev, R. Z., Blamont, J., Galeev, A. A., Moroz, V. I., Shapiro, V. D., et al. 1986a. *Nature* 321: 259–62

Sagdeev, R. Z., Szabo, F., Avanesov, G. A., Cruvellier, P., Szabo, L., et al. 1986b. *Nature* 321: 262–66

Scherb, F., Roessler, F. L., Magee, K., Harlander, J., Reynolds, R. J. 1986. Paper presented at COSPAR Symp., 26th, Toulouse, Fr.

Schleicher, D. G., et al. 1986. Paper presented at the ESA/ESTEC Halley Conf., Heidelberg

Schloerb, F. P., Kinzel, W. M., Swade, D. A., Irvine, W. M. 1986. *Ap. J. Lett.* In press

Schlosser, W., Schulz, R., Cocret, P. 1986. Paper presented at the ESA/ESTEC Halley Conf., Heidelberg

Schwarz, G., et al. 1986. Paper presented at the ESA/ESTEC Halley Conf., Heidelberg

Sekanina, Z., Larson, S. M. 1986a. *Astron. J.* 92: 462–82

Sekanina, Z., Larson, S. M. 1986b. *Nature* 321: 357–61

Slavin, J. A., et al. 1986. *Geophys. Res. Lett.* In press

Smith, B. A., Terrile, R. J. 1984. *Science* 226: 1421–24

Smith, E. J., Tsurutani, B. T., Slavin, J. A., Jones, D. E., Siscoe, G. L., Mendis, D. A. 1986. *Science* 232: 383–85

Smoluchowski, R. 1981. *Ap. J. Lett.* 244: L31–34

Spinrad, H. 1982. *Publ. Astron. Soc. Pac.* 94: 1008–16

Spinrad, H., Belton, M. J. S., Wehinger, P. A., Wyckoff, S. 1986a. In *Asteroids, Comets, Meteors II*, ed. C.-I. Lagerkvist, B. A. Lindblad, H. Lundstedt, H. Rickman, pp. 491–96. Uppsala: Uppsala Univ.

Spinrad, H., McCarthy, P., Strauss, M. A. 1986b. Paper presented at the ESA/ESTEC Halley Conf., Heidelberg

Spinrad, H., Stauffer, J., Newburn, R. L. 1979. *Publ. Astron. Soc. Pac.* 91: 707–11

Stewart, A. I. F. 1986. Paper presented at the ESA/ESTEC Halley Conf., Heidelberg

Strauss, M. A., McCarthy, P. J., Spinrad, H. 1986a. *Geophys. Res. Lett.* 13: 389–92

Strauss, M. A., McCarthy, P. J., Spinrad, H. 1986b. *Bull. Am. Astron. Soc.* 18(3): 793

Tokunaga, A. T., Hanner, M. S. 1985. *Ap. J. Lett.* 296: L13–16

von Rosenvinge, T. T., Brandt, J. C., Farquhar, F. W. 1986. *Science* 232: 353–56

Wannier, P. G. 1980. *Ann. Rev. Astron. Astrophys.* 18: 399–437

Weaver, H. A., Feldman, P. D., Festou, M. C., A'Hearn, M. F. 1981. *Ap. J.* 251: 809–19

Weaver, H. A., Mumma, M. J. 1984. *Ap. J.* 276: 782–97

Weaver, H. A., Mumma, M. J., Larson, H. P., Davis, D. S. 1986. *Nature* 324: 441–44

Wehinger, P. A., Peterson, B. A., Wyckoff, S., Lindholm, E., Festou, M. 1986. Paper presented at the ESA/ESTEC Halley Conf., Heidelberg

Weissman, P. R. 1982. In *Comets*, ed. L. L. Wilkening, pp. 637–58. Tucson: Univ. Ariz. Press

Weissman, P. R. 1985. In *Protostars and Planets II*, ed. D. C. Black, M. S. Matthews, pp. 895–919. Tucson: Univ. Ariz. Press

Weissman, P. R. 1986. In *Asteroids, Comets, Meteors II*, ed. C.-I. Lagerkvist, B. A. Lindblad, H. Lundstedt, H. Rickman, pp. 197–206. Uppsala: Uppsala Univ.

Weissman, P. R., Kiefer, H. H. 1981. *Icarus* 47: 302–11

West, R. M., Pederson, H. 1984. *Astron. Astrophys.* 138: L9–10

Whipple, F. L., Huebner, W. F. 1976. *Ann. Rev. Astron. Astrophys.* 14: 143–72

Williams, T. B., Caldwell, N., Schommer, R. A. 1984. *Ap. J.* 281: 579–84

Wisniewski, W., Fay, T. D. 1985. *IAU Circ. No. 4041*

Wisniewski, W., Fay, T. D., Gehrels, T. 1986. In *Asteroids, Comets, Meteors II*, ed. C.-I. Lagerkvist, B. A. Lindblad, H. Lundstedt, H. Rickman, pp. 337–40. Uppsala: Uppsala Univ.

Woods, T. N., Feldman, P. D., Dymond, K. F., Sahnow, D. J. 1986. *Nature* 324: 436–38

Wyckoff, S. 1982. In *Comets*, ed. L. L. Wilkening, pp. 3–55. Tucson: Univ. Ariz. Press

Wyckoff, S., Wagner, M., Wehinger, P. A., Schleicher, P. G., Festou, M. C. 1985. *Nature* 316: 241–42

Yamamoto, T. 1985. *Astron. Astrophys.* 142: 31–36

Yeomans, D. 1986. Paper presented at the ESA/ESTEC Halley Conf., Heidelberg

ROTATION AND MAGNETIC ACTIVITY IN MAIN-SEQUENCE STARS[1]

Lee W. Hartmann and Robert W. Noyes

Harvard-Smithsonian Center for Astrophysics, 60 Garden Street, Cambridge, Massachusetts 02138

1. INTRODUCTION

Magnetic fields play an important role in many astrophysical phenomena. The physical processes involved in generating magnetic fields are complicated and poorly understood at present. Theoretical advances are dependent upon further observations of nearby astronomical objects, which can provide detailed information on the generation and dissipation of magnetic fields in environments not realizable in the laboratory.

Observations of roughly cyclic magnetic fields on the Earth and Sun led to the development of dynamo theories of magnetic-field generation, in which magnetic activity results from the interaction of rotation and convection. In recent years, improvements in observational techniques have made it possible to study magnetic dynamo activity on stars. Although stellar observations provide far less detailed information than solar measurements, an understanding of the dependence of magnetic-field generation on global parameters can be obtained only from the systematic study of many objects. In addition, observations of stellar magnetic activity indicate how the magnetic activity of the Sun has evolved over the course of time, and they may provide insights into the solar-terrestrial environment at early epochs.

In this review we discuss recent observational results that shed light on how rotation interacts with convection to produce stellar magnetic fields. We concentrate on the activity of main-sequence stars, for which a large

---

[1] The US Government has the right to retain a nonexclusive royalty-free license in and to any copyright covering this paper.

body of magnetic activity and rotation results have become available. In Section 2 we sketch the evolution of stellar rotation from pre-main-sequence to main-sequence phases. Some implications of this rotational evolution for internal velocity fields, and therefore for the dynamo processes, are noted. In Section 3 we discuss surface manifestations of magnetic activity, thought to be generated by the magnetic dynamo acting within rotating, convecting stars, and how this activity depends on mass, rotation, and age. Finally, in Section 4 we summarize the present state of our understanding of rotation and magnetic activity in main-sequence stars and outline some promising areas for future work.

Many topics involved in solar and stellar magnetic activity have been reviewed recently [see, e.g., Baliunas & Vaughan (1985) and Rosner et al. (1985) for a summary of the more recent results]. For this reason we have concentrated on results particularly relevant to magnetic-field generation rather than attempting an encyclopedic review of all related literature. More detailed discussion of various aspects can be found in the above-mentioned reviews.

2. ROTATIONAL EVOLUTION OF LOW-MASS STARS

Overview

In the magnetic dynamo model, the rotation of a star is an important parameter affecting the rate of magnetic-field generation. It follows that determinations of stellar rotation are important in testing the dynamo theory and in general for understanding the magnetic activity of stars.

For some time it was thought that solar-type stars (main-sequence objects with $M \lesssim 1 \ M_\odot$) exhibited a monotonic decline of rotation with time between the ages of $\sim 7 \times 10^7$ and 4.5×10^9 yr (Kraft 1967, Skumanich 1972). This decline of surface rotation with age is accompanied by a reduction in surface magnetic fields, as indicated by various tracers of magnetic activity (Wilson 1963, 1966; cf. Section 3). Stellar spin-down is generally attributed to the torque of winds magnetically coupled to the stellar surface. The form of the rotational slowdown with age ($v_{\rm rot} \approx t^{-1/2}$; Skumanich 1972) implies that the wind torque is smaller at lower rotational velocities.

These observational results can be explained by the following scenario. As a star ages, the wind torque decreases the stellar rotation, which in turn causes the dynamo generation of magnetic fields to decrease, consistent with observations of magnetic-field tracers. The decreased magnetic activity produces less coronal heating and/or a smaller Alfvén (corotation) radius, which results in less spin-down torque. The result is a feedback mechanism, in which stellar wind torque is largest at the fastest rotation rates and declines for slower rotational periods. The $v_{\rm rot} \approx t^{-1/2}$ "Skumanich law" implies such a feedback. One might expect that stars of a given mass will eventually settle into similar tracks of rotation and activity with age, independent of their initial angular momenta.

In recent years the number of measurements of stellar rotation and magnetic activity has increased dramatically. The periodic modulation of starlight produced by magnetic areas rotating across the disk has been exploited to provide precise rotation rates for many main-sequence stars and some pre-main-sequence stars. In addition, improvements in detector technology on large telescopes have led to the spectroscopic measurement of rotational line broadening for many faint stars, permitting the exploration of stellar rotation among very young stars.

The new observations have modified the old picture of stellar rotational evolution substantially. We now know that low-mass stars either are born slowly rotating or are spun down very quickly at an early age. These stars subsequently spin up dramatically during contraction to the main sequence. A rapid phase of main-sequence spin-down then follows; the age at which this spin-down occurs is mass dependent, occurring later for lower mass stars.

This complicated behavior could well produce differences between internal and surface rotation, with important consequences for stellar magnetic activity. The rotation effects important for dynamo action incorporate not only the mean surface rotation rate, but also the differential rotation both with latitude and depth. Because the spin-down torque is probably directly applied only to the outermost layers, it is conceivable that the interior will lag behind, depending upon the rates of internal diffusion of angular momentum. The difference between surface and internal rotation may be magnified if the surface undergoes rapid variations in rotation.

Recent data from helioseismology suggest little differential rotation in the solar interior, but do not rule out larger shears at earlier evolutionary stages. As we show below, it seems easier to explain the observed dependence of rotation on mass in young clusters if only the outer convective envelope is initially spun down.

Here we outline the new observations of rotation during the pre-main-sequence and main-sequence phases, concentrating on aspects that may be relevant to the problem of magnetic-field generation. Further details are given in the observational review by Stauffer & Hartmann (1986).

Evolution of Rotation of Main-Sequence Stars

ROTATION OF PRE-MAIN-SEQUENCE STARS ($t \lesssim 10^6$ YR)   We begin by considering the rotation of solar-type stars when they are in the T Tauri phase (ages $\sim 10^6$ yr, $R \sim 3$–4 $R_\odot$). Most of the data available come from $v \sin i$ measurements (Bouvier et al. 1986, Hartmann et al. 1986, 1987), although in a few cases photometric rotational periods have been measured (Kappelmann & Mauder 1981, Schaefer 1983, Rydgren & Vrba 1983, Vrba et al. 1984, Rydgren et al. 1985, Bouvier et al. 1986). The equatorial rotational velocities estimated from photometric modulation periods are generally in good agreement with $v \sin i$ measurements when both are available, and this correspondence suggests that the assigned radii of many T Tauri stars (and hence their placement in the H-R diagram) are reasonably correct. The assignment of masses, however, is far less certain and is quite sensitive to details of stellar evolution models and calibration problems. We adopt the H-R diagram positions and evolutionary tracks of Cohen & Kuhi (1979) and present the data in a single mass bin.

The recent spectroscopic results for low-mass T Tauri stars are summarized in Figure 1a. In this figure we have included rotational velocities for weak-emission stars recently discovered by Herbig et al. (1986). Walter

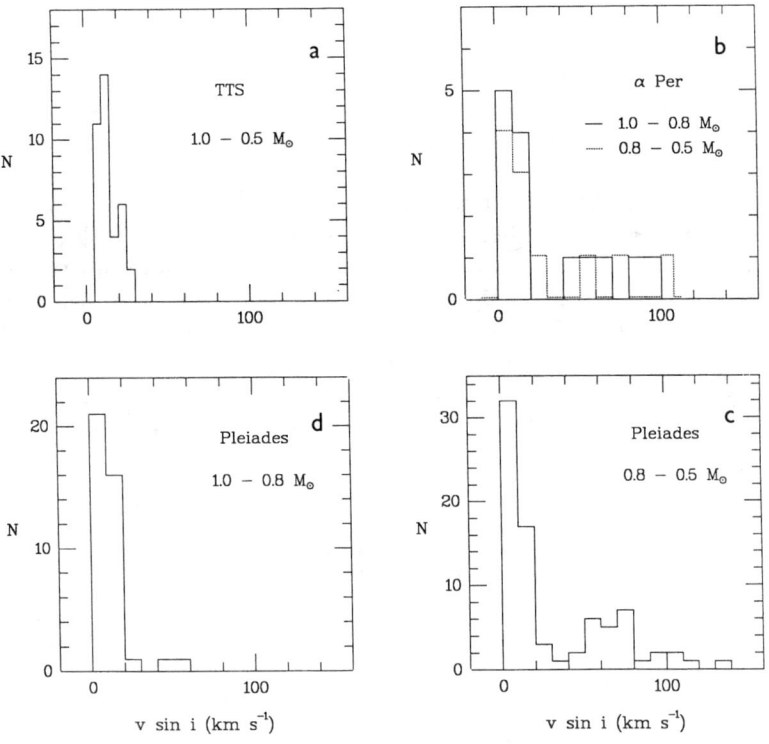

Figure 1 Observed rotational velocity distributions for low-mass T Tauri stars and for young cluster stars, as described in the text. For α Per, the two histograms are superimposed, not additive.

et al. (1987) have found similar results for other pre-main-sequence stars. The observed low rotational velocities are consistent with the earlier upper limits of Vogel & Kuhi (1981).

Figure 1a shows that at an early age (10^6 yr), the rotational velocities of low-mass stars are remarkably small in comparison with breakup velocity (~ 200 km s^{-1}). Either the angular-momentum problem is solved very efficiently during star formation, or else an early phase of rapid spin-down as a young star takes place. [It should be pointed out that no rotational velocities are available for the strong-emission objects, whose continua show little trace of the photospheric absorption lines necessary for spectroscopic velocity analysis. These objects constitute perhaps 10–20% of the total sample in typical magnitude-limited surveys (Hartmann et al. 1986).]

The appearance of the rotational velocity distribution shown in Figure 1a differs significantly from that which would be expected if all stars had the same rotational velocity modified by a random distribution of inclination i. In the latter case, the rotational velocities would be skewed to large values of $v \sin i$. Attempts to correct the $v \sin i$ distribution for differences in mass and age are model dependent and therefore uncertain. However, if one assumes that angular momentum loss in the T Tauri phase is minimal, as discussed below, the data suggest that young low-mass stars of the same mass and age exhibit a significant spread in their angular momentum (Hartmann et al. 1986).

SPIN-UP DURING CONTRACTION TO THE MAIN SEQUENCE: THE α PERSEI CLUSTER ($t \sim 5 \times 10^7$ YR) At present we have no further information about the evolution of low-mass stellar rotation until stars approach the zero-age main sequence. The youngest cluster for which a number of rotational velocities are available is α Per, which has a nominal upper-main-sequence turnoff age of 5×10^7 yr. The distribution of rotation shown in Figure 1b (Stauffer et al. 1985) exhibits remarkable skewness, possessing both a concentration of stars at low velocities and a broad high-velocity tail.

Comparison between Figures 1a and 1b implies that at least some low-mass stars increase their rotational velocities as they approach the main sequence. Such an effect is quite consistent with the standard picture of pre-main-sequence stellar evolution. Adopting the evolutionary tracks of Cohen & Kuhi (1979), typical T Tauri stars shown in Figure 1a contract in radius by a factor of three by the time they arrive on the main sequence, leading to a reduction in the moment of inertia by a factor of 10 to 20. Detailed calculations (Hartmann et al. 1986) show that if one assumes solid-body rotation and no angular momentum loss during pre-main-sequence contraction, then one can comfortably account for the range in rotation shown in Figure 1*b*, with some stars approaching velocities of 200 km s^{-1}. The most rapidly rotating T Tauri stars known can lose no more than half of their angular momentum and still match the rapid rotators in α Per after contraction to the main sequence. Slightly more angular momentum loss may be needed to produce the slow rotators in α Per. These limits on angular momentum loss do not prevent the rotational angular frequency from increasing somewhere between the pre-main-sequence and main-sequence phases.

At first sight the inference of spin-up is surprising. It is thought that magnetically coupled stellar winds carrying away angular momentum cause the spin-down of solar-type stars (Schatzman 1962). T Tauri stars are known to have strong winds, and straightforward application of the Alfvén-wave wind theory for these stars results in relatively short spin-down times (Hartmann 1985). However, it is likely that realistic modifications of magnetic-field structure can reduce the angular momentum loss from wind models to levels consistent with spin-up (Roxburgh 1983, Mestel 1984). Furthermore, magnetic braking during pre-main-sequence phases must overcome the rapidly decreasing stellar moment of inertia, while main-sequence spin-down takes advantage of longer evolutionary time scales.

RAPID SPIN-DOWN ON THE MAIN SEQUENCE: THE PLEIADES CLUSTER ($t \sim 7 \times 10^7$ YR) The surprising aspect of rotation in this slightly older cluster is the extreme dependence of $v \sin i$ on spectral type or mass. Many K–M dwarfs (Figure 1*c*) rotate rapidly, at rates comparable to α Per, whereas the G dwarfs (Figure 1*d*) do not (van Leeuwen & Alphenaar 1982, Soderblom et al. 1983, Stauffer et al. 1984, van Leeuwen et al. 1986, Stauffer & Hartmann 1987). The transition in spectral type between the slow- and rapid-rotator regimes is fairly sharp (Stauffer et al. 1984, Stauffer & Hartmann 1987). By comparison with the rotational velocities of stars in α Per, one infers that the Pleiades G dwarfs have spun down rapidly on the main sequence, presumably as a result of magnetic wind braking (Schatzman 1962).

Rapid spin-down for 1-$M_\odot$ stars had been predicted theoretically by Endal & Sofia (1981; hereafter ES). In the ES model, a stellar wind coupled to the surface by magnetic fields removes angular momentum at the stellar surface. Transfer of angular momentum was assumed to be very efficient in convecting regions, whereas the diffusion of angular momentum occurred much more slowly in radiative zones. As a result, the convective envelope of the ES model is nearly in solid-body rotation, whereas the radiative zone is far less effectively coupled and spins down on a much longer time scale. Since the convective envelope of a 1-$M_\odot$ star contains only about 10% of the total stellar moment of inertia, the surface rotation is reduced much faster for a given angular momentum loss than would occur if the star were uniformly rotating.

Qualitative application of the ES ideas for internal diffusion of angular momentum suggests that surface spin-down will take longer in low-mass main sequence stars because the fraction of the total moment of inertia in the convective envelope increases (Stauffer et al. 1984; see also Bohigas et al. 1986). In this way one can explain the sharp difference in surface rotation between G and K dwarfs in the Pleiades without having to postulate an ad hoc turnoff of angular momentum loss at $\sim$K2.

Stauffer & Hartmann (1987) showed that if angular momentum loss is relatively independent of rotation and spectral type, then one can reproduce the rapid change in rotational velocity distribution as a function of spectral type seen in the Pleiades simply from the scaling of the envelope moment of inertia with mass. While this assumption about angular momentum loss is fairly arbitrary and not provable at this point, it does receive some support from the lack of any strong dependence upon (rapid) rotation for tracers of magnetic activity in Pleiades stars, as discussed below.

The other striking aspect of the distributions of rotational velocities shown in Figure 1c is the surprisingly large range of angular momenta compared with that seen in older clusters such as the Hyades (see below). The possibility that rapidly rotating stars are members of close binary systems, where tidal synchronization enforces rapid rotation, is ruled out by the absence of measurable radial velocity variations (Stauffer et al. 1984, Stauffer & Hartmann 1987).

One would like to know whether the spread in rotational velocities indicates a corresponding range in the time interval of star formation, or whether a real spread of angular momentum is present in a coeval cluster that gets damped out later by the wind-braking feedback. The answer to this question is not known at present. The idea that the slow rotators may simply be older objects that have had extra time to spin down is supported by the results of Duncan & Jones (1983), who found a range in lithium abundances among Pleiades dwarfs. Since lithium is destroyed in late-type dwarfs as they age, Duncan & Jones interpreted their results in terms of a large age spread ($\gtrsim 10^8$ yr), which would clearly allow enough time for different amounts of spin-down. Additional evidence for variations in Li abundances has been found by Butler et al. (1987).

However, an age spread of 10^8 yr in this compact cluster is difficult to understand theoretically. This age interval is equivalent to nearly 10^2 free-fall or sound-crossing times. One might expect that energy and momentum input from T Tauri winds would either trigger star formation on much shorter time scales or eject gas from the cluster over periods of 10^6 yr.

This second possibility is especially likely given the low escape velocity (~ 1.5 km s^{-1}) from the cluster (van Leeuwen 1983).

The alternative idea—that rotation in the Pleiades and α Per reflects a broad initial distribution of angular momentum—has the problem that the observed range of $v \sin i$ seems to be proportionately larger in these clusters than among T Tauri stars. However, the data for T Tauri stars contain upper limits to $v \sin i$, and not all T Tauri stars have measured rotation, so this possibility cannot be ruled out.

A third possibility is that an initial spread in angular momentum (or a spread due to differing ages) may be amplified by the nature of the angular momentum loss mechanism. Stauffer & Hartmann (1987) showed that the observed distributions of rotational velocities in these two young clusters could be produced from the "initial" distribution of T Tauri rotation if angular momentum loss is independent of rotation. This causes an increase in the dispersion of angular momentum because the slow rotators lose a larger fraction of their rotation. While rotation-independent braking differs from the Skumanich-type law, Stauffer & Hartmann note that emission activity in the Pleiades, in α Per, and among T Tauri stars does not show the strong dependence on rotation seen in older, more slowly rotating solar-type stars. Thus, the pattern of magnetic activity differs from those objects for which the Skumanich braking was derived. The absence of a strong rotation-activity correlation among very young stars suggests that magnetic braking may also not be strongly rotation dependent. At present we do not have enough information to sort out these possibilities.

OLD CLUSTERS: THE HYADES AND OTHERS   From the work of Kraft (1967) it was known that late-F and early-G dwarfs in the Hyades were slower rotators than comparable stars in the Pleiades. This result was confirmed in more detail by Benz et al. (1984). Measurements of rotational modulation (Lockwood et al. 1984, Radick et al. 1986) now provide direct period determinations without the ambiguity of $\sin i$ for a number of objects in the Hyades. These data indicate a remarkably small spread in rotation ($\sim 20\%$ or so) at a given spectral type (consistent with the $v \sin i$ analysis of Benz et al. 1984). Clearly, there is some strong feedback mechanism operating in the angular momentum loss process for solar-type stars between the ages of the Pleiades and the Hyades, as envisaged by the original Skumanich (1972) relation.

Recently, Stauffer et al. (1987) found that some M dwarfs in the Hyades rotate considerably faster (10–20 km s^{-1}) than do both G and K dwarfs (5 km s^{-1}). This result suggests a distribution of rotation with spectral type qualitatively similar to that observed in the Pleiades but with a turn-on of rapid rotation at a much lower mass. It seems implausible that rapid angular momentum loss should arbitrarily "turn on" at progressively later ages for lower mass stars. The model in which only the convective envelope is spun down at first naturally predicts the observed sequence of rotational velocity distributions. Lower mass stars spin down more slowly because they have larger fractional moments of inertia in their convective zones, which causes the turnup in rotation to occur at progressively lower masses in older clusters.

Benz et al. (1984) also report $v \sin i$ measurements for two other clusters (Coma Ber and Praesepe) of age comparable to the Hyades, obtaining similar rotational velocities. Their results indicate that the braking of rotational velocities between the ages of the Pleiades and 6×10^8 yr is slower than indicated by the Skumanich $t^{-1/2}$ relation. Interpolation between the Hyades and the Sun indicates a spin-down rate slightly faster than the Skumanich relation.

STILL OLDER STARS: THE FIELD Studies of older stars suffer from the lack of nearby clusters with appropriate ages. Other indicators of age, such as kinematic properties or lithium abundances, must be used. In addition, the smaller rotational velocities of these stars make it difficult to measure $v \sin i$ accurately.

So far, the evidence suggests a Skumanich-type spin-down ($v_{\rm rot} \approx t^{-1/2}$; Smith 1979, Soderblom 1983). The Sun appears to have a rotational velocity typical for stars of its age (Soderblom 1983, Gray 1984b). The spread of rotational velocities at a given mass and age is not well constrained by presently available data.

The Mount Wilson program has produced accurate rotational periods for many older main-sequence stars, as reviewed by Baliunas & Vaughan (1985). Although these data do not directly constrain spin down as a function of age, they have permitted detailed analyses of the rotation-activity connection (see below).

Soderblom (1985) used empirical calibrations of the relation between chromospheric activity and rotation (see next section) to infer additional rotational velocities of field stars from Ca II emission measurements. He showed that the distribution of rotation among field stars is also reasonably consistent with the Skumanich spin-down law if one assumes a nearly constant birthrate in the solar neighborhood.

Summary

In Figure 2 we summarize observations of low-mass stellar rotation as a function of age. For comparison we show the results for surface spin-down for one Endal & Sofia (1981) evolutionary model. The ES model was

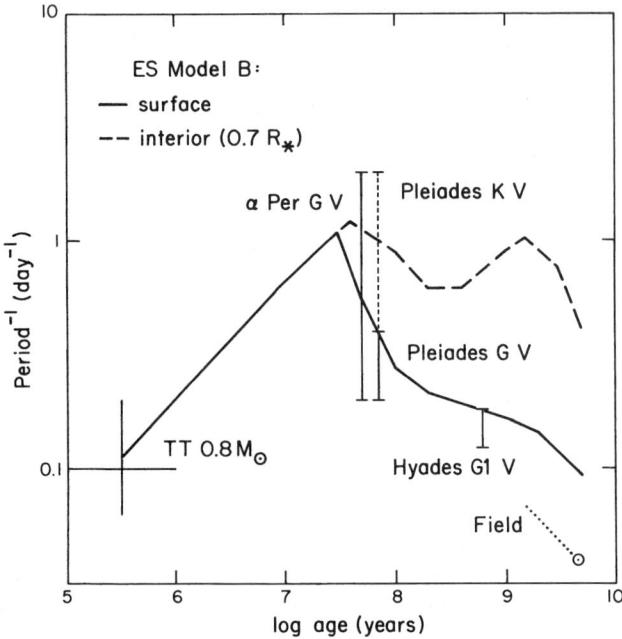

Figure 2 Surface rotational velocities of solar-type stars as a function of age. The rotational evolution of solar Model B of Endal & Sofia (1981) is shown for comparison, both at the surface (solid line) and at an interior level close to the radiative core–convective envelope interface (dashed line). Dotted line is the "Skumanich law" fit to the Sun.

started at an age of 10^7 yr; we have extrapolated back to approximate T Tauri ages by assuming conservation of angular momentum and solid-body rotation, which should not be too bad an approximation for nearly completely convective objects.

The ES models do a remarkably good job in reproducing the qualitative behavior of surface rotation at early ages. They fail in later main-sequence evolution, where the rotation predicted near the bottom of the convective envelope far exceeds the possible interior rotation of the Sun from limits on its gravitational quadrupole moment (as pointed out by ES) or from the presently available helioseismological data that indicate that rotation throughout the convection zone is only slightly slower than the surface rate (Duvall & Harvey 1984, Brown 1985, Libbrecht 1986, Brown & Morrow 1987). This probably means that both surface braking and internal diffusion of angular momentum have been underestimated in late stages.

The evolution of rotation as sketched in Figure 2 indicates that solar-type stars reach their maximum rotational velocities at a transition age of 3×10^7 to 1×10^8 yr and undergo rapid spin-down thereafter. This rapid spin-down may cause large differential rotation between the radiative core and the convective envelope (ES). Some theories suggest that the site of dynamo magnetic field generation is deep in the convective envelope, near the boundary with the radiative core (Parker 1975, Galloway & Weiss 1981, Golub et al. 1981, Spruit & van Ballegooijen 1982). If this is the case, and if the ES picture of early spin-down is qualitatively correct, then the magnetic-field generation of stars during the early phases of surface spin-down may be influenced by the strong radial gradient of differential rotation. The magnetic activity of stars in this phase of spin-down may well exhibit qualitative differences from the activity observed in pre-main-sequence stars, which are more completely convective, and from activity in older main-sequence stars, in which internal redistribution of angular momentum has decreased the radial differential rotation over time.

3. MAGNETIC ACTIVITY DEPENDENCE ON STELLAR PARAMETERS

Overview

It is generally thought that stellar magnetic activity derives from the interaction of stellar rotation and convection through a magnetic dynamo process. Although it is difficult to measure stellar magnetic fields directly, other indicators of magnetic activity can be measured, and observations of one such indicator (Ca II emission) suggest periodic and/or quasi-periodic behavior reminiscent of the solar magnetic activity cycle (Wilson 1978, Baliunas & Vaughan 1985). The idea that rotation plays a crucial role in the generation of stellar magnetic activity has been amply supported by observations [beginning with those of Wilson (1963, 1966), Kraft (1967), and Skumanich (1972)] that various tracers of magnetic activity are enhanced in more rapidly rotating main-sequence stars [see, e.g., Baliunas & Vaughan (1985) and Rosner et al. (1985) for a summary of more recent results].

The relation between rotation and activity, and the rate of decline of both with age, might be expected to depend also on stellar mass or composition because these parameters determine the structure and dynamical behavior of the stellar interior. In particular, the depth of the convection zone and the characteristic time scale of convection, both of which may play a critical role in the magnetic dynamo process, are strongly dependent on stellar mass and more weakly dependent on composition. If, for simplicity, we concern ourselves only with stars of nearly solar composition, we would then expect the empirical rotation-activity-age relation to show a clear mass dependence.

Separating the effects of rotation from the influence of other stellar parameters is an important step toward understanding dynamo generation of magnetic fields. With vastly increased data bases of rotation and magnetic activity measurements, it is now possible to investigate this problem in some detail. The principal difficulty in assessing the importance of various parameters is that the theory is not sufficiently advanced to tell observers what to look for. As Schüssler (1983) has said, "Instead of fitting parameters for ad hoc models to get 'reasonable agreement,' observations should be used as a guide for the development of physical concepts." Unfortunately, this is hard to do because stellar rotation is generally correlated with other parameters (radius, convection-zone properties, age) that might affect magnetic-field generation. In addition, the most easily measured tracers characterizing activity—chromospheric or coronal emission—bear a complex relation to the underlying magnetic activity. There are also uncertainties in our understanding of the internal structure of even main-sequence stars and difficulties in interpreting the observations properly. The result is that several approaches to separating the effects of rotation from internal stellar properties on observed magnetic activity have been devised, each internally consistent and each dependent on initial assumptions that are difficult or impossible to prove at present.

Despite these difficulties, some progress has been made in understanding stellar magnetic-field generation. We consider below a variety of tracers of magnetic activity, each with its own set of observational problems and difficulties of interpretation.

Direct Measurements of Magnetic Fields

Conventional Zeeman polarization measurements have been generally unsuccessful in detecting magnetic fields on late-type dwarfs (Boesgaard et al. 1975, Vogt 1980, Brown & Landstreet 1981). The expected bipolar nature of the stellar fields probably tends to cancel out the circular polarization. (We note that for the Sun seen as a star, the near cancellation of regions of opposite magnetic polarity leads to a net field strength of order 1 G, even though the field strength in spatially resolved magnetic regions generally exceeds 10^3 G.) Borra et al. (1984) found a marginal detection for one dwarf, χ Boo A, on one night with a mean net longitudinal field strength of roughly 25 G. Since other measurements discussed below indicate much larger field strengths, Borra et al. conclude that the magnetic geometry must be extremely complex, composed of several hundred regions with opposite polarities. Tinbergen & Zwaan (1981) found evidence for linear polarization in some late-type stars that they attribute to magnetic fields.

The direct measurement of magnetic fields by means of Zeeman broadening appears to be a more promising technique in that it avoids the field-cancellation effect. Studies of optical spectral lines have been conducted by Robinson et al. (1980), Marcy (1981, 1984), Marcy & Bruning (1984), Gray (1984a), Saar et al. (1986), and Saar & Linsky (1987). For optical lines the excess line broadening due to a magnetic field of 2 to 3 kg covering an appreciable fraction of the stellar surface is extremely modest and shows up principally in the line wings.

Marcy (1984) and Gray (1984a) have produced a number of Zeeman-broadening observations. Marcy observed the magnetically sensitive Fe I line at 6173 Å in a sample of 29 main-sequence stars, reporting detections for 19 objects. These detections are interpreted to result from magnetic fields typically in the range 1500–2000 G and covering factors of 0.3–0.8 of the stellar surface. A general increase in magnetic field strengths and areal coverage was found in later spectral types, a result that agreed qualitatively but not quantitatively with the dynamo prediction of Durney & Robinson (1982).

Marcy observed a correlation of magnetic flux with Ca II emission consistent with the predictions of Ulmschneider & Stein (1982) for chromospheric heating by slow-mode waves. Marcy also found a correlation between the magnetic flux ϕ and stellar parameters of the form $\phi \approx T_{\rm eff}^{-2.8} V_{\rm rot}^{0.55}$. Surprisingly, as Marcy noted, the Sun deviates badly from this relation.

Gray (1984a), using Fourier analysis of several spectral lines, measured Zeeman broadening in the spectra of 7 out of a sample of 18 late-type dwarfs. He detected fields only in stars later than G6, in general agreement with the trend toward larger fields in cooler stars found by Marcy. Remarkably, Gray (1985) showed that magnetic-field measurements from a variety of observers using a variety of lines and modeling techniques suggest a nearly constant value for the total magnetic flux.

Saar & Linsky (1987) reported preliminary broadening measurements for 20 late-type stars. They found a correlation between magnetic-field strength and photospheric gas pressure, as might be expected. They also found only weak correlations between rotation and field strength but observed a strong correlation between rotation and the fraction f of the stellar surface covered by magnetic fields. An even stronger correlation was found between f and the ratio of convective overturn time to rotational period, the latter being an important parameter of dynamo theory (see below). They found no evidence for Gray's (1985) constant magnetic-flux relation.

In deriving the field strengths and areas from such observations, one assumes that the surface brightness of the magnetic regions is the same as that of the nonmagnetic photosphere. This is surely a simplification of the true situation, but the data are not capable of supporting more detailed modeling (Gray 1984a). Typical starspot models show that the spots are likely to be quite "black" in the optical spectral region; thus, these magnetic-field measurements are more likely to refer to plage or facular areas and probably miss spot magnetic fields entirely.

It is surprising that these "plage" areas, with magnetic-field strengths approaching sunspot-field strengths, cover a large fraction (up to 80%) of the visible surface of these stars, whereas any dark spots present cover very small areas on many stars, given the observed photometric variations of $\lesssim 2\%$ for many objects (Dorren & Guinan 1982, Radick et al. 1983). It is surprising that the "plage" areas, with fields large enough to perturb photospheric conditions, do not produce larger disk-averaged photometric variability directly. One might argue that the absence of observed photometric rotational modulation implies substantial surface homogeneity, but this is difficult to justify given the observed rotational modulation of Ca II emission (cf. Baliunas & Vaughan 1985).

The subtlety of the Zeeman-broadening effect not only requires high signal-to-noise data, but also demands extreme care in the interpretation of the results. For example, numerical experiments indicate that rotational broadening of spectral lines begins to compromise the detection of Zeeman effects at reported levels for $v \sin i > 5$ km s^{-1}. At 10 km s^{-1}, Zeeman broadening at usual levels can be difficult to detect, which perhaps explains why some fairly rapidly rotating, active G dwarfs do not show any Zeeman signal (Marcy 1984, Gray 1984a).

Another problem is that the optical Zeeman signal is typically a small perturbation in the line wings. Because the vast majority of lines studied for Zeeman broadening are saturated, accurate field measurements require a careful calculation of the line profile, including the effects of atmospheric structure, turbulent velocities, and line strengths. Some uncertainties in line profile modeling can be minimized by differential analysis, in which comparisons are made between lines with roughly similar strengths from the same multiplet but with different magnetic sensitivities (Marcy 1984, Gray 1984a, Saar et al. 1986). However, saturated lines of different strengths do not have precisely similar shapes. Modest adjustments of poorly known damping constants used in the Voigt profile can introduce large differences in the derived field strengths. Normalizing to the saturated cores, as in Marcy's (1984) technique, will necessarily leave differences in the line wings, where the weak optical Zeeman signal is observed. Often the lines chosen for their strongest Zeeman sensitivity also have the largest *gf* values and so have the deepest wings from saturation effects alone. More recent analyses treat this effect more carefully (Saar & Linsky 1987, Basri & Marcy 1986). Because of these possible systematic problems of interpretation, one should treat the results from these difficult measurements with some caution.

Probably the least ambiguous measurements of magnetic fields have been made in the infrared spectral region, where the Zeeman splitting is larger relative to other line-broadening processes (Giampapa et al. 1978, Gondoin et al. 1985, Saar & Linsky 1985). For reasonable field strengths, one can observe the Zeeman triplet as individual components in the infrared, not just as slightly deeper wings superimposed on a normal line profile. Saar & Linsky (1985) found fields of ~ 3–4 kG with covering factors of 70–80% for the dMe star AD Leo. Unfortunately, with present infrared equipment not many stars can be observed in this way.

Emission Tracers of Magnetic Activity

The process responsible for heating the tenuous outer atmospheres of late-type stars are not understood at present, but solar observations show a clear correlation of high-temperature emission with magnetic activity. A number of emission lines and continua formed at various temperatures higher than the stellar effective temperature can thus be used as tracers of magnetic fields. While the calibration of these tracers in terms of magnetic properties is difficult or poorly known, there is a huge literature on the subject, principally due to the ease of detection of such tracers.

Many lines and continua formed between temperatures of 6000 K to $\gtrsim 10^7$ K can be used as magnetic tracers. Coronal stellar X-ray emission has been reviewed recently by Rosner et al. (1985), and recent results on ultraviolet chromospheric and transition-region emission have been discussed by Linsky (1983, 1985), Jordan et al. (1984), and Zwaan (1986). In general these tracers correlate well with age and rotation, confirming the trend expected for magnetic activity.

At present, observations of the Ca II H and K chromospheric emission lines offer advantages over observations of any other type of diagnostic for determining the effect of stellar parameters on magnetic activity. Surveys in the X-ray and ultraviolet regions are substantially incomplete owing to limitations of satellite observing time, telescope size, or detector efficiency. A sample in which many of the data consist of upper limits is said to be censored, a situation that obviously makes quantitative study of stellar activity more difficult [see Rosner et al. (1985) for a discussion of the problems created by substantial censoring of the X-ray data]. In contrast, censoring of the Ca II HK data is minimal if attention is restricted to stars of spectral type G or later. Moreover, a huge data base of consistent measurements has been accumulated at Mount Wilson, begun by O. C. Wilson in the late 1960s and continued by A. H. Vaughan and collaborators up to the present (see the review by Baliunas & Vaughan 1985).

Solar data indicate that Ca II emission is an extremely sensitive diagnostic of magnetic fields (e.g. Skumanich et al. 1975). Ca II emission is also directly relatable to other types of emission produced by magnetic activity, such as Mg II h and k, far-ultraviolet lines from chromospheric and transition-zone species like C II, H Lα, He II, C IV, N V, etc., and coronal X rays. Many researchers (e.g. Ayres et al. 1981, Oranje et al. 1982, Schrijver 1983, 1986, Hartmann et al. 1984a, Marilli & Catalano 1984, Zwaan 1986, Marilli et al. 1986, Basri 1987) have found that there is a tight correlation between activity-related emission in these lines both for dwarfs and for evolved stars of different masses and rotation rates. After correcting for any emission component not related to magnetic activity (important for chromospheric lines such as Ca II H and K, as described below), Schrijver (1983, 1986) found, for example, that various chromospheric and transition-zone lines are essentially proportional, while coronal X-ray emission scales approximately as the 3/2 power of chromospheric emission indicators. This means that the structure of magnetically heated features in stellar atmospheres has a (perhaps surprising) similarity for stars of very different effective temperatures, surface gravities, and photospheric convective motions. The results also have the practical consequence that many of the data needed to survey stellar activity may be acquired by observing Ca II emission.

MOUNT WILSON CA II HK EMISSION MEASUREMENTS, AND THE VAUGHAN-PRESTON "GAP" For a number of years, beginning with O. C. Wilson's first measurements of stellar Ca II H and K emission in the early 1960s, there has been an ongoing program at Mount Wilson of studying this emission in many stars. An important milestone of this project was the publication (Wilson 1978) of evidence for activity cycles in other stars (to be discussed below). A second milestone was the publication (Vaughan & Preston 1980) of a survey of emission levels in over 400 late-type main-sequence stars in the solar neighborhood. The results of this survey are shown in Figure 3.

In their paper, Vaughan & Preston (1980) called attention to the appearance of two concentrations of stars in the $B-V$ range 0.4–1.0, rather well separated by their mean level of HK emission (see Figure 3). From kinematic arguments they showed that the strong-emission stars were younger, extending the results of Kraft (1967) and Wilson (1963). The results suggested a rapid decay in emission in stars of intermediate ages and possibly even a discontinuous drop in activity level between the two groups (Vaughan & Preston 1980, Durney et al. 1981, Knobloch et al.

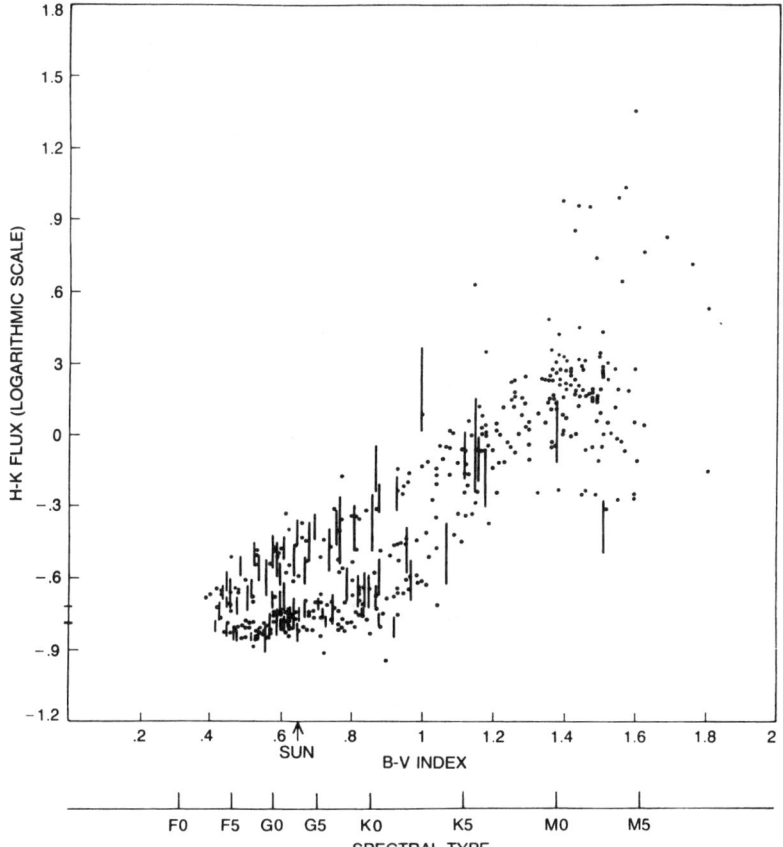

Figure 3 The survey of Ca II emission in solar neighborhood stars by Vaughan & Preston (1980). S is an index proportional to the line-center emission relative to the Ca II line wings (see text). The solid lines indicate the excursion of activity index S for stars monitored by Wilson (1978) for long-term activity variations.

1982, Middelkoop 1982). This interpretation of the Vaughan-Preston data led to suggestions of a change in dynamo behavior between young and old G stars (e.g. Knobloch et al. 1982, Durney et al. 1981). Originally this was supported by suggestions that the young stars did not exhibit the long-term cyclic behavior observed in many older objects (Vaughan 1980), but evidence for cyclic periods has now been found among young survey stars (Baliunas & Vaughan 1985).

Before discussing the origin of the two concentrations of stars in the Vaughan-Preston diagram, or alternatively the cause of the apparent gap in the emission activity, it is important to understand how the data were taken. The Mount Wilson HK photometer is a four-channel chopping spectrometer (Vaughan et al. 1978) that yields the value of the Ca II flux index S, proportional to the total equivalent width of the Ca II H and K emission reversals and photospheric light in 1-Å bands centered on H and K. In Figure 4 we show the bandpass of the HK photometer near the Ca II K line, superimposed upon a quiet Sun spectrum and the spectrum of a plage region. By an Eddington-Barbier argument, one expects that the damping line wings are formed mostly in the upper photosphere, whereas the central emission and the final absorption reversal are formed in the chromosphere. Comparison between the plage and quiet Sun spectra provides evidence for this schematic division; the central emission increases by a much larger factor than the wing emission for magnetically active regions.

Thus, the Mount Wilson HK photometer measurement for stars like the Sun is substantially "contaminated" by photospheric emission. The photospheric contribution results in a nonzero measured HK flux for a star even if it has *no* chromosphere. This photospheric contamination is larger for hotter stars, making it impossible to detect Ca II chromospheric emission in stars hotter than about middle F and rendering its detection in low-activity late-F and early-G stars difficult.

Hartmann et al. (1984b) showed that in spite of the apparent discontinuity represented by the Vaughan-Preston gap, the data could be modeled with a smoothly varying decay of Ca II emission with age. The pileup of points in the lower left of the diagram is produced by the photospheric "floor" discussed above. The upper concentration of points can be approximately reproduced by assuming that the chromospheric emission exponentially saturates to a constant level in young stars, a result originally suggested by Barry et al. (1981) from lower resolution data. Hartmann et al. showed that this distribution was consistent with fragmentary unpublished observations by A. H. Vaughan for the Pleiades. Barry et al. (1984) found a similar dependence of emission on age. X-ray observations of the Pleiades similarly show a tendency toward saturation (Caillault & Helfand 1985). Neither the fragmentary Pleiades data nor the Vaughan-Preston survey are consistent with the Skumanich law (Ca II emission $\sim t^{-1/2}$) for stars 10^8 yr old or less.

Even with the effect of saturation of emission, Hartmann et al. could not quite reproduce the concentration of strong-emission stars. However, plausible variations in the local stellar birthrate of a factor of two on time scales of a few times 10^8 yr could provide agreement with observation. The analysis also suggested that there must be a quantitative change in the *rate* of emission decay with time, even if there is no discontinuity in emission

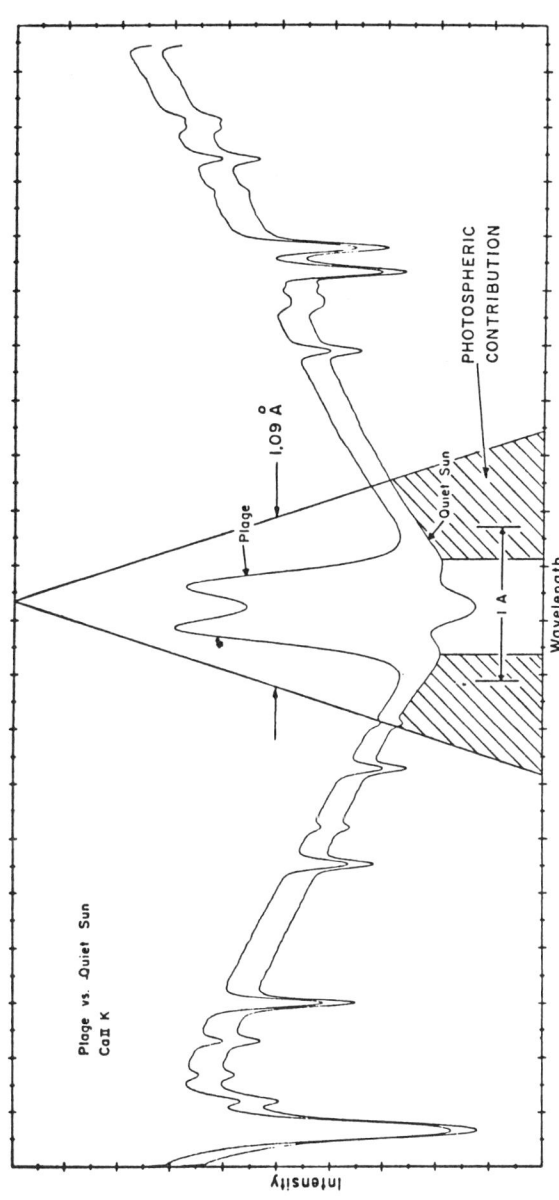

Figure 4 The K-line bandpass of the instrument used to produce the Vaughan-Preston (1980) survey and other Mount Wilson Ca II HK flux measurements, superimposed on both quiet Sun and plage spectra (Hartmann et al. 1984b).

levels. The decay must be slow at first in order to populate the upper branch of stars in the Vaughan-Preston diagram, accelerating later so that there is a real underdensity of stars below this upper branch. Under these circumstances the apparent near-absence of stars in the "gap" can be no more than a statistical fluctuation, i.e. the data do not require that there be a sharp decrease in magnetic activity generation to produce the gap.

A slow decay in activity at ages of $\sim 10^8$ yr, followed by a later stage of rapid decay with age, has also been inferred from studies using other tracers (e.g. Caillault & Helfand 1985, Simon et al. 1985, Herbig 1985). In Figure 5 we plot the variation of several emission tracers of magnetic activity. Pre-main-sequence stars exhibit a wide range of emission fluxes, and Simon et al. (1985) have suggested that the decay of emission activity in T Tauri stars is different from the slow weakening of activity observed in main-sequence stars. Many tracers indicate a "plateau" of emission between the ages of the Pleiades and the Hyades. One explanation of this behavior is that the fractional stellar surface area covered by magnetic fields is approaching unity (Saar & Linsky 1987) and thus is producing a saturation in activity. Another possibility is that core rotation is still rapid

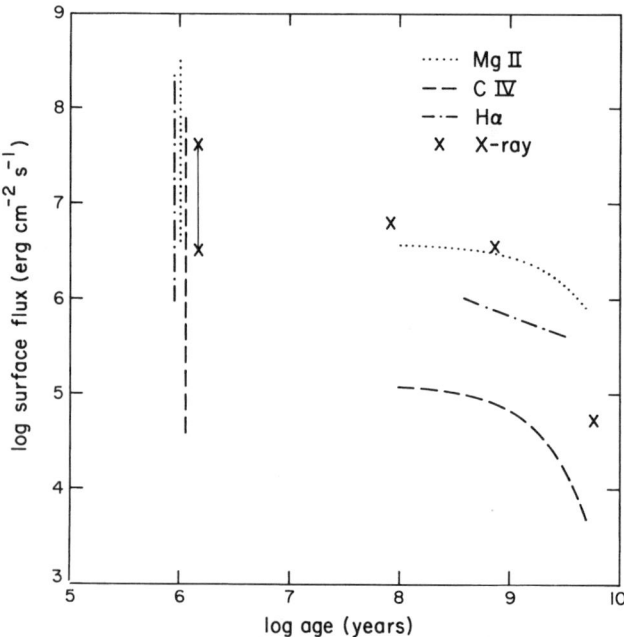

Figure 5 Variation of emission levels in solar-type stars with age for a variety of tracers of magnetic activity (from Herbig 1985, Caillault & Helfand 1985, Simon et al. 1985).

in the "plateau" phase (see Figure 2), and so magnetic fields generated near the convective envelope–radiative core interface do not weaken directly with surface spin-down.

The correction process used by Zwaan and colleagues for the "basal" HK flux (e.g. Zwaan 1986) led Rutten & Schrijver (1986) to conclude that there is a "knee" in the relation between excess surface flux $\Delta F_{\rm HK}$ and rotation period $P_{\rm rot}$ that occurs at a value of surface flux located near the top of the Vaughan-Preston gap. This knee is much sharper than the gradual change in the rate of emission decay found by Hartmann et al. (1984b), and the rapid decay of emission helps to produce the gap. However, the basal component must be added to the Rutten & Schrijver emission decay relations in order to model the original Vaughan-Preston data. Adding the basal flux results in an emission law equivalent to that used by Hartmann et al. (1984b).

Rutten & Schrijver (1986) make the important point that the Vaughan-Preston "gap" is not apparent for $B-V > 1.0$ (see Figure 3), which suggests a qualitative difference in the decay of emission for stars with $M \lesssim 0.8\,M_\odot$. One wonders whether this change in emission activity might be related to the change in convective-zone properties proceeding down the main sequence, or whether instead it is the result of the different rotational histories of G dwarfs in comparison with the rotational spin-down of K–M stars, as discussed in Section 2.

THE EMISSION-ROTATION RELATION: THE ROSSBY PARAMETER   Mean-field dynamo theories generally are characterized by a dimensionless parameter, the dynamo number $N_{\rm D}$ (Parker 1979). Models generally require a minimum value of $N_{\rm D}$ for magnetic-field amplification, and the resultant fields are expected to be enhanced for larger values of $N_{\rm D}$ (cf. Durney & Robinson 1982, Robinson & Durney 1982). In general terms, we have

$$N_{\rm D} = \frac{\alpha \Omega' d^4}{\eta^2},$$

where α is the product of the mean helicity of convection $\langle \mathbf{v} \cdot (\nabla \times \mathbf{v}) \rangle$ and its characteristic convective overturn time $\tau_{\rm c}$, Ω' is the radial gradient of angular velocity, d is the characteristic length scale of the convection, and η is a turbulent magnetic diffusivity. Noting that α may scale as Ωd and making the plausible (but not necessarily correct) assumption that Ω' scales like Ω/d (cf. Durney & Latour 1978), we find that $N_{\rm D} \sim (\Omega \tau_{\rm c})^2 \sim N_{\rm R}^{-2}$, where $N_{\rm R}$ is the Rossby number, or the ratio of the rotation period to the convective overturn time.

In an attempt to explore whether the Rossby number is indeed a valid way to parameterize magnetic activity, Noyes et al. (1984a) compared mean chromospheric activity levels observed with the Mount Wilson HK photometer with estimates of the Rossby number derived from the observed rotation period and calculated convective overturn times τ_c. [An analysis along similar lines was independently carried out by Mangeney & Praderie (1984).] Following Gilman (1980), τ_c was set equal to the convective overturn time one scale height above the bottom of the convective zone. The ratio $Ro \equiv P_{rot}/\tau_c$ is used to characterize the Rossby number N_R in the deep convection zone, where dynamo generation of magnetic fields is thought to occur. Because this ratio involves the measured rotation at the surface and a theoretical convective overturn time deep in the convection zone and therefore does not strictly equal the Rossby number, we denote Ro the Rossby parameter.

An uncertain term in this analysis is the ratio of mixing length to pressure scale height, L/H (assumed to be the same for all stars along the lower main sequence); larger values of L/H (that is, larger convective efficiency) translate into deeper convective zones, with longer convective overturn times at their base. It was found that a choice of $L/H = 1.9$ minimizes the scatter of R'_{HK} versus Ro, and in fact the scatter is slightly less than in an alternative plot of F'_{HK} versus P_{rot} (Figure 6). This by itself does not indicate that the former is preferable to the latter, since the addition of an extra free parameter (L/H) should of course decrease the scatter.

In principle, the choice of L/H is not completely free; independent arguments from helioseismology and from stellar interior calculations (summarized by Noyes et al.) suggest that $L/H \approx 1.5$–2 is reasonable.

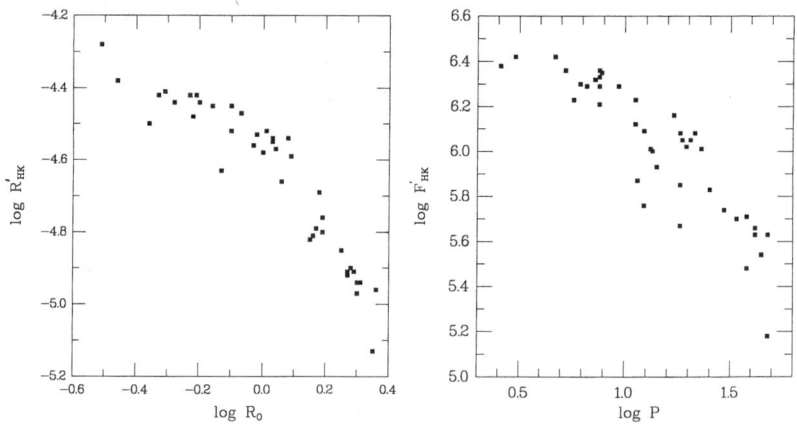

Figure 6 The dependence of the normalized Ca II chromospheric flux R'_{HK} (corrected for the line-wing contribution) on the Rossby parameter Ro, and the dependence of Ca II surface flux F'_{HK} on stellar rotational period P (from Noyes et al. 1984a).

However, the problem of determining the "correct" characteristic time scale that should enter into the Rossby parameter goes beyond the selection of L/H. It requires an assumption of where in the convection zone the dynamo process is operating, since the convective overturn time is strongly depth dependent; Gilman's time scales were for one scale height above the bottom of the convection zone, under the assumption that the dynamo process was localized in this region (as opposed, for example, to being distributed throughout the convection zone).

The problem of determining from observations alone whether a Rossby parameter or the rotation period itself best parameterizes stellar activity is compounded by uncertainty over whether the emission ratio R'_{HK}, the surface flux per unit area F'_{HK}, the HK emission luminosity L'_{HK}, or some other emission indicator best characterizes magnetic activity. It turns out that the ratio $R'_{HK}/F'_{HK} = (\sigma T^4)^{-1}$ has roughly the same dependence on spectral type as τ_c for G and K stars if $L/H \simeq 1.9$; thus, the question of whether the "correct" dependence is one of R'_{HK} on Ro, F'_{HK} on P_{rot}, or some other relation is difficult to answer based simply on the criterion of minimal observational scatter in an empirical activity-rotation relationship (Figure 6). Zwaan (1986), Rutten & Schrijver (1986), and Basri (1986, 1987) suggest that the surface flux F is the best variable to parameterize stellar activity because the relations between different chromospheric, transition-zone, and coronal emissions, when plotted for stars of different masses and luminosity, show less scatter if the emission is expressed in terms of F rather than the flux ratio R or emission luminosity L.

Theory also provides little support in finding the appropriate parameters with which to characterize magnetic activity. Theories of the generation of acoustic and magnetic waves, which may heat the chromosphere, suggest that the output flux should be dependent upon stellar effective temperature (cf. Ulmschneider & Stein 1982). The exact temperature behavior, however, depends upon the mode being excited and the (unknown) field strengths present. Furthermore, the theory deals only with the generation of waves in the deep photosphere. The exact amount of wave energy dissipated to produce observable emission in the outer atmosphere may well have additional dependencies on mass or (equivalently) effective temperature.

Different ways of correcting for the fraction of emission within the HK photometer bandpass which is not due to magnetic activity (see, e.g., Linsky & Ayres 1978) also lead to different conclusions about the nature of the mass-dependent term in the activity-rotation relationship. While Noyes et al. found that the activity indicator R'_{HK} depends only on Ro, Zwaan and colleagues (using basically the same data) found it impossible to fit any activity indicator to Ro independent of color.

The reasons for these differing conclusions are worth outlining in some detail to illustrate the problems of interpreting the emission-rotation relation. Noyes et al. (1984a) subtracted off only the photospheric wing flux from the HK measurements (Figure 4) and found similar shapes of the R'_{HK} relation for stars of differing spectral types. On the other hand, Schrijver (1983, 1986), Oranje & Zwaan (1985), Rutten (1986), and Rutten & Schrijver (1986) all subtracted off a larger flux component (the "basal" flux), which is largest for F and G dwarfs and rapidly declines for later spectral types. The residual "magnetic" Ca II emission then falls off more rapidly with rotation or age for G dwarfs (where the correction is larger) than for K dwarfs (where the correction is smaller).

A primary justification for applying this "basal" flux correction is that it produces the tightest correlations between Ca II excess flux and other indicators (principally X-ray emission). Schrijver (1983, 1986) verified this through a common-factor analysis of Ca II and various chromospheric, transition-region, and X-ray emission indicators. His results depend critically on the assumed power-law form of the correlation. However, there is presently no theoretical reason why the ratio of coronal to chromospheric emission should have any particular form, and since the basal flux minimization technique involves a fit using two free parameters, one wonders whether other parameterizations with two free parameters might similarly reduce the scatter from the fit.

Given the many uncertainties in determining the appropriate indicator of magnetic activity–induced emission and the uncertainties in characterizing stellar rotation and convection, the case for considering the Rossby parameter as the basic determinant of stellar magnetic activity is not airtight. A primary argument in its favor remains the key role that such a parameter plays in mean-field dynamo theory, coupled with the rather tight observed relations between emission ratios and Ro.

It should be noted that the rotation-activity relations discussed above do not appear to hold for all rotating main-sequence stars with convection zones. At the early end of the lower main sequence, Wolff et al. (1986) studied C IV $\lambda 1549$ emission and He I $\lambda 5876$ for F stars and noted that activity sets in about at F0 ($B-V \sim 0.28$, $T_{\rm eff} \sim 7300$ K). However, for the early-F stars ($7300 > T_{\rm eff} > 6600$ K) the activity is quite high and shows no dependence on rotation or on Ro, whereas for late-F stars ($6600\ {\rm K} > T_{\rm eff}$) they find the same relation between C IV emission and Ro as that found by Noyes et al. (1984a). Wolff et al. argue that the physical mechanism responsible for chromospheric and coronal emission is different for stars hotter and cooler than $T_{\rm eff} = 6600$ K, and they suggest that acoustic heating could play the dominant role in the heating of early-F star atmospheres. [They also note, following Walter (1983), that primordial magnetic fields, rather than dynamo-generated fields, could be producing the activity in early-F stars.]

At the late end of the lower main sequence, Doyle (1987) shows that there is a good correlation between Mg II $\lambda 2800$ emission and Rossby parameter Ro for stars as late as M4.5. For even later stars, departures might be expected from such an activity-rotation relationship. For example, if the magnetic dynamo is a shell dynamo, operating at the interface between the convection zone and the radiative zone, magnetic activity should cease beyond spectral type $\sim$ M5, when stars become fully convective. Nevertheless, recent data by Giampapa & Liebert (1986) indicate that at least some stars later than M5 show Hα emission characteristic of a nonradiatively heated chromosphere, which implies that regenerative dynamo action does not require the existence of a convectively stable boundary at the base of the convection zone.

One way to sort out the effects of rotation and stellar structure is to include evolved stars in the analysis, obtaining wider ranges in various stellar parameters (e.g. Basri 1986, 1987). Gilliland (1985) calculated the variation of convective overturn times as stars evolve off the main sequence. He found that for stars more massive than the Sun, these evolutionary effects can have important observable effects on stellar activity and spin-down. For example, an evolutionary increase of τ_c could explain the surprisingly large rotation and activity of the F7 subgiant ι Vir, given its calculated age of 4.5×10^9 yr. The models indicate that ι Vir has spent most of its lifetime with a much smaller convective overturn time than at present, and thus until recently the star has been much less active and has experienced correspondingly less spin-down. Unfortunately, uncertainties in stellar interior structure are greater for giants than for main-sequence stars. Also, many evolved stars where activity has been studied are members of binary systems, whose surface layers at least have been spun up by tidal interaction with a binary companion. The single stars might well have outer convective envelopes that are more slowly rotating than their radiative cores, whereas the situation could be reversed in an evolved binary, thus leading to qualitative changes in activity.

Magnetic Cycle Periods

Perhaps the most direct contact between observation and dynamo theory is the determination of magnetic cycle periods. The data require no theory relating emission levels to magnetic-field strengths to determine how long-term relative variations in activity level relate to stellar properties. Although a variety of techniques have been employed to search for stellar magnetic cycles, the most complete, accurate, and consistent record comes from the initial observations of Ca II emission by Wilson (1978), supplemented by the later monitoring from Mount Wilson (Vaughan et al. 1981, Baliunas et al. 1983). In this case, as for disk-integrated sunlight, the Ca II H and K resonance-line emission serves as a tracer for the spot cycle. The H and K data on stellar activity cycles have recently been thoroughly reviewed by Baliunas & Vaughan (1985), so we do not discuss them further here.

Initially, Vaughan (1980) argued from visual inspection that active, rapidly rotating main-sequence stars did not possess solar-type cycles. This finding was interpreted to mean that rapid rotation excites high-order dynamo modes (Durney et al. 1981) or that a change in convection-zone patterns in rapidly rotating stars substantially alters the dynamo (Knobloch et al. 1982). However, this observational result did not stand up to additional data that increased the time baseline, as well as to more objective measures of periodicity (Baliunas et al. 1987). It now appears that cycles are present among stars with varying levels of activity.

A number of attempts have been made to determine how activity cycle periods should scale with P_{rot} and with spectral type. Since τ_c near the bottom of the convection zone is uniquely related to spectral type, this either implicitly or explicitly amounts to a scaling with the two parameters P_{rot} and τ_c. Belvedere et al. (1980) calculated a series of five dynamo models for main-sequence stars between F5 and M0 and found that the period of the dynamo oscillation increased with advancing spectral type, with the increase between F5 and G5 being more than a factor of 10. They attributed this increase to the increasing depth of the convection zone with advancing spectral type. A more gradual increase, by a factor of 4, from G5 to M0 was attributed to the increase of magnetic diffusivity with advancing spectral type.

Robinson & Durney (1982; see also Durney & Robinson 1982) arrived at just the opposite conclusion: that cycle periods should decrease with advancing spectral type for stars of a given rotation period P_{rot}. Their result was based on the consideration that magnetic amplification is possible only if the magnetic amplification time t_A is equal to or shorter than the buoyant rise time t_R of magnetic fields. Calculations based on this consideration predicted that for later-type stars having the same P_{rot}, the surface fields should be stronger and the amplification time t_A (assumed to be proportional to the cycle period) shorter. Also, for a given spectral type, the cycle period is shorter for shorter rotation periods.

Noyes et al. (1984b) analyzed the Mount Wilson data for the Sun and 12 other slowly rotating field dwarfs of spectral type G2 to K7, all of which had been found in Wilson's (1978) data to show quasi-cyclical behavior like that of the Sun. They found for this limited sample that the measured cycle periods, for stars of similar spectral type, varied as P_{rot}^n, where

$n = 1.25 \pm 0.5$. For stars of a given rotation period, P_{cyc} decreases with increasing $B-V$, in agreement with the prediction of Robinson & Durney (1982) but counter to the results of Belvedere et al. (1980). When all spectral types were considered together, it was found that the cycle period varied as Ro^n with essentially the same exponent n. This behavior was compared with predictions of analytical nonlinear dynamo wave models developed by N. Weiss and collaborators: If the saturation of the dynamo is attained by magnetic freezing of the differential rotation or of the α-affect, thereby quenching the dynamo action, then the dynamo period is independent of P_{rot} (so that $n = 0$); however, if the saturation occurs through buoyant loss of magnetic flux, then the dynamo period is directly proportional to P_{rot} (so that $n = 1$). The observations appear to be compatible with the second mechanism, but not the first.

To the extent that activity cycle periods can be shown to vary with the Rossby parameter Ro, independent of spectral type, this would provide strong support for the idea that the Rossby number, rather than the rotation rate, is the fundamental determinant of stellar activity. Unfortunately, there are a number of uncertainties in the analyses of Noyes et al (1984b), including the fact that the number of stars sampled is very small, and their period estimates are based only on a number of years of data less than twice the derived cycle periods. Furthermore, there may be significant selection effects [only slowly rotating stars are considered, and the relation may not hold for more rapidly rotating stars (see Baliunas et al. 1987)].

Baliunas and colleagues (see Baliunas 1986) have brought the original survey of Wilson (1978) up to date, extending the period of analysis for 99 lower main-sequence stars to 18 yr. The data, and their analysis through periodogram studies, clearly show that long-term activity variations cannot be described by one simple formula. About one eighth of the stars in the survey show essentially no long-term variation of the HK index S, even though they may not be the weakest chromospheric emitters in their class. Some show large-amplitude long-term variability but without any clearly dominant period in the periodogram analysis. Others show a more or less steady decrease or increase in S, which implies that if there is a dominant period, it must be decades long. The remaining stars (about three fifths of the total) are cyclic or likely so, with apparent periods ranging from 2.6 yr through about 12 yr. Some stars show indications of multiple periodicity in their periodograms. Cyclic variability is present in stars as early as F2 and as late as M0, but because these stars are near the extremes of spectral type in the sample, this does not imply a cutoff of cyclic behavior at these spectral types. Selection effects are doubtless important also in the longest periods measured, which approach the duration of the monitoring period, and there is no reason to doubt the existence of significantly longer activity cycles. In this connection it may be noted that an apparent cyclical variation with about a 60-yr period is present in the photometric data on the dK5e star BD+26°730 (Hartmann et al. 1981), presumably owing to a starspot cycle.

It remains to be seen whether or not the continuous monitoring of HK variability in Wilson's original list of stars (and other stars that may be added) will verify or refute the relation between cycle period and Rossby parameter found by Noyes et al. (1984b). It is already clear that the analysis problems are complex owing to the very nonperiodic nature of activity cycles, the small ratio of observing time span to cycle periods, and the significant noise introduced by stellar active-region growth and decay. It may require a number of additional years of monitoring before definitive conclusions are possible.

4. CONCLUSIONS AND FUTURE PROSPECTS

The tight correlations observed between stellar emission activity and rotation show that magnetic-field generation in stars is a deterministic process controlled by global stellar parameters. The well-defined activity-rotation relations resulting from improved observational techniques hold out hope that satisfactory theories of magnetic-field generation can ultimately be constructed.

It appears that we are beginning to detect the effects of convection-zone properties and rotation on stellar magnetic activity. Theory suggests that the ratio of rotation period to convective overturning time may be a crucial parameter for magnetic-field generation. There is some observational support for this idea, but other interpretations of the data are possible. Extensions of activity surveys to include stars with a wider range of internal properties will help to sort out the various possibilities. Accurate measurements of surface magnetic fluxes would help calibrate emission tracers, removing ambiguities of interpretation or even eliminating the need for indirect measurements. The availability of such measurements may depend strongly on improvements in near-infrared spectroscopic techniques.

The detection of roughly 10-yr cycles in old solar-type stars and the measurement of short cycles in some young stars suggest that cycle period measurements may also be an important clue to understanding magnetic dynamos. Longer baselines of observation are needed. Although improving cycle data will be a lengthy task, advances in small-telescope automation suggest that the necessary data can be acquired relatively painlessly.

Eventually better constraints on stellar interior structure and velocity fields will be needed for realistic dynamo theories. While the necessary stellar oscillation data may not be acquired soon, it is possible that recent observations of mass-dependent rotational evolution can be exploited to place limits on internal differential rotation and to provide the foundations for a better understanding of magnetic-field generation in nearby astrophysical systems.

ACKNOWLEDGMENTS

We acknowledge useful conversations with Sallie Baliunas, John Stauffer, and Aad van Ballegooijen, and assistance in preparing figures from Rob Hewett and Bob Donahue.

Literature Cited

Ayres, T. R., Marstad, N. C., Linsky, J. L. 1981. *Ap. J.* 247: 545
Baliunas, S. L. 1986. In *Cool Stars, Stellar Systems, and the Sun*, ed. M. Zeilik, D. Gibson, p. 3. Berlin: Springer-Verlag
Baliunas, S. L., Donahue, R. A., Duncan, D. K., Frazer, J., Gilliland, R., et al. 1987. In preparation
Baliunas, S. L., Vaughan, A. H. 1985. *Ann. Rev. Astron. Astrophys.* 23: 379
Baliunas, S. L., Vaughan, A. H., Hartmann, L., Middelkoop, F., Mihalas, D., et al. 1983. *Ap. J.* 275: 752
Barry, D. C., Cromwell, R. H., Hege, K., Schoolman, S. A. 1981. *Ap. J.* 247: 210
Barry, D. C., Hege, K., Cromwell, R. H. 1984. *Ap. J. Lett.* 277: L65
Basri, G. 1986. In *Cool Stars, Stellar Systems, and the Sun*, ed. M. Zeilik, D. Gibson, p. 184. Berlin: Springer-Verlag
Basri, G. 1987. Preprint
Basri, G., Marcy, G. 1986. *Bull. Am. Astron. Soc.* 18: 984
Belvedere, G., Paterno, L., Stix, M. 1980. *Astron. Astrophys.* 91: 328
Benz, W., Mayor, M., Mermilliod, J. C. 1984. *Astron. Astrophys.* 138: 93
Boesgaard, A. M., Chesley, D., Preston, G. W. 1975. *Publ. Astron. Soc. Pac.* 87: 353
Bohigas, J., Carrasco, L., Torres, C. A. O., Quast, G. R. 1986. *Astron. Astrophys.* 157: 278
Borra, E. F., Edwards, G., Mayor, M. 1984. *Ap. J.* 284: 211
Bouvier, J., Bertout, C., Benz, W., Mayor, M. 1986. *Astron. Astrophys.* 165: 110
Brown, D. N., Landstreet, J. D. 1981. *Ap. J.* 246: 899
Brown, T. M. 1985. *Nature* 317: 591
Brown, T. M., Morrow, C. A. 1987. Submitted for publication
Butler, R. P., Cohen, R. D., Duncan, D. K., Marcy, G. W. 1987. In preparation

Caillault, J.-P., Helfand, D. J. 1985. *Ap. J.* 289: 279
Cohen, M., Kuhi, L. V. 1979. *Ap. J. Suppl.* 41: 743
Dorren, J. D., Guinan, E. F. 1982. *Astron. J.* 87: 1546
Doyle, J. D. 1987. *MNRAS*. In press
Duncan, D. K., Jones, B. F. 1983. *Ap. J.* 271: 663
Durney, B. R., Latour, J. 1978. *Geophys. Astrophys. Fluid Dyn.* 9: 241
Durney, B. R., Mihalas, D., Robinson, R. D. 1981. *Publ. Astron. Soc. Pac.* 93: 537
Durney, B. R., Robinson, R. D. 1982. *Ap. J.* 253: 290
Duvall, T. L., Harvey, J. W. 1984. *Nature* 310: 19
Endal, A. S., Sofia, S. 1981. *Ap. J.* 243: 625
Galloway, D. J., Weiss, N. O. 1981. *Ap. J.* 243: 945
Giampapa, M. S., Golub, L., Worden, S. P. 1978. *Ap. J. Lett.* 268: L121
Giampapa, M. S., Liebert, J. 1986. *Ap. J.* 305: 784
Gilliland, R. 1985. *Ap. J.* 299: 286
Gilman, P. 1980. In *Stellar Turbulence, IAU Colloq. No. 51*, ed. D. Gray, J. Linsky, p. 19. New York: Springer-Verlag
Golub, L., Rosner, R., Vaiana, G. S., Weiss, N. O. 1981. *Ap. J.* 243: 309
Gondoin, Ph., Giampapa, M. S., Bookbinder, J. A. 1985. *Ap. J.* 297: 710
Gray, D. F. 1984a. *Ap. J.* 277: 640
Gray, D. F. 1984b. *Ap. J.* 281: 719
Gray, D. F. 1985. *Publ. Astron. Soc. Pac.* 97: 719
Hartmann, L. 1985. *Sol. Phys.* 100: 587
Hartmann, L., Baliunas, S. L., Duncan, D. K., Noyes, R. W. 1984a. *Ap. J.* 279: 778
Hartmann, L., Bopp, B., Dussault, M., Noah, P. V., Klimke, A. 1981. *Ap. J.* 249: 662

Hartmann, L., Hewett, R., Stahler, S., Mathieu, R. D. 1986. *Ap. J.* 309: 275
Hartmann, L., Soderblom, D. R., Noyes, R. W., Burnham, N. 1984b. *Ap. J.* 276: 254
Hartmann, L., Soderblom, D. R., Stauffer, J. R. 1987. *Astron. J.* In press
Herbig, G. H. 1985. *Ap. J.* 289: 269
Herbig, G. H., Vrba, F. J., Rydgren, A. E. 1986. *Astron. J.* 91: 575
Jordan, C., Judge, P., Johansson, S. 1984. In *Ultraviolet and X-Ray Spectroscopy of Astrophysical and Laboratory Plasmas, IAU Colloq. No. 86*, ed. G. A. Doschek, p. 51. Washington, DC: Nav. Res. Lab.
Kappelmann, N., Mauder, H. 1981. *ESO Messenger* 23: 18
Knobloch, E., Rosner, R., Weiss, N. O. 1982. *MNRAS* 197: 45P
Kraft, R. P. 1967. *Ap. J.* 150: 551
Libbrecht, K. G. 1986. *Nature* 319: 753
Linsky, J. L. 1983. In *Solar and Stellar Magnetic Fields: Origins and Coronal Effects, IAU Symp. No. 102*, ed. J. O. Stenflo, p. 313. Dordrecht: Reidel
Linsky, J. L. 1985. *Sol. Phys.* 100: 333
Linsky, J. L., Ayres, T. R. 1978. *Ap. J.* 220: 619
Lockwood, G. W., Thompson, D. T., Radick, R. R., Osborn, W. H., Baggett, W. E., et al. 1984. *Publ. Astron. Soc. Pac.* 96: 714
Mangeney, A., Praderie, F. 1984. *Astron. Astrophys.* 130: 143
Marcy, G. W. 1981. *Ap. J.* 245: 624
Marcy, G. W. 1984. *Ap. J.* 276: 286
Marcy, G. W., Bruning, D. H. 1984. *Ap. J.* 281: 286
Marilli, E., Catalano, S. 1984. *Astron. Astrophys.* 133: 57
Marilli, E., Catalano, S., Trigilio, C. 1986. *Astron. Astrophys.* 167: 297
Mestel, L. 1984. In *Cool Stars, Stellar Systems, and the Sun*, ed. S. L. Baliunas, L. Hartmann, p. 49. Heidelberg: Springer-Verlag
Middelkoop, F. 1982. *Astron. Astrophys.* 107: 31
Noyes, R. W., Hartmann, L. W., Baliunas, S. L., Duncan, D. K., Vaughan, A. H. 1984a. *Ap. J.* 279: 763
Noyes, R. W., Weiss, N., Vaughan, A. H. 1984b. *Ap. J.* 287: 769
Oranje, B. J., Zwaan, C. 1985. *Astron. Astrophys.* 147: 265
Oranje, B. J., Zwaan, C., Middelkoop, F. 1982. *Astron. Astrophys.* 110: 30
Parker, E. N. 1975. *Ap. J.* 198: 205
Parker, E. N. 1979. *Cosmical Magnetic Fields: Their Origin and Their Activity.* Oxford: Clarendon. 860 pp.
Radick, R., Duncan, D., Baggett, W. E., Thompson, D. T., Lockwood, G. W. 1986. In preparation
Radick, R. R., Worden, S. P., Wilkerson, M. S., Africano, J. L., Klimke, A., et al. 1983. *Publ. Astron. Soc. Pac.* 95: 300
Robinson, R. D., Durney, B. R. 1982. *Astron. Astrophys.* 108: 322
Robinson, R. D., Worden, S. P., Harvey, J. W. 1980. *Ap. J. Lett.* 236: L155
Rosner, R., Golub, L., Vaiana, G. S. 1985. *Ann. Rev. Astron. Astrophys.* 23: 413
Roxburgh, I. W. 1983. In *Solar and Stellar Magnetic Fields: Origins and Coronal Effects, IAU Symp. No. 102*, ed. J. O. Stenflo, p. 449. Dordrecht: Reidel
Rutten, R. G. M. 1986. *Astron. Astrophys.* 159: 291
Rutten, R. G. M., Schrijver, C. 1986. In *Cool Stars, Stellar Systems, and the Sun*, ed. M. Zeilik, D. Gibson, p. 120. Berlin: Springer-Verlag
Rydgren, A. E., Vrba, F. J. 1983. *Ap. J.* 267: 191
Rydgren, A. E., Vrba, F. J., Chugainov, P. F., Shakhovskaya, N. I. 1985. *Bull. Am. Astron. Soc.* 17: 556
Saar, S. H., Linsky, J. L. 1985. *Ap. J. Lett.* 299: L47
Saar, S. H., Linsky, J. L. 1987. *Proc. COSPAR Meet., 26th.* In press
Saar, S. H., Linsky, J. L., Beckers, J. M. 1986. *Ap. J.* 302: 777
Schaefer, B. 1983. *Ap. J. Lett.* 266: L45
Schatzman, E. 1962. *Ann. Astrophys.* 25: 18
Schrijver, C. J. 1983. *Astron. Astrophys.* 127: 289
Schrijver, C. J. 1986. PhD thesis. Univ. Utrecht, Neth.
Schüssler, M. 1983. In *Solar and Stellar Magnetic Fields: Origins and Coronal Effects, IAU Symp. No. 102*, ed. J. O. Stenflo, p. 213. Dordrecht: Reidel
Simon, T., Herbig, G. H., Boesgaard, A. M. 1985. *Ap. J.* 293: 551
Skumanich, A. 1972. *Ap. J.* 171: 565
Skumanich, A., Smythe, C., Frazier, E. N. 1975. *Ap. J.* 200: 747
Smith, M. A. 1979. *Publ. Astron. Soc. Pac.* 91: 737
Soderblom, D. R. 1983. *Ap. J. Suppl.* 53: 1
Soderblom, D. R. 1985. *Astron. J.* 90: 2103
Soderblom, D. R., Jones, B. F., Walker, M. F. 1983. *Ap. J. Lett.* 274: L37
Spruit, H. C., van Ballegooijen, A. A. 1982. *Astron. Astrophys.* 106: 58
Stauffer, J. R., Hartmann, L. 1986. *Publ. Astron. Soc. Pac.* 98: 1233
Stauffer, J. R., Hartmann, L. 1987. *Ap. J.* In press
Stauffer, J. R., Hartmann, L., Burnham, N., Jones, B. 1985. *Ap. J.* 289: 247
Stauffer, J. R., Hartmann, L., Latham, D., Stefanik, R. 1987. Submitted for publication
Stauffer, J. R., Hartmann, L., Soderblom,

D. R., Burnham, N. 1984. *Ap. J.* 280: 202
Tinbergen, J., Zwaan, C. 1981. *Astron. Astrophys.* 99: 173
Ulmschneider, P., Stein, R. F. 1982. *Astron. Astrophys.* 106: 9
van Leeuwen, F. 1983. PhD thesis. Leiden Univ., Neth.
van Leeuwen, F., Alphenaar, P. 1982. *ESO Messenger* 28: 15
van Leeuwen, F., Alphenaar, P., Meys, J. J. M. 1987. *Astron. Astrophys. Suppl.* In press
Vaughan, A. H. 1980. *Publ. Astron. Soc. Pac.* 92: 392
Vaughan, A. H., Baliunas, S. L., Middelkoop, F., Hartmann, L., Mihalas, D., et al. 1981. *Ap. J.* 250: 276
Vaughan, A. H., Preston, G. W. 1980. *Publ. Astron. Soc. Pac.* 92: 385
Vaughan, A. H., Preston, G. W., Wilson, O. C. 1978. *Publ. Astron. Soc. Pac.* 90: 267
Vogel, S. N., Kuhi, L. V. 1981. *Ap. J.* 245: 960
Vogt, S. S. 1980. *Ap. J.* 240: 567
Vrba, F. J., Rydgren, A. E., Zak, D. S., Chugainov, P. F., Shakhovskaya, N. I. 1984. *Bull. Am. Astron. Soc.* 16: 998
Walter, F. M. 1983. *Ap. J.* 274: 794
Walter, F. M., Brown, A., Mathieu, R., Myers, P., Vrba, F. 1987. In preparation
Wilson, O. C. 1963. *Ap. J.* 138: 832
Wilson, O. C. 1966. *Ap. J.* 144: 695
Wilson, O. C. 1978. *Ap. J.* 226: 379
Woolf, S. C., Boesgaard, A. M., Simon, T. 1986. *Ap. J.* 310: 360
Zwaan, C. 1986. In *Cool Stars, Stellar Systems, and the Sun*, ed. M. Zeilik, D. Gibson, p. 19. Berlin: Springer-Verlag

THE LOCAL INTERSTELLAR MEDIUM

Donald P. Cox and Ronald J. Reynolds

Department of Physics, University of Wisconsin, Madison, Wisconsin 53706

The local interstellar medium (LISM) has been an active area of investigation for at least 15 years, but for the first several of those it was uncommon to appreciate what was being learned in this area by others. The story began as many widely dispersed rivulets, which collected and converged on Madison, Wisconsin in June 1984 for IAU Colloquium No. 81 on the Local Interstellar Medium. Many of the principals in LISM work were present, and an adequate history could be assembled by tracing back through the references in the proceedings of that meeting (62). In making that history, however, it will make no more sense to pick a particular stream as primary than it does to legislate just where in La Salle's Louisiana one finds the origin for the Mississippi River. Since IAU Colloquium No. 81 occurred only two years ago, was well organized, and the proceedings are nicely edited into major subject groupings, nearly all but the most recent developments can be found conveniently referenced there (62). The proceedings (43) of a second LISM meeting, Symposium 7 of COSPAR XXVI in Toulouse, July 1986, provide some updated discussions, introduce additional principals, and add further topics.

GLOSSY OVERVIEW

For the purposes of this review, the local interstellar medium will be taken to consist of material within the nearest 10^{19} H atoms cm^{-2}. The distance to that column density is surprisingly large, ranging from roughly 30 pc in the Galactic plane to as much as 200 pc within 20° of the north Galactic pole.

Within that column-density contour there is an irregularly shaped volume (see Figure 1), wasp waisted in the plane and very extended in the north, for which the most prevalent conditions are a temperature of 10^6 K, a density of about 5×10^{-3} cm^{-3}, and therefore a thermal pressure of about 10^4 cm^{-3} K. The volume of this region is essentially equal to that of a sphere of radius 100 pc. The hot gas within it generates the soft X-ray background via its normal thermal excitation processes.

Early in this review, we consider whether the Sun occupies a typical location in the interstellar medium (ISM). Although an entirely satisfactory answer cannot be given, we can at least be reasonably confident that X-ray-emitting gas does not continuously pervade a large volume fraction of interstellar space. More particularly, the X-ray-emitting gas right around us appears to be limited in actual extent to the region we see, rather than our ability to see somewhat more distant gas being limited by absorption. This is not a conclusion easily reached, owing to a strong negative correlation between the distributions of large column densities of hydrogen (e.g. several times 10^{20} cm^{-2}) and the surface brightness of the soft X-ray background (SXRB). The effect looks very much like absorption; we review some of the evidence showing that it almost certainly is not.

Having convinced ourselves that the soft X-ray background derives from a limited region, we can comfortably name that region the Local Bubble and consider local events and processes involved in its origin or reason for being. Such causes are obliged to provide the cavity occupied by the hot gas, as well as the hot gas itself. We note that although the thermal energy present now within the Local Bubble could have been contributed by a single supernova explosion, that explosion could not also have generated the observed cavity in an originally average (possibly distinct from typical or probable) interstellar environment.

The low density in the Local Bubble has at least two possible interpretations. If a large fraction of the general interstellar medium is occupied by a phase of typical density $n \lesssim 10^{-2}$ cm^{-3}, then there is no further need for local cavity production. On the other hand, a cavity of the necessary scale can also be blown in an interstellar medium of average density by the winds and explosions of a moderate number of stars acting in concert. Presumably such bubble blowing will leave behind association B stars that are further down the main sequence from those that did the work. The nearest of these associations is Sco-Cen, which at a distance of $\simeq 210$ pc is outside the Local Bubble and more likely responsible for another hot bubble nearby, circumscribed by Loop I. If association bubble blowing is responsible for the Local Bubble, then the latter is almost surely of great age. It may be a feeble remnant of a major event some 2×10^7 yr ago that has been postulated from time to time to help understand the larger scale distributions of material and velocities (and even cosmic rays) in the solar neighborhood. It must be remembered, however, that the Local Bubble is extremely small compared with Gould's Belt and the realm of phenomena previously discussed in this regard.

In addition to there being at least two possible reasons for the LISM's low density, there are also at least two reasons for the high temperature. One possibility, rather tightly constrained, is that the hot gas is bounded by an active shock wave, today still raising interstellar material to 10^6 K. For the model to work, the ambient density must have been roughly 4×10^{-3} cm^{-3} to 10^{-2} cm^{-3}, the explosion energy must have been about that of one supernova, and the explosion must have occurred about 10^5 yr ago somewhere near the Sun. The other serious possibility is that the Local Bubble consists of pressure-confined hot gas and that the last supernova occurrence(s) within it was at least 10^6 yr ago (more likely nearly 10^7 yr ago or longer). Such hot gas has a roughly 10^7-yr lifetime against radiative losses at ISM pressures; the Local Bubble has a total luminosity of roughly 10^{36} erg s^{-1}. (Its volume is sufficiently large that this luminosity is conceivably available from steady energy input from low-energy cosmic rays, stellar winds, or other nonexplosive phenomena that we disregard.)

The essential beauty of the possibility that the Local Bubble is a quiescent region of hot gas is that such regions can be produced by the same sort of explosion needed for the active blast-wave model, but they last as much as 100 times longer. This increases our odds of being within such a region, by about the same factor of 100. Alternatively, the explosion of perhaps a dozen supernovae in a region of average interstellar density roughly 10^7 yr ago could have both created the cavity and left the required hot interior. These pressure-confined models would have essentially four parameters for explosion production: the ambient preexplosion density, the explosion energy, the ambient pressure with which the explosion-produced hot bubble will equilibrate, and the age serve as such a set. Two of these parameters (say, the explosion energy and ambient density) collapse to determine one observable, the bubble radius. After formation, the bubble has a rapid early evolution driven by conduction within the hot gas, a long period of nearly constant appearance, and then a rapid demise with the onset of significant radiative cooling. Because of the slow evolution of the bubble over most of its life, its age parameter is essentially irrelevant so long as it falls within the bubble duration. There are therefore two dominant parameters—the ambient pressure p and the bubble radius R—together with a relation for the required explosion energy versus density. Since the ISM pressure and the cavity scale are known without reference to the X-ray observations, it is encouraging that the bubble temperature T predicted to accompany p and R is in fact the bubble temperature one observes via the soft X-ray emission. Then, given p, R, and T, the soft X-ray brightness is automatically of the magnitude observed. One waxes euphoric before recalling that in ISM work, there are far too many coincidences and little evidence.

Pressure-confined bubble models, though attractive, are precarious. A standard distribution of diffuse clouds inside the bubble, undergoing thermal evaporation at the theoretical rate, would both shorten the lifetime and lower the characteristic temperature to the point that the SXRB would have to derive from a young hot phase, returning us to the active blast-wave regime. Alternatively, the fact that the quiescent bubbles (without clouds) have lifetimes 100 times longer and should be correspondingly more common than their young active counterparts is a little embarrassing when one has estimated the probability of being within the latter to be about 10% (24). In short, the supernova rate in the solar neighborhood must be lower than the Galaxy-wide average in order for long bubble durations to have validity. Without such a reduction, the Local Bubble would have supernova energy input greatly in excess of its apparent radiative loss rate in X rays.

Taking a small side trip, let us suppose that the supernova rate is not small, but that the Local Bubble does include a theoretician's cloud population. Radiation in evaporative boundaries of the embedded clouds could then provide a match between bubble heating (by the supernovae) and cooling. In such a system, however, the bubble properties are independent of the bubble scale. A bit of searching for consistency soon leads essentially to a McKee & Ostriker (73) ISM model.

As it happens, however, the supernova rate locally is likely to be small enough not to endanger the quiescent bubble time scale, which leaves cloud evaporation as the most serious problem. It thus becomes necessary to look within the Local Bubble to see what things are found there. What stars, what clouds, what other preying denizens are in fact within the X-ray emitting volume? In order to aid in the study of this question, we discuss the best estimated bubble geometry, based on the brightness distribution of the SXRB. With this geometry, we can avoid being misled in discussions about what nearby interstellar material is actually inside the bubble.

If the LISM includes the average density of interstellar clouds and these clouds are spherical, then the number of clouds inside the Local Bubble should be of order 2000. If the clouds are sheetlike or are arranged in sheetlike collections of ~ 100-pc coherence, then roughly zero should be within the Bubble, with the ones expected being its walls. From such remarks it is clear that observations rather than theory are necessary to determine the local cloud population, even given that it is "average." Observations, in fact, show that at high latitude there is considerably less H I than average, that what hydrogen there is, is smoothly distributed rather than being concentrated in standard diffuse clouds, and that quite close to the Sun there are none of the standard diffuse clouds that might have been expected (with perhaps one exception). It is possible that the formation of the Local Bubble cleared the region of clouds, or that clouds have sheetlike distributions in the ISM as a whole, or both. In any case, no standard or denser cloud (except, possibly, the feature seen in α Oph) has yet been found within the Local Bubble as defined in Figure 1. We expect that in most directions the Bubble so defined will have $N_H \lesssim 10^{19}$ cm^{-2}, appreciably less than the typical column density of a single cloud.

We are not saying that there are no clouds of any sort within the Local Bubble. There are low-density features, which if they were numerous could evaporatively endanger quiescent bubbles. Perhaps more interesting are how such features can themselves survive in the bubble environment and what their properties can tell us about that environment.

There are indications of the existence of these low-density clouds (or fluff) within the Local Bubble from a variety of observations. Interpre-

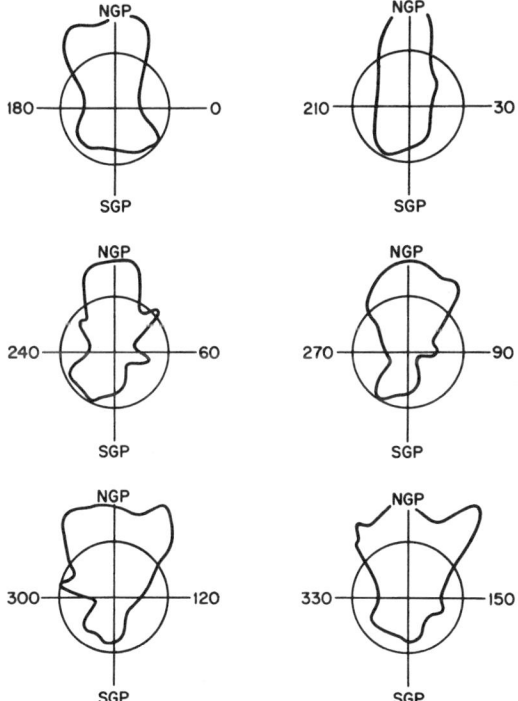

Figure 1 Preliminary maps in polar slices of the X-ray-emitting cavity. Slices follow longitudes shown. Circles have radii of 100 pc [from (97)].

tation of backscattered solar UV radiation shows the Sun to be inside such a cloud, in relative motion at about 20 km s^{-1}, with $n \sim 10^{-1}$ cm^{-3} and $T \sim 10^4$ K. Lines of sight to neighboring stars with measured H I column densities indicate that this very diffuse cloud extends roughly 3 to 5 pc in several directions. Lines of sight to more distant stars tend to show average H I densities that are much lower (so long as the stars are within the Bubble), and it is not uncommon for the $\sim 10^{18}$ cm^{-2} of the Local Fluff to be the entire N_H seen. There are, however, indications of a larger concentration of material ($N_H \sim 10^{19}$ cm^{-2}) of large angular extent in the general direction of the Galactic center roughly 5 to 15 pc from the Sun. Such a cloud could provide a region to test the Local Bubble model of the SXRB if very soft X-ray observations can be obtained there.

The ionization state of the Local Fluff is of considerable interest. A number of observations suggest that the gas has a modest to substantial ionization fraction ($n_e/n \sim 0.3$–0.7), but there is no observed UV source in the solar vicinity that could maintain that ionization. The anticipated level from the observed sources is roughly $n_e/n \simeq 0.09$. The ionization could be out of equilibrium, showing exposure to a more vigorous UV flux roughly 10^5 to 10^6 yr ago, a point that is occasionally used in favor of the recent and still active local supernova models. There could also be a local but otherwise unobserved UV source (for example, a conduction boundary on the outside of the Local Fluff). Or, perhaps, the indirect indicators of significant ionization are incorrect.

Results of attempts to measure the velocity structure in the Local Fluff have varied between suggestions that there is overall coherent motion, with gradual change in direction with increasing scale, to recent observations showing three separate velocity components in stars less than 5 pc distant. Each of these components, however, is found to show coherence over a very large solid angle. There is thus a tantalizingly complex velocity structure on the smallest scale. At a larger scale, the observed velocities indicate that the geometry proposed in Figure 1 is necessarily transient on a time scale of 10^6 to 10^7 yr. Simple views of the velocities involved and their relations to the quiescent bubble possibility are considered at the close of this review and provide an example of the tentativeness of much of our understanding of this field.

QUESTIONS

Three major questions are addressed in this review:

1. What is the LISM like?
2. Why is it like that?

3. Is the LISM typical of the ISM as a whole?

Consider first question 3. Quite different replies are commonly given to it, less because of our ignorance of properties of the LISM than because of our disagreements over those in more distant regions. The following responses span the range of possibilities:[1]

3a. The LISM is typical, with a normal distribution among phases and no real discontinuity distinguishing it from neighboring regions (25, 56).
3b. The LISM is somewhat typical. Its low density is like that of the intercloud medium elsewhere, but its properties resemble those in that small percentage of the general ISM that has recently experienced supernova reheating. Thus, the local value of the intercloud temperature is abnormally high (22–24, 73).
3c. The LISM is distinctly unusual, dominated by the presence of a large cavity of exceedingly low density and high temperature, as compared with interstellar norms (24, 26, 32, 51, 97).

Without the correct answer to question 3, there is no hope of knowing why the LISM has its properties (question 2). Are we trying to understand interstellar normalcy or a gross departure from normalcy? What should we model? What should be considered ambient? and what should be taken as the perturbation? Depending on which answer is correct, the Sun could be in a fairly quiescent interstellar environment, or within the bounds of a rapidly expanding young (1.5×10^5 yr old) supernova remnant, or even within the heavily evolved zone of a major interstellar event some 2×10^7 yr ago. Presumably all three could be true. The picture is, in any case, full of the raging of interstellar storms and the great churning engine that has so dramatically participated in giving us being.

We turn now to properties of the LISM as we seem to know them, with a brief introduction to major attributes, concepts, and quarrels.

PROPERTIES OF THE LOCAL BUBBLE

The Vacancy

Although it has always been difficult to detect local interstellar matter, the absence of detection was stared at for quite some time before it inverted, to the detection of an absence. With the development of UV, EUV, X-ray, and very sensitive optical astronomy, local interstellar material has been

---

[1] The three responses are all opinions that have been held by one of us (DPC), in the order listed, starting about 1973.

found and the situation has clarified. Yet the recurring surprise has been the extreme rarity of matter in much of nearby interstellar space.

A simple picture that seems so far to be consistent with the data is that on a scale of several tens of parsecs, the Sun is surrounded by an irregularly shaped region whose most probable density is only about 4×10^{-3} cm^{-3}. The pressure in this region is not particularly unusual by interstellar standards (thermal $p/k \approx 10^4$ cm^{-3} K), so that the low-density material has a temperature of about 10^6 K and is observable via its soft X-ray emission (62, 93, 98).

The suggestion above, that the interstellar pressure normally has $p/k \approx 10^4$ cm^{-3} K, is not universally accepted, chiefly because a number of measurements of *thermal* pressures in clouds have tended to imply that $p/k \sim 3000$ cm^{-3} K. But vertical integration of the weight distribution of interstellar material yields a total $p/k \gtrsim 22{,}000$ cm^{-3} K (e.g. 6a, 26). Of this, roughly 10^4 cm^{-3} K is supplied by the observed cosmic rays and cloud motions, while the large-scale average field is estimated to contribute about 2000 cm^{-3} K. Thus, about 10^4 cm^{-3} K are needed to complete this budget. This is consistent with the estimated thermal pressure in the X-ray-emitting gas around us but is two to three times larger than the thermal pressure within clouds, as measured by most techniques (64a). This discrepancy exists independent of Local Bubble conditions. It may be that magnetic pressure in H I regions has been underestimated (a 5-μG field solves the problem), and some of the error may derive from the large-amplitude fluctuations of the field that are expected to be associated with waves in the clouds (22, 23). See (64a) for a more complete discussion of the data underlying this problem.

The Boundary Type

There has been considerable controversy over whether the Local Bubble of hot gas has an actual boundary beyond which there is no further X-ray emission, or whether instead it is part of a widespread interstellar phase that has denser clouds of X-ray absorbing material interspersed within it, limiting our vision to the local portion (56, 57, 68, 69, 93). For brevity we refer to these interpretations as the emission- and absorption-bounded options.

A major piece of evidence is that the brightness of the soft X-ray background shows significant negative correlation with the measured hydrogen column density along the line of sight (13, 14, 68, 69). This should be characteristic of absorption-bounded visibility at high Galactic latitude if much of the X-ray emission lies beyond most of the absorbing material (14). Such a view is attractive, given the strong pole brightening of the observed X rays, particularly in the north. (Absorption of the relevant energies is mainly photoelectric, from hydrogen and helium.) This negative correlation has been compared with a variety of models, including one in which the soft X-ray background at high latitudes was largely cosmological in origin (14), one in which it derived mainly from a hot corona of the Milky Way (e.g. 68), and one dominated by the hot-phase emission of the ISM in a distribution somewhat thicker than that of the H I (56).

There are other pieces of evidence, however, that strain the credibility of absorption models. The negative correlation of the SXRB with N_H is only about half as great as it should be, given the known absorption cross section (13). If the emissions were largely extragalactic or coronal, this would require that the N_H be clumped into optically thick clouds on a scale smaller than the X-ray measurements. The average transparency would then be increased because absorbing atoms are self-shielded, with areas of higher transmission between clouds. On the other hand, if the emission were interstellar, then intermixing of absorbing clouds with the emission also reduces the average absorption (29, 38, 56). It is possible to mock up the observed behavior by adjustments to the scale heights of absorbing and emitting regions and to the degree of clumpiness of the ISM. In the process, one also fits the observed degree of scatter in the SXRB vs. N_H relation (56).

But the absorption cross section is highly energy dependent, so that the softer X rays (e.g. B band) should be noticeably more sensitive to absorption than the harder ones (C band). [See McCammon et al. (69) for a general survey of SXRB topics.] In simple configurations a negative correlation introduced by absorption should thus be accompanied by an additional negative correlation between the X-ray intensity and its spectral hardness. Except for the North Polar Spur, which clearly does show absorptive behavior, this second expectation is *not* confirmed.

Various distributions of emitting and absorbing material that lessen or avoid this predicted effect of absorption have been considered. A venerable possibility is that *all* of the absorbing material is clumped to rather high optical depth ($N_H \gtrsim 2 \times 10^{20}$ cm^{-2}), making absorption gray. But careful studies of the high-latitude distribution of interstellar material have shown conclusively that it is not clumped to anything like the degree required (31, 54, 55).

Further complication is provided by recent measurements of the SXRB at yet lower energies (Be band) over about 10% of the sky (for which there is about a factor of two to three range in the count rate at higher energies) (10a). The ratio of Be to B band is amazingly constant, given that the Be band would be significantly absorbed by $N_H \sim 10^{19}$ cm^{-2}. The implications are somewhat model dependent. Clumping models are made even more unreasonable, since the smooth H I component is so heavily constrained. It is also fairly certain that the *variations* in any N_H lying entirely between the X-ray emission and us cannot exceed roughly 2×10^{18} cm^{-2} over the broad swath of sky examined thus far. It is very tempting to infer that the Be band counts arise from emission closer than the nearest 5×10^{18} cm^{-2} of hydrogen, and that by implication the B band counts also arise from this limited region, even though the B band quite clearly anticorrelates with hydrogen column densities exceeding 10^{20} cm^{-2}. Another possibility that bears consideration is one in which the absorbing and emitting material are intermixed to large optical depth (61). On average, this scenario also provides a spectrum that is independent of total optical depth. It allows for variations in intensity in different directions through variations in the average absorber density (or volumetric emission rate). It suffers, however, from requiring these directional variations while also requiring a considerable uniformity along each line of sight (after averaging over a fairly large solid angle). Roughly speaking, the ratio of emissivity to absorption density would have to be the same over the nearest 10^{19} cm^{-2} along the path as it is over the nearest 7×10^{19} cm^{-2}. (The total column densities for the Be band–sampled directions ranged from 2 to 8×10^{20} cm^{-2}.) An additional problem is that although embedded absorber models can be found that provide either the global and N_H variation of the C (and B) band maps (although suffering from unreasonable clumping requirements) or the constant intensity ratio of Be to B bands (although requiring possibly unreasonable line-of-sight uniformity), no embedded absorber model so far proposed can do both.

There are probably schemes that would allow the SXRB vs. N_H anticorrelation to be attributed to absorption, but no effect of absorption other than the negative correlation with N_H has been found, after years of careful searching. If absorption is responsible, it has been playing a diabolical game.

An alternative view is that absorption has little or nothing to do with the SXRB because nearly all of the X-ray emission occurs closer than any significant amount of absorber ($N_H \lesssim 5 \times 10^{18}$ cm^{-2}). In this view, the Local Bubble is emission bounded, and one's only difficulty is understanding the N_H anticorrelation.

Such an alternative has long gone under the name "displacement" (63, 93). The basic idea is that around the solar location there is an extensive but irregular ISM cavity. This cavity is of such a scale that it "displaces" a significant fraction of the hydrogen that would normally be found at high latitudes, with the amount displaced being roughly proportional to the extent of the cavity. The X-ray emission is imagined to arise entirely within the cavity, with the brightness proportional to the cavity extent (constant volume emissivity). Clearly, when the cavity extent differs between two directions, the direction of greater extent has brighter X-ray emission and lesser hydrogen column density. Although this sounds somewhat ad hoc, it has recently been shown that the introduction of an X-ray-emitting cavity of the observed scale[2] into a gaseous disk with the observed average vertical distribution of hydrogen, $n(z)$, would produce exactly the observed anticorrelation (97, 98). There are no free parameters. The model also reproduces the observed local paucity of high-latitude hydrogen compared with what would be expected based on measurements at high z elsewhere. The fits are almost equally good, no matter whether the displaced hydrogen is assumed to be moved radially to the cavity boundary or is somehow removed entirely. The latter picture works slightly better. In the displacement model the SXRB and N_H actually each vary separately with cavity extent rather than through a direct connection, and the model works quantitatively to relate the observed extents, densities, and anticorrelation. There are, of course, no constraints implied on any X-ray emission that might occur behind opaque screens.

Stated somewhat more carefully, we can see from column-density measurements to stars that there is a large cavity in the local H I distribution (39, 81), and this cavity is found reflected in local reductions of N_H seen via 21-cm emission at high latitudes (66). From comparison with the soft X-ray background, we find the cavity to be apparently filled with hot gas and containing very little absorptive material. The total N_H is beyond the cavity boundary and is modulated, as is the X-ray intensity, by the cavity extent. See (18) for further evidence in this regard.

Thus, let us provisionally assume that the Local Bubble is emission bounded. The hot gas seen in the soft X rays does not extend beyond the distance to which we can see it. The LISM is not sufficiently typical to warrant answer 3a.

The Remaining Possibilities

The chief difference between answers 3b and 3c is whether the general ISM has a pervasive intercloud component with density less than, say, 10^{-2} cm^{-3}. If it does, then a single local supernova could have made the temperature near us high enough to emit soft X rays. If, however, the pervasive low-density phase does not exist, then something must have caused the density near us to be abnormally low.

It is useful to note at the outset that even within these two possibilities there are significant subdivisions and even some blurring of the divisions. For later clarity, let us list the possibilities. Under 3b, the Local Bubble

---

[2] Observed via onset of significant N_H measurements in stars beyond the cavity boundary.

could be (*a*) the actively growing remnant of a supernova explosion. But because the interstellar pressure is large, a successful model could also be made in which (*b*) the explosion has completed its rapid growth, leaving behind a pressure-confined bubble of hot gas that is gradually cooling off (51).

Similarly, with regard to possibility 3c (that the low density seen locally is a cavity), the X-ray emission could be due to (*c*) a young remnant reheating the diffuse material within the cavity, (*d*) an older event that has reheated the entire cavity, leaving a confined hot bubble of diffuse gas with denser surroundings, or even (*e*) a single explosion of sufficient magnitude to have both created the local cavity and left the interior hot.

Possibilities (*a*) and (*c*) both portray the Local Bubble as being a young active remnant and can be discussed together. Possibilities (*b*) and (*e*) are examples of old pressure-confined bubbles that are gradually cooling, while (*d*) is similar but inertially confined by the dense cavity wall.

Finally, in such an extensive enumeration we should probably include caveats. Perhaps a diffuse ISM energy source will be found that can drive regions of low density to coronal conditions. Perhaps the tunnel networking of the general coronal phase picture can have local hot spots due not so much to a recent reheating as to an absence of surfaces on which to dump heat. There must be other possibilities that will make our current attempts at understanding seem naive and/or contrived.

The Local Bubble as an Active Supernova Remnant

It may be incorrect to assume that the X-ray emissivity in the Local Bubble is fairly uniform per unit volume. Consider those models in which the bubble represents a somewhat typical region of the ISM, but one in which the lowest density material has been recently reheated by a supernova explosion. Then, depending on the interactions between the remnant and its embedded clouds and on the initial distribution near the explosion site, the X-ray emissivity can be anywhere from very heavily edge brightened to centrally peaked within the remnant (the latter for remnants that have been actively involved in thermal evaporation of clouds).

In these scenarios, it is less clear how to regard the notion of displacement. One is required to have the remnant surface brightness correlate with distance to the remnant boundary, which in turn presumably correlates negatively with the distribution of mean density so far encountered in each direction. In short, running into less causes more rapid growth, and in the Local Bubble it must also result in *brighter* X rays.

The situation is complicated, but as a rule the expected correlation is in the other direction. This is quite apparent, for example, in the Cygnus Loop. X rays are brightest and the spectrum softest where the growth has been slowed (18a, 48, 64). On the other hand, for nonevaporative remnants it has been shown (32) that if the postshock temperature is roughly 0.8×10^6 K in directions of great extent, then in regions of lesser extent the emission enhancement is shifted into the EUV. The X-ray emission can then be correlated positively with the radial extent of the cavity (negatively with N_H), while the total emission correlation is opposite. Although this model sounds very unlikely, it has the same basic temperature requirement as is inferred from the X-ray spectrum. It is therefore the behavior expected for the Bubble as seen (32). This model holds another surprise. One would expect that the spectral variation with surface brightness should be obvious, since the temperature is lower in directions of lesser extent (hence the shift to EUV). This shift does not appear in the more detailed models, however, because the cooler edge material radiates so rapidly that it cools off and disappears, leaving only hotter, dimmer, more interior parcels to be seen. Thus, although the naive expectation is that this form of displacement would have a spectral variation with surface brightness opposite to that predicted for absorption-bounded models, the best (albeit crude) models suggest that the spectral variation may be too slight to notice. Such variation, in fact, does not appear in the data. The latter show no first-order association between brightness and spectrum.

Models in which recent supernova reheating is invoked to explain the locally high temperature are rather heavily constrained (5, 24, 32). Their observed temperatures constrain the remnant expansion rate (shock velocity). Their N_H anticorrelation sets the radial scale. The radius to shock velocity ratio fixes the age. The X-ray surface brightness—together with the assumption that elemental abundances are solar—determines the mean density. From density and temperature one has the pressure, and with radius the explosion energy required. The overall result is that the explosion of 1 supernova ($\sim 5 \times 10^{50}$ erg) roughly 10^5 yr ago not far from the solar location, in a medium with mean density about 4×10^{-3} cm^{-3}, could provide pretty much what is seen. Details depend on the preexplosion density distribution and cloud interaction mechanisms. The remnant's internal pressure (i.e. that in the solar vicinity and presumably in the Local Fluff) is about $p/k \sim 2 \times 10^4$ cm^{-3} K. The edge pressure is about a factor of three higher in the nonevaporative models for which the calculations have been made.

Because this model is so tightly constrained, it makes some firm predictions that may be at odds with O^{+5} absorption-line measurements, medium-energy X rays, etc. (24). The essential reason for the heavy constraint, however, is strict adherence to the idea that the Local Bubble supernova remnant will continue evolving rapidly as it ages, never again looking as it does now.

If, on the other hand, the Local Bubble had some sort of walls, rather than just an end to being hot, then the supernova explosion within the preexisting cavity would have reheated it entirely in roughly 2×10^5 yr, after which the hot gas would be confined by its surroundings. Evolution would slow immensely, taking place on time scales associated with moving the wall material and cooling of the hot gas. There would not be an active shock at the outer boundary of the hot gas. The three versions mentioned earlier of these stagnant, or confined, bubbles are considered collectively in the next section.

There are various other interesting aspects of the possibility that a supernova explosion occurred within perhaps 15 pc of the Sun roughly 10^5 yr ago. Such an extraordinary event could conceivably have had an evolutionary impact on Earth. In addition, the resulting remnant would today have properties similar to those commonly advocated for cosmic-ray acceleration, with the result that being inside we could be sampling a perturbed cosmic-ray sample (20, 75, 102). The supernova ejecta could also be responsible for the observed ^{26}Al gamma rays (20).

Confined Bubbles

If the most recent supernova within the Local Bubble occurred between roughly 2×10^5 and 10^6 yr ago, the system would be extremely difficult to model. It would probably have strong reflected shocks driven into the denser walls and a very important interaction with the ambient pressure. The outgoing wave would be degenerating to a sound wave. The interior pressure would very likely undershoot the ambient value and then rebound as pressure equilibrium was gradually established.

Because we have seen no evidence for an outflow of material from the solar neighborhood, as might be incited by the interaction of the remnant during this period with the H I in its surroundings, we doubt that the Local Bubble's last supernova falls in this age range. The quoted upper end of the range is roughly the sound crossing for $R \sim 100$ pc at 10^6 K, although complete pressure relaxation could take even longer than 10^7 yr, the time scale for wall motion of 100 pc with characteristic interstellar speed of 10 km s^{-1}.

For LISM reheating events longer ago than roughly 10^6 yr, one might expect the residual presence of a bubble of hot gas, roughly in pressure equilibrium with its surroundings (51). Interpreting the soft X-ray observations in this fashion returns the results we originally quoted: radial scale varying from 30 to 200 pc, temperature of 10^6 K, and thermal pressure $p/k \simeq 10^4$ cm^{-3} K. Since the inferred pressure is in complete agreement with the value expected from pressure equilibrium, a suitable model can be constructed (26).

In the discussion to follow, we seek the parameters of a pressure-confined explosion that would have left us with conditions such as those in the Local Bubble. Some of these results could also be applied to inertially confined explosions, i.e. explosions that have reheated the entire volume of a preexisting cavity with dense walls. On the other hand, since we do not see much evidence for such walls and the motions that might be induced within them, and because the pressure magnitude makes pressure confinement plausible, we do not consider wall confinement further.

Models

PRESSURE-CONFINED BUBBLES   We consider two quite separate possibilities (b and e in the earlier list) in the same general discussion as proposed by Innes & Hartquist (51). In each, the Local Bubble is the pressure-confined residual of an explosion. In one case the explosion took place in an already extremely low-density intercloud phase ($n \lesssim 10^{-2}\,\mathrm{cm}^{-3}$), while in the other case the explosion was in a much higher density medium ($n \sim 0.1$ to 1 cm^{-3}) and has itself been obliged to create the Local Bubble cavity as well as leave the interior able to generate the soft X-ray background. In the second case, the explosion energy inferred is much greater than that of a single supernova.

We explore an extremely idealized model, assuming a point explosion in a homogeneous medium, followed by adiabatic decompression of the heated gas to the ambient pressure; thermal conduction and simple radiative effects are included only after the decompression. This model can be used profitably to locate the basic parameter regimes that could possibly explain the soft X-ray background.

The adiabatic expansion of a strong point explosion of energy E_0 in a homogeneous medium of mass density ρ_0 is described by "the Sedov solution" (94), whose shock radius R_s after time t is given by

$$R_s = \left\{\frac{2.025 E_0 t^2}{\rho_0}\right\}^{1/5}.$$

Including the number density and pressure jump conditions for a strong shock $[n_s = 4n_0,\ p_s = 0.75\rho_0 v_s^2$ with $v_s = dR_s/dt$ and $p_s = \chi n_s kT_s$, $\chi = (n+n_e)/n$, where n is the density of nuclei (ions and neutrals) and n_e is the electron density], we can evaluate the adiabatic constant κ for a parcel as a function of the previously shocked mass M_s:

$$\kappa = \frac{p_s}{(n_s)^{5/3}} = \frac{8.1\pi}{25(4)^{5/3}} \frac{m}{n_0^{2/3}} \frac{E_0}{M_s},$$

where $m = \rho/n$, and M_s is the total mass of shocked gas ($= 4\pi R_s^3 \rho_0/3$).

If a remnant's interior remains adiabatic during the time required to bring it to pressure equilibrium, then the adiabatic constant of a parcel determines its final density; when $p = p_0$, $n = 4\rho_0(p_0/p_s)^{3/5}$. The fact that $n(M) \propto M^{3/5}$ implies that $n \propto R^{9/2}$, $T \propto R^{-9/2}$ in the adiabatic cavity. The radius at which temperature T is found is given by

$$R(T) = \left\{\frac{3}{5} \frac{2.025}{2(4)^{5/3}} \frac{E_0}{(\chi k n_0 T)^{2/3} p_0^{1/3}}\right\}^{1/3}$$

$$= (36.1 \text{ pc}) \frac{E_{51}^{1/3}}{(n_0 T_6^{2/9} p_{-12}^{1/9})},$$

where $E_0 = E_{51} \times 10^{51}$ erg, $p = p_{-12} \times 10^{-12}$ dyn cm^{-2}, $T = T_6 \times 10^6$ K (24).

Following the adiabatic production of this bubble, we suppose that, thanks to thermal conduction, isothermality is approached over a gradually increasing portion of its central region. When this equilibration has reached radius R, the temperature interior to R is approximately the average of the adiabatic values, 5/2 of the adiabatic temperature at R. Replacing T in the $R(T)$ equation above by 0.4 $\bar{T}$, the radius of a bubble with smoothed temperature $\bar{T}$ should be

$$R(\bar{T}) = (44.25 \text{ pc}) \frac{E_{51}^{1/3}}{(n_0 \bar{T}_6)^{2/9} p_{-12}^{1/9}}.$$

In order for such a bubble to provide the soft X-ray background, we must have $\bar{T}_6 \approx 1$, in which case the Wisconsin carbon-band count rate should be (69)

$$C = 5 \times 10^4 n^2 R_{pc}/\phi = 80 p_{-12}^2 (R/100 \text{ pc})/\phi,$$

where ϕ is a factor of order unity (described below). Taking $R \approx 100$ pc as a typical dimension from the anticorrelation and 160 as a typical C-band count rate, we require

$$p_{-12} \approx \sqrt{2\phi}.$$

By using this and taking $R(\bar{T})$ above to be equal to 100 pc, we find the energy requirement for the explosion to be

$$E_{51} = 12.9 n_0^{2/3} \phi^{1/6},$$

or

$$n_0 = 7.5 \times 10^{-3} \text{ cm}^{-3} (2E_{51})^{3/2}/\phi^{1/4}.$$

The factor ϕ has been introduced above to show the effects of uncertainty in the atomic rates used to derive the conversion between emission measure and carbon-band count rate. Presently the rates are, in some quarters, considered to be more uncertain than has previously been appreciated, with the current range being $1 \lesssim \phi \lesssim 5$ (M. Arnaud, private communication).

THERMAL CONDUCTION AND RADIATIVE COOLING We take the classical thermal conductivity of a plasma to be $\kappa_c = \beta T^{5/2}$ (99), where $\beta = 6 \times 10^{-7}$ cgs, and use an approximate cooling function $L = \alpha T^{-1/2}$ (86), where $\alpha = 1.3 \times 10^{-19}$ cgs. We can expect saturated thermal conduction at high temperatures and in the steep gradient regions near the edge (21). We should, however, be able to use the classical formula to estimate the rate of energy redistribution over the interior whenever $\kappa_c T/R \ll pc$, where c is the sound speed. With $R = R_{100} \cdot 100$ pc, this criterion is

$$T_6 < 2(p_{-12} R_{100})^{1/3}.$$

Since we are estimating the time scale to equilibrate when the temperature is roughly 10^6 K or a little less, the classical estimate should suffice.

The time scale for achieving approximate isothermality at temperature T and pressure p within radius R is then

$$\tau_I \sim \frac{5}{12} \frac{pR^2}{\beta T^{7/2}}$$

$$\sim 2 \times 10^6 \text{ yr } p_{-12}(R_{100})^2/T_6^{7/2}.$$

Inverting this to $T(t)$ provides a reasonable estimate of the interior temperature of the bubble and the parcel to which isothermality has reached.

With the approximate cooling function chosen, the radiative cooling time for an otherwise adiabatic parcel of gas is independent of its pressure environment (60). The volume emissivity is $Ln^2 = \alpha T^{-1/2} n^2$, from which

$$\frac{d}{dt}\left[\frac{p}{n^{5/3}}\right] = -\frac{2}{3}\frac{Ln^2}{n^{5/3}} = -\frac{2\alpha(\chi k)^{1/2}}{3[p/n^{5/3}]^{1/2}},$$

or

$$\frac{d}{dt}\left[\frac{p^{3/2}}{n^{5/2}}\right] = -\alpha(\chi k)^{1/2}.$$

The adiabatic constant reaches zero after a time

$$\Delta t_{\text{cool}} = p^{3/2}/[\alpha(\chi k)^{1/2} n^{5/2}] = \frac{p}{Ln^2} = \frac{\chi kT}{Ln}$$

$$\approx 7 \times 10^4 \text{ yr } T_6^{3/2}/n$$

$$\approx 2 \times 10^7 \text{ yr } T_6^{5/2}/p_{-12}.$$

If we now consider our bubble, which initially had a strong negative temperature gradient, we find that thermal conduction is rapidly leveling the interior temperature while radiative cooling is quickly steepening the gradient at the outer edge. The two processes eventually "meet" to provide a sharp-edged isothermal bubble (or crude approximation thereof) when $\Delta t_{cool} \sim \tau_I$; this condition is met when the bubble parameters are given by

$$T_B \sim 8 \times 10^5 \text{ K} \left[\frac{p_{-12} R_{100}}{1.4}\right]^{1/3},$$

$$t_B \sim 7 \times 10^6 \text{ yr } (R_{100})^{5/6} \left(\frac{1.4}{p_{-12}}\right)^{1/6}.$$

This time scale for thermal shaping of the bubble equals $11 R/c$ (11 sound-crossing times), or 49 times the age of an active adiabatic remnant with the same radius and temperature.

The evolution of the explosion bubble should take, overall, roughly two t_B, one to reach the T_B state and one more to die radiatively from there. The temperature T_B should be the characteristic temperature over most of the central time period. Higher temperatures are short lived owing to thermal conduction; lower temperatures are short lived due to radiative processes.

We wish to stress that the temperature T_B depends only on the ambient pressure of the system and the radius, not separately on the explosion energy or the ambient density. In addition, the ambient pressure observed via the measurements of the interstellar matter weight distribution (26) combines with the radial scale of the bubble observed via the N_H anticorrelation and interstellar absorption-line studies to provide a value of T_B equal to the temperature actually measured in the soft X-ray background. The predicted X-ray surface brightness is also then correct. We have to conclude that what we actually see around us is the most likely thing to be there, given only that we are within tens of parsecs and 10^7 yr of the site of an explosion sufficiently energetic to create a 100-pc radius hot cavity in whatever ambient density was present. As quoted previously, the energy requirement for the explosion is $E_{51} \sim 13 n_0^{2/3} \phi^{1/6}$.

EVAPORATION OF CLOUDS WITHIN THE BUBBLE The presence of evaporable clouds of interstellar material within the Local Bubble could seriously shorten the characteristic bubble evolution time. Conditions are such that classical conduction provides a reliable rate estimate (21), in which case the steady mass-loss rate from a spherical cloud of radius R in a medium of asymptotic temperature T_F and conductivity $\kappa_F = \beta T_F^{5/2}$ is given by

$$\dot{m} = \frac{16\pi\kappa_F m R \phi_G}{25\chi k},$$

where ϕ_G is a conduction suppression coefficient arising from magnetic field geometry. From this rate, the evaporation time scale for a cloud as a whole is

$$t_{evap} \sim 3 \times 10^5 \frac{R_{pc}^2 n}{\phi_G T_6^{5/2}} \text{ yr} = \frac{5}{\phi_G}\left(\frac{R_{cl}}{R_B}\right)^2 \frac{n}{n_F} \cdot \tau_I.$$

Here R_{pc} and R_{cl} refer to the cloud radius (the former in parsecs), R_B the bubble radius, n the cloud internal density, and n_F the asymptotic density, where $T = T_F = T_6 \times 10^6$ K. The corresponding time scale for appreciable density enhancement within the bubble is

$$t_\rho = \frac{\rho_F}{\dot{\rho}} = \frac{n_F}{f_{cl} n} t_{evap} = \frac{5}{\phi_G f_{cl}}\left(\frac{R_{cl}}{R_B}\right)^2 \cdot \tau_I.$$

A reasonable upper limit to the "cloud"-filling factor within the Local Bubble is $f_{cl} \lesssim 0.1$, while $R_B/R_{cl} \sim 50$. Hence the cloud presence could shorten the evaporative time scale by a factor of $(50\phi_G)^{-1}$. By equating t_ρ with Δt_{cool} once again to find the modified bubble temperature and evolution time scale, we obtain

$$T'_B = 4 \times 10^5 \text{ K} \left(\frac{p_{-12}}{1.4}\frac{R_{pc}}{2}\right)^{1/3}\left(\frac{0.1}{\phi_G f_{cl}}\right)^{1/6},$$

$$t'_B = 1.5 \times 10^6 \text{ yr } (R_{pc}/2)^{5/6}(1.4/p_{-12})^{1/6}\left(\frac{0.1}{\phi_G f_{cl}}\right)^{5/12}.$$

With the estimates $R_{pc} \sim 2$, $p_{-12} \sim 1.4$, $f_{cl} \sim 0.1$, and $\phi_G \sim 1$, the bubble would have both a significantly shortened lifetime and a factor of 2 lower characteristic temperature. The duration of higher temperatures (before evaporative energy dilution) is $t \sim t'_B(T'_B/T)^{7/2}$, or only about 6×10^4 yr for $T \gtrsim 10^6$ K.

Hence, the presence of a substantial population of evaporable clouds lowers the typical bubble temperature, shortens its lifetime, and pushes the X-ray emissive era back toward or into the regime of active blast-wave evolution. Both the comforting coincidence between T_B and the X-ray emission temperature and the satisfyingly long period over which the bubble should look as it does look evaporate along with the clouds (if, that is, there are such clouds, and if they do evaporate as modeled).

There are several factors that could reduce the cloud evaporation rate. Insulation via magnetic field geometry has been mentioned in introducing ϕ_G. The inclusion of time-dependent effects could be important (7). Changing to filamentary or planar cloud geometry eliminates the steady-state evaporation case altogether, with the result that transition layers steadily thicken until they become radiative. These effects combine; conduction limited to flux tubes causes linear cloud distension until the gradients are too weak for conduction to beat radiation.

If the Local Bubble is a hot spot within a general low-density "coronal" intercloud phase, then another possibility is that there were few if any clouds near us initially, and that essentially all of these were rapidly evaporated long ago. This is easily possible given the short evaporation time scale quoted above. Conversely, if the Local Bubble constitutes a cavity blown into the normally much denser ISM by some grand explosion, then residual clouds within the cavity are not expected—except possibly for extraordinarily dense ones that are too uncommon to affect this discussion.

What we find is that the presence or absence of a moderate filling factor of low-density H I regions (e.g. 0.1–1 cm^{-3}) within the cavity could have a major impact on our understanding of the bubble origin and evolution. We have already noted that the bubble shows up as a local paucity of H I, a sign that such clouds are at least uncommon; however, the solar system is inside such a cloud, which implies that the absence may not be so complete as some models may require.

It has become somewhat fashionable to invoke "hydrodynamic stripping" of mass off of clouds (e.g. 46) rather than thermal evaporation as the mechanism for mass transfer from clouds to the hot phase in active supernova remnants. Such a mechanism necessarily operates only during active expansion of a remnant, not during a quiescent pressure-confined afterglow.

The essential trio of questions raised by this section are whether there are today clouds suitable for evaporation within the Local Bubble, whether evaporation acts on such clouds as we previously estimated, and whether there are sufficient numbers of such clouds to cause the rapid demise of pressure-confined bubbles. We postpone further discussion until later but note that the tentative answers are probably, maybe, and perhaps.

TOWARD A GENERAL ISM MODEL, AND THEN BACKING OFF   Rapid cloud evaporation threatens the propriety of models using long-lived pressure-confined bubbles. Models with quiescent versions of such bubbles can also be made untenable by frequent supernovae. If the recurrence rate within a bubble exceeds t_B^{-1}, the bubble radiates less rapidly than the rate at which energy is deposited within it. A quick estimate shows the importance of this possibility. Assuming the Local Bubble to have an effective radius of $R_{100} \times 100$ pc compared with a Galactic radius of perhaps 15 kpc, the bubble subtends $5 \times 10^{-5} R_{100}^2$ of the area of the disk. If that fraction of the Galactic supernovae occurs within the Bubble, and if the Galactic rate is 1 per 50 yr, then the rate within the bubble is 1 per $10^6/R_{100}^2$ yr. For $R_{100} \sim 1$, this is clearly a high rate compared with the roughly 10^7-yr evolutionary time scale of a cloud-free bubble. We have two choices: to either cut the supernova rate, or to change the quiescent bubble model. If the latter choice is taken, we then require a more rapid bubble evolution rate consistent with that provided by cloud evaporation. But a general model for arbitrary supernova rate and cloud population requires that the interstellar pressure be allowed to float as a free parameter. Forcing self-consistency then provides a lovely picture long known as the MO (McKee & Ostriker) model of the interstellar medium. Having taken that approach, the Local Bubble is then the local active remnant of a recent supernova explosion (73).

But the opposite tack is also still open. The effective bubble area could well be somewhat less than is implied by $R_{100} = 1$. The occurrence of Type II supernovae within spiral arms suggests that our bubble will have zero rate for those (47). The high scale height for older stars that may be progenitors of Type I supernovae implies that our bubble will not capture the high-z ones in its subtended column (47). The average supernova rate may follow the stellar density, concentrating somewhat closer to the Galactic center (47). The combination of these effects could easily push the supernova recurrence time for our bubble into the 10^7-yr regime. But bubbles elsewhere will not be so lucky. Statistics aside, we know of no star within the Local Bubble volume defined below that is a candidate supernova progenitor, due to detonate within the next 10^7 yr.

Cavity Geometry

We have thus far concentrated on only two aspects of the Local Bubble—its low density and high temperature. We have found that a quiescent bubble model has some nice features. Parenthetically, some of these bubble models are also consistent with evidence of a local downflow in overlying H I (e.g. 4), as would be found about 2×10^7 yr after a large local explosion (since this is the time scale for reversal of upward motion in the Galactic gravitational potential). But it is increasingly clear that other things within the bubble must be examined before we can hope to know its origin or significance. Other cavity residents with stories to tell include the stellar population (searched for supernova progenitors, studied for mass return rate, and examined for remnants of recent planetary nebula formation), the cloud population (studied for phases present, sizes, degrees of ionization, abundance anomalies, and signatures of an evaporative boundary), the radiation field (particularly in ionizing photons), and, perhaps critically, the cosmic-ray population at low energy (studied in particular for signs of recent local acceleration and peculiarities). It would also be satisfying to witness spectral signatures of the ions thought to be present in the X-ray emitting gas.

In order to carefully carry out such a study, however, it is necessary to have a fairly accurate model for the bubble geometry. With it, we can avoid such misconceptions as the notion that having abundant H I within 100 pc of the Sun is equivalent to having it within the bubble. To that end, we consider a proposed bubble geometry that derives from the brightness distribution of the soft X-ray background (26, 97, 98). This geometry assumes that there is a linear scaling between the X-ray brightness (specifically in the B band) and the cavity extent.

The two scaling parameters were chosen by arranging for the cavity to exclude all stars whose known N_H exceeds 6×10^{18} cm^{-2} and by optimizing the agreement between the predicted and observed magnitudes for the locally missing N_H at high latitudes, while maintaining consistency with models of a pressure-confined bubble with thermal $p/k \approx 10^4$ cm^{-3} K. In short, it is a picture that is already consistent with the SXRB, the local hydrogen distribution, and the anticipated pressure for a quiescent bubble.

The resulting geometrical structure, shown in Figure 1, is proposed specifically as a map of the extent to which X-ray-emitting gas exists. It is not associated with any particular contour in N_H, apart from the requirement that $N_H \leq 6 \times 10^{18}$ cm^{-2} in all directions. In some directions, the N_H levels jump suddenly to high values just beyond the Local Bubble boundary, so that one can think meaningfully of there being a cavity wall (39, 81). In other directions, the total column density remains low to much greater distances than the boundary of the hot gas. (This is particularly noticeable in the third quadrant.) Regions of warm H II apparently occupy substantial volumes beyond the hot cavity in those directions. (Since X-ray emissivity occurring for a considerable distance would be easily visible to us, the modesty of the X-ray surface brightness in such directions was one of the earliest and most discouraging indications that an X-ray-emitting tunnel network does not pervade the interstellar medium.)

To be a bit more explicit, the Local Bubble cavity model based on the soft X-ray background brightness indicates that in the plane of the third quadrant, the cavity extends only about 50 pc. Below the plane, toward β CMa (at $l = 226°$, $b = -14°$), the cavity extent increases to about 70 pc.

Observations of β CMa (44, 81, and references therein) clearly show very little material of any detectable sort along the entire path, a distance of 140 to 270 pc. The cavity of X-ray-emitting gas probably occupies only half the line of sight or less. The column density of neutral hydrogen is only about 10^{18} cm^{-2}, all of which may belong to the local cloud around the Sun, occupying perhaps 3 pc. There is also absorption due to ions characteristic of diffuse H II, with an ionized hydrogen column density estimated at 2×10^{19} cm^{-2} (or 7 cm^{-3} pc) from measured S II and S III column densities (44). By means of the C$^+$ fine-structure excitation, the local electron density was measured to be approximately 0.1 cm^{-3}, which implies an occupation pathlength of roughly $65 f^{-1}$ pc of the line of sight to the star, where f is the filling fraction of the gas. The thermal pressure is a reasonably acceptable $p/k \sim 2000$ cm^{-3} K, a value comparable to that of the very local interstellar medium.

The brightness of the diffuse Hα background near β CMa, measured with the Wisconsin Fabry-Perot spectrometer (R. J. Reynolds, unpublished), indicates an Hα intensity of about $15R$ ($1R = 10^6/4\pi$ photon cm^{-2} s^{-1} sr^{-1}), which corresponds to a total emission measure $\simeq 41\ T_4^{0.9}$ cm^{-6} pc. (T_4 is the temperature in units of 10^4 K.) The Hα line widths [$\simeq 30$ km s^{-1} (FWHM)] limit the temperature to $<20{,}000$ K.

Probably only a small fraction of this Hα emission originates from the warm H$^+$ in front of β CMa. The Hα emission and H$^+$ column density, taken together, would require a total occupation pathlength of 1 pc at a density of 6 cm^{-3}. A high-pressure H II region would be needed to produce those figures, which are in disagreement with the electron density derived from the C$^+$ fine-structure excitation as well as the large angular scale of the Hα emission. More likely, the Hα emission is produced over a much larger pathlength, extending well beyond the distance of β CMa. Without any other phase to occupy the volume in this direction, we have to believe that it is accomplished by the warm H$^+$. Perhaps, however, it will turn out that a short-lived bubble of $\sim 10^5$-K gas will be found in this direction, radiating copiously in the EUV.

The general features of the geometry shown in Figure 1 are that the bubble extent in the plane is small (30–60 pc), and that it is very elongated perpendicular to the plane, particularly in the north where it extends as far as 200 pc. It is distinctly wasp waisted.

Considering evidence from slightly harder (M band) X rays, there is reasonably convincing evidence that the Local Bubble has a neighbor, generally referred to as the Loop I (or North Polar Spur) Bubble from associated features (52). This neighboring bubble is large, and is so close that it looms over about one fourth of the sky. But at latitudes below roughly 40°, there is clearly a wall of neutral hydrogen separating the Loop

I Bubble from ours. At higher latitudes, the two bubbles may well be connected (97). Figure 2 shows a rough sketch of the possible relationship between these two. We note in passing that the gas within Loop I must be hotter than that within the Local Bubble, since it is so bright in M-band X-rays.

The challenge then, which is only partly taken up in the remainder of this review, is to study the distributions of gas-phase material, stars, and other denizens within the postulated bubble geometry, looking for insight into both time scales and processes as well as inconsistencies with the quiescent bubble scenario.

Clouds Within the Local Bubble

The standard "diffuse clouds" explored via extinction studies have been characterized statistically with a column density $N_H \sim 3 \times 10^{20}$ cm^{-2} and a interception rate in the disk of 6.2 kpc^{-1} (100). Their mean contribution to the midplane hydrogen density is 0.7 cm^{-3}. With a nominal density of 40 cm^{-3} (i.e. $nT \sim 3000$ cm^{-3} K at $T \sim 80$ K), their mean thickness is 2.7 pc, and they occupy 1.7% of interstellar space. If such clouds were spheri-

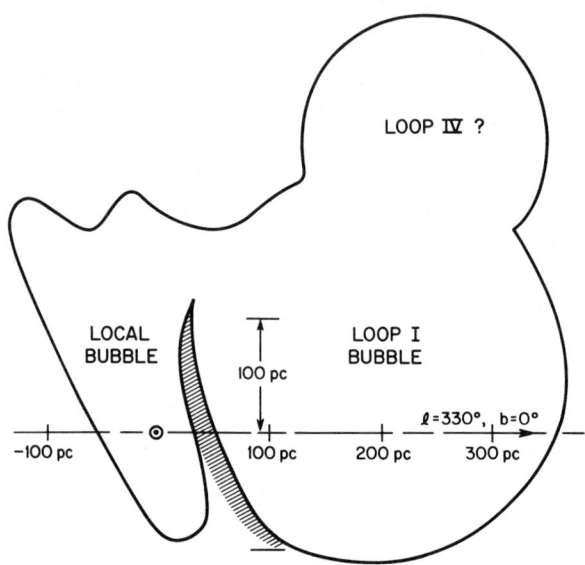

Figure 2 Schematic representation of the relationship between the Local and Loop I Bubbles and the Sun, including the intervening wall of H I. The column density of this wall decreases with latitude. The distance scale of the Loop I center is uncertain and was chosen a bit on the large side to make this figure sensible [combined views of (97) and (52)].

cal, their radii would be 2 pc. There would be about 2000 of them in a region with the volume of the Local Bubble.

A considerably more complete description of the cloud distribution function is now available from optical absorption-line (48a) and 21-cm studies [reviewed in (64a)]. The number of clouds per kiloparsec with column densities in excess of N_H is 5.7 $(N_H/10^{20}$ cm$^{-2})^{-0.8}$, with $3 \times 10^{19} < N_H < 2 \times 10^{20}$ cm^{-2}. From temperature information available in 21 cm, and an assumed thermal pressure of 3000 cm^{-3} K, the above statistic would be produced by a population of spherical clouds having radii $R \sim 0.8$ pc $(N_H/10^{20}$ cm$^{-2})^{1/3}$ and the differential number density $dn_c/dR = 2.4 \times 10^{-3} (R/1$ pc$)^{-5.4}$ pc^{-4}. The number of such clouds having radii greater than R, in a volume equaling that of a sphere with radius 100 pc, is 2300 $(1$ pc$/R)^{4.4}$, where $1 \gtrsim R \gtrsim 0.5$ pc expresses the nominal size range. The number of interstellar clouds expected within the Local Bubble is very sensitive to the cloud size cutoff chosen, but surely it exceeds the 2000 "standard clouds" estimated above.

The low hydrogen column densities N_H at high galactic latitudes are consistent with such clouds being entirely absent from the Local Bubble (66). Furthermore, the H I that is found at high latitudes appears to be very smoothly distributed, with no holes or lumps of high contrast (55). It could be entirely in a fairly uniform distribution of the "warm neutral medium" component.

Attempts to locate nearby H I clouds that could provide opaque screens for low-energy X-ray-shadowing experiments have been fruitless (38). With the above distribution of spherical clouds, there should have been a score near enough to have angular radii greater than 6°. None seem to have been found. This is the cloud population that would a priori have been expected to provide the evaporation that would shorten the life of the Local Bubble as a hot quiescent region.

It is not actually necessary to have missing clouds to explain the disparity between those expected and those seen. By assuming that "diffuse clouds" have sheetlike rather than spherical geometry, or that small clouds are grouped in sheets with coherence over several tens of parsecs, the number expected within the Local Bubble can be reduced from 2000 to zero. The expected clouds are then those actually seen in the higher density H I regions beyond the hot-gas boundary. On a fairly large scale, it may yet be true that the ISM has a spongy "tunnel" structure, or even perhaps a cellular one.

The notion of spherical diffuse clouds dies hard. But the clouds in question are pressure confined and in a hostile environment. They probably have no significant surface tension or self-gravity attempting to minimize their area or radial irregularities. Unless the magnetic fields are tangled or twisted, they very likely have negligible cohesion, let alone any restoring force countering geometrical distortion. In short, there is no advantage in either roundness or smoothness. On the other hand, the magnetic pressure is appreciable, thermal conduction tends to be confined to flux tubes, and instabilities in relative motion tend to wrap those tubes into flux ropes. Linear structures thus have good reason for existence. The embedding of dense H I within the warm H I, and the tendency of major interstellar disturbances to push material toward sheetlike geometry, would then seem to promote a structure similar to mats of angel hair,[3] rather like what is actually seen in reflection of the Pleiades.

We have, however, fallen off the track. The H I observed above the Local Bubble contains no appreciable dense H I structures. Close to the Sun, dense H I is sparse. From the lack of Be to B band variation in soft X rays, we estimate that within the proposed cavity, the maximum effective[4] distributed H I column density from Earth to edge is 10^{19} cm^{-2}. (This conclusion is firm only over the $\sim 10\%$ of the sky that has so far been studied in the Be band.) Thus the maximum consistent *mean* density of absorbers varies from about 0.1 cm^{-3} in the plane (30-pc pathlengths) down to 0.02 cm^{-3} at the northern extent of 200 pc. (In the plane, one must take care to leave room for the emission, however.) Such low densities are, as we have already noted, consistent with the sparse information available for H I column densities to stars within the cavity (with the possible exception of α Oph), but they cannot otherwise be regarded as observed. They are what is necessary, however, for consistency between observations and our present picture of the origin of the SXRB. This absorber density limit is sufficiently low that it requires a significant reduction in the mean density of even the warm H I component. This requirement is not so stringent, however, that it cannot be made to agree with the limited data on very local H I column densities, as we shall see. But given the low-density Local Fluff, near the Sun and clearly inside the Local Bubble, one is somewhat anguished to understand its survival.

PROPERTIES OF THE LOCAL CLOUDS

Solar Backscatter Observations

The existence of a cloud or clouds of interstellar gas (Local Fluff) within a few parsecs of the Sun has been revealed by a variety of observations. The first opportunities to study the interstellar gas in the immediate vicinity of the Sun were provided by the detection of backscattered sunlight [see

---

[3] Spun glass fiber wool used in Christmas decorations.

[4] The effective cross-section ratio is taken to be that of a gas with both hydrogen and helium neutral.

review by Fahr (33)]. Over the last 15 years, sounding rockets, satellites, and interplanetary spacecraft have been used to study solar H Lα 1216 Å and He 584 Å emission backscattered from interstellar H I and He I atoms flowing through the solar system. The observations include photometric intensity measurements and high-resolution spectroscopic (hydrogen absorption cell) measurements over the sky. The data basically reveal that there is a relative velocity of about 20 km s^{-1} between the Sun and a surrounding medium that has a density of $\sim 10^{-1}$ cm^{-3} and a temperature of $\sim 10^4$ K. Thus, the Sun is located in a region or cloud with a significantly higher density and lower temperature than that associated with the surrounding Local Bubble. The properties of this cloud may provide important clues about the recent history of the local interstellar medium and interactions between the hot gas and embedded cooler clouds. Also, the presence of this inflowing gas provides a unique opportunity to measure the properties of an interstellar cloud at a "point" rather than as the usual line-of-sight averages.

An accurate determination of the properties of the very local interstellar medium (VLISM, a "point" sample of the Local Cloud) from the solar backscatter data depends critically upon understanding how the density and velocity distributions of the inflowing atoms are modified near the Sun. The effects of solar gravity, radiation pressure, photoionization, and interactions with the solar wind and outer heliopause all must be taken into account in order to interpret properly the 1216-Å and 584-Å data. Global fits of solar interaction models to some of the most recent and best data available have been made by Dalaudier et al. (28), Bertaux et al. (10), and Chassefière et al. (19). Their analyses yield the following characteristics for the VLISM: $n_{\text{H I}} \simeq 0.03$–$0.065$ cm^{-3}, $n_{\text{He I}} \simeq 0.01$–$0.02$ cm^{-3}, $T = 8000 \pm 1000$ K, and a flow speed of 16 ± 1 km s^{-1} with respect to the local standard of rest (LSR) toward $l = 124° \pm 3°$, $b = +4° \pm 3°$. The sample lengths are a few $\times 10^{-5}$ pc, which is the distance covered by the flow over the time interval ($\simeq 1$–2 yr) of the observations.

A particularly interesting aspect of the derived properties is the apparent noncosmic value for the $n_{\text{H I}}/n_{\text{He I}}$ density ratio. Figure 3 summarizes the densities derived for H I and He I in the VLISM from the various backscatter experiments. Values of $n_{\text{H I}}$ are represented by filled symbols, whereas values of $10 \times n_{\text{He I}}$ are represented by open symbols. The mission, date, and reference for each value is given in Table 1. Figure 3 shows a clear, consistent deviation of the derived H I and He I densities from the "cosmic" abundance ratio of 10:1; specifically, $n_{\text{H I}} \lesssim 5n_{\text{He I}}$. This may have important implications for the properties of the cloud. For example, Meier (74), Weller & Meier (106), and Bertaux et al. (10) have all suggested that the observed H I/He I density ratios might be due to a substantial

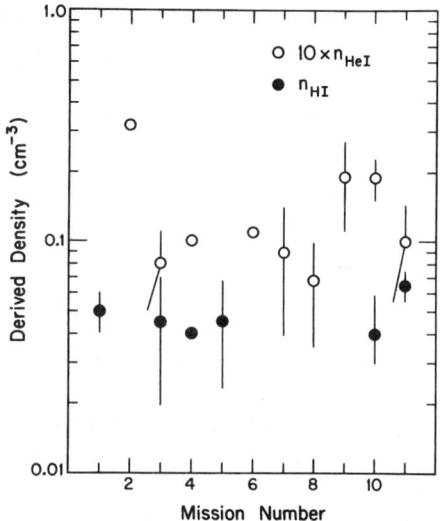

Figure 3 A summary of H I densities and He I densities derived from 11 experiments (see Table 1) measuring solar Lα 1216 Å and He 584 Å backscatter from the interplanetary medium. For He I, the value plotted is 10 times the derived He I density [from a compilation by Chassefière et al. (19)].

ionization fraction of the local hydrogen. If this interpretation is correct, then the data in Figure 3 indicate a fractional ionization of hydrogen $\simeq 50$–70% and a total hydrogen (H I + H II) density ≈ 0.1–0.2 cm^{-3}. The total thermal pressure would then be in the range 1300–3000 cm^{-3} K, which is nearer the traditional interstellar medium value than that (≈ 500 cm^{-3} K) provided by the directly observed neutral atoms.

Table 1 Summary of VLISM He I and H I observations

Number	Date	Mission	References
1.	Mar. '71	OGO-5	103
2.	Feb. '73	Rocket	82
3.	Dec. '73–Jan. '74	Mariner 10	2, 3, 15
4.	'73–'78	Pioneer 10	108
5.	Apr. '75	Copernicus	1
6.	July '75	OSO-8	105
7.	July '75	Apollo-Soyouz	37
8.	June '76	Rocket	34
9.	Nov. '76–Feb. '77	Solrad 11B	106
10.	Dec. '76–Jan. '78	Prognoz 5/6	10, 28
11.	Mar. '78–Mar. '80	Venera 11/12	19

It must be reiterated, however, that the H I and He I densities have been derived by fitting the observations with a solar interaction model; and if the model is incomplete, then systematic errors could be present in the derived parameters. One interaction that has not been included in the above models is the possible modification of the inflowing gas by its passage through the heliospheric interface. Very little is known about the nature of the transition that must exist between the heliosphere, dominated by the solar wind and solar magnetic field, and the unperturbed interstellar medium. Ripken & Fahr (89, 90) have considered some transition models and have explored the effects of charge exchange between protons and H I atoms in the interstellar gas as it approaches the heliosphere. They conclude that a nonequilibrium state between the distribution functions of the protons and the H I atoms, due to the perturbation of the interstellar magnetic field ahead of the heliosphere, could result in a significant charge exchange–induced depletion of H I atoms flowing into the solar system. If this process is important, then the H I/He I abundance ratio discrepancy of a factor of two could be accounted for by an interstellar gas that is only 10–20% ionized (depending upon the heliospheric model) rather than 50% ionized. The corresponding VLISM thermal pressure is then $p/k \sim 1000$–2000 cm^{-3} K.

This process apparently cannot account for the H I/He I abundance ratio if the hydrogen in the VLISM is less than about 10% ionized (i.e. electron or proton density $\lesssim 0.01$ cm^{-3}), since interstellar protons are needed to deplete the inflowing H I by charge exchange. Conversely, an ionization fraction much above 20% would produce too much depletion of the H I (i.e. an observed He I/H I ratio larger than that observed). Thus, to achieve the observed ratio there is only a narrowly allowed range of ionization.

In steady state the ionization rate per atom, Γ_H, required to maintain a hydrogen ionization fraction x is given by

$$\Gamma_H \approx \alpha_H \frac{x^2}{1-x} n_H,$$

where n_H is the total (H I + H II) hydrogen density, and α_H is the hydrogen recombination coefficient (3.1×10^{-13} cm^3 s^{-1} at 8000 K; case B). Therefore, the minimal 10% ionization fraction would in steady state require $\Gamma_H \geq 3.4 \times 10^{-16}$ $(n_H/0.1)$ s^{-1}. A lower limit to the ionization rate present within the gas can be calculated from the observed ionizing radiation from nearby hot white dwarfs (50, 80) and the soft X-ray background (10a, and references therein) and from the inferred rate due to cosmic rays (104a). The hydrogen-ionizing flux detected from HZ 43 (50) provides a rate of

$\Gamma_H \approx 7.7 \times 10^{-17}$ s^{-1}, and the combined flux from the other known EUV sources—F24, G191–B2B, and GD 153—increases the total to about 1.5×10^{-16} s^{-1}. The X-ray background ($hv \geq 90$ eV) and cosmic rays provide an additional 2.6×10^{-17} s^{-1} and 7×10^{-17} s^{-1}, respectively, for a total $\Gamma_H \simeq 2.5 \times 10^{-16}$ s^{-1}. This value must be considered a lower limit to the typical rate within the Local Fluff, since optical depths to the solar location are nonnegligible and a significant fraction of the ionizing radiation must be absorbed in portions of the cloud nearer to the sources [e.g. a column density of 10^{18} cm^{-2} (see below) provides an optical depth of unity for 25-eV photons]; there also may be additional, as yet undetected sources of Lyman continuum photons. Therefore, in the VLISM the observed radiation and observed neutral density combine to imply that $x \gtrsim 0.085$, corresponding to an electron (proton density) ≥ 0.009 cm^{-3} in the circumsolar environment. Since this value is close to the minimum ionization requirement for the Ripken & Fahr mechanism, we regard that requirement as essentially satisfied.

According to Ripken & Fahr (89, 90), this heliospheric interaction would have two other effects on the inflowing gas: (a) The temperature of hydrogen would be raised about 1500 K above that of helium (which is not significantly affected by this process due to the low cross section for helium-proton charge exchange); and (b) the depletion of O I, which does have a very large charge exchange cross section with protons, should be much larger than that of H I. Observations designed to search for these effects would be important in resolving this ambiguity in the interpretation of the backscatter data. An accurate determination of the hydrogen temperature (8000 ± 1000 K; 10) has been made in the upwind hemisphere, where effects of a heliospheric interaction would be most easily detected. However, all temperature determinations for helium have been made downwind, where solar wind interactions may have increased the temperature (10, 35). The helium temperature is determined from photometric observations that measure the density and lateral spread of the gravitationally induced helium density enhancement along the downwind line behind the Sun. More direct, high-spectral-resolution measurements of the 584-Å line have not yet been successful because of (apparently surmountable) technical difficulties in constructing a helium absorption cell. The available temperature measurements indicate that T_H (downwind) $\geq T_{He}$ (downwind) $> T_H$ (upwind), which is not the temperature relationship predicted by a heliospheric interaction that heats the upwind hydrogen to a higher temperature than the helium. Bertaux (9) has pointed out that these temperature relationships are a strong indication that deviations from the models are more related to an interaction (possibly solar wind heating) near the Sun than near the heliosphere boundary. However, an additional heliospheric interaction has not been ruled out, and accurate measurements of both the hydrogen and the helium temperatures upwind would be important. Other model complications are also beginning to be investigated (see *Résumés des Communications*, 10th Session, XXVI COSPAR 86).

In quick summary, the VLISM is inferred to have a neutral helium density of 0.01–0.02 cm^{-3} and a neutral hydrogen density exceeding 0.03 cm^{-3} and perhaps exceeding 0.05 cm^{-3}. Assuming that the hydrogen to helium abundance ratio is 10 by number, that the helium density has been correctly estimated (the modelers are reviewing this point), and that helium is completely neutral, the total hydrogen density must be 0.1–0.2 cm^{-3}. If heliospheric attenuation is negligible, the difference between neutral and total density implies an ionization fraction of roughly 0.5 to 0.7. On the other hand, with heliospheric attenuation of neutral H at the Ripken & Fahr (89) rate, the observations require the inflowing plasma to be roughly 10 to 20% ionized. Too little ionization would offer too few charge-exchanging ion targets, and too much would allow far too little neutral H to penetrate the solar system. Thus both mechanisms proposed so far depend on there being partial ionization. The required ionization rates differ by about a factor of 30 between the two, with $\Gamma_H \sim 5 \times 10^{-16}$ s^{-1} for 10% ionization of hydrogen with total density 0.15 cm^{-3}, and $\Gamma_H \sim 1.7 \times 10^{-14}$ s^{-1} for 50% ionization. The rate available from known sources is 2.5×10^{-16} s^{-1}; other possibilities are discussed later.

The models above agree about the VLISM density, disagree about the VLISM thermal pressure by roughly a factor of 1.4 (i.e. the electron contribution), and disagree about the ionization fraction by about a factor of 5 and the ionization rate by a factor of 30.

Other Observations

Interstellar absorption-line observations of nearby stars are another important source of information about the Local Fluff. McClintock et al. (70, 71) derived H I column densities from Copernicus Lα observations toward nearby ($d = 1.3$–14 pc) stars and concluded that the Local Cloud in which the Sun is embedded extends about 3.5 pc from the Sun with $n_{H\,I} \sim 0.1$ cm^{-3}. The derived b-values (≈ 11–15 km s^{-1}) for the hydrogen toward the four closest stars ($1.3 \leq d \leq 3.5$ pc) correspond to temperatures of 7000–13,000 K. A larger volume around the Sun subsequently has been mapped by Frisch & York (39) and Paresce (81), who collected hydrogen column density data for 140 stars at distances 10–3000 pc from the Sun. The data show an asymmetry with respect to the Sun in the distribution of H I. In the third quadrant (180° to 270° in Galactic longitude) H I column densities are less than 5×10^{18} cm^{-2} out to at least 200 pc, while in the first and fourth quadrants, near the Galactic plane, column densities of 5×10^{18} cm^{-2} are reached within about 20–50 pc. This asymmetry, which is also revealed in the amount of interstellar reddening within a few hundred parsecs of the Sun (67), suggests a concentration of interstellar matter near the Sun in the general direction of the Galactic center.

There are two independent pieces of evidence for an appreciable ($N_H > 10^{19}$ cm^{-2}) cloud within about 17–35 pc of the Sun in the general direction of the Galactic center. Tinbergen (104) measured the optical polarization of about 180 stars within 35 pc and found evidence for a region of uniform, enhanced polarization about 30–60° in diameter centered near $l = 0°$, $b = -20°$. This polarization is observed in stars with distances ranging from 5 to 35 pc. Tinbergen concluded that this result is not spurious and that it implies the existence of an interstellar dust cloud within 20 pc of the Sun with $E_{B-V} \approx 0.003$ mag, corresponding to $N_H \approx 1-2 \times 10^{19}$ cm^{-2} at a standard gas-to-dust ratio (12). A substantial hydrogen column density is also inferred from a variety of interstellar absorption-line observations toward α Oph at $l = 36°$, $b = +23°$, which is believed to be at a distance ≈ 17 pc (41). Column densities of Ti II, Ca II, Na I, Fe II, and Mg II all indicate that $2.5 \times 10^{19} \lesssim N_H \lesssim 4 \times 10^{20}$ cm^{-2} toward this star (38a, 78). The radial velocity of the absorbing gas appears to be different from the star and in fact is consistent with the general flow of gas within 100 pc found by Crutcher (27). Paresce (78) has suggested that the VLISM flow observed at the Sun is the warm, partially ionized edge of this cloud à la McKee & Ostriker. The direction of the VLISM flow suggests that the Sun is just entering this cloud. It is interesting, however, that α Oph does not have an enhanced polarization (104) and thus is not included in the Tinbergen "patch." A satisfactory map of the distribution of small-column-density features in the region very close to the Sun does not yet exist; it would be very helpful in promoting a better understanding of the Local Bubble.

In principle, observations of interstellar absorption lines[5] toward nearby stars could provide information on the ionization state of the gas. For example, if ionization equilibrium is assumed, then the column density ratio Mg I/Mg II gives the electron density

$$n_e = \frac{\Gamma}{\alpha} \frac{N(\text{Mg I})}{N(\text{Mg II})} \text{ cm}^{-3}, \qquad 1.$$

[5] Emission-line studies of the Local Fluff appear to be ruled out by the exceedingly small emission measure of the gas. An ionization fraction of 50% implies an emission measure of $\sim 10^{-2}$ cm^{-6} pc, which is approximately two orders of magnitude beyond the capabilities of current hydrogen recombination-line detection techniques for warm gas (88).

where Γ is the Mg I photoionization rate, α the Mg II recombination coefficient, and N(Mg I) and N(Mg II) the column densities of the first two ionization states of magnesium through the cloud. In practice, the value for α is uncertain, and the derivation of N(Mg II) from the equivalent widths of the 2800-Å line are very difficult to obtain owing to line saturation. An acceptable value for Γ appears to be 4×10^{-11} s^{-1} [derived by Bruhweiler et al. (17) from the radiation field of Gondhalekar et al. (42), whereas a value of 8×10^{-11} s^{-1} would be consistent with the radiation field of Witt & Johnson (107) and the analysis of de Boer et al. (30)]. The recombination coefficient α is a steep function of temperature due to the dominance of dielectronic recombination at temperatures of $\sim 10^4$ K. A parameterization of the work of Jacobs et al. (53) by Shull & Van Steenberg (96) gives values for α that range from 7.9×10^{-13} to 2.2×10^{-12} cm^3 s^{-1} for temperatures between 7000 and 9000 K, respectively (the range of hydrogen temperatures consistent with the backscatter data); at 10,000 K, $\alpha = 3.1 \times 10^{-12}$ cm^3 s^{-1}. Therefore, depending upon the temperature, we have the relation $\Gamma/\alpha \approx 13$ to 50 cm^{-3}; there is an additional uncertainty of unknown magnitude due to our incomplete knowledge of the atomic physics.

Based on Copernicus and IUE data for three nearby stars—α Lyr (8 pc), α Gru (20 pc), and α Eri (40 pc)—Bruhweiler et al. (17) adopted a value for the column density ratio N(Mg I)/N(Mg II) $\approx 2 \times 10^{-3}$ in the Local Cloud. If this is true, then Equation 1 implies $n_e \approx 0.026$–0.10 cm^{-3}, which suggests a moderate to high fractional ionization of the interstellar hydrogen that spans the range required by models discussed earlier. However, another nearby star, α PsA (7 pc), listed by Bruhweiler et al. (17) has N(Mg I)/N(Mg II) $\lesssim 4.5 \times 10^{-4}$, which implies that $n_e \lesssim 0.02$ cm^{-3}. This may reflect the significant variations expected in the physical conditions within the cloud due to strong differentials in optical depths for ionizing photons, or else inaccuracies in the derived values of N(Mg II) due to an incomplete understanding of line formation.

Rather than deriving N(Mg II) from the equivalent width of the 2800-Å line, Frisch et al. (40) have used the fact that Mg II should represent virtually all of the magnesium in the gas phase [see Murray et al. (76)]. In this case, the electron density can be represented by the relation

$$n_e = 7.7 \times 10^9 \frac{\Gamma}{\alpha} \frac{W_{2852}}{N_H(\text{Mg/H})} \text{ cm}^{-3}, \qquad 2.$$

where N_H is the total hydrogen column density through the cloud, Mg/H is the gas-phase magnesium to hydrogen abundance ratio, and W_{2852} is the equivalent width of the Mg I λ2852 line (in mÅ). In low-density regions magnesium appears to be nearly undepleted. Studies by Murray et al. (76) and Jenkins et al. (58) suggest a depletion of about a factor of two (corresponding to a Mg/H abundance ratio of $\approx 2 \times 10^{-5}$) in regions having hydrogen densities ≈ 0.1 cm^{-3}. Absorption-line measurements indicate a hydrogen column density through the Local Cloud of $\approx 10^{18}$ cm^{-2} [e.g. McClintock et al. (70)], although this value may depend upon direction (see above). A column density of 10^{18} cm^{-2} is also supported by the analyses of the soft X-ray and EUV spectra of the hot DA white dwarf HZ 43 (63 pc) at $l = 54°$, $b = +84°$, which is a line of sight about 90° from the Galactic center direction. An H I column density of $(1–1.3) \times 10^{18}$ cm^{-2} was derived from the EXOSAT spectrum (77), which is primarily shortward of the helium ionization edge at 504 Å and sensitive to He I absorption. On the other hand, an H I column density of $(2–6) \times 10^{17}$ cm^{-2} has been derived from the Voyager spectrum (50), which is primarily longward of 504 Å and thus sensitive to H I absorption. [Interestingly, Paerels et al. (77) point out that the discrepancy between the two measurements could be reconciled if a significant fraction of the hydrogen along the line of sight to HZ 43 is ionized.]

For an assumed $N_H = 10^{18}$ cm^{-2} and a Mg/H abundance ratio of 2×10^{-5} for the Local Cloud, Equation 2 becomes

$$n_e \approx (0.005-0.02)\, W_{2852}\, \text{cm}^{-3}, \qquad 3.$$

where the range in the coefficient reflects the range in Γ/α for temperatures between 10,000 and 7000 K, respectively. The relation is appropriate only for those stars far enough away to be outside the local 10^{18} cm^{-2}, but not so far to have any further N_H. This contrasts with the approach of Frisch et al. (40), who adopt a mean hydrogen density of 0.12 cm^{-3} and leave the pathlength L through the cloud a free parameter; they also assume that there is no depletion of magnesium and use a value for Γ/α that corresponds to a temperature of 11,500 K. Detections of the Mg I line are reported by Bruhweiler et al. (17) toward α Lyr (8 pc) and α Gru (20 pc), for example, which give values for W_{2852} of 22 and 12.5 mÅ, respectively. These results, together with Equation 3, suggest an electron density of ~ 0.1 cm^{-3}. On the other hand, the 1.5-mÅ upper limit for W_{2852} toward α CMa (2.7 pc) implies an electron density that is much lower (i.e. $n_e \lesssim 0.01$ cm^{-3}). Since we know that the observed EUV flux at the Sun provides $n_e \gtrsim 0.009$ cm^{-3}, we expect a minimum Mg I equivalent width $W_{2852} \gtrsim 0.43-1.7$ mÅ for stars sampling the Local Cloud. Current observations do not have sufficient sensitivity to verify this expected lower limit. While the available data suggest significant variations in the local electron density, it must be kept in mind that this interpretation may be negated by possible variations in hydrogen column densities and gas temperatures along different lines of sight through the cloud. The two Mg I detections cited above are within 70° of the Galactic center direction, whereas α CMa samples a very different region of the sky.

Support for the possibility that the properties of the Local Cloud may not be homogeneous is provided by high signal-to-noise observations of Ca II profiles toward 20 nearby stars (65). The profiles appear to indicate three separate, coherent motions for the gas within 5 pc of the Sun! The mean velocity dispersion of the components is about 10 km s^{-1}. Lallement et al. (65) suggest that such complex motions within such a small region could be the result of shocks. Whatever the origin, if the structure exists, then it is likely that physical conditions within the Local Cloud vary rapidly with position. Therefore, owing to the steep dependence of $\alpha_{Mg\ II}$ on temperature, even if the atomic parameters are known precisely and accurate measurements of Mg I and Mg II column densities are obtained toward many nearby stars, accurate measurements of the temperature would still be needed in order to map unambiguously the properties of the Local Cloud.

Possible Sources of Ionization

As the above discussion indicates, the degree of ionization within the Local Cloud is uncertain, and unambiguous measurements probably will be very difficult to obtain. Since a number of observations can be interpreted in terms of a high ($\gtrsim 50\%$) fractional ionization of hydrogen, at least in certain locations within the cloud, it is useful to consider possible sources of ionization. The evidence for ionization mentioned above includes the H I and He I densities derived from the solar backscatter data, the difference in the values for the H I column density toward HZ 43 derived from the EUV and soft X-ray portions of the spectrum, and the derived magnesium column densities toward some stars. In addition, a high fractional ionization of hydrogen toward α Aur (14 pc) has been proposed by Bobroff et al. (11) in order to explain the absence of what was expected to be a measurable amount of He II 304 Å emission from the star; however, their interpretation depends very much on the accuracy of the stellar atmosphere model (80).

Strong evidence for significant ionization of hydrogen in the LISM is provided by the dispersion measure of a nearby (130 ± 15 pc) pulsar. Gwinn et al. (45) used VLBI techniques to measure the parallax of PSR 0950+08, which has a dispersion measure of 2.97 cm^{-3} pc, corresponding to an electron column density $N_e = 9.16 \times 10^{18}$ cm^{-2}. This pulsar is located at $l = 229°$, $b = +44°$, which appears to place it within the low H I column density cavity in the third quadrant as mapped by Frisch & York (39) and Paresce (81) and defined by H I column density contours $<5 \times 10^{18}$ cm^{-2}

and $<1 \times 10^{19}$ cm^{-2}, respectively. (The Local Bubble, as defined in Figure 1, appears to extend about 80 pc in this direction.) The dispersion measure data therefore suggest that $N_e > N_{H\,I}$ along the line of sight to the pulsar. No O or early B-type stars are located near this line of sight; note also that the hot medium, with a density $\approx 5 \times 10^{-3}$ cm^{-3}, could account for only about 20% of the dispersion measure if it occupied 100% of the line of sight. Of course, the locations of the ionized clouds that are contributing to the dispersion measure are not known; they could be located anywhere along the line of sight to the pulsar. In fact, not more than about 5–10% of the electron column density is probably associated with the Local Cloud (an electron density of 0.05–0.10 cm^{-3} over a distance of 3 pc). However, this pulsar observation implies that ionized clouds *are* present. Within these ionized regions the mean electron density $n_e = 0.023\, f^{-1}$ cm^{-3}, where f is the fraction of the line of sight occupied by the ionized clouds. Salter et al. (91) have derived a (parallax) distance of about 50 pc for another pulsar, PSR 1929+10 at $l = 47°$, $b = -4°$, which has a dispersion measure of 3.2 cm^{-3} pc (9.9×10^{18} cm^{-2}). However, the validity of this distance is in some doubt (6).

If the Local Fluff had a total hydrogen density of 10 n_{He} and hydrogen ionization fraction $x \gtrsim 0.50$, then in steady state a hydrogen ionization rate $\Gamma_H \gtrsim 1.6 \times 10^{-14}$ s^{-1} would be required. This rate is much larger than that due to cosmic rays or the soft X-ray background (see above). An examination of available data on B stars and known hot white dwarf stars in the solar neighborhood reveals that the total photon flux from these sources also is not capable of accounting for such an ionization rate at the Sun (87). Thus a significant stellar source of ionizing photons appears to be ruled out observationally. However, a diffuse flux of $\gtrsim 5 \times 10^3$ photon cm^{-2} s^{-1} is possible. A very long integration time spectrum of the 500–1100 Å band with the Voyager spectrometer has placed an upper limit of $\lesssim 10^5$ photon cm^{-2} s^{-1} on the EUV flux in this band near the north Galactic pole (49). At lower Galactic latitudes Sandel et al. (92) have indicated a possible detection on Voyager spectra of a continuum flux at photon energies $\gtrsim 16.5$ eV. This flux appears to vary significantly with direction; in fact, if this flux is real and extends over an energy band of a few electron volts, then the source need occupy only ~ 0.1 sr in order to provide significant ionization of the local hydrogen. The source of the Sandel et al. flux (if it exists) is not clear; however, an ionizing flux of 5×10^3 photon cm^{-2} s^{-1} could be produced by a 1-pc-thick slab of isothermal plasma at a temperature of 3×10^5 K (emission measure of 2×10^{-4} cm^{-6} pc in collisional and pressure equilibrium at $p/k = 10^4$ cm^{-3} K; 84). The tentative detection of far-UV emission lines near the north Galactic pole by Feldman et al. (36) has been interpreted by Paresce et al.

(83) as emission from $\approx 2 \times 10^5$ K plasma with an emission measure of ≈ 0.1 cm^{-6} pc. If a small amount ($<1\%$) of this plasma were located in the immediate vicinity of the Local Cloud, then a substantial ionization of the cloud could result. Furthermore, Slavin (95) has calculated the ionizing photon flux that would be emitted by a conducting interface (with equilibrium ionization) between the 10^6-K gas of the Local Bubble and the 10^4-K Local Cloud. Preliminary results yield an ionizing flux of $\approx 5 \times 10^3$ photon cm^{-2} s^{-1} and a hydrogen ionization rate of $\approx 1.5 \times 10^{-14}$ s^{-1} near the surface of the cloud, a rate that would cause ionization of half of the hydrogen if the total density were 0.1 cm^{-3}. This point is discussed further below.

Time-dependent ionization processes may also have affected the Local Fluff. The cooling time scale for gas at a density of 0.1 cm^{-3} and a temperature of 10^4 K is at least 4×10^5 yr (59) and may be as long as 10^8 yr (64a); the corresponding recombination time scale from a fully ionized state is $\approx 10^6$ yr. Thus, the supernova event discussed above could be responsible for a significant residual ionization in the Fluff. Although the low-velocity (16 km s^{-1} with respect to the LSR) of the VLISM suggests that it has not interacted recently with a strong shock, intense radiation from the supernova or its remnant could have affected the local gas. Thus, a determination of the ionization state of the Local Cloud, while difficult, should be given a high priority, since it may have important implications for the properties of cloud–hot gas interfaces and the recent history of the local interstellar medium.

CLOUD EXISTENCE IN THE LOCAL BUBBLE

We have considered the possible impact that a large population of clouds within the Local Bubble would have on the bubble evolution. We find observationally that standard diffuse clouds are conspicuously absent from the Bubble, but that there are low-column-density clouds (with the Sun inside one, and several possibly within ~ 5 pc) and indications of only one larger column density feature ($\geq 2 \times 10^{19}$ cm^{-2}) nearby. We cannot be sure whether there are sufficient numbers of the lower column density wisps (fluff) within the Bubble that their evaporation could significantly affect the lifetime of the Bubble.

On the other hand, the very presence of these clouds must be something of a surprise. If the bubble were created by a large explosion, sweeping the region clean, it would not leave behind such low-N_H pieces. In addition, we earlier calculated the cloud lifetime against unimpeded evaporation. Taking $R = 3$ pc, $T_\mathrm{F} = 10^6$ K, and $n_\mathrm{F} = 5 \times 10^{-3}$ cm^{-3} (so that $p_\mathrm{F} = 1.4 \times 10^{-12}$ dyn cm^{-2}) as parameters representative of the VLISM cloud and the surrounding hot gas, the evaporation time scale is about $3 \times 10^6 n$ yr, where n is the mean density of the cloud. If indeed $n \sim 0.15$ cm^{-3}, as implied by the helium backscatter measurements, then the cloud lasts only about 5×10^5 yr.

This lifetime estimate is almost surely wrong, since it depends upon an idealized spherical steady-state model for thermal conduction. This is the same model for which the ionizing photon flux (from the conduction interface) was calculated and found marginally capable of contributing an appreciable ionization near the cloud boundary (95). In fact, the ionizing flux depends on the temperature structure around the cloud, which is to first order independent of the actual magnitude of the conductivity. As a consequence, it may be possible with a suppressed conductivity to have both an ionizing flux of order 5000 cm^{-2} s^{-1} and an evaporative lifetime appreciably greater than $3 \times 10^6 n$ yr.

An interesting test is a calculation of the O^{+5} column density expected from the boundary of this same cloud. For a boundary with equilibrium ionization structure, $N(\mathrm{O}^{+5})$ is small and is given by

$$N(\mathrm{O}^{+5}) \sim 2 \times 10^{17} \, A_0 R_{\mathrm{pc}} n_{\mathrm{F}} / T_6^{3/2} \text{ cm}^{-2}$$

where A_0 is the oxygen abundance relative to hydrogen (72). If we include the effect of nonequilibrium ionization, the column density is directly proportional to the evaporation rate (since this rate governs the number of ions passing through O^{+5}) and to the mean time scale for ionization (8). An approximate calculation suggests that

$$N(\mathrm{O}^{+5}) \sim \frac{10^{15}}{R_{\mathrm{pc}}} \frac{A_0}{n_{\mathrm{F}}} T_6^{3/2} \text{ cm}^{-2},$$

or roughly 5×10^{13} cm^{-2} for the Local Cloud. This estimate should be appropriate when it exceeds the equilibrium values. Now, if the thermal conductivity is for some reason simply reduced in magnitude, the cloud evaporation rate and $N(\mathrm{O}^{+5})$ decrease in direct proportion. Hence the product of the evaporation time scale and $N(\mathrm{O}^{+5})$ remains constant:

$$\tau_{\mathrm{evap}} N(\mathrm{O}^{+5}) \sim 3 \times 10^{20} \frac{R_{\mathrm{pc}} A_0 n}{n_{\mathrm{F}} T_6} \text{ cm}^{-2} \text{ yr},$$

which is $10^{20} n$ cm yr for the Local Cloud. Since we can be reasonably certain that the Local Cloud boundary contributes at most an O^{+5} column density of 10^{13} cm^{-2} (57a), and certainly not even that in all directions (21a), the evaporation rate is apparently reduced by at least a factor of about 5, which increases the evaporative lifetime to at least $10^7 n$ (cm^{-3}) yr $\sim 2 \times 10^6$ yr. Perhaps a restudy of O VI, looking specifically for a local component, together with a more careful calculation of the anticipated column density will extend this lifetime even further, into the 10^7 yr or beyond regime. Of course, we cannot be entirely certain that the cloud evaporates at all until we actually see the outflowing ions. Perhaps when the anticipated column densities of C IV, S IV, and other species are calculated, it will be found that existing observations severely limit evaporation at any significant rate. See (21a) for further discussion.

So maintenance of the Local Cloud against the tendencies of thermal conduction may be something it knows how to do by itself. We can still wonder why such clouds are within the Local Bubble in the first place. One suggestion is that the VLISM and the Local Fluff are part of the larger wall of material that seems to lie between the Local Bubble and Loop I (16). This wall is shown schematically in Figure 2. The velocities involved are of about the same magnitude, similar in direction, and generally consistent with their being due to Loop I expansion. With a Loop I centered roughly on the Sco-Cen association at 200 pc, a current expansion velocity of 20 pc Myr^{-1}, and an assumed constant r^3v as in momentum-conserving pressure-free snowplow expansion, we have an explosion age $T \sim r/(4v) \sim 3 \times 10^6$ yr for Loop I. The principal difficulty with this interpretation is one of detail. The soft X-ray background in the Galactic center direction, toward the near face of the Loop I sphere, is not particularly dim. The cavity extent shown in Figure 1 is shown as roughly 30 pc in order to get the necessary emission measure. It is conceivable that small wisps of material could have gotten somewhat ahead of the main shell, or that the material in the shell is shredded into thin filaments interspersed in the X-ray-emitting gas. In this case, it could be necessary to rethink the possibilities of embedded absorber models for this particular direction. But this picture is presently just conjecture in need of study and observational support. (The Loop I expansion time scale, however, is probably about right, independent of whether or not the Local Fluff is part of the shell. In addition, in another million or two years, the denser shell should be upon us.)[6] It has also been suggested that some very diffuse clouds within the Local Bubble could be the expanded remnants of individual planetary nebula shells (26). With some detective work, it might be possible to associate particular clouds with their corresponding white dwarfs. At the very least one must be careful in making assumptions regarding the anticipated relative abundances of hydrogen and helium or the normalcy of the gas-to-dust ratio in these clouds. They may individually represent large excursions from a well-mixed interstellar medium. The Local Cloud, for example, has a probable total mass of only about 0.5

---

[6] The LSR and solar-centered velocities have about the same magnitude.

$M_\odot$ and could have expanded to its current extent in a few hundred thousand years. We hope, in closing on this note, that one can appreciate the level of challenging ignorance that still permeates and stimulates efforts in studying the LISM. For those seeking a survey of what is known, it is found in the opening section.

ACKNOWLEDGMENTS

The authors would like to thank P. Frisch, F. Bruhweiler, J. Raymond, J. Slavin, R. Lallement, and F. Paresce for valued discussions, as well as C. McKee, C. Heiles, D. McCammon, and C. Goebel for critical readings that greatly improved the manuscript.

This work was supported in part by the National Aeronautics and Space Administration under Grant NAG5-629 and the National Science Foundation under Grant AST8511808.

Literature Cited

1. Adams, T. F., Frisch, P. C. 1977. *Ap. J.* 212: 300
2. Ajello, J. M. 1978. *Ap. J.* 222: 1068
3. Ajello, J. M., Witt, N., Blum, P. W. 1979. *Astron. Astrophys.* 73: 260
4. Albert, C. E. 1983. *Ap. J.* 272: 509
5. Arnaud, M., Rothenflug, R., Rocchia, R. 1984. In *The Local Interstellar Medium, IAU Colloq. No. 81*, ed. Y. Kondo, F. C. Bruhweiler, B. D. Savage, p. 301. *NASA CP-2345*
6. Backer, D. C., Sramek, R. A. 1982. *Ap. J.* 260: 512
7. Balbus, S. A. 1986. *Ap. J.* 304: 787
8. Ballet, J., Arnaud, M., Rothenflug, R. 1986. *Astron. Astrophys.* 161: 12
9. Bertaux, J. L. 1984. In *The Local Interstellar Medium, IAU Colloq. No. 81*, ed. Y. Kondo, F. C. Bruhweiler, B. D. Savage, p. 3. *NASA CP-2345*
10. Bertaux, J. L., Lallement, R., Kurt, V. G., Mironova, E. N. 1985. *Astron. Astrophys.* 150: 1
10a. Bloch, J. J., Jahoda, K., Juda, M., McCammon, D., Sanders, W. T., Snowden, S. L. 1986. *Ap. J. Lett.* 308: L59
11. Bobroff, N., Nousek, J., Garmire, G. 1984. *Ap. J.* 277: 678
12. Bohlin, R. C., Savage, B. D., Drake, J. F. 1978. *Ap. J.* 224: 132
13. Bowyer, C. S., Field, G. B. 1969. *Nature* 223: 573
14. Bowyer, C. S., Field, G. B., Mack, J. E. 1968. *Nature* 217: 32
15. Broadfoot, A. L., Kumar, S. 1978. *Ap. J.* 222: 1054
16. Bruhweiler, F. C. 1984. In *The Local Interstellar Medium, IAU Colloq. No. 81*, ed. Y. Kondo, F. C. Bruhweiler, B. D. Savage, p. 39. *NASA CP-2345*
17. Bruhweiler, F. C., Oegerle, W., Weiler, E., Stencel, R., Kondo, Y. 1984. In *The Local Interstellar Medium, IAU Colloq. No. 81*, ed. Y. Kondo, F. C. Bruhweiler, B. D. Savage, p. 64. *NASA CP-2345*
18. Burrows, D. N., McCammon, D., Sanders, W. T., Kraushaar, W. L. 1984. *Ap. J.* 287: 208
18a. Charles, P. A. Kahn, S. M., McKee, C. F. 1985. *Ap. J.* 295: 456
19. Chassefière, E., Bertaux, J. L., Lallement, R., Kurt, V. G. 1986. *Astron. Astrophys.* 160: 229
20. Clayton, D. D., Cox, D. P., Michel, F. C. 1986. In *The Galaxy and the Solar System*, ed. R. Smoluchowski, J. N. Bahcall, M. S. Mathews, p. 129. Tucson: Univ. Ariz. Press
21. Cowie, L. L., McKee, C. F. 1977. *Ap. J.* 211: 135
21a. Cowie, L. L., Songaila, A. 1986. *Ann. Rev. Astron. Astrophys.* 24: 499
22. Cox. D. P. 1979. *Ap. J.* 234: 863
23. Cox. D. P. 1981. *Ap. J.* 245: 534
24. Cox. D. P., Anderson, P. R. 1982. *Ap. J.* 253: 268
25. Cox, D. P., Smith, B. W. 1974. *Ap. J. Lett.* 189: L105
26. Cox, D. P., Snowden, S. L. 1986. *Adv. Space Res.* 6: 97
27. Crutcher, R. M. 1982. *Ap. J.* 254: 82
28. Dalaudier, F., Bertaux, J. L., Kurt, V. G., Mironova, E. N. 1984. *Astron. Astrophys.* 134: 171
29. Davidson, A., Shulman, S., Fritz, G.,

Meekins, J. F., Henry, R. C., Friedman, H. 1972. *Ap. J.* 177: 629
30. de Boer, K. S., Koppenaal, K., Pottasch, S. R. 1973. *Astron. Astrophys.* 28: 145
31. Dickey, J. M., Salpeter, E. E., Terzian, Y. 1978. *Ap. J. Suppl.* 36: 77
32. Edgar, R. J. 1986. *Ap. J.* 308: 389
33. Fahr, H. J. 1974. *Space Sci. Rev.* 15: 483
34. Fahr, H. J., Lay, G., Wulf-Mathies, C. 1977. *Adv. Space Res.* 18: 393
35. Fahr, H. J., Nass, H. U., Rucinski, D. 1984. In *The Local Interstellar Medium, IAU Colloq. No. 81*, ed. Y. Kondo, F. C. Bruhweiler, B. D. Savage, p. 32. *NASA CP-2345*
36. Feldman, P. D., Brune, W. H., Henry, R. C. 1981. *Ap. J. Lett.* 249: L51
37. Freeman, J., Paresce, F., Bowyer, S. 1979. *Ap. J. Lett.* 231: L37
38. Fried, P. M., Nousek, J. A., Sanders, W. T., Kraushaar, W. L. 1980. *Ap. J.* 242: 987
38a. Frisch, P. C. 1981. *Nature* 293: 377
39. Frisch, P. C., York, D. G. 1983. *Ap. J. Lett.* 271: L59
40. Frisch, P. C., York, D. G., Fowler, J. R. 1987. Preprint
41. Gliese, W. 1969. *Catalogue of Nearby Stars*. Heidelberg: Veroff
42. Gondhalekar, P. M., Phillips, A. P., Wilson, R. 1980. *Astron. Astrophys.* 85: 272
43. Gry, C., Wamsteker, W., eds. 1986. *Proceedings of COSPAR Symposium on UV Space Astronomy: Physical Processes in the Local Interstellar Medium. Advances in Space Research*, Vol. 6. In press
44. Gry, C., York, D. G., Vidal-Madjar, A. 1985. *Ap. J.* 296: 593
45. Gwinn, C. R., Taylor, J. H., Weisberg, J. M., Rawley, L. A. 1984. In *The Local Interstellar Medium, IAU Colloq. No. 81*, ed. Y. Kondo, F. C. Bruhweiler, B. D. Savage, p. 281. *NASA CP-2345*
46. Hartquist, T. W., Dyson, J. E., Pettini, M., Smith, L. J. 1986. *MNRAS* 221: 715
47. Heiles, C. 1987. *Ap. J.* 315: 555
48. Hester, J. J., Cox, D. P. 1986. *Ap. J.* 300: 675
48a. Hobbs, L. M. 1974. *Ap. J.* 191: 395
49. Holberg, J. B. 1987. Preprint
50. Holberg, J. B., Sandel, B. R., Forrester, W. T., Broadfoot, A. L., Shipman, H. L., Barry, D. C. 1980. *Ap. J. Lett.* 242: L119
51. Innes, D. E., Hartquist, T. W. 1984. *MNRAS* 209: 7
52. Iwan, D. 1980. *Ap. J.* 239: 316
53. Jacobs, V. L., Daves, J., Rogerson, J. E., Blaha, M. 1979. *Ap. J.* 230: 627
54. Jahoda, K., McCammon, D., Dickey, J. M., Lockman, F. J. 1985. *Ap. J.* 290: 229
55. Jahoda, K., McCammon, D., Lockman, F. J. 1986. *Ap. J. Lett.* 311: L57
56. Jakobsen, P., Kahn, S. M. 1986. *Ap. J.* 309: 682
57. Jakobsen, P. 1986. *Adv. Space Res.* In press
57a. Jenkins, E. B. 1978. *Ap. J.* 219: 845
58. Jenkins, E. B., Savage, B. D., Spitzer, L. Jr. 1986. *Ap. J.* 301: 355
59. Kafatos, M. 1973. *Ap. J.* 182: 433
60. Kahn, F. D. 1976. *Astron. Astrophys.* 50: 145
61. Kahn, S. M., Jakobsen, P. 1986. Preprint
62. Kondo, Y., Bruhweiler, F. C., Savage, B. D., eds. 1984. In *The Local Interstellar Medium, IAU Colloq. No. 81. NASA CP-2345*. See Foreword, p. v
63. Kraushaar, W. L. 1977. *Soft X rays and the interstellar medium*. Presented at Am. Astron. Soc. Meet., 149th, Honolulu, Hawaii
64. Ku, W. H.-M., Kahn, S. M., Pisarski, R., Long, K. S. 1984. *Ap. J.* 278: 615
64a. Kulkarni, S. R., Heiles, C. 1987. In *Galactic and Extragalactic Radio Astronomy*, ed. K. I. Kellermann, G. L. Verschuur, Chap. 3. 2nd ed. In press
65. Lallement, R., Vidal-Madjar, A., Ferlet, R. 1986. *Astron. Astrophys.* 168: 225
66. Lockman, F. J. 1986. *Proc. NRAO Workshop on Gaseous Halos of Galaxies*, ed. J. N. Bregman, F. J. Lockman, p. 63. Charlottesville, Va: NRAO
67. Lucke, P. B. 1978. *Astron. Astrophys.* 64: 367
68. Marshall, F. J., Clark, G. W. 1984. *Ap. J.* 287: 633
69. McCammon, D., Burrows, D. N., Sanders, W. T., Kraushaar, W. L. 1983. *Ap. J.* 269: 107
70. McClintock, W., Henry, R. C., Linsky, J. L., Moos, H. W. 1978. *Ap. J.* 225: 465
71. McClintock, W., Henry, R. C., Moos, H. W., Linsky, J. L. 1976. *Ap. J. Lett.* 204: L103
72. McCray, R., Snow, T. P. Jr. 1979. *Ann. Rev. Astron. Astrophys.* 17: 213
73. McKee, C. F., Ostriker, J. P. 1977. *Ap. J.* 218: 148
74. Meier, R. R. 1980. *Astron. Astrophys.* 91: 62
75. Morfill, G. E., Hartquist, T. W. 1985. *Ap. J.* 297: 194
76. Murray, M. J., Dufton, P. L., Hibbert, A., York, D. G. 1984. *Ap. J.* 282: 481
77. Paerels, F. B. S., Bleeker, J. A. M.,

Brinkman, A. C., Gronenschild, E. H. B. M., Heise, J. 1986. *Ap. J.* 308: 190
78. Paresce, F. 1983. *Nature* 302: 806
79. Deleted in proof
80. Paresce, F. 1984. In *The Local Interstellar Medium, IAU Colloq. No. 81*, ed. Y. Kondo, F. C. Bruhweiler, B. D. Savage, p. 169. *NASA CP-2345*
81. Paresce, F. 1984. *Astron. J.* 89: 1022
82. Paresce, F., Bowyer, S., Kumar, S. 1974. *Ap. J. Lett.* 188: L71
83. Paresce, F., Monsignori Fossi, B. C., Landini, M. 1983. *Ap. J. Lett.* 266: L107
84. Raymond, J. C. 1984. Private communication (an update of Ref. 85)
85. Raymond, J. C., Smith, B. W. 1977. *Ap. J. Suppl.* 35: 419
86. Raymond, J. C., Cox, D. P., Smith, B. W. 1976. *Ap. J.* 204: 290
87. Reynolds, R. J. 1986. *Astron. J.* 92: 653
88. Reynolds, R. J. 1987. *Ap. J.* 315: 234
89. Ripken, H. W., Fahr, H. J. 1983. *Astron. Astrophys.* 122: 181
90. Ripken, H. W., Fahr, H. J. 1984. In *The Local Interstellar Medium, IAU Colloq. No. 81*, ed. Y. Kondo, F. C. Bruhweiler, B. D. Savage, p. 28. *NASA CP-2345*
91. Salter, M. J., Lyne, A. G., Anderson, B. 1979. *Nature* 280: 477
92. Sandel, B. R., Schemansky, D. E., Broadfoot, A. L. 1979. *Ap. J.* 227: 808
93. Sanders, W. T., Kraushaar, W. L., Nousek, J. A., Fried, P. M. 1977. *Ap. J. Lett.* 217: L87
94. Sedov, L. I. 1959. *Similarity and Dimensional Methods in Mechanics.* New York: Academic
95. Slavin, J. 1986. Private communication
96. Shull, J. M., Van Steenberg, M. 1982. *Ap. J. Suppl.* 48: 95
97. Snowden, S. L. 1986. PhD thesis. Univ. Wis., Madison
98. Snowden, S. L., Cox, D. P., Kraushaar, W. L., Sanders, W. T. 1986. In preparation
99. Spitzer, L. Jr. 1962. *Physics of Fully Ionized Gases.* New York: Interscience. 170 pp.
100. Spitzer, L. Jr. 1985. *Phys. Scr.* T11: 5
101. Spitzer, L. Jr., Jenkins, E. B. 1975. *Ann. Rev. Astron. Astrophys.* 13: 133
102. Streitmatter, R. E., Balasubrahmanyan, V. K., Protheroe, R. J., Ormes, J. F. 1985. *Astron. Astrophys.* 143: 249
103. Thomas, G. E., Krassa, R. F. 1971. *Astron. Astrophys.* 11: 218
104. Tinbergen, J. 1982. *Astron. Astrophys.* 105: 53
104a. van Dishoeck, E. F., Black, J. N. 1986. *Ap. J. Suppl.* 62: 109
105. Weller, C. S., Meier, R. R. 1979. *Ap. J.* 227: 816
106. Weller, C. S., Meier, R. R. 1981. *Ap. J.* 246: 386
107. Witt, A. N., Johnson, M. W. 1973. *Ap. J.* 181: 363
108. Wu, F. M., Judge, D. L. 1979. *Eos, Trans. Am. Geophys. Union* 60: 365

CEPHEIDS AS DISTANCE INDICATORS

M. W. Feast and A. R. Walker

South African Astronomical Observatory, P.O. Box 9, Observatory 7935, South Africa

1. INTRODUCTION

The crucial importance of Cepheids as distance indicators in both galactic and extragalactic research has long been recognized. At the present time, for instance, they appear to be the most satisfactory indicators of relative and absolute distances to nearby galaxies and to be the foundation for extensions of the distance scale to cosmologically important distances. The early history of the Cepheid period-luminosity (PL) and period-luminosity-color (PLC) relations and their calibration has been covered in a comprehensive way by Fernie (1969). A well-known series of investigations of the Cepheid distance scale by Sandage & Tammann was summarized by Sandage (1972). Another investigation that has been much referred to by later workers is that of de Vaucouleurs (1978). Among the many other recent studies are those of van den Bergh (1977), Fernie & McGonegal (1983), and Caldwell (1983). Stothers (1983) assembled all the estimates available to him. We have attempted in this review to give a detailed assessment of the current situation, taking into account the very considerable improvements on both the observational and the theoretical side since these earlier studies.

2. THE PROBLEMS OF REDDENINGS AND ABSORPTIONS

The methods of calibrating and using Cepheid distances are sensitive in varying degrees to the effects of interstellar reddening and absorption. Fortunately, it is possible to correct Cepheid magnitudes and colors individually for these effects. While suitably calibrated spectral type–color relations can in principle be used for this purpose, the most satisfactory method is to use multicolor photometry. UBV photometry is unsuitable, essentially because the reddening and intrinsic lines are nearly parallel (cf. Cogan 1979). However, the BVI system (Dean et al. 1978 = DWC), the Walraven system (Pel 1978), and the DDO system (Dean 1981) have all been applied successfully to this problem. It appears (Feltz & McNamara 1980) that satisfactory results can also be obtained in the Strömgren system. The application of any of these methods requires the adoption of an intrinsic line in the relevant two-color diagram. The slope of this line can be determined from the loci of individual Cepheids around their cycles. The zero point of the line can be fixed from Cepheids of known reddening. These are generally taken to be ones in clusters and associations.

The position of the appropriate intrinsic line will depend on metallicity. If we assume that the lines derived as above, using (nearby) clusters and associations, refer at least in the mean to Cepheids of solar metallicity, the displacement necessary for other metallicities can be estimated from theory. This has been done in the case of the BVI intrinsic line by Caldwell & Coulson (1985 = CCI) using the model atmosphere calculations of Bell & Gustafsson (1978) and of Kurucz (1979). These corrections are of importance, for instance, in deriving the reddenings of Magellanic Cloud Cepheids.

The accuracy that can be achieved by these methods depends on observational errors (including any effects due to binary companions to the Cepheids), width in the adopted intrinsic "line" (due for instance to gravity effects), uncertainty in the adopted abundances, and errors in the zero point of the intrinsic line and the theoretical abundance corrections to it. The width of the intrinsic line in the BVI system due to gravity effects is likely to be small, since the loops made by individual Cepheids are narrow in a direction perpendicular to this line. Some idea of the internal scatter in reddening determinations can be obtained from Magellanic Cloud Cepheids. CCI find that values of $E(B-V)$ derived from BVI photometry have dispersions of 0.037 mag and 0.026 mag in the Large Magellanic Cloud (LMC) and Small Magellanic Cloud (SMC), respectively. Obviously some of this scatter is due to real reddening differences from Cepheid to Cepheid. CCI estimate upper limits to the observational uncertainties of 0.032 mag (LMC) and 0.019 mag (SMC). Thus, relative values of E_{B-V} (and A_V) can be derived with high accuracy from BVI photometry, and similar accuracy is possible in other color systems. The zero point obviously depends on the calibrators used. Pel (1978) used an intrinsic line (Walraven system) that depends on nearby Ib and Iab (nonvariable) supergiants, theoretical results, S Nor (in a cluster), and β Dor (in the foreground of the LMC). His reddenings average ~ 0.05 mag less in E_{B-V} than those of DWC. So far as distance determinations are concerned, errors in the reddening zero point cancel out provided that the same set of Cepheids (in clusters and associations) is used to provide both the reddening and the absolute magnitude calibration. There are 12 stars in common between the reddening calibrations of DWC (their Table 2) and our Table 2 of absolute magnitude calibrators. The difference between the adopted reddenings, E_{B-V}, for these stars is

$$\Delta \text{ (DWC}-\text{Present)} = 0.00 \pm 0.01 \text{ (12 Cepheids)}.$$

Thus the absolute magnitudes to be discussed in Section 7.1 can be used consistently in conjunction with the DWC BVI reddening scale.

Uncertainty in the ratio of total to selective absorption can introduce uncertainty into the absorptions derived from reddenings. Unless otherwise stated, we have adopted $\mathscr{R} = A_V/E_{B-V} = 3.3$ as the appropriate value for stars with intrinsic colors in the Cepheids' range (cf. de Vaucouleurs 1978). Caldwell & Coulson (1986 = CCII), on the other hand, use $\mathscr{R} \sim 3.5$, while earlier investigators often used $\mathscr{R} \sim 3.0$. If $\mathscr{R}$ is regarded as being uncertain by ~ 0.2, then this may introduce an appreciable error in the distances of heavily reddened Cepheids. It will not be of importance in most extragalactic applications and is quite negligible in the case of the LMC and SMC distances (cf. CCII).

3. THE FORMS OF THE PL AND PLC RELATIONS (BVI)

The distances of Cepheids are derived using a PL or a PLC relation. These relations were first found to exist, empirically, in the Magellanic Clouds, and the Clouds (especially the LMC) are still the best place to establish the relations, apart from their zero points. Martin et al. (1979 = MWF) showed that when LMC Cepheids were corrected individually for reddening using BVI photometry, the resulting $\langle V_0 \rangle$–log P relation was well defined but had a significant scatter ($\sigma = 0.27 \pm 0.04$). This scatter could be reduced to $\sigma = 0.14 \pm 0.02$ (i.e. to the level of their observational errors) by using a PLC relation of the form

$$\langle M_v \rangle = \alpha \log P + \beta (\langle B_0 \rangle - \langle V_0 \rangle) + \phi. \qquad 1.$$

Since observational error is present in more than one variable, it is necessary to solve for the coefficients of the PLC relation using maximum-likelihood techniques (cf. Balona 1983). MWF found the following relation:

$$\langle M_v \rangle = -3.80 \log P + 2.70(\langle B_0 \rangle - \langle V_0 \rangle) + \phi \text{(PLC2)}. \qquad 2.$$

Suggestions have occasionally been made that the PLC relation does not exist at all or that β (in Equation 1) is radically different from the value given in Equation 2 (e.g. that $\beta = 0$ or $\beta > 5$). Such suggestions have been discussed in detail in the literature (see summaries by Feast 1984a, 1985) and have been found to be unsubstantiated (see also Section 5). In principle, the coefficients in Equation 1 depend on the chemical abundances. However, because the metal deficiency of the LMC Cepheids is slight (a factor of ~ 1.4 times less than solar), the differences between the values of α and β given in Equation 2 and those appropriate to a metal-normal Cepheid are very small. The effect on the zero point ϕ is of greater significance. (Abundance effects are discussed in Section 4.)

More recently, CCII have analyzed BVI data for Cepheids in both Magellanic Clouds, taking into account the effects of metallicity (which become more significant for the SMC). They find the following relation for metal-normal Cepheids:

$$\langle M_v \rangle = -3.53 \log P + 2.13(\langle B_0 \rangle - \langle V_0 \rangle) + \phi(\text{PLC1}). \qquad 3.$$

CCII tabulate the changes expected in the various coefficients as a function of metallicity. These were derived from the theoretical predictions of Iben & Tuggle (1975) and Becker et al. (1977) (cf. Iben & Renzini 1984).

The rather large differences in α and β between Equations 2 and 3 are primarily due to the different ways in which the data were analyzed statistically. In fact, the two solutions are, for a variety of reasons (cf. CCII), at about the two extremes possible. These extremes in the maximum-likelihood solutions correspond to two possible least-squares solutions with in one case, all the errors in the colors, and in the other with all the errors in the magnitudes. One may reasonably suppose that the actual values lie between these two extremes. In much of what follows we give results derived using both equations. The distances obtained using either of the two equations will in general differ little from one another. The methods of calibration, in effect, ensure that the values of $\langle M_v \rangle$ derived from Equations 2 and 3 are the same for Cepheids on the PL and period-color (PC) ridge lines. The range in $(\langle B_0 \rangle - \langle V_0 \rangle)$ at a given period is about 0.32 (CCII) for Cepheids of a given composition. This will result in a difference of 0.09 in the predicted $\langle M_v \rangle$ values from Equations 2 and 3 for a Cepheid at the edge of the PC strip. For groups of Cepheids, the mean difference between the predictions of the two equations will be smaller than this (generally much smaller). Here and throughout, we have omitted the standard errors of α and β (though these are generally given in the original papers). Since changes in the two constants are correlated, their individual uncertainties are not of direct relevance for present purposes. In using the PLC (or PL) relations, the main uncertainty one faces is in ϕ (see Section 7).

Theory predicts that β is a function of period (cf. van Genderen 1983). However, the empirical evidence in the Magellanic Clouds (CCII) suggests that over the range of periods considered, a constant value is adequate. It should be noted that throughout we have excluded from consideration, in discussions of the PL and PLC relations, the long-period Cepheids ($P \geq 100$ days). These important stars are in some ways anomalous and are discussed separately in Section 8.

For certain purposes it is useful to define a PLC relation in V and $(V-I)$. CCII give the following relation:

$$\langle M_v \rangle = -3.66 \log P + 3.71 \overline{(V-I)_0} + \phi(\text{PLC3}), \qquad 4.$$

where $\overline{(V-I)_0}$ is a magnitude mean.

The bolometric PL relation should be nearly free of abundance effects (cf. Becker et al. 1977), and this should also be approximately so for the relation in $\langle M_v \rangle$, since the bolometric corrections to V are very small (cf. Pel 1985). For the Magellanic Clouds, CCII find that

$$\langle M_v \rangle = -2.78 \log P + \phi(\text{PL}V). \qquad 5.$$

It has sometimes been suggested that despite the greater real scatter of the PL relation compared with the PLC relation, the former is more useful because of its insensitivity to abundance effects. However, it should be remembered that the PL relation in V (unlike the PLC relation in $V, B-V$) is very sensitive to errors in the reddening corrections. Thus in any case, unless adequate estimates of the reddening can be made by means not involving observations of the Cepheid itself, we require multicolor observations of the Cepheids, as well as estimates of metallicity in order to derive reddenings, before a reliable PL distance can be obtained.

4. ABUNDANCE EFFECTS

In order to apply the procedures of the last section, some estimate of a Cepheid's metallicity is necessary. In many cases (especially in extragalactic work), it will probably be necessary to rely on estimates from H II regions in the general vicinity of the Cepheids concerned. This is likely to be quite adequate in many cases, since a metal deficiency of a factor of 1.4, which is clearly established in the case of the H II regions in the LMC (cf. CCII), corresponds to a metallicity correction to the modulus of 0.14 (PLC in V, $B-V$). So in practice it should be possible to keep the errors in the moduli well below this value. There are, however, a number of methods of estimating metallicities from the Cepheids themselves (in addition to high-dispersion spectroscopy, which is applicable only to the brighter Cepheids). These methods as applied in the Magellanic Clouds are listed in Table 1.

Harris (1981a) has also applied the Washington photometric system to determinations of abundances of Cepheids within our own Galaxy. His results are consistent with a galactic abundance gradient of $\Delta[Fe/H] = -0.07$ kpc^{-1}, as are the available results of high-dispersion spectroscopy (Giridhar 1986, Harris & Pilachowski 1984). It is important to take this gradient into account in deriving accurate distances for galactic Cepheids (Caldwell & Coulson 1987).

Eggen (1985a,b,c) has made extensive observations of galactic Cepheids using the Strömgren system, and it may be anticipated that these data contain valuable information on metallicities. A recent discussion (Feast 1987) suggests that the observed spread in the Strömgren metallicity index $[M_1]$, at a given period, largely reflects a spread in temperatures. Thus, means will have to be devised to take this into account in deriving abundances from the Strömgren photometry.

CCII show that because of the different metallicity dependences of the PLC, the PL, and the reddening relations in the BVI system, it is possible to estimate the metallicities in the Magellanic Clouds by forcing the PLC and PL moduli together. Their results are shown in Table 1. Despite the intrinsic scatter in the PC relation, the displacement of this relation with metallicity can also be used (together with theoretical models) to estimate metallicity for a group of Cepheids. The results obtained in this way by CCII are also in Table 1. It is clear that the results of these last two estimates are in good agreement with more direct methods. This is a useful indication that the methods being used to correct the various relations for metallicity are satisfactory.

Laney & Stobie (1986a = LS) have used their JHK infrared photometry of Magellanic Cloud Cepheids (see Section 5) to show that the log T_e–log P relation for the SMC is displaced 210 ± 80 K hotter than that for the

Table 1 Abundances $[A/H]$ of Magellanic Cloud Cepheids

Method	LMC	SMC	Reference
Washington four-color photometry	-0.09	-0.65	Harris 1981b, 1983
Walraven five-color photometry	-0.2	-0.6	Pel 1984, Pel et al. 1981
Curve of growth (low-dispersion spectra)	-0.06	-0.50	C. D. Laney, in preparation
PL, PLC, and BVI reddening relations	-0.20	-0.49	CCII
PC relations	-0.15	-0.53	CCII
Adopted	-0.15	-0.60	
Depletion factor (D)	1.4	4	

LMC. They combine this with theory as summarized by Iben & Renzini (1984) to obtain a metallicity difference between the two Clouds of $\Delta Z = 0.014 \pm 0.006$, with the SMC being more metal poor than the LMC. This result agrees well with those in Table 1, where the adopted depletions correspond to $\Delta Z = 0.012$.

In deriving the necessary metallicity corrections, CCII assume that the helium abundance varies with metal abundance according to the relation $\Delta Y = 2.8\, \Delta Z$ (Lequeux et al. 1979), although there is considerable uncertainty in this ratio (cf. Pagel 1986). However, this does not introduce any large uncertainty into the PLC relation. Thus, if in the Magellanic Clouds we adopt $\Delta Y = 0$, the PLC moduli would decrease by 0.034 (LMC) and 0.078 (SMC). The PL moduli would be unchanged.

5. INFRARED PL AND PLC RELATIONS

McGonegal (1982) pointed out that in deriving distances to Cepheids, infrared (JHK) photometry has a number of advantages over optical photometry. Specifically, (a) the infrared magnitudes and colors are much less sensitive to interstellar reddening, (b) the effective temperature range at a given period introduces less scatter in a PL relation as one goes to longer wavelengths, (c) the differences in chemical abundance between Cepheids have little effect on the infrared colors, and (d) the light amplitudes of Cepheids are lower at longer wavelengths and this makes the estimation of mean magnitudes easier when few observations are available. In view of these advantages, infrared photometry of Cepheids in our own and other galaxies has been carried out by these workers and their colleagues and by others.

As with the optical work, the slope of the infrared period-luminosity relation is best established in the Magellanic Clouds; LS (who give references to earlier work) find that

$$\langle M_H \rangle = -3.42 \log P + \phi(\text{PL}H), \qquad 6.$$

with a scatter of $\sigma = 0.13$.

This scatter is, as anticipated, considerably less than that about Equation 5, and Equation 6 will be satisfactory for many practical purposes in distance estimation. Nevertheless, LS show that by using good quality photometry of LMC and SMC Cepheids, one can find a PLC relation even in the infrared, namely

$$\langle M_H \rangle = -3.64 \log P + 3.28(\langle J-K \rangle_0) + \phi(\text{PLC}H). \qquad 7.$$

Using standard transformations (cf. LS), we find that the color coefficient in Equation 7 corresponds to a value of $\beta = 2.60 \pm 0.19$ (Equation 1), in satisfactory agreement with the results of Section 3.

As already noted, the bolometric PL relation should be virtually free of metallicity effects. However, LS point out that for magnitudes at a given wavelength, a metallicity effect is introduced through the bolometric correction, because the mean effective temperature at a given period is a function of metallicity. They show, for instance, that the displacement between the log T_e–log P relations for the two Magellanic Clouds (as derived from the displacement in their infrared color–log P relations) leads to a difference in zero points, $\Delta\phi$, given by

$$\Delta\phi = \phi(\text{PL}H)(\text{LMC}) - \phi(\text{PL}H)(\text{SMC}) = -0.10. \qquad 8.$$

A qualitatively similar effect must be present at all wavelengths. However, in V, where the bolometric corrections are themselves very small, the effect is unlikely to be of significance in most practical cases.

A useful check on the accuracy of Cepheids as relative distance indicators can be obtained by comparing the *BVI* and *JHK* moduli of the two Magellanic Clouds. Because of the large depth of the SMC (cf. Section 9), it is necessary to define the SMC centroid used, in the same way for both the optical and infrared data. One then finds the following relations:

BVI: $\Delta\text{Mod}(\text{SMC–LMC}) = 0.30 \pm 0.06$ (CCII),

JHK: $\Delta\text{Mod}(\text{SMC–LMC}) = 0.32 \pm 0.04$ (LS).

The effects of metallicity and reddening are rather different in the two wavelength ranges. Thus the agreement shown above indicates that the manner in which these effects are being taken into account is satisfactory for distance determination purposes.

6. PL AND PLC RELATIONS AT OTHER WAVELENGTHS AND SHORTCUT METHODS

In any large-scale application of Cepheids to distance-scale problems, it is obviously important to consider carefully the most economical method of proceeding, taking into account problems connected with reddening corrections, metallicity corrections, mean magnitude and color estimates, period determination, and intrinsic spread in the relations used (e.g. the PL relation). The method adopted will depend on the accuracy required in the derived distances.

Visvanathan (1985) and Mathewson et al. (1986) have made Cepheid observations in the Magellanic Clouds at 1.05 μm (*IV* band), where the advantages of the farther infrared are partially retained. As applied by

Visvanathan, the method is a "shortcut." It uses the existing, extensive photoelectric and photographic data on Magellanic Cloud Cepheids to reduce random-phase IV observations approximately to mean light. Average reddenings are adopted from existing multicolor optical photometry. The intrinsic spread in the $\langle IV \rangle$–log P relation will presumably be somewhat greater than in the farther infrared, but it is probably sufficiently small to make the 1.05-μm region relatively attractive.

The form of the PL relation derived by Visvanathan is

$$\langle M_{IV} \rangle = -3.140 \log P + \phi(\text{PL}IV). \qquad 9.$$

Visvanathan's general methods would appear to be useful in cases such as the Magellanic Clouds, where much data on light curves, reddenings, and metallicities are already available. It does indeed seem very desirable to find ways of using the vast body of photographic work to cut down the amount of new data needed.

It may be useful in certain circumstances to write the PLC relation in I and $(R-I)$ (Eggen 1977). However, the way in which this particular equation was derived may have led to biased coefficients (Cogan 1980). Whether for this or other reasons, applying the equation to Eggen's LMC data (following his procedure) leads to a strong dependence of the derived LMC modulus on period.

Until very recently, the main observational difficulty in the photometry of Cepheids in distant galaxies at any wavelength has been the problem of carrying out accurate photometry in crowded fields. Conventional aperture photometry is very difficult in these circumstances. The work of McAlary & Madore (1984) on photometry of the Cepheids in NGC 2403 at J clearly shows how the crowded nature of the field limits the accuracy of results in the infrared. The availability of panoramic detectors has therefore revolutionized the situation, at least in the wavelength range covered by the BVR and I bands, where the commonly available detectors are sensitive. These detectors make observations such as those outlined in Section 3 for BVI photometry feasible in galaxies out to distances of several megaparsecs. Thus, in principle, full PLC studies such as those already carried out in the Magellanic Clouds are now possible in a considerable number of other galaxies. Since the PL relation becomes narrower at longer wavelengths, it may be attractive for some purposes to derive distances from an $\langle R_0 \rangle$ or $\langle I_0 \rangle$–log P relation with reddenings from BVI or other multicolor work. In view of this, it would be useful to set up definitive $\langle R_0 \rangle$ and $\langle I_0 \rangle$–log P relations in the Magellanic Clouds. To do this may require more observing at R and I. Freedman et al. (1985) show that the width of the PL relation at B, V, or I is roughly the same as the scatter of random-phase observations at the respective wavelength. Thus it may be more satisfactory to observe large numbers of Cepheids in a galaxy a few times, rather than to observe a few Cepheids very intensively. However, complete light curves will be needed for the most accurate results. [Note that the two Cepheids recently discovered in M101 (Cook et al. 1986) have R amplitudes of ~ 1.0 mag.] Freedman (1985) has suggested that if, in a given galaxy, Cepheids are observed in $BVRI$, then the reddening can be obtained by comparing the apparent moduli derived from the four PL relations. It will be interesting to see whether in practice this method is superior to determining the reddening from the BVI diagram (which is clearly possible from the same data if the multicolor observations are nearly simultaneous).

An approximate method of correcting random-phase observations (in V) to mean light was suggested by Madore (1985a), but Moffett & Barnes (1986) find that the underlying assumptions of the method are false and that it can lead to errors in the moduli of galaxies of several tenths of a magnitude.

The large amount of effort necessary to identify Cepheids and to derive their periods is a real constraint on the wider use of these variables as distance indicators. Automated plate searching and measuring has considerably relieved this situation. Nevertheless, a full analysis still requires large amounts of plate material. Madore & Freedman (1985) have suggested that it might be possible to identify Cepheids and to determine approximate periods for them from a small number of observations. Extensive tests of actual data along the lines they suggest would help establish the strengths and limitations of their approach. An attempt to apply their method to the two newly discovered Cepheids in M101 was apparently not successful (Cook et al. 1986).

7. PL AND PLC ZERO POINTS

In the following subsections we discuss in detail the absolute calibration of the PL and PLC relations.

7.1 *Clusters, Associations, Geometrical Determinations*

Earlier work (e.g. Sandage & Tammann 1969, van den Bergh 1977, Fernie & McGonegal 1983, Caldwell 1983) on the PL and PLC zero points from Cepheids in galactic clusters and associations can now be revised using the improved distances recently obtained for several of the clusters. Walker and coworkers (Walker 1985c,d, Walker & Laney 1986) have given charge-coupled device (CCD) color-magnitude diagrams for Lyngå 6, NGC 6067, and NGC 6649; Turner (1986) and Pel (1985) have made new observations of NGC 6087 and M25; and Pedreros et al. (1984) have published new photoelectric and photographic photometry for NGC 7790. All this work gives Cepheid absolute magnitudes, and cluster distances, based on the fitting of a zero-age main sequence (ZAMS) to unevolved stars in the clusters. The fiducial ZAMS used here is that of Turner (1976a, 1979), adjusted to a Pleiades modulus of 5.57 ± 0.08 derived by van Leeuwen (1983) from a fit of the Pleiades main sequence to field stars of known parallax, or else the published ZAMS fit has been adjusted to this modulus. This Pleiades modulus corresponds to a Hyades modulus of 3.27, with a metallicity correction for the latter of 0.22 (cf. Pel 1985). Use of a Pleiades ZAMS answers some of the criticisms of the ZAMS fitting method (discussed by Schmidt 1984). However, Walker & Laney (1986) note that good quality data are necessary [$\sigma(M_v) < 0.3$, which implies $\sigma(B-V) < 0.05$] if the spurious displacement of the main sequence due to unrecognized binaries is to be kept well below 0.1 mag.

Cluster distances can also be found from Strömgren-Hβ photometry of the OB stars in the clusters by using a suitable luminosity calibration for these stars. The results of Schmidt (1984) for eight clusters have been rediscussed by Balona & Shobbrook (1984) in the context of an improved luminosity calibration. The Strömgren luminosity calibration used by Eggen (e.g. 1983) will also need changing if it is to be put on the system of Balona & Shobbrook. Turner (1986) criticizes the Strömgren-Hβ method for finding the luminosity of OB stars, and he concludes that the neglect of rotation leads to incorrect evolutionary corrections.

The list of calibrating Cepheids given in Table 2 totals 28 stars, consisting of 17 stars in 14 clusters, 8 stars in associations, 2 stars illuminating gas or dust, and 1 double star system. Only 6 of the clusters can be considered to have available an accurate photometric study of a well-populated main sequence; this situation is due to many of the clusters being extremely sparse. As a consequence, some of the stars have absolute magnitudes that are much better determined than others; in particular, the membership of the longer period stars in the assigned associations is not very secure.

Detailed notes are given elsewhere (Walker 1987b) for each calibrator and also for several stars that at times have been included in lists of calibrating Cepheids. Recently, searches have been made for new calibrating stars; in particular, van den Bergh et al. (1982, 1983, 1985) have searched for associations near long-period Cepheids, while Turner (1985) lists several Cepheids that may lie in cluster coronae.

The ZAMS fitting assumes implicitly that $[Fe/H]_{cluster} = [Fe/H]_{Pleiades}$. There is little direct evidence that the calibrating clusters, the Cepheids they contain, or nearby Cepheids in general have [Fe/H] significantly different from this (on this last point, cf. Section 4 and Feast 1987). When zero points for each calibrating Cepheid were recalculated assuming a

Table 2 Cepheids in clusters, etc.

Star	Period	μ_v	E_{OB}	$\langle V \rangle$	$\langle B \rangle - \langle V \rangle$	μ_0	$\langle M_v \rangle$	$\langle B_0 \rangle - \langle V_0 \rangle$	ϕ_{PL}	ϕ_{PLC1}	ϕ_{PLC2}	Type[a]
SU Cas	1.95	7.96	0.29	5.97	0.70	7.08	−1.98	0.43	−1.18	−1.87	−2.03	G
EV Sct	3.09	12.70	0.60	10.14	1.16	10.88	−2.53	0.60	−1.17	−2.09	−2.31	C
α UMi	3.97	5.19	0.00	1.99	0.59	5.19	−3.20	0.59	−1.54	−2.34	−2.52	B
SZ Tau	4.03	9.97	0.39	6.53	0.84	8.79	−3.43	0.47	−1.74	−2.30	−2.41	C
CEb Cas	4.48	14.34	0.64	10.99	1.12	12.39	−3.33	0.52	−1.52	−2.14	−2.26	C
CF Cas	4.87	14.34	0.64	11.14	1.18	12.39	−3.17	0.59	−1.26	−2.00	−2.14	C
HD 144972	5.10	12.19	0.35	8.87	0.91	11.13	−3.31	0.59	−1.34	−2.06	−2.20	C
CEa Cas	5.14	14.34	0.64	10.92	1.20	12.39	−3.39	0.61	−1.42	−2.18	−2.33	C
UY Per	5.37	14.77	0.99	11.35	1.48	11.74	−3.38	0.56	−1.35	−1.99	−2.12	C
CV Mon	5.38	13.60	0.77	10.30	1.29	11.25	−3.27	0.58	−1.24	−1.91	−2.04	C
V Cen	5.49	10.07	0.31	6.82	0.88	9.13	−3.24	0.59	−1.18	−1.89	−2.03	C
V367 Sct	6.29	15.42	1.35	11.56	1.83	11.27	−3.80	0.58	−1.58	−2.21	−2.32	C
U Sgr	6.74	10.49	0.48	6.70	1.10	9.03	−3.77	0.66	−1.46	−2.25	−2.40	C
DL Cas	8.00	12.85	0.53	8.97	1.15	11.24	−3.86	0.66	−1.35	−2.08	−2.21	C
S Nor	9.75	10.38	0.19	6.42	0.94	9.81	−3.95	0.77	−1.20	−2.09	−2.26	C
TW Nor	10.79	15.54	1.34	11.67	2.00	11.42	−3.79	0.78	−0.92	−1.81	−1.97	C
CPD-537400	11.22	12.19	0.35	8.37	1.18	11.13	−3.80	0.86	−0.88	−1.93	−2.14	C
VY Car	18.93	12.43	0.28	7.46	1.16	11.58	−4.95	0.91	−1.40	−2.38	−2.55	A
RU Sct	19.70	14.70	1.03	9.49	1.63	11.55	−5.16	0.68	−1.56	−2.05	−2.09	C
RZ Vel	20.40	12.25	0.35	7.09	1.13	11.19	−5.14	0.81	−1.50	−2.25	−2.36	A
WZ Sgr	21.83	12.92	0.56	8.03	1.39	11.22	−4.86	0.89	−1.13	−2.02	−2.16	C
SW Vel	23.47	13.35	0.41	8.12	1.15	12.11	−5.21	0.78	−1.40	−2.03	−2.10	A
T Mon	27.02	11.69	0.20	6.13	1.17	11.09	−5.55	0.99	−1.57	−2.61	−2.78	A
KQ Sco	28.69	15.36	1.00	9.81	1.94	12.30	−5.47	1.05	−1.42	−2.57	−2.78	A
RS Pup	41.39	13.01	0.57	7.01	1.43	11.28	−5.96	0.92	−1.47	−2.21	−2.30	G
SV Vul	45.01	13.29	0.49	7.14	1.45	11.80	−6.12	1.01	−1.52	−2.44	−2.57	A
GY Sge	51.07	16.56	1.30	10.23	2.23	12.56	−6.22	1.08	−1.47	−2.50	−2.65	A
S Vul	67.60	15.90	0.89	9.00	1.86	13.18	−6.83	1.07	−1.74	−2.66	−2.77	A

[a] Abbreviations: A = association, B = binary, C = cluster, G = geometrical.

galactic metallicity gradient of $\Delta[\text{Fe/H}] = -0.1$ kpc^{-1}, the change in Cepheid luminosities was insignificant. The calibrators are well distributed in space around the Sun, and there is no significant difference in mean galactocentric distance for the short-period stars in clusters and the long-period stars in associations. It is, however, clear that improved information on the metallicities of the calibrating clusters and of Cepheids in general would be of considerable interest.

The input data for calculating M_v are the apparent V modulus of the cluster, μ_v, the OB star reddening (at the location of the Cepheids), $E(\text{OB})$, and the intensity mean magnitudes $\langle V \rangle$ and $\langle B \rangle - \langle V \rangle$ for the Cepheid itself. We use the apparent V modulus (corrected for differential extinction and referenced to a Pleiades modulus of 5.57) because, as pointed out by J. A. R. Caldwell (private communication), the published true moduli contain the dispersion caused by the various values of $\mathscr{R}$ that have been used. We find μ_v (the apparent modulus) by adding $\mathscr{R}(\text{OB}) \cdot E(\text{OB})$ to the published μ_0 [where the $\mathscr{R}(\text{OB})$ value adopted is that used in the cluster study] and then correct to the adopted Pleiades modulus. In most cases, we have preferred to use the reddening determined from the cluster stars rather than that for the Cepheid itself, observing that for the short-period stars, which here are those with the best cluster data, the Cepheid reddening is not as well determined from the BVI photometry as it is for longer period stars.

In addition, the available I band photometry for the calibrating stars is on both the Johnson system (northern stars) and the Kron-Cousins system (southern stars), and there are difficulties in transforming from one system to the other (cf. Taylor 1986). However, in the majority of cases, the Cepheid reddening agrees with that for the cluster to within the errors.

We use

$$\mathscr{R} = 3.06 + 0.25(B-V)_0 + 0.05\, E(B-V), \qquad \qquad 10.$$

where the coefficients of the color and reddening terms are from Olson (1975). The zero point is chosen (Laney & Stobie 1986b) so that the value for $\mathscr{R}$ agrees with the mean value found for 51 galactic clusters by Turner (1976b).

Then, we have

$$\mu_0 = \mu_v - \mathscr{R}(\text{OB}) \cdot E(\text{OB}),$$

$$M_v = V - \mathscr{R}(\text{Cepheid}) \cdot E(\text{Cepheid}) - \mu_0,$$

where $E(\text{Cepheid}) = E(\text{OB}) \cdot [0.98 - 0.09(\langle B_0 \rangle - \langle V_0 \rangle)_{(\text{Cepheid})}]$ (Schmidt-Kaler 1982). The absolute magnitude M_v is nearly independent of both the zero point (3.06) in Equation 10 and μ_0. The values of ϕ are then calculated for Equations 2, 3, and 5 above. Rather than assign individual weights, we have preferred to group the stars in various ways and to calculate mean zero points for these groups (see Table 3). These means give very similar zero points, except that the association stars appear to be a little brighter and redder than expected. These are, in fact, the stars for which membership is the most doubtful.

The preferred solution is that in which the Cepheids in clusters are weighted 1 and all other Cepheids weighted 0.5, as indicated in Table 3.

The small range of zero points for the various solutions suggests that the adopted values are accurate to ± 0.10, not including the uncertainty (± 0.08) in the Pleiades distance modulus. This in turn suggests that the standard error of a modulus determination should be ~ 0.15, or perhaps less. The changes in zero points are 0.12 mag (PLC) and 0.06 mag (PL) over those used by CCII in their discussion of the distance of the Magellanic Clouds (our distance scale being shorter).

Table 3 also shows the Balona & Shobbrook (1984) recalibration of Schmidt's (1984) intermediate-band photometry, adjusted to a Pleiades modulus of 5.57. We conclude that there is no significant difference between the present calibration and that derived from intermediate-band photometry when they are both referred to the same Pleiades distance.

Table 3 BVI zero points[a] from clusters, etc.

	No.	$\phi(PLV)$	$\phi(PLC1)$	$\phi(PLC2)$
(1) All data: unit weight per star	28	-1.37 ± 0.04	-2.17 ± 0.04	-2.31 ± 0.04
(2) As (1) without associations	20	-1.32 ± 0.05	-2.06 ± 0.03	-2.21 ± 0.03
(3) As (1) but $P < 10$ days	15	-1.37 ± 0.04	-2.09 ± 0.04	-2.24 ± 0.04
(4) As (1) but clusters only	17	-1.31 ± 0.05	-2.06 ± 0.03	-2.20 ± 0.03
(5) Clusters only: unit weight per cluster	14	-1.31 ± 0.06	-2.06 ± 0.04	-2.19 ± 0.04
(6) All data: cluster stars unit weight, others weight 0.5	28	-1.35 ± 0.04	-2.13 ± 0.04	-2.27 ± 0.04
(7) As (1) but $P > 10$ days	13	-1.38 ± 0.07	-2.26 ± 0.08	-2.40 ± 0.08
(8) As (1) but associations only	8	-1.50 ± 0.04	-2.40 ± 0.07	-2.54 ± 0.08
(9) Recalibration of β photometry of 6 clusters	—	-1.25 ± 0.11	—	-2.31 ± 0.06
(10) Recalibration of β photometry of 5 clusters[b]	—	-1.34 ± 0.08	—	-2.29 ± 0.05
Adopted [= solution (6)]	—	-1.35	-2.13	-2.27

[a] All the standard errors in this table are internal (that is, they represent the scatter among the individual values of Table 2).

[b] This omits NGC 7790, for which the reddening from intermediate-band photometry (Schmidt 1984) is very different from that adopted by Pedreros et al. (1984).

7.2 Statistical Parallaxes

The most recent work on the luminosities of Cepheids from statistical parallaxes is that of Clube & Dawe (1980), who used the proper motions and radial velocities listed by Wielen (1974) for 45 Cepheids. Wielen lists distances for these Cepheids, calculated with a set of adopted magnitudes and reddenings and using the PLC relation of Tammann (1970, Equation 9). Clube & Dawe express their result as a correction to this distance scale and find it to be $\langle \Delta M_v \rangle = 0.0 \pm 0.3$. For 42 of the stars, $(B-V)$ colors and reddenings are listed by DWC. For these stars, distances were calculated on the basis of the PLC relation (Equation 2) above. The ratios of these distances to those of Wielen, together with the Clube & Dawe result, then lead to the zero point $\phi(\text{PLC2}) = -2.32 \pm 0.3$. Within the errors, this value is the same as that derived from clusters and associations (Table 3).

7.3 Pulsation Parallaxes

The method for determining the linear radius of a pulsating star, suggested originally by Baade (1926) and modified in an important way by Wesselink (1946; see also Becker 1940), can be used to derive distances (and thus absolute magnitudes) provided we have some means of measuring or estimating the angular diameter of the variable. We consider first the linear radius determination and then its conversion to absolute magnitude.

7.3.1 LINEAR RADIUS The assumptions implicit in the Baade-Wesselink method are probably most clearly set out by Balona (1977).

The effective temperature T_e is defined by the equation

$$M_{\text{Bol}} = -5 \log R - 10 \log T_e + M_{\text{Bol}\odot} + 10 \log T_{e\odot}, \qquad 11.$$

where R is the radius in solar units and $\odot$ refers to the Sun. Provided we can write linear equations of the form

$$\log T_e = a_0 + a_1 (B-V)_0, \qquad 12.$$

$$M_{\text{Bol}} = M_v + b_0 + b_1 (B-V)_0, \qquad 13.$$

we obtain

$$M_v = A(B-V)_0 - 5 \log (R_0 + r) + C, \qquad 14.$$

where

$$A = -(10a_1 + b_1), \qquad 15.$$

$$C = M_{\text{Bol}\odot} + 10 \log T_{e\odot} - (10a_0 + b_0), \qquad 16.$$

and $R_0 + r = R$ (R_0 is the mean stellar radius, r the displacement from the mean). If $\mathscr{R}$ is the ratio of total to selective absorption, E the color excess, and D the distance modulus, and linearizing the term in $R_0 + r$, we find

$$V = A(B-V) - (5 \log_{10} e/R_0)r + B,\qquad 17.$$

where

$$B = (\mathscr{R} - A)E - 5 \log R_0 + C + D.\qquad 18.$$

Here r can be obtained by integrating the radial velocity curve, namely

$$r = -kP \int V_r \, d\theta,\qquad 19.$$

where P is the period, V_r the radial velocity, θ the phase, and k the limb darkening and projection correction. Balona (1977) shows how to obtain R_0, A, and B from Equations 19 and 17 using a maximum-likelihood method.

This method of formulation shows clearly that provided $\log T_e$ and M_{Bol} can be linearized as in Equations 12 and 13, we do not need to adopt a priori any specific values for the coefficients in these equations (since A is determined by the solutions). It is also clear that the results are independent of interstellar reddening. Evidently, we can substitute any other magnitude and color for V and $(B-V)$. Most modern radius determinations adopt a formulation like Balona's, although with variations in detail as to how it is applied and using different color indices. For instance, we could adopt values of a_1 and b_1 from calibrations of nonvariable supergiants. Since A does not appear to be correlated with period, intrinsic color, light amplitude, etc. (Thompson 1975, Balona 1977), it might be thought advantageous to use a mean of all the derived A values in obtaining final values of R_0 for a group of Cepheids. Balona (1977) found that this, in fact, increased the scatter in the final $\log R_0$–$\log P$ relation. He suggested that by making A the same for all stars, we throw all the errors due to phase mismatching, presence of companions, etc., on R_0.

The "surface brightness method" is, in principle, the same as that just described, except that as applied by Barnes et al. (1977) $(V-R)$ is used as the color index and a mean value of A is adopted. This is evident from the definition of surface brightness F_v, namely

$$F_v = \log T_e + 0.1(M_{\text{Bol}} - M_v),\qquad 20.$$

and from the fact that in actual application, F_v is linearized to

$$F_v = F + G(V-R)_0,\qquad 21.$$

with G determined from the Cepheids themselves (Barnes 1980, Barnes &

Moffett 1985). Thus, writing the earlier equations in $(V-R)$, we have $A \equiv 10G$.

In the early literature it was usual to avoid linearizing Equations 12, 13, or 20 by comparing magnitudes and displacements at pairs of phases of the same color. Balona (1977) suggests, however, that in the presence of random errors, the higher-order terms in the color would lead to biased estimates of the radii. It might be valuable to look into this more closely.

Coulson et al. (1986) have shown how to correct R_0 for the error introduced by linearizing Equation 14. This effect is generally small. Of more potential importance is the implicit assumption that the F_λ-color relation (or the T_e and bolometric correction versus color relations separately) are independent of gravity or other phase-related effects, such as turbulence (Benz & Mayor 1982). This is a cause of uncertainty whether or not the equations are linearized. The relation in $(V-R)$ appears to be less gravity dependent than that in $(B-V)$ (Barnes & Evans 1976, Barnes et al. 1976), although this conclusion is not entirely certain, since the empirical value of G for Cepheids is different from that for nonvariable supergiants (Barnes 1980). Barnes also reports that the F_v–$(V-R)$ relation is the same on the rising and falling branches of Cepheids, but this conclusion is based on radii derived on the assumption of no gravity effects in $(B-V)$. The systematically larger radii found by Coulson et al. (1986) using $(V-I)$ rather than $(B-V)$ from otherwise identical data and methods may be due to gravity effects being more important in $B-V$.

Two ways of improving radius determinations have recently been suggested. One is to work in the infrared (say at K), where the surface brightness changes little around the cycle of a Cepheid (e.g. Welch et al. 1985b, Fernley et al. 1985). Since this latter effect is due to the opposing effects of bolometric correction and effective temperature changes, it is not clear whether the smaller value of A that results in Equation 14 when M_v is replaced by M_K is more accurate. However, use of a K, $V-K$ system would probably be advantageous. The other possibility is to take $\log g$ into account by using multicolor photometry (cf. Caccin et al. 1981 = CORS method).

The lack of simultaneous photometry and the radial velocities used in Baade-Wesselink solutions has often been a cause of concern. Recent programs have sought to remedy this situation. [A number of these are summarized in Madore (1985b).]

Balona (1977) finds that the scatter about his $\log R_0$–$\log P$ relation is $\sim 10\%$ in R (~ 0.2 in M_v). Thus random errors can be reduced to negligible proportions by observing ~ 40 stars. Systematic errors arise through gravity effects and from the assumption that the derived radial velocities measure the displacement of the photosphere. These uncertainties are difficult to quantify.

7.3.2 ANGULAR DIAMETERS Using the occultation angular diameter of ζ Gem [1.81 ± 0.31 milliarcsecs (Ridgway et al. 1982)], the Baade-Wesselink data of Scarfe (1976), $\langle V \rangle = 3.89$ (Schaltenbrand & Tammann 1971), and $E_{B-V} = 0.03$ (DWC), we obtain $M_v = -3.88$, with an uncertainty of 0.37 from errors in the angular diameter alone. This result corresponds to ϕ(PLC2) $= -2.19(\pm 0.4?)$.

Other angular diameters must be derived using an estimated zero point for the F_λ-color relation [$= -(10a_0 + b_0)$] of Equation 16. This is derived either directly from measured angular diameters of nonvariable stars or by combining such measures with theory and discussing bolometric corrections and T_e values separately. The discussion by Pel (1985) shows that while bolometric corrections are unlikely to be appreciably in error, T_e values could be uncertain by ~ 250 K and thus lead to an uncertainty of ~ 0.2 in M_v.

7.3.3 APPLICATION TO THE ZERO-POINT DETERMINATIONS Fernie (1984) counted 594 Baade-Wesselink–type determinations of radii for 182 Cepheids up to May 1982. In view of current work in this field, it is sufficient here to consider three representative bodies of data:

1. The radii determined by Balona (1977) using radial velocities and BV photometry. These were used in a zero-point determination by MWF, who adopted a_0 and b_0 from Parsons (1971) and Kraft (1961) ($C = 3.40$).
2. Gieren's (1986) tabulation of results obtained using $(V-R)$ colors with a surface brightness zero point and slope from Barnes & Moffett (1985).
3. The luminosities derived by Sollazzo et al. (1981) using the CORS method.

We obtain values for ϕ(PLC2) of -2.42 ± 0.05, -2.32 ± 0.05, and -2.52 ± 0.05 (all errors being internal) for the data in (1.), (2.), and (3.), respectively. For comparison, our adopted value from clusters and associations is -2.27, with an external error of 0.10–0.15 (internal error 0.04).

7.4 Binaries

The study of Cepheid binaries in the far-ultraviolet is now possible. Of 21 Cepheids with known or suspected blue companions, Böhm-Vitense & Proffitt (1985) detected 13 companions using IUE observations. Böhm-Vitense (1985) derived luminosities for five of these stars. Kurucz models were used to derive effective temperatures, and the luminosities were obtained by assuming the companions lay on the ZAMS.

The resulting absolute magnitudes of the Cepheids are shown in Table

4 together with the differences from those predicted by PLC and PL relations using our adopted zero points [Equations 3 and 5, with $\phi(\text{PLC1}) = -2.13$ and $\phi(\text{PL}V) = -1.35$ (Table 3)]. The Böhm-Vitense zero points would be ~ 0.5 fainter than ours. Feast (1986a) suggested that this difference could be due to inadequacies in the models or to evolution of the companion off the ZAMS. Recent work (Evans & Ferro 1986) suggests that the analysis of the IUE data is a major source of uncertainty. Evans & Ferro give data from blue companions to three Cepheids, two of which are in common with those studied by Böhm-Vitense. As can be seen in Table 4, there are large differences in the results of the two investigations, with the absolute magnitudes of Evans & Ferro being brighter than our predictions. Ferro & Madore (1985) also obtained a bright absolute magnitude for the low-amplitude Cepheid HD 129708 from its blue companion.

Taken together, these results from binaries show no evidence for a significant deviation from our adopted distance scale, although the scatter is very large.

7.5 Magellanic Clouds and M31

The Cepheid distance scale could obviously be calibrated if we had an independent distance to any galaxy containing Cepheids.

Magellanic Cloud moduli can be derived from (a) RR Lyrae variables, (b) horizontal branches of old clusters, (c) Mira variables, (d) Type II Cepheids, (e) fits to the giant clump in intermediate-age clusters, and (f) main sequence fits.[1]

Table 4 Absolute magnitudes for Cepheid binaries[a]

	ΔM_1	ΔM_2		ΔM_1	ΔM_2
Böhm-Vitense (1985)			Evans & Ferro (1986)		
SU Cyg	0.25	0.14	SU Cyg	-0.46	-0.57
V636 Sco	0.70	0.79	SU Cas[b]	-0.71	-0.78
U Aql	0.67	0.70	W Sgr	-0.23	-0.35
η Aql	0.47	0.37	Ferro & Madore (1985)		
W Sgr	0.61	0.49	HD 129708[c]	—	$-0.59 \, (= \pm 0.3)$

[a] ΔM_1 is the derived $\langle M_v \rangle$ minus predicted (PLC), ΔM_2 the derived $\langle M_v \rangle$ minus predicted (PL) (see text for details).
[b] SU Cas is treated as an overtone pulsator (otherwise $\Delta M_1 = -1.17$, $\Delta M_2 = -1.14$).
[c] No ΔM_1 for HD 129708, since $(B-V)$ affected by companion.

---

[1] OB stars may also be used (Crampton 1979, Crampton & Greasley 1982), but we have not considered them here. The calibration of OB giants and supergiants seems rather uncertain [e.g. the calibration of Balona & Shobbrook (1984) is ~ 0.3 mag fainter than that of Balona & Crampton (1974)].

The current calibration of the luminosities of RR Lyrae variables has been reviewed recently (Feast 1987), and $\langle M_v(\text{RR})\rangle = 0.6\pm0.2$ was the value adopted (at all metallicities). In practice, methods (a), (b), (c), and (d) all depend on this result, since other means of calibrating the Miras and Type II Cepheids are less reliable than that depending on a galactic globular cluster scale with $\langle M_v(\text{RR})\rangle = 0.6$.

Table 5 summarizes the available data on the Magellanic Clouds, together with the distance moduli of the Clouds from the Cepheids with our adopted zero point (Section 7.7). The moduli that depend on $\langle M_v(\text{RR})\rangle = 0.6$ are all quite consistent with the adopted Cepheid scale. There is a suggestion that main sequence fits to intermediate-age clusters may give somewhat lower moduli, at least when the fit is made to a theoretical sequence. The problems of systematic errors in main sequence fitting were mentioned briefly in Section 7.1 and have been discussed by Walker & Laney (1986). Their numerical experiments using Pleiades data suggest that where the fitting of an empirical ZAMS is made to main sequences showing a large scatter (as in the currently available Magellanic Cloud data), a systematic error of ~ 0.1 in the modulus (in the sense of the derived modulus being too small) is quite possible. If, instead of an empirical ZAMS, one fits to models (that is, to isochrones), the situation becomes worse. Isochrones may be considered as true lower envelopes, and their use as ridge lines through scattered data can give an additional 0.05–0.1 systematic underestimation of the modulus. In addition, the fitting of main sequences to theory is sensitive to the effects of convective overshooting on the upper part of the main sequence, as discussed by Chiosi & Pigatto (1986), and neglect of this effect can also lead to an underestimation of the modulus.

Main sequence fitting in the Magellanic Clouds is of obvious importance to the distance-scale problem. However, taking into account the above considerations, such fittings would not appear to be giving results in serious conflict with our adopted Cepheid scale.

The detection of RR Lyraes in M31 (Pritchet 1986) allows an additional check to be made on the Cepheid scale. For 10 RR Lyraes, Pritchet (1986) finds $\langle B\rangle = 25.52\pm0.12$. He takes $\overline{M_v} = 0.76\pm0.14$ from statistical parallaxes (Hawley et al. 1986), $\langle (B-V)_0\rangle = 0.26$, and $A_B = 0.32$ [from H I (Burstein & Heiles 1982)] to get a true modulus of 24.18. Correcting for the fact that the Hawley et al. magnitudes are not true intensity means will partially cancel the effect of increasing the absorption by ~ 0.1 mag to place it on the Sturch system, which is the appropriate system to use with these statistical parallaxes (cf. Feast 1987). This will also make the adopted value of A_B the same as that used in discussing the M31 Cepheids (Section 9). Then, by using $\langle M_v(\text{RR})\rangle = -0.6$, $\langle (B-V)_0\rangle = 0.26$, and $A_B = 0.43$,

Table 5 Magellanic Cloud distance moduli

Method	Modulus[a]
Large Magellanic Cloud (LMC)	
1. *Variables*	
Cepheids (this review)	18.47 ± 0.15
RR Lyraes in NGC 2210, $\langle M_v(RR) \rangle = 0.6$	
(Walker 1985b)	$18.42 \pm (0.10 \text{ Int})$
Miras [$\langle M_v(RR) \rangle = 0.6$ (Feast 1987)]	$18.48 \pm (0.06 \text{ Int})$
Type II Cepheids in infrared, $\langle M_v(RR) \rangle = 0.6$	
(C. D. Laney, private communication)	18.45
2. *Old clusters* (*Horizontal Branch at* $M_v = 0.6$)	
NGC 2210 (Walker 1984) (as RR's above)	(18.42)
Hodge II (Andersen et al. 1984)	$18.4 \pm (0.1 \text{ Int})$
3. *Intermediate-age clusters*	
NGC 2162/NGC 2190 (Schommer et al. 1984)	
(*a*) Main sequence (Hyades) (cf. Feast 1984b)	18.3 ± 0.2
(*b*) Main sequence (theory)	18.1
H4 (Mateo & Hodge 1986)	
(*a*) Fit to giant clump	18.1 ± 0.3
(*b*) Main sequence (theory)	18.1 or 18.7
ESO121–SCO3 (Mateo et al. 1986)	
(CMD)	18.2 marginally better than 18.7
NGC 1978 (Christian et al. 1985)	
CMD comparison with NGC 2158	18.2 ± 0.2
(data do not warrant arguing for change from 18.6)	
NGC 2213/NGC 1651 (Seidel et al. 1986)	
Fit to clump (theory) (evidence uncertain)	18.2 or 18.7
4. *Young clusters* (*main sequence*)	
NGC 1866 (Walker 1985a)	18.4 or 18.5 ± 0.2
Depends on [Fe/H]	
Small Magellanic Cloud (SMC)	
1. *Variables*	
Cepheids (this review)	18.78
RR Lyraes (18.42 ± 0.39) (cf. Feast 1987)	18.8:
2. *Clusters*	
NGC 411 (Da Costa & Mould 1986)	
CMD of cluster and adjacent field	18.8 better than 19.3
NGC 411, L113, NGC 121 (Seidel et al. 1986)	
Giant clump (theory)	18.8 not 19.3

[a] Int = internal error.

we get a true modulus of 24.23, in excellent agreement with the Cepheid results (Section 9).

7.6 Theoretical Estimates of Zero Points

In principle it should be possible to obtain theoretical PLC and PL relations for Cepheids by combining computed evolutionary tracks with pulsation theory. The results of Iben & Tuggle (1975), Becker & Iben (1980), and Becker et al. (1977) are summarized by Iben & Renzini (1984). Equation 24 of the last reference gives, for $Z = 0.02$, $Y = 0.28$,

$$M_v^{Th} = -3.731 \log P - 12.873 \log T_e + 47.433 - BC. \qquad 22.$$

To apply this equation, we must adopt values for the bolometric corrections (BC) and the effective temperatures. The problems regarding these quantities were set out in Section 7.3. The discussion of Pel (1985, especially his Figure 3) suggests that errors in BC will be small (probably less than ~ 0.05 mag). The heavy line of his Figure 3 may be approximated by

$$BC = 2.239 \log T_e - 8.427, \qquad 23.$$

which substituted into Equation 22 gives

$$M_v^{Th} = -3.731 \log P - 15.112 \log T_e + 55.860. \qquad 24.$$

Figure 2 of Pel (1985) shows various possible calibrations of log T_e in terms of $(B-V)_0$. For four of these calibrations we estimate a_0 and a_1 (see Equation 12 above) from this figure. Table 6 shows the values of $M_v^{Th} - M_v^{Obs}$ calculated at a few representative periods using these values of a_0 and a_1 and taking M_v^{Obs} from Equation 3 using our adopted zero

Table 6 Comparison between theory and observation

1. PLC relation

T_e relation used	log P			
	0.5	1.0	1.5	2.0
		$M_v^{Th} - M_v^{Obs}$		
Kraft	−0.58	−0.59	−0.59	−0.59
Pel	−0.58	−0.47	−0.37	−0.24
Flower I	−0.40	−0.28	−0.13	0.00
Böhm-Vitense I	−0.24	−0.09	+0.16	+0.22

2. PL relation

log P	$\Delta M_{Bol} + BC$	ΔM_{Bol}
0.5	−0.35	−0.37
1.0	−0.52	−0.44
1.5	−0.69	−0.51
2.0	−0.86	−0.58

point $\phi(\text{PLC1}) = -2.13$ and with $(B-V)_0$ from the galactic ridge-line relation (CCII)

$$(B-V)_0 = 0.412 \log P + 0.310. \qquad 25.$$

It might be supposed from Table 6 that the theoretical results could be made to agree with observations by taking a cool enough calibration of log T_e. However, consideration of the PL relation suggests that this is not necessarily the basic problem. Equation 18 of Becker et al. (1977) can be written

$$M_{\text{Bol}}^{\text{Th}} = -3.125 \log P - 1.525. \qquad 26.$$

The difference between this and Equation 5 with $\phi(\text{PL}V) = -1.35$ (i.e. $\Delta M_{\text{Bol}} + BC$) is also shown in Table 6. The bolometric corrections will be small. We show in Table 6 the values of ΔM_{Bol} found by using bolometric corrections from Pel (1985) (the Flower I relation of his Figure 2 together with his Figure 3 and our Equation 25). It follows that there is a discrepancy of ~ 0.5 mag, with the theoretical results being brighter.

7.7 *Summary of Zero Points*

The PLC and PL zero points that we have adopted from the discussion on clusters and associations (Section 7.1) are given at the bottom of Table 3. We adopt these as our primary values. Other values that we adopt are either derived from these or come from adopting $\mu_0(\text{LMC}) = 18.47$, a value that depends on the primary values (see Section 9.1).

The other evidence on zero points discussed above may be summarized as follows:

1. Statistical parallaxes are consistent with our adopted zero point, but with a large uncertainty.
2. Pulsation parallaxes range from close to our adopted zero point to ~ 0.25 mag brighter.
3. Results from binaries show a large scatter.
4. Other types of variables in the Magellanic Clouds and M31 show good agreement with our zero point if $\langle M_v(\text{RR}) \rangle = 0.6$.
5. Intermediate-age clusters in the LMC may indicate a zero point 0.2–0.4 mag fainter than our adopted one. However, the cluster moduli may have been systematically underestimated (see above).
6. Theoretical PL and PLC zero points are brighter than our adopted values (by ~ 0.5 mag).

The theoretical work and the pulsation parallaxes might lead us to believe that our zero point is somewhat too faint. However, any significant brightening of the adopted zero point would imply that the LMC cluster moduli obtained from main sequence fits are quite seriously in error and that $\langle M_v(\text{RR})\rangle$ is significantly brighter than 0.6. In view of the above, we have preferred to retain our adopted values while recognizing that there are still a number of uncertainties.

Preliminary calibrations of the infrared PL (and PLC) relations have been carried out (Welch et al. 1985a, LS) using infrared photometry of Cepheids in some of the calibrating clusters and associations of Table 2. These calibrations are currently somewhat uncertain because (a) they rely on incomplete data and (b) the results depend somewhat on the adopted value of A_V/E_{B-V} (LS). It seems best at present to calibrate Equations 6 and 7 from Magellanic Cloud Cepheids using distances implied by our adopted zero points above (see Section 9). In a similar way, we use the Magellanic Clouds to calibrate Equations 9 and 4. These zero points are:

$\phi(\text{PL}H) = -2.14,$

$\phi(\text{PLC}H) = -3.59,$

$\phi(\text{PL}IV) = -1.88,$

$\phi(\text{PLC3}) = -3.13.$

For extragalactic applications we also require the $\langle M_B\rangle$–log P relation. From Equations 3 and 25 with our adopted zero point, we obtain

$$\langle M_B\rangle = -2.24 \log P - 1.16. \qquad 27.$$

8. LONG-PERIOD CEPHEIDS

It has long been recognized that the 100–200 day Cepheids in the Magellanic Clouds lie below a linear extrapolation of the PL and PLC relations in various colors (e.g. see MWF and CCII for BVI, and LS for JHK). These Cepheids also show a considerable scatter about any mean relation. Circumstellar extinction could only explain these effects if it were nearly neutral over the wavelength range from B to H (see LS). Van Genderen (1983) suggests that these stars are a mixture of first and second crossing Cepheids. Grieve et al. (1985) have proposed that two low-amplitude variables in the LMC with $P \sim 250$ days may be regarded as Cepheids. LS raise the possibility of a period, $\langle H_0\rangle$, amplitude relation for long-period Cepheids. It would obviously be very important to understand these objects, since they are potentially important as distance indicators. At present they can only be used with considerable caution.

9. CEPHEID DISTANCES TO GALAXIES

9.1 *Magellanic Clouds*

Equations 3 and 5 were used by CCII to determine the moduli of the Magellanic Clouds. With our revised zero points (Table 3) and the abundance model adopted by CCII ($D = 1.4Y$), their results for the LMC now become

μ_0(PL)(LMC) = 18.51,

μ_0(PLC)(LMC) = 18.50.

Walker (1987a) has obtained *BVI* photometry of the short-period (~ 3 days) Cepheids in the LMC cluster NGC 1866 and has analyzed this together with data on other short-period Cepheids ($\log P < 1$) in the LMC. This period range has the advantage that it is the range in which many of the best calibrators (Section 7.1) lie. It has the disadvantage that the reddening corrections are less precise in the color range of these stars than in that of longer period ones. Also, for some of the stars, the data are less extensive than for longer period Cepheids. With the same zero point and model as above, the Cepheids considered by Walker give

μ_0(PL)(LMC) = 18.44,

μ_0(PLC)(LMC) = 18.41.

We adopt a Cepheid mean μ_0(LMC) = 18.47.

In view of the discussion in Section 5, this choice leads to

μ_0(SMC) = 18.47+0.31 = 18.78.

The Magellanic Clouds also provide two excellent examples of the power of Cepheids as relative distance indicators. Accurate photometry by Gascoigne & Shobbrook (1978) shows that the eastern edge of the LMC is nearer than the western, consistent with the tilted-plane model of de Vaucouleurs (1960). The difference amounts to 0.14 in the modulus over an angular separation of 6.5. Work on early-type stars has suggested that the SMC is a complex structure of considerable extent in the line of sight. This conclusion has been firmly established by the Cepheid work of CCII and Mathewson et al. (1986). The depth of the SMC is at least 0.6 in the modulus, and probably more. The infrared work on SMC Cepheids (LS) confirms this large depth.

9.2 *More Distant Galaxies*

The general problems of photographic photometry of faint stars in galaxies are well known (e.g. background contamination, the need to extend photoelectric sequences, etc.). Most photographic observations of Cepheids in galaxies beyond the Magellanic Clouds have been made in the *B* band or something approximating it. The large spread in the PL relation in *B* together with a magnitude limit can then lead to a bias in the slope and position of the observed PL relation. In addition, the reddenings have generally been estimated by indirect methods, and this can result in considerable uncertainties in the appropriate A_B values.

The advantages of infrared photometry have already been discussed (Section 5). Background problems and crowding impose severe limitations on the aperture photometry currently employed. These restrictions will be considerably eased when infrared arrays become generally available. Cepheid light amplitudes are much reduced in the infrared, and as a consequence it has often been thought adequate to obtain only one (or perhaps a few) observations per star. However, the observations of Magellanic Cloud Cepheids (Laney & Stobie 1986b) show that many (especially at the longer periods) have considerable amplitudes (e.g. ~ 0.5 mag at H; see also Section 6).

Much work with CCD arrays is currently in progress, although few detailed reports have been published. Work is under way in M31, the galaxies of the Sculptor group, NGC 5236 (Centaurus group), NGC 3109, and IC 5152, among others. Feast (1985) lists galaxies in which Cepheids (Type I or II) have been reported.

Table 7 lists Cepheid distance moduli obtained by a variety of techniques (CCD, infrared, and photographic). Where detailed results have been published, we have reevaluated them using the following relations, which come from our previous discussions and imply $\mu_0(\text{LMC}) = 18.47$:

$\langle M_v \rangle = -1.35 - 2.78 \log P,$

$\langle M_B \rangle = -1.16 - 2.24 \log P,$

$\langle M_H \rangle = -2.14 - 3.42 \log P,$

$\langle M_J \rangle = -1.83 - 3.32 \log P.$

The last relation is based on LS and is used for NGC 2403 only. The adopted values of $E(B-V)$ for each galaxy are also shown in the table. These come from the references cited and generally refer to galactic foreground reddening only. Detailed notes on the data in Table 7 are given by Walker (1987c).

In the case of the moduli in parentheses in Table 7 the detailed observations have not yet been published, and the absorptions adopted are not known. We have, however, corrected them to a scale defined by $\mu_0(\text{LMC}) = 18.47$.

Table 7 Extragalactic Cepheid distances[a] (excluding LMC, SMC)

Galaxy	Adopted $E(B-V)$[b]	μ_0 (CCD)	μ_0 (IR)	μ_0 (pg)	References[c]
NGC 6822	0.3	(23.37)	23.32	23.25	1, 2, 3
IC 1613	0.03	(23.97)	24.07	24.3	3, 4, 5
M33	0.2	24.07	24.19	24.1	6, 7, 8, 9
M31	0.1	—	24.3	24.2	10, 11
WLM	0.03	—	—	24.8	12
Sextans A, B	0.03	—	—	25.5	13, 14
NGC 300	0.04	(25.87)	—	26.0	3, 15
NGC 3109	0.05	—	—	26.1	16
NGC 2403	0.06	(27.47)	27.9	27.3	3, 17, 18
M81	0.06	—	—	(28.1)	19
M101	0.0	29.27	—	—	20

[a] We have used $\mathscr{R} = 3.3$ as being applicable to the Cepheids. This is consistent with Equation 10. The neglect of variation of $\mathscr{R}$ with color and reddening introduces negligible error in the distance moduli.

[b] For all galaxies, except M33 and M31, the $E(B-V)$ adopted is from consideration of foreground absorption only and is obtained from the cited references. Only for NGC 6822 is the foreground absorption significant. For M33 the adopted absorption is that given by Freedman (1985); the Madore et al. (1985) value is similar. In M31 the photographic observations are of Cepheids in Baade Field IV, for which various measurements confine $A_v = 0.2$–0.4. The infrared measurements by Welch et al. (1986) also include some stars in Baade Field III, for which the absorption is expected to be greater. The adopted $E(B-V)$ in the table is a compromise value.

[c] References: 1. Kayser 1967, 2. McAlary et al. 1983, 3. Freedman 1986, 4. Sandage 1971, 5. McAlary et al. 1984, 6. Sandage 1983, 7. Sandage & Carlson 1983, 8. Madore et al. 1985, 9. Freedman 1985, 10. Baade & Swope 1963, 11. Welch et al. 1986, 12. Sandage & Carlson 1985a, 13. Sandage & Carlson 1985b, 14. Walker 1986, 15. Graham 1984, 16. Demers et al. 1985, 17. Sandage 1984a, 18. McAlary & Madore 1984, 19. Sandage 1984b, 20. Cook et al. 1986.

The general consistency of the various results for a given galaxy in Table 7 is good. The most discrepant case is NGC 2403, which shows an infrared modulus that stands off from the others. This discrepancy may be at least partly due to the fact that the infrared method was being pushed to its practical limit with current instrumentation and that a very broadband filter was used in the work.

The way in which more general distance indicators can be established on the basis of Cepheid distances to nearby galaxies is beyond the scope of this review. There are many remaining uncertainties in this step to greater distances, and in the estimation of the Hubble constant H_0. Nevertheless, following Aaronson et al. (1980) and Aaronson & Mould (1983) and calibrating the infrared Tully-Fisher relation on the basis of M31 and M33 alone (using the moduli of Table 7), we obtain $H_0 = 100$ km s^{-1} Mpc^{-1}. However, if we base our calculations entirely on M81 and NGC

2403 (the two other galaxies in Table 7 with relevant data in Aaronson et al. 1982), we obtain $H_0 = 77$ km s^{-1} Mpc^{-1}. Or, using the three galaxies Ho IX, NGC 5585, and NGC 5204 in the M101 group (at the Cepheid distance of M101) for the Tully-Fisher zero point, we find $H_0 = 73$ km s^{-1} Mpc^{-1}. The above assumes the slope of the infrared Tully-Fisher relation to be $b = 10$ (as used by Aaronson & Mould 1983). As pointed out by Sandage & Tammann (1984), the best current value is $b = 11.5$, and a change to this value can affect the results somewhat. In fact, assuming that results can be compared at log $\Delta V_{20}^c(0) = 2.5$ (the log H I line width), we find, from all seven galaxies mentioned in this paragraph, Tully-Fisher zero points that scale the Aaronson-Mould H_0 value to $H_0 = 81$ km s^{-1} Mpc^{-1} ($b = 10$) and $H_0 = 77$ km s^{-1} Mpc^{-1} ($b = 11.5$). The standard error of a determination is probably ~ 10 km s^{-1} Mpc^{-1} (Aaronson & Mould 1983). These values of H_0 must clearly be regarded as provisional determinations, from one line of attack only.

10. CONCLUSIONS

Recent years have seen substantial improvements in the calibration and application of the Cepheid distance scale. This is largely due to the availability of improved instrumentation (CCD arrays, infrared detectors, multicolor photometric systems, radial velocity spectrometers, etc.). Work to take further advantage of these instrumental developments is now in progress. When two-dimensional infrared detectors become readily available, it will be possible to take full advantage of this wavelength region for Cepheid distance determinations. It is anticipated that space astrometry will be of great importance in providing both improved proper motions of galactic Cepheids and also parallaxes of nearby main sequence stars on which the Cepheid zero point derived from star clusters currently depends. An improved astrometric distance for the Hyades will also be important for the Cepheid problem (cf. Tammann 1979, Feast 1986b). The present uncertainty in the calibration of the basic scale appears to be $\sim 7\%$ (in distance), or perhaps slightly less.

It is now possible to substantially improve the quality and quantity of Cepheid data in nearby galaxies. The discovery of Cepheids has already been pushed as far as M101 (~ 7 Mpc), and it is expected that the Hubble Space Telescope will be used in attempts to find Cepheids in more distant galaxies, possibly including some in the Virgo cluster (cf. Aaronson & Mould 1986). Thus, in the next few years we can expect a further refinement in the calibration of the Cepheid zero point and in its application to the precise determinations of galactic and extragalactic distances.

Acknowledgments

We are grateful to the many astronomers who sent us preprints and reprints of their work. These have been of great value. We are particularly indebted to Drs. Balona, Caldwell, Coulson, and Laney as well as to other colleagues at SAAO for helpful discussions on Cepheids and on distance-scale problems in general, and also, in some cases, for access to unpublished work.

Literature Cited

Aaronson, M., Mould, J., Huchra, J. 1980. *Ap. J.* 237: 655–65
Aaronson, M., Huchra, J., Mould, J., Tully, R. B., Fisher, J. R., et al. 1982. *Ap. J. Suppl.* 50: 241–62
Aaronson, M., Mould, J. 1983. *Ap. J.* 265: 1–17
Aaronson, M., Mould, J. 1986. *Ap. J.* 303: 1–9
Andersen, J., Blecha, A., Walker, M. F. 1984. *MNRAS* 211: 695–705
Baade, W. 1926. *Astron. Nachr.* 228: 359–62
Baade, W., Swope, H. H. 1963. *Astron. J.* 68: 435–70
Balona, L. A. 1977. *MNRAS* 178: 231–43
Balona, L. A. 1983. In *Statistical Methods in Astronomy*, ESA SP-201, pp. 187–89. Paris: ESA
Balona, L. A., Crampton, D. 1974. *MNRAS* 166: 203–17
Balona, L. A., Shobbrook, R. R. 1984. *MNRAS* 211: 375–90
Barnes, T. G. 1980. In *Highlights of Astronomy*, ed. P. A. Wayman, 5: 479–82. Dordrecht: Reidel
Barnes, T. G., Dominy, J. F., Evans, D. S., Kelton, P. W., Parsons, S. B., Stover, R. J. 1977. *MNRAS* 178: 661–74
Barnes, T. G., Evans, D. S. 1976. *MNRAS* 174: 489–502
Barnes, T. G., Evans, D. S., Parsons, S. B. 1976. *MNRAS* 174: 503–12
Barnes, T. G., Moffett, T. J. 1985. See Madore 1985b, pp. 53–55
Becker, S. A., Iben, I. 1980. *Ap. J.* 237: 111–29
Becker, S. A., Iben, I., Tuggle, R. S. 1977. *Ap. J.* 218: 633–53
Becker, W. 1940. *Z. Astrophys.* 19: 289–303
Bell, R. A., Gustafsson, B. 1978. *Astron. Astrophys. Suppl.* 34: 229–40
Benz, W., Mayor, M. 1982. *Astron. Astrophys.* 111: 224–28
Böhm-Vitense, E. 1985. *Ap. J.* 296: 169–74
Böhm-Vitense, E., Proffitt, C. 1985. *Ap. J.* 296: 175–84

Burstein, D., Heiles, C. 1982. *Astron. J.* 87: 1165–89
Caccin, B., Onnembo, A., Russo, G., Sollazzo, C. 1981. *Astron. Astrophys.* 97: 104–9
Caldwell, J. A. R. 1983. *Observatory* 103: 244–48
Caldwell, J. A. R., Coulson, I. M. 1985. *MNRAS* 212: 879–88 (Erratum: *MNRAS* 214: 639 (CCI)
Caldwell, J. A. R., Coulson, I. M. 1986. *MNRAS* 218: 223–46 (CCII)
Caldwell, J. A. R., Coulson, I. M. 1987. *Astron. J.* In press
Chiosi, C., Pigatto, L. 1986. *Ap. J.* 308: 1
Christian, C. A., Heasley, J. N., Janes, K. A. 1985. *Ap. J.* 299: 683–94
Clube, S. V. M., Dawe, J. A. 1980. *MNRAS* 190: 591–610
Cogan, B. C. 1979. *MNRAS* 188: 297–300
Cogan, B. C. 1980. *Ap. J.* 239: 941–52
Cook, K. H., Aaronson, M., Illingworth, G. 1986. *Ap. J. Lett.* 301: L45–48
Coulson, I. M., Caldwell, J. A. R., Gieren, W. P. 1986. *Ap. J.* 303: 273–79
Crampton, D. 1979. *Ap. J.* 230: 717–23
Crampton, D., Greasley, J. 1982. *Publ. Astron. Soc. Pac.* 94: 31–35
Da Costa, G. S., Mould, J. R. 1986. *Ap. J.* 305: 214–27
Dean, J. F. 1981. *MNRAS* 197: 779–90
Dean, J. F., Warren, P. R., Cousins, A. W. J. 1978. *MNRAS* 183: 569–83 (DWC)
Demers, S., Kunkel, W. E., Irwin, M. J. 1985. *Astron. J.* 90: 1967–81
de Vaucouleurs, G. 1960. *Ap. J.* 131: 265–81
de Vaucouleurs, G. 1978. *Ap. J.* 223: 351–63
Eggen, O. J. 1977. *Ap. J. Suppl.* 34: 1–31
Eggen, O. J. 1983. *Astron. J.* 88: 379–85
Eggen, O. J. 1985a. *Astron. J.* 90: 1297–1319
Eggen, O. J. 1985b. *Astron. J.* 90: 1278–96
Eggen, O. J. 1985c. *Astron. J.* 90: 1260–77
Evans, N. R., Ferro, A. A. 1986. In *New Insights in Astrophysics (Eight Years of Ultraviolet Astronomy with IUE)*. ESA SP-263, pp. 403–4. Paris: ESA

Feast, M. W. 1984a. In *Structure and Evolution of the Magellanic Clouds, IAU Symp. No. 108*, ed. S. van den Bergh, K. S. de Boer, pp. 157–70. Dordrecht: Reidel

Feast, M. W. 1984b. *MNRAS* 211: 51P–55P

Feast, M. W. 1985. See Madore 1985b, pp. 157–65

Feast, M. W. 1986a. In *Galaxy Distances and Deviations from Universal Expansion*, ed. B. F. Madore, R. B. Tully, pp. 7–14. Dordrecht: Reidel

Feast, M. W. 1986b. In *Highlights of Astronomy*, ed. J. P. Swings, 7: 59–62. Dordrecht: Reidel

Feast, M. W. 1987. In *The Galaxy*, ed. G. Gilmore, R. Carswell. Dordrecht: Reidel. In press

Feltz, K. A., McNamara, D. H. 1980. *Publ. Astron. Soc. Pac.* 92: 609–47

Fernie, J. D. 1969. *Publ. Astron. Soc. Pac.* 81: 707–31

Fernie, J. D. 1984. *Ap. J.* 282: 641–49

Fernie, J. D., McGonegal, R. 1983. *Ap. J.* 275: 732–36

Fernley, J. A., Jameson, R. F., Sherrington, M. R. 1985. See Madore 1985b, pp. 58–62

Ferro, A. A., Madore, B. F. 1985. *Observatory* 105: 207–9

Freedman, W. L. 1985. See Madore 1985b, pp. 225–27

Freedman, W. L. 1986. Paper presented at NATO Conf. on Galaxy Distances and Deviations from Universal Expansion, Hawaii

Freedman, W. L., Grieve, G. R., Madore, B. F. 1985. *Ap. J. Suppl.* 59: 311–21

Gascoigne, S. C. B., Shobbrook, R. R. 1978. *Proc. Astron. Soc. Aust.* 3: 285–86

Gieren, W. P. 1986. *Ap. J.* 306: 25–29

Giridhar, S. 1986. *J. Astrophys. Astron.* 7: 83–98

Graham, J. A. 1984. *Astron. J.* 89: 1332–42

Grieve, G. R., Madore, B. F., Welch, D. L. 1985. *Ap. J.* 294: 513–16

Harris, H. C. 1981a. *Astron. J.* 86: 707–18

Harris, H. C. 1981b. *Astron. J.* 86: 1192–99

Harris, H. C. 1983. *Astron. J.* 88: 507–17

Harris, H. C., Pilachowski, C. A. 1984. *Ap. J.* 282: 655–66

Hawley, S. L., Jeffreys, W. H., Barnes, T. G., War, L. 1986. *Ap. J.* 302: 626–31

Iben, I., Renzini, A. 1984. *Phys. Rep.* 105: 329–406

Iben, I., Tuggle, R. S. 1975. *Ap. J.* 197: 39–54

Kayser, S. E. 1967. *Astron. J.* 72: 134–48

Kraft, R. P. 1961. *Ap. J.* 134: 616–32

Kurucz, R. L. 1979. *Ap. J. Suppl.* 40: 1–340

Laney, C. D., Stobie, R. S. 1986a. *MNRAS* 222: 449–72 (LS)

Laney, C. D., Stobie, R. S. 1986b. *S. Afr. Astron. Obs. Circ. No. 10*, pp. 51–81

Lequeux, J., Peimbert, M., Rayo, J. F., Serrano, A., Torres-Peimbert, S. 1979. *Astron. Astrophys.* 80: 155–66

Madore, B. F. 1985a. *Ap. J.* 298: 340–44

Madore, B. F., ed. 1985b. *Cepheids: Theory and Observations, Proc. IAU Colloq. No. 82.* Cambridge: Cambridge Univ. Press. 300 pp.

Madore, B. F., Freedman, W. L. 1985. *Astron. J.* 90: 1104–12

Madore, B. F., McAlary, C. W., McLaren, R. A., Welch, D. L., Neugebauer, G., Matthews, K. 1985. *Ap. J.* 294: 560–66

Martin, W. L., Warren, P. R., Feast, M. W. 1979. *MNRAS* 188: 139–57 (MWF)

Mateo, M., Hodge, P. 1986. *Ap. J. Suppl.* 60: 893–937

Mateo, M., Hodge, P., Schommer, R. S. 1986. Preprint

Mathewson, D. S., Ford, V. L., Visvanathan, N. 1986. *Ap. J.* 301: 664–74

McAlary, C. W., Madore, B. F. 1984. *Ap. J.* 282: 101–5

McAlary, C. W., Madore, B. F., Davis, L. E. 1984. *Ap. J.* 276: 487–90

McAlary, C. W., Madore, B. F., McGonegal, R., McLaren, R. A., Welch, D. L. 1983. *Ap. J.* 273: 539–43

McGonegal, R., McLaren, R. A., McAlary, C. W., Madore, B. F. 1982. *Ap. J. Lett.* 257: L33–36

Moffett, T. J., Barnes, T. G. 1986. *Ap. J.* 304: 607–16

Olson, B. I. 1975. *Publ. Astron. Soc. Pac.* 87: 349–51

Pagel, B. E. J. 1986. ESO Sci. Preprint No. 463

Parsons, S. B. 1971. *MNRAS* 152: 121–31

Pedreros, M., Madore, B. F., Freedman, W. L. 1984. *Ap. J.* 286: 563–72

Pel, J. W. 1978. *Astron. Astrophys.* 62: 75–94

Pel, J. W. 1984. See discussion following Feast 1984a

Pel, J. W. 1985. See Madore 1985b, pp. 1–16

Pel, J. W., Genderen, A. M., van Lub, J. 1981. *Astron. Astrophys.* 99: L1–L4

Pritchet, C. J. 1986. In *Galaxy Distances and Deviations from Universal Expansion*, ed. B. F. Madore, R. B. Tully, pp. 35–38. Dordrecht: Reidel

Ridgway, S. T., Jacoby, G. H., Joyce, R. R., Siegel, M. J., Wells, D. C. 1982. *Astron. J.* 87: 680–84

Sandage, A. 1971. *Ap. J.* 166: 13–35

Sandage, A. 1972. *Q. J. R. Astron. Soc.* 13: 202–21

Sandage, A. 1983. *Astron. J.* 88: 1108–25

Sandage, A. 1984a. *Astron. J.* 89: 630–35

Sandage, A. 1984b. *Astron. J.* 89: 621–29

Sandage, A., Carlson, G. 1983. *Ap. J. Lett.* 267: L25–28

Sandage, A., Carlson, G. 1985a. *Astron. J.* 90: 1464–73

Sandage, A., Carlson, G. 1985b. *Astron. J.* 90: 1019–26
Sandage, A., Tammann, G. A. 1969. *Ap. J.* 157: 683–708
Sandage, A., Tammann, G. A. 1984. *Nature* 307: 326–29
Scarfe, C. D. 1976. *Ap. J.* 209: 141–45
Schaltenbrand, R., Tammann, G. A. 1971. *Astron. Astrophys. Suppl.* 4: 265–314
Schmidt, E. G. 1984. *Ap. J.* 285: 501–14
Schmidt-Kaler, Th. 1982. In *Landolt-Börnstein Numerical Data and Functional Relationships in Science and Technology*, New Series, Group 6, Vol. 2b, ed. K. Schaifers, H. H. Voigt, p. 12. Berlin: Springer-Verlag
Schommer, R. A., Olszewski, E. W., Aaronson, M. 1984. *Ap. J. Lett.* 285: L53–57
Seidel, E., Da Costa, G. S., Demarque, P. 1986. Preprint
Sollazzo, C., Russo, G., Onnembo, A., Caccin, B. 1981. *Astron. Astrophys.* 99: 66–72
Stothers, R. B. 1983. *Ap. J.* 274: 20–30
Tammann, G. A. 1970. In *The Spiral Structure of Our Galaxy, IAU Symp. No. 38*, ed. W. Becker, G. Contopoulos, pp. 236–45. Dordrecht: Reidel
Tammann, G. A. 1979. In *European Satellite Astrometry*, ed. C. Barbieri, P. L. Bernacca, pp. 271–80. Padova: Antoniana
Taylor, B. J. 1986. *Ap. J. Suppl.* 60: 577–99
Thompson, R. J. 1975. *MNRAS* 172: 455–63
Turner, D. G. 1976a. *Ap. J.* 210: 65–75
Turner, D. G. 1976b. *Astron. J.* 81: 1125–33
Turner, D. G. 1979. *Publ. Astron. Soc. Pac.* 91: 642–47
Turner, D. G. 1985. See Madore 1985b, pp. 209–11
Turner, D. G. 1986. *Astron. J.* 92: 111–18
van den Bergh, S. 1977. In *Décalages Vers le Rouge et Expansion de l'Univers, Colloq. UAI No. 37*, ed. C. Balkowski, B. E. Westerlund, pp. 13–41. Paris: CNRS
van den Bergh, S., Brosterhus, E. B. F., Alcaino, G. 1982. *Ap. J. Suppl.* 50: 529–49
van den Bergh, S., Younger, P. F., Brosterhus, E. B. F., Alcaino, G. 1983. *Ap. J. Suppl.* 53: 765–90
van den Bergh, S., Younger, P. F., Turner, D. G. 1985. *Ap. J. Suppl.* 57: 743–50
van Genderen, A. M. 1983. *Astron. Astrophys.* 124: 223–35
van Leeuwen, F. 1983. PhD thesis. Leiden Univ., Neth.
Visvanathan, N. 1985. *Ap. J.* 288: 182–86
Walker, A. R. 1984. *Mon. Notes Astron. Soc. S. Afr.* 43: 89–96
Walker, A. R. 1985a. *MNRAS* 217: 13P–15P
Walker, A. R. 1985b. *MNRAS* 212: 343–52
Walker, A. R. 1985c. *MNRAS* 213: 889–97
Walker, A. R. 1985d. *MNRAS* 214: 45–53
Walker, A. R. 1986. *MNRAS*. In press
Walker, A. R. 1987a. *MNRAS*. In press
Walker, A. R. 1987b. *S. Afr. Astron. Obs. Circ.* In press
Walker, A. R. 1987c. *S. Afr. Astron. Obs. Circ.* In press
Walker, A. R., Laney, C. D. 1986. *MNRAS* 224: 61–74
Welch, D. L., Evans, N. R., Drukier, G. 1985b. See Madore 1985b, pp. 51–52
Welch, D. L., McAlary, C. W., Madore, B. F., McLaren, R. A., Neugebauer, G. 1985a. *Ap. J.* 292: 217–21
Welch, D. L., McAlary, C. W., McLaren, R. A., Madore, B. F. 1986. *Ap. J.* 305: 583–90
Wesselink, A. J. 1946. *Bull. Astron. Inst. Neth.* 10: 91–99
Wielen, R. 1974. *Astron. Astrophys. Suppl.* 15: 1–33

PHYSICAL CONDITIONS, DYNAMICS, AND MASS DISTRIBUTION IN THE CENTER OF THE GALAXY

R. Genzel

Max-Planck-Institut für Physik und Astrophysik,
Institut für Extraterrestrische Physik, D-8046 Garching,
West Germany

C. H. Townes

Department of Physics, University of California, Berkeley,
California 94720

1. INTRODUCTION

This review emphasizes recent investigations of the central 10 pc of the Galaxy and conclusions on energetics, dynamics, and mass distribution derived from them. New information on the Galactic center comes from X- and gamma-ray measurements and from infrared and microwave studies, especially from spectroscopy, high-resolution imaging, and interferometry. Section 2 is an overview of the phenomena seen in the Galactic center. In Section 3 we cover the questions of energetics and luminosity, followed by a discussion of the distribution, excitation, and dynamics of the neutral and ionized interstellar matter in Sgr A (Section 4). Our current knowledge about the mass distribution and the evidence for and against a massive black hole is analyzed in Section 5, while in Section 6 we discuss the nature of the central 0.1 pc of the Galaxy. We conclude in Section 7 with a summary. There are other recent reviews on the Galactic center region, and those by Oort (1977), Lacy et al. (1982), Townes et al. (1983), Brown & Liszt (1984), Lo (1986), Genzel (1987), and Serabyn (1987) are especially relevant for the present discussion. We also refer the reader to the proceedings of workshops on the Galactic center held at Caltech in 1982 (Riegler & Blandford 1982) and in Berkeley in 1986 (Backer 1987).

For convenience and easy comparison with other discussions, we refer all quantities to a Sun–Galactic center distance of $R_0 = 10$ kpc and give scale factors in units of $R_0(10) = R_0/10$ kpc. The most likely value for R_0 is between 7 and 9 kpc. The distance recommended by the International Astronomical Union is 8.5 kpc, but more recent investigations of Miras, Cepheids, RR Lyrae variables, and globular clusters point to a value near or below 8 kpc (cf. Feast 1987, Frenk & White 1982, Glass & Feast 1982, Blanco & Blanco 1985). From the velocity field of H II regions in the outer Galaxy, Blitz (1987) finds a distance of 7.6 kpc. A direct determination of the distance to the molecular cloud Sgr B2 has recently been made for the first time with Very Long Baseline Interferometry from the statistical parallax of proper motions of H_2O maser features (Reid et al. 1987). The value derived from this independent technique is 7.1 ± 1.2 kpc, where the error bars are mainly determined by the statistical uncertainty (1σ) due to the finite number of maser features whose motions could be determined. Sgr B2 ($l^{II} = 0.5°$) is almost certainly within a few hundred parsecs of the Galactic center.

2. OVERVIEW OF THE PHENOMENA FOUND IN THE GALACTIC CENTER

2.1 *The Stellar Cluster*

Because of strong foreground extinction by interstellar dust ($A_v \approx 30$; Becklin et al. 1978a), the Galactic center stellar cluster can only be observed at $\lambda \gtrsim 1$ μm. Observations at 2 μm by Becklin & Neugebauer (1968, 1975), Allen et al. (1983), and Matsumoto et al. (1982) have shown the presence of an extensive stellar cluster with a brightness distribution scaling approximately as $R^{-0.75}$ from $\sim 1°$ (180 pc) to about 0.5" (0.025 pc; see also Bailey 1980). Beyond $R \sim 6°$, the surface brightness distribution steepens and scales approximately like R^{-2} (Matsumoto et al. 1982). The stellar light is peaked within a few arcseconds of a complex of infrared sources called IRS 16, which is, therefore, commonly interpreted to be the center of the Galaxy. Most of these stars probably represent M and K giants of surface temperature about 4000 K. Allen et al. (1983) have interpreted the central cusp of the brightness distribution as an indication of a small core radius of the cluster [$R_{core} \lesssim 0.03$ pc $R_0(10)$]. However, the central parsec of the Galaxy happens to lie in a minimum of line-of-sight extinction (Rieke & Lebofsky 1987). Furthermore, a number of the 2-μm sources, including the bright supergiant IRS 7 and the "blue" IRS 16 complex itself, are not part of the general red giant population, which thus artificially decreases the observed core radius. With allowance for these effects, current estimates for the cluster's core radius range between 0.05 pc $R_0(10)$ (Allen 1987) and 1 pc $R_0(10)$ (Rieke & Lebofsky 1987).

On scales less than about 3 pc, the 2-μm brightness distribution on the sky appears "near-spherical" (Allen et al. 1983). On larger scales the flux distribution is aligned with the Galactic plane and has a minor to major axis ratio of about 0.4 on scales of tens of parsecs (Becklin & Neugebauer 1968, Sanders & Lowinger 1972), and of about 0.75 on scales of several hundred parsecs (Matsumoto et al. 1982). The latter value is similar to the flattening of the RR Lyrae distribution (Oort & Plaut 1975) or of the OH-IR stars (Habing 1987). This flattening may indicate that the stellar distribution is not spherical, but spheroidal, as has been assumed by Sanders & Lowinger (1972) in their modeling of the mass distribution. The 2-μm brightness distribution, however, is strongly influenced by patchy foreground absorption that can in part be identified with known molecular cloud complexes (Lebofsky 1979, Güsten et al. 1981, Hiromoto et al. 1984). Any detailed conclusions about the intrinsic flattening of the cluster from the surface brightness distribution alone are, therefore, uncertain. The near-infrared stellar luminosity of the central 2-pc radius region, obtained by extrapolating the measured near-infrared flux with $T \sim 4000$ K and allowing for reddening due to intervening dust, is $L_{\rm IR} \approx 6.5 \times 10^6 \, L_\odot$ $R_0^2(10)$ (Becklin & Neugebauer 1968).

2.2 Radio Continuum Emission and Interstellar Clouds

The thermal and nonthermal radio continuum emission in the vicinity of the Galactic center has been investigated by a number of groups and has been discussed in detail in the last article in this series on the Galactic center (Brown & Liszt 1984). Apart from the Sgr A complex (that is, the Galactic center itself) the most striking structure is the "arc," consisting of a "spur" perpendicular to the Galactic plane at $l^{\rm II} \approx 0.18°$ and a "bridge" connecting the spur with the halo of Sgr A. High-resolution VLA mapping by Yusef-Zadeh et al. (1984) shows that the spur and bridge largely consist of filamentary or sheetlike structures. The morphology of nonthermal emission in the straight threads of the spur is especially unusual and may be due to large-scale magnetic field structures aligned perpendicular to the Galactic plane (Yusef-Zadeh et al. 1984, Yusef-Zadeh 1986, Seiradakis et al. 1985, Tsuboi et al. 1986). A corresponding spur at negative Galactic longitudes has been found by Sofue & Handa (1984) and Liszt (1985). Sofue & Handa (1984) propose that both spurs belong to the same structure and are the result of a recent explosion ($E_{\rm tot} \geq 10^{51}$ erg). Uchida et al. (1985) suggest that the lobes are due to magnetically accelerated gas streaming away from the Galactic plane. The arched filaments in the bridge, tilted about 25° relative to the Galactic plane, are emitting thermally in the radio continuum and also contain dust emitting in the far-infrared (Dent et al. 1982).

Molecular gas is associated with the bridge and Sgr A complex, and possibly with parts of the spur (Fukui et al. 1977, Güsten et al. 1981, Sandqvist 1982, Liszt et al. 1985, Bally et al. 1986a, Güsten et al. 1987b). The giant molecular clouds within a few hundred parsecs of Sgr A are warmer ($T_{gas} \sim 50-100$ K), denser ($n_{H_2} \sim 10^4$ cm^{-3}; CS and NH$_3$ emission is seen throughout the clouds), and of larger local line width ($\Delta v_{FWHM} \sim 20-40$ km s^{-1}) than their counterparts in the Galactic disk at a few kiloparsecs from the center (e.g. Güsten & Downes 1980, Güsten et al. 1981, Armstrong & Barrett 1985, Bally et al. 1986a,b). The relative distribution of ionized (H II), photodissociated [C$^+$ (Lugten et al. 1986)], and molecular gas (Liszt et al. 1985, Güsten et al. 1987b) suggests that these clouds are ionized by external sources. The high gas temperature (Güsten et al. 1985) and relatively low dust temperature ($T_d \sim 20-30$ K; Gatley et al. 1978, Hildebrand et al. 1978, Mezger et al. 1986) indicate that gas heating is not due to dust-gas collisions but rather to a direct mechanism, such as cosmic rays or (more likely) shock dissipation of the turbulent motions in the clouds (Güsten et al. 1985). With the exception of a few compact H II regions near the centers of the $+20$ km s^{-1} and $+40$ km s^{-1} clouds (Ekers et al. 1983, Ho et al. 1985), there is no evidence for current, massive star formation in the Galactic center clouds. The $+20$ and $+40$ km s^{-1} clouds and the clouds associated with the bridge (Güsten et al. 1987b) are likely located within 20 to 100 pc of the Galactic center, where the cosmic-ray and UV energy density is high, and where tidal shearing might efficiently convert rotational energy around the Galactic center into local turbulent motions.

From CS $J = 2 \to 1$ mapping with the IRAM 30-m millimeter telescope, Güsten et al. (1987b) conclude that the negative velocity gas associated with the bridge [$V_{LSR} = -20$ to -80 km s^{-1} (Yusef-Zadeh 1986)] connects with a massive cloud within 5' [15 pc $R_0(10)$] of Sgr A, which on morphological grounds is likely at a galactocentric distance comparable to the projected distance. Hence, the bridge—and possibly the spur—are probably located in the Galactic center region. Since these gas clouds have negative velocities that are opposite to the direction of Galactic rotation, they probably have significant radial velocities. From this fact and the morphology of the gas distribution, Güsten et al. (1987b) propose that the clouds are being tidally disrupted and stream inward toward the center at a rate of a few tenths of a solar mass per year. Because of that large mass flow rate and corresponding energy ($\sim 10^{52}$ erg), they consider outflow less likely. A scenario of highly eccentric orbits could also be attractive for the +20 and +40 km s^{-1} clouds (Bally et al. 1986b).

2.3 The Sgr A Complex

The Sgr A complex, making up the central 10 pc of the Galaxy, is a strong emitter at infrared and radio wavelengths. From radio continuum observations, the region has been traditionally divided into the nonthermal, shell-like source Sgr A East and the thermal source Sgr A West (Ekers et al. 1975, 1983, Sandqvist 1974), separated by about 1.5 arcmin. The H II region Sgr A West, coincident with the peak of the 2-μm stellar emission, is also prominent at mid- and far-infrared wavelengths [5 to 100 μm (Becklin & Neugebauer 1975, 1982, Harvey et al. 1976, Rieke et al. 1978)]. More recently, Yusef-Zadeh et al. (1984), Yusef-Zadeh (1986), and Yusef-Zadeh et al. (1986) have drawn attention to a third component of partly thermal, partly nonthermal radio continuum emission that forms a "halo" of scale size 10' (30 pc) around Sgr A West and East. As shown in Figure 1, taken from Yusef-Zadeh (1986), the halo resolves into a plethora of streamers, protrusions, and filamentary loops that seem to emerge from the nuclear region mostly perpendicular to the Galactic plane. Yusef-Zadeh et al. (1986) also find a narrow ridge of nonthermal 160-MHz emission emerging from the center and stretching about 30 pc approximately perpendicular to the Galactic plane (see also Kassim et al. 1987). Yusef-Zadeh (1986) and Morris & Yusef-Zadeh (1987) interpret the streamers and protrusions in the Sgr A halo as indicators of outflow of ionized gas along poloidal magnetic field lines (see also the model by Heyvaerts et al. 1987).

The physical relationship between Sgr A East, traditionally interpreted as a young supernova remnant (age $\gtrsim 10^2$–few $\times 10^3$ yr), and Sgr A West has been a matter of intense discussion (for an overview, see Brown & Liszt 1984). In part because of the close correlation between the western edge of Sgr A East and the Sgr A West H II region, a consensus has emerged that Sgr A East is located near the Galactic center.

An overview of the central 8 pc of the Galaxy is given in Figure 2. The spatial distribution of the thermal radio emission in Sgr A West has the morphology of a minispiral (Figure 2; Ekers et al. 1983, Lo & Claussen 1983). Velocity measurements of the ionized gas show, however, that the motions of gas in the minispiral correspond to a combination of an approximately circular ring, streamers that are indeed somewhat spirallike in morphology, and individual clouds (Lo & Claussen 1983, Serabyn & Lacy 1985, van Gorkom et al. 1987). Some of the streamers appear to connect with features in the ring. Near the center of the minispiral lies a compact, nonthermal radio source [Sgr A* (Lo et al. 1985; Figures 2, 3)].

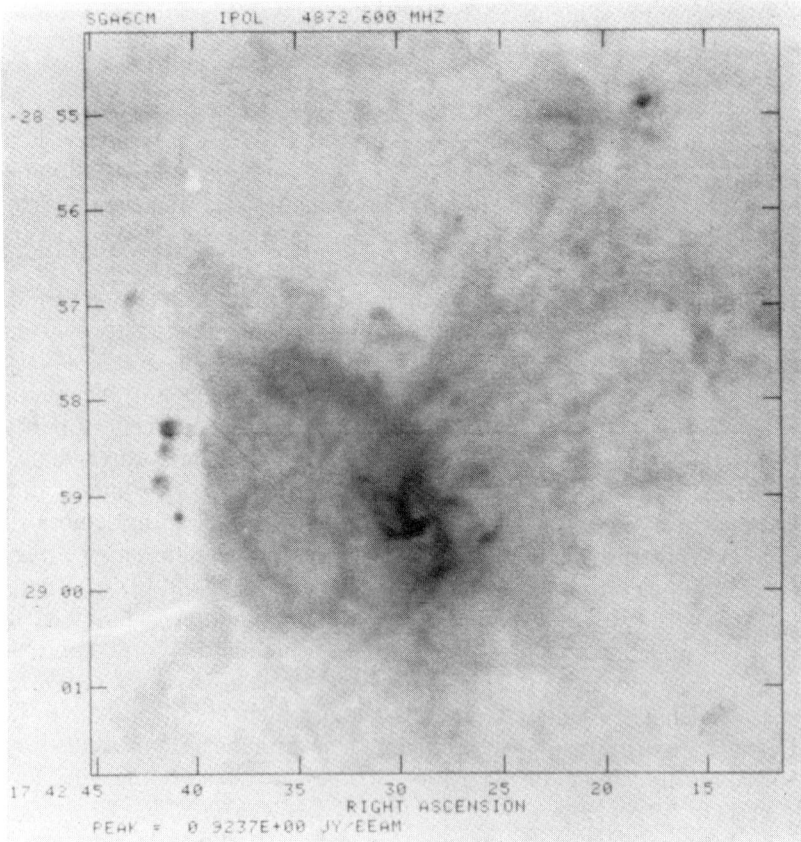

Figure 1 VLA radio continuum map (beam size 3.4" × 2.9") at 6 cm of Sgr A, showing the Sgr A East nonthermal shell, the Sgr A West thermal "minispiral," and the protrusions in the Sgr A halo (from Yusef-Zadeh 1986).

At high spatial resolution (Figure 3; Lo 1986), the structure of the minispiral breaks up further into thin, somewhat long filaments or arcs (diameter ≤ 0.1 pc) with embedded knots. The filaments and arms appear to converge in the "bar" region a few arcseconds south of IRS 16/Sgr A*. Yusef-Zadeh (1986) interprets the southern part of the western arc and the northern arm as physically connecting to the protrusions in the halo (Figure 1).

The mid- and far-infrared continuum emission from warm dust grains is also centered on Sgr A West. The 5–30 μm emission follows closely the radio continuum distribution in the minispiral (Becklin & Neugebauer

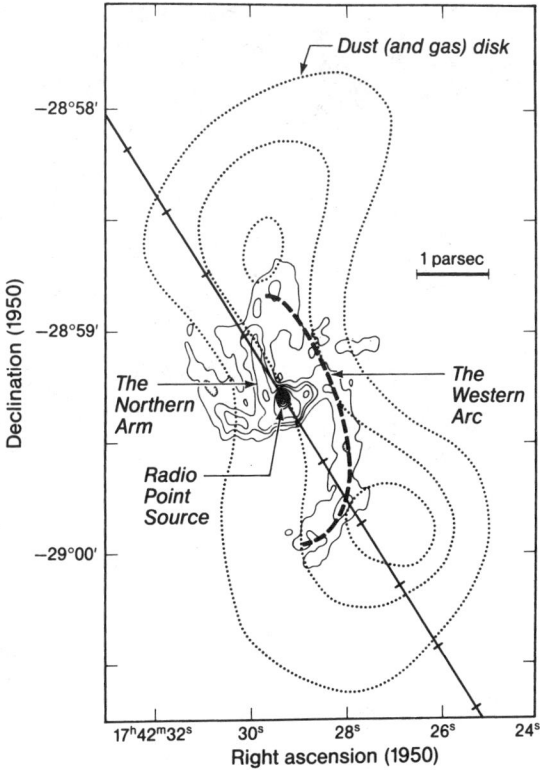

Figure 2 "Road map" of the central 8-pc diameter region (from Crawford et al. 1985), showing the radio continuum "minispiral" at 2 cm (from Ekers et al. 1983), the compact radio source Sgr A*, and the 60–100 μm dust lobes (Becklin et al. 1982). The Galactic plane is indicated by a diagonal line, and some of the ionized streamers referred to in the text are marked.

1975, Rieke et al. 1978, Gatley 1982, Gezari et al. 1985). The $\lambda \gtrsim 50$ μm emission, however, forms a double-lobed structure (Figure 2) with two emission peaks at $\sim 1.7\,R_0(10)$ pc located on either side of IRS 16/Sgr A* and aligned approximately along the Galactic plane (Becklin et al. 1982). Becklin et al. (1982) conclude from these observations that centered on the nucleus is a dust ring of inner radius ~ 1.7 pc whose plane is close to the line of sight. Just inside of this radius lies some of the ionized gas. At still smaller radii, there is very little dust and gas except the ionized streamers. Hence the region inside of 1.7 pc can be regarded as a central cavity. Beyond the dust lobes there is partially ionized (C^+), atomic (H I,

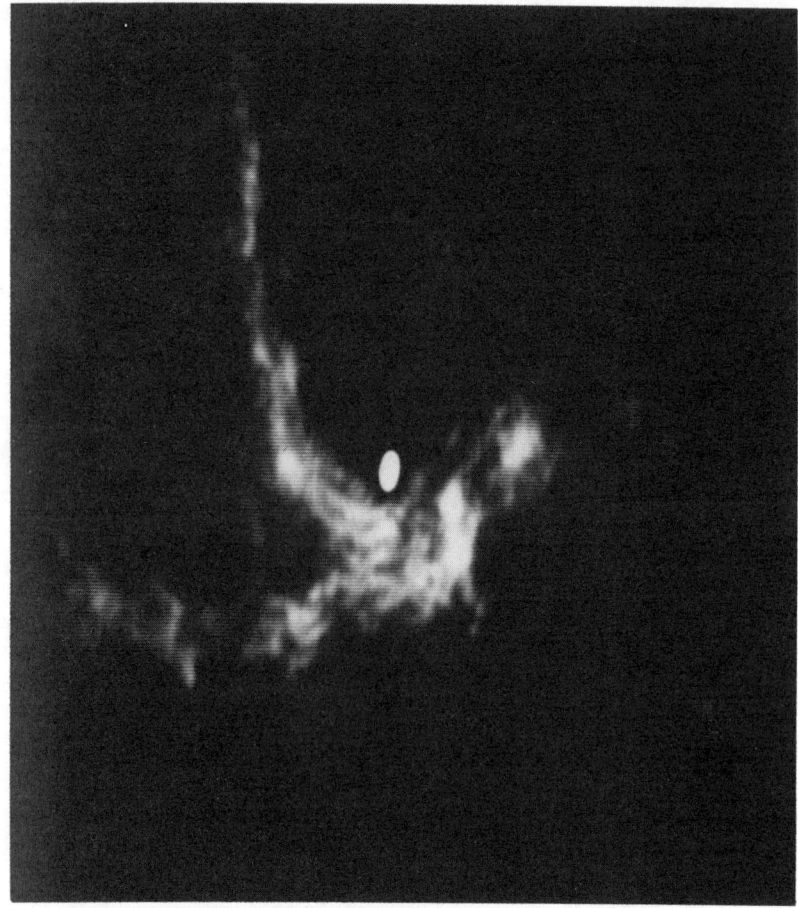

Figure 3 VLA radio continuum map at 6 cm of the central 40″ of Sgr A West at 0.3″ × 0.6″ (FWHM) resolution (Lo 1986). The radio point source Sgr A* is the bright oval in the center of the image.

O I), and molecular (H_2, CO, OH, HCO^+, CS, HCN, NH_3, etc.) gas. The neutral and ionized gas is discussed in more detail below.

2.4 X- and Gamma-Ray Emission

Figure 4 is a superposition of the 0.5–4 keV Imaging Proportional Counter (IPC) picture (1′ resolution) from the Einstein satellite (Watson et al. 1981) on the 20-cm radio continuum map. Within the positional uncertainties of the X-ray map, Sgr A West is coincident with a "compact" ($\Theta \leq 1.5′$) hard ($T \approx 10^8$ K) X-ray source. Because of large connections for foreground

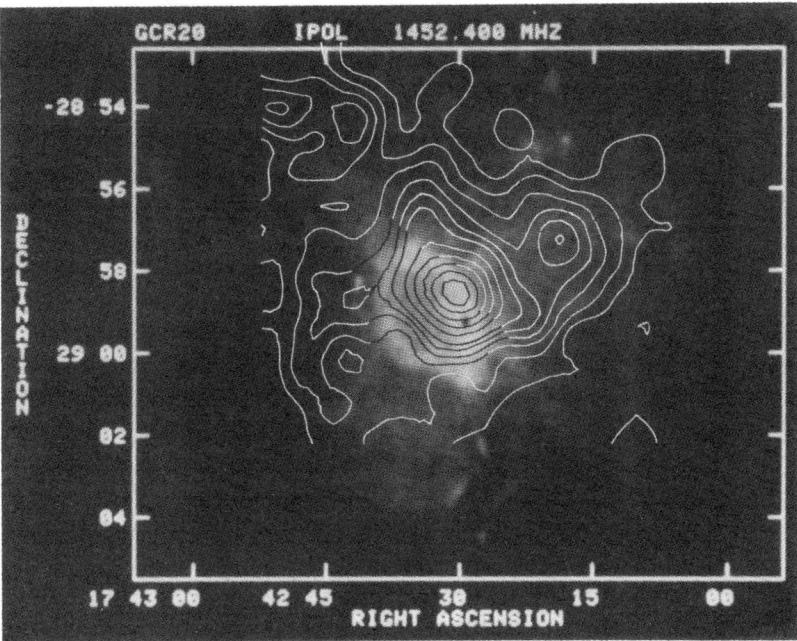

Figure 4 Contours of the 2–4 keV Einstein (HEAO-2) IPC X-ray map of Sgr A [$\approx 1'$ resolution (Watson et al. 1981)], superposed on the 20-cm VLA radio continuum map of the Sgr A complex (from Yusef-Zadeh 1986).

absorption ($N_H \sim 2$–20×10^{22} cm^{-2}), the estimated 2–6 keV luminosity [1.5×10^{35} erg s^{-1} $R_0^2(10)$] and derived effective temperature are quite uncertain (Watson et al. 1981). During the period of the Einstein mission, the Galactic center X-ray source did not exhibit significant variability. More recent data from a SPARTAN mission (Kawai et al. 1987) in the 1–15 keV range, however, do not show any evidence for a source at the Galactic center stronger than about one fourth of the strength reported by Watson et al. (1981). Yusef-Zadeh (1986) suggests that weaker X-ray emission may be associated with the filamentary protrusions, northwest of Sgr A West (Figure 4). The origin of the soft X-ray emission is uncertain. According to Watson et al. (1981), stellar sources (late-type dwarfs and early-type stars) could account for about half of the flux. If all of the emission reported by Watson et al. (1981) were thermal bremsstrahlung from interstellar gas of temperature $\approx 10^8$ K, the derived rms electron density of the hot medium would be $\langle n_e^2 \rangle^{1/2} \approx 5[f/0.1]^{-1/2}$ cm^{-3}, where f is the volume filling factor. However, the results of Kawai et al. (1987) indicate that the X-ray source at the Galactic center is variable and, hence, that much of its flux must be very localized.

Observations at 10 keV to 10 MeV from HEAO-1, HEAO-3, and balloons with spatial resolutions between 2° and 40° show the presence of a luminous [L (10 keV to 10 MeV) $\sim 2 \times 10^{38}$ erg s^{-1}] source of relatively small size [$\leq 0.5°$ (10–100 keV), $\leq 5°$ (100 keV–10 MeV)] and hard spectrum ($T_{\text{eff}} = 10^9$–10^{10} K) in or near the Galactic center (e.g. Matteson 1982, and references therein). The hard X-ray source is highly variable, which suggests that much of the flux comes from a source or a few sources with diameters $\lesssim 0.1$ pc. Because of the poor spatial resolution, there are currently few constraints on the emission mechanism or the character of the source(s) (but see discussion in Rees 1987).

One of the most remarkable discoveries has been the detection of a strong and time-variable source of 511-keV e^+–e^- annihilation-line radiation within a few degrees of the Galactic center. A series of 12 balloon and satellite experiments dating back to 1970 and involving six different groups has established the following facts (for reviews, see Jacobson 1982, Leventhal & MacCallum 1982, Lingenfelter & Ramaty 1982, Ramaty & Lingenfelter 1987, Leventhal 1987). The line radiation (peak luminosity $\sim 10^{37}$ erg s^{-1}) was strong in the 1970s but decreased or "turned off" rather abruptly ($\tau \lesssim 10^7$ s) at the beginning of this decade (Riegler et al. 1981). The source has remained in an "off" or "low" state through the last observation of 1984. The narrow width of the annihilation line [$\Delta E \sim 2$ keV (Riegler et al. 1981)] implies gas velocities of $\lesssim 700$ km s^{-1}. The probable detection of a 3-photon continuum at $E < 511$ keV implies that about 80% of all incident positrons decay via positronium formation (Brown & Leventhal 1987). The uncertainty on the positronium fraction— which is a strong constraint on the medium—unfortunately is large, and even 0% cannot be excluded at the 3σ level. A large fraction of positronium formation, together with the narrow line width, would imply that the annihilations take place in a reasonably cold and dense molecular or atomic medium [$n_{\text{H}+2\text{H}_2} \gtrsim 10^5$ cm^{-3} (Brown & Leventhal 1987)]. Alternatively, the annihilations may occur in a warm ($T \lesssim$ few $\times 10^4$ K), partially ionized medium [$n_e > 10^3$ cm^{-3} (Bussard et al. 1979)], or on dust grains in a hot H II region (Zurek 1985). The variability implies a source size of ≤ 0.3 pc.

A gamma-ray line at 1.8 MeV caused by the decay of ^{26}Al into ^{26}Mg has also been detected toward the Galactic center (Mahoney et al. 1982, 1984, Share et al. 1985a,b). The interpretation of this emission has been that ^{26}Al is produced in explosive nucleosynthesis in novae and supernovae or in mass loss from red giants, O, or Wolf-Rayet stars. With the relatively low spatial resolution measurements on the HEAO-C and SMM satellites, the distribution of the 1.8-MeV line emission is in good agreement with the spatial distribution of these objects in the Galactic plane. More recently, von Ballmoos et al. (1987) reobserved the 1.8-MeV line emission with better spatial resolution with a balloon-borne Compton telescope and found that their measurements are best fit by an unresolved ($\Theta < 10°$) source located at the center. The extended distributions of novae or supernovae give poorer fits, but they cannot be excluded by the data with certainty. In contrast, MacCallum et al. (1987) find a much lower central 1.8-MeV flux than von Ballmoos et al. (1987), despite a similar beam size. MacCallum et al. (1987) conclude that their measurements are consistent with a diffuse nature of this gamma-ray emission. The β-decays of the ^{26}Al nuclei emitting the 1.8-MeV emissions also lead to emission of positrons. Webber et al. (1986) have pointed out that the positron production rate estimated from the flux of the 1.8-MeV line is consistent with the time-averaged strength of the 511-keV line. Hence, *if* both gamma lines originate in the central few parsecs, they might be physically connected and could suggest that within the last 10^5–10^6 yr a large number of supernovae or novae ($\gtrsim 10^3$), or alternatively a supermassive object [$\gtrsim 5 \times 10^5$ $M_\odot$ (Hillebrandt et al. 1986, 1987)], have exploded in the Galactic center. However, since the actual degree of concentration of the ^{26}Al-line emission toward the center is still uncertain, it is not clear that an "exotic" explanation of this type is required.

3. ENERGETICS AND PHYSICAL CONDITIONS IN THE CENTRAL 4 pc: STAR BURST OR CENTRAL SOURCE?

Figure 5 shows a composite infrared and radio spectrum of the central 4-pc diameter region of the Galaxy, assembled from a number of different observations. Several important parameters can be derived. First, the observed far-infrared continuum luminosity from thermally emitting dust grains, estimated from the total flux in Figure 5, is $\sim 3 \times 10^6$ $L_\odot$ $R_0^2(10)$. The *total* far-infrared luminosity of Sgr A may be 3 to 10 times larger, as a result of photons escaping to larger radii (Becklin et al. 1982). Second, the number of Lyman continuum photons emitted from this region, as estimated from the total radio continuum flux density (Lacy et al. 1982, Ekers et al. 1983, Mezger & Wink 1986), is about 2 to 4×10^{50} s^{-1} $R_0^2(10)$, again taking into account that a significant fraction ($\geq 50\%$) of Lyman continuum photons may escape to $R \gtrsim 2$ pc and be absorbed in lower density gas. Third, the intensities of infrared fine-structure lines in high-ionization stages (S IV, O III, Ar III) are weak (Lacy et al. 1980, Watson et al. 1980). This indicates that the ultraviolet (UV) radiation field has a low effective temperature [$T_{\text{eff}} \sim 35,000$ K (Lacy et al. 1980, Serabyn

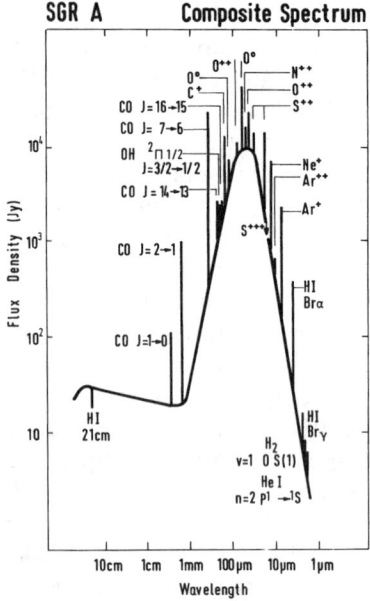

Figure 5 Composite line and continuum spectrum of the central 4-pc diameter region of the Galactic center. Note that line flux *densities*, and not the fluxes, are given. As with the continuum radiation, this means that to get line fluxes and luminosities in various lines, the flux densities have to be multiplied by $\Delta v = v(\Delta v/c)$, making the shorter wavelength lines relatively more important for the energetics. The line and continuum data are taken from the literature and referenced in the text (from Genzel et al. 1985).

& Lacy 1985)]. A low effective temperature is also consistent with the ratio of $L\alpha$ luminosity (as estimated from the mid-infrared dust emission in the ionized region) to total far-infrared luminosity (Becklin et al. 1982). The number of Lyman continuum photons and the effective temperatures of the UV field and of the far-infrared luminosity are consistent with each other and indicate that the intrinsic UV luminosity of the Galactic center is about $10^7 L_\odot R_0^2(10)$. Most of this UV radiation is absorbed by dust grains in the central 10 pc and is converted to far-infrared radiation.

Is the ultraviolet radiation produced by a cluster of OB stars, or does it come from a central "exotic" object? The presently available data from luminosity and stellar measurements do not give a clear answer to this important question and probably allow both scenarios.

On the one hand, the UV radiation could indeed come from a cluster of hot stars if special conditions are assumed. The low effective temperature of the ultraviolet field excludes early-O stars as the primary source of UV photons, but a cluster of approximately 10^2 late-O and early-B stars would be consistent with the measurements. Rieke (1982), Lebofsky et al. (1982a), and Lacy et al. (1982) propose that the absence of early-O stars could be due to a burst of star formation about 10^7 yr ago. The early-O stars formed in this star-formation event would by now have left the main sequence and have become red supergiants, and only B and lower mass stars would still be on the main sequence. Six late-type stars within about 2 pc of the center are intense enough for their infrared spectra to have been studied. Lebofsky et al. (1982a; see also Wollmann et al. 1982) classified four of these as supergiants, thus supporting the star-formation burst model. However, spectral work by Sellgren et al. (1987) indicates that only one, IRS 7, has the spectrum of a supergiant, whereas the others have spectra similar to late-M-type giants. The conclusion from the latter work is that there may have been no recent major star-formation event within this small radius about the center. Still more recently, Rieke & Lebofsky (1987), from a comparison of the dereddened 2-μm brightness of stars near the Galactic center with bulge stars in Baade's window and assuming similar populations, conclude that stars near the Galactic center are more luminous, and hence that star formation has taken place there within the last 10^8 yr.

On the other hand, there is no evidence for *current* star formation in the central 10 pc of the center, and the observed characteristics of the radiation field are also consistent with a single central source. The characteristics of the Sgr A West H II region are quite different from those of other star-formation regions in the Galaxy. There is little dense molecular gas in the central region, and the observations, discussed in more detail in Section 4.2, suggest that the dense cloudlets of ionized gas are not heated by embedded early-type stars (e.g. Gatley 1987).

Much of the luminosity may originate within a few arcseconds of IRS 16/Sgr A*. The dust temperature increases steadily inward from $R \sim 3$ pc to $R \sim 0.5$ pc (Becklin et al. 1982, Gatley 1982, 1987). Assuming that most of the 5–20 GHz continuum emission in the bar is thermal, van Gorkom et al. (1985) conclude that the electron temperature in the "bar" region is at least as high as 12,000 K, significantly higher than in the western arc and northern arm ($T_e \approx 5000$ K) and indicative of a radial gradient in electron temperature. However, Wright et al. (1987) find evidence for a significant nonthermal component in the bar from a comparison of continuum maps between 1.6 and 88 GHz and cast some doubt on the magnitude of the electron temperature gradient inferred by van Gorkom et al. (1985). Geballe et al. (1984) and Wade et al. (1987) find that the flux of the He I $n = 2$ $^1P-^1S$ line at 2.06 μm near IRS 16/Sgr A* (Hall et al. 1982) is consistent with the proposition that Lyman continuum flux originating near the IRS 16 complex ionizes all of Sgr A West. A single central source of ionization with $T_{\text{eff}} \approx 35,000$ K and $L_{\text{UV}} \approx 10^7$ $L_\odot$ $R_0^2(10)$ is also consistent with the ionization state of the gas in the central 3 pc, the bolometric and Lyman continuum luminosity, and also the near-infrared (2 μm) flux density of individual sources in the IRS 16 complex (Henry et al. 1984, Serabyn & Lacy 1985). This supersedes an earlier conclusion (Lacy et al. 1982) that the 2-μm flux density is too low to be consistent with the effective temperature and luminosity.

4. INTERSTELLAR MATTER: CIRCUMNUCLEAR RING, IONIZED STREAMERS, RELATIVISTIC AND HOT GAS IN THE CAVITY

4.1 *The Neutral Circumnuclear Disk (R = 2 to 8 pc): Spatial Distribution and Physical Conditions*

Following the detection of the 2-pc dust lobes surrounding the Galactic center by Becklin et al. (1982; see also Rieke et al. 1978, Harvey et al. 1981), the corresponding gas was found in the neutral oxygen fine-structure line at 63 μm (Lester et al. 1981, Genzel et al. 1982, 1984), in the 2-μm $S(1)$ line of vibrationally excited molecular hydrogen (Gatley et al. 1984), in the 21-cm atomic hydrogen line, and in the 2.6-mm CO $J = 1 \rightarrow 0$ line (Liszt et al. 1983). Sandqvist (1987) notes that some evidence for the neutral circumnuclear gas may have been present in earlier lunar occultation studies of the 18-cm OH lines (Sandqvist 1974). These and other observations by a number of groups show the existence of a $\sim$2-pc radius, double-lobed structure of warm, atomic and molecular gas (Genzel et al. 1985, Harris et al. 1985, Lugten et al. 1986, Gatley et al. 1986) approximately along the Galactic plane, surrounded by cooler, lower excitation gas (Liszt et al. 1983, Fukui et al. 1983, Sandqvist et al. 1985, 1987, Lo 1986, Serabyn et al. 1986, Fukui & Churchwell 1987).

Recent high-spatial-resolution observations now demonstrate that these lobes represent in fact an essentially complete, 2-pc radius ring of gas that forms the inner edge of a thin, disklike structure extending to at least 8 pc from the center (Gatley et al. 1986, Lugten et al. 1986, Serabyn et al. 1986, Kaifu et al. 1987, Güsten et al. 1987a, Sandqvist et al. 1987, Jackson et al. 1987). The ring/disk structure is inclined by 60–70° with respect to the line of sight. The major axis at $R \sim 2$ pc is near the Galactic plane, but the large scale disk is tilted by $\sim 20°$ relative to the Galactic plane. Inside its sharp inner edge, there is a thin ring of ionized material (Lo & Claussen 1983, Serabyn & Lacy 1985) but very little other interstellar gas. The mean density of neutral gas at $R \leq 1.7$ pc is at least a factor of 10 lower than that in the neutral ring. As an example, Figure 6 shows a velocity-integrated map of the 3-mm HCN $J = 1 \rightarrow 0$ transition, obtained with the Berkeley Hat Creek interferometer at 10″ resolution, together with typical spectra. Figure 7 shows the circumnuclear ring in different tracers. The HCN line samples warm, dense ($n_{H_2} \gtrsim 10^5$ cm^{-3}) molecular gas; the C$^+$ 158-μm line shows the distribution of gas partially ionized by the far-ultraviolet radi-

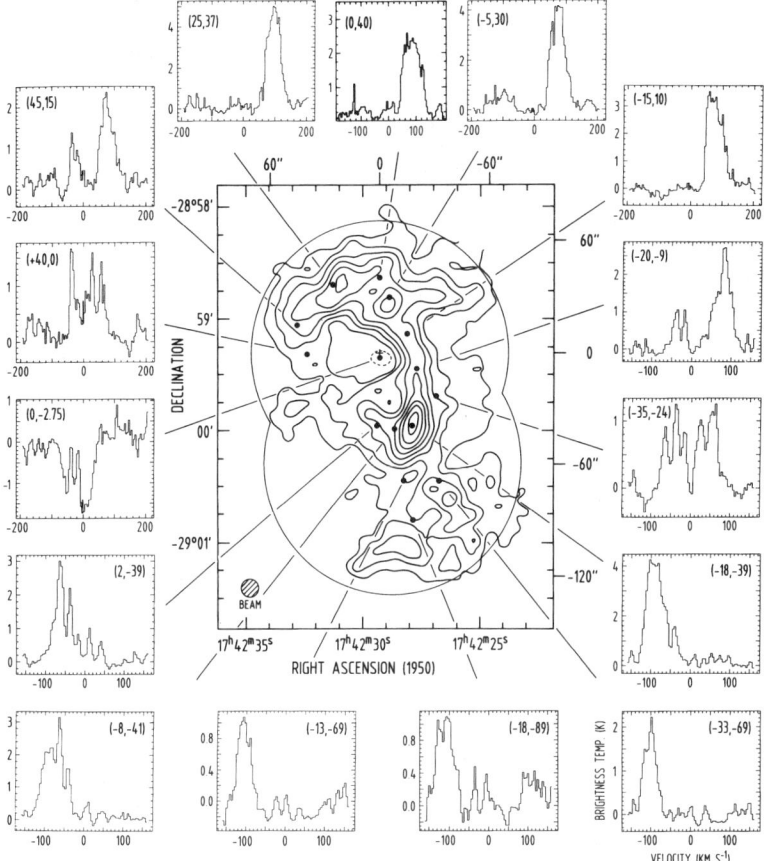

Figure 6 A 10" resolution cleaned map of the 3-mm HCN $J = 1 \to 0$ transition toward Sgr A, together with typical spectra at 4 km s^{-1} resolution (from Güsten et al. 1987).

ation from the center; the CO $J = 1 \to 0$ emission traces somewhat cooler, lower density gas at larger distances from the center; and the 2-μm emission from vibrationally excited molecular hydrogen samples hot molecular gas excited by shock waves or ultraviolet radiation.

The measurements clearly show that the atomic and molecular gas is unusually warm and dense and that the excitation and energy density or pressure of the gas fall rapidly with increasing radius. The unusual characteristics of the neutral circumnuclear interstellar matter reflect the large energy density and concentration of energy sources in the central few parsecs. Genzel et al. (1985), using the far-infrared dust opacity and H I and CO column densities, estimate the total amount of the warm, circum-

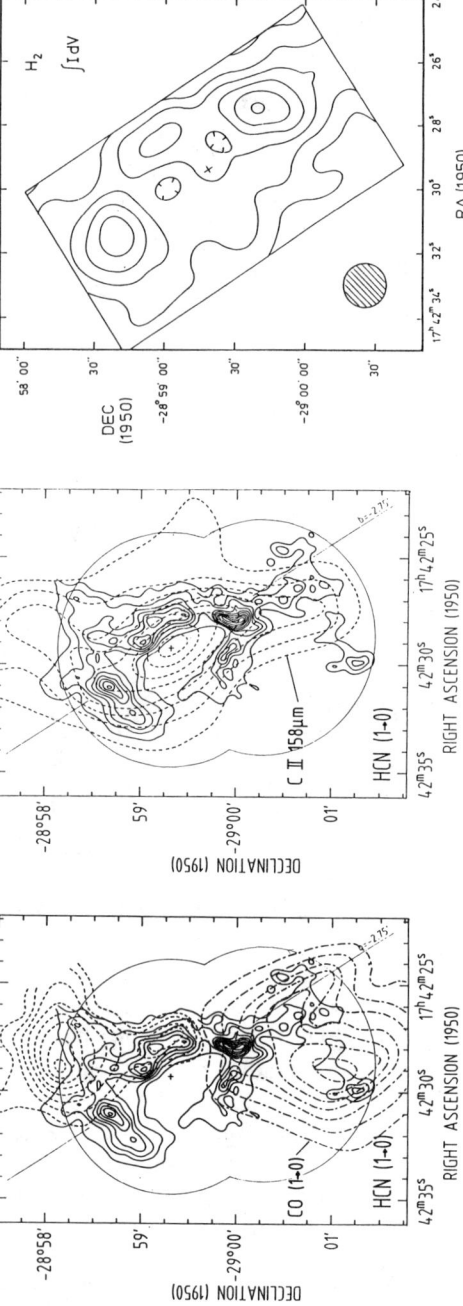

Figure 7 The circumnuclear disk/ring in different tracers. (*Left*) A 5″ resolution MEM velocity-integrated 3-mm HCN map [solid contours (Güsten et al. 1987a)] and 2.6-mm CO $J = 1 \rightarrow 0$ map at $80 \leq |v_{LSR}| \leq 110$ km s^{-1} [22″ beam, dashed-dotted (Serabyn et al. 1986)]. (*Middle*) An HCN map and 158-μm $^2P_{3/2}-^2P_{1/2}$ C$^+$ velocity-integrated map [55″ beam, dashed (Lugten et al. 1986)]. (*Right*) A 2-μm H$_2$ $v = 1 \rightarrow 0$ $S(1)$ map [18″ beam (Gatley et al. 1986)].

nuclear neutral gas between 2 and 5 pc to range between one and a few times 10^4 $M_\odot$. Figure 8 displays measurements of far-infrared, submillimeter, and millimeter emission lines of CO at two different positions in the circumnuclear disk (Harris et al. 1985, Lugten et al. 1987). At the inner edge of the ring, the CO gas is dominated by a component of temperature $\gtrsim 400$ K and hydrogen density 5×10^4 cm^{-3}, while temperature and density fall to $\lesssim 300$ K and 1.5×10^4 cm^{-3} at $R = 4.5$ pc. The warm atomic gas, sampled by the O I and C$^+$ far-infrared fine-structure lines, is coexistent with the warm molecular gas and is also characterized by similar physical parameters [$n_{H_2} \sim 10^5$ cm^{-3}, $T \sim 350$ K (Genzel et al. 1985)]. Even hotter gas at the boundary of the ring ($T \approx 2000$ K) is emitting in the 2-μm ro-vibrational lines of H$_2$ (Gatley et al. 1984, 1986). The gas pressure estimated from the relative intensities of the far-infrared and submillimeter lines decreases by a factor of 4 ± 1 between 2 and 5 pc from the center. A similar excitation gradient, and gas densities of approximately 10^5 cm^{-3} are also derived from observations of several CS rotational lines in the millimeter range (Güsten et al. 1987b, Evans 1987). Finally, the dust temperature decreases from $T \sim 100$ K at 1 pc to $\lesssim 60$ K at 4 pc (Becklin et al. 1982, Drapatz et al. 1985, Lester et al. 1987).

Direct as well as indirect evidence indicates that the circumnuclear ring is very clumpy, with a volume filling factor of about 0.05 to 0.1. Direct evidence comes from the high-resolution maps of molecular gas emission

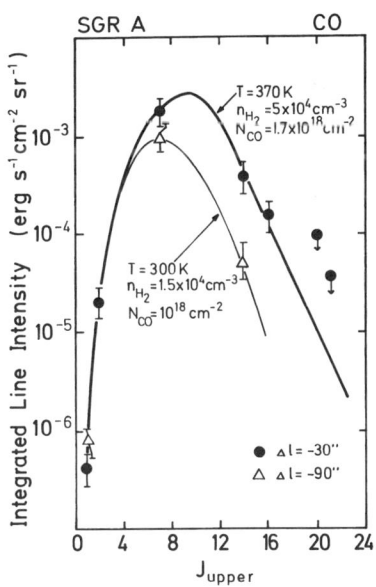

Figure 8 Intensities of different submillimeter and far-infrared CO lines at offsets $-30''$ and $-90''$ from the position of Sgr A* along the Galactic plane, together with best-fit models (Lugten et al. 1987, Harris et al. 1985).

(Figures 6, 9), which show clumps of diameter 0.2 to 0.5 pc. Indirect evidence comes from a comparison of the local hydrogen volume density needed for collisional excitation of the CO and O I/C$^+$ emission lines ($n_H \sim 10^5$ cm^{-3}) and the average hydrogen volume density estimated from the column density ($N_{H\,I+2H_2} \sim 2$–5×10^{22} cm^{-2}) and scale size of the region. Similar clumpiness is seen in the ionized gas ring (cf. Figure 2) at the inner edge of this neutral region. The low volume filling factor means that UV radiation emerging from the center can penetrate relatively far into the disk.

4.2 Gas Dynamics and Nature of the Circumnuclear Material

The available infrared and microwave spectroscopic measurements give detailed information on the neutral gas dynamics. The dominant large-scale motion of the majority of the neutral gas is rotation about the center. Several investigations find that any systematic, large-scale radial motion of the ring as a whole must be less than about 30 km s^{-1}, although a few clouds do show strong deviations from pure circular motion (Lugten et al. 1986, Serabyn et al. 1986, Güsten et al. 1987a). On the other hand, Gatley et al. (1986) and Kaifu et al. (1987) conclude that systematic radial motions

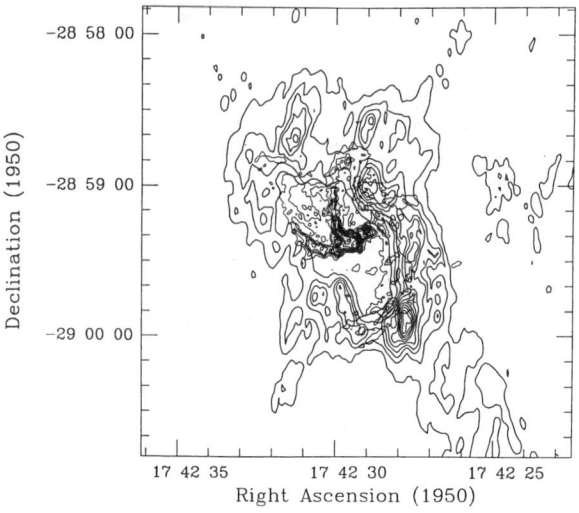

Figure 9 A 3-mm HCN $J = 1 \rightarrow 0$ aperture synthesis map at 3" resolution (MEM; Wright et al. 1987) combined with the 6-cm radio continuum map from Lo & Claussen (1983; 1" resolution).

as large as 50 km s^{-1} are characteristic of the ring as a whole. The derived rotation velocity as a function of distance from the center, corrected for inclination ($i \approx 70°$), is approximately constant between 2 and 4 pc ($v_{\rm rot} \approx 110$ km s^{-1}; Figure 10).

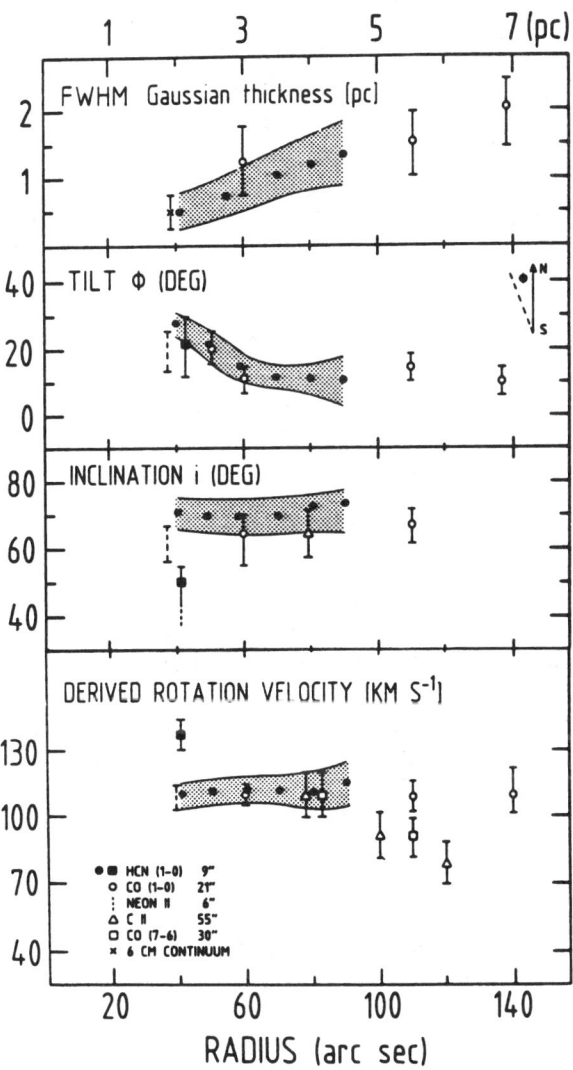

Figure 10 Parameters of the neutral circumnuclear disk as derived from a number of measurements discussed in the text (from Güsten et al. 1987a).

Beyond 4 pc the velocity field is more complex. Profiles of HCN, CS and CO millimeter emission lines in the southeastern part of the disk show a splitting into two velocity components [-110 and -80 km s^{-1} (R. Güsten, personal communication, Serabyn et al. 1986)], and the observed velocity centroids of the high excitation C$^+$ and CO $J = 7 \to 6$ lines decrease from -100 to about -70 km s^{-1} between 4 and 6 pc (Harris et al. 1985, Lugten et al. 1986, Güsten et al. 1987a). It is presently not clear which if any of the two components are representative of the rotation velocity, and both are plotted in Figure 10.

While a circulating disk is a good first-order description of the neutral gas dynamics, the rotation pattern is strongly perturbed. Large line widths *everywhere* in the disk ($\Delta v_{\text{FWHM}} \sim 50$–$60$ km s^{-1} in a 5–10″ beam) indicate large, *local* perturbations in the general circular motion. Spectra of the ionized gas at the inner edge of this region also show irregular variations in average velocity of as much as 20 km s^{-1} between regions separated by only 5″ (Serabyn & Lacy 1985). A bright cloud near the minor axis at $R \sim 2$ pc west of the nucleus, prominent in H$_2$, CO, and HCN emission, does not fit into the general rotation pattern, and it is clearly an example where the radial velocity of a particular cloud is comparable to its circular velocity. Furthermore, the spatial distribution and velocity field of the HCN gas at the inner edge of the disk indicate that not all of the material is in one plane, but rather that the inner disk is somewhat warped or kinked, or that the inner disk consists of a number of gas streamers with slightly different orientations (Güsten et al. 1987a). Finally, the disk's position angle on the sky appears to change with distance from the center. The plane of rotation is within 5° of the Galactic plane at the inner edge of the disk, but is tilted by about 20 to 25° relative to the plane at radii of 3 to 5 pc (Figure 10).

The large local turbulence makes up about 10–15% of the rotational energy and may support the thickness of the disk. The FWHM thickness increases from about ~ 0.5 pc at 2 pc to about 2 pc at 7 pc, approximately consistent with hydrostatic equilibrium (Figure 10; Güsten et al. 1987a, Serabyn et al. 1986). Cloud-cloud shocks driven by the large turbulence and the ultraviolet radiation from the center via the photoelectric effect may be the dominant heating mechanisms for the warm molecular and atomic gas components, respectively (Genzel et al. 1985, Harris et al. 1985, Serabyn & Güsten 1986). The clumpiness of the disk may explain how warm atomic gas (C$^+$, O I), which probably comes from cloud surfaces photodissociated by the UV radiation from the center, can exist at large distances from the center and can coexist with molecular gas. Mass outflow from the center (see Section 4.5) may also play a role for the excitation of the hot molecular hydrogen at the inner edge of the disk, which has a total luminosity of $L_{H_2} \sim 10^3 \, L_\odot \, R_0^2(10)$ (Gatley et al. 1984). However, outflow alone probably cannot account for the larger amount of warm CO, O I, and C$^+$ gas emitting in the far-infrared because of its large spatial extent throughout the disk and the energy requirements [$L_{CO+O\,I} \sim 5 \times 10^4 \, L_\odot \, R_0^2(10) \approx 0.005 \, L_{UV}$ within $R = 4$ pc $R_0(10)$]. The ratio of inferred total luminosity in the near-infrared H$_2$ and far-infrared CO lines is about two orders of magnitude lower than that found in the shocked region of Orion, which perhaps indicates low-velocity ($v_s \leq 10$ km s^{-1}) shocks or high magnetic fields (Draine & Roberge 1985). Furthermore, some of the molecular hydrogen appears to be excited by ultraviolet radiation rather than by shocks alone (Gatley et al. 1986).

The present distribution and physical parameters of the circumnuclear disk cannot represent a static equilibrium configuration. Although the neutral gas clouds are subject to tidal disruption [$n_H(R = 2 \text{ pc}) \ll n_{Roche} \approx 10^7$ cm^{-3}] and dispersal because of substantial velocity differentials, the disk is nevertheless very clumpy and has a well-defined, sharp inner edge. Radiation pressure, thermal pressure from the H II region, or hot gas in the cavity (see Section 4.6) are not sufficient to maintain the sharp edge, and magnetic fields $\gtrsim 10$ mG would be required to counterbalance the turbulent pressure of the neutral clouds (cf. Güsten et al. 1987a). Furthermore, the apparent warping of the disk at 2 pc and the radial change of the disk's position angle on the sky also cannot be maintained in steady state. The time scale for dispersal of the clumpy clouds or for the warping is no more than a few rotation periods, or a few $\times 10^5$ yr. The time scale for dissipation of the disk's turbulent energy ($\sim 2 \times 10^{50}$ erg) by shock cooling, as estimated from gas mass and turbulent velocity dispersion and the total infrared and submillimeter molecular-line cooling ($\sim 2 \times 10^4 L_\odot$), is about $8.5 \times 10^4 \, \eta$ yr, where $\eta \leq 1$ is the conversion efficiency from the dissipation rate of kinetic energy into line radiation. Finally, unless there are forces pressurizing the central cavity, the lower velocity gas at the inner edge of the ring will fall inward, eliminating the sharp density discontinuity on a time scale of $\sim 10^5$ yr.

A possible dynamical equilibrium explanation of the circumnuclear disk might be the conversion of rotational energy into turbulence by "turbulent" or magnetic friction, leading to a slow contraction of the disk. In this scenario, the mass accretion rate necessary to maintain the present line radiation from the disk is about $10^{-2} \, M_\odot$ yr^{-1}. Over a period of a few million years, however, the central cavity would fill in with gas unless there is a mechanism for its removal, and mass has to be replenished from farther out, possibly from the clouds within 10 to 100 pc of the center (Section 2.2). Another significant source of influx of gas into this region of the Galaxy may be mass loss from asymptotic giant branch stars as measured by IRAS and 18-cm OH maser emission (Habing 1987). An alternative and perhaps simpler scenario is an explosion (Sgr A East?) or burst of mass outflow with a required energy of $10^{51} R_0^2(10)$ erg not longer than about 10^5 yr ago. Such temporary, intense activity could also explain the central cavity.

4.3 Ionized Gas Streamers ($R \leq 1.7$ pc): Spatial Distribution and Physical Conditions

Inside of 1.7 pc, the gas is predominantly ionized. Some evidence for neutral atomic gas in the central region comes from mapping of the [O I] 63-μm line (Genzel et al. 1985). Although this evidence is not entirely conclusive because of the relatively low spatial resolution (30″) of the far-infrared measurements, it may suggest the presence of a few tens of solar masses in atomic material at $R \leq 1.7$ pc. This is to be compared with about 10^2 $M_\odot$ of ionized gas, as derived from the radio free-free flux density and an assumed average electron density of about 10^4 cm^{-3}.

Lacy et al. (1979) first examined in some detail the distribution and velocities of ionized gas in the central region. While suggesting that the ridge now identified as the northern arm and part of the western arc might be either falling through or in orbit about the center, they concentrated primarily on the central region and concluded that there were bloblike clouds that might result from stellar-stellar collisions (Lacy et al. 1980, 1982). However, detailed mapping with the VLA (e.g. Brown et al. 1981, Ekers et al. 1983, Lo & Claussen 1983) brought out the continuity and filamentary character of much of the gas, especially those parts more distant from the center.

On the basis of the filamentary morphology of the ionized gas (Section 2.3; Figures 2, 3), Lo & Claussen (1983), Ekers et al. (1983), and Brown (1982) have suggested that the arcs represent coherent streamers of gas. This impression is strengthened if, in addition to the spatial morphology, the velocity field of the gas is considered. Figure 11 gives an overview of the ionized gas velocities, as derived from mapping of the 12.8-μm [Ne II] fine-structure line (Lacy et al. 1980, Serabyn & Lacy 1985), from VLA mapping of the 76α H I recombination line (van Gorkom et al. 1985, 1987), and from mapping of the broad, near-infrared lines of He I and H I near IRS 16 (Geballe et al. 1987). It is apparent that the velocities change more or less continuously along the individual arcs, as expected for gas streamers.

There is a large range of electron densities in the ionized gas. The rms density in the filaments and knots, as derived from the radio free-free emission measure and the ratio of [Ne II] to infrared H I recombination lines, is as high as 10^4 to 10^5 cm^{-3} (Lacy et al. 1980, Brown et al. 1981,

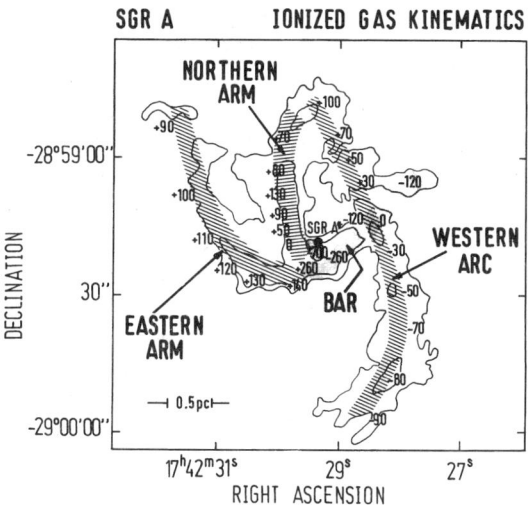

Figure 11 Overview of the ionized gas velocities, superposed on the Lo & Claussen (1983) 6-cm map. The measurements in the western arc, bar, and northern and eastern arms are from 12.8-μm [Ne II] spectroscopy by Lacy et al. (1980) and Serabyn & Lacy (1985), and from 76α H I recombination-line measurements by van Gorkom et al. (1985, 1987). The ± 700 km s^{-1} broad-line emission region is from 2-μm He I/H I spectroscopy by Hall et al. (1982) and Geballe et al. (1984, 1987).

Serabyn & Lacy 1985). Measurements of average electron density with beams covering the central 20–30″ diameter region come from the ratio of the two far-infrared [O III] fine-structure lines [$n_e = 8000^{+\infty}_{-5000}$ cm^{-3} (Genzel et al. 1984)] and the two infrared [S III] lines [$n_e = 1800 \pm 500$ cm^{-3} (Herter et al. 1984)]. The [O III] line intensities also show that any other material of similar temperature but clearly outside the detected ionized material has a density less than about 40 cm^{-3} (Watson et al. 1980). The difference between these numbers and the higher values derived from the interferometric radio measurements may be due to a substantial amount of lower density material outside of the high-density knots and filaments (cf. Mezger & Wink 1986).

4.4 *Nature of the Ionized Gas Clouds*

Detailed information about ionized clouds in the Galactic center has been made available by high-resolution radio mapping, mapping of the mid-infrared thermal dust emission, and rather fine-scale mapping of velocities of the 12.8-μm [Ne II] fine-structure line and of the hydrogen and helium recombination lines. The nature of the ionized clouds has been discussed in detail in a number of publications (Rieke et al. 1978, Becklin et al.

1978b, 1982, Lacy et al. 1980, 1982, Brown et al. 1981, Gatley 1982, Rieke 1982, Lo & Claussen 1983, Gezari et al. 1985, Allen & Sanders 1986). Most authors have stressed the differences between the Sgr A West H II region and "normal" H II regions: In the former the mean dust temperature is high ($T_d \sim 200$–300 K), the $\lambda \geq 50$ μm flux from the central region is relatively small, the total far-infrared luminosity is only slightly greater than the Lα luminosity, there is a lack of intrinsic silicate absorption, the ratio of 10-μm to radio continuum flux is very large, and the variations in 8–12 μm color temperatures of the different compact cloudlets are small. The differences with normal compact H II regions are further supported by the high-spatial-resolution maps and the velocity measurements discussed above. Most of the cloudlets on the radio continuum and 4–30 μm maps appear to be just regions of additional brightness—that is, probably density enhancements in large, coherent gas structures rather than regions of increased temperature. This suggests that the ionized clouds are externally photoionized by the general UV radiation field near the center, and that the 4–30 μm emission is due to thermal emission from dust grains within the ionized clouds, which are heated by trapped Lα photons (e.g. Becklin et al. 1982, Lo & Claussen 1983, Gezari et al. 1985). The dust column densities through the knots are low ($A_v < 1$), so that most of the nonionizing ultraviolet and visible luminosity is not absorbed there. A contrasting view, proposed by Rieke (1982) and Allen & Sanders (1986), is that internal heating sources are nevertheless required, since the same sources emitting at 10 μm are also apparent on 3–4 μm maps (implying the presence of hot dust grains) and since the source sizes appear to increase with wavelength. Gatley (1982, 1987) notes that these facts are not necessarily in contradiction to the external heating model, since trapped Lα heating produces a range of dust temperatures, and hotter dust grains are expected in density concentrations because of the higher recombination rate, and hence higher Lα energy density, there.

Further information on the nature of the ionized gas comes from a comparison with the neutral interstellar matter, which is shown in Figure 11. From a comparison of spatial distributions and velocity fields, Güsten et al. (1987a) emphasize the close physical relationship between the two. The western arc (Figure 2) tracks closely the inner edge of the neutral ring, with detailed correspondences of individual peaks in either distribution apparent at the southern and northern ends of the arc. The velocity fields are also similar. The ionized gas velocities change approximately linearly with position offset from IRS 16/Sgr A* (e.g. along Galactic longitude) and approach ± 100 km s^{-1} LSR (local standard of rest) at the northern and southern ends of the arc, respectively (Serabyn & Lacy 1985, van Gorkom et al. 1985, 1987). This behavior is characteristic of a ring of rotating gas at constant radius of 1.7 pc $R_0(10)$. Hence, the western arc almost certainly is the inner, photoionized surface of the neutral ring, rotating at about 110 km s^{-1} and inclined by about 60–70° relative to the line of sight (Serabyn & Lacy 1985, Genzel et al. 1985). However, this model raises the question of the eastern and southern counterparts of the western arc, which are not apparent on the radio maps by Lo & Claussen (1983) or Ekers et al. (1983). Serabyn & Lacy (1985) have proposed that the northern arm and bar intercept and absorb most of the ionizing UV radiation, thus decreasing the Lyman continuum flux impinging on the eastern part of the ring. The recent high-resolution maps of the neutral ring (Kaifu et al. 1987, Güsten et al. 1987a; Figure 11), however, show that the molecular emission on the eastern side of the ring is also weaker by a factor of about 2 to 3 than on the western side. It is more likely, therefore, that the eastern counterpart of the ionized western arc is just somewhat less dense on average and for that reason weaker than the lowest contour of the present 5–15 GHz radio continuum maps. Van Gorkom et al. (1985, 1987) and Serabyn & Lacy (1985) find line emission along the southeastern portion of the ring at the appropriate velocities, in support of this scenario. Further evidence for the presence of the southeastern portion of the ring in the ionized gas comes from continuum maps at 23 and 88 GHz (Ishiguro et al. 1986, Wright et al. 1987).

While the northern and eastern arms do not have a counterpart in molecular gas, a physical relationship to the neutral ring is suggested by the fact that both arms terminate at the ring and seem to trail from there inward to the center. In both cases, the velocities of the ionized arms at the points closest to the ring agree with the neutral gas velocities there. Furthermore, the hydrogen densities, large local line widths, and short lifetimes of the ionized gas clouds (Lacy et al. 1980) are similar to those found later in the adjacent neutral ring.

4.5 *Inflow vs. Outflow*

These findings suggest the following scenario, first proposed by Lo & Claussen (1983) and Ekers et al. (1983), at least for much of the ionized gas inside the cavity. Low-angular-momentum material sporadically falls toward the center as a result of collision of gas clouds in the inner part of the neutral ring or of inward-streaming gas from farther out. This material becomes ionized and is stretched to long streamers by the tidal forces during the fall (time scale a few $\times 10^4$ yr). Velocity measurements generally support such a model and clarify which streamers are related (Lo & Claussen 1983, Serabyn & Lacy 1985). In further support of this model for the northern and eastern arms there is a relative lack of neutral gas on the ring near their end points. Both arms share the same direction of angular momentum, which also may be the direction of angular momentum of the disk (see discussion in Güsten et al. 1987a). Depending on the assumed average ionized gas density, a mass-inflow rate of 10^{-3}–10^{-2} $M_\odot$ yr^{-1} is derived from the present amount of material in the central cavity (Lacy et al. 1979, 1982, Lo & Claussen 1983, Ekers et al. 1983), not inconsistent with the possible accretion rate in the neutral disk (Güsten et al. 1987a). While there are some bloblike clouds in the central region that are difficult to explain by this type of infall, an origin of any large fraction of the ionized clouds in stellar collisions (Lacy et al. 1982) now appears unlikely.

The possibility of the streamers representing outflow is discussed below. Assuming that they flow inward, what would be the fate of the infalling gas? Is most of it propelled out after streaming through the center on defined orbits, or do collisions with other clouds, friction with distributed gas, or interaction with magnetic fields lead to gas collecting near the dynamic center? Estimates indicate that the clouds would collide with each other on a time scale of about 10^4 yr, or they may disperse due to their internal velocity dispersion in about the same time (Lacy et al. 1980). A sharp bend in the northern arm near IRS 1 (Figure 3; Serabyn 1985, Serabyn et al. 1987) may be an example of such a cloud-cloud collision. The apparent braiding of the filamentary structures in Figure 3 (Lo 1986) and the detection of intrinsic, coherent 10-μm dust polarization in the northern arm (Aitken et al. 1986) may indicate magnetic field strengths of at least 10 mG $(n_c/10^5 \text{ cm}^{-3})^{1/2}$ (Aitken et al. 1986, Dyck et al. 1974, Capps & Knacke 1976, Lebofsky et al. 1982b). The energy density of a $\geq$ 10-mG field is $\geq 4 \times 10^{-6}$ erg cm^{-3} and the kinetic energy density of the cloud motions in the northern arm is $\sim 1.4 \times 10^{-5}$ erg cm^{-3} $(n_c/10^5 \text{ cm}^{-3})$ $(v/130$ km s$^{-1})^2$, where n_c is the electron density inside the ionized clumps and v is their velocity. Hence, magnetic forces could play a significant role in the ionized gas dynamics. Quinn & Sussman (1985) have proposed a model in which the western arc and northern arm together form a streamer of gas that originates in the ring and spirals inward as a result of friction by distributed, hot gas. Their particular model does not fit the measured velocities (Serabyn 1985); however, does friction play some role? Quinn & Sussman (1985) note that to be dynamically important in a single orbit, such gas must have a density $\gtrsim n_c/50$. They discuss the possible nature of such gas and some evidence against frictional effects of the magnitude required. Section 4.6 below provides some details on observational limits for this gas density and indicates that only for the lower ($n_c \sim 10^3$–10^4 cm^{-3}) cloud densities (for example, in the northern arm) could gas generally distributed in the cavity be of dynamic importance and not be in conflict with current observations. In summary, frictional effects, cloud-cloud collisions, and magnetic forces are all likely to have some dynamic impact on the gas in the central cavity. It appears that a substantial fraction of the gas would fall inward, possibly onto a massive object (should one exist) or to form a reservoir of gas leading to a future star-formation event (cf. Lacy et al. 1982, Lo & Claussen 1983). The transient nature of the ionized gas in the central cavity suggests that the inferred accretion has not been a steady process. The current infall must have been preceded by events on the order of 10^4–10^5 yr ago that strongly affected the gas in this region and eliminated a large fraction of the gas from the central 2 pc.

The observed radial velocities of the ionized gas clouds are consistent with outward as well as inward motion. Yusef-Zadeh (1986) and van Gorkom et al. (1987) have noted that the northern arm appears to connect to the large-scale protrusions discussed in Section 2.3 (Figure 1), and Morris & Yusef-Zadeh (1987) suggest that there is outflow from the center, channeled and directed by the circumnuclear ring and poloidal magnetic fields. The density valley in the ring (Figures 7 and 9) may mark where the outflow has punctured through the dense gas ring. However, the presence of ionized gas beyond the inner HCN ring at $R \sim 2$–3 pc may instead be explained by Lyman continuum photons penetrating deeper into the disk north of the center *because* of the lower gas densities there. Heyvaerts et al. (1987) suggest that the gas streamers in the cavity, as well as the protrusions and larger scale radio filaments discussed above, represent gas trapped in expanding magnetic loops generated by a central source. The bar is structurally more complex; it appears to contain more than one streamer as well as one or two more isolated clouds. A considerable part of the gas may be extensions of the northern and eastern arms (Lo & Claussen 1983, Lo 1986, Serabyn et al. 1987, Lacy et al. 1987). Quinn & Sussman (1985) have proposed that the bar represents outflow aligned approximately perpendicular to the plane of the neutral gas disk. Rees (1987) suggests that the ionized streamers are formed as "vapor trails" as a result of interaction of material in the cavity with ejecta from stars disrupted by a central, massive black hole. He also proposes that the ejecta from successive stellar disruptions could maintain the central cavity. Hall et al. (1982) and Geballe et al. (1984) conclude that the most likely explanation for the high-velocity (± 700 km s^{-1}) He I and H I gas immediately around IRS 16 is an outflow or wind from a star or stars at the center. The inferred mass outflow rate is $\sim 4 \times 10^{-3}$ $M_\odot$ yr^{-1} (Geballe et al. 1987).

An outflow of this magnitude would have a significant impact on velocities in the central region and could explain the presence and luminosity of the shocked molecular hydrogen gas at the inner edge of the molecular ring (Gatley et al. 1984, 1986). However, the impact of an isotropic central outflow does not appear to explain the spatial distribution and velocity field of the ionized streamers. The streamers converge in a region $\sim 5''$ south of the broad-line region and are not centered on IRS 16. Velocities show a marked decrease with distance from IRS 16, rather than the constant or slowly increasing velocities expected in the case of acceleration in an outflow. The proposed outflow also does not appear to have enough momentum or energy to explain either the excitation of most of the molecular gas in the ring or the existence of a central cavity, given the large turbulent ram pressure in the neutral ring (Harris et al. 1985, Güsten et al. 1987a).

The problem of producing outflow of the character and amount required is discussed in Section 6.

4.6 Relativistic or Hot Gas in the Central Cavity Not Directly Observed

Evidence for the presence of relativistic electrons in the central cavity comes from a comparison of the extended ($\sim 1'$ scale) emission in the 1.7- and 5-GHz VLA maps by Ekers et al. (1983) with the distribution of neutral gas (Figure 11; Wright et al. 1987). There is a sharp increase of extended radio continuum emission within the central cavity, tracking more or less exactly the inner edge of the neutral ring. This emission is predominantly nonthermal, as indicated by the spectral index map of Ekers et al. (1983; called the "diffuse" Sgr A West component there). With the standard assumptions (see, e.g., Miley 1980), an energy density of relativistic electrons of 10^{-9}–10^{-8} erg cm^{-3} and an equipartition magnetic field of ~ 300 μG are derived. Additional evidence for an enhanced density of relativistic electrons in the northern arm and bar comes from a comparison of 1.7–80 GHz radio continuum maps and flux densities by Wright et al. (1987).

As mentioned in Section 2.4, one suggested interpretation of the X-ray emission from Sgr A West is that hot gas is present [$T \sim 10^7$–10^8 K (Watson et al. 1981)]. The recent measurements by Kawai et al. (1987) set a limit on this radiation that is four times less than what was previously reported; from this time variation we can conclude that the radiation must have originated in a localized source, and that any X-ray radiation in the 1–15 KeV range resulting from a broad distribution of hot gas in the Galactic center must have a luminosity $\lesssim 4 \times 10^{34}$ erg s^{-1}.

Gas broadly distributed throughout the "cavity" and having a temperature $\lesssim 10^5$ K would fall into the center if it does not have additional velocity of motion $\gtrsim 100$ km s^{-1}. Gas of temperature $\gtrsim 2 \times 10^6$ K would evaporate rapidly in directions out of the Galactic plane where there is little restraining gas, since the rms proton velocity at this temperature is about 225 km s^{-1} (cf. Quinn & Sussman 1985). Ionized gas at intermediate temperatures cools by radiation so rapidly that providing a sufficient input of energy to it is very demanding (Dalgarno & McCray 1972). In addition to these general considerations, a number of observations provide information on the amount of gas that may be present in the "cavity." If the gas were in a low state of ionization, so that Ne II would be present, the maximum intensity found in certain directions between known clouds (Lacy et al. 1980) sets an upper limit on the density $n \lesssim 700$ cm^{-3}. Actually, the known UV flux should further ionize such tenuous gas, in which case the intensity ratio of [O III] lines (Watson et al. 1980) sets a more restricted upper limit of 40 cm^{-3}, as noted in Section 4.3. For gas at higher temperatures, possible energy sources need be considered that could provide for its rate of radiation. These include a high-velocity wind from the center, shocks due to motion of the known ionized clouds through the medium, possibly relativistic particles, and varying magnetic fields. The wind proposed by Geballe et al. (1987), 4×10^{-3} $M_\odot$ yr^{-1} at an average velocity of 650 km s^{-1} (which appears to be an upper limit), could provide an energy flow of about 5.3×10^{38} erg s^{-1}.

The moving clouds could adequately heat all the gas only by moving as much as about one full orbit around the center. Hence, we take the time required to heat most of the gas by this mechanism as greater than 10^4 yr. The relativistic particles lose their energy (radiatively) at times longer than about 10^5 yr in any likely gas density, and if their energy density is $< 10^{-8}$ erg cm^{-3} as estimated, this is equal to the energy density of a 10^7-K gas of only 5 particles cm^{-3}. Magnetic fields need to be considered, but no mechanism for providing as much energy as does the proposed wind has been suggested without the presence of an unknown and exotic object.

Table 1 shows the gas densities filling the cavity that would radiate at a total power less than or equal to that from the above wind (5.3×10^{38} erg s^{-1}) and those that would cool by radiation in a time greater than 10^4 yr. This severely limits the gas density at any temperature between 10^4 and 10^7 K. While very hot gas ($\gtrsim 2 \times 10^6$ K) would evaporate, it is still of interest to examine the maximum gas densities that might fill the cavity and not produce X rays in excess of the luminosity reported by Watson et al., as modified by the results of Kawai et al. (1987) to $<4 \times 10^{34}$ erg s^{-1}. Results of such a calculation are also shown in Table 1, where it can be seen that the X-ray observations put a severe limit on densities at any temperature above 4×10^6 K. While the various estimates (e.g. those based on X-ray intensities or winds) are only approximate, the density limits are not very

Table 1 Upper limits to densities of gas broadly distributed in the Galactic center "cavity"

Temperature (K)	n_{max} allowed by X-ray observations (cm^{-3})	Radiative cooling time T_c (yr)	n_{max} allowed if $T_c > 10^4$ yr (cm^{-3})	n_{max} if power radiated is $\leq 5.3 \times 10^{38}$ erg s^{-1} (cm^{-3})[b]
10^4	—	$700/n$	0.1	70
10^6	—	$1.4 \times 10^5/n$	15	100
2×10^6	1500	$1.4 \times 10^6/n$	150	200
4×10^{6}[a]	80	$3 \times 10^6/n$	300	200
10^{7}[a]	12	$7 \times 10^6/n$	700	200

[a] Gas at these temperatures would evaporate rapidly (see text).
[b] Power that might be provided by a central wind.

sensitive to errors, since in all criteria except cooling time they depend on the square root of the experimentally determined quantities. Hence, it appears that no gas of density greater than a few hundred per cubic centimeter is likely to be widespread throughout the "cavity" of radius about 1.7 pc around the center. These considerations also give an upper limit to the thermal pressure of about 10^8 K cm^{-3} (a few $\times 10^{-8}$ erg cm^{-3}), which would be marginally sufficient to confine some, but not adequate to limit expansion of most, of the clouds in the "cavity"; these clouds have equivalent pressures up to several hundred times this value due to their high densities and velocity dispersions. Any pressure within the "cavity" can have an effect on the innermost edge of the ring surrounding it. However, this hot gas pressure, if present, could not counteract gravitational forces deep enough into the ring to be very important on the scale size of most of the measurements.

5. MASS DISTRIBUTION: EVIDENCE FOR A MASSIVE BLACK HOLE?

Determination of the mass distribution in the center of the Galaxy is of obvious importance to understand the dynamics of the system and as a direct test of whether or not there is a (massive) central object. We consider three approaches to this fundamental question. First, the stellar surface brightness distribution can be converted to a mass distribution if a constant mass-to-luminosity ratio can be adopted. Second, the velocities of interstellar gas clouds can probe the mass distribution if the gas clouds rotate about the dynamical center, if the velocities can be interpreted in terms of other well-defined orbits in the center's gravitational field, or if the virial theorem can be applied in a statistical sense. Third, the mass distribution can be determined from statistics of the stellar velocities, provided that the stellar distribution has a relatively simple symmetry (such as spherical or spheroidal). We now discuss the observations in these three categories in turn.

5.1 Stellar Light Distribution

The 2-μm light distribution, discussed in Section 2.1, suggests that the mass of stars emitting in the near-infrared (that is, presumably of K and M giants) increases somewhat faster than linearly with radius [$(M_*(R) \propto R^{1.2})$] between about 1 and a few hundred parsecs (Becklin & Neugebauer 1968, Allen et al. 1983, Matsumoto et al. 1982). Inside of the cluster's core radius, which may range between 0.1 and 1 pc, mass increases much faster [$(M_*(R) \sim R^2$ to $R^3)$]. Beyond about 1 kpc, the mass inferred from the light distribution increases much slower than linearly with radius (Matsumoto et al. 1982). To assign to the so-derived distribution an absolute scale, Becklin & Neugebauer (1968) adopted a mass to total near-infrared luminosity ratio equal to that in the nucleus of M31 ($M/L \sim 3 M_\odot/L_\odot$), reasoning that the 2-$\mu$m distribution and total near-infrared luminosity in the M31 nucleus are quite similar to those in the Galactic center. On the basis of Becklin & Neugebauer's data, Sanders & Lowinger (1972) and Oort (1977) constructed more refined models, taking into account deviations from spherical symmetry as represented by the flattening of the cluster's light distributions.

5.2 Interstellar Gas Dynamics: 100 to 1000 pc

Historically, the rotating "nuclear disk" of neutral atomic hydrogen on a scale of a few hundred parsecs to about 1 kpc was the other early indicator of the mass distribution near the center. The mass distribution derived from the rotation curve of the atomic gas along the Galactic plane (Rougoor & Oort 1960, Oort 1977, Sinha 1978), with the assumption of spherical symmetry, is shown in Figure 12. From the mass derived from the 21-cm H I rotation curve, an improved M/L ratio for the stellar cluster can be derived (Oort 1977). In their best fit of the 2-μm data to the H I rotation curve, Matsumoto et al. (1982) find $M/L_{IR} = 2 M_\odot/L_\odot$. A further correction comes from the finding (Burton & Liszt 1978, Liszt & Burton 1980) that the "nuclear disk" is actually tilted with respect to the Galactic plane and has significant noncircular motions. The observed terminal velocities along the plane thus are strict upper limits to the equilibrium rotation velocities. Burton & Liszt's (1978) analysis suggests that radial and tan-

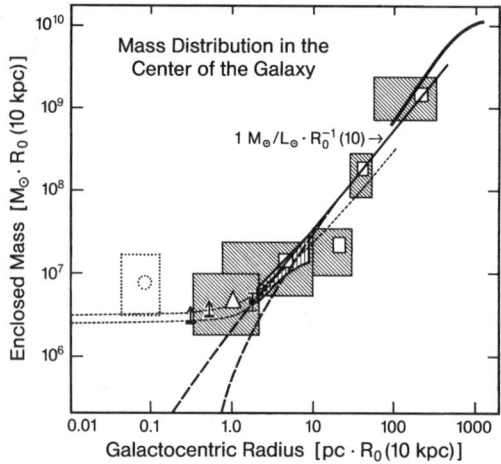

Figure 12 Composite mass distribution in the Galactic center. The heavy solid curve is the observed H I rotation along the Galactic plane (Rougoor & Oort 1960, Oort 1977, Sinha 1978). The vertically striped region near the center of the figure is the rotation curve in the neutral circumnuclear ring or disk. The black dot and the lower limits represent mass estimates from the ionized streamers at $R \leq 1.7$ pc (Serabyn & Lacy 1985, Serabyn et al. 1987). The dotted box represents a mass estimate from the fast ionized gas near IRS 16 if it is in orbit about a localized mass. The thin solid line is the mass distribution derived from the 2-μm surface brightness distribution (Becklin & Neugebauer 1968, Matsumoto et al. 1982) with an M/L ratio of 1 $M_\odot/L_\odot$ $R_0^{-1}(10)$, calibrated on the mass at a galactocentric radius of about 300 pc derived from the H I rotation curve [$v_{rot} = 180$ km s^{-1}, corrected for noncircular motions (Liszt & Burton 1980)]. The long-dashed continuations of that curve at $R < 3$ pc are for core radii of 1 and 0.1 pc. The large boxes with central quadrangles represent mass estimates from the velocity dispersion of OH-IR stars (Winnberg et al. 1985, Habing et al. 1983). The large box with central triangle represents the mass derived from the velocity dispersion of six stars emitting at 2 μm (Sellgren et al. 1987). The upper thin-dashed curve on the left represents a mass model with a central point mass of 3.0×10^6 $M_\odot$, a stellar cluster of core radius 0.1 pc, and a mass distribution derived from the 2-μm surface brightness distribution with $M/L = 0.6$ $M_\odot/L_\odot$. The lower thin-dashed curve represents a model with a 2.5×10^6 $M_\odot$ point mass, a core radius of 1 pc, and a mass distribution derived from 2-μm observations with $M/L = 1$ $M_\odot/L_\odot$ (cf. Serabyn et al. 1987). It merges into the thin solid line.

gential velocities in the "nuclear disk" or "nuclear bar" are comparable. With $v_{rot} \approx v_{tan} = 180$ km s^{-1}, the mass-to-luminosity ratio calibration at a few hundred parsecs gives $M/L_{IR} \sim 1$ $M_\odot/L_\odot$ R_0^{-1} (10). The thin solid line in Figure 12 marks the mass distribution derived from the 2-μm light distribution with this calibration of M/L_{IR} and under the assumption that M/L_{IR} is constant with radius.

5.3 10 to a Few $\times 10^2$ pc

Between about 10 and a few hundred parsecs radius, there is little certain information on the mass distribution from interstellar gas dynamics, although there are many massive molecular clouds in this region. However, these clouds do not show a strong signature of rotation and have large (50–100 km s^{-1}) noncircular velocities. In their recent analysis of extensive mapping of ^{13}CO $J = 1 \rightarrow 0$ line emission in the central few degrees (Bally et al. 1986a), Bally et al. (1986b) conclude that the rotation curve between 10 and 10^2 pc from the center is approximately flat.

5.4 *1.7 to 10 pc*

The assumption of predominantly circular motion appears to be moderately well justified for the neutral circumnuclear disk between 1.7 and 8 pc. The derived mass distribution shown in Figure 12 [$M(R) \approx 2.8 \times 10^6$ R (pc) $M_\odot$ between 2 and 4 pc] is in good agreement with several different data sets (Section 4.1), including the western arc of ionized gas at its inner edge. It also agrees well with the stellar light distribution and $M/L_{IR} \sim 0.8$–1.0 $M_\odot/L_\odot$ $R_0^{-1}(10)$. As discussed above (Section 4.1), the velocity field of the neutral material at $R > 4$ pc is more complex and shows a velocity component $<|100$ km s$^{-1}|$. If this dropoff is representative of the equilibrium rotation curve, the mass distribution outside 4 pc must increase more slowly with radius than is indicated by the 2-μm light distribution or by an isothermal cluster.

5.5 *0.1 to 1.7 pc*

Figure 11 clearly shows that inside of about 1.7 pc, the average absolute velocities increase toward the center, from about 100 km s^{-1} at $R = 1.7$ pc $R_0(10)$, to 150 km s^{-1} at 0.7 pc, to 260 km s^{-1} at 0.3 pc, and to 700 km s^{-1} at 0.1 pc. If this increase can be interpreted in terms of the virial theorem, a mass of 1–4 $\times 10^6$ $M_\odot$ $R_0(10)$ must be concentrated within 0.1 pc of IRS 16/Sgr A* (Crawford et al. 1985, Serabyn & Lacy 1985, Mezger & Wink 1986). This conclusion confirms the earlier results of Lacy et al. (1980). However, since the velocity field is dominated by a few coherent gas streamers, an orbital analysis rather than a statistical approach may be more appropriate. The western arc has already been discussed above. Other streamers do not appear to represent circular orbits. As a demonstration, Figure 13 gives the velocity centroids of the 12.8-μm [Ne II] lines along the northern arm, plotted against declination offset from the bright cloud IRS 1 (from Serabyn & Lacy 1985). The mean velocity gradient of the northern arm is about four times larger than that of the western arc, which suggests that the former comes much closer to the center ($R_{\min} \sim 0.5$ pc). There are also clear deviations from an average linear slope, indicating that the gas is moving radially. Serabyn & Lacy (1985) have demonstrated that the spatial distribution and velocity field of the northern arm can be well fitted by *simple*, single particle orbits in gravitational fields of various

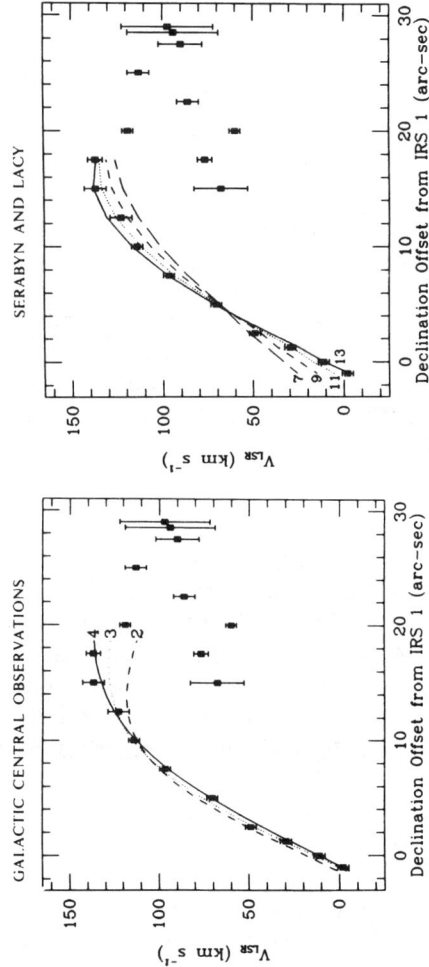

Figure 13 [Ne II] velocities in the northern arm (squares). Also shown are best-fit parabolic orbits about a point mass (models marked in units of $10^6 M_\odot$) centered 1.7″ west and 0.7″ north of IRS 1 (*left*) and best-fit orbits in an R^{-2} (isothermal) stellar cluster (*right*). Models are labeled in terms of mass (in units of $10^6 M_\odot$) enclosed within 1.7 pc of the center (from Serabyn & Lacy 1985).

forms. This streamer has a substantially smoother variation in velocity than does the western arc, and it is well represented by a parabolic or elliptical orbit in the field of a 2–4 × 10^6 $M_\odot$ $R_0(10)$ point mass (Figure 13, *left*). More specifically, Serabyn & Lacy show that a mass of at least 3.3 × 10^6 $M_\odot$ $R_0(10)$ must be contained within about 0.5 pc, the pericenter distance of the northern arm. Gravitational fields due only to a stellar cluster with isothermal ($\rho_* \propto R^{-2}$) or shallower distributions do not fit orbits of both the western arc and northern arm. The mass density of such a cluster that gives an adequate acceleration to fit the northern arm also gives a mass contained within 1.7 pc of $\geq 12 \times 10^6$ $M_\odot$, which is not consistent with the mass derived from the western arc ($4.7 \pm 1 \times 10^6$ $M_\odot$). To fit the data without a central point mass, any stellar cluster model has to have a density distribution $\rho_*(R) \propto R^{-\alpha}$, with $\alpha > 2.7$. In a similar analysis for the eastern arm and central ridge, Serabyn et al. (1987) find that a mass of at least 2–3 × 10^6 $M_\odot$ $R_0(10)$ must be contained within $\lesssim 0.2$ to 0.3 pc from the center (Figure 12). Again, isothermal or shallower cluster models do not fit the measurements. Analysis of the eastern arm and an apparent extension toward the center shows similar results, but the continuity of this streamer is not certain.

Since a number of the innermost [Ne II] clouds do not fit into clear extended portions of orbits, a convincing orbital analysis for them and the He I/H I broad-line zone is not available, although there is some evidence for short streamers even in the innermost region of the bar (Serabyn et al. 1987, Lacy et al. 1987). For these inner and less extended clouds, a statistical treatment in accordance with the virial theorem is tempting, in which case it immediately leads to the presence of a large mass concentrated near the center because of the many high velocities found there (260 km s^{-1} and greater). A more conservative, and probably more reasonable, approach is to assume that this high-velocity material has fallen toward the center from some distance and to examine the mass distribution that might give it such a large kinetic energy.

A test mass of initially zero or near-zero radial velocity falling directly inward toward the center from a distance R_0, where the rotational velocity is v_0, to R_1 would acquire a kinetic energy corresponding to velocity

$$v_0 \left[2 \frac{R_0 - R_1}{R_1} \right]^{1/2}$$

in the field of a point mass, to velocity

$$v_0 \left[2 \ln \frac{R_0}{R_1} \right]^{1/2}$$

if it falls in a stellar cluster of density distribution R^{-2}, and to velocity

$$v_0\sqrt{10}\left[1-\left(\frac{R_1}{R_0}\right)^{1/5}\right]^{1/2}$$

in a stellar cluster of density distribution $R^{-1.8}$, which is indicated for the Galactic center. According to these expressions, infall from 2 pc, where $v_0 \approx 110$ km s^{-1}, to only 0.25 pc (5") from the center would produce velocities of 411, 224, and 203 km s^{-1}, respectively, for these three cases. Thus, a stellar cluster alone would only marginally explain an observed velocity of 260 km s^{-1}. Furthermore, for this case the recently discovered [Ne II] clouds with velocities of ± 400 km s^{-1}, or even possibly 600 km s^{-1}, about 5" or more from IRS 16 (Lacy et al. 1987) (and of course the 700 km s^{-1} velocity of He I and H I lines) would require infall from distances that are so large ($\gtrsim 60$ pc for 400 km s^{-1}) that it is difficult to understand how the clouds could have both avoided loss of energy due to collision and have kept their observed relatively compact form rather than being quite extended. A central massive object provides these high velocities rather naturally, although some of them require a modest infall.

A stellar cluster with a core radius of $r_c \approx 1$ pc, as suggested by Rieke & Lebofsky (1987), provides still more severe problems for the observed velocities if a central mass is not present. First, for a standard King model with this core radius, the rotational velocity of 110 km s^{-1} at 1.7 pc implies a rotational velocity of 175 km s^{-1} at 10 pc, which is quite different from the observed values (Figure 10). Secondly, infall from a radius of 2 pc, where the rotational velocity would be 110 km s^{-1}, to 0.25 pc would give a velocity due to the gravitational potential change of only 131 km s^{-1}. Hence, the occurrence of velocities as high as 400 km s^{-1} or even 260 km s^{-1} due to infall from the disk seems quite improbable with this core radius unless, of course, there is also a large central mass.

It should be noted that the above arguments on the mass distribution also hold at least approximately if the ionized streamers represent *outward* and not inward motion. The main reason is that the *change* of velocity along a given path is used to measure the shape of the gravitational field. This change, of course, is independent of the direction of motion, as long as gravitation is the dominant force and single particle orbits (one set of initial conditions) are viable approximations to the streamers' motions.

5.6 *Stellar Dynamics*

Measurement of stellar velocities in the Galactic center come from a sample of about 55 OH-IR stars in the central ± 200 pc from work by Habing et al. (1983) and Winnberg et al. (1985), and from a sample of 5 M giants and the supergiant IRS 7 in the central 2 pc of the Galaxy from work by Sellgren et al. (1987). The OH-IR stars probably sample mass-losing asymptotic giant branch stars of mass $\geq 1\ M_\odot$ and age 10^8–10^9 yr (Habing 1987). There is some evidence that the OH-IR stars at distances ≥ 50 pc from the center [the Habing et al. (1983) sample] represent somewhat older stars in the bulge, whereas the OH-IR stars at 1–50 pc galactocentric radius in the Winnberg et al. sample (which are more concentrated toward the Galactic center and the Galactic plane) are younger and somewhat more massive.

The enclosed masses at different radii derived from the OH-IR stars by calculation of the velocity dispersion and application of the virial theorem [$M(R) = \beta \sigma_v^2 R/G$, with $\beta \sim$ 2–3 for a spherical distribution) are shown as the large boxes in Figure 12. The mass estimated from the Sellgren et al. sample is indicated by an open triangle. The surrounding boxes give the uncertainties, as estimated from the statistical error bars and by the application of different mass estimators discussed by Sellgren et al. (1987). While the stars within a few parsecs on either side of the center show no rotation and a line-of-sight velocity dispersion of about 80 km s^{-1}, the stars beyond about $R = 3$ pc show a radially increasing component of rotation in the same sense as Galactic rotation. McGinn et al. (1987) find a rotation velocity of about 50–90 km s^{-1} for the M stars at ~ 4 pc. For the OH-IR stars, Winnberg et al. (1985) and Habing et al. (1983) derive $v_{\rm rot} \sim (270 \pm 77)l$ km s^{-1} for distances between a few parsecs and 300 pc, where l is Galactic longitude (in degrees).

5.7 *Discussion of Indicators of the Mass Distribution*

The different estimators of mass shown in Figure 12 appear to give a very consistent determination of the mass distribution in the central few hundred parsecs of the Galaxy. While individual stars do not have the same velocities as do gas clouds along the same line of sight, the statistical distribution of velocities agree very well within the uncertainties with the velocities derived from the gas dynamics.

Overall, the enclosed mass decreases approximately linearly, or steeper with decreasing galactocentric radius from a few hundred to a few parsecs, representing a stellar cluster with $M/L_R \sim 1\ M_\odot/L_\odot\ R_0^{-1}(10)$. There is some (although not conclusive) evidence from low gas and stellar velocities that the mass distribution decreases less steeply than linear between about 30 and 5 pc and more steeply between a few hundred and 30 pc; such a distribution suggests that there are different populations in the central region [Galactic center cluster vs. bulge (Winnberg et al. 1985, Rieke & Lebofsky 1987)]. Inside of about 2 pc, the enclosed mass decreases more slowly with decreasing radius and may approach a constant value. The measurements are most consistent with a central point mass of 2–3 × 10^6 $M_\odot\ R_0(10)$, in combination with a near-isothermal cluster of mass 1.5–

2.5×10^6 R (pc) $M_\odot$ outside of its core radius ($0.1 \leq R_c \leq 1$ pc). Suitable models are shown as dashed lines in Figure 12. The gas and stellar dynamics measurements are also consistent with only a stellar cluster and no central point source if the stellar density increases with decreasing radius as fast or faster than $R^{-2.7}$ inside of ~ 2 pc and about as R^{-2} between 2 and 4 pc. The 2-μm stellar surface brightness distribution, however, does not show such a distribution, which makes it unlikely that a cluster with such a density distribution dominates the mass. Furthermore, even for this steep stellar distribution, the presence of gas clouds of velocity $\gtrsim 400$ km s^{-1} is difficult to understand.

The mass distribution of the stellar cluster may have to be scaled for deviations of the stellar distribution from spherical symmetry. Evidence for such deviations comes from the distribution of OH-IR stars [minor to major axis ratio $c/a = 0.7$ at $R \leq 50$ pc (Winnberg et al. 1985) and $c/a = 0.6$ for the bulge at $R \sim$ a few $\times 10^2$ pc (Habing 1987)] and probably from the 2-μm stellar surface brightness distribution (see Section 2.1). Further evidence comes from the fact that the cluster appears to have a significant amount of angular momentum (Winnberg et al. 1985, McGinn et al. 1987). For $v_{\text{rot}}/\sigma_v \sim 0.8$ at 4 pc (McGinn et al. 1987), the calculations by Ostriker & Peebles (1973) and Binney (1981) give $c/a \sim 0.7$. The corresponding correction factor for the mass enclosed within a sphere of a given radius compared with the spherical case is about 0.87. This factor is derived from the relationship between rotation velocity and mass in nonhomogeneous spheroids given by Schmidt (1965).

Therefore, if the 2.2-μm light distribution is a measure of the total mass distribution of all stars in the center, the central mass concentration could be in the form of a supermassive star (should such an object be stable) or, more likely, in the form of a massive black hole. The presence of a $\sim 10^6$-$M_\odot$ black hole is *by far sufficient but not necessary* to explain the current UV luminosity of the Galactic center. The required mass accretion rate for a massive black hole is about 10^{-5} $M_\odot$ yr^{-1} $R_0^2(10)$, a value that is small compared with the estimates of current gaseous accretion in the central parsec [10^{-3}–10^{-2} $M_\odot$ yr^{-1} $R_0^2(10)$], and probably (Ozernoy 1979, Gurzadyan & Ozernoy 1981, Lacy et al. 1982) but not necessarily (Rees 1982) small compared with the estimated average tidal disruption rate of stars. Ozernoy (1987) stresses that if the 511-keV line comes from a black hole, then the mass could not be greater than a few hundred solar masses.

The estimates of the mass distribution in the Galactic center *could be misleading* for the following reason. Each estimator of mass requires certain basic assumptions that may be challenged. The conversion of surface brightness to mass not only requires calibration of the M/L ratio by other methods (note that the M/L_{IR} ratio discussed above is strictly an upper limit to M/L because of additional radiation in the visible, UV, or far-IR region), but also relies on a reasonably constant M/L ratio. This assumption may be hazardous because of the possible population changes mentioned above, and because there seem to be sources of 2-μm radiation in the very center that are not part of the M and K giant population. To obtain the apparent concentration of mass toward the center from stars alone, it appears that a substantial component of the stellar population would need to be both heavier and less luminous than average. In the case of the dynamics measurements, the assumption of gravitation as the dominant force is essential. This assumption is more likely to hold for the stars than the gas, but a straightforward analysis of the stellar velocities also requires that the stellar distribution function is not explicitly a function of angular momentum (as would be the case for a triaxial distribution, for example). For the gas dynamics, forces other than gravitation may be important. This concern is especially pertinent for the motions of the ionized gas streamers, of the bar region, and of the high-velocity He I/H I gas, which are the most important for the black hole question. While the broad-line zone near IRS 16 would be the most definite indication of a massive point source if the motions were known to be due to rotation or inflow of gas, the current data do not indicate rotation (Geballe et al. 1987) and the gas may be outflowing (Geballe et al. 1984, Ozernoy 1984a). Outflow also must be considered as an alternative possibility for the dynamics of the ionized gas streamers. This possibility not only would change the conclusions about the mass distribution, but also frictional effects would be more misleading than in the case of infall. For the [Ne II] clouds at $R < 2$ pc, magnetic forces could strongly affect the "single-orbit" analysis of Serabyn & Lacy (1985), and they could also possibly produce the observed high-velocity gas (Heyvaerts et al. 1987). The possibility of such magnetic effects needs to be studied and further analyzed.

6. IRS 16 AND Sgr A*: THE CENTRAL 0.1 pc

Figure 14 is a 2.23-μm map (Forrest et al. 1986) of the central 10" (0.5 pc) centered on the IRS 16 complex and the compact radio source Sgr A* (filled box). As discussed in Section 3, most of the ionizing luminosity of Sgr A West may originate from this 0.5-pc diameter region, and the dynamical center of the gas motions is also centered within a few arcseconds of IRS 16 (Serabyn & Lacy 1985). On the map in Figure 14 [and on similar maps by Storey & Allen (1983) and Allen & Sanders (1986)], IRS 16 is resolved into about three or four individual components. Becklin et al. (1987) report from recent lunar occultation measurements that IRS

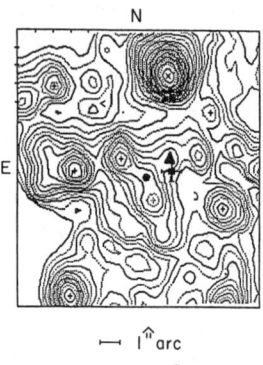

Figure 14 A plot of the equal-surface-brightness contours from the 2.23-μm image (from Forrest et al. 1986). Contours are shown every factor of 1.20 (0.2 mag) in surface brightness, with every other contour dotted. The display has been rectified to portray the true appearance on the sky, with tick marks 1″ apart shown in the upper left corner. An "×" marks IRS 7, and "+"s mark various 2-μm sources. The filled box represents the calculated position of the Sgr A* compact radio source; the ±0.5″ error bars are an estimate of the positional uncertainty. A heavy dot marks IRS 16 Center, and a filled triangle denotes IRS 16 NW.

16 SW and Center are compact ("star"-like), while IRS 16 NE is resolved at the 0.5″ scale.

The primary candidate for a possible massive black hole is the radio point source, analogous to the cases of extragalactic compact radio sources (Lynden-Bell & Rees 1971). Its linear size is ≤ 20 AU (Lo et al. 1985), and its radio luminosity is about 2×10^{34} erg s^{-1}, which makes it a unique object within the Galaxy (unlike pulsars, binary stellar sources, or young supernovae). Lo et al. (1985) also report that the compact source is elongated approximately east-west, with a major to minor axis ratio of about 2:1. Its characteristics are very similar to the nuclear sources in M81 and M104 (Lo 1986). Backer & Sramek (1982, 1987) have measured the proper motion of Sgr A* with respect to extragalactic background quasars. They find that Sgr A* shows the proper motion expected from the solar motion about the center. They find that any intrinsic motion of Sgr A* within the center is less than 40 km s^{-1} for $R_0 = 8$ kpc, which makes it likely that Sgr A* is near the Galactic center and is a relatively massive object. The obvious question then is, Does an infrared counterpart of the radio source exist, and if so, does it satisfy the characteristics of an accreting $\approx 10^6$-$M_\odot$ black hole of luminosity $\approx 10^7 L_0$? The luminosities of IRS 16 Center (dot in center of Figure 14) and of IRS 16 NE and SW, as estimated from the extrapolation of their dereddened infrared flux distribution into the ultraviolet, are consistent with high intrinsic luminosities [$L \gtrsim 10^6 L_0$ (Henry et al. 1984, Serabyn & Lacy 1985)]. However, it is now clear that Sgr A* is not coincident with IRS 16 Center, NE, or SW, but rather is displaced about 1″ west of IRS 16 Center (Figure 14; Storey & Allen 1983, Allen & Sanders 1986). On the basis of this result, Allen & Sanders (1986) conclude that Sgr A* cannot be a massive black hole. They also state that all other intrinsic characteristics of the radio point source, as well as the characteristics of the 511-keV emission (assuming that the positron-annihilation radiation is associated with Sgr A*), can easily be accounted for by a small ($\approx 10^2 \, M_\odot$) black hole (see also Ozernoy 1984b, 1987). Allen & Sanders (1986) and Rieke (1982) propose that IRS 16 is a compact cluster of late-O and early-B stars formed in a burst $\lesssim 10^7$ yr ago.

An alternate view has been proposed by Stein & Forrest (1986), who suggest that Sgr A* is close to and may coincide with another blue source that they call IRS 16 NW (triangle in Figure 14). IRS 16 NW appears coincident within the error bars, with one or two faint sources visible on red (0.9-μm) CCD images (Storey & Allen 1983, Biretta et al. 1982, Ricker et al. 1982). Dereddening the flux distribution of IRS 16 NE then leads to the conclusion that this source, and hence Sgr A*, could be very luminous, a result that would be consistent with the massive black hole hypothesis (Stein & Forrest 1986). However, a still more recent determination of the position of Sgr A* from matching Bα and radio images of the central region places this source at a position 0.3" farther south than the position shown in Figure 14, and not coincident with any of the near-infrared sources (Forrest et al. 1987).

An argument against Sgr A* being at the center of mass is that the highest velocity Ne II clouds appear to be centered on a region about 4" south of Sgr A* (Lacy et al. 1979; cf. with maps of Lacy et al. 1987). A large point mass this far away would allow the small measured transverse velocity of Sgr A* noted above to occur with some ($\sim 15\%$) probability, even though it is not at the center. It is also clear that the nature of the IRS 16 complex is far from being understood. Whether this source and its surrounding gas could come from a star recently disrupted by the possible black hole is probably worth consideration.

Mass loss from a star or stars has been suggested as a source of the high-velocity ($\gtrsim 700$ km s^{-1}) gas near IRS 16 [or its acceleration (Geballe et al. 1987)], of the origin and dynamics of the ionized clouds and filaments, or of excitation in some of the surrounding neutral material (Gatley et al. 1984, 1986). Qualitatively, such effects appear attractive and may occur. For the majority of the gas, it is difficult to explain on this basis either the morphology or the amounts of material and accelerations observed as due to outflow from presently *known* types of stars. Stars that emit gas with very high velocities are generally very hot and accelerate the gas by means of radiation pressure. Wolf-Rayet stars are perhaps the most powerful example; they and other hot stars emit as much as a few times $10^{-5} \, M_\odot$ yr^{-1} (Chiosi & Maeder 1986). Hence, to produce the outflow of $\gtrsim 10^{-3} \, M_\odot$ yr^{-1} necessary to explain the gas around IRS 16 or to replenish the gas dissipated in the cavity, a total of 30 to 100 such stars would be needed.

Stars of this type would also produce enough radiation to ionize the observed clouds and to dominate the ultraviolet luminosity. Their UV radiation temperature could be higher than the maximum of 35,000 K allowed by the observed state of ionization. Such a collection of stars would emit gas almost isotropically and more or less continuously, even if a single star emitted in a more jetlike pattern or in short bursts. This model would then not produce the distribution of cloud shapes nor the rather well-defined streamers observed.

A further argument against general acceleration of clouds by a central source of wind is that ionized cloud velocities are systematically higher near the center than farther from it, which is the reverse of what is expected from a central source of accelerating wind. It is not easy to attribute to friction the observed slowing down of clouds if they are moving outward (under continued acceleration by winds from the center). Since the cloud dimensions are about one tenth of the distance from the center to the outer radius of the cavity, the density of the material that is to provide frictional deceleration must be at least about 10% that of the clouds (or about 10^3 cm^{-3}) to have an important effect. Section 4.6 indicates that a wide distribution of gas of the required density is quite unlikely. Since there does not appear to be enough intercloud gas to provide the required friction, the high cloud velocities close to the center and the lower ones farther away would be inconsistent with acceleration by an outflowing wind.

Outflow and additional effects from as yet unknown stars or other central "engines" may, of course, be a possible explanation for the activity at the Galactic center, and several recent scenarios have been proposed by Rees (1987), Ozernoy (1984b, 1987), and Heyvaerts et al. (1987).

7. SUMMARY

STARS The density of stars emitting near 2 μm increases with $R^{-1.8}$ down to a radius of about 1 to 0.1 pc: At the latter radius, the stellar density would be $>10^6$ $M_\odot$ pc^{-3} (for a spherical distribution). If these stars are primarily M and K giants, their mass-to-luminosity ratio is ≈ 1 $M_\odot/L_\odot$ $R_0^{-1}(10)$. In addition, radiation of $1-3 \times 10^7$ $L_\odot$ $R_0^2(10)$ emerges as thermal dust emission in the far-infrared, but it probably originates as ultraviolet emission. The UV radiation could emerge from a compact cluster of late-O and early-B stars centered on or near the IRS 16 complex or from a single exciting source. In the former case, the mass-to-luminosity ratio of the whole cluster is ≈ 0.2 $M_\odot/L_\odot$, which suggests that a burst of star formation took place $\lesssim 10^7$ yr ago.

INTERSTELLAR MATTER   Most of the neutral interstellar gas and dust in the central 10 pc originates in a thin, highly clumped disk whose thickness is probably supported by its large turbulence. The disk rotates about the center, appears to be warped, and shows a change of the position angle of its major axis with radius. It has a well-defined central hole of radius ~ 2 pc, and its inner edge appears photoionized by the UV radiation from the center. The mean gas and dust density in the central cavity is about 10 to 100 times lower than in the surrounding circumnuclear disk. The large amount ($\sim 10^4$ $M_\odot$) of unusually warm ($T = 300$–2000 K) and dense ($n \approx 10^4$–10^5 cm^{-3}) molecular and atomic gas may be due to (*a*) shock heating resulting from frequent cloud-cloud collisions and from mass outflow from the center and/or (*b*) photoelectric heating by the UV radiation from the center.

Perhaps the most likely explanation of the ionized gas streamers inside the central cavity forming the "minispiral" on radio continuum maps is that they are gas filaments falling from the edge of the neutral disk toward the center. The streamers are probably externally heated and photoionized by the general UV radiation field in the center. Magnetic fields could be important for the morphology and dynamics of the streamers. In addition, there may be hot ($\sim 10^6$ K) gas of density $\lesssim 10^2$ cm^{-3} and relativistic electrons filling the cavity. One remarkable feature is the "broad-line" region of velocity width 1500 km s^{-1} and of diameter ≈ 0.1 pc centered on the IRS 16/Sgr A* complex. These large velocities could be due to infall onto a massive object or an outflow of about 3×10^{-3} $M_\odot$ yr^{-1} from an unusual star (or stars) near the center. The ≥ 10-pc scale radio continuum protrusions emanating from the center and possibly the ≥ 200-pc radio lobes perpendicular to the Galactic plane may also be evidence for gas expanding from the center.

The distribution and physical state of interstellar matter in the Galactic center is a nonequilibrium configuration of short ($t \lesssim 10^5$ yr) lifetime. The central cavity and the appearance of the neutral circumnuclear disk may have been caused by a violent explosion ($E \approx 10^{51}$ erg) about 10^5 yr ago. The clumpiness, turbulence, and line emission of the ionized streamers and of the circumnuclear disk may also be explained by a dynamical equilibrium in which rotational energy is fed into turbulent energy. Neutral and ionized gas should currently accrete toward the center at a rate of 10^{-3}–10^{-2} $M_\odot$ yr^{-1}. The Galactic center thus may be temporarily in a "quiescent" state of accretion.

THE CASE FOR AND AGAINST A MASSIVE BLACK HOLE   The current evidence for a massive ($\approx 10^6$ $M_\odot$) black hole from the observed radiation phenomena and the gas and stellar dynamics is substantial but not fully convincing.

Radiation phenomena indicating a black hole include the presence of the compact radio source Sgr A* and the unusual, time-variable 511-keV electron-positron annihilation radiation. However, these phenomena and the possible UV radiation associated with the hole's accretion disk probably do not require a very large mass ($\approx 10^2\ M_\odot$). Furthermore, the UV characteristics of the Galactic center can be accounted for by less exotic explanations, such as time-variable stellar formation. The case for a *massive* black hole is mainly supported by the gas dynamics, which shows an increase of velocities inside of ≈ 2 pc. Both the magnitude of gas velocity and the apparent orbital motions make a strong case for a large central mass concentration, unless a mechanism or mechanisms other than the gravitational field can be found that provide an appropriate pattern of acceleration for the gas.

ACKNOWLEDGMENTS

We are grateful to D. Backer, E. Becklin, T. Geballe, R. Güsten, M. Gwinn, A. Harris, W. Hillebrandt, D. Jaffe, J. Lacy, M. Leventhal, K. Y. Lo, J. Lugten, J. Middleditch, M. Morris, G. Rieke, K. Sellgren, E. Serabyn, G. Stacey, J. Stutzki, V. Schönfelder, J. van Gorkom, M. Wright, and F. Yusef-Zadeh for permission to cite and display data prior to publication. We also thank E. Becklin, J. Lugten, L. Ozernoy, G. Rieke, K. Sellgren, and E. Serabyn for helpful comments and discussions.

This work has been supported in part by NASA grants no. NAG 2-208 and NGR 05-003-272, and by the Presidential Young Investigators Award to R. Genzel (NSF grant no. AST83-51381 and Rockwell International).

Literature Cited

Aitken, D. K., Roche, P. F., Bailey, J. A., Briggs, G. P., Hough, J. H., Thomas, J. A. 1986. *MNRAS* 218: 363
Allen, D. A. 1987. See Backer 1987. In press
Allen, D. A., Hyland, A. R., Jones, T. J. 1983. *MNRAS* 204: 1145
Allen, D. A., Sanders, R. M. 1986. *Nature* 319: 191
Armstrong, J. T., Barrett, A. H. 1985. *Ap. J. Suppl.* 57: 535
Backer, D., ed. 1987. *Proceedings of the Symposium on The Galactic Center in Honor of Charles H. Townes.* New York: Am. Inst. Phys. In press
Backer, D., Sramek, R. 1982. *Ap. J.* 260: 512
Backer, D., Sramek, R. 1987. See Backer 1987. In press
Bailey, M. E. 1980. *MNRAS* 190: 217
Bally, J., Stark, A. A., Wilson, R. W., Henkel, C. 1986a. *Ap. J. Suppl.* In press
Bally, J., Stark, A. A., Wilson, R. W.,

Henkel, C. 1986b. *Ap. J.* In press
Becklin, E. E., Neugebauer, G. 1968. *Ap. J.* 151: 145
Becklin, E. E., Neugebauer, G. 1975. *Ap. J. Lett.* 200: L71
Becklin, E. E., Matthews, K., Neugebauer, G., Willner, S. P. 1978a. *Ap. J.* 219: 121
Becklin, E. E., Matthews, K., Neugebauer, G., Willner, S. P. 1978b. *Ap. J.* 220: 831
Becklin, E. E., Gatley, I., Werner, M. W, 1982. *Ap. J.* 258: 134
Becklin, E. E., et al. 1987. In preparation
Binney, J. 1981. In *The Structure and Evolution of Normal Galaxies,* ed. S. M. Fall, D. Lynden-Bell, p. 55. Cambridge: Cambridge Univ. Press
Biretta, J. A., Lo, K. Y., Young, P. J. 1982. *Ap. J.* 262: 578
Blanco, V. M., Blanco, B. M. 1985. *Mem. Soc. Astron. Ital.* In press
Blitz, L. 1987. *Proc. NATO Summer School*

on *The Galaxy*, ed. R. Carswell, G. Gilmore. Dordrecht: Reidel. In press

Brown, B. L., Leventhal, M. 1987. *Ap. J.* In press

Brown, R. L. 1982. *Ap. J.* 262: 110

Brown, R. L., Liszt, H. S. 1984. *Ann. Rev. Astron. Astrophys.* 22: 223

Brown, R. L., Johnston, K. J., Lo, K. Y. 1981. *Ap. J.* 250: 155

Burton, W. B., Liszt, H. S. 1978. *Ap. J.* 225: 815

Bussard, R. W., Ramaty, R., Drachman, R. J. 1979. *Ap. J.* 228: 928

Capps, R. W., Knacke, R. F. 1976. *Ap. J.* 270: 76

Chiosi, C., Maeder, A. 1986. *Ann. Rev. Astron. Astrophys.* 24: 329

Crawford, M. K., Genzel, R., Harris, A. I., Jaffe, D. T., Lacy, J. H., et al. 1985. *Nature* 315: 467

Dalgarno, A., McCray, R. A. 1972. *Ann. Rev. Astron. Astrophys.* 10: 375

Dent, W. A., Werner, M. W., Gatley, I., Becklin, E. E., Hildebrand, R. H., et al. 1982. See Riegler & Blandford, p. 33

Draine, B. T., Roberge, W. G. 1985. *Ap. J.* 282: 491

Drapatz, S., Haser, L., Hofmann, R., Oda, N., Wilczek, R. 1985. *Adv. Space Res.* 5(3): 7

Dyck, H. M., Capps, R. W., Beichman, C. A. 1974. *Ap. J. Lett.* 188: L103

Ekers, R. D., Goss, W. M., Schwarz, U. J., Downes, D., Rogstad, D. H. 1975. *Astron. Astrophys.* 43: 159

Ekers, R. D., van Gorkom, J. H., Schwarz, U. J., Goss, W. M. 1983. *Astron. Astrophys.* 122: 143

Evans, N. J. 1987. See Backer 1987. In press

Feast, M. W. 1987. *Proc. NATO Summer School on The Galaxy*, ed. R. Carswell, G. Gilmore. Dordrecht: Reidel. In press

Forrest, W. J., Pipher, J. L., Stein, W. A. 1986. *Ap. J. Lett.* 301: L49

Forrest, W. J., Shore, M. A., Pipher, J. L., Woodward, C. E. 1987. See Backer 1987. In press

Frenk, C. S., White, S. D. 1982. *MNRAS* 198: 173

Fukui, Y., Churchwell, E. 1987. See Backer 1987. In press

Fukui, Y., Iguchi, T., Karifu, N., Chicada, Y., Morimoto, M., et al. 1977. *Publ. Astron. Soc. Jpn.* 29: 643

Fukui, Y., Ogawa, H., Deguchi, S. 1983. *Ap. J. Lett.* 275: L55

Gatley, I. 1982. See Riegler & Blandford 1982, p. 25

Gatley, I. 1987. See Backer 1987. In press

Gatley, I., Becklin, E. E., Werner, M. W., Harper, D. A. 1978. *Ap. J.* 220: 822

Gatley, I., Jones, T. J., Hyland, A. R.,
Beattie, D. H., Lee, T. J. 1984. *MNRAS* 210: 565

Gatley, I., Jones, T. J., Hyland, A. R., Wade, R., Geballe, T. R., Krisciunas, K. L. 1986. *MNRAS* 222: 299

Geballe, T. R., Krisciunas, K. L., Lee, T. J., Gatley, I., Wade R., et al. 1984. *Ap. J.* 284: 118

Geballe, T. R., Wade, R., Krisciunas, K. L., Gatley, I., Bird, M. C. 1987. *Ap. J.* In press

Genzel, R. 1987. *Proc. NATO Summer School on The Galaxy*, ed. R. Carswell, G. Gilmore. Dordrecht: Reidel. In press

Genzel, R., Watson, D. M., Townes, C. H., Lester, D. F., Dinerstein, H., et al. 1982. See Riegler & Blandford, p. 72

Genzel, R., Watson, D. M., Townes, C. H., Dinerstein, H. L., Hollenbach, D., et al. 1984. *Ap. J.* 276: 551

Genzel, R., Watson, D. M., Crawford, M. K., Townes, C. H. 1985. *Ap. J.* 297: 766

Gezari, D. Y., Tresch-Fienberg, R., Fazio, G. G., Hoffman, W. F., Gatley, I., et al. 1985. *Ap. J.* 299: 1007

Glass, I. S., Feast, M. W. 1982. *MNRAS* 198: 199

Gurzadyan, V. G., Ozernoy, L. M. 1981. *Astron. Astrophys.* 95: 39

Güsten, R., Downes, D. 1980. *Astron. Astrophys.* 87: 6

Güsten, R., Walmsley, C. M., Pauls, T. 1981. *Astron. Astrophys.* 103: 197

Güsten, R., Walmsley, C. M., Ungerechts, H., Churchwell, E. 1985. *Astron. Astrophys.* 142: 381

Güsten, R., Genzel, R., Wright, M. C. H., Jaffe, D. T., Stutzki, J., Harris, A. I. 1987a. *Ap. J.* In press

Güsten, R., Serabyn, E., Evans, N. J., Downes, D. 1987b. Submitted for publication

Habing, H. J. 1987. *Proc. NATO Summer School on The Galaxy*, ed. R. Carswell, G. Gilmore. Dordrecht: Reidel. In press

Habing, H. J., Olnon, F. M., Winnberg, A., Matthews, H. E., Baud, B. 1983. *Astron. Astrophys.* 128: 230

Hall, D. N. B., Kleinmann, S. G., Scoville, N. Z. 1982. *Ap. J. Lett.* 262: L53

Harris, A. I., Jaffe, D. T., Silber, M., Genzel, R. 1985. *Ap. J. Lett.* 294: L93

Harvey, P. M., Campbell, M. F., Hoffmann, W. F. 1976. *Ap. J. Lett.* 205: L69

Harvey, P. M., Campbell, M. F., Hoffmann, W. F. 1981. *Ap. J. Lett.* 241: L183

Henry, J. P., DePoy, D. L., Becklin, E. E. 1984. *Ap. J. Lett.* 285: L27

Herter, T., Houck, J. R., Shure, M., Gull, G. E., Graf, P. 1984. *Ap. J. Lett.* 287: L15

Heyvaerts, J., Pudritz, R. E., Norman, C. 1987. See Backer 1987. In press

Hildebrand, R. H., Whitcomb, S. E.,

Winston, R., Stiening, R. F., Harper, D. A., Moseley, S. H. 1978. *Ap. J.* 219: L101
Hillebrandt, W., Mair, G., Ziegert, W. 1986. *Proc. IAP Rencontre Nucl. Astrophys.*, 2nd, ed. J. Audouze, M. Cassé, J. P. Chièze. Dreux, Fr: Ed. Frontières
Hillebrandt, W., Thielemann, F. K., Langer, N. 1987. *Ap. J.* In press
Hiromoto, N., Maihara, T., Mizutani, M. K., Takami, H., Shibai, H., Okuda, H. 1984. *Astron. Astrophys.* 139: 309
Ho, P. T. P., Jackson, J. M., Barrett, A. H., Armstrong, J. T. 1985. *Ap. J.* 288: 575
Ishiguro, M., Fomalont, E., Morita, K. I., Kasugo, T., Kanzawa, T., et al. 1986. Preprint
Jackson, J. M., Ho, P. T. P., Barrett, A. H. 1987. See Backer 1987. In press
Jacobson, A. S. 1982. See Riegler & Blandford 1982, p. 123
Kaifu, N., Hayashi, M., Inatami, J., Gatley, I. 1987. See Backer 1987. In press
Kassim, N. E., Erickson, W. C., La Rosa, T. N. 1987. See Backer 1987. In press
Kawai, N., Fenimore, E. E., Middleditch, J., Cruddace, R., Fritz, G., et al. 1987. Preprint
Lacy, J. H., Baas, F., Townes, C. H., Geballe, T. R. 1979. *Ap. J. Lett.* 227: L17
Lacy, J. H., Townes, C. H., Geballe, T. R., Hollenbach, D. J. 1980. *Ap. J.* 241: 132
Lacy, J. H., Townes, C. H., Hollenbach, D. J. 1982. *Ap. J.* 262: 120
Lacy, J. H., Lester, D. F., Arens, J. F., Peck, M. C., Gaalema, S. 1987. See Backer 1987. In press
Lebofsky, M. J. 1979. *Astron. J.* 84: 324
Lebofsky, M. J., Rieke, G. R., Tokunaga, A. T. 1982a. *Ap. J.* 263: 736
Lebofsky, M. J., Rieke, G. R., Deshpande, M. R., Kemp, J. C. 1982b. *Ap. J.* 263: 672
Lester, D. F., Werner, M. W., Storey, I. W. V., Watson, D. M., Townes, C. H. 1981. *Ap. J. Lett.* 248: L109
Lester, D. F., Joy, M., Harvey, P. M., Ellis, H. B. 1987. See Backer 1987. In press
Leventhal, M. 1987. *Proc. Tex. Symp. Relativistic Astrophys., 13th.* In press
Leventhal, M., MacCallum, C. J. 1982. See Riegler & Blandford 1982, p. 132
Lingenfelter, R. E., Ramaty, R. 1982. See Riegler & Blandford 1982, p. 148
Liszt, H. S. 1985. *Ap. J. Lett.* 293: L65
Liszt, H. S., Burton, W. B. 1980. *Ap. J.* 236: 779
Liszt, H. S., van der Hulst, J. M., Burton, W. B., Ondrechen, M. P. 1983. *Astron. Astrophys.* 126: 341
Liszt, H. S., Burton, W. B., van der Hulst, J. M. 1985. *Astron. Astrophys.* 142: 237
Lo, K. Y. 1986. *Science* 233: 1394
Lo, K. Y., Claussen, M. J. 1983. *Nature* 306: 647
Lo, K. Y., Backer, D. C., Ekers, R. D., Kellermann, K. I., Reid, M. J., Moran, J. M. 1985. *Nature* 315: 124
Lugten, J. B., Genzel, R., Crawford, M. K., Townes, C. H. 1986. *Ap. J.* 306: 691
Lugten, J. B., Harris, A. I., Stacey, G. J., Genzel, R., Townes, C. H. 1987. See Backer 1987. In press
Lynden-Bell, D., Rees, M. J. 1971. *MNRAS* 152: 461
MacCallum, C. J., Huters, A. F., Stang, P. D., Leventhal, M. 1987. *Ap. J.* In press
Mahoney, W. A., Ling, J. C., Jacobson, A. S., Lingenfelter, R. E. 1982. *Ap. J.* 262: 742
Mahoney, W. A., Ling, J. C., Wheaton, W. A., Jacobson, A. S. 1984. *Ap. J.* 286: 578
Matsumoto, T., Hayakawa, S., Koizumi, H., Murakami, H., Uyama, K., et al. 1982. See Riegler & Blandford 1982, p. 48
Matteson, J. L. 1982. See Riegler & Blandford 1982, p. 109
McGinn, M. T., Sellgren, K., Becklin, E. E., Hall, D. N. B., Gatley, I. 1987. See Backer 1987. In press
Mezger, P. G., Wink, J. E. 1986. *Astron. Astrophys.* 157: 252
Mezger, P. G., Chini, R., Kreysa, E., Gemünd, H. P. 1986. *Astron. Astrophys.* In press
Miley, G. 1980. *Ann. Rev. Astron. Astrophys.* 18: 165
Morris, M., Yusef-Zadeh, F. 1987. See Backer 1987. In press
Oort, J. H. 1977. *Ann. Rev. Astron. Astrophys.* 15: 295
Oort, J. H., Plaut, L. 1975. *Astron. Astrophys.* 41: 71
Ostriker, J. P., Peebles, P. J. E. 1973. *Ap. J.* 186: 467
Ozernoy, L. M. 1979. In *The Large-Scale Characteristics of the Galaxy, IAU Symp. No. 84*, ed. B. Burton, p. 395. Dordrecht: Reidel
Ozernoy, L. M. 1984a. *Astron. Tsirk.* 1342: 1
Ozernoy, L. M. 1984b. *Astron. Tsirk.* 1349: 1
Ozernoy, L. M. 1987. See Backer 1987. In press
Quinn, P. J., Sussman, G. J. 1985. *Ap. J.* 288: 377
Ramaty, R., Lingenfelter, R. E. 1987. See Backer 1987. In press
Rees, M. 1982. See Riegler & Blandford 1982, p. 166
Rees, M. 1987. See Backer 1987. In press
Reid, M. J., Schneps, M. H., Moran, J. M., Gwinn, C. R., Genzel, R., et al. 1987. In *Star Forming Regions, IAU Symp. No. 115*, ed. M. Peimbert. Dordrecht: Reidel
Ricker, G. R., Bautz, M. W., DePoy, D. L., Meyer, S. S. 1982. See Riegler & Blandford 1982, p. 97
Riegler, G. R., Blandford, R. D., eds. 1982.

The Galactic Center, AIP Conf. Proc. No. 83. New York: Am. Inst. Phys.
Riegler, G. R., Ling, J. C., Mahoney, W. A., Wheaton, W. A., Willett, J. B., et al. 1981. *Ap. J. Lett.* 248: L13
Rieke, G. H. 1982. See Riegler & Blandford 1982, p. 194
Rieke, G. H., Lebofsky, M. J. 1987. See Backer 1987. In press
Rieke, G. H., Telesco, C. M., Harper, D. A. 1978. *Ap. J.* 220: 556
Rougoor, G. W., Oort, J. H. 1960. *Proc. Natl. Acad. Sci. USA* 46: 1
Sanders, R. H., Lowinger, T. 1972. *Ap. J.* 77: 292
Sandqvist, Aa. 1974. *Astron. Astrophys.* 33: 413
Sandqvist, Aa. 1982. See Riegler & Blandford 1982, p. 12
Sandqvist, Aa. 1987. See Backer 1987. In press
Sandqvist, Aa., Wooten, A., Loren, R. B. 1985. *Astron. Astrophys.* 152: L25
Sandqvist, Aa., Karlsson, R., Whiteoak, J. B., Gardner, F. F. 1987. See Backer 1987. In press
Schmidt, M. 1965. In *Galactic Structure (Stars and Stellar Systems,* Vol. 5), ed. A. Blaauw, M. Schmidt, p. 513. Chicago: Univ. Chicago Press
Seiradakis, J. H., Lasenby, A., Yusef-Zadeh, F., Wielebinski, R., Klein, U. 1985. *Nature* 317: 697
Sellgren, K., Hall, D. N. B., Kleinmann, S. G., Scoville, N. Z. 1987. *Ap. J.* In press
Serabyn, E. 1985. PhD thesis. Univ. Calif., Berkeley
Serabyn, E. 1987. *Proc. NATO School on Astrophysical Jets and Their Engines.* Dordrecht: Reidel. In press
Serabyn, E., Güsten, R. 1986. *Astron. Astrophys.* 161: 334
Serabyn, E., Lacy, J. H. 1985. *Ap. J.* 293: 445
Serabyn, E., Güsten, R., Wink, J. E., Walmsley, C. M., Zylka, R. 1986. *Astron. Astrophys.* 169: 85
Serabyn, E., Lacy, J. H., Townes, C. H. 1987. Submitted for publication
Share, G. H., Kinzer, R. L., Kurfess, J. D., Forrest, D. J., Chupp, E. L., Rieger, E.1985a. *Ap. J. Lett.* 292: L61
Share, G. H., Kinzer, R. L., Chupp, E. L., Forrest, D. J., Rieger, E. 1985b. *Proc. Int. Cosmic Ray Conf., 19th, La Jolla, Calif.* 1: 353
Sinha, R. P. 1978. *Astron. Astrophys.* 69: 227
Sofue, Y., Handa, T. 1984. *Nature* 310: 568
Stein, W. A., Forrest, W. J. 1986. *Nature* 323: 232
Storey, J. W. V., Allen, D. A. 1983. *MNRAS* 204: 1153
Townes, C. H., Lacy, J. H., Geballe, T. R., Hollenbach, D. J. 1983. *Nature* 301: 661
Tsuboi, M., Inoue, M., Handa, T., Tabara, H., Kato, T., et al. 1986. *Astron. J.* 92: 819
Uchida, Y., Shibata, K., Sofue, Y. 1985. *Nature* 317: 699
van Gorkom, J. H., Schwarz, U. J., Bregman, J. D. 1985. In *The Milky Way Galaxy, IAU Symp. No. 106,* ed. H. van Woerden, W. B. Burton, R. J. Allen. Dordrecht: Reidel
van Gorkom, J. H., Schwarz, U. J., Bregman, J. D. 1987. In preparation
von Ballmoos, P., Diehl, R., Schönfelder, V. 1987. *Ap. J.* In press
Wade, R., Geballe, T. R., Krisciunas, K., Gatley, I., Bird, M. C. 1987. *Ap. J.* In press
Watson, D. M., Storey, J. W. V., Townes, C. H., Haller, E. E. 1980. *Ap. J. Lett.* 241: L43
Watson, M. G., Willingale, R., Grindlay, J. E., Hertz, P. 1981. *Ap. J.* 250: 142
Webber, W. R., Schönfelder, V., Diehl, R. 1986. *Nature* 323: 692
Winnberg, A., Baud, B., Matthews, H. E., Habing, H. J., Olnon, F. M. 1985. *Ap. J. Lett.* 291: L45
Wollman, E., Smith, H. A., Larson, H. P. 1982. *Ap. J.* 258: 506
Wright, M. C. H., Genzel, R., Güsten, R., Jaffe, D. T. 1987. See Backer 1987. In press
Yusef-Zadeh, F., Morris, M., Chance, D. 1984. *Nature* 310: 557
Yusef-Zadeh, F., Morris, M., Slee, O. B., Nelson, G. J. 1986. *Ap. J. Lett.* 300: L47
Yusef-Zadeh, F. 1986. PhD thesis. Columbia Univ., New York, N.Y.
Zurek, W. H. 1985. *Ap. J.* 289: 603

EXISTENCE AND NATURE OF DARK MATTER IN THE UNIVERSE

Virginia Trimble

Astronomy Program, University of Maryland, College Park, Maryland 20742, and Department of Physics, University of California, Irvine, California 92717

1. HISTORICAL INTRODUCTION AND THE SCOPE OF THE PROBLEM

The first detection of nonluminous matter from its gravitational effects occurred in 1844, when Friedrich Wilhelm Bessel announced that several decades of positional measurements of Sirius and Procyon implied that each was in orbit with an invisible companion of mass comparable to its own. The companions ceased to be invisible in 1862, when Alvan G. Clark turned his newly-ground $18\frac{1}{2}''$ objective toward Sirius and resolved the 10^{-4} of the photons from the system emitted by the white dwarf Sirius B. Studies of astrometric and single-line spectroscopic binaries are the modern descendants of Bessel's work.

A couple of generations later, data implying nonluminous matter on two very different scales surfaced almost simultaneously. First, Oort (498, 499) analyzed numbers and velocities of stars near the Sun and concluded that visible stars fell shy by 30–50% of adding up to the amount of gravitating matter implied by the velocities. Then, in 1933, Zwicky (777) concluded that the velocity dispersions in rich clusters of galaxies required 10 to 100 times more mass to keep them bound than could be accounted for by the luminous galaxies themselves. The former result was taken much more seriously than the latter by contemporary and succeeding astronomers (being dignified by the name "the Oort limit"), which is perhaps more a statement about the personalities of Oort and Zwicky than about anything else.

The next decades were by no means devoid of relevant ideas and investigations (346, 484). The beginning of the modern era of dark-matter research can, however, be dated to 1974, when Ostriker, Yahil & Peebles (506) and Einasto, Kraasik & Saar (193) tabulated galaxy masses as a function of the radius to which they applied and found M increasing linearly with R out to at least 100 kpc and $10^{12}\,M_\odot$ for normal spirals and ellipticals.

Since then, a mainstream astronomer who seriously doubted that we are somehow not readily seeing 90% or more of the stuff in the Universe has found himself in the position of having to justify his discordant views. The low-mass torch, upheld for a time by Burbidge (109) and Woltjer (755), has recently been refueled by Valtonen (706–708).

Because dark matter has been invoked in many different objects and on many different scales, a very large fraction of astronomical research bears in some way on the issue. Necessarily, then, many aspects are given rather short shrift here.

First, nothing is said about the value of the Hubble constant, though it enters in powers from -2 to $+2$ into various determinations of mass and luminosity of distant objects and is arguably the largest single uncertainty in these determinations (268). Hodge's (301) 1981 conclusion that an impartial choice of value for H_0 would be both difficult and unprofitable remains regrettably correct. Besides, like Hodge, I have friends in both camps.

Next, several other relevant topics that have recently been reviewed in this series are somewhat neglected. These include evolution of galaxies in clusters (186), models of the Milky Way (36), the contribution of low-mass stars to the local mass density (415), constraints on dark matter in globular clusters (196), and properties of the Milky Way spheroid (229). In general, little is said about how calculations were done, except where methods have been substantially criticized. References cited for results generally explain how they were obtained, and, for the researcher desiring to acquire a thorough knowledge of methods, the standard starting place is Chandrasekhar's *Stellar Dynamics* (132). For many of the cases considered, $GM/R \sim V^2$ is all the physics needed.

Finally, the literature search approached completeness only for English-language journals received by University of California, Irvine, and/or University of Maryland between May 1984 and July 1986 and catalogued under Library of Congress designation QB (astronomy and astrophysics). Some relevant work appearing earlier, in books, and in non-English or nonastronomical journals has undoubtedly been missed. Much of the pre-1979 literature can be accessed through Faber & Gallagher's (200) fine review of masses of galaxies.

For the objects and systems discussed here, masses and luminosities will normally be given in solar units. That is, $M/L_B = 10$ means 10 solar masses of gravitating material for every one solar luminosity in the blue band. If no wavelength band is indicated, either bolometric luminosity is meant or the number has sufficiently large error bars that the wavelength makes no difference.

There exist a modest number of objects and systems where little or no dark matter can be present. These either have the dynamical mass equal to the luminous mass, or larger (but there exist more attractive explanations than dark matter), or seemingly smaller (so that some kind of energy input is needed to prevent collapse). The solar system, having been stable for some 4.5×10^9 yr, can be used to put rather stringent limits on nearest permitted approaches of black holes, substellar mass objects, and other hypothetical forms of dark matter (299, 325), including a possible substellar companion to the Sun (746).

Other objects under this heading include high-latitude molecular clouds (seemingly unbound, but perhaps confined magnetically or by hot coronal gas pressure; 366, 433), Galactic plane molecular clouds (which require nonthermal support or continuous energy input; 621, 638), and the radio-emitting lobes of extragalactic radio sources [for which gravitational confinement by a massive black hole has been proposed (108), but pressure confinement is now more widely accepted]. One intergalactic H I cloud that looks unbound (594) may be gravitationally attached to a pair of galaxies (593) or be a genuinely transient configuration (559, 734).

Star clusters, both open (440, 441, 747) and globular (158, 159, 196, 269, 524, 535), display dynamical properties fully explicable by the mass in visible stars and predictable stellar remnants. The very large proper motions, leading to $M = 10^8$ $M_\odot$ for one cluster (486), need to be confirmed. Most measurements pertain only to brighter parts of clusters and would permit very extended, massive dark halos (520), but at least in M15, the velocity dispersion declines from 15 km s^{-1} in the core to less than 1 km s^{-1} in the outskirts [see work by P. Seitzer & K. C. Freeman reported in (231)], suggesting that the light edge is also the mass edge.

2. SINGLE GALAXIES

Mass estimates applying to individual galaxies can come from velocity dispersions or rotation curves of the stars and gas (including X-ray-emitting gas) making up the galaxy itself or from positions and velocities of test particles like globular clusters and satellite galaxies. Where several methods can be applied to the same part of the same galaxy, results are frequently, but not always, in reasonable agreement. In the conventional terminology, once a given amount of gravitating mass has been found on one scale, it no longer counts as "dark" or "missing" on larger scales. Thus, for instance, $M/L = 3$ is considered proper for inner disks of spirals in clusters of galaxies—because it is what we measure for the solar neighborhood—even though a third or more of the M required may be in unknown form.

2.1 Milky Way

2.1.1 SOLAR NEIGHBORHOOD   Later versions of the analysis of stellar velocities and distributions perpendicular to the Galactic plane pioneered by Oort (498, 499) have been quite consistent over the years. Bahcall (35), for instance, reported $\rho(\text{dark})/\rho(\text{luminous}) = 0.5\text{--}1.5$, with the dark matter confined to a scale height less than 0.7 kpc, rather like the old disk population. The total disk M/L is then about 3 and the local density of dark matter 0.1 $M_\odot$ pc^{-3} or 30 $M_\odot$ pc^{-2} in a cylinder perpendicular to the plane.

Most studies have, however, used rather similar tracer populations—main-sequence F stars and K giants. For each, owing to effects of evolution and metallicity, stellar brightnesses may have been overestimated and distances underestimated (268), resulting in an overestimate of the local mass density. This happens because we measure the velocity distribution only locally, but count tracer stars far from the plane. Thus, if a given velocity carries a star farther from the plane than you thought, the amount of mass holding it back is less than you thought.

The dynamics of a younger population, confined more closely to the Galactic disk, provides little evidence for dark matter (392), which could mean that the distribution is that of a thick disk, or that there is a problem with the analysis (37). A new, independent data set, extending to about 2 kpc from the plane, apparently requires little or no dark matter in either a thin or thick disk, the stellar populations in these and other components of the Galaxy providing all the necessary gravitating mass (G. Gilmore & K. Kuijken, personal communication, 1986). The existence of a separate dark component belonging specifically to the Milky Way disk should probably, therefore, not be accepted without reservation (268).

2.1.2 MILKY WAY ROTATION CURVES AND THE DISK MASS   If our Galaxy were a spherically or even radially symmetric configuration, then the circular orbit speeds of test particles at known distances from the center could be turned into a curve of $M(R)$ with only modest uncertainties arising from the difference between the gravitational potentials of spheres, ellipsoids, and thin disks.

There are several stumbling blocks along this seemingly smooth path, including the problem of determining accurate distances to objects whose velocities are known (or conversely) and that of converting heliocentric velocities to galactocentric ones. The former has been solved by using stars closely associated with gas clouds as the distance indicators (82, 390) or by using velocities of bright stars themselves (587). The main uncertainty in the velocity conversion comes from uncertainties in the values of our own distance from the Galactic center (R_0) and rotation speed (V_c) (81). Traditional values of 10 kpc and 250 km s^{-1} are in the process of being replaced by 8.5 kpc and 220 km s^{-1} (365), but the residual uncertainty is enough that masses mentioned in this section should be interpreted as containing a factor $(R_0/10)(V_c/250)^2$.

The traditional R_0 and V_c imply $1.5 \times 10^{11} M_\odot$ and $M/L_B = 10$ for the inner disk. Since L scales as R_0^2 (713), the ratio M/L is better known than either separately. The Galactic infrared luminosity contributes another $1-2 \times 10^{10} L_\odot$ (281), so that M/L_{bol} is closer to 5 than 10.

Out beyond R_0, rotation curves determined from gas continue to rise (81, 138, 376, 390, 527), while that dervied from stars remains flat at $V = V_c$ (31, 587) out to about 2 R_0. The implied masses are 4.3 and $3.0 \times 10^{11} M_\odot$ for the traditional rotation constants (or 3 and $2 \times 10^{11} M_\odot$ for the new ones), and $M/L_B \sim 15-20$. Some of the discrepancy probably arises from noncircular gas motions (144, 279), but there must also be some problem in at least one of the distance scales and, therefore, a real uncertainty in the mass. That $L(R)$ does not continue to rise is in no doubt, since we can see individual bright stars (or, rather, the relative deficiency of them) out to many kiloparsecs and are not dependent on deconvolutions of surface photometry as is the case for most other galaxies (Section 2.2.1).

Although these rotation curves use tracers in the disk, several lines of argument strongly suggest that much of the nonluminous mass is in a spherical or spheroidal component, especially outside R_0. These arguments include (*a*) the flaring of the Galactic H I disk (390); (*b*) detailed deconvolutions of both the rotation curve (36, 280) and surface photometry (384) into several Galactic components, including nucleus, bulge, thin Population I disk, thick old Population I disk, Population II spheroid, and dark halo; and (*c*) the need for a spheroidal or hot component (not necessarily dark) to stabilize bar modes in a thin cold disk (474, 503, 605).

Thus we have more or less talked ourselves into a relatively dark halo with $M \gtrsim M$(disk) out to 1-2 R_0 without being able to say whether it is merely an unexpectedly large mass associated with the known Population II spheroid or a physically distinct component. Discussion of observations within the Galactic disk that constrain the nature (as opposed to the amount) of dark matter appears in Section 6.3.

2.1.3 THE HALO OF OUR GALAXY Charting the velocity field for the Galactic halo presents the same problems as for the disk in identifying a sufficient number of objects whose distances and velocities can both be measured. The uncertainties scaling with R_0 and V_c^2 persist for most techniques. An additional one arises because there is no reason to expect halo orbits to be circular, and a given position/velocity pair "belongs" to an orbit of larger semimajor axis and, therefore, of larger central mass if the orbit is nearly a circle than if it plunges straight in. The common assumption of locally isotropic velocity distributions need not be correct. The reward that compensates for these additional difficulties is that halo objects can be found out to 5–10 R_0 (vs. 2 R_0 for disk ones), enabling us to probe $M(R)$ further out. Other properties of halo objects, including ages and chemical compositions, that may be correlated with dynamics have been elegantly reviewed by Carney (124).

Commonly used probes include RR Lyraes and other field stars, globular clusters, and satellite galaxies. The latter two can be treated in two different ways, estimating the mass interior to their positions either from their velocities or from the assumption that their observed sizes represent tidal limits set by the Galaxy as a whole. Because stars leak gradually away from the edges of clusters and satellites, rather than vanishing instantly, tidal radii fitted to star counts or photometry are probably underestimates by factors of about 1.4 (501), and masses derived from them should be scaled up by about $(1.4)^3$. Masses assumed for the satellite dwarf galaxies (Section 2.4) also enter about linearly into calculations based on them, and the Milky Way comes out very massive only if they are.

Table 1 summarizes the results of recent investigations. Features to be noted are the upper envelope, which has $M(R)$ gradually increasing with radius to at least $10^{12}\ M_\odot$ at 100 kpc (implying $M/L \gtrsim 30$), and fluctuations due to different choices of velocity distribution, the argument in favor of circular orbits for the outer objects being that they are so diffuse that they would not have survived even one close passage of the Galactic center (523). The methods used in most of the velocity-based estimates are described clearly by Lynden-Bell et al. (429). Two recent results (489 and, especially, 676) are considerably lower than average despite making use of roughly the usual data base and techniques. The reasons remain to be elucidated, but if the results are correct, most of the mass of the Milky Way is actually within a few R_0 and the total is considerably less than $10^{12}\ M_\odot$.

Significant reduction of the numbers in Table 1 can be achieved in two ways. First, a nonstandard modeling of the gravitational effects of the Magellanic Clouds (707) says that cluster and satellite velocities do not currently probe the gravitational potential very well. Second, the most

Table 1 Determinations of Galactic mass as a function of galactocentric distance

R (kpc)	Method	Mass ($M_\odot$)	Reference
12–17	RR Lyraes toward the LMC (isotropic velocities)	$2.6–2.9 \times 10^{11}$	148
17	Globular cluster velocities	2×10^{11}	106
20	Globular clusters (circular orbits)	$0.3–0.8 \times 10^{11}$	487
	(isotropic orbits)	2×10^{11}	
44	Globular cluster and satellite galaxy tidal radii	$8.9 \pm 2.6 \times 10^{11}$	333
50	Halo star and cluster velocities	4.4×10^{11}	489
65	Escape velocity of 3 RR Lyraes	$10–30 \times 10^{11}$	282
50–100	Globular clusters (isotropic velocities)	$5 \pm 2 \times 10^{11}$	497
	(no radial orbits)	10×10^{11}	
50–100	Globular cluster velocities	2×10^{11}	676
100	Globular cluster and satellite velocities (if all bound)	9×10^{11}	471
100	Globular cluster isotropic velocities	10×10^{11}	523
118	One globular cluster	$\leq 10 \times 10^{11}$	654
Total	Escape velocities of field halo stars	$\geq 5 \times$ mass to R_0	125
Total	Globular cluster and satellite galaxy tidal radii	10×10^{11}	138

extreme members of the stellar and cluster populations may not currently be bound to the Milky Way. Some stars with high positive velocities could be runaways from disrupted binary systems, but the large negative velocities [-161 km s^{-1} for a globular cluster at 84 kpc (523) and -465 km s^{-1} for one RR Lyrae at 64 kpc (282)], even if the objects are in free-fall toward us for the first time, require large Milky Way masses. The only alternative is a large velocity acquired nongravitationally, through a runaway process in some nearby galaxy, or from mutual containment in the potential of some larger, more distant mass. The ad hoc explanations will come to seem quite improbable if the samples of high-velocity stars and clusters continue to increase, and mutual containment only shoves the dark-matter problem further away without removing it.

Existing data on halo velocity distributions are rather insensitive to the difference between flat and spheroidal mass distributions, but the round shape strongly suggests round equipotentials (473).

Despite the uncertainties and reservations expressed in the preceding two sections, it seems safe to conclude that (a) within R_0, there is about as much mass in a spheroidal (mostly dark) halo as within the luminous disk; and (*b*) outside R_0, there is at least 2 and probably 3–10 times as much matter as inside. The alternative is that the Good Lord is a good deal more *raffiniert*, and the average astronomer considerably less so, than we would like. Because the dark matter of Section 2.1.1 is confined at least to a thick disk and that of Sections 2.1.2 and 2.1.3 necessarily roughly spherical, they are in some sense two separate components, though they could perhaps (Section 6.3) consist of physically similar objects, whether faint stars or suitably chosen black holes.

2.2 Other Spiral Galaxies

The data (liberally interpreted) base pertinent to dark matter in single spiral galaxies consists of (*a*) a large number of rotation curves for galaxies seen more or less edge on, determined from optical (322, 562, 564) or 21-cm (11, 93, 259) emission lines; (*b*) some observations bearing on the issue of whether most of the gravitating mass is in the disk or in some more spheroidal component; and (*c*) a grab bag of ideas about disk stability, barred and ring galaxies, gravitational lenses, and the visible spheroidal component.

In comparison with studies of the Milky Way, we have the advantage of a larger sample and of not having to assume a local rotation speed, but we lose out in angular resolution and in having the poorly known distance scale parameter H_0 with an uncertainty of $\pm 50\%$ (centered around 75 km s^{-1} Mpc^{-1}) rather than R_0 with an uncertainty of only $\pm 20\%$. Where authors have explained which value of H_0 they assumed, values of M and M/L will sometimes include a parameter h, meaning $(H_0/100)$.

2.2.1 ROTATION CURVES AND DISKS The rotation of M31 was first detected by Slipher (628) and followed well out into the disk by Babcock (33) as part of his PhD dissertation. His rotation curve was still rising at the last measured point and (scaling to the modern distance) implied $M = 3 \times 10^{11}\ M_\odot$ and $M/L = 17$ out to 18 kpc. He remarked upon the difference from the Milky Way rotation curve as then understood and on the high M/L compared to solar neighborhood values. Freeman (230) was among the first to notice that such non-Keplerian rotation curves were a widespread phenomenon and to deduce that there might be considerable gravitating mass outside the observed regions. Ten years later, the entire astronomical community was dashing madly in the direction (which was, therefore, by definition, forward) of M/L increasing monotonically with R, only to trip over Kalnajs (349), who pointed out that some of the existing rotation curves, at least the optical ones, could be equally well fit by a disk of constant M/L, especially if it had a sharp edge not far outside the last observed point.

There has ensued a new round of data collection and, probably, a new consensus (with a few caveats) along the following lines (11, 39, 121, 361):

Many rotation curves remain flat or even rise to large radii. Once one accepts the possibility of many components with different central densities, M/L's, and exponential or power-law scale lengths, these cannot be uniquely interpreted. As far out as optical data extend, one can get away with little or no halo (provided the disk properties are chosen carefully), but most of the likely fits have at least as much mass in spheroid as in disk, independent of Hubble type (Sa to Sd), and the importance of the halo increases as we look farther out with H I rotation curves. NGC 3198 (11), for instance, has been traced to 22.5 h^{-1} kpc (11 disk scale lengths), by which point the integral M/L_B has risen to 24 h and the average halo/disk ratio to at least 4. A tentative point at 37.5 h^{-1} kpc has $M/L_B =$ 33 h, close to values for binary galaxies (Section 3.2).

Work on multiplicity of spiral arms as a function of halo/disk mass ratio provides independent support for values near one (27, 28, 674). Disk masses, analogous to the Oort limit, can be extracted from stellar and gas velocity dispersions as a function of radius in face-on spirals (386). The average old disk M/L_B for six galaxies is 6 ± 2. Clearly, we do not know how many low-mass main-sequence stars contribute to this, but the disk colors are consistent with an initial-mass function much like the local one, so that it is at least self-consistent to conclude that disks of other spirals contain $50 \pm 20\%$ dark matter (386).

Now, about the caveats. First, one worries about noncircular gas velocities due to perturbations by companions or outlying gas that has not had time to come into dynamical equilibrium. These can be mapped and are sometimes as small as 3 km s^{-1} for very regular spirals (387), but they can be as large as 45 km s^{-1} or 30% of V_c^{max} for others (48, 564). It is some consolation that visible companions and crowding within clusters tend to correlate with rotation curves that remain flat or turn down rather than rising (111, 770), suggesting that this sort of noncircular velocity effect, at least, causes us to derive low masses, not high ones.

Second, there are anomalies in both directions. Some galaxies with low M/L have experienced a recent burst of star formation [Mkn 348 (624a)] or may just have been placed at excessive distances (450), but for NGC 3992 the conclusion that M/L from the inner rotation curve and from some very distant companion galaxies is the same (corresponding to a total mass of $2 \times 10^{11} M_\odot$) is independent of distance (260). Anomalously large values can occur because star formation has been depressed for a long time (96, 237), because the circular velocity rises as high as 500 km s^{-1} (251), or because the rotation curve remains flat out to 100 kpc or more (555).

Finally, if the background sky brightness around spiral galaxies has been overestimated, $L(R)$ may also continue to increase (75, 385), so that M/L remains near 10 even for very massive galaxies.

2.2.2 THE SHAPE OF THE DARK MATTER DISTRIBUTION, OR HOW ROUND WAS MY HALO For external spiral galaxies as well as for the Milky Way, evidence favoring a spheroidal rather than flat dark component comes from H I flaring and warps and considerations of disk stability. Polar ring galaxies provide independent support for this picture, and a bit can be said about higher moments of the distribution.

H I flaring M31 (103) and other galaxies (17) share with the Milky Way H I disks that widen at large radii. Since the gas velocity dispersion perpendicular to galactic disks does not increase with radius (387), this flaring must mean that the equipotentials are becoming rounder far out, requiring an increasing fraction of the total mass to be outside the visible thin disk. Although a spheroidal component seems most likely, a thick disk of at least several kiloparsecs will also work (414, 572). The phenomenon is not universal (94).

H I warps Disks whose outer regions look as if their rotation axes were tilted relative to the rest of the galaxy are also shared by the Milky Way, M31 (103), and other galaxies (93). To prevent differential rotation from destroying these requires a somewhat special potential shape. Massive spherical halos are one possibility (682), but truncated disks and companion galaxies are also possibilities (635). An interesting counterargument is that, since the most remote gas in spirals may have been recently acquired through cannibalism, the fact that it finds its way into a plane quickly indicates the existence of a (dark) flat disk potential at large radii (573).

Disk stability The bar modes of a thin (cold) stellar disk can be damped by a comparable amount of mass in a thick (hot) halo (503, 605, 769). This halo need not be dark, and the visible spheroid contributes significantly (29). The stability consideration can at least be regarded as an argument against putting much dark matter in the inner disks of spirals, although, for a gaseous disk, a spheroid may not be a stabilizing influence at all (553).

Spindle or polar ring galaxies These are a dozen or two otherwise-normal S0s, seen edge on and ringed by annuli of gas more or less perpendicular to the main galactic disks (26). Both models for the formation and, more important, measured velocities of the gas rings require a roughly spherical mass distribution (588, 599, 744).

Core radius and triaxiality   Because the inner, visible parts of spirals have much of their mass in disks, even the best existing velocity data and photometry turn out to justify only a lower limit to the halo core radius, typically in the range of 10–20 kpc (G. Lake, personal communication, 1986). This vitiates phase-space arguments that would constrain the mass of fermions contributing to dark matter if the core radii were small (516, 677). Binney (76) has compiled the arguments in favor of triaxial halos that come from numerical simulations of their formation and from detailed models of the driving and sustaining of warps, polar rings, and ripples. He mildly favors aligned spheroids for spiral galaxies and tilted ones for ellipticals. On the other hand, an antiparallel halo is an alternative cause of warps (636), and a tilted one could drive spiral arm formation where other mechanisms fail (442).

2.2.3   BARRED SPIRALS, GRAVITATIONAL LENSES, AND THE VISIBLE SPHEROID

Bars and rings   Since massive halos stabilize bar modes and make more difficult the formation of rings at the outer Lindblad resonance and of those attributed to galaxy collisions (104, 134a, 673), one might expect galaxies with these features not to show evidence for large halos. The SB NGC 3992 (261) and one barred Seyfert (58) have low M/L's as expected.

Gravitational lenses   Two cases where QSOs seem to be lensed, but no galaxy is seen, require about 10^{12} $M_\odot$ with $M/L_B \geq 21$ h for 2016+112 (591) and as much as 10^{13} $M_\odot$ with M/L in excess of 100 for 2345+007 (696). Conversely, some cases where we see a foreground galaxy or QSO, but no lensed image of the background one produced by it (334, 695), set upper limits of 2×10^{12} $M_\odot$ for Q1548+114A,B and 2×10^{11} $M_\odot$ within a radius of 50 kpc for a number of galaxies. A proposed loosening of the latter limit (528) has probably mistaken too many of the background galaxies for dwarfs in the foreground (694) and requires rethinking. Serious worry about either the large lower limits or the small upper ones should probably be saved until gravitational lensing in general is better understood (110, 479).

The visible spheroid   Velocity dispersions in the cores of a number of S0s yield M/L ratios of a few, exactly as expected from population synthesis (530) and the dominance of disk mass at small radii. Velocities of the globular clusters of M31 (278), if assumed isotropic, imply a mass of 3×10^{11} $M_\odot$ out to 20 kpc, very similar to the Milky Way result. This should be redone with the large sample now available (319). In a few early-type spirals, the spheroid is bright enough for photometry out to many kiloparsecs, most notably in the Sombrero galaxy (109a), for which the spheroidal M/L_B is 5–6 out to 20 kpc, the disk value 10, and the halo/disk mass ratio about $4:1 \times 10^{11}$ $M_\odot$. In such cases, the spheroid could be purely stellar (171a).

2.3 Elliptical Galaxies

The gravitational potentials of elliptical galaxies can be probed with velocities of stars, globular clusters, companion galaxies, and thermal and X-ray gas, none of which can be expected to show essentially circular motion like that in spiral disks. Most ellipticals, even highly flattened ones, are not rotationally supported (73). A few exceptions can be modeled as rotating structures within massive dark halos for which $M \propto R$ is, at any rate, a consistent solution (411). They are dynamically perhaps not very different from S0s. For the rest, the choice among isotropic, mostly radial, and mostly circular velocity distributions contributes significantly to uncertainties in most of the mass measurements addressed in the following sections.

2.3.1 NUCLEI AND BINARY NUCLEI The central black holes invoked in most models of active galactic nuclei (80) will constitute a form of dark matter once accretion ceases but make a negligible contribution to the mass inventory of the Universe. The subset of ellipticals (relatively common at the centers of rich clusters) that display two or more bright nuclei give the impression that they or their clusters must have very large M/L's because the velocity differences between nuclei are often large. But this seems not to be the case (669–671). Stellar velocity profiles for these galaxies and their several nuclei are indistinguishable from those of other Es and lead to M/L_B in the range 10–20.

2.3.2 STELLAR VELOCITY DISPERSIONS Normally, absorption-line profiles are deconvolved into $\sigma_v(r)$ using some standard model for $\rho(r)$ (73). The deconvolution is typically not unique, even for very good data. For instance, the central cD in Abell 2029, whose line widths increase with R, can be fit either by isotropic velocities and M/L increasing linearly outward or by a preponderance of circular orbits and constant M/L (667). Similar caveats apply to other galaxies (128). Not all conceivable distributions of orbit shapes result in self-consistent gravitational potentials within which those orbits are stable, but the difficult process of finding such self-consistent orbit distributions has only just begun (149, 598). With less than perfect line profiles, even at many radii, errors of 40% or more are easily possible (360, 551). The strongest probably safe statement is that the visible parts of most normal ellipticals have M/L of 7 to 20 (34, 340).

The globular clusters have a larger velocity dispersion than the field stars at the same (two-dimensional) radius, at least in NGC 5128 (297) and M87 (475a). The difference between the two in radial distribution, plus projection effects, can account for this (264), but additional dynamical differences are also possible.

2.3.3 THERMAL GAS Most ellipticals have very little gas, and cases showing a coherent gas velocity field are even rarer. A rotation curve for the H I in NGC 4278 (370) yielded $M/L_B = 39\ h$ out to two Holmberg radii. Two recent measurements with higher angular resolution suggest that M/L rises to these values from smaller central ones, as would be expected in the presence of an extended dark halo. Population synthesis to match line profiles in E cores (530) is also consistent with relatively small central M/L.

The galaxies concerned are the small, gas-rich NGC 5666, for which a preliminary VLA map (G. Lake, personal communication, 1986) shows something like a normal rotation curve, with M/L increasing about a factor of two from center to outskirts, and the larger NGC 7097, which has enough optical gas to trace out to $1.5\ h^{-1}$ kpc. By this point, M/L has risen to $7\ h$ from $2\ h$ at the center (116). Comparison of gas and stellar velocity dispersions in the latter indicates a preponderance of radial orbits. Thus, a central M/L derived assuming isotropy will be too large. Poor angular resolution in the absorption-line spectroscopy may also contribute to the discrepancy between these low central values and the larger ones of the previous section. It is important to look for color and other gradients associated with the changing M/L to distinguish population changes with radius from the effects of a hypothetical dark component.

2.3.4 SHELLS A dark halo, about as massive as that of a spiral with the same luminosity, but rather more compact, is a definite asset in modeling the shells and ripples seen in the outer regions of some normal and peculiar ellipticals (436). If the material contributing the shell light has come from a low-angular-momentum encounter with a smaller galaxy, then test particles representing it experience phase wrapping in the potential well of the elliptical and oscillate back and forth in radius. A sharp crest at the turnaround radius is seen as a shell, and multiple shells result from a spread in initial particle energies (191, 294, 776). There are other models for these shells which are not helped by a dark halo (203, 751).

2.3.5 PLANETARY NEBULAE, GLOBULAR CLUSTERS, AND COMPANION GALAXIES Emission-line velocities for the planetary nebulae in M32 yield a total mass of $8 \times 10^8\ M_\odot$ and $M/L = 3\text{--}4$ for the visible parts of the galaxy (488). To within the statistical errors, the nebular velocities are distributed isotropically, suggesting to the authors that M/L does not increase with radius. M32 could probably not retain an independent halo in any case, given its close promixity to M31.

The most thoroughly analyzed elliptical cluster system is that of M87, the central dominant galaxy in the Virgo cluster (316). The cluster velocity dispersion is larger than that of the stars further in, suggesting a preponderance of circular orbits. Assuming this to be correct and taking a distance of 16 Mpc, Huchra & Brodie (316) then find a mass of 6×10^{12} $M_\odot$ and $M/L_B = 150$ out to 18 kpc. This rises to nearly 10^{13} $M_\odot$ for an isotropic velocity distribution.

The velocities of companions of M87 probe much the same region and imply similarly large mass and M/L, though with somewhat lower statistical significance, since fewer objects are involved (485). Because M87 is a cluster-center galaxy, one cannot be quite sure whether it is the galaxy or the cluster potential being traced out.

NGC 720, on the other hand, is a relatively isolated elliptical with six faint companions. Comparison of their velocity dispersion (353 km s^{-1}) and average radial position (23.2') with those of the main galaxy (214 km s^{-1} and 86") suggests that there is 44^{+40}_{-20} times as much gravitating mass in the large volume as in the small one (189).

Some other central cD's seem to have a semi-bound population of companion galaxies with characteristic velocity dispersions of 250 km s^{-1} [vs. 1400 km s^{-1} for the clusters as a whole (151)], from which mass information could probably be extracted.

2.3.6 RADIO AND X-RAY SOURCES Of various models proposed for the confinement of radio jets and lobes, there is one (704) that works only with the potential of a massive halo and one (71) that works only without it.

A subset of X-ray clusters show emission strongly concentrated toward a central dominant galaxy, e.g. a core radius for Virgo of 50 kpc vs. 250 kpc for typical rich clusters. Other members of the subset are Centaurus [NGC 4696 (443)], Perseus [NGC 1275 (204)], and several poorer clusters (381, 430, 434). The strong central concentration suggests that the X-ray gas may belong mostly to the central galaxies rather than to the clusters as a whole.

The standard model of gas shed from stars, radiatively cooling and flowing toward the galactic center, and ways of using these X-ray cooling flows to measure galactic masses are comprehensively reviewed by Sarazin (578, 579). The derived mass is, unfortunately, extremely sensitive to the form of temperature gradient assumed, from which arises most of the present controversy, as well as an earlier one pertaining to M87 (77).

Given the measured Einstein temperature gradient for M87, its total halo mass amounts to $3-10 \times 10^{12}$ $(r/300$ kpc$)$ $M_\odot$ at a distance of 20 Mpc (382, 652, 727). The material concerned is quite dark; beyond 100 kpc, the total M/L_B value is at least 150 and the local value greater than 500. The circular velocity at 300 kpc corresponding to this mass is about 600 km s^{-1}, double the 300 km s^{-1} of the central galaxy (581). For M87, then, the combination of X-ray data with that from companion galaxies, globular clusters, and stellar velocity dispersions implies $M(R)$ rising linearly with R over the whole range from 0.3 to 300 kpc (579).

Results for the other central dominant galaxies mentioned have larger error bars, but are similarly suggestive of total masses near 10^{13} $M_\odot$ and M/L's $\gtrsim 100$. Some of this mass may, however, belong properly to the clusters rather than to the individual central galaxies.

Cooling-flow-type X-ray emission has also been seen from some non-central cluster galaxies, isolated ellipticals, and early-type spirals (227, 490, 680). These are all fainter than M87 by factors of about 100 and are at least as far away, so that even the average temperatures, let alone the gradients, are poorly known and the error bars on the masses very large. Isothermal gas at 10^7 K (227, 659) or gas with reasonable constraints on the temperature and pressure gradients (206) leads to typical masses of $1-10 \times 10^{12}$ $M_\odot$ and M/L's of 60 ± 20 (for $H_0 = 50$). The X-ray gas itself is then not a major contributor toward the total mass. This is distinctly the majority view. It is supported by models of cooling flows as a function of the shape of the gravitational potential confining the gas (580), which predict X-ray surface brightnesses more centrally peaked than the observed ones unless the parent galaxies have heavy extended halos. These halos are hotter than the stellar systems they contain.

A minority view (681) permits temperatures to rise or fall toward galaxy centers and so finds possible M/L ranges that overlap values derived from stellar velocity dispersions. At least one Sa galaxy, the Sombrero, has both a rotation curve (47a) and X-ray emission measured. The lower X-ray mass fits smoothly onto an extension of the rotation curve, while the higher conventional one falls far above it (J. van Gorkom, personal communication, 1986; G. Fabbiano, personal communication, 1986). In this case also, the standard massive halo implied by isothermal X rays would be much hotter than the visible part of the galaxy.

2.4 Dwarf Galaxies

The dwarf galaxies are particularly interesting because they have the potential for telling us the smallest configuration that can have a dark halo and thereby constraining the minimum particle mass possible in the halos (677). The situation is somewhat different for dwarf irregulars and dwarf spheroidals, which are discussed separately without any attempt to resolve the vexing issue of whether there is an evolutionary, environmental, or other close relationship between them. It is worth remembering, however, that if you turn off star formation in the SMC for some billions of years, it will surely become very faint and tend toward $M/L_B = 20$, and that any set of things assumed about halo masses, tidal or wind-driven stripping of gas, chemical evolution, and changes in star-formation rates and M/L had better be self-consistent.

2.4.1 IRREGULAR GALAXIES The irregular galaxies (321) trail continuously downward from the latest spiral types (Sd) and clumpy irregulars with $M_B \leq -20$, through the Sm's or Magellanic types with detectable arms and rotational velocity fields, to low-surface-brightness, truly irregular Irr's and slowly rotating dIrr's (dwarf irregulars) with $M_B \geq -13$.

For representative objects brighter than $M_B = -14.5$, H I rotation curves extend far enough out to demonstrate the existence of large and rising M/L's (dark matter) outside the bright optical regions (120, 121, 231, 259, 389), though as for normal spirals, the optical data alone can be fit by constant, relatively low M/L (661, 662). In the range $M_B = -14.5$ to -12, there is still clearly rotation, but it has been traced out only to the optical edge, and the relatively small M/L's pertain to the core and do not constrain dark matter one way or the other. VLA mapping of Irr's in this magnitude range (G. Lake et al., personal communication, 1986) now in progress may reveal whether they share the halo/disk ratio of about one found for brighter Irr's. Finally, below $M_B = -12$, velocity fields show little evidence of rotation, but H I line widths are still large enough to say that M/L is large and the visible stars and gas are probably not self-gravitating. One caveat comes from NGC 1705 at -15, which has double optical emission lines split by the H I velocity width. This could plausibly be bipolar outflow rather than bound gas moving in the potential of a dark halo (231).

The blue compact dwarf galaxies are a subset dominated by bright star-forming nuclei but possessing also considerable gas and an underlying old stellar population. Existing data (660–662; T. X. Thuan, personal communication, 1986) pertain largely to the visible regions, and so the fact that the stars and gas account for the Virial mass is not surprising given the similar results for other irregular and spiral galaxy cores.

2.4.2 DWARF SPHEROIDAL GALAXIES These are the most bitterly argued of all. They have such low surface brightnesses that integrated velocity dispersions cannot be measured, and the data comprise only individual velocities for 10–20 stars per galaxy and the six globular clusters of Fornax, plus measurements of tidal radii. Because the data pertain to the central, visible parts, any necessary halos must put a good deal of mass there, so that phase-space constraints suffice to rule out neutrinos or anything else of mass $\lesssim 100$ eV (417). These galaxies are so faint that we have data only for the seven companions of our own Galaxy, but Andromeda has three or more, and they are undoubtedly the commonest sort, at least in small groups.

The tidal M/L's (201) are $\gtrsim 40$ for Carina and Ursa Minor, $\lesssim 10$ for Sculptor and Draco. Perversely, the qualitative argument for large mass coming from the ability of some dSph's to retain their own globular clusters in the presence of the parent galaxy applies only to Fornax (and the M31 companions NGC 147 and 185), whose dynamical M/L proves to be low; but it is the brightest of the lot. This qualitative argument and masses derived from tidal radii both depend upon the mass assumed for the Milky Way. Thus, if our Galaxy has no massive dark halo, neither have the dwarf ellipticals from this point of view (468a).

The first velocity dispersions came from carbon stars, whose optical velocities are unreliable (344) at about the 6 km s^{-1} level of the dispersions found. Much additional hard work (1-3, 22, 146, 158a, 604) has broadened the data base to include K giants and the Fornax clusters while eliminating radial velocity variables. Velocity dispersions of 6-10 km s^{-1} persist for five of the companion galaxies (Leo I and II remain to be studied) and imply (158a, 231) M/L's near 2 for Fornax, 5 for Sculptor, 10 for Carina (but 15-20 if it had as few young stars as the others), 40-60 for Draco, and 80-100 for Ursa Minor. The largest M/L ratios belong to the least luminous galaxies, and, if they persist in larger data samples, then we have learned something important about the smallest dark configurations and what they can be made of.

3. GALAXIES IN BINARIES AND SMALL GROUPS

The existence of binary *stars* was proposed by Michell (463) on the statistical basis that close pairs of stars in the sky were far too common to represent accidental superpositions. The conclusion was, however, not widely accepted by natural philosophers of the time until 1803-4, when Herschel succeeded in tracing out portions of several orbits, showing them to be consistent with Newtonian gravitation. Binary galaxies were similarly discovered in a statistical way (311), and confirmation from measured orbits can be expected about the year 1,000,001,987.

In the meantime, galaxies in pairs and small groups present the problem that any particular system for which we might want to attempt a mass determination (including, unfortunately, the Local Group) could turn out to be only a brief encounter or an optical double. Even worse, the parameters used to select a sample of nominally bound systems (separations in space and velocity, R and V) are exactly the same ones that dominate calculated masses, in the form mass proportional to RV^2. Inevitably, then, the larger the range of separations admitted in a sample, the larger the masses found.

3.1 *The Local Group*

The mass of our own small cluster of galaxies has been estimated in three ways, using the Virial theorem (320, 470), the fact that the Milky Way and Andromeda are approaching each other in an expanding universe (293, 346, 418, 428, 574), and the effects of the Local Group on the velocity field of nearby galaxies (574). The first leads to a large, but quite uncertain, value of M/L, as do models of the Magellanic Stream (419).

The second consideration was pioneered by Kahn & Woltjer (346), who required that the self-gravitation of the pair suffice to have turned around their initial Hubble separation velocity by the present age of the Universe. The calculation is a pleasant application of classical mechanics, the resulting equations being given in convenient form by Lynden-Bell (428) and by Sandage (574). The chief uncertainties are the amount of time allowed for the turnaround (6–20 × 10^9 yr, depending on choice of H_0 and q_0) and the present approach velocity. This is the 300 km s^{-1} heliocentric speed of Andromeda (561) minus 80% (geometrical projection factor) of the local Galactic rotation speed of 184–294 km s^{-1} (365, 418, 619), or 65–153 km s^{-1}. The most probable range of derived masses is 2–7 × 10^{12} $M_\odot$, and M/L is about 100. Appreciable quantities of dark matter can be avoided only if the local circular speed is as improbably high as 290 km s^{-1} (293). If it were, the inner Milky Way, with a flat rotation curve out to 20 kpc, would itself contain 4 × 10^{11} $M_\odot$ and have $M/L = 16$.

Sandage (574) has applied the third method and found that the Local Group does, indeed, perturb the nearby velocity field, but only slightly, consistent with a total mass of 4 × 10^{11} $M_\odot$ and M/L about 5. The mass thus found is, inevitably, sensitive to choices of H_0, q_0, circular velocity, and the mass distribution within the Local Group, but is invariably less than the binding mass for the same set of assumptions. That two bright galaxies should be this close together by chance is not horribly improbable (62, 267), but the large relative velocity still has to be accounted for, either through nongravitational processes or through the gravitational effect of some other mass concentration. Random velocities of 100 km s^{-1} could be contained by the potential of the Virgo supercluster at a distance of 18 Mpc if the total mass is as large as 4 × 10^{13} $M_\odot$. This does not eliminate the dark-matter problem but only moves it elsewhere!

A bound Local Group of relatively small mass is possible only if one follows Arp (24) in attributing an appreciable fraction of the blueshift of Andromeda to nonvelocity effects, but the absolute minimum for two test particles falling toward each other inside a single central potential well is only $7 \times 10^{11} M_\odot$.

3.2 Binary Galaxies

The measurement of binary galaxy masses using velocity differences and projected separations was pioneered and persistently pursued by Page (511–513). He recognized most of the problems that are still with us: collecting a large enough sample that the need to average over projection angles does not produce enormous error bars; being sure that the sample includes only bound systems; and deciding the intrinsic eccentricity of the orbits. In addition, his photographically determined velocities typically had errors comparable with the velocity differences being sought, which may be the reason that his pairs of spiral galaxies yielded much smaller values of M/L (3 ± 2) than those in later investigations. Curiously, his results for Es and S0s (90 ± 40) do not disagree.

There now exist a number of additional data sets with much more accurate optical or 21-cm velocity values (79, 354, 400, 525, 600, 688, 742). Once differences in assumed values of H_0, orbital eccentricity, and waveband for L have been reconciled, nearly all of these are consistent with $M/L_B = 70\pm20\ h$ within radii of 100 h^{-1} kpc, assuming circular orbits (500). This drops immediately to 30 ± 10 if the orbital velocities are assumed to be distributed isotropically, and still lower for radial orbits (742).

The velocity separations are, on average, larger for nonisolated systems, which might reflect the effects of potentials that belong to larger groups rather than the binary pairs, but N-body simulations of galaxy clustering (197) suggest that this is not a problem, implying that galaxies forming in dense regions really are more massive (285). The binary galaxy ΔVs are not much correlated with luminosity (742), while line widths of individual galaxies clearly are. This should not surprise us if the former probe primarily the dark halos and the latter the visible disks of galaxies.

Worrisomely, the one discordant sample (354) with $M/L_B = 7\pm1$ (the same as the rotation masses for the same galaxies singly) is the one that most nearly meets a strict criterion for containing only bound systems (706). One cannot, unfortunately, resolve the issue by considering only pairs that show signs of interaction. These indeed yield lower M/L's (21, 706), but the members must be very close together, and so will be inside of, and not probe, most of their shared dark halo according to the conventional view.

3.3 Small Groups

Early cataloguers of groups containing three to tens of comparably bright galaxies (107, 711) typically applied the Virial theorem to their data, finding that masses near 10^{13} $M_\odot$ and M/L's near 100 were necessary for binding, while at the same time they expressed severe reservations about both the permanence of the configurations and the appropriateness of the technique.

The most extensive published data set (238) includes 90 groups in the Center for Astrophysics redshift survey and yields a mean M/L of 170 h (317). A number of other observational and theoretical investigations essentially concur (51, 52, 102, 128, 197, 273, 289, 685). Once again, a dissenting analysis of triples (356) finds a mean M/L of only 10, and a test using redshift asymmetries suggests that many of the systems with apparently large M/L's are, in fact, unbound (706), as earlier suggested by Materne & Tammann (439a). For 36 small groups velocity-mapped in H I, the masses implied by the individual galaxy rotation curves are enough to bind the systems if the orbits are circular (594a). These groups have not been statistically tested for boundedness.

Two special cases are poor cD clusters and compact groups. The former (189) typically have velocity dispersions of the companion galaxies not much larger than those of the central galaxies, implying that much of the mass is fairly near the center. The velocity-dispersion masses (60) are sometimes rather larger than the X-ray masses (434), unlike the case of single ellipticals. For the Hickson compact groups, even the visible mass implies awkwardly short dynamical times (750), although once one allows for accidental alignments (437) and more sophisticated dynamical evolution (51, 53, 736), the number of merger products may just about agree with the number of galaxies that could have formed that way (53). The rotation curves of the individual Hickson galaxies typically turn over (i.e. no massive halos) and show other peculiarities (V. C. Rubin, personal communication, 1986).

The situation for all the systems mentioned in the preceding three sections seems to be that dark matter, exceeding that implied by rotation curves and other local measurements by about a factor of 10, is needed to guarantee their stability over a Hubble time, but that there is in every case some alternative possible interpretation of the data. Unfortunately, the only discriminants so far suggested are likely to be believed only by those who already believed a particular answer anyway. A signature for bound but noninteracting systems is badly needed!

4. RICH CLUSTERS, SUPERCLUSTERS, AND GLOBAL CONSIDERATIONS

4.1 *Cluster Cores*

The clustering of the spiral nebulae was recognized long before their extragalactic nature was universally accepted (296, 754), and a dynamical study of Virgo (777) was among the first indicators of dark matter in the Universe. Nearly all modern studies concur with this pioneering one in finding cluster mass-to-light ratios considerably larger than those of individual galaxies, even ones with extensive rotation curves. Readers who agree with Fasenko (212) and Zabrenowski (766) that clusters and larger entities merely reflect the effects of nonuniform absorption in our own Galaxy can skip the rest of this section.

Published M/L values range from 30 to 1000, but this narrows considerably when normalized to a single value of H_0, face-on galaxies, and a unique waveband for L. By way of reminder, $M/L(H = 50) = 1/2 M/L(H = 100)$, $M/L_V \approx 0.7 \, M/L_B$, and, probably, $M/L_{bol} \approx 1/2 M/L_V$. There is still another factor of two between the Virial mass $(2T+U = 0)$ and the just-bound mass $(T+U = 0)$, And, finally, M/L drops by yet another factor two when corrected for absorption in our Galaxy and normalized to face-on external galaxies (217, 218; J. E. Felten, personal communication, 1986). It is not always very easy to determine which set of parameters was used in a particular study, but virtually all the very large values found in the literature drop to 100–200 for face-on galaxies, the correct choice of solar L_V, and $H = 50$ km s^{-1} Mpc^{-1} (J. E. Felten, personal communication, 1986).

Virial M/L's of 100 or more (in the sense of the previous paragraph) have been derived for Virgo (253, 318, 686), Coma (178, 362, 469, 657), Perseus (363), and a number of other rich clusters (95, 239, 507), including one at a redshift of 0.39 (592). Estimators other than the Virial theorem yield concordant, but typically slightly larger, masses (289, 631). Some uncertainty attaches to these numbers owing to faint galaxies in the clusters which cannot be counted directly and are included using an average luminosity function (469, 558). Light from between obvious galaxies or from their extreme outskirts is also rather poorly measured, but almost certainly less than that seen in the galaxies (664). M/L could be changed by a factor of 2–4.

More uncertainty comes from cluster dynamics. If, for instance, M/L is not constant throughout Coma, acceptable fits to measured velocities can be achieved with masses from $6 \times 10^{14} \, M_\odot$ and circular orbits in the outskirts to $5 \times 10^{15} \, M_\odot$ and radial orbits far out (461, 658), corresponding to $M/L_B = 40$ to 300 ($H_0 = 50$). If outer parts of clusters are not relaxed, Virial theorem masses could be too large by factors of 3–5 (130, 481). An extreme version of this idea (708) associates dark matter only with central dominant galaxies, the others acting like unrelaxed test particles. Finally, some clusters are certain to be contaminated by foreground and background galaxies that will push up the velocity dispersion and so M/L (219, 239, 522) or to contain substructures that can have the same effect (426).

Even Virgo may be a victim of this latter phenomenon (657, 710). All these considerations serve primarily to widen the error bars on M/L, though perhaps more in the direction of low values than high ones.

Potential wells of rich clusters can also be probed via X-ray emission of hot gas in them. The standard assumptions (578) of hydrostatic, isothermal gas lead to M/L values of 100 or more (208, 698), with the usual noise coming from different choices of distance, luminosity scale, etc. Once this noise is eliminated, results from analyses that assume different gas distributions and temperature structures (152) are not terribly discordant, with $M/L \sim 200\ h$, although the X-ray gas itself contributes as much as 30% of the mass in this case.

One very large value, $M/L = 10^{3.3}$ for the cluster that lenses the QSO 0857+561 (262, 554), is subject to the same caveat as single galaxy lenses—that the process is not yet very well understood (110, 479).

How is this dark matter distributed? Clearly, 100-kpc halos must have blended in cluster cores where the intergalactic distance is small. In addition, halos attached to the individual galaxies would make them hefty enough to segregate, the most massive ones falling to the center. While morphological segregation in clusters is conspicuous (252), luminosity segregation is much less so (118, 236, 306, 758), leading to a majority conclusion that $M/L \le 30$ for individual cluster galaxies (557, 571, 738). A similar limit comes from the requirement that gravitational drag heating of X-ray gas not be so large as to prevent cooling flows where they are seen (468). The amount of dark matter in rich clusters could, however, easily be the same as would have been associated with the individual galaxies in a less dense environment (227). The likelihood of this depends mostly on how you think galaxies formed, but, in any case, most of the dark matter must now belong communally to the cluster, not individually to the galaxies.

4.2 Superclusters

These larger scale structures, catalogued by Abell (6) and others (46, 609, 614, 709), provide another opportunity for measuring mass-to-luminosity ratios. Results on this and larger scales are, however, usually given in terms of the ratio Ω of density in a particular component to the density needed to close the Universe, $\rho_c = 3H^2/8\pi G = 2 \times 10^{-29}\ h^2$ g cm^{-3}. So defined, Ω coming out of most dynamical analyses is independent of H. Since the average luminosity density of the Universe is about $2.4 \times 10^8\ h$ $L_\odot$ Mpc^{-3} (217), a component with a particular value of $(M/L)h$ will contribute $\Omega = (M/L)h/1000h$ toward closure.

Necessarily, the values of Ω found from these large-scale structures can pertain only to volumes where there is at least some luminous material! In addition, many, though not all, of the cluster and supercluster analyses assume a constant ratio of dark to luminous mass throughout the volume considered. The effect of both these limitations is probably to make derived values of Ω lower limits to the real value, though the latter could, in principle, go either way.

Abell (7) first noted that binding typical superclusters would require 10^{16-17} $M_\odot$ in a 50-Mpc radius. More recent studies of our own Virgo supercluster are countably infinite in number (4, 187, 416, 656, 687, 715, etc.). The measurements are often stated in terms of a "Virgo-centric infall." This does not mean that we are actually moving toward the Virgo cluster, but only that we are moving away less rapidly than in a Hubble velocity field unperturbed by the large mass concentration. Most determinations fall in the range 250 ± 50 km s^{-1}, implying $\Omega = 0.2 \pm 0.1$ if all superclusters have the same M/L.

Other superclusters, if bound structures, also tell us that $\Omega = 0.2 \pm 0.1$ out to scales of 30–50 Mpc (172, 226, 276, 521). Smaller values have been claimed for a few structures (263, 656), and some things that look like superclusters may not be bound (47), but the main remaining controversy seems to be whether we can rule out $\Omega = 1$ only at the 1-σ level (113, 456) or with considerably more confidence (304, 410, 521).

Structure on scales of clusters to superclusters can, as an alternative to looking at particular objects, be probed via the correlation function

$$\xi(r) = \xi(r/r_0)^{-n}, \qquad 1.$$

which expresses the excess probability of finding an object a distance r away from an index object relative to the probability in a random distribution. The method was pioneered by Peebles (517), and the probability of its use shows a large excess at small distances from Princeton. Calculations of Ω from the correlation function (208a) are sometimes dignified by the name "Cosmic Virial Theorem" and sometimes undignified by the phrase "finding Ω written on the sky" (mostly by people who say they didn't).

There is reasonable agreement that $n = 1.8$ for galaxies and about 2.0 for clusters, that r_0 is about 10 Mpc for galaxies and considerably larger for clusters, and that there are real differences among kinds of galaxies and among the catalogues that have been analyzed (44, 142, 167, 194, 403, 531, 568, 617). Some disagreement persists about the shape of the function at large distances (353, 534, 560, 651) and the amount of clustering of clusters and superclusters (42, 44, 348, 597). These differences and disagreements should be kept in mind when considering a particular model of galaxy formation that claims to match or not to match "the" observed $\xi(r)$.

Determinations of Ω from the form of $\xi(r)$ may not be particularly meaningful, in the sense that a number of different initial conditions can relax to what we see by the present time (582). Where the attempt has been made, however, the best fit is about 0.2 (168, 519), the same as found from the superclusters considered as separate entities. In so far as Ω is written in the sky, it is written small.

4.3 Very Large Scale Structures and Streaming (VLSS)

Structures on scales still larger than superclusters provide an opportunity for probing a larger fraction of the total mass density, and the difficulty of forming them without simultaneously introducing detectable lumpiness into the 3-K microwave background radiation puts very severe constraints on the properties of dark matter participating in galaxy formation.

The rich clusters themselves apparently extend out to 30 h^{-1} Mpc and so include about a quarter of all galaxies (43, 249). Abell superclusters are clumped on scales up to 150 h^{-1} Mpc (57). And a number of recent investigations (224, 287, 313, 459) have pointed out bridges, filaments, and connections among well-known superclusters, including Virgo, Perseus-Pisces, Hydra-Centaurus, and Abell 569, 2634, and 2666. The current record is held by Tully's (684) joining of Virgo to Coma to Hydra-Centaurus and beyond, for a structure at least 279 h^{-1} Mpc across. The most famous hole in the distribution of galaxies, the great void in Bootes (368), is at least 60 h^{-1} Mpc across (491).

The density contrast between voids (or the largest lumps) and the Universe as a whole is an important boundary condition for formation processes. The Bootes void definitely contains a few low-luminosity, emission-line galaxies (491) and perhaps some gas clouds (105). Other voids in the KOSS (368) and other surveys (377, 404) approach the size of the Bootes one and are not completely empty either. The density contrast could be as low as $\delta\rho/\rho = 2$–3 (easily produced in models of biased galaxy formation) but is more probably about 10, presenting severe difficulties (171).

The topology of the largest structures has been described as chains (342), cells (341), pancakes (772), filaments (254), disks (683), and bubbles (404). The critical question is whether it is the high- or the low-density regions that form a connected network, and one recent statistical investigation (258) finds that both do, so that the topology is most like that of a sponge.

Velocity deviations from uniform Hubble flow on scales of about 100 h^{-1} Mpc and with amplitudes of 400 km s^{-1} or more showed up more than a decade ago (563) and were initially greeted with some scepticism, partly because the implied direction of motion for the Milky Way was roughly orthogonal to that given by the dipole anisotropy of the 3-K radiation, though the magnitude was about the same (749). Additional work on both spiral (5, 147) and elliptical (112) galaxies has reproduced roughly the magnitude and direction of the early result, but the way of describing the motion has changed. Rather than as deviant motion of the Local Group, it is now perceived (112) as coherent streaming at 600–700 km s^{-1} of the galaxies in a 100-Mpc diameter region relative to the microwave background rest-frame. The vectorial sum of this streaming and our motion relative to the sample galaxies (= Local Group velocity + motion of Milky Way in LG + rotation of MW at position of Sun) is then the solar motion measured relative to the 3-K radiation.

It is not clear that these large-scale motions have much to do with dark matter, except to suggest the physical reality of very large systems and, therefore, perhaps, of density enhancements 100 Mpc or more across. They are, however, very nearly impossible to account for within the currently population-biased cold dark-matter models of galaxy formation (726a, 739) and could, if confirmed, force the rejection of such models and, therefore, of cold dark matter in general.

In addition, several attempts have been made to calculate Ω directly from the largest scale clustering. Lahav (397) finds 0.3 from 15,000 catalogued optical galaxies at an average distance of 50 h^{-1} Mpc. This differs little (but in the "right" direction) from the 0.2 given by supercluster scales. Finally, two analyses of the distribution of IRAS galaxies (449, 762) yielded $\Omega = 0.5$ and 0.85 ± 0.15, respectively. The main difference between the IRAS and optical samples is that the former has an average, baseline number density of bright galaxies twice that of the latter (760), so that a given fractional density enhancement contains more total mass. Which is "right" cannot be certainly determined until more of the IRAS galaxies have measured distances (166), but, in the meantime, it is just possible that something very close to the closure density has already been detected, though its form and distribution remain largely unknown.

4.4 *The Global Value of* Ω

Given that the known mass-energy in the Universe comes within a factor of about five of the closure density, and that redshifts and apparent magnitudes of distant galaxies limit the effective Ω to at most four (257), one might well guess $\Omega = 1$ exactly. The motivation for this is strengthened when it is recognized (180) that a value within a factor 10 of one now can be achieved only if Ω fell within about 1 part in 10^{15} of unity during the epoch of nucleosynthesis ($T \approx 1$ MeV) and 1 part in 10^{49} at the time of the GUT phase transition ($T \approx 10^{14}$ GeV).

The case for $\Omega = 1$ is enhanced by a model for the evolution of the very early Universe called inflation (100, 271). The general idea is that an early period of exponential expansion, triggered by a high-temperature phase transition, guarantees that widely separated parts of the Universe were once in communication (and so can reasonably have the same density and temperature now) and $\Omega = 1$ to very high precision [though there are alternative versions of inflation that have most of the standard virtues but permit $\Omega < 1$ (256, 424)]. The whole subject of the relationship between particle physics and cosmology and its implications for values of H_0, Ω, and the cosmological constant Λ is an exceedingly active one, with archival journal papers often lagging current research by three or four ideas even when they first appear. The interested reader is advised to start with recent conference proceedings (375, 607) and then to haunt the preprint department of his local library.

The other main argument for a critical-density universe arises from the difficulties of forming galaxies and larger structures if $\Omega < 1$ without simultaneously introducing into the 3-K radiation larger inhomogeneities than the present upper limit, $\Delta T/T \leq$ few $\times\ 10^{-5}$, on the angular scales that correspond to galaxies at the epoch ($Z \sim 10^3$) when matter and radiation decouple (701).

Qualitatively, to make $\Delta\rho/\rho \approx 1$ now requires $\Delta\rho/\rho$ of about 10^{-3} at decoupling. (Lumps grow at most linearly with Z in an expanding universe.) This, in turn, imposes fluctuations of $\Delta T/T = (1/3)\ \Delta\rho/\rho$, which we will still see if the Universe has remained optically thin. These are too big even if $\Omega = 1$. The situation is worse if $\Omega < 1$, so that the galaxies, etc., we see correspond to $\Delta\rho/\rho \gg 1$. On the other hand, $\Delta T/T$ can be greatly reduced if there is a background of nonbaryonic dark matter that does not interact with radiation at $T \lesssim 1$ GeV. Then lumps in it can start growing early and reach sizable amplitudes without perturbing the radiation. The baryons begin to follow the dark matter only after decoupling, eventually (because they are dissipative) becoming more clumped than the dark background.

Quantitative versions of this argument are very much more complex, but lead to the same conclusion that galaxy formation is enormously simplified if the Universe is closed with at most weakly interacting dark matter (83, 89, 169, 427, 723, 775), though the very large scale streaming velocities are still a problem (726a, 739). This subject also is in a phase of rapid development and most easily followed through conferences and preprints.

5. INTERMISSION

Most of the authors cited have been convinced of the existence of significant amounts of nonluminous mass. This is not entirely a bandwagon effect; in addition, if you work hard on something, you want it to be important. Table 2 summarizes the current majority view, but the dark matter needed on many scales can be reduced or possibly eliminated if you are willing to accept some of the following alternative hypotheses.

1. Solar neighborhood: tracer stars brighter than assumed.
2. Rotation curves of Milky Way and other spirals: outer gas in noncircular and probably impermanent orbits owing to effects of recent arrival, companions, etc.; luminosity at large radius underestimated because sky background brightness overestimated.
3. Velocities of globular clusters, companion galaxies, and outlying stars of Milky Way and other galaxies: outer high-speed objects not in permanent bound orbits.
4. X-ray emission from elliptical galaxies: gas temperature distribution that declines steeply toward galaxy centers.
5. Dwarf spheroidals: not tidally distorted by Milky Way because Galaxy mass small; stellar velocity dispersion due partly to binary and pulsating stars.
6. Velocity dispersion of stars and globular clusters in galaxies and galaxies in clusters: preponderance of circular orbits at large radius.
7. Binary galaxies: preponderance of radial orbits or isotropic distribution.
8. Small groups: many unbound or bound only as part of larger structures (including Local Group).
9. Rich clusters: not yet relaxed; interacting subsystems; X-ray gas polytropic rather than isothermal; dynamics dominated by central massive core.

There is no way of avoiding sizable quantities of nonluminous mass if the advantages of inflation or some other argument incline you to favor

Table 2 Amounts of mass on various scales implied by "mainstream" investigations

Scale	$\langle M/L \rangle$	Ω
Visible stars and clusters	1	0.001
Visible parts of galaxies	10	0.01
Binary galaxies and groups	10–100	0.01–0.1
Rich clusters and superclusters	100–300	0.2 ± 0.1
Largest scale coherent structures	700 ± 150?	0.5–1.0
Inflationary scenario	$1000\,h$	1.0

$\Omega = 1$. The final section is devoted to a brief discussion of possible kinds of dark matter and observational constraints on them.

6. OBSERVATIONAL CONSTRAINTS ON THE NATURE OF DARK MATTER

If one accepts the evidence for nonluminous mass on many scales, the next obvious question is, what is it made of? A single candidate would be the most elegant solution, but it is asking a great deal of any one sort of thing that it be "cold" enough to settle down into the Milky Way disk and simultaneously "hot" enough to remain much less clustered than the superclusters over the age of the Universe.

Curiously ordinary, baryonic matter may come closest to meeting this stringent condition, since it is known to exist in both dense stars and very diffuse X-ray-emitting gas. The objection to $\Omega = 1$ in baryons derived from nucleosynthetic considerations and possible ways around it are addressed in Section 6.2. Other candidates currently under consideration include black holes, neutrinos, and a whole zoo of (mostly hypothetical) particles predicted by various branches of theoretical high-energy physics. Section 6.4 discusses these.

6.1 *The Nonstarters*

A few dark-matter candidates (stars and gas) would be so conspicuous that they can be ruled out easily, and a few others (gravitational radiation and primordial black holes) would be so inconspicuous that little can be said about them.

True stars, which derive most of their energy from nuclear reactions, extend down to $0.085 \, M_\odot$ and $\lesssim 10^{-3} \, L_\odot$ (345). Dim though they are, they could be seen individually in our own Galaxy (161, 248) and collectively in other galaxies (626) if they were responsible even for the $M/L = 5$–10 implied by spiral rotation curves.

Gas is similarly well inventoried in all possible forms in the Milky Way and other galaxies. On intergalactic scales, a closure density of cold or warm gas would produce conspicuous emission and/or absorption lines which we do not see (515). Very hot gas would radiate X rays, and indeed we see a highly isotropic X-ray background well matched by 40 ± 5 keV thermal bremsstrahlung (439), which if produced by hot intergalactic gas would require something close to the closure density (266). There are two objections. First, heating the gas requires more than 10% of all available nuclear energy in the Universe, and there are no very obvious sources (220, 266). Second, after removing the contributions of galaxies, quasars, and clusters that are known X-ray sources, the remaining spectrum can no longer be fit by thermal radiation (244).

Gravitational radiation of many but not all possible wavelengths could close the Universe without having yet been detected (19, 556). Clearly it cannot be clustered or clumped, but it can mimic dark matter in clusters in some ways (543). Primordial black holes could also dominate Ω (126, 480), and the extent to which they might cluster depends upon their (unknown) velocities. Because PBHs must form very early, if at all, they do not count as baryonic matter in the context of nucleosynthesis and are not conspicuous in any other way either.

6.2 Baryons: The Nucleosynthesis Problem

The ability of the conventional hot big bang to account for about 25% helium in the matter expanding and cooling from it (284) is generally regarded as one of its great triumphs. Simultaneous minor reactions can also yield small amounts of H^2, He^3, and Li^7, consistent with the quantities observed (728). But the initial conditions must be chosen very carefully. If the baryon density is too high, or the number of neutrino species larger than three (or several other if's), then too much He and not enough H^2 come out.

The largest Ω(baryon) consistent with deuterium production depends on the present abundance [which varies at least from $D/H = 3 \times 10^{-6}$ to 2×10^{-5} (717)], on how much has been destroyed by passage through stars [perhaps as much as two thirds of the original supply (143)], and on a number of details of nuclear physics and cosmology (86). Given conventional physics, a standard hot big bang, and generous error bars, $0.015 \lesssim \Omega h^2 \lesssim 0.15$ (86, 170).

This nucleosynthesis limit overlaps the Ω determined from clusters and superclusters, which could, therefore, consist entirely of baryonic matter. It can also be stretched toward one in a variety of ways, most simply by lowering H_0 to $\lesssim 25$ km s^{-1} Mpc^{-1} (610), in which case clusters are bound by their X-ray gas. A second, more drastic modification is nonzero lepton number, which changes the equilibrium n/p ratio and can therefore either raise or lower H^2 and He production at a given baryon density (228, 703). Third, various inhomogeneities in density, temperature (569a), and n/p ratio can also push production up or down, possibly enough to permit $\Omega = 1$ in baryons. Variations in n/p arising because neutrons can drift out of dense regions in the presence of a magnetic field, but protons cannot, look particularly promising (20, 307). Finally, one might abandon big-bang nucleosynthesis completely and produce deuterium, helium, and perhaps the photon background elsewhere. Although early efforts at this were not terribly successful (549), considerably recent effort has focused on pregalactic stars, especially supermassive ones of $10^{6\pm2}$ $M_\odot$ that have long since become black holes (90, 126, 359, 398, 540, 616).

In light of these possibilities, it may be premature to rule out baryonic dark matter, even at the level of $\Omega = 1$. Conversely, the idea that dark matter at least up to the scale of galaxies must be baryonic to account for the near-constancy of $\rho(\text{dark})/\rho(\text{luminous})$ (39) is subject to the reservation that formation of the luminous disks might have been directly controlled by the dark halos, or the halos by the disks (202, 209).

6.3 Baryonic Dark Matter: The Candidates (Brown Dwarfs, White Dwarfs, Black Holes)

It is not absolutely certain that we need baryonic dark matter even in the Galactic disk. The process of disk formation could draw inward enough of the hypothetical nondissipative matter of the halo to account for the Oort limit (54, 140). But baryons are clearly a possibility, at least up to $\Omega \sim 0.1$–0.2. How might they be assembled? Normal stars and gas were among the nonstarters in Section 6.1. In addition, even the Oort limit in neutron stars and stellar-remnant black holes can probably be ruled out. First, ongoing accretion of interstellar gas would result in the radiation of more X rays than we see. Second, the stellar evolution and demise that formed them would have scattered into the interstellar gas far more heavy elements than are present (288, 447). The additional upper limit of 2 $M_\odot$ on disk dark objects based on the long-term survival of wide binary systems (40) is probably a bit less stringent than the two already mentioned (733).

Brown dwarfs are substellar objects whose only energy source is contraction. It is not certain that any (apart from the Jovian planets) have ever been detected (627). A number of searches have identified no candidates (345). On the other hand, at least one survey (283) found interesting numbers of very red (presumably cool, faint) images, and at least one set of models for the evolution of low-mass stars and substars (163) implies that the known faint stars are part of a population whose number is still rising as $M^{-1.68}$ at 0.1 $M_\odot$. In addition, a separate population of objects $\lesssim 0.08$ $M_\odot$ as old as the Galaxy could not yet have been seen in any way, and might be expected to arise from star formation in gas under pressure (205) or in pregalactic objects of 10^{6-8} $M_\odot$ (359). The upturn of star numbers in M13 for star masses below 0.5 $M_\odot$ (427a) is interesting in connection with the possibility of a separate, low-mass population.

For other galaxies, we can say very little about the initial mass function of small stars (171a, 585), and it could continue to rise smoothly, putting large amounts of material into faint stars and substars. Brown dwarfs in sufficient numbers to be dynamically important might show up in deep exposures taken with the Wide Field/Planetary Camera of the Hubble

Space Telescope (640) or in the infrared flux from galaxies at moderate redshift (639).

White dwarfs are the normal remnants of $0.5-8 \pm 2$ $M_\odot$ stars and can fade below detectability in less than the age of the Galaxy (163). Even so, the ones remaining from stars produced at constant initial mass function and formation rate over the history of the Galaxy contribute at most 0.007 $M_\odot$ pc^{-3} locally (163, 221), only about 5% of the Oort limit. One model of galactic chemical evolution does, however, posit an early generation of intermediate-mass stars whose white-dwarf remnants might account for the dynamical mass in the Milky Way disk and, probably, the halo as well (406, 493). This early generation would have had to be even more prominent to contribute significantly on the scale of clusters and superclusters. Since these very old degenerate dwarfs could be as cool as the hypothetical brown dwarfs and will surely be much smaller, they will be exceedingly difficult to detect or rule out.

Black holes of stellar masses have already been noted as nonstarters, and very small primordial ones are essentially unconstrainable. This leaves massive ones that might have formed in the early Universe (126, 127, 398). If these make up the halo dark matter, then, in addition to having contributed (perhaps) helium and photons in the past, they will now be stirring up the disk stars and contributing to the known increase of stellar velocity dispersion with age that is generally blamed on giant molecular clouds (393) or spiral arms (122). Black holes that are too massive will be too efficient at this, but a 10^{12} $M_\odot$ halo consisting of 10^6 $M_\odot$ objects is just right (336, 350, 394). If similar black holes make up the dark matter in dwarf ellipticals, then they must be clustered toward the center, predicting an outward decrease of stellar velocity dispersion as a test of the model (394). In addition, 10^6 $M_\odot$ black holes in halos should reveal themselves by gravitationally lensing radiation from objects behind them (504, 509). Resolution of radio components may be possible with (VLBI), but optical observations will have to wait for the space optical interferometer.

By way of summary, baryonic matter definitely cannot be ruled out and even has some advantages at least up to the $\Omega \sim 0.15$ level, consistent with conventional nucleosynthesis. If one takes very seriously the nucleosynthetic lower limit on Ω_b or the need for dissipative material in galactic disks, then at least some of the nonluminous mass must be baryonic.

6.4 *Nonbaryonic Dark Matter*

One's first impression is that this category includes a countably infinite number of indistinguishable, hypothetical entities. There is some truth in this. But a count of words mentioned in three recent reviews of particle physics and cosmology (492, 532, 690) uncovered only 36 names, not quite all of which in fact designate physically distinct entities. These can be classified in terms of their predicted masses and the theories that predict them and by their astrophysical contribution as dark-matter mimics, hot dark matter (relativistic when galaxies form and so promoting large-scale structure), or cold dark matter [nonrelativistic when galaxies form and so promoting small-scale structure (89)].

Table 3 attempts such a classification. For most of the entities, the requirement that $\Omega(\text{baryon})/\Omega(\text{dark}) \sim 0.1$ constitutes a new form of fine tuning, replacing that required to get Ω close to but not exactly one. The ratio may arise naturally for quark nuggets (753), but it must otherwise require a particular value of the energy scale of a symmetry breaking or some other process (690).

6.4.1 THE DARK MATTER MIMICS—$G(R)$ AND Λ A gravitational coupling constant that increases monotonically with separation or varies suitably with acceleration would mean that the amount of matter present in luminous galaxies and parts of galaxies could act like a larger amount of dynamical mass and suffice to account for flat rotation curves, large velocity dispersions, etc. (466, 576, 666). There are theoretical (245) and even experimental (482) indications that G may indeed vary, though probably on distance scales much smaller than those required to simulate dark matter. Although many details of such models remain to be worked out (216, 577), preliminary results are that they do not match observations as well as dark-matter models (188, 295). Their main virtue is that of producing the required effect over a range of distance scales.

A nonzero cosmological constant Λ, on the other hand, acts like the zero-point energy of a quantum field theory (771) and contributes homogeneously. It also permits the curvature of space k and the deceleration parameter q_0 to become independent parameters, so that most of the standard cosmological tests (apparent magnitude or angular diameter vs. redshift, source counts, etc.) do not really tell us Ω, but only some combination of Ω and Λ or k or q_0, even when authors (422) hope they are measuring Ω. To make $q_0 = 1/2$ or $k = 0$ with $\rho < \rho_c$ takes Λ of order $\pm 10^{-35}$ s^{-2}. This is also a sort of fine-tuning problem, at least in an inflationary universe, where Λ is briefly enormous (100). Astrophysicists have debated the likelihood of nonzero Λ for years (339, 446, 665) without reaching any definite conclusion. It shares with most "real" dark matter the virtue of making galaxy formation a bit easier (724), but cannot solve the dark-matter problem on all scales, being unable to cluster. This objection need not apply to the zero-point energy of a real field theory (J. Weber, personal communication, 1986).

Table 3 Summary of nonbaryonic dark matter candidates[a]

Candidate/particle	Approximate mass	Predicted by	Astrophysical effects
$G(R)$	—	Non-Newtonian gravitation	Mimics DM on large scales
Λ (cosmological constant)	—	General relativity	Provides $\Omega = 1$ without DM
Axion, majoron, goldstone boson	10^{-5} eV	QCD; PQ symmetry breaking	Cold DM
Ordinary neutrino	10–100 eV	GUTs	Hot DM
Light higgsino, photino, gravitino, axino, sneutrino[b]	10–100 eV	SUSY/SUGR	Hot DM
Para-photon	20–400 eV	Modified QED	Hot/warm DM
Right-handed neutrino	500 eV	Superweak interaction	Warm DM
Gravitino, etc.[b]	500 eV	SUSY/SUGR	Warm DM
Photino, gravitino, axino, mirror particle, simpson neutrino[b]	keV	SUSY/SUGR	Warm/cold DM
Photino, sneutrino, higgsino, gluino, heavy neutr.no[b]	MeV	SUSY/SUGR	Cold DM
Shadow matter	MeV	SUSY/SUGR	Hot/cold (like baryons)
Preon	20–200 TeV	Composite models	Cold DM
Monopoles	10^{16} GeV	GUTs	Cold DM
Pyrgon, maximon, perry pole, newtorites, Schwarzschild	10^{19} GeV	Higher-dimension theories	Cold DM
Supersymmetric strings	10^{19} GeV	SUSY/SUGR	Cold DM
Quark nuggets, nuclearites	10^{15} g	QCD, GUTs	Cold DM
Primordial black holes	10^{15-30} g	General relativity	Cold DM
Cosmic strings, domain walls	$10^{8-10}\,M_\odot$	GUTs	Promote galaxy formation, but cannot contribute much to Ω

[a] Abbreviations: DM, dark matter; QCD, quantum chromodynamics; PQ, Peccei & Quinn; GUTs, grand unified theories; SUSY, supersymmetric theories; SUGR, supergravity; QED, quantum electrodynamics.
[b] Of these various supersymmetric particles predicted by assorted versions of supersymmetric theories and supergravity, only one, the lightest, can be stable and contribute to Ω, but the theories do not at present tell us which one it will be or the mass to be expected.

6.4.2 HOT DARK MATTER AND GALAXY FORMATION All constraints derived from models of galaxy formation should be accepted with some caution, because a number of additional, unrelated parameters and processes undoubtedly contribute, and their effects are not always separable. Among these are (a) the spectrum of the initial perturbations; (b) the possibility that galaxies do not trace mass but form only at 2–3 σ peaks [biasing (50, 546, 623, 757)]; (c) gravitational clustering and relaxation after formation (155, 655), some kinds of which can mimic biasing (305); and (d) nongravitational mechanisms for galaxy formation including the selective shadowing of radiation pressure by dust (309) and gas-dynamical effects of exploding stars, supermassive objects, or active nuclei that pile up gas (330). All of these complicate the extrapolation back from observed galaxies to the nature of the underlying dark matter.

Nevertheless, the hot/cold distinction is an important one because in an expanding, cooling universe dominated by particles of mass m_i, the Jeans mass is

$$M_J = 3 \times 10^{18} \, M_\odot/m_i \, (\text{eV}) \qquad \qquad 2.$$

(92, 773). Thus, with a dominant particle of 10–100 eV (HDM), supercluster-sized structures will acquire their identity first and later fragment into galaxies, while masses of MeV, GeV, or more will lead to galaxies or smaller structures forming first and larger things being built up by gravitational clustering (CDM). Observations of amount of clustering as a function of redshift ought eventually to be able to tell us which happened (343) but have not yet done so.

Hot dark matter was the first to be considered and has several virtues. First, one sort is actually known to exist—the neutrinos and antineutrinos of electron, muon, and (presumably) tau flavors—which in most modern pictures of symmetry breaking should have some rest mass (690). Second, small extrapolations back in time of the hot big-bang conditions known from nucleosynthesis permit a fairly precise calculation of their number density at about 100 cm^{-3} for each species. Thus, the rest masses must be 10–100 eV if they are to add up to $\Omega = 1$ (154, 243). Other (hypothetical) particles in the same mass range do not share these virtues, but will behave in the same way during galaxy formation.

Three different experimental groups have reported evidence for neutrino rest masses in the cosmologically interesting range (423, 550, 625). Each has in turn been firmly doubted (16, 64, 85, 164, 421), and the case must currently be regarded as not proven. One of the (many) suggested resolutions of the solar neutrino problem (70) invokes a rest mass that might also fall in the interesting range.

Simulations of galaxy formation in a universe dominated by HDM have

DARK MATTER   459

been carried out by many groups, first enthusiastically (596), next with the realization that they make galaxies too late ($Z \lesssim 3$) and with velocity dispersions in the small-scale structure much larger than we see (135, 233, 740), and then in a spirit of "save the HDM" (455, 612). The current situation can probably be summarized by saying that hot dark matter is exceedingly useful in producing large-scale structure, including perhaps even the very large-scale streaming (153, 183, 234, 457, 700), and may have some part in scenarios with decaying dark matter or two "tooth fairies" (see Sections 6.4.4 and 6.4.5), but that it cannot be the only nonbaryonic component present at the time of galaxy formation.

6.4.3 COLD DARK MATTER A large fraction of the candidates listed in Table 3 come under this heading and are collectively called "ino's," WIMPs (weakly interacting massive particles), or just cold dark matter. Axions are formed cold (690) and so belong here despite their small masses.

It was recognized from the beginning that CDM scenarios would require something like biasing to provide correlations over large scales (270). If galaxies form only at several-sigma density peaks, then they will be more clustered than the underlying dark matter, because most of these peaks will be small fluctuations sitting atop larger scale, smaller amplitude ones. The biased cold dark matter program has now been explored in considerable detail and has achieved a number of successes in accounting for the observed properties of galaxies and clusters (50, 83, 84, 123, 169, 328, 539, 548, 570, 586, 634, 743, 765). There are minor disagreements about how best to do the calculations and interpret them (629), but the chief problem is in accounting for the largest scale voids and superclusters and, especially, for the very large-scale streaming motions (725, 726a, 739).

Because HDM gets into trouble with small-scale phenomena and CDM with large-scale ones, a natural thought is to try to combine their virtues in some way. This has been attempted through the assumption of massive particles that decay to relativistic ones (Section 6.4.4) and through the positing of two or more kinds of dark matter (Section 6.4.5).

6.4.4 DECAYING DARK MATTER The general idea here is that the Universe came out of its inflationary epoch with $\Omega = 1$ in some unstable WIMP that helped to make galaxies and then decayed away, leaving $\Omega = 1$ total in the decay products but only 0.2 in bound structures. There must be no photons in the decay products, thus the WIMPs cannot be photinos (495), but gravitinos and heavy neutrinos are possibilities. Calculations of this type are numerous (185, 240, 303, 326, 496, 583, 647, 691, 703a) and reasonably successful in making galaxies without disturbing the 3-K background. They require, however, fine tuning of the decay epoch, make the look-back age of the Universe uncomfortably short (since $R \propto t^{1/2}$ in a radiation-dominated universe), and must leave at least half the initial dark mass of a spiral galaxy in the halo in order to reproduce flat rotation curves (223). The advantages, therefore, probably do not outweigh the disadvantages.

6.4.5 SCENARIOS WITH TWO TOOTH FAIRIES By way of explanation, when an American child loses a milk (baby) tooth, he is supposed to put it under his pillow at bedtime. The tooth fairy comes during the night, takes away the tooth, and leaves a suitable coin in its place. The hope in these scenarios (595) is that we can put dark matter under our computers at quitting time and find that a good fairy has left galaxies by morning, except that in most of the schemes currently under investigation, it really takes two of them.

Biased CDM belongs to this category, unless the biasing comes automatically out of the way the perturbations grow (548). Percolation of explosive galaxy formation plus HDM is another possible combination (133). More popular, however, are the combination of (a) one hot and one cold entity, which together with the baryons add up to $\Omega = 1$ (9, 195, 211, 608, 699, 716), (b) one dark matter candidate plus nonzero cosmological constant (532), or (c) a dark matter background in which the seeds for galaxy formation are quark nuggets, primordial black holes, or cosmic strings (774). The strings, at least, automatically introduce large-scale correlations or biasing and are the 1986 "best-buy" model (12, 68, 255, 308, 547, 548, 643, 692, 718). It is, however, a safe bet that, by the time you read this in 1987 or later, some other combination will seem at least as likely to leave realistic galaxies under the computer.

6.4.6 DETECTION OF DARK MATTER CANDIDATES—ASTROPHYSICAL METHODS Existing or proposed observations can constrain, or possibly provide evidence for, the presence of several possible kinds of DM. Clearly the two are not entirely distinct; an effect attributable, for instance, to photino decay either proves the existence of photinos or sets an upper limit to them, depending on the likelihood of other causes of the same effect.

The excluded candidates are those with combinations of mass, lifetime, and cross section that would more than close the Universe, spoil nucleosynthesis, or produce a larger background of photons than we see in some waveband (86, 91, 268, 374, 590, 624, 649, 726). No whole class can be eliminated in this way.

Prospects for future detection include baryons (gas, faint galaxies, supermassive black holes, etc.) in voids whose presence would favor biasing and whose absence would favor strings as the mechanism for producing large-scale coherent structure. Gamma rays from WIMP decays or annihilations in the Milky Way halo should also eventually be measurable (565, 637).

Of the observations that could indicate detection of DM candidates has already occurred, some gravitational lenses and a feature near the center of our own Galaxy have been blamed on strings (141, 510), a possible 1667-Å feature in the ultraviolet radiation background has been attributed to light neutrino decay (644), and photons at higher energies attributed to decay or annihilation of more massive WIMPs (601–603). These could also be responsible for the unexpectedly high flux of low-energy cosmic-ray antiprotons (646); and sneutrinos might cool white dwarfs faster than conventional energy-transport mechanisms (472).

The most far-reaching of these "we may already have seen them" proposals is the simultaneous use of WIMPs to close the Universe, reduce the solar neutrino flux to the observed value, and adjust the frequencies of solar normal modes to match data (215, 241, 246, 532, 650). The particles must necessarily be trapped in cores of other stars as well and influence their structure and evolution, in ways that remain to be worked out but could be either good (J. Faulkner, personal communication, 1986) or bad (A. Renzini, personal communication, 1986) from the point of view of bringing theory and observation into accord.

6.4.7 DETECTION OF DARK MATTER CANDIDATES—LABORATORY METHODS
As in the astrophysical case, some volumes of parameter space can be ruled out on the basis of existing experiments (10, 177, 236a). All positive results so far reported—neutrino rest mass (550), axions (756), and monopoles (119)—have been questioned and need further work for their confirmation.

Future prospects include both the use of existing proton-decay apparatuses to look for high-energy neutrinos from WIMP decay (379) and the construction of new, dedicated apparatus. Some of these, like a proposed large, high Q microwave cavity to detect axions through their propensity to convert to photons in the presence of a strong magnetic field (620), carry price tags in the modern high-energy physics range. Two others, germanium or silicon spectrometers (32, 732) and bolometers (115, 190, 232), that detect WIMPs through their propensity to interact with nuclei and deposit energy in superconducting grains, could register roughly one count per day with a 10-kg detector and a cost in the individual PI grant range. These are, of course, enormously exciting possibilities, although we confess that we have not yet started urging our friends and relations to build suitable detectors.

7. L'ENVOI

Is there a dark-matter problem? Not necessarily, though there clearly are a number of astrophysical problems to which different kinds of dark matter are among the possible solutions. There are also astrophysical problems [e.g. the Great Red Spot on Jupiter (19a)] to which dark matter is not a solution.

No one kind of dark matter, with the remotely possible exception of ordinary baryons, can solve all the problems at once. On the other hand, to invoke five or six different kinds to match the properties of spiral disks and halos, ellipticals, dwarf galaxies, clusters, superclusters, and galaxy formation is probably also the wrong strategy. How are we betting? At least evens on baryons on scales up to superclusters, and maybe one chance in four on baryons all the way (either closing the Universe or managing to exist in an open one). The remaining probability spreads rather uniformly over a very large number of candidates at the moment. Several different events (confirmation of neutrino oscillations, laboratory detection of 10-GeV photinos, or whatever) could collapse the wave function considerably.

Finally, there are well-defined, finite, observational, theoretical, and experimental programs that can be expected to improve our understanding of the amount, distribution, and nature of nonluminous mass. These range from studying the brightness of K giants as a function of metal abundance and looking for baryonic matter in cosmic voids to pursuing higher dimensional theories of particle physics and building superconducting microwave cavities. There is surely at least one task suitable for every scientist who is interested in the problem. Failing this, feel free to keep in mind this thought from James Russell Lowell ("A Fable for Critics"): "Nature fits all of her children with something to do. He who would write, but can't write, can surely review."

Acknowledgments

A truly remarkable number of colleagues responded to a request for preprints, reprints, advice, and counsel. Sincere (and alphabetical) thanks, therefore, to Marc Aaronson, John and Neta Bahcall, Joshua Barnes, Timothy Beers, Ed Bertschinger, James Binney, Roger Blandford, Albert Bosma, Bruce Carney, Bernard Carr, Leo Connolly, Ramanath Cowsik, Marc Davis, Alan Dressler, Andrew Fabian, Giuseppina Fabbiano, S. Michael Fall, James Felten, Holland Ford, Carlos Frenk, John Gallagher, Gerard Gilmore, Riccardo Giovanelli, J. Richard Gott, Richard Harms, Michael Hawkins, Martha Haynes, Mark Henriksen, Jack Hills, Yehuda Hoffman, Satoru Ikeuchi, James Ipser, Bernard and Janet Jones, Alexander Kashlinsky, Stephen Kent, Robert Kirshner, George Lake, Richard Larson, Brian Lewis, V. N. Lukash, Richard Matzner, Adrian Melott, David Merritt, Mordehai Milgrom, Richard Miller, R. M. Muradian,

Richard Mushotzky, William Newman, Igor Novikov, Keith Olive, Jeremiah Ostriker, Ruth Peterson, Tsvi Piran, William Press, Joel Primack, Peter Quinn, Martin Rees, Herbert Rood, Edwin Salpeter, Renzo Sancisi, Allan Sandage, Robert Sanders, William Saslaw, Paul Schechter, David Schramm, Dennis Sciama, Jacob Shaham, Paul Shapiro, Ed Shaya, Joseph Silk, Michael Skrutskie, Floyd Stecker, Gary Steigman, Curtis Struck-Marcell, T. X. Thuan, John Tonry, Michael Turner, Mauri Valtonen, Gerard de Vaucouleurs, Ira Wasserman, Simon White, Amos Yahil, and W. H. Zurek.

Literature Cited

1. Aaronson, M. 1987. See Ref. 198. In press
2. Aaronson, M., Olszewski, E. 1986. *Astron. J.* 92: 580
3. Aaronson, M., Olszewski, E. 1986. See Ref. 369, p. 153
4. Aaronson, M., et al. 1982. *Ap. J.* 258: 64
5. Aaronson, M., et al. 1986. *Ap. J.* 302: 536
6. Abell, G. O. 1958. *Ap. J. Suppl.* 3: 211
7. Abell, G. O. 1961. *Astron. J.* 66: 607
9. Achilli, S., et al. 1985. *Ap. J.* 299: 577
10. Ahlen, S. P., et al. 1987. Preprint CfA No. 2292
11. Albada, T. S. van, et al. 1985. *Ap. J.* 295: 305
12. Albrecht, A. 1987. See Ref. 198. In press
16. Altzitzoglou, T., et al. 1985. *Phys. Rev. Lett.* 55: 799
17. Allen, R. S., Allen, R. J. 1979. *Astron. Astrophys.* 74: 73
19. Anderson, J. D., Mashoon, B. 1985. *Ap. J.* 290: 445
19a. Antipov, S. V., et al. 1986. *Nature* 323: 238
20. Applegate, J. H., et al. 1986. *Phys. Rev. D* 35: 1151
21. Arkhipova, V. P., et al. 1986. *Sov. Astron. AJ* 30: 11
22. Armandroff, T. E., Da Costa, G. 1986. *Astron. J.* 92: 777
24. Arp, H. C. 1986. *Astron. Astrophys.* 156: 207
25. Athanassoula, E., ed. 1983. *Internal Kinematics and Dynamics of Galaxies, IAU Symp. No. 100.* Dordrecht: Reidel
26. Athanassoula, E., Bosma, A. 1985. *Ann. Rev. Astron. Astrophys.* 23: 147
27. Athanassoula, E., et al. 1986. See Ref. 369, p. 133
28. Athanassoula, E., Bosma, A. 1986. *Astron. Astrophys.* In press
29. Athanassoula, E., Sellwood, J. 1986. *MNRAS* 221: 213
31. Avedisova, A. S. 1985. *Sov. Astron. Lett.* 11: 378
32. Avignone, F. T., et al. 1986. In *Weak and Electromagnetic Interactions in Nuclei*, ed. H. V. Klapdor, p. 676. Berlin: Springer-Verlag
33. Babcock, H. W. 1939. *Lick Obs. Bull.* 19: 41
34. Bacon, R., et al. 1985. *Astron. Astrophys.* 152: 315
35. Bahcall, J. N. 1984. *Ap. J.* 276: 169; also 287: 926
36. Bahcall, J. N. 1986. *Ann. Rev. Astron. Astrophys.* 24: 577
37. Bahcall, J. N. 1986. See Ref. 369, p. 17
39. Bahcall, J. N., Castertano, S. 1985. *Ap. J.* 293: L3
40. Bahcall, J. N., et al. 1985. *Ap. J.* 290: 15
42. Bahcall, N. 1984. *Adv. Space Res.* 3: 367
43. Bahcall, N. 1986. *Ap. J. Lett.* 302: L41
44. Bahcall, N. 1987. See Ref. 198. In press. Also, *Comments Astrophys.* 11: 263
46. Bahcall, N., Soneira, R. 1984. *Ap. J.* 277: 27
47. Bahcall, N., et al. 1986. *Ap. J.* 311: 15
47a. Bajaja, E., et al. 1984. *Astron. Astrophys.* 141: 309
48. Ball, R. 1986. *Ap. J.* 307: 453
50. Bardeen, J., et al. 1986. *Ap. J.* 304: 15
51. Barnes, J. 1984. *MNRAS* 208: 873
52. Barnes, J. 1985. *MNRAS* 215: 517
53. Barnes, J. 1986. Talk at Santa Cruz Workshop, Nearly Normal Galaxies, 1986
54. Barnes, J. 1987. See Ref. 198. In press
57. Batusky, D. J., Burns, J. O. 1985. *Astron. J.* 90: 1413
58. Beckmann, J. E., et al. 1985. *Astron. Astrophys.* 157: 48
60. Beers, T. C., et al. 1984. *Ap. J.* 283: 33

62. Bergh, S. van den 1971. *Astron. Astrophys.* 11: 154
64. Bergqvist, K. 1985. *Phys. Lett.* 154B: 224
68. Bertschinger, E. 1987. *Ap. J.* 316: In press
70. Bethe, H. 1986. *Phys. Rev. Lett.* 56: 1305
71. Bicknell, G. V. 1986. *Ap. J.* 305: 109
73. Binney, J. 1982. *Ann. Rev. Astron. Astrophys.* 20: 399
75. Binney, J. 1986. *Philos. Trans. R. Soc. London Ser. A* 320: 431
76. Binney, J. 1986. See Ref. 369, p. 303
77. Binney, J., Cowie, L. L. 1981. *Ap. J.* 247: 464
79. Blackman, C. P., van Morsel, G. A. 1984. *MNRAS* 208: 91
80. Blandford, R. D. 1986. See Ref. 654a, p. 359
81. Blitz, L. 1983. See Ref. 25, p. 43
82. Blitz, L., et al. 1980. In *Interstellar Molecules, IAU Symp. No. 87*, ed. B. Andrews, p. 213. Dordrecht: Reidel
83. Blumenthal, G., et al. 1984. *Nature* 311: 405
84. Blumenthal, G., et al. 1986. *Ap. J.* 301: 27
85. Boehm, F., Vogel, P. 1984. *Ann. Rev. Nucl. Part. Sci.* 34: 125
86. Boesgaard, A. M., Steigman, G. 1985. *Ann. Rev. Astron. Astrophys.* 23: 319
89. Bond, J. R., Szalay, A. 1983. *Ap. J.* 274: 443
90. Bond, J. R., Carr, B. J., Arnett, W. D. 1983. *Nature* 304: 514
91. Bond, J. R., Carr, B., Hogan, C. 1985. Fermilab Preprint 85-115
92. Bond, J. R., et al. 1980. *Phys. Rev. Lett.* 45: 1980
93. Bosma, A. 1981. *Astron. J.* 86: 1721, 1825
94. Bottema, R., et al. 1986. *Astron. Astrophys.* 167: 34
95. Bothun, G., et al. 1983. *Ap. J.* 268: 47
96. Bothun, G., et al. 1985. *Astron. J.* 90: 2487
100. Brandenberger, R. 1985. *Rev. Mod. Phys.* 57: 1
102. Briggs, H. 1986. *Ap. J.* 300: 613
103. Brinks, E., Burton, W. B. 1984. *Astron. Astrophys.* 141: 195
104. Brosch, N. 1985. *Astron. Astrophys.* 153: 199
105. Brosch, N., Gondalekhar, P. 1986. *Ap. J. Lett.* 140: L49
106. Brosche, P., et al. 1985. *Astron. J.* 90: 2033
107. Burbidge, E. M., Burbidge, G. R. 1961. *Astron. J.* 66: 541
108. Burbidge, G. R. 1967. *Nature* 216: 1287
109. Burbidge, G. R. 1975. *Ap. J. Lett.* 196: L7
109a. Burg, G. van den, Shane, W. W. 1986. *Astron. Astrophys.* 168: 49
110. Burke, B. 1986. See Ref. 654a, p. 517
111. Burstein, D., et al. 1986. *Ap. J. Lett.* 305: L11
112. Burstein, D., et al 1986. See Ref. 432
113. Bushouse, H., et al. 1985. *MNRAS* 217: 7P
115. Cabrera, B., et al. 1985. *Phys. Rev. Lett.* 55: 25
116. Caldwell, R., et al. 1986. *Ap. J.* 305: 136
118. Capelato, H. V., et al. 1985. *Astrophys. Space Sci.* 108: 363
119. Caplin, A. D., et al. 1986. *Nature* 321: 402
120. Carignan, C. 1985. *Ap. J.* 299: 59
121. Carignan, C., Freeman, K. C. 1985. *Ap. J.* 294: 494
122. Carlberg, R., Sellwood, J. 1985. *Ap. J.* 292: 79
123. Carlberg, R., et al. 1986. *Ap. J. Lett.* 300: L1
124. Carney, B. W. 1984. *Publ. Astron. Soc. Pac.* 96: 841
125. Carney, B. W., Latham, D. 1986. See Ref. 369, p. 39
126. Carr, B. J. 1985. In *Observational and Theoretical Aspects of Relativistic Astrophysics and Cosmology*, ed. J. L. Sanz, L. J. Goicoecha, p. 1. Singapore: World Scientific
127. Carr, B. J., et al. 1984. *Ap. J.* 277: 445
128. Carter, D., et al. 1985. *MNRAS* 212: 471
130. Cavaliere, A., et al. 1986. *Ap. J.* 309: 651
132. Chandrasekhar, S. 1957. *Stellar Dynamics.* New York: Dover
133. Charlton, J. C., Schramm, D. N. 1986. *Ap. J.* 310: 26
134a. Chatterjee, T. K. 1984. *Astrophys. Space Sci.* 106: 309
135. Chau, H.-Y., et al. 1984. *Ap. J.* 281: 560
138. Chini, R., Wink, J. E. 1984. *Astron. Astrophys.* 139: L5
140. Chubb, T. A. 1986. *Ap. J.* 305: 609
141. Chudnovsky, E. M., Field, G. B., Spergel, D., Vilenkin, A. 1986. Preprint
142. Ciardullo, R., et al. 1985. *Ap. J.* 293: 69
143. Clayton, D. D. 1985. *Ap. J.* 290: 428
144. Clemens, D. P. 1985. *Ap. J.* 295: 422
146. Cohen, J. G. 1983. *Ap. J. Lett.* 270: L41
147. Collins, C. A., et al. 1986. *Nature* 320: 506
148. Connolly, L. 1985. *Ap. J.* 299: 728
149. Contopoulos, G., Magnenat, P. 1986. *Celest. Mech.* 37: 387

151. Cowie, L. L., Hu, E. M. 1986. *Ap. J. Lett.* 305: L39
152. Cowie, L. L., et al. 1987. *Ap. J.* In press
153. Cowsik, R., Ghosh, P. 1987. *Ap. J.* 317: In press
154. Cowsik, R., McClelland, J. 1972. *Phys. Rev. Lett.* 29: 669
155. Crane, P., Saslaw, W. C. 1986. *Ap. J.* 301: 1
158. Cudworth, K. 1986. *Astron. J.* 92: 348
158a. Da Costa, G. 1987. See Ref. 265. In press
159. Da Costa, G., Freeman, K. C. 1985. In *IAU Symp. 113*, p. 69
161. Dahn, C., et al. 1986. *Astron. J.* 91: 621
163. D'Antona, F., Mazzitelli, I. 1986. *Astron. Astrophys.* 162: 80
164. Datar, V. M., et al. 1986. *Nature* 318: 547
166. Davis, M. 1986. See Ref. 369, p. 97
167. Davis, M., Djorgovski, S. 1985. *Ap. J.* 299: 15
168. Davis, M., Peebles, P. J. E. 1982. *Ap. J.* 267: 265
169. Davis, M., et al. 1985. *Ap. J.* 292: 371
170. Dearborn, D. S., et al. 1986. *Ap. J.* 302: 35
171. Dekel, A. 1987. See Ref. 198. In press. Also, *Comments Astrophys.* 11: 275 (1986)
171a. Dekel, A., Shaham, J. 1978. *Astron. Astrophys.* 74: 186
172. Dekel, A., Shaham, J. 1980. *Astron. Astrophys.* 85: 154
177. De Rújula, A., et al. 1986. *Nature* 320: 38
178. Des Forêts, G., et al. 1984. *Ap. J.* 280: 15
180. Dicke, R. H., Peebles, P. J. E. 1979. In *General Relativity: An Einstein Centenary Review*, ed. S. W. Hawking, W. Israel, p. 504. Cambridge: Cambridge Univ. Press
183. Doroshkevich, A. G. 1984. *Sov. Astron. AJ* 28: 253
185. Doroshkevich, A. G., et al. 1985. *Sov. Astron. Lett.* 11: 201
186. Dressler, A. 1984. *Ann. Rev. Astron. Astrophys.* 22: 185
187. Dressler, A. 1984. *Ap. J.* 281: 512
188. Dressler, A., Lecar, M. 1983. Preprint
189. Dressler, A., et al. 1986. *Astron. J.* 91: 1059
190. Drukier, A., et al. 1986. *Phys. Rev. D* 33: 3495
191. Dupraz, C., Combes, F. 1986. *Astron. Astrophys.* 166: 53
193. Einasto, J., et al. 1974. *Nature* 250: 309
194. Einasto, J., et al. 1986. *MNRAS* 219: 457
195. Einasto, J., Einasto, M., Gramann, M., Saar, E. 1986. Submitted for publication
196. Elson, R., Hut, P., Inagaki, S. 1987. *Ann. Rev. Astron. Astrophys.* 25: 565
197. Evrard, A. E., Yahil, A. 1985. *Ap. J.* 296: 299, 310
198. Faber, S. M., ed. 1987. *Nearly Normal Galaxies*. Springer-Verlag. In press
200. Faber, S. M., Gallagher, J. S. 1979. *Ann. Rev. Astron. Astrophys.* 17: 135
201. Faber, S. M., Lin, D. C. 1983. *Ap. J. Lett.* 266: L17
202. Fabian, A. C., et al. 1986. See Ref. 369, p. 201
203. Fabian, A. C., et al. 1980. *Nature* 287: 613
204. Fabian, A. C., et al. 1981. *Ap. J.* 248: 47
205. Fabian, A. C., et al. 1986. *Ap. J.* 305: 9
206. Fabian, A. C., et al. 1986. *MNRAS* 221: 1049
208. Fabricant, D., et al. 1986. *Ap. J.* 308: 530
208a. Fall, S. M. 1975. *MNRAS* 172: 23P
209. Fall, S. M., Efstathiou, G. 1981. *MNRAS* 186: 133
211. Fang, L. Z., et al. 1985. *Astrophys. Space Sci.* 115: 99
212. Fasenko, B. 1985. *Astrofizika* 20: 495
215. Faulkner, J., et al. 1986. *Nature* 321: 226
216. Felten, J. 1984. *Ap. J.* 286: 1
217. Felten, J. 1986. *Comments Astrophys.* 11: 53
218. Felten, J. 1986. See Ref. 369, p. 111
219. Ferney, J. A., Bhavsar, S. P. 1985. *MNRAS* 210: 883
220. Field, G. B., Perrenod, S. C. 1977. *Ap. J.* 215: 717
221. Fleming, T. A., et al. 1986. *Ap. J.* 308: 176
223. Flores, R., et al. 1986. *Nature* 323: 781
224. Focardi, P., et al. 1986. *Astron. Astrophys.* 161: 217
226. Ford, H. C., et al. 1981. *Ap. J. Lett.* 245: L53
227. Forman, W., et al. 1985. *Ap. J. Lett.* 293: 102
228. Fowler, W. A. 1971. *Quad. Accad. Naz. Lincei* 157: 115
229. Freeman, K. C. 1987. *Ann. Rev. Astron. Astrophys.* 25: 603
230. Freeman, K. C. 1970. *Ap. J.* 160: 811
231. Freeman, K. C. 1987. See Ref. 198. In press
232. Freese, K. 1987. See Ref. 198. In press
233. Frenk, C. S., et al. 1983. *Ap. J.* 271: 417
234. Fry, J. N. 1986. *Ap. J.* 306: 358
236. Fujishima, F., Tosa, M. 1985. *Publ. Astron. Soc. Jpn.* 36: 333
236a. Gaisser, T. K., Steigman, G. 1986. *Phys. Rev. D* 34: 2260

237. Gallagher, J. S., Bushouse, H. 1983. *Astron. J.* 88: 55
238. Geller, M., Huchra, J. 1983. *Ap. J. Suppl.* 52: 61
239. Geller, M., et al. 1984. *Astron. J.* 89: 319
240. Gelmini, G., et al. 1984. *Phys. Lett.* 146B: 311
241. Gelmini, G. C., Hall, L. J., Lin, M. J. 1986. Preprint HUTP 86/A042
243. Gershtein, S. S., Zeldovich, Ya. B. 1966. *JETP Lett.* 4: 175
244. Giaconni, R., Zamorani, G. 1987. *Ap. J.* 313: 20
245. Gibbons, G. W., Whitting, B. F. 1981. *Nature* 291: 636
246. Gilliland, R. L., et al. 1986. *Ap. J.* 306: 703
248. Gilmore, G., et al. 1985. *MNRAS* 213: 257
249. Giovanelli, R., Haynes, M. 1985. *Astron. J.* 90: 2445
251. Giovanelli, R., et al. 1986. *Ap. J. Lett.* 301: L1
252. Giovanelli, R., et al. 1986. *Ap. J.* 300: 77
253. Giraud, E. 1986. *Astron. Astrophys.* 167: 41
254. Gott, J. R. 1983. In *Early Evolution of the Universe and Its Present Structure, IAU Symp. No. 104*, ed. G. Chincarini, G. Abell, p. 235. Dordrecht: Reidel
255. Gott, J. R. 1985. *Ap. J.* 288: 422
256. Gott, J. R. 1985. See Ref. 375. In press
257. Gott, J. R., Gunn, J. E., Schramm, D. N., Tinsley, B. M. 1974. *Ap. J.* 194: 543
258. Gott, J. R., et al. 1986. *Ap. J.* 306: 341
259. Gottesman, S. T., Hawarden, T. G. 1986. *MNRAS* 219: 759
260. Gottesman, S. T., et al. 1983. See Ref. 25, p. 93
261. Gottesman, S. T., et al. 1984. *Ap. J.* 286: 471
262. Greenfield, P. E., et al. 1985. *Ap. J.* 293: 379
263. Gregory, S. A., Thompson, L. A. 1984. *Ap. J.* 286: 422
264. Grillmair, C., et al. 1986. *Astron. J.* 91: 1328
265. Grindlay, J., Philip, A. G. D., eds. 1987. *Globular Cluster Systems in Galaxies, IAU Symp. No. 126*. Dordrecht: Reidel. In press
266. Guilbert, P. W., Fabian, A. C. 1986. *MNRAS* 220: 439
267. Gunn, J. E. 1974. *Comments Astrophys. Space Sci.* 6: 7
268. Gunn, J. E. 1986. See Ref. 369, p. 537
269. Gunn, J. E., Griffin, R. F. 1979. *Astron. J.* 84: 712
270. Gunn, J. E., et al. 1978. *Ap. J.* 223: 1015
271. Guth, A. H. 1984. *Ann. NY Acad. Sci.* 422: 1
273. Hammer, F., Notale, L. 1986. *Astron. Astrophys.* 155: 420
276. Harms, R., et al. 1981. *Ann. NY Acad. Sci.* 375: 178
278. Hartwick, F. D. A., Sargent, W. L. W. 1974. *Ap. J.* 190: 283
279. Haud, U. 1984. *Astrophys. Space Sci.* 104: 337
280. Haud, U., et al. 1985. In *The Milky Way, IAU Symp. No. 106*, ed. H. van Woerden, W. B. Burton, R. J. Allen, p. 85. Dordrecht: Reidel
281. Hauser, M., et al. 1984. *Ap. J.* 285: 74
282. Hawkins, M. R. S. 1984. *MNRAS* 206: 433
283. Hawkins, M. R. S. 1987. *MNRAS* 223: 845 (and personal communication)
284. Hayashi, C. 1950. *Progr. Theor. Phys.* 5: 224
285. Haynes, M. 1987. See Ref. 198. In press
287. Haynes, M., Giovanelli, R. 1986. *Ap. J. Lett.* 306: L55
288. Hegyi, D. J., et al. 1986. *Ap. J.* 300: 492
289. Heisler, J., et al. 1985. *Ap. J.* 298: 8
293. Herbst, E. 1975. *Publ. Astron. Soc. Pac.* 87: 827
294. Hernquist, L., Quinn, P. J. 1987. *Ap. J.* 312: 1
295. Hernquist, L., Quinn, P. J. 1987. *Ap. J.* 312: 11
296. Herschel, W. H. 1784. In *Collected Works*, 1: 157
297. Hesser, J., et al. 1986. *Ap. J. Lett.* 303: L51
299. Hills, J. G. 1986. *Astron. J.* 92: 595
301. Hodge, P. 1981. *Ann. Rev. Astron. Astrophys.* 19: 357
303. Hoffman, Y. 1986. *Ap. J. Lett.* 305: L1
304. Hoffman, Y., Shaham, J. 1985. *Ap. J.* 297: 16
305. Hoffman, Y., et al. 1982. *Ap. J.* 262: 413
306. Hoffman, Y., et al. 1985. *Ap. J. Lett.* 289: L15
307. Hogan, C. 1987. See Ref. 198. In press
308. Hogan, C., Rees, M. J. 1985. *Nature* 311: 109
309. Hogan, C., White, S. D. M. 1986. *Nature* 321: 575
311. Holmberg, E. 1937. *Lund Obs. Ann.* 6: 173
313. Hopp, U., Materne, J. 1985. *Astron. Astrophys. Suppl.* 61: 93
316. Huchra, J., Brodie, J. 1987. *Astron. J.* 93: 779
317. Huchra, J., Geller, M. 1982. *Ap. J.* 257: 423
318. Huchra, J., et al. 1985. In *Clusters and Groups of Galaxies*, ed. F. Mardirossian et al., p. 79. Dordrecht: Reidel
319. Huchra, J., et al. 1980. *Ap. J. Lett.* 259: L57

320. Humason, M. L., Wahlquist, H. D. 1955. *Astron. J.* 60: 254
321. Hunter, D., Gallagher, J. S. 1984. *Ann. Rev. Astron. Astrophys.* 22: 37
322. Hunter, D., et al. 1986. *Astron. J.* 91: 1086
325. Hut, P., Tremaine, S. D. 1985. *Astron. J.* 90: 1548
326. Hut, P., White, S. D. M. 1984. *Nature* 310: 637
328. Ikeuchi, S. 1986. *Astrophys. Space Sci.* 118: 500
330. Ikeuchi, S., Ostriker, J. P. 1986. *Ap. J.* 301: 522
333. Innanen, K. A., et al. 1983. *Astron. J.* 88: 338
334. Iovino, A., Shaver, P. A. 1986. *Astron. Astrophys.* 166: 119
336. Ipser, J., Semenzato, R. 1985. *Astron. Astrophys.* 149: 408
339. Islam, J. N. 1983. *Phys. Lett.* 97A: 239
340. Jarvis, B. J., Freeman, K. C. 1985. *Ap. J.* 295: 246
341. Joeveer, M., Einasto, J. 1978. In *The Large Scale Structure of the Universe, IAU Symp. No. 79*, ed. M. Longair, J. Einasto, p. 241. Dordrecht: Reidel
342. Joeveer, M., et al. 1978. *MNRAS* 185: 357
343. Jones, B. J. T., Palmer, P. L. 1986. In *Axisymmetric Systems, Galaxies, and Relativity*, ed. M. A. H. MacCallum. Cambridge: Cambridge Univ. Press. In press
344. Jura, M. 1986. *Astron. J.* 91: 539
345. Kafatos, M., Harrington, R. S., eds. 1986. *Brown Dwarfs*. Cambridge: Cambridge Univ. Press
346. Kahn, F. D., Woltjer, L. 1959. *Ap. J.* 130: 105
348. Kalinikov, M., Kuneva, I. 1986. *MNRAS* 218: 49P
349. Kalnajs, A. J. 1983. See Ref. 25, p. 87
350. Kamahori, O., Fujimoto, M. 1986. *Publ. Astron. Soc. Jpn.* 38: 151
353. Karachentsev, I. D. 1984. *Sov. Astron. Lett.* 10: 77
354. Karachentsev, I. D. 1985. *Sov. Astron. AJ* 29: 243
356. Karachentsev, V. T., Karachentsev, I. D. 1982. *Astrofizika* 18: 1
359. Kashlinsky, A., Rees, M. J. 1983. *MNRAS* 205: 955
360. Katz, N., Richstone, D. O. 1985. *Ap. J.* 299: 331
361. Kent, S. M. 1986. *Astron. J.* 91: 1301
362. Kent, S. M., Gunn, J. E. 1982. *Astron. J.* 87: 945
363. Kent, S. M., Sargent, W. L. W. 1983. *Astron. J.* 88: 697
365. Kerr, F. J., Lynden-Bell, D. 1986. *MNRAS* 221: 1027
366. Keto, E. R., Myers, P. C. 1986. *Ap. J.* 304: 466
368. Kirschner, R. P., Oemler, A., Schechter, P. L., Shectman, S. 1981. *Ap. J. Lett.* 248: L57 (KOSS)
369. Knapp, G. R., Kormendy, J. F., eds. 1986. *Dark Matter in the Universe, IAU Symp. No. 117*. Dordrecht: Reidel
370. Knapp, G. R., et al. 1978. *Ap. J.* 222: 808
374. Kolb, E. W., et al. 1986. *Phys. Rev. D* 34: 2197
375. Kolb, E. W., et al., eds. 1986. *Inner Space/Outer Space*. Chicago: Univ. Chicago Press
376. Kolesnik, I. G., Yurevich, L. V. 1985. *Astrofizika* 22: 272
377. Koo, D., Kron, R., Szalay, A. 1986. Unpublished redshift survey
379. Krauss, L. M., et al. 1986. *Phys. Rev. D* 33: 2079
381. Kriss, G. A., et al. 1983. *Ap. J.* 272: 439
382. Krol, V. A. 1986. *Astrofizika* 23: 499
384. Kruit, P. van der 1986. *Astron. Astrophys.* 157: 230
385. Kruit, P. van der 1986. See Ref. 369, p. 415
386. Kruit, P. van der, Freeman, K. C. 1984. *Ap. J.* 278: 81
387. Kruit, P. van der, Shostak, G. S. 1984. *Astron. Astrophys.* 134: 258
389. Krumm, N., Burstein, D. 1984. *Astron. J.* 89: 1312
390. Kulkarni, S. R., et al. 1982. *Ap. J. Lett.* 259: L63
392. Kuzmin, G. G. 1955. *Tartu Astron. Obs. Publ.* Cited in Ref. 369, p. 243
393. Lacey, C. G. 1984. *MNRAS* 208: 687
394. Lacey, C. G., Ostriker, J. P. 1985. *Ap. J.* 299: 633
397. Lahav, O. 1987. *MNRAS* 225: 213
398. Lahav, O. 1986. *MNRAS* 220: 259
400. Lake, G., Schommer, R. 1984. *Ap. J. Lett.* 279: L19
403. Lapparent, V. de, et al. 1986. *Ap. J.* 304: 585
404. Lapparent, V. de, et al. 1986. *Ap. J. Lett.* 302: L1
406. Larson, R. B. 1986. *MNRAS* 218: 400
410. Lee, H., et al. 1986. *Ap. J. Lett.* 304: L1
411. Levison, H. F., Richstone, D. O. 1985. *Ap. J.* 295: 340
414. Lewis, B. M. 1987. *Ap. J. Suppl.* 63: 515
415. Liebert, J., Probst, R. G. 1987. *Ann. Rev. Astron. Astrophys.* 25: 473
416. Lilje, P. B., et al. 1986. *Ap. J.* 307: 91
417. Lin, D. C., Faber, S. M. 1983. *Ap. J. Lett.* 266: L21
418. Lin, D. C., Lynden-Bell, D. 1977. *MNRAS* 181: 7
419. Lin, D. C., Lynden-Bell, D. 1982. *MNRAS* 198: 707

421. Lindhard, J., Hansen, P. G. 1986. *Phys. Rev. Lett.* 57: 965
422. Loh, E. D., Spillar, E. J. 1986. *Ap. J. Lett.* 301: L1
423. Lubimov, V. A., et al. 1980. *Phys. Lett.* 94B: 266
424. Luccin, F., et al. 1986. *Astron. Astrophys.* 162: 13
426. Lucey, J. R., et al. 1986. *MNRAS* 221: 453
427. Lukash, V. N. 1986. See Ref. 369, p. 379
427a. Lupton, R. H., Gunn, J. E., Griffin, R. 1986. Preprint
428. Lynden-Bell, D. 1983. In *Kinematics, Dynamics, and Structure of the Milky Way*, ed. W. H. L. Shuter, p. 349. Dordrecht: Reidel
429. Lynden-Bell, D., et al. 1983. *MNRAS* 204: 87P
430. Maccagni, O., et al. 1987. *Ap. J.* 316: In press
432. Madore, B. F., Tully, R. B., eds. 1986. *Galaxy Distances and Deviations from Universal Expansion.* Dordrecht: Reidel
433. Magnani, L., et al. 1985. *Ap. J.* 295: 402
434. Malamuth, E. M., Kriss, G. A. 1986. *Ap. J.* 308: 10
436. Malin, D. F., Carter, D. 1983. *Ap. J.* 274: 534
437. Mamon, G. 1986. *Ap. J.* 307: 426
439. Marshall, R. E., et al. 1980. *Ap. J.* 235: 4
439a. Materne, J., Tammann, G. 1974. *Astron. Astrophys.* 37: 383
440. Mathieu, R. D. 1987. See Ref. 265. In press
441. Mathieu, R. D., Latham, D., Griffin, R. F., Gunn, J. E. 1986. *Astron. J.* 92: 1100
442. Mathur, S. D. 1984. *MNRAS* 211: 901
443. Matilsky, T., et al. 1985. *Ap. J.* 291: 621
446. McCrea, W. H. 1971. *Q. J. R. Astron. Soc.* 12: 140
447. McDowell, J. 1986. *MNRAS* 217: 77
449. Meiksin, A., Davis, M. 1986. *Astron. J.* 91: 191
450. Meisel, A. 1984. *Astron. Astrophys.* 145: 135
455. Melott, A. 1985. *Ap. J.* 289: 2
456. Melott, A. 1986. *Phys. Rev. Lett.* 57: 257
457. Melott, A. 1986. Preprint (Some Like It Hot)
459. Merigh, R., et al. 1986. *Astron. Astrophys.* 160: 298
461. Merritt, D. 1987. *Ap. J.* 313: 121
463. Michell, J. 1767. *Philos. Trans. R. Soc.* 1767: 234
466. Milgrom, M. 1986. *Ap. J.* 306: 9
468. Miller, L., et al. 1986. *MNRAS* 220: 713
468a. Miller, R. H. 1986. *Astron. Astrophys.* 167: 41
469. Millington, S. J. C., Peach, J. V. 1986. *MNRAS* 221: 15
470. Misra, R. 1985. *MNRAS* 212: 163
471. Miyamoto, M., et al. 1980. *Astron. Astrophys.* 90: 215
472. Mochkovich, R., Olive, K. A., Silk, J. 1986. Preprint
473. Monet, D. G., et al. 1981. *Ap. J.* 245: 454
474. Morozov, A. 1983. *Sov. Astron. Lett.* 9: 370
475a. Mould, J., Oke, J. B., Nemec, J. M. 1986. *Astron. J.* 93: 53
479. Narayan, R. 1986. See Ref. 654a, p. 529
480. Nasel'skii, P. N., Poharëv, A. C. 1985. *Sov. Astron. AJ* 29: 487
481. Navarro, J. F., et al. 1986. *Astrophys. Space Sci.* 123: 117
482. Newman, R. 1983. *Proc. Marcel Grossman Meet. General Relativ.*, 3rd, ed. Hu Ning. Amsterdam: North-Holland
484. Neyman, J., Page, T., Scott, E., eds. 1961. *Astron. J.* 66: 533
485. Nieto, J.-L., et al. 1984. *Astron. Astrophys.* 139: 464
486. Ninkovich, S. 1984. *Astrofizika* 20: 150
487. Ninkovich, S. 1984. *Astrophys. Space Sci.* 110: 379
488. Noltheníus, R., Ford, H. 1986. *Ap. J.* 305: 600
489. Norris, J. 1986. *Ap. J. Suppl.* 61: 667
490. Nulsen, P. E., et al. 1984. *MNRAS* 208: 165
491. Oemler, A. 1987. See Ref. 198. In press
492. Olive, K. A. 1986. See Ref. 375
493. Olive, K. A. 1986. *Ap. J.* 309: 210
495. Olive, K. A., et al. 1985. *Nucl. Phys. B* 255: 495
496. Olive, K. A., et al. 1985. *Ap. J.* 291: 1
497. Olszewski, E. W., et al. 1986. *Ap. J. Lett.* 302: L45
498. Oort, J. 1932. *Bull. Astron. Inst. Neth.* 6: 249
499. Oort, J. 1960. *Bull. Astron. Inst. Neth.* 16: 45
500. Ostriker, J. P. 1986. See Ref. 369, p. 85
501. Ostriker, J. P. 1987. See Ref. 265. In press
503. Ostriker, J. P., Peebles, P. J. E. 1973. *Ap. J.* 186: 467
504. Ostriker, J. P., Vietri, M. 1986. *Ap. J.* 300: 68
506. Ostriker, J. P., et al. 1974. *Ap. J. Lett.* 193: L1
507. Pacheco, J. A. de F. 1984. *Astrophys. Space Sci.* 105: 393
509. Paczyński, B. 1986. *Ap. J.* 304: 1
510. Paczyński, B. 1986. *Nature* 319: 567

511. Page, T. L. 1952. *Ap. J.* 116: 63
512. Page, T. L. 1962. *Ap. J.* 136: 685
513. Page, T. L. 1975. In *Galaxies and the Universe* (*Stars and Stellar Systems*, Vol. 9), ed. A. Sandage, M. Sandage, J. Kristian, p. 541. Chicago: Univ. Chicago Press
515. Peebles, P. J. E. 1971. *Cosmology*. Princeton, NJ: Princeton Univ. Press
516. Peebles, P. J. E. 1979. In *Physical Cosmology*, ed. R. Balian, J. Audouze, D. N. Schramm, p. 213. Amsterdam: North-Holland
517. Peebles, P. J. E. 1980. *The Large Scale Structure of the Universe*. Princeton, NJ: Princeton Univ. Press
519. Peebles, P. J. E. 1984. *Ap. J.* 284: 439
520. Peebles, P. J. E. 1984. *Ap. J.* 277: 470
521. Peebles, P. J. E. 1986. *Nature* 321: 27
522. Perea, J., et al. 1985. *MNRAS* 219: 511; 222: 49 (1986)
523. Peterson, R. C. 1985. *Ap. J.* 297: 309
524. Peterson, R. C., Latham, D. 1986. *Ap. J.* 305: 645
525. Peterson, S. D. 1979. *Ap. J.* 232: 20
527. Petrovskaya, I. V., Terrikorpi, P. 1986. *Astron. Astrophys.* 163: 39
528. Phillips, S. 1986. *Nature* 314: 721
530. Pickles, A. 1985. *Ap. J.* 296: 340
531. Postman, M., et al. 1986. *Astron. J.* 91: 1267
532. Primack, J. R. 1986. See Ref. 607
534. Pritchet, C., et al. 1986. *Astron. J.* 91: 1
535. Pryor, C., et al. 1986. *Astron. J.* 91: 546
539. Quinn, P. J., et al. 1986. *Nature* 322: 329. Also, see Ref. 369, p. 316
540. Ramadurai, S., Rees, M. J. 1985. *MNRAS* 215: 53P
543. Rees, M. J. 1971. *MNRAS* 154: 187
546. Rees, M. J. 1985. *MNRAS* 213: 75P
547. Rees, M. J. 1986. *MNRAS* 222: 27P
548. Rees, M. J. 1987. See Ref. 198. In press
549. Reeves, H. 1974. *Ann. Rev. Astron. Astrophys.* 12: 437
550. Reines, F., et al. 1980. *Phys. Rev. Lett.* 45: 1307
551. Richstone, D. O., Tremaine, S. D. 1984. *Ap. J.* 286: 27
553. Robe, H., Leruth, L. 1984. *Astron. Astrophys.* 133: 369
554. Roberts, D. H., et al. 1985. *Ap. J.* 293: 356
555. Roelfsema, P. R., Allen, R. J. 1985. *Astron. Astrophys.* 146: 213
556. Romani, R., Taylor, J. H. 1983. *Ap. J. Lett.* 265: L35
557. Rood, H. 1965. PhD thesis. Univ. Mich., Ann Arbor
558. Rood, H. 1980. *Rep. Progr. Phys.* 44: 1078
559. Rood, H., Williams, B. A. 1985. *Ap. J.* 288: 535
560. Rowan-Robinson, M., Needham, G. 1986. *MNRAS* 222: 611
561. Rubin, V. C., D'Odorico, S. 1969. *Astron. Astrophys.* 2: 484
562. Rubin, V. C., et al. 1980. *Ap. J.* 238: 471
563. Rubin, V. C., et al. 1976. *Astron. J.* 81: 687, 719
564. Rubin, V. C., et al. 1985. *Ap. J.* 289: 81
565. Rudaz, S. 1986. *Phys. Rev. Lett.* 56: 2128
568. Sadler, E., Sharp, N. 1985. *Ap. J.* 287: 80
569a. Sale, K. E., Mathews, G. J. 1986. *Ap. J. Lett.* 309: L1
570. Salmon, J., Hogan, C. 1986. *MNRAS* 221: 93
571. Salpeter, E. E. 1986. See Ref. 432
572. Sancisi, R. 1986. Remark at Santa Cruz Workshop, Nearly Normal Galaxies
573. Sancisi, R., van Albada, T. S. 1986. See Ref. 369, p. 67
574. Sandage, A. 1986. *Ap. J.* 307: 1
576. Sanders, R. H. 1986. *Astron. Astrophys.* 154: 135
577. Sanders, R. H. 1986. *MNRAS* 223: 539
578. Sarazin, C. 1986. *Rev. Mod. Phys.* 58: 1
579. Sarazin, C. 1987. In *Structure and Dynamics of Elliptical Galaxies, IAU Symp. No. 127*, ed. T. de Zeeuw. Dordrecht: Reidel. In press
580. Sarazin, C., White, S. D. M. 1986. Preprint
581. Sargent, W. L. W., et al. 1978. *Ap. J.* 221: 731
582. Saslaw, W. C. 1985. *Ap. J.* 297: 49
583. Sato, K., et al. 1986. *MNRAS* 218: 637
585. Scalo, J. M. 1986. *Fundam. Cosmic Phys.* 11: 1
586. Schaeffer, R., Silk, J. 1985. *Ap. J.* 292: 315
587. Schechter, P. 1987. See Ref. 198. In press
588. Schechter, P., et al. 1984. *Astron. J.* 89: 618
590. Scherrer, R., Turner, M. S. 1986. *Phys. Rev. D.* In press
591. Schneider, D. P., et al. 1986. *Astron. J.* 91: 991
592. Schneider, D. P., et al. 1986. *Astron. J.* 92: 523
593. Schneider, S. E. 1985. *Ap. J. Lett.* 288: L33
594. Schneider, S. E., et al. 1983. *Ap. J. Lett.* 273: L1
594a. Schneider, S. E., et al. 1986. *Astron. J.* 92: 742
595. Schramm, D. N. 1985. Talk at Tex. Symp. Relativ. Astrophys., 12th, Jerusalem

596. Schramm, D. N., Steigman, G. 1981. *Ap. J.* 243: 1
597. Schramm, D. N., Szalay, A. 1985. *Nature* 314: 718
598. Schwarzschild, M. 1979. *Ap. J.* 232: 236
599. Schweizer, F., et al. 1983. *Astron. J.* 88: 909
600. Schweizer, L. 1987. *Ap. J. Suppl.* In press
601. Sciama, D. W. 1982. *MNRAS* 198: 1P
602. Sciama, D. W. 1983. *Phys. Lett.* 121B: 119
603. Sciama, D. W. 1984. *Phys. Lett.* 137B: 169
604. Seitzer, P., Frogel, J. A. 1985. *Astron. J.* 90: 1796
605. Sellwood, J. 1985. *MNRAS* 217: 127
607. Setti, G., Hove, L. van, eds. 1986. *ESO/CERN Symp. Cosmology, Astronomy, and Fundamental Physics, 2nd.* Munich: ESO
608. Shafi, Q., Stecker, F. 1984. *Phys. Rev. Lett.* 53: 1292
609. Shane, C. D., Wirtanen, C. A. 1954. *Astron. J.* 59: 285
610. Shanks, T. 1985. *Vistas Astron.* 28: 595
612. Shapiro, P., Struck-Marcell, C. 1985. *Ap. J. Suppl.* 57: 205
614. Shapley, H. 1933. *Proc. Natl. Acad. Sci. USA* 19: 591
616. Shchekinov, Yu. A., Vainer, B. V. 1986. *Astrophys. Space Sci.* 123: 103
617. Shectman, S. 1985. *Ap. J. Suppl.* 57: 77
619. Shuter, W. H. L. 1981. *MNRAS* 194: 851
620. Sikivie, P. 1983. *Phys. Rev. Lett.* 51: 1415
621. Silk, J. 1985. *Ap. J. Lett.* 292: L71
623. Silk, J. 1985. *Ap. J.* 297: 1
624. Silk, J., Bloemen, H. 1987. *Ap. J. Lett.* 313: L47
624a. Simkin, S. M., et al. 1987. *Science* 235: 1367
625. Simpson, J. 1985. *Phys. Rev. Lett.* 54: 1891
626. Skrutskie, M. F., et al. 1985. *Ap. J.* 299: 303
627. Skrutskie, M. F., Forrest, W. J., Shure, M. A. 1986. *Ap. J. Lett.* 312: L55
628. Slipher, V. M. 1914. *Lowell Obs. Bull.* 2: 65
629. Smith, B. F., Miller, R. H. 1986. *Ap. J.* 309: 522
631. Smith, H. 1984. *Ap. J.* 285: 16
634. Solovevo, L. V., Starobinski, A. A. 1985. *Sov. Astron. AJ* 29: 367
635. Sparke, L. 1984. *Ap. J.* 280: 117
636. Sparke, L. 1984. *MNRAS* 211: 911
637. Srednicki, M., et al. 1986. *Phys. Rev. Lett.* 56: 263
638. Stahler, S. W. 1984. *Ap. J.* 281: 209
639. Stahler, S. W., Falla, F., Salpeter, E. E. 1986. *Ap. J.* 308: 697
640. Staller, R. F. A., Jong, T. de 1981. *Astron. Astrophys.* 98: 140
643. Stebbins, A. 1986. *Ap. J. Lett.* 303: L21
644. Stecker, F. 1980. *Phys. Rev. Lett.* 45: 1460
646. Stecker, F., et al. 1985. *Phys. Rev. Lett.* 55: 2622
647. Steigman, G. 1984. *Nucl. Phys. B* 252: 73
649. Steigman, G., Turner, M. 1984. *Nucl. Phys. B* 253: 378
650. Steigman, G., et al. 1978. *Astron. J.* 83: 1050
651. Stevenson, P. R. F., et al. 1985. *MNRAS* 213: 953
652. Stewart, G. C., et al. 1984. *Ap. J.* 278: 536
654. Suntzeff, N., et al. 1985. *Astron. J.* 90: 1481
654a. Swarup, G., Kapahi, V. K., eds. 1986. *Quasars, IAU Symp. No. 119.* Dordrecht: Reidel
655. Szalay, A. S., Schramm, D. N. 1985. *Nature* 314: 718
656. Tammann, G., Sandage, A. 1985. *Ap. J.* 294: 81
657. Tanaka, K. I. 1985. *Publ. Astron. Soc. Jpn.* 37: 481
658. The, L. S., White, S. D. M. 1986. *Astron. J.* 92: 1248
659. Thomas, P. A. 1986. *MNRAS* 220: 949
660. Thuan, T. X. 1983. *Ap. J.* 268: 667
661. Thuan, T. X. 1985. *Ap. J.* 299: 885
662. Thuan, T. X. 1986. *Ap. J. Suppl.* In press
664. Thuan, T. X., Kormendy, J. 1977. *Ap. J.* 205: 696
665. Tinsley, B. M. 1977. *Ann. NY Acad. Sci.* 302: 423
666. Tohline, J. E. 1983. See Ref. 25, p. 205
667. Tonry, J. 1983. *Ap. J.* 266: 58
669. Tonry, J. 1984. *Ap. J.* 279: 13
670. Tonry, J. 1985. *Astron. J.* 90: 2431
671. Tonry, J. 1985. *Ap. J.* 291: 45
673. Toomre, A. 1978. In *The Large Scale Structure of the Universe, IAU Symp. No. 79,* ed. M. Longair, J. Einasto, p. 109. Dordrecht: Reidel
674. Toomre, A. 1981. In *Structure and Evolution of Normal Galaxies,* ed. S. M. Fall, D. Lynden-Bell, p. 111. Cambridge: Cambridge Univ. Press
676. Tremaine, S. D. 1987. See Ref. 198. In press
677. Tremaine, S. D., Gunn, J. E. 1979. *Phys. Rev. Lett.* 42: 407
680. Trinchieri, G., Fabbiano, G. 1985. *Ap. J.* 296: 447
681. Trinchieri, G., Fabbiano, G., Canizares, C. 1986. *Ap. J.* 310: 637
682. Tubbs, A. D., Sanders, R. H. 1979. *Ap. J.* 230: 736

DARK MATTER 471

683. Tully, R. B. 1983. In *Early Evolution of the Universe and Its Present Structure*, *IAU Symp. No. 104*, ed. G. Chincarini, G. Abell, p. 239. Dordrecht: Reidel
684. Tully, R. B. 1986. *Ap. J.* 303: 251
685. Tully, R. B. 1986. Talk at Santa Cruz Workshop, Nearly Normal Galaxies
687. Tully, R. B., Shaya, E. 1984. *Ap. J.* 281: 31
688. Turner, E. L. 1976. *Ap. J.* 208: 304
690. Turner, M. S. 1986. See Ref. 369, p. 445
691. Turner, M. S., et al. 1984. *Phys. Rev. Lett.* 52: 2090
692. Turok, N., Brandenberger, R. H. 1986. *Phys. Rev. D* 33: 2175, 2182
694. Tyson, J. A. 1987. See Ref. 198. In press
695. Tyson, J. A., et al. 1984. *Ap. J. Lett.* 281: L59
696. Tyson, J. A., et al. 1986. *Astron. J.* 91: 1274
698. Ulmer, M. P., et al. 1985. *Ap. J.* 290: 551
699. Umemura, M., Ikeuchi, S. 1985. *Ap. J.* 299: 583
700. Umemura, M., Ikeuchi, S. 1986. *Astrophys. Space Sci.* 119: 243
701. Uson, J. M., Wilkinson, D. T. 1984. *Ap. J. Lett.* 277: L1
703. Vainer, B. V. 1985. *Sov. Astron. Lett.* 11: 275
703a. Vainer, B. V., et al. 1986. *Astrofizika* 23: 733
704. Valtonen, M. J. 1980. *Ap. J.* 236: 750
706. Valtonen, M. J., Byrd, G. G. 1986. *Ap. J.* 303: 523
707. Valtonen, M. J., et al. 1985. *Astrophys. Space Sci.* 107: 209
708. Valtonen, M. J., et al. 1985. *Astron. Astrophys.* 143: 182
709. Vaucouleurs, G. de 1953. *Astron. J.* 58: 30
710. Vaucouleurs, G. de 1961. *Ap. J. Suppl.* 6: 213
711. Vaucouleurs, G. de 1965. In *Galaxies and the Universe (Stars and Stellar Systems*, Vol. 9), ed. A. Sandage, M. Sandage, J. Kristian, p. 557. Chicago: Univ. Chicago Press (Publ. 1975)
713. Vaucouleurs, G. de 1982. *The Cosmic Distance Scale and the Hubble Constant*, p. 75. Canberra: Mt. Stromlo & Siding Spring Obs.
715. Vaucouleurs, G. de, Peters, W. L. 1985. *Ap. J.* 297: 27
716. Veldarnini, R., Bonometto, S. 1985. *Astron. Astrophys.* 146: 237
717. Vidal-Madjar, V., Gry, C. 1984. *Astron. Astrophys.* 138: 235
718. Vilenkin, A. 1985. *Phys. Rep.* 121: 263
723. Vittorio, N., Silk, J. 1984. *Ap. J. Lett.* 285: L39
724. Vittorio, N., Silk, J. 1985. *Ap. J. Lett.* 297: L1
725. Vittorio, N., Silk, J. 1985. *Ap. J. Lett.* 293: L1
726. Vittorio, N., Silk, J. 1985. *Phys. Rev. Lett.* 54: 2269
726a. Vittorio, N., et al. 1986. *Nature* 323: 132
727. Volkov, E. V. 1985. *Sov. Astron. AJ* 29: 262
728. Wagoner, R. V., Fowler, W. A., Hoyle, F. 1967. *Ap. J.* 148: 3
732. Wasserman, I. 1986. *Phys. Rev. D* 33: 2071
733. Wasserman, I., Weinberg, M. D. 1986. *Ap. J.* 312: 390
734. Watson, W. D., Degeuchi, S. 1984. *Ap. J. Lett.* 281: L5
736. Weinberg, M. D. 1986. *Ap. J.* 300: 93
738. White, S. D. M. 1978. *MNRAS* 184: 185
739. White, S. D. M. 1987. See Ref. 198. In press
740. White, S. D. M., et al. 1984. *MNRAS* 209: 27P
742. White, S. D. M., et al. 1983. *MNRAS* 203: 701
743. White, S. D. M., Frenk, C. S., Davis, M., Efstathiou, G. 1987. *Ap. J.* 313: 505
744. Whitmore, B. D., McElroy, D. B., Schweizer, F. 1987. *Ap. J.* 314: 439
746. Whitmore, D. P., Jackson, A. A. 1984. *Nature* 308: 713
747. Wielen, R. 1975. In *Dynamics of Stellar Systems*, *IAU Symp. No. 69*, ed. A. Hayli, p. 119. Dordrecht: Reidel
749. Wilkinson, D. T. 1983. In *Early Evolution of the Universe and Its Present Structure*, *IAU Symp. No. 104*, ed. G. Chincarini, G. Abell, p. 143. Dordrecht: Reidel
750. Williams, B. A., Rood, H. 1986. *Ap. J. Suppl.* 63: 265
751. Williams, R. E., Christiansen, W. A. 1984. ESO Preprint 322
753. Witten, E. 1984. *Phys. Rev. D* 30: 272
754. Wolf, H. 1906. *Astron. Nachr.* 170: 21
755. Woltjer, L. 1975. *Astron. Astrophys.* 42: 109
756. Wong, C.-Y. 1986. *Phys. Rev. Lett.* 56: 1047
757. Xiang, S. P. 1985. *Astrophys. Space Sci.* 111: 171
758. Yabushita, S., Allen, A. J. 1985. *MNRAS* 213: 117
760. Yahil, A. 1986. See Ref. 432
761. Yahil, A. 1987. See Ref. 198. In press
762. Yahil, A., et al. 1986. *Ap. J. Lett.* 301: L1
765. Zabotin, N. A., Nazelsky, P. D. 1985. *Sov. Astron. AJ* 29: 239
766. Zabrenowski, M. 1986. *Astrophys. Space Sci.* 117: 179

769. Zasov, A. V. 1985. *Sov. Astron. Lett.* 11: 307
770. Zasov, A. V., Kyazumov, G. A. 1983. *Sov. Astron. AJ* 27: 384
771. Zeldovich, Ya. B. 1968. *Sov. Phys. Usp.* 11: 381
772. Zeldovich, Ya. B., Novikov, I. D. 1983. *Relativistic Astrophysics*, Vol. 2. Chicago: Univ. Chicago Press
773. Zeldovich, Ya. B., Sunyaev, R. A. 1980. *Pisma Astron. Zh.* 6: 457
774. Zeldovich, Ya. B., et al. 1975. *JETP* 67: 401
775. Zeldovich, Ya. B., et al. 1982. *Nature* 300: 407
776. Zurek, W. H., et al. 1986. See Ref. 369, p. 316
777. Zwicky, F. 1933. *Helv. Phys. Acta* 6: 110

VERY LOW MASS STARS

James Liebert

Steward Observatory, University of Arizona, Tucson, Arizona 85721

Ronald G. Probst

Kitt Peak National Observatory, National Optical Astronomy Observatories, Tucson, Arizona 85726

> *The commonplace I sing.*
> Walt Whitman

1. INTRODUCTION

Very low mass (VLM) stars, which we define here somewhat arbitrarily as those with masses $\lesssim 0.3\ M_\odot$, pose some of the more interesting problems in stellar astrophysics. The physics of their envelopes and atmospheres are more complicated and their structures less well modeled than their more massive counterparts. The properties of stellar configurations near the hydrogen-burning mass limit, $\sim 0.08\ M_\odot$, are particularly uncertain. The likely onset of complete interior convection for stars in this mass range raises questions as to how applicable is the solar model in describing their surface activity—photospheric, chromospheric, and coronal.

While they dominate numerically the stellar population in the solar neighborhood, their low luminosities and cool temperatures have made them difficult objects for detailed study. However, new observing techniques from both the ground and space have afforded much recent progress, now including results on the halo population (subdwarfs). Still to come are the first observations of the nearest globular clusters with the Hubble Space Telescope, which may spatially resolve tens of thousands of M subdwarfs down to near the main-sequence mass limit (cf. Bahcall & Schneider 1987). Also ahead of us are final answers to such questions of broad astronomical impact as (*a*) how many stars of very low mass there really are in our Galactic disk and in other stellar populations, (*b*) whether substellar-mass "brown dwarfs" really exist in substantial numbers, (c) what are the properties of individual substellar objects, and (d) whether the sum of low-mass stellar and substellar objects can account for the dynamical disk or halo mass in our Galaxy and others. In this review, we discuss several theoretical and observational topics involved in discovering and analyzing very low mass stellar objects below about 0.3 $M_\odot$, as well as their likely extension into the substellar range. We hereafter refer to these two classes of objects as VLM stars and brown dwarfs,[1] respectively; collectively, they can be called VLM objects. Section 2 outlines recent theoretical work on low-mass stellar interiors and atmospheres, the determination of the hydrogen-burning mass limit, important dynamical evidence bearing on the expected numbers of such objects, and the expectations for such objects from star-formation theory. The emphasis of our discussion of brown dwarfs is on the properties of substellar objects near the stellar mass limit. We do not attempt to discuss the interior and atmospheric structures of configurations closer to Jovian mass (see, instead, Stevenson 1982a, Hubbard & Stevenson 1984, Lunine et al. 1986) or the formation and subsequent evolution of such objects in binaries or planetary systems (see Stevenson 1982b). Section 3 summarizes the observational techniques used to discover and analyze VLM objects. In Section 4 we compare the stellar parameters derived from observations with the theory. In Section 5, the chromospheric and coronal activity of VLM stars is discussed, with emphasis on how these properties might be used to derive stellar ages and population parameters. Finally, in Section 6 we conclude with a discussion of the luminosity and mass functions for VLM disk and halo stars in the solar neighborhood.

2. THEORY OF VERY LOW MASS STARS

2.1 *Interior Models and the Mass Limit*

To first order, stars of the lowest masses are among the easiest to model, since they may be completely convective throughout. Limber (1958) showed that they are well represented by $n = 3/2$ polytropes and calculated models down to 0.09 $M_\odot$, with approximate outer boundary conditions. Kumar (1963a) extended this work to the mass range 0.04–0.09 $M_\odot$, taking into account nonrelativistic degeneracy in the equation of state. He realized

---

[1] Substellar-mass objects unbound to stars are usually referred to as brown dwarfs or, alternatively, black dwarfs in the literature. Here we use the former term (Tarter 1975, 1986), which is useful for emphasizing two points: (a) Substellar-mass objects do emit radiation and hence are not "black," but the best wavelength range or "color" for detecting them depends on their stage of evolution (Section 2.2); (b) the recent analysis of Lunine et al. (1986) suggests that their atmospheres will turn out to be cloudy and dusty.

that the onset of degeneracy could prevent the onset of thermonuclear reactions below a certain mass limit. Kumar (1963b) used similar polytropic configurations to calculate the Kelvin lifetimes for gravitationally contracting objects below the limit. The subsequent phase of cooling of the degenerate hydrogen configurations was later explored in more detail by Stevenson (1978). Hoxie (1970) and Straka (1971) incorporated gray and nongray model atmospheres, respectively, for their outer boundary conditions to polytropic interiors.

The development of Henyey codes allowed researchers to discard the polytropes and other analytical relations in favor of detailed equation of state (EOS) and opacity tables. The first models with more sophisticated physics for masses near the limit were presented in a series of papers by Grossman, Graboske, and coworkers (Grossman et al. 1974; hereafter GHG). Their treatment included nonideal gas effects, electron screening, degeneracy, Los Alamos opacity tables, and nongray model atmospheres for surface boundary conditions. They found that the lower mass limit to the main sequence lies near 0.085 $M_\odot$.

The more recent model grids of VandenBerg et al. (1983) cover the low-mass star range more completely (0.1–0.75 $M_\odot$) for a wide range of metallicity ($Z = 10^{-5}$ to 0.02). These models incorporate at low temperatures the opacities of Alexander (1975), which treat in detail molecular and particulate matter as well as atomic absorption processes and therefore increase substantially the total opacities. Note, however, that the more recent tables of D. R. Alexander et al. (1983) have somewhat smaller opacities. Since the primary purpose was to compare with the derived stellar parameters for metal-poor subdwarf M stars, VandenBerg et al. did not incorporate some of the EOS refinements that are more important closer to the stellar mass limit.

Cox et al. (1981) have challenged the early assumption that VLM stars with $M \lesssim 0.3$ $M_\odot$ should be completely convective throughout. Noting that the empirical determinations show somewhat higher luminosities for a given mass near the hydrogen-burning mass limit, these authors were able to fit the empirical data by choosing l/H, the ratio of the mixing length to the scale height in the theoretical treatment of the convective envelope, more than an order of magnitude smaller than the values near unity usually chosen. The resulting adiabat leads to higher central temperatures, higher resulting luminosities, and in some cases radiative cores. The persistence of radiative cores could be relevant to the occurrence of active chromospheres and coronae in stars well below 0.3 $M_\odot$ (see Section 5). However, the issue is apparently not whether stars near the main-sequence mass limit are completely convective, but rather at what mass does this transition occur.

2.2 Recent Interior Models Near and Below the Main Sequence Mass Limit

The emphasis of recent research on low-mass interiors, however, continues to be on the physics at and below the stellar mass limit. In a series of papers, D'Antona & Mazzitelli (1985, 1986a,b, and references therein) have explored the consequences of incorporating further refinements in the EOS and the opacities. In particular, their EOS (Magni & Mazzitelli 1979) treats the envelope regions of partial dissociation and ionization in more detail. They also explore the use of both "standard" Los Alamos opacities (Cox & Stewart 1970) and the more recent opacity tables of D. R. Alexander et al. (1983), which treat molecules. The principal differences between these models and those of GHG are that (a) configurations near the mass limit (~ 0.08 $M_\odot$) stabilize at substantially lower luminosities, and (b) there exists a range of transition masses—0.06–0.08 $M_\odot$, but depending on the composition—for which an object can sustain appreciable luminosities through hydrogen burning for 10^9–10^{10} yr before finally failing to stabilize as a main-sequence star. Long-lived, marginal hydrogen-burning configurations are also predicted in the models of Stringfellow (1986), which used similar opacities. The accuracy of any of the current model sets is probably limited by the precision of the opacities. It may prove easier in the future to make modest improvements in the EOS, as is being undertaken by Saumon & Van Horn (1987). See also the review by Van Horn (1986).

Brown dwarfs evolve through at least three distinct phases: (a) hydrodynamic collapse with decreasing radius R, increasing central temperature T_c, but essentially constant surface effective temperature T_e; (b) deuterium burning (slowly decreasing T_e and R); and (c) degenerate cooling (decreasing T_e and T_c, slowly decreasing R). Configurations with masses below about 0.015 $M_\odot$ do not undergo significant deuterium burning (Lunine et al. 1986). During phases (a) and (b), brown dwarfs are detectable, as are VLM stars, at red/near-infrared wavelengths ($T_e \sim 2000$–3000 K). When they reach phase (c), the IRAS far-infrared bandpasses are more appropriate (~ 500 K).

The degenerate cooling phase is similar to that occurring in white dwarfs, except primarily for the importance of thermal effects in the EOS, the relatively low densities, and of course the hydrogen-rich (solar) composition. The effective polytropic index is near $n = 3/2$ but drops toward $n = 1$ for configurations well below the stellar mass limit. As the cooling proceeds toward $T_c = 0$, the brown dwarf interior may undergo significant differentiation in the chemical elements, as has been modeled for Jovian planets. The maximum radius for the zero-temperature EOS (Zapolsky &

Salpeter 1969) occurs at about twice the Jovian mass (0.001 $M_\odot$), with a value for solar composition at about the radius of Jupiter (0.1 $R_\odot$).

Stevenson (1978, 1986) continues with a simple philosophy for calculating configurations in the regime of hydrogen degeneracy, at least for masses $>0.05~M_\odot$. He argues that the pressure can still be approximated by the polytropic formula to within 10%, so that for many purposes the complicated EOS tables are not worth bothering with. He recognizes more serious problems with the treatment of envelope and atmospheric opacities, which remain uncertain by much more than 10% at many temperatures and densities, no matter whether one uses tables of detailed calculations or analytical expressions. These affect in particular the emergent luminosity of the configuration and the observed energy distribution or spectrum, as has been explored in detail by Lunine et al. (1986) (see Section 2.3). The assumption of a simpler EOS is also a feature of the recent series of calculations by Nelson et al. (1985, 1986), who explored in particular the dependencies upon various physical assumptions and parameters in the codes.

2.3 Atmospheres

In addition to their relevance to the outer boundary conditions for interior models, reasonable stellar atmosphere calculations are necessary in order to determine temperatures and luminosities from the observations for comparison with theory. These calculations remain at such a rudimentary state that the energy distributions of the truly VLM stars are generally fit with blackbodies! Of course, the preponderance of theoretical work on cool stellar atmospheres has dealt with red giants. In some respects, the atmospheres of dwarf stars are easier to calculate. For example, convection is much more efficient, and complications due to extended envelopes and mass loss are minimal. However, at these high pressures, the formation of numerous molecules and their complicated opacities pose more formidable problems than for giant stars.

Following the first exploratory radiative models (Tsuji 1966, Gingerich et al. 1966), the most important molecules including H_2O and TiO were incorporated into radiative atmospheres (Tsuji 1968, Carbon et al. 1968). The next steps were to include both molecular blanketing, convection (Auman 1968, Mould 1975), and detailed atomic-line blanketing (Mould 1976a). Mould covered the temperature range 3000–4250 K for metal abundances $[M/H] = 0$ to -2. While he succeeded in fitting both photometry and spectroscopic features (see also Hartwick et al. 1984, Berriman & Reid 1987), it was evident that there were large uncertainties in the molecular blanketing and in the derived temperatures for stars near 3000 K (or about spectral type M3, near our VLM boundary; Persson et al. 1977, Reid & Gilmore 1984). New calculations by M. S. Bessell & R. Wehrse (in preparation) are intended to cover a similar range of metal abundances down to around 2000 K.

The ground-breaking paper of Lunine et al. (1986) presents the first complicated atmospheres for brown dwarf masses (0.001–0.08 $M_\odot$). They explore a range of possible atmospheric opacities, including condensed clouds of refractory materials, which may appear for masses below 0.08 $M_\odot$. The dominant CO molecule may give way to CH_4 for masses below about 0.03 $M_\odot$. Water may be very important or negligible, depending on the atmospheric carbon-to-oxygen ratio. However, provision for possible enhancements of heavier elements is applicable primarily to brown dwarfs (planets) formed in binary systems. Lunine et al. present preliminary emergent energy distributions for a range of opacity possibilities; in particular, it appears that infrared spectra of brown dwarfs will have important diagnostic roles. It will be important to determine observationally if low-luminosity objects form large grains and a cloud deck; these could change the atmospheric opacities enough to drastically distort the shape of the observed luminosity function (Stevenson 1986).

2.4 *Dynamical Constraints on the Numbers of VLM Objects*

The PhD thesis of Tarter (1975) inspired much theoretical and observational interest in the possibility that substellar-mass brown dwarfs and VLM stars might be responsible for "missing mass" inferred dynamically for galactic halos and clusters of galaxies. Other solutions now seem more plausible on these scales (see Trimble 1987). In the solar neighborhood, however, there is a similar, long-standing problem. It appears that the local mass density derived from star counts in the direction of the Galactic poles is larger by at least 0.06 $M_\odot$ than the mass density accounted for by visible stars, gas, and dust (Oort 1960, Bahcall 1984, 1986). The existence of wide binaries in the Galactic disk allows the inference that this component should consist of individual entities each of mass <2 $M_\odot$ (Bahcall et al. 1985). Since the local (disk) missing mass is also required to be dissipative, attention has been confined to two kinds of very faint stellar components— (*a*) remnant stars, especially cool white dwarfs; and (*b*) VLM main-sequence stars/brown dwarfs. Larson (1986) advocates the former in the context of a bimodal star-formation picture, but most attention has been focused on the latter possibility. Obliquely related to the general problem may be the possible existence of a solar brown dwarf companion (cf. Davis et al. 1984).

In a comprehensive review primarily of the observational evidence, Scalo (1986) concludes that the stellar initial-mass function (IMF) reaches a peak near 0.3 $M_\odot$, then declines at lower masses, but remains very uncertain near the main-sequence mass limit. On the other hand, D'Antona & Mazzitelli (1986b) conclude that the slope is still increasing down to 0.1 $M_\odot$. There is little doubt, at least, that the local IMF slope is less than the Salpeter (1955) value (2.35) in the VLM range. However, our assessment (Section 6) is that the uncertainty in the theoretical mass-luminosity relation is too great to be confident of the IMF's behavior or to be certain of the existence of a peak in the mass function near 0.3 $M_\odot$.

As pointed out by numerous authors, even a modestly rising IMF slope below 0.3 $M_\odot$ would provide little to the local mass density in the form of VLM stars plus brown dwarfs. The necessary conclusion, if VLM objects are to add substantially to the local mass density, is that there must be a sharp increase in the mass function near or below the stellar mass limit (see Hawkins 1986b, Reid 1987; see also Section 6.4). This would appear to require an additional "low-mass mode" of star formation. It is therefore important to ask (a) if there is a lower limit to the masses of stellar fragments that can form, and (b) if there are valid theoretical or observational reasons to expect that very large numbers of such small-mass fragments can be made in the Galactic disk (or the halo). These questions are addressed in Section 2.5. The mode of formation for low-mass stellar and substellar binary companions may be largely a different subject, less relevant to the resolution of the disk mass problem.

One further dynamical consequence of a population of brown dwarfs sufficiently large to solve the mass accountability problem discussed above might be detectable effects on the outer solar system. Hills (1985) shows that the observed low eccentricity of planetary orbits indicates that no object more massive than 0.02 $M_\odot$ has passed inside of the orbits of the outer planets during the lifetime of our solar system. The probability of an interstellar brown dwarf interloper passing this close to the Sun is small, even at the very high densities required to account for the remaining disk mass. However, this does lead to a strong constraint on the mass of a solar companion. Using Table 1 of Nelson et al. (1986), a companion with $M < 0.02\ M_\odot$ and an age of 5.5×10^9 yr should have $\log L/L_\odot < -6.36$ and $T_e < 450$ K. Optical searches for such a cool, dim solar companion would be futile, although further analysis of the IRAS data base is apropos.

2.5 *The Formation of VLM Objects*

Hoyle (1953) first outlined the concept of hierarchical star formation, by which massive gas clouds may fragment into progressively smaller masses. If the fragmentation is characterized by simple, stochastic processes, a log-normal distribution similar to observed field and cluster IMFs may be produced, even with no input physics in the model (Larson 1973, Elmegreen & Mathieu 1983). The fragmentation problem for the disk must, however, be considered in the context of our modern understanding of giant molecular clouds (GMCs), with typical masses of order 10^5 $M_\odot$; these clouds are the sites for the bulk of star formation in the Galactic disk (see the review by Shu et al. 1987). Each generation of star formation in these long-lived entities uses only a small fraction of the total cloud mass (cf. Duerr et al. 1982). The initial stage is the formation of molecular cloud cores, probably via ambipolar diffusion (cf. Mestel & Spitzer 1956, Mouschovias & Paleologou 1981, Shu 1983). The collapse, fragmentation, and subsequent behavior of these can now be calculated with increasing degrees of sophistication.

Unfortunately, the outcome of the final stages of fragmentation and other processes may depend on the detailed thermal behavior and the velocity and/or magnetic fields of each collapsing cloudlet, while the time-dependent terms in the energy equation must be treated accurately where the optical depths are becoming large. Bodenheimer (1978) and Larson (1985) showed that fragmentation may be enhanced for rotating clouds, which collapse to flat disks. With similar assumptions, Zinnecker (1984) reproduced in his fragmentation spectrum a (possible) observed peak in the IMF near 0.3 $M_\odot$ and a decline toward lower masses. Other observational results may suggest that such a peak should occur. In particular, Larson (1985) argues that the plot of observed temperatures vs. densities of interstellar clouds (and comparison with semiempirical model curves) suggests that minimum temperatures of ~ 5 K occur at densities of 10^{-18} g cm^{-3} (or lower). This corresponds to a limiting Jeans mass around 0.3 $M_\odot$. In particular, since the observed temperatures of clouds having higher densities appear to be relatively higher, these may correspond to conditions that are becoming opaque, so that further fragmentation to masses much below 0.3 $M_\odot$ will not occur often. Larson notes that various environmental heating effects (associated with rapid formation of massive stars) may increase both the minimum cloud temperature and the fragment masses. However, no mechanism is identified that would tilt the fragmentation mass spectrum to much lower masses.

A number of authors have attempted to calculate more specifically the minimum mass at which the fragmentation should stop. For dust-grain opacities and physical conditions applicable to dark molecular clouds of solar composition gas, Low & Lynden-Bell (1976) suggest a minimum fragment mass of 0.007 $M_\odot$ and Silk (1977) estimates 0.01 $M_\odot$. Boss (1986, and references therein) has calculated three-dimensional models for the collapse of rotating protostellar clouds. He finds that fragmentation stops with the formation of binary components of 0.01 $M_\odot$. The limits can be considerably larger—approaching 0.1 $M_\odot$—if dust grains are totally lacking in a metal-free environment, possibly applicable to the initial collapse of a galactic spheroid population. Thus, these calculations provide little basis for believing that "unseen" massive halos around galaxies consist of low-mass stellar or substellar objects (Tohline 1980; see, however, Zinnecker 1986).

On the other hand, Shu (1985) does not regard the question of the minimum mass of condensations as a fragmentation problem at all. The collapse of individual cores immersed in GMCs is accompanied by slow accretion from the surroundings (Stahler et al. 1980). Given the surrounding mass that is available, there is no reason for the buildup in the stellar mass to stop until the star reverses the process by giving off a wind. This in turn may occur when a convective envelope forms and a magnetic dynamo begins converting rotational energy into surface activity and mass outflow. Shu argues that the onset of convection, which requires the establishment of a strong outward temperature gradient, is driven by the appearance of thermonuclear energy generation in the core due to deuterium burning. Accordingly, GMC cores must therefore build up to at least the minimum stellar mass for deuterium burning ($\sim 0.015~M_\odot$; Lunine et al. 1986). A comprehensive exposition of these and related star-formation ideas is given in Shu (1987).

In summary, there appears to be every likelihood that some stellar objects should form with masses down to of order $0.01~M_\odot$. However, while we are not overly impressed with the ability of such calculations to predict detailed IMF shapes, no mechanism has yet been identified that would result in a second mode of star formation producing large numbers of VLM/substellar-mass objects, as required to satisfy the Oort-Bahcall mass requirement or as suggested by the results of Hawkins (1986a,b).

3. OBSERVATIONAL METHODS AND RESULTS

3.1 Searches and Surveys

The most fruitful means of identifying VLM stars has been astrometric: photographic proper-motion surveys followed by trigonometric parallaxes. The first systematic proper-motion survey with large areal coverage, faint limiting magnitude, and small limiting motion was carried out for the Southern Hemisphere by W. J. Luyten. It provided material for the first determination of the luminosity function for VLM stars (Luyten 1938). A review of this and earlier work is given by Luyten (1963a). Since that review, a survey of the northern sky has been completed at Lowell Observatory by Giclas (1958 et seq.; Giclas et al. 1971) and extended southward around the South Galactic Pole (Giclas et al. 1972, 1978). Together, the Luyten and Giclas surveys cover the whole sky to limits of

$m_{pg} \lesssim 16.5$, $\mu \gtrsim 0.2$ arcsec yr^{-1}. The use of blue-sensitive plates and an onset of incompleteness at $m_{pg} \sim 15$ cause these surveys to be strongly magnitude limited for inclusion of VLM stars. Faint, high-motion stars are still being found in the Southern Hemisphere in deeper plate material (Ruiz et al. 1986).

By far the most ambitious proper-motion survey to date has been conducted by Luyten using plates from the 1.2-m Palomar Schmidt (Luyten 1963b et seq.). While previous surveys used the human eye, the Luyten Palomar survey was largely carried out by a specially constructed automatic machine. The plate-matching algorithm was limited to $\mu \lesssim 2.5$ arcsec yr^{-1} except for selected polar regions, and fields near the Galactic plane were not surveyed because of the very high star densities. The combination of red plates, a very faint limiting magnitude ($m_R \sim 20$), and a small motion limit ($\mu \gtrsim 0.1$ arcsec yr^{-1}) provides for both penetration of the substellar regime in luminosity (Probst 1983a) and inclusion of a substantial number of late halo dwarfs and cool white dwarfs in addition to late disk stars (Liebert et al. 1979). A catalog and finding charts have been issued for the ~ 3600 stars with $\mu \geq 0.5$ arcsec yr^{-1} (Luyten 1979a, Luyten & Albers 1979). Dawson (1986) finds this portion of the survey 90% complete for $m_R < 18$ over 65% of the sky. A catalog only is available for the $\sim 50,000$ stars with $0.2 \leq \mu < 0.5$ arcsec yr^{-1} (Luyten 1979b).

Selection of stars by proper motion introduces a bias that may strongly affect results concerning the sample kinematics. Recognizing this, Vyssotsky (1963) conducted a large-scale objective-prism survey to identify red dwarfs spectroscopically, independent of motion. Upgren et al. (1972) surveyed the Southern Hemisphere in a similar fashion. The use of blue spectral criteria and bright limiting magnitudes largely prevented the discovery of VLM stars. Several investigations have attempted to extend this approach to significantly later types and fainter magnitudes by very low dispersion spectra in the far-red (600–6000 Å mm^{-1}; 5800–9000 Å). Since a dwarf-giant luminosity discriminant is unavailable at this dispersion, these surveys have been limited to high Galactic latitudes.

Results of this work fueled a considerable controversy over the existence of a substantial population of VLM stars with small transverse velocities. Sanduleak (1965, 1976), using photographic $V-I$ colors of spectrum-selected stars, claimed to find a large space density of late dwarfs. Similar results were obtained by Smethells (1974) using spectra only, and by Staller (1979) from spectra and $R-I$ colors. These results raised considerable difficulties both dynamically and in terms of consistency with the nearby star sample (Schmidt 1974, Luyten 1976), although theoretical explanations were soon forthcoming (Biermann 1974, Staller 1975). Subsequent investigations initially focused on kinematics (Murray & Sanduleak 1972,

Jones 1972, Gliese 1974, Koo & Kron 1975), but the crux of the matter was the accuracy of the spectral subtyping and the photographic photometry. The former has been found to be systematically too late in type, and the latter to suffer from errors in both photometry and calibration (Upgren 1972, Weistrop 1976a,b, 1977, 1980, Pesch 1976, Pesch & Dahn 1982, Reid 1982, Smethells 1983, Reid & Gilmore 1982). Subsequent surveys have abandoned spectral subtyping (Pesch & Sanduleak 1978, Robertson 1983, 1984). In Robertson's (1984) survey, only 10% of the late dM stars found spectroscopically lacked prior proper-motion data.

Purely photometric techniques provide an alternative bias-free survey mode. Direct photography can be used to cover large areas to deep limiting magnitudes quickly. Such surveys are also limited to high Galactic latitudes to avoid contamination by giants. Giclas et al. (1971, 1980) have published lists of very red stars identified by color in South Galactic Pole fields in the course of their survey. However Thé & Staller (1974) found these stars to be of relatively early type, while Warren (1976) determined that the sample was not motion independent. Staller et al. (1981) identified 2600 very red stars at the South Galactic Pole by blinking red and blue Palomar Sky Survey plates and claimed an enhanced space density. Alcaino & Thé (1982) concluded that these were late-M dwarfs from photographic $B-V$ colors. Besides the inappropriate color choice, this work had the potential for systematic errors in that the calibrating photoelectric sequences did not include stars as red or as faint as the blink sample. Reid & Gilmore (1982) found that photographic $V-I$ colors for 600 of these stars indicated they were of much earlier type, deflating the space density.

Weistrop (1972) conducted a more quantitative survey with iris photometry of B and V plates in a 13.5 deg^2 field at the North Galactic Pole. She found a much higher space density and significantly smaller scale height for dwarfs of $M_V \leq 12$ than had been determined from motion-selected samples. The choice of passbands was not well suited to red stars, and her conclusions were invalidated by systematic errors in both the photographic and calibrating photoelectric photometry (Weistrop 1976c, Faber et al. 1976).

The availability of sensitive emulsions, automated measuring machines, considerable computer power, and testable models has led to a revival of interest in number-magnitude-color counts for Galactic structure studies. Bahcall (1986) reviews recent work. Reid & Gilmore have applied this formidable apparatus to VLM stars in a series of major papers. Using deep $BVRI$ plates of 18 deg^2 at the South Galactic Pole, Reid & Gilmore (1982) obtained $V-I$ and $I-K$ colors to define a complete sample of stars with $M_V \leq 19$ and so determine a bias-free luminosity function. The data reduction and screening task is impressive: $\sim 10^5$ images per plate yielded a final sample of 89 stars within 100 pc, 7 with $M_V > 14.5$. Subsequent work has extended the sample size in this field with brighter stars (Gilmore & Reid 1983) and the sample volume with a second field (Gilmore et al. 1985). Other fields are in progress. Hawkins (1986a,b) has used the same techniques with R and I plates in an intermediate-latitude field, obtaining a greater limiting distance for the faintest stars. His results are in general agreement with those of Reid & Gilmore, but with an indication of an upturn in the space density for the least luminous stars.

Great care has been taken in this recent work to properly calibrate the faint images, which ultimately carry high weight in the final luminosity function. However, history shows that systematic errors can be insidious, e.g. when photoelectric sequences are photographically extended to fainter and redder stars. The use of faint CCD sequences is very helpful. Hawkins found a large, nonlinear color effect in the $R_{pg} \to R_{pe}$ transformation that had been missed in an earlier investigation using bright photoelectric standards (Blair & Gilmore 1982). The automated image identification and measuring techniques can also introduce systematic effects. Some of Hawkins' original sample had been previously found and discarded as red compact galaxies but were retained by him because they showed measurable proper motion (Hawkins 1986a). Remeasurement of some images by an alternative technique on a different machine indicated that the faintest stars were initially measured as too bright (Hawkins 1986b). Although Hawkins' survey was basically colorimetric, he also obtained proper motions for his sample, and these contributed significantly to the discussion of his results.

Very deep optical surveys in small fields have been done with CCD detectors. For VLM stars this allows sampling well beyond 100 parsecs for the latest disk dwarfs and deep into the halo for cool subdwarfs (Boeshaar et al. 1986). However, only a small total volume has been surveyed so far. It is noteworthy that none of these photometric surveys has found any stars significantly redder than VB 10, which suggests that they are indeed rare.

All-sky infrared surveys for cool brown dwarfs are being conducted with the IRAS data bases (Low 1986). Null results thus far do not significantly constrain the brown dwarf population owing to a lack of sensitivity. The coadded survey data may lead to more interesting results. Future ground- or space-based infrared surveys will be very powerful for probing this portion of the (sub)stellar population (Low 1986).

What is the future of large-scale optical surveys for VLM stars? The nearby-star sample is statistically complete to $r \simeq 13$ pc for $M_V \lesssim 14$ (Upgren et al. 1986). Interest now centers on extending a well-determined luminosity function to very faint limits. The question of kinematic bias, which prompted objective-prism surveys, has been answered in the negative for the Vyssotsky dwarfs (Upgren 1978) and for the smaller samples of later types (e.g. Smethells 1983). While densitometric methods of classification (Kelly et al. 1982) offer the possibility of automated, objective surveys, the systematic effects noted by Robinson (1984) are intrinsic to the emulsion. This approach seems to have little future application to VLM star surveys. The Palomar Sky Survey is being repeated to significantly fainter limiting magnitudes, and a short-interval proper-motion survey could detect or meaningfully constrain the field brown dwarf population (Probst 1986). Deep photographic or other photometric surveys have a bright future in Galactic structure studies, with continued application to VLM star discovery as a side benefit if appropriate colors and very careful calibrations are employed. Reid (1987) has outlined the benefits of such surveys for discovery of brown dwarfs. Proper-motion determinations in photometric fields will be a useful supplement.

3.2 Classification and Quantification

Trigonometric parallaxes provide the fundamental datum of distance, from which luminosities, masses, kinematics, and secondary distance indicators are derived. Recent photographic parallax programs have emphasized stars of astrophysical interest (VLM stars, white dwarfs), fainter limiting magnitudes to reach intrinsically faint objects, and new methods of measurement and computation. These new methods result in greatly decreased internal and external errors, giving the resulting physical information high weight. Upgren (1985) discusses accidental and systematic errors in parallaxes in a historically oriented review. Van Altena (1983) includes a discussion of future developments in parallaxes in a broader overview of astrometry in general.

Largely because of the convergence of the US Naval Observatory (USNO) 1.5-m astrometric reflector and the Giclas survey, the known stellar population within 20 pc of the Sun increased by 25% over a 10-yr period (Gliese 1969, Gliese & Jahreiss 1979). This was due more to an increase in limiting magnitude than to a decrease in proper-motion limit for parallax program stars. Attention is now turning to the stars of smaller motion (Weis 1984, 1986) and to the faintest Luyten stars for VLM candidates (Dahn 1985, Probst 1985). The latter are too faint for photographic work but well within reach of red-sensitive CCDs. Initial results with solid-state detectors have been positive (Monet & Dahn 1983, Dahn 1985), and a full-scale CCD parallax program has been implemented at the USNO. An important limitation is the paucity of reference stars in the small CCD fields at high Galactic latitudes. This problem should be alleviated by the development of larger format CCDs.

Broadband photoelectric photometry in the red and infrared has been an invaluable tool for work on VLM stars. The original R, I system was defined by Kron (Kron & Smith 1951, Kron et al. 1953, 1957) to provide a better temperature index than $B-V$ for cool stars. It has been redefined with a different photodetector and used extensively by Eggen (1975, 1978, 1982); variant systems have also been used for VLM stars (Kron & Mayall 1960, Weistrop 1975). The Kron-Cousins system is now standard in the Southern Hemisphere (Cousins 1976, 1980a,b, Menzies et al. 1980, Bessell 1979). An equatorial network defined on this system has been established by Landolt (1983), and he is presently incorporating some late red dwarfs into it (private communication). Properties of M dwarfs in the Cousins system are discussed by Thé et al. (1981, 1984). The Cousins R filter has a redward tail that causes its effective wavelength to shift into the I band for cool stars (Bessell 1979). This tail does not reproduce photographically, which causes difficulties in calibration (Hawkins 1986b). Bessell (1986) suggests a redefinition of the filter to avoid these difficulties.

An alternative R, I system was defined by Johnson and used for M dwarf studies (Johnson 1965a, Johnson et al. 1966). The I band is quite different from that of Kron & Cousins (Johnson 1965b). Recent work has attempted to refine the Johnson system to higher precision with fainter stars; a critical discussion is given by Bessell (1983). There is an extensive literature on transformations between various systems (Eggen 1971a, Fernie 1983, Cousins 1980c,d, 1984, and other citations above). These transformations can be nonlinear, especially for very red stars.

M dwarfs show a tight correlation between $R-I$ color and absolute magnitude (Upgren 1975, Weistrop 1977, 1980), and a linear relation between M_I and M_{bol} (Eggen 1969, 1971b). $R-I$ is not a pure temperature index because of the effect of TiO blanketing in the R band (Mould 1976a,b). A cleaner index can be found in the 0.8–1.2 μm region (Mould & McElroy 1978). Blanketing effects can be used to discriminate between dwarfs and subdwarfs with $B-R$, $R-I$ (Hartwick 1977) or $B-V$, $V-I$ (C. C. Dahn et al., in preparation).

The bulk of the energy from VLM stars is radiated between 0.8 and 3.5 μm. Johnson (1964) defined the first broadband infrared system with bands J, K, L, and M (1.2, 2.2, 3.5, and 5.0 μm). An H band (1.6 μm) was later interpolated (e.g. Glass 1974). The original Johnson filters are broader than those used presently, and his standards contain large random errors relative to currently attainable precision. His system has been largely supplanted by a variety of natural systems (Engels et al. 1981, Jones & Hyland 1982, Allen & Cragg 1983, Elias et al. 1982, Joyce et al. 1984 and unpublished, Sinton & Tittemore 1984, Carter 1984). Transformations between systems are given by Glass (1983, 1985), Elias et al. (1983), and

Koornneef (1983a) in addition to the other citations above. The need for a single, generally accepted, standard system has grown with the improvement in photometric precision. Koornneef (1983a,b) has attempted to define a refined version of Johnson's system by transforming data on a variety of other systems. This procedure has been criticized for *UBV* photometry (Manfroid & Heck 1985).

While optical bandpasses can be chosen from astrophysical considerations, the location of infrared bands is determined by terrestrial water vapor absorption. This complicates the measurement of the primary absorbers in cool stellar atmospheres at these wavelengths, H_2O and CO, and biases the determination of stellar fluxes (Reid & Gilmore 1984). Low-level terrestrial absorption across the bandpasses introduces systematic effects in photometric reductions and absolute flux calibrations (Manduca & Bell 1979). Glass (1985) provides a compact introduction to many aspects of infrared photometry.

$J-H$, $H-K$ colors provide information on abundance and gravity (Mould 1976a, Mould & Hyland 1976). Photometry at 1–5 μm is necessary for the determination of bolometric luminosity and effective temperature. A long-baseline color such as $V-K$ is an excellent temperature index, while M_K is well correlated with bolometric magnitude (see Section 4 below).

Spectral classifications of red dwarfs are usually on one of three different systems—those of Kuiper (1942), Joy (1947), and the MK system (Johnson & Morgan 1953), which is identical to Kuiper's for types M0.5 and later. All use TiO bands in moderate-dispersion, blue-visual photographic spectra. The Mount Wilson (Joy) system has been essentially redefined by a classification of 426 M dwarfs by Joy & Abt (1974). Boeshaar (1976) has derived a new classification scheme using yellow-red photographic spectra, which is used in a redefinition of the MK system by Keenan & McNeil (1976). She finds CaOH more useful than TiO for late-M dwarfs. Bidelman (1985) has collated the ~ 3200 classifications by Kuiper and corrected them to this revised MK system. Additional classifications for substantial numbers of proper-motion and/or late-type dwarfs are given by Bidelman & Lee (1975), Cowley & Hartwick (1982, using their own criteria), Cowley et al. (1967), Walker (1983), Lee (1984), and P. C. Boeshaar (Dahn et al. 1983 and unpublished). Thus, while spectral types are available for a large number of VLM stars, they are on a variety of systems and have been obtained with widely varying wavelength coverage and resolution on both photographic and digital detectors.

Wing (1973) takes a different approach by using an eight-color, narrow-band photometric system to measure molecular band strengths and continuum slopes for M dwarfs. Much of this work remains unpublished;

classifications for nearby stars appear in Wing & Dean (1983). Jones and coworkers have defined a four-color system to separate giants and dwarfs and to obtain photometric parallaxes (Jones 1973, Jones et al. 1981, J. B. Alexander et al. 1983). While the giant-dwarf separation is clear, the parallaxes are not entirely reliable (cf. Ianna & Bessell 1986).

Photographic spectroscopy of these faint red stars has been supplanted by spectrophotometry with red-sensitive detectors. Representative digital spectra for VLM stars are given by Gilmore & Hewett (1983), Pettersen et al. (1985), Liebert et al. (1984), and Dahn et al. (1986), among others. These data have been used for comparative studies of abundance and temperature effects and their correlation with chromospheric activity and kinematics (Section 5). Qualitative physical arguments can be based on the spectral energy distribution (Probst & Liebert 1983), but the lack of model atmospheres precludes quantitative analyses.

Radial velocity work on VLM stars has seen changing scientific rationales. Determination of bias-free kinematics motivated Dyer (1954a,b) and Wilson (1967). Dyer emphasized the correlation between proper motion and radial velocity; a sample biased in transverse motion shows bias in radial motion as well. A recent use of this material in kinematic studies is the work of Upgren (1978). The massive-halo hypothesis provided the impetus for Cowley & Hartwick (1982) and Hartwick et al. (1984). Interest in surface activity and its correlation with kinematics is driving current work (Section 5). Jahreiss & Gliese (1985) point out the lack of radial velocities for many of the fainter stars in the nearby-star sample.

The identification of spectroscopic binaries among dM stars has been a poor stepchild of radial velocity surveys. The activity-duplicity hypothesis for dMe stars (A. R. Upgren, private communication) and the search for brown dwarfs (Marcy et al. 1986) are producing more deliberate programs. H. A. Abt (private communicaion) is repeating his duplicity survey of F-G V stars (Abt & Levy 1976) with significantly higher precision and expects to find more VLM companions. Discovery of such companions to solar-type stars may also benefit from the search for extrasolar planetary systems (Campbell et al. 1986) and from studies of atmospheric mass motions (Dravins 1985).

3.3 *Binaries*

Binaries among VLM stars provide the fundamental data on stellar masses and the calibration of the mass-luminosity law. Statistics of the distribution of masses and orbital parameters may eventually constrain models of star (and possibly planet) formation. Observed luminosity functions must be corrected for undetected duplicities to obtain accurate mass densities. Contrariwise, if the mass distribution of companions can be shown to be similar to that of the field for stellar objects, searches directed at substellar binary companions could determine or constrain the field mass density of brown dwarfs.

Large-scale visual surveys of field stars for duplicity have been carried out repeatedly over the last 160 years. These surveys have been generally unsuited for the detection of VLM systems or companions because of bright limiting magnitudes. More productive surveys for this purpose have targeted previously identified nearby or low-luminosity stars. Kuiper (1934, 1936) observed stars within 25 pc and found many close, rapid pairs that have since yielded masses. Worley (1962) examined 700 M dwarfs, including many observed by Kuiper, and found several more short-period systems. Both programs were carried out on large refractors with limiting magnitudes $m_V \sim 14$–15. The increase in the known stellar population within 20 pc since Worley's survey, and the time-varying probability of detection for a given system, suggest that a repeat survey with very large aperture would be fruitful. Visual observations, principally with the filar micrometer, have remained a major source of data for VLM companions discovered by this and other means. Worley (1983) discusses the extent and precision of the visual data base. The future of this classic technique is in doubt, owing to the declining number of observers and the development of new instrumental methods.

Van Biesbroeck (1944, 1961) applied proper-motion techniques in a search for faint, widely separated companions to proper-motion stars. This elegant approach netted several benchmark VLM companions; VB 10 remained the least luminous nondegenerate star known for almost forty years after its discovery. J. Irwin (private communication) has taken plate pairs at Las Campanas Observatory for an extension of Van Biesbroeck's work to the Southern Hemisphere.

Photometric measurements of binary components are fundamental, but they are very difficult at small separations. Various visual techniques, reviewed by Wallenquist (1954), have provided the available magnitude differences Δm for many close pairs. Rakos (1965, 1974) invented the photoelectric area scanner for application to this problem. Results from this device are given by Rakos et al. (1982) and recent developments by Rakos (1983a). Atwood & Curott (1975; Atwood 1975) built a similar device and obtained data for a few VLM binaries. The limit of separation for reliable photometry with scanners is about 0.5 arcsec. Photometric measurements for much closer pairs will be possible with the Hubble Space Telescope (Rakos 1983b).

Optical speckle interferometry has not yet significantly contributed to VLM binary studies. Speckle observation offers the possibility of reliable photometry at separations an order of magnitude smaller than those that are reached with scanners. The combination of speckle and spectroscopic data has been very fruitful for more massive systems (McAlister 1985), and the recent availability of high-precision radial velocity measurements for M dwarf spectroscopic binaries may yield similarly synergistic results. However, even on 4-m class telescopes the limiting magnitude is $m_V \sim 15$ for binary companions whose existence and approximate position are already known (H. A. McAlister, private communication). Thus, optical speckle interferometry does not yet appear to be a suitable discovery technique for new close, faint companions.

The extension of speckle techniques to infrared wavelengths, via a single detector and scanning slit, was demonstrated by Sibille et al. (1979). McCarthy (1983, 1986) has applied the method to VLM binaries in a very productive program. His work has included the resolution of known astrometric binaries for new or improved mass determinations and a general survey of VLM stars for new companions. Observation at 2.2 μm greatly reduces the luminosity ratio between primary and secondary for VLM stars, compared with that of optical wavelengths, and allows penetration quite far down the main sequence. However, the reality of the least luminous object detected to date, a possibly substellar companion to VB 8, remains controversial (McCarthy et al. 1985, McCarthy 1986, Harrington et al. 1983, Skrutskie et al. 1987, Perrier & Mariotti 1987). Speckle observations yield 2.2-μm magnitude differences ΔK that have twofold utility: They are easily related to ΔV for application to astrometric results, and the individual deconvolved K magnitudes yield M_{bol} directly with little scatter (cf. Section 4). The present limit on the technique is $\Delta K \lesssim 4$. Sorokin & Tokovinin (1985) suggest a method that could extend the detectable magnitude difference if applied in the infrared.

Direct imaging at infrared wavelengths is also a promising approach for detection of brown dwarfs (Probst 1983a). Jameson et al. (1983) used raster scanning in an unsuccessful search for substellar companions to nearby stars. Skrutskie et al. (1986) have conducted a more extensive but also unsuccessful search with an infrared CCD camera. These null results do not yet seriously constrain the brown dwarf mass function because of the small search areas (Probst 1986). However, G. H. Rieke & M. J. Lebofsky (private communication, 1987) have nearly finished a search for faint companions around all known stars within 8 pc of the Sun, north of $-20°$ declination and fainter than $M_V = 11$. The survey has a detection limit of $K = 15.5$, corresponding to at least 6 mag fainter in M_K than VB 10, and covers fields of radius >100 AU around each star. No plausible brown dwarf candidates have been found.

The use of infrared photometry to detect unresolved stellar binaries by anomalous colors has occasionally been suggested (Heintz 1969, 1983).

Displacement in a color-magnitude diagram due to duplicity has long been recognized and applied to open clusters (Haffner & Heckmann 1936). The effect has been applied to nearby field stars using a long-baseline $V-K$ color with only modest success (Probst 1981a,b). Probst also investigated broadband color-color relations, but no combination appeared successful for field VLM binaries owing to the similarity of spectral energy distributions and the dispersion in intrinsic colors. The approach may still prove viable if one looks for anomalies in the energy distribution of individual stars with higher spectral resolution and a very long baseline (Berriman & Reid 1987, Backman et al. 1986). A visual-infrared color method has been successfully used in a search for VLM companions to white dwarfs (Probst 1983b,c), a circumstance in which the energy distributions are quite different. This approach has been extended to a search for white dwarf–brown dwarf pairs by Shipman (1986), by Kumar (1985, 1986), and by Zuckerman & Becklin (preprint).

Direct imaging of binaries against a background reference frame is the only observational technique that in itself yields all the information needed to determine individual masses. Historically this has been done photographically, and the minimum separation for reliable results ($\sim$ few arcseconds; van de Kamp 1967) implies periods of decades or longer, so results are slow in coming. Direct astrometric photography has also been the medium for detection of VLM companions by the effect of mass rather than light, i.e. perturbations. The first photographically discovered astrometric companion was found by Reuyl (1936) from a parallax series delayed long enough for the perturbation to become apparent (van de Kamp 1986). Other perturbations have since been found in parallax series, many with the USNO astrometric reflector. Very small effects can be well established if the period is short (e.g. Harrington 1971), and longer period perturbations are detectable if geometrical conditions are favorable (Christy 1978). While such discoveries are valuable, they remain largely accidental. Parallax series with short time spans (3–5 yr) are generally not well suited for detection of perturbations. A substantial portion of a nonlinear effect can be absorbed into the linear (proper-motion) component of a parallax solution, even with a full period covered (Black & Scargle 1982).

Recognizing both the significance of Reuyl's find and the conditions that led to it, van de Kamp (1939) established a long-term systematic program at Sproul Observatory for detection of low-amplitude, long-period perturbations. Program methods and results are reviewed by Lippincott (1978). The benefit of unbroken coverage of selected stars for decades is illustrated by her Table II, with definite results achieved only after 30 yr or more in many cases. Since the revival of long-focus astrometry in the 1960s, other observatories have instituted or resuscitated similar long series.

A single resolved observation of separation and Δm allows the determination of individual masses for an astrometric binary. Lacking this, plausible values can be estimated from a mass-luminosity relation. Small perturbations usually yield two limiting cases: nearly equal masses, or a very low mass secondary. If the former case is doubtful for other reasons, the inference is a VLM stellar or substellar companion (Harrington 1986). Known VLM astrometric systems have been resolved photographically (Lippincott 1955), visually (Lippincott & Hershey 1972), with photometric scanners (Atwood & Curott 1975), by optical speckle imaging (Balega et al. 1984[2]; H. A. McAlister, in Harrington 1986), and by infrared speckle (McCarthy 1986). Remaining unresolved systems include several with possibly substellar mass companions that are frequently mentioned as likely targets for the Hubble Space Telescope (e.g. Rakos 1983b), although McCarthy's (1986) nondetections at $\Delta K \simeq 4$ (inferring $\Delta V \sim 8$) already indicate exceedingly low luminosities for the companions if the perturbations are real.

Some small-amplitude perturbations are well established by independent results, e.g. Wolf 1062 (Harrington 1977, Lippincott 1977). The reality of others (chiefly Sproul results) has been challenged. Those most familiar with the material have taken opposing positions. Heintz (1976, 1978, 1985) strongly urges the presence of systematic errors due to inadequate modeling of telescope imaging characteristics; high internal precision may then be deceptive. Lippincott & Hershey (1983; also Hershey & Lippincott 1982, Hershey 1982), on the other hand, maintain from internal and external comparisons of long plate series that companions down to a few Jovian masses could be detected around the nearest stars from Sproul data. The history and status of the most interesting case, Barnard's star, has been recently reviewed by its chief protagonist (van de Kamp 1981, 1986). Heintz (1978) gives an opposing view. Recent McCormick Observatory and USNO results do not support the perturbation, with the McCormick data over 12 yr being essentially flat (P. Ianna, private communication). Quantitative, impersonal methods for deciding the reality of a perturbation or fitting an orbit should see increasing use (e.g. Monet 1979, McNamara et al. 1987). But sophisticated analyses that are too far removed from the data can easily produce spurious results (e.g. Jensen & Ulrych 1973). The figures in van de Kamp (1981, Chap. 14; also van de Kamp 1982) provide a useful graphical introduction to the reliability of perturbation analyses as a function of noise level and orbital characteristics.

---

[2] The Balega et al. detection appears unrelated to Reuyl's (1943) perturbation.

The increasing interest in the detection of extrasolar planetary systems has produced suggestions for very high precision astrometry with new methods (Gatewood 1976, Gatewood et al. 1980, 1986, Black & Scargle 1982). Gatewood is implementing a new photoelectric technique on an essentially new telescope in a dedicated perturbation search program (Gatewood 1983, Gatewood et al. 1985). This work requires duplication of effort, lest we find ourselves two decades hence with another collection of potentially exciting results at the instrumental noise level and without independent confirmation.

4. DERIVED STELLAR PARAMETERS

4.1 Radiometric Properties

Determination of bolometric luminosity L requires complete photometric coverage, a parallax, and a flux calibration. Photometry or spectrophotometry can now be routinely obtained to a few-percent precision from 0.3 to 5 μm. Due to the limiting magnitudes of source surveys, known VLM stars are usually very nearby, so the parallaxes and absolute magnitudes are well determined. The status of optical and infrared flux calibrations is reviewed by Hayes (1985). Optical ($\lambda < 1$ μm) absolute calibrations compare Vega directly with laboratory sources. Multiple independent determinations yield a mean flux calibration and energy distribution accurate to better than 2% (Hayes 1985). Three methods have been widely used to derive infrared calibrations at 1–5 μm: extrapolation of the optical calibration of Vega via a model atmosphere (e.g. Strecker et al. 1979), use of the solar absolute flux via solar-analog stars (e.g. Campins et al. 1985), and direct comparison of Vega to terrestrial sources (citations in Hayes 1985; also Mountain et al. 1985). As Hayes (1985) points out, infrared calibrations are still few in number and have significant systematic errors; but recent results agree to within 3%.

Determination of effective temperature T_e for VLM stars can proceed independent of parallaxes, but another difficulty intrudes: the lack of model atmospheres for very cool, high-gravity stars (Section 2.3). Until reliable models become available, T_e for VLM stars is estimated by simple blackbody fitting of the energy distribution.

Pioneering work in (L, T_e) determinations for M dwarfs was done by Johnson (1965a). However, his photometry did not include the H band near the peak of the energy distributions, and T_e was determined from a broadband color-T_e relation derived from giants. As the atmospheric structure and opacity sources are quite different in dwarfs and giants, there is no reason to believe that the same such relations will apply to both (Bessell 1979). Greenstein et al. (1970) devised a procedure for blackbody fitting in which the optical blanketing is constrained by a graybody model. Veeder (1974) used this method in his analysis of 145 red dwarfs with infrared photometry. He found strong correlations between M_{bol} and M_K, and between log T_e and both $V-K$ and $R-I$. Veeder lacked data at J but argued that compensating errors were at work in this region. His data included the L band, which showed some depression below the blackbody fit from H_2O absorption. Mould & Hyland (1976) used Mould's (1976a) models for a more quantitative determination of (L, T_e). The resulting luminosities are comparable to Veeder's, but the T_e scale is considerably cooler at its 2800-K cool end, probably from overestimated H_2O opacity.

Reid & Gilmore (1984) attacked the T_e problem for the coolest stars with *BVRIJHK* photometry of a substantial sample. They used infrared spectrophotometry to assess systematic errors due to the placement of the broadbands in minima of the stellar atmospheric absorption. Blackbodies were fit by taking the K-band flux to represent the continuum level. Tight correlations were again noted between M_{bol} and M_K, and between log T_e and various colors. The temperature scale is systematically hotter than Veeder's scale, an effect attributed to proper inclusion of the J-band flux. However, the most recent results (Berriman & Reid 1987) indicate that the Reid & Gilmore work erred by approximating the flux longward of K with a blackbody fit to shorter wavelengths. Spectrophotometry and broadband photometry show a substantial flux depression at 3 μm from H_2O bands. Fitting blackbodies through the K point to the complete 0.4–5 μm range gives temperatures and luminosities in good agreement with those of Veeder for the coolest stars. It is encouraging that these two much different methods of blackbody fitting now give the same results for L and T_e. This gives greater credence to the results from this necessarily crude approach.

4.2 Masses

The number of reasonably well-determined masses for VLM stars has increased steadily over the last decade. The Sproul Observatory staff have been particularly productive in the determination of orbits and mass ratios for visual and photographic pairs, while McCarthy's (1983, 1986) work on astrometric binaries has added several to the ranks of resolved systems. We have identified binaries with reliable orbits in the Worley & Heintz (1983) catalog and with high-quality parallaxes in Gliese (1969) or Gliese & Jahreiss (1979), and have computed the masses for VLM components given in Table 1. Additional sources for parallax, orbital, or photometric data are given in the notes to Table 1. Our mass errors are smaller than Popper's (1980) results for stars in common because we typically use somewhat smaller errors for the parallax and semimajor axis.

The photometric results in Table 1 were obtained by deconvolving joint

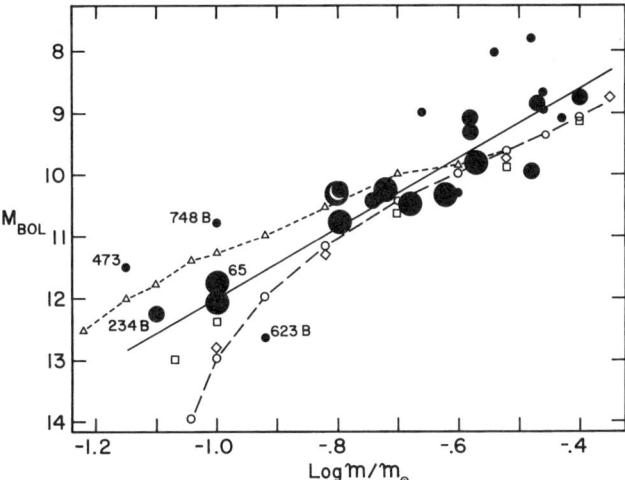

Figure 1 Mass-luminosity diagram for the binary components in Table 1. ●, data from Table 1, with larger symbols indicating higher weight; --△--, 1 × 10^8 yr points from D'Antona & Mazzitelli (1985, DM); —○—, 5 × 10^9 yr points from DM; □, ZAMS from Grossman et al. (1974); ◇, ZAMS from VandenBerg et al. (1983); *solid line*, empirical fit from Smith (1983). A few empirical points are labeled with Gliese numbers.

photometry. We use measurements of component magnitude difference Δm and the well-determined slopes $\Delta m/\Delta V$ for the lower main sequence.[3] Finally, we use (M_{bol}, M_K) and (log T_e, color) relations fit to the data in Berriman & Reid (1987) to derive luminosities and temperatures. We take $M_{bol}(\odot) = +4.70$ from the solar V magnitude of Hayes (1985) and the bolometric correction of Buser & Kurucz (1978). This data set forms the basis for the next section.

4.3 Comparison With Theoretical Models

Figure 1 is the mass-luminosity diagram for our binary sample. We do not derive an empirical *M-L* relation. Our data are not all of the requisite quality, and Smith's (1983) fit to Popper's (1980) data is a good fit to ours as well, as shown in the figure. Model results are plotted from Grossman et al. (1974), VandenBerg et al. (1983), and D'Antona & Mazzitelli (1985; hereafter DM). The agreement with observational data is good for $M \geq 0.2\ M_\odot$, but the zero-age main-sequence (ZAMS) models become systematically more underluminous with decreasing mass below this value. Since three different codes, with differing treatments of the constitutive

---

[3] Probst (1981a) finds $\Delta I_K/\Delta V = 0.67$ and $\Delta K/\Delta V = 0.53$ from the overall slope of H-R diagrams and values for 11 photometrically resolved binaries of widely varying ΔV.

Table 1 Masses and radiometric properties of VLM binary components

Name	Gliese No.	M_K	$V-K$	M_{bol}	$\log T_e$	$M/M_\odot$	Notes
μ Cas B	53B	7.5 ±0.1	4.9 ±0.2	10.2 ±0.2	3.497 ±0.015	0.19 ±0.02	Russell & Gatewood 1984
L726-8 A	65A	8.77 0.04	6.51 0.03	11.7 0.1	3.440 0.013	0.10 0.01	C. E. Worley et al., in preparation; Δm from Rakos et al. 1982
L726-8 B	65B	9.22 0.04	6.90 0.05	12.2 0.1	3.426 0.013	0.10 0.01	
L726-8 A	65A	8.85 0.04	6.55 0.05	11.7 0.1	3.438 0.013	0.10 0.01	Alternate solution with Δm from Harrington & Behall 1973
L726-8 B	65B	9.12 0.04	6.78 0.07	12.1 0.1	3.430 0.013	0.10 0.01	
+41 328 B	67B	6.5 0.2	4.6 0.5	9.1 0.3	3.51 0.02	0.4 0.1	Lippincott et al. 1983
o^2 Eri C	166C	7.55 0.03	5.20 0.03	10.3 0.1	3.486 0.013	0.16 0.01	D. Backman, private communication
Ross 614 A	234A	7.67 0.04	5.44 0.04	10.4 0.1	3.478 0.013	0.18 0.03	D. W. McCarthy, private communication
Ross 614 B	234B	9.3 0.1	7.0 0.3	12.2 0.1	3.42 0.02	0.08 0.01	
+67 552 B	310B	—	—	10.3 0.3	3.47 0.02	0.25 0.08	Lippincott 1973; D. W. McCarthy, private communication
−12 2918	352AB	6.4 0.2	4.51 0.02	9.0 0.2	3.511 0.013	0.22 0.07	Components assumed equally massive
Wolf 424	473AB	8.61 0.06	6.45 0.02	11.5 0.1	3.442 0.013	0.07 0.02	Components assumed equally massive
+48 2108	508B	—	—	8.8 0.2	3.539 0.014	0.34 0.06	Eggen 1974
Ci 20, 986 B	623B	9.6 0.2	7.0 0.3	12.6 0.2	3.42 0.02	0.12 0.03	McCarthy & Henry 1987

Name	ID						Reference
CM Dra	630.1Aa	7.6 ± 0.1	5.08 ± 0.04	10.3 ± 0.1	3.498 ± 0.014	0.24 ± 0.01	Lacy 1977
CM Dra	630.1Ab	7.7 ± 0.1	5.19 ± 0.04	10.5 ± 0.1	3.498 ± 0.014	0.21 ± 0.01	
Wolf 630	644AB	6.2 ± 0.1	4.60 ± 0.01	8.8 ± 0.1	3.508 ± 0.013	0.40 ± 0.06	Heintz 1984a; components assumed equally massive
Wolf 630	644A	6.4 ± 0.1	4.41 ± 0.02	9.0 ± 0.1	3.514 ± 0.013	0.24 ± 0.04	Weis 1982; alternate solution with B very overmassive
+45 2505	661A	6.5 ± 0.1	4.5 ± 0.1	9.1 ± 0.1	3.512 ± 0.013	0.26 ± 0.03	Heintz & Borgman 1984
+45 2505	661B	6.7 ± 0.1	4.7 ± 0.1	9.3 ± 0.1	3.505 ± 0.013	0.26 ± 0.03	
+29 3029	677A	—	—	7.8 ± 0.4	3.594 ± 0.013	0.3 ± 0.1	Lippincott 1982
+29 3029	677B	—	—	8.0 ± 0.4	3.586 ± 0.013	0.3 ± 0.1	
μ Her B	695B	6.1 ± 0.1	4.68 ± 0.06	8.6 ± 0.2	3.505 ± 0.013	0.35 ± 0.10	Components assumed equally massive
μ Her C	695C	6.4 ± 0.1	4.9 ± 0.1	8.9 ± 0.2	3.496 ± 0.013	0.35 ± 0.10	
+59 1915	725A	6.74 ± 0.04	4.45 ± 0.02	9.4 ± 0.1	3.513 ± 0.013	0.5 ± 0.1	Hershey 1982, Rakos et al. 1982; assume 15% uncertainty in the mass sum
+59 1915	725B	7.27 ± 0.04	4.71 ± 0.02	10.0 ± 0.1	3.504 ± 0.013	0.3 ± 0.1	
Wolf 1062	748A	6.6 ± 0.1	4.63 ± 0.03	9.2 ± 0.1	3.508 ± 0.013	0.4 ± 0.2	Lippincott 1977; D. W. McCarthy, private communication
Wolf 1062	748B	8.0 ± 0.1	5.8 ± 0.2	10.8 ± 0.2	3.47 ± 0.02	0.10 ± 0.04	
+56 2783	860A	7.13 ± 0.05	4.72 ± 0.04	9.8 ± 0.1	3.503 ± 0.013	0.27 ± 0.02	Eggen 1956; D. Backman, private communication
+56 2783	860B	7.98 ± 0.06	5.5 ± 0.1	10.8 ± 0.1	3.477 ± 0.013	0.16 ± 0.01	
+19 5116	896B	7.5 ± 0.1	6.0 ± 0.2	10.2 ± 0.1	3.46 ± 0.02	0.16 ± 0.03	Heintz & Borgman 1984, Heintz 1984b, Kuiper 1943, Baize 1966; assumed 15% uncertainty in mass sum

physics and atmospheric opacities, give the same general result, it may have real astrophysical significance.

At the lowest masses, the luminosities decline precipitously with age in the DM models, illustrated by two epochs in Figure 1. The increasing slope of their M-L relation for $M < 0.3\ M_\odot$ allows a *declining* luminosity function to translate into a *rising* mass function (Section 6.2). Scalo (1986) dismisses this result on the grounds that the ZAMS M-L relation of DM does not fit the data for VLM stars. We view this dismissal as premature. Selection effects, chiefly apparent magnitude limits, bias the few known binaries at the lowest masses toward the most luminous and hence youngest systems (e.g. van de Kamp 1986, p. 305). Available age indicators lend some support to this idea. Gliese 65, 234, and 473 are emission-line flare stars with young disk kinematics. Gl 65 and 473 are both members of the Hyades supercluster according to Eggen (1984), although their relative placement in Figure 1 is not consistent with equal age. Gl 65 and 234 are X-ray emitters (Johnson 1983). The somewhat underluminous Gl 623B has old disk kinematics and is not a known emission-line or flare star. The mass uncertainty for this star derives almost equally from errors in parallax, photocentric semimajor axis, and observed separation. But Gl 748B, possibly quite luminous for its mass, is lacking in chromospheric or coronal activity and has a large (~ 100 km s^{-1}) space velocity. Its mass error is dominated by the precision of the photocentric semimajor axis. Resolution of several well-known astrometric binaries with the Hubble Space Telescope could verify the DM model results if some prove to have components of stellar mass at very low luminosity.

Figure 2 is the observational HR diagram for this same sample. Model results are transformed to (M_K, $V-K$) using our fits to Berriman & Reid's (1987) data; their stars that define the transformations are included in Figure 2.[4] By otherwise limiting the observational data to binaries, we retain mass as an implied parameter. The mean observed values for $M = 0.3, 0.2$, and $0.1\ M_\odot$ are shown for comparison to the models. Figure 1 shows good agreement in L for $M \geq 0.2\ M_\odot$, and the discordance in Figure 2 for these masses is almost entirely in the temperatures (e.g. the leftward progression of models at $0.3\ M_\odot$). Since L is determined primarily by the central temperature, whereas T_e is photospheric, this mismatch suggests continuing problems with atmospheric opacities but no serious difficulties with interior physics in the models.

The least luminous stars are of greatest interest. Evolutionary tracks from DM for $M \lesssim 0.1\ M_\odot$ show a considerable stretchout in L and T_e at

---

[4] Gl 866 has been found to be binary (Leinert et al. 1986, McCarthy et al. 1986), and we have corrected the Berriman & Reid data for this.

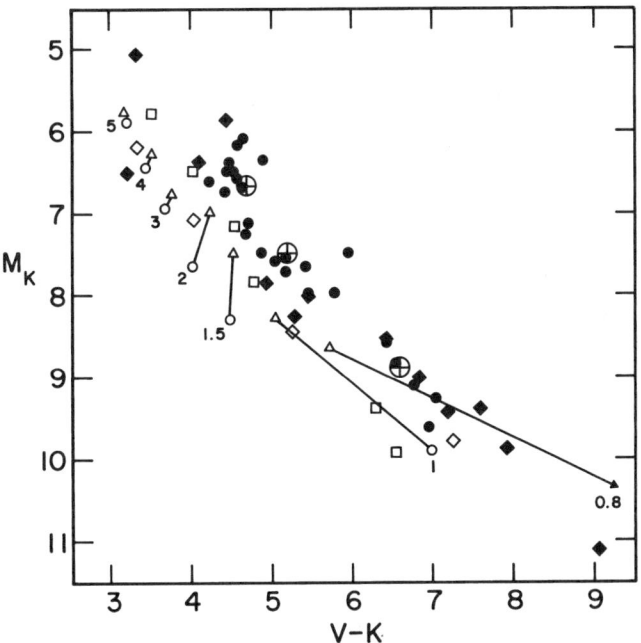

Figure 2 Observational H-R diagram for entries in Table 1 plus late dwarfs from Berriman & Reid (1987). Symbols as in Figure 1, plus ⊕, mean observational points for 0.3, 0.2, and 0.1 $M_\odot$; ◆, Berriman & Reid data. DM values for different epochs at the same mass are connected by lines and labeled by the mass in 0.1-$M_\odot$ units.

a given mass, which is qualitatively duplicated by our binaries: Compare Gl 65AB, 623B, and 748B at 0.1 $M_\odot$, or Gl 234B and 473AB at 0.07 $M_\odot$. The T_e scale of Reid & Gilmore (1984) placed their least luminous stars very close to the theoretical tracks, and they argued from this that all of these objects are stellar. The revised (Berriman & Reid 1987) scale used here moves these stars rightward from the tracks and revives the possibility that known objects include some genuine brown dwarfs. The least luminous object in Figure 2, LHS 2924, shows no chromospheric activity, which suggests that it is an old star (Probst & Liebert 1983, Liebert et al. 1984). Another object of similar color and luminosity, LHS 2397a, has strong Hα emission and may be quite young (Liebert & Dahn 1986). All these results indicate that this portion of the H-R diagram may contain objects of very different age; hence, the search for brown dwarfs must seek either much cooler objects or those with a well-determined age for unambiguous identification and interpretation. Young clusters such as the Pleiades, or "field" stars identifiable with young superclusters (Eggen

1985), offer promising possibilities. Even here, however, a mass-dependent age spread may cause difficulties (Eggen 1982, Adams et al. 1983).

4.4 The Local Subdwarf M Population and Derived Metal Abundances

By definition, subdwarf M stars lie significantly below the main sequence, as defined by solar neighborhood stars of the old disk population. Stellar structure theory predicts that the main sequence for stars of lower mean metallicity (Z) than those representative of the bulk of the solar neighborhood population should lie below the old disk main sequence in a theoretical H-R diagram. Two aspects of this phenomenon should be understood, however. First, the predicted shift is primarily to higher surface temperatures and smaller radii; in fact, a metal-poor star of a given mass is actually more luminous than its normal Z counterpart. The second point is that a composition difference may also affect the transformation from the observed fluxes to such theoretical quantities as T_e, i.e. a metal-poor object looks "bluer" in certain colors anyway, regardless of whether it truly has a higher effective temperature. The inclusion of infrared colors to specify most of a star's energy distribution permits effective temperature estimates that are somewhat less sensitive to these blanketing problems. These show that the subdwarfs truly lie to the left of the main sequence for normal metallicity (cf. Mould & Hyland 1976).

In practice, the M subdwarfs are an ill-defined population with several recognizable properties besides subluminosity that are assumed to be correlated. These include the following:

1. *spectra* characterized by relatively weak TiO bands for a given blanketing-insensitive color, but relatively strong hydride bands and atomic lines (cf. Ake & Greenstein 1980, Bessell 1982);
2. *broadband colors* due either to the reduced blanketing in a metal-poor atmosphere, such as in a $B-R$, $R-I$ diagram (Hartwick 1977), or to the higher pressures in such an atmosphere, such as the $J-H$, $H-K$ plot of Mould & Hyland (1976);
3. *high total space motions*, which identify halo stars in the solar neighborhood.

The first two properties are of course directly attributable to the reduced metal content of the atmospheres; the third is a consequence of the fact that halo stars generally have low metallicity. However, any metal-poor stars of the old disk population may not have large space motions. Indeed, it is difficult to establish how to use these subdwarf characteristics of subluminosity, colors, spectra and kinematics to define the same stellar population. This is because the metallicities of the low-mass subdwarfs are poorly determined compared with those of more massive stars, and because the low-mass subdwarfs in the solar neighborhood represent a substantial (but unknown) range and distribution of metallicity.

Very few M dwarfs have been analyzed in sufficient detail for absolute abundances of several heavy elements to be determined independently. Only weak lines of elements not affected by molecular association are easily interpreted, and only a few M stars have been bright enough for spectroscopy of the required dispersion (although the advent of CCD detectors is allowing this list to be expanded). Hartmann & Anderson (1977) derived sets of abundances close to solar values for several early-M dwarfs; Mould (1976c) found a mean $[M/H] = -0.55$ for the high-velocity subdwarf Kapteyn's star. Mould (1976b) used photometric measurements of the continuum color, TiO, and CaH blanketing to estimate mean metallicities for a large number of stars classified sdM, and the sample was later extended by Hartwick et al. (1984). The derived metallicities extended down only to about one tenth of the solar values. Given the comparatively larger volume that has been searched for metal-deficient stars of higher luminosities, as well as the extreme rarity of field halo stars with very low metallicity, the small abundance span of these samples is perhaps not surprising.

The most metal-deficient main-sequence star yet analyzed is the unique subdwarf G77−61, which has a carbon abundance greater than that for oxygen; the estimated mass is 0.3 ± 0.1 $M_\odot$. Its carbon enrichment, but not its metal deficiency, may be attributed to a binary mass transfer history (Dahn et al. 1977, Dearborn et al. 1986). The heavy-element abundance distribution (Gass 1985) shows that low-mass halo stars extend to metallicities comparable with the most extreme halo giants—$[M/H] < -5$ for iron. It seems fortunate to find even a single such representative in the solar neighborhood, especially one as bright as the 13th magnitude. Or could there be other examples among the generally fainter subdwarfs having more normal ratios of carbon to oxygen? Bessell (1982 and unpublished) would use the few low-luminosity subdwarfs with bright, well-analyzed companions having low metallicity to calibrate subdwarf spectra onto an approximate $[M/H]$ scale for the unanalyzed VLM region. However, the new model atmosphere calculations by R. Wehrse & M. S. Bessell (in preparation) may make this scale more direct and quantitative.

5. STELLAR SURFACE ACTIVITY AND DERIVED POPULATION PARAMETERS

The theory and observation of activity near and above the surfaces of late-type dwarf stars has become a huge subject, driven by detailed solar work, by the IUE and Einstein Observatories and other space experiments, and by recent advances in ground-based instrumentation. These now permit relatively rudimentary, semiempirical modeling of solar-type activity in other stars. The subject thus encompasses time-dependent phenomena near the photospheres and in the overlying chromospheres and coronae, the manner in which nonradiative energy is deposited at these layers, and the way in which such energy is produced (i.e. by rotational braking due to a magnetic dynamo).

The fundamental thesis that has emerged in recent years is that the active phenomena, the variations observed from the outer envelope to the corona, are somehow a direct consequence of strong, variable magnetic fields (Linsky 1983). The working model for the behavior of other dwarf stars is that magnetic flux tubes like those in the solar atmosphere are the basic structures responsible for atmospheric inhomogeneity, are the regions of enhanced nonradiative heating, and are the locations of time-variable phenomena on various time scales. Diverse observations of M dwarf stars in fact suggest that the level of surface activity may be orders of magnitude higher in these stars than in the Sun, especially in the coronae. For example, Saar & Linsky (1985) recently reported the first positive measurement—by modeling Zeeman-broadened absorption lines—of a mean magnetic field well above the kilogauss level in an active M dwarf star.

While magnetic activity is therefore identified as the direct stimulus for the release of nonthermal radiative and mechanical energy at stellar surfaces and above, the existence of surface magnetic fields for late-type stars appears to depend on the presence of an extensive, outer convective envelope. Moreover, the *amount* of activity—especially coronal X rays—appears to be most directly related to the stellar rotation rate, although it is by no means clear that the loss of rotational energy (spin-down) is the direct source of energy dissipated in the surface, chromospheric, and coronal activity.

For the VLM dwarf stars, far down in mass from the Sun, it is not clear how fully the solar model applies. As discussed in Section 2.1, VLM stars may lack the radiative cores that some have argued may be necessary to anchor the magnetic flux (Galloway & Weiss 1981, Schmitt et al. 1984). It is also more difficult, for these faint stars, to obtain such relevant observations as the profiles of absorption lines from which rotational velocities or magnetic field strengths can be estimated.

In the next sections we document the evidence for a direct dependence of coronal and chromospheric activity on the stellar rotation period for the relatively small samples of M dwarf stars—mostly above VLM masses—having measured rotation periods or velocities. Unfortunately, the increase of rotation period with age is not yet as well calibrated for late-M dwarfs as it is for solar masses (cf. Simon et al. 1985), and binary stars may behave differently. However, the spin-down may cause a decline in surface activity, as measured in various ways. Surface activity would be expected then to correlate strongly with other indicators of age, such as stellar kinematics (Section 5.3). Thus there is the potential of using the surface activity measures as broad indicators of stellar age, population type, and multiplicity.

Extensive reviews of general aspects of stellar envelope physics and surface activity include Hartmann & Noyes (1987), Baliunas & Vaughan (1985), and Rosner et al. (1985). Various conference proceedings may also be consulted. Especially useful is *Activity in Red-Dwarf Stars* (Byrne & Rodonò 1983), although the bulk of the discussions necessarily pertain to more massive M dwarfs. Even Stauffer & Hartmann's (1986b) nice review of "Rotation in Low Mass Stars" barely overlaps the VLM range.

5.1 *Coronal Activity and Stellar Rotation*

The level of coronal activity may be directly measured by X-ray as well as radio observations. Recent attention has been focused on the X-ray results: Observations of over 100 mostly early/middle-M dwarfs within 25 pc of the Sun were obtained using the Einstein Observatory (cf. Johnson 1981, 1983, 1986, Vaiana et al. 1981, Rucinski 1984, Bookbinder 1985, Rosner et al. 1985). Only a handful of Einstein targets were (presumably) single VLM stars, although such stars are included in numerous known binary targets where the origin of the detected X rays is ambiguous. By cross-checking against the catalogues of nearby stars, Johnson (1986) found only a few known M dwarf stars with $M_V > +15$ having positions inside of the Einstein fields (but which were not the targets of the observations).

The X-ray luminosity function for M dwarf stars from these samples (Rosner et al. 1981, Johnson 1986) covers at least four orders of magnitude, from $L_x > 10^{30}$ erg s^{-1} to $L_x < 10^{26}$ erg s^{-1}, compared with a value of about 10^{27} for the quiet Sun. Moreover, within the M dwarf range, there is nearly this wide a dispersion at a given M_V. Such a wide dispersion is also characteristic of G and K stars. For the latter stars it is believed that L_x depends most directly on approximately the square of the rotation rate (Pallavicini et al. 1981, Golub 1983), while the mean L_x does not vary much with stellar mass among late-type dwarfs. The M dwarfs with observed rotational periods or velocities are also consistent with the direct decline of L_x with rotation rate. However, the mean L_x for M dwarfs declines modestly with decreasing mass (Golub 1983), in contrast with its behavior for the more massive main-sequence star; possible implications of this mass dependence are discussed in Section 5.4.

5.2 Chromospheric Activity and Rotation

The stellar chromospheres are more amenable to a variety of synoptic observational projects from ground-based and space ultraviolet telescopes. The characteristic line emission and ultraviolet continua are the best diagnostics of chromospheres. Traditionally, the focus has been on Ca II H&K line emission in solar-type stars (cf. Wilson 1967, Vaughan & Preston 1980) as the primary measure of nonradiative, chromospheric heating. Fortunately, for those interested in very cool, dim stars, Herbig (1985) has shown that quantitative measures of Hα equivalent widths in G stars correlate well with the Ca II H&K line data, and thus these widths might be used in a similar manner. The use of Hα equivalent widths as a diagnostic of the physical conditions in M dwarf chromospheres has been discussed by Cram & Mullan (1979), Zarro (1983), and Giampapa (1985). High-resolution spectroscopic observations covering Hα for a few dM and dMe stars have been discussed by Kraft et al. (1964), Worden et al. (1981), Zarro & Rodgers (1983), and Young et al. (1984). The chromospheric models and predicted Balmer-line profiles of Cram & Mullan (1979) permit estimates of chromospheric densities and temperatures for observed stars, both during flare activity and in quiescence.

The first comprehensive project to study the mean chromospheric properties in this way for a large sample of M dwarfs is that of Stauffer & Hartmann (1986a; hereafter SH). Unfortunately, the brightness limit of the SH study excluded all but a few stars in the VLM range. However, the general conclusions are an important starting point for present and future studies of lower luminosity objects.

SH find that their sample of several hundred stars culled from Gliese (1969) divides into two groups that show Hα in absorption and emission, respectively. The emission-line (dMe) objects generally include the stars for which cluster membership and/or kinematics indicate relative youth (Section 5.3), as well as binary stars. Stars that are suggested by the indicators to be old, including metal-poor subdwarfs, show absorption lines. The absorption strength generally declines with decreasing temperature, as expected for a photospheric origin. However, the chromospherically active stars include those with enhanced absorption (cf. Cram & Mullan 1979, Giampapa 1985) as well as those showing emission. SH argue that chromospheric emission is in fact triggered by a minimum rotation rate (see Bopp et al. 1981). About half of the emission-line stars show broadened line profiles, interpreted as rotation rates ($V \sin i$) above the detection limit of 10 km s^{-1}. Further discussion of the case for a direct dependence of chromospheric activity on rotation is found in Stauffer & Hartmann (1986b).

Extensive photometric and spectroscopic studies have shown that chromospheric activity is often highly variable in active stars owing to the occurrence of "flares" (both large events and continuous microflaring). For further discussion, the reader is referred to the reviews by Pettersen (1983) and Mirzoyan (1984), and to their extensive papers. A catalogue of flare star data is available in Pettersen (1976), and a discussion of derived physical parameters for this set of active stars can be found in Pettersen (1980). Cram & Woods (1982) discuss semiempirical models of flare activity.

5.3 Ages and Kinematics

The levels of both coronal and chromospheric activity in M dwarf stars clearly decline with age, as we expect from the hypothesis that the activity levels depend most directly on the stellar rotation rate. Close binary stars, which can maintain higher rotation rates into old age, are often clear exceptions. Both X-ray flux levels and the indicators of chromospheric activity may vary widely as a result of the occurrence of stellar flares in the active objects. On the other hand, it is not clear whether the age-rotation relations derived for more massive stars apply to M dwarfs—in particular, to VLM stars; age estimates for groups of these stars may also be obtained from the stellar kinematic properties (cf. Wielen 1977).

The PhD thesis of Bookbinder (1985) included an examination of the statistical behavior of X-ray emission, stellar ages and color dependences for 243 K and M dwarf stars within 25 pc of the Sun. He found that stars showing old disk kinematics exhibit an average level of coronal (X-ray) emission an order of magnitude less than do young disk stars. The same trend is found in the investigations of Johnson (1983, 1986). Again, we caution that there is very little evidence as to whether X-ray activity correlates in the same manner with stellar age for masses below 0.3 $M_\odot$.

The correlation of chromospheric activity with measures of stellar age is well demonstrated in the large sample of SH and in their previous work on members of young clusters such as the Pleiades and Hyades. The only attempt to extend this kind of investigation to the VLM stellar regime is that of Giampapa & Liebert (1986; hereafter GL). Despite the use of the echelle spectrograph on the Multiple Mirror Telescope (4.5-m effective aperture), only a subset of the known VLM stars culled from proper-motion catalogues may be observed successfully at high spectral resolution, and these at a signal-to-noise ratio far poorer than that of the SH sample.

There are clear indications from the work of GL that chromospheric activity correlates in a similar manner with age and the stellar kinematics at masses below 0.3 $M_\odot$. VLM objects with young disk space motions generally show strong Hα emission. Stars with halo space motions down to $M_V \sim +14$–15 show Hα in absorption. A majority of the VLM stars with old disk motions lacked strong Hα emission—see also Liebert et al. (1979) and Pettersen et al. (1985). These correlations carry other implications, which have been investigated for M dwarfs of higher masses. The oldest, inactive dwarfs with larger space motions are also more likely to be metal deficient and to lie accordingly in the H-R diagram below the main sequence of solar composition stars (Section 6.3).

5.4 Surface Activity Near the Stellar Mass Limit

The GL investigation did not suggest that the age/kinematics dependence of chromospheric activity differs qualitatively between VLM stars (whose interiors are likely to be completely convective) and stars of higher masses. Indeed, of the stars near the stellar mass limit, many are flare stars of high flare frequency. Certainly, the majority of known objects near the main-sequence mass limit—i.e. with $M_V \gtrsim +17$—do show strong Hα emission. Bessell (1986) finds variability in the $V-I$ and $R-I$ colors for several such objects, probably due to chromospheric activity. However, the proper-motion samples may be badly biased near the stellar mass limit; GL certainly found stars lacking Hα emission at least as faint as $M_V = +16$ among those with generally higher space motions.

On the other hand, evidence has been accumulating that M dwarfs show a different dependence of coronal activity on stellar mass than do G and K stars. The mean X-ray luminosity (L_x) shows a small, steady decline through the M dwarf mass range (Golub 1983). Indeed, Bookbinder (1985) concludes that VLM stars later than M5 spectral type were strongly under-represented among the detected targets of his optically selected sample. This effect was argued to be a direct result of lower activity levels in such stars, rather than due to a bias in the sample of M dwarf targets.

The meaning of a decline of the mean L_x in the M dwarf mass range depends on the origin of the X-ray energy: If the coronal energy were derived primarily from the loss of rotational energy, only a small mass dependence might be expected, assuming that the efficiency of rotational braking were not a function of mass. However, if the energy were drawn ultimately from the thermonuclear reservoir, a more proper measure of coronal activity might be L_x/L_{bol} (Rucinski 1984); the mean value of this ratio does not clearly change with mass in the M star range, although Bookbinder (1985) still finds a small increase with decreasing mass. Rucinski (1984) argues that the coronal energy losses alone are so large that the rotational energy of M dwarfs would be expended on a time scale far too short (10^6–10^7 yr) to account for the observed relationships of activity vs. age.

It is apparent that VLM stars may have very active chromospheres and active coronae, and that these may show qualitatively similar declines with increasing stellar age or decreasing rotation as those for more massive stars. This leaves unclear what effect the disappearance of the radiative core has on the driving of surface activity; an exploration of alternative dynamo mechanisms is beyond the scope of this discussion. Two possible applications are worthy of note: First, this result lends little support to the hypothesis that a drastic decline in surface activity (and therefore in magnetically induced angular momentum loss) should occur in the mass-losing secondary stars of cataclysmic variables as they are stripped down to the 0.2–0.3 $M_\odot$ appropriate to the 2–3 hr "period gap" (cf. Rappaport et al. 1983, Spruit & Ritter 1983). Note that such secondary stars must necessarily be rapid rotators. Second, if it can be assumed that at least some level of surface activity may necessarily occur in young, VLM stellar objects, delineation of the low-luminosity limit below which active chromospheres or coronae are not found is a potential empirical way of identifying necessarily old, visible stars of a given composition and therefore of defining the main-sequence mass limit. Thus one also expects detectable substellar mass objects to exhibit "young disk" space motions.

6. THE LUMINOSITY AND MASS FUNCTIONS

Because of the history of discovery of low-mass stars, the luminosity function (LF) of these stars is usually expressed first as the number density of stars per unit magnitude vs. either M_{pg} or M_V. For the motivations discussed in Section 2, one wishes to use this information to derive the aggregate initial-mass function (IMF), which for low-mass stars with main-sequence lifetimes much longer than the age of the Galactic disk is equal simply to the observed present-day mass function (PDMF)—see Scalo (1986). The path from the LF to the IMF for low-mass stars is especially tortuous, as we discuss below. Reid (1987) gives an extensive review of this subject, which permits us to be somewhat briefer; see also Scalo (1986).

We may express the initial-mass function $\psi(M)$—that is, the number of stars per unit mass, per cubic parsec—in this simple case as

$$\psi(M) = \Phi(M_V) \times (dM_V/dL) \times (dL/dM).$$

The first factor on the right-hand side of the equation is the observed LF per unit M_V, the second factor is the bolometric correction, and the third is the mass-luminosity relation. Since most observed samples of low-mass stars lie within one disk scale height, we pose the problem in terms of volume rather than surface densities.

6.1 The Luminosity Function

The history of this subject is too long for review here. In the early work, however, the enormous contributions of Luyten (1938, 1968) produced a basically correct LF, consistent with more recent determinations down to the VLM interval. It is now established that the number density reaches a peak near $M_V = 12$–13; this maximum shows in analyses of proper-motion catalogues (Luyten 1968, Schmidt 1983), nearby-star samples (cf. Wielen et al. 1983, Dahn et al. 1986), photometric surveys (Gilmore et al. 1985, Hawkins 1986a,b), and spectroscopic surveys (see Reid 1982). Past this peak in the VLM interval ($M_V \gtrsim 14$), there is considerable uncertainty in the shape of the distribution, as is discussed below.

The distribution vs. absolute visual magnitude becomes increasingly nonlinear at the lower luminosities, with a given M_V interval occupying a smaller and smaller M_{bol} interval. Although the bolometric corrections are accurately enough known for an M_V distribution to be transformed into an LF (cf. Reid 1984), it is easier to compare the various empirical LF determinations (which employ differing color systems) after the conversion to M_{bol}.

Reid (1987) displays four of the most important recent determinations of the LF in his Figure 4. These empirical distributions show peaks generally near $M_{bol} \sim 10$, or $\log L/L_\odot \sim -2.1$. However, it is not the existence of a peak in the LF per se that reveals the qualitative nature of the IMF, but rather its location and shape. Even a steeply increasing Salpeter mass function ($\psi(M) \sim M^{-2.35}$) maps back into an LF with a broader peak near $M_{bol} = +11$–12 and $M_V = +14$–15 (Figure 3).

Two important differences among the best empirical samples are critical to the shape of the derived IMF. First, the photometric LFs of Gilmore et al. (1985) and Hawkins (1986b) show sharper declines at lower luminosities past the peak than do the nearby/proper-motion star samples of Wielen et al. (1983; hereafter WJK), C. C. Dahn et al. (in preparation; called LHSb in Reid 1987), and especially the sample of very nearby stars (within 5.2 pc) presented in Dahn et al. (1986). Figure 3 shows the averaged LF adopted by Reid (1987) and the WJK nearby-star LF. The use of a linear, rather than logarithmic, scale for number density and the inclusion of error bars emphasize the uncertainties in even the best available determinations. In Figure 3, the Reid averaged LF shows a sharp decline in comparison with the WJK nearby-star sample alone. Consideration of the well-studied stars of the immediate solar neighborhood (Dahn et al. 1986) suggests a reason for this difference. This is the only sample in which we may expect that the vast majority of close binaries have been recognized and counted separately, and its statistical significance and likely com-

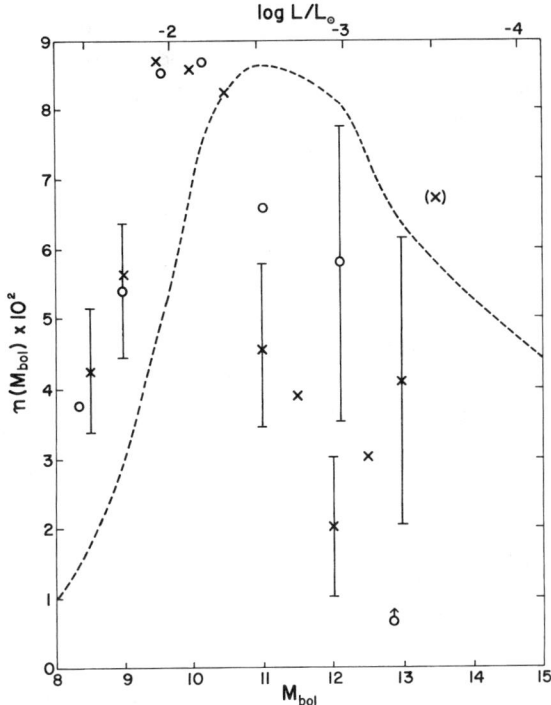

Figure 3 The luminosity function (LF), shown as the number of stars per unit absolute bolometric magnitude (M_{bol}) in a complete volume of radius 20 pc. vs. M_{bol}. The top abscissa axis is log $L/L_\odot$. ○, adopted LF of Reid (1987); X, data taken from Wielen et al. (1983, WJK), after subtraction of white dwarfs, with representative error bars. The Salpeter power law IMF is transformed back into an LF using the M-L relation of DM, and it is shown as the dashed curve.

pleteness are competitive with the other samples at $M_{bol} > 11$. The resolved binary components add substantially to the low end of the LF. While different effects are predicted to dominate by the ad hoc modeling of Reid (1987), the simple census of the nearby stars in this sample and the WJK sample suggests that a proper accounting for low-luminosity binary companions in samples of more distant, astrometrically unstudied stars would produce a true LF with a flatter decline at luminosities lower than the peak. The possible significance for the mass function is discussed in the next section.

A second possible difference among the empirical LFs is that the photometric samples show evidence for an upturn in the LF near $M_{bol} = +13$, as first reported by Hawkins (1986a,b)—note the LF of Reid (1987) in Figure 3. Hawkins' (1986b) proper-motion estimates also suggest the small tangential velocities for these faint stars that one expects if they are young brown dwarfs. While the statistical significance of an upturn is open to some question, and although the result also depends on the plate calibrations (see Section 3), this finding would certainly require an increasing IMF at $\lesssim 0.1\ M_\odot$. The nearby/proper-motion star samples do not suggest comparably large numbers of stars at $M_{\rm bol} \gtrsim 13$. However, the uncertainties, particularly in the completeness, are large enough that their LFs are not necessarily inconsistent with the possible existence of an upturn.

6.2 The Initial-Mass Function

The conversion from an LF to an IMF is even more perilous and controversial. For example, D'Antona & Mazzitelli (1986b) employ their theoretical (D'Antona & Mazzitelli 1985; DM) relation of mass to luminosity (M-L); Scalo (1986) relies upon an empirical relation. We show in Figure 4, and discuss below, how these lead to vastly different IMFs for

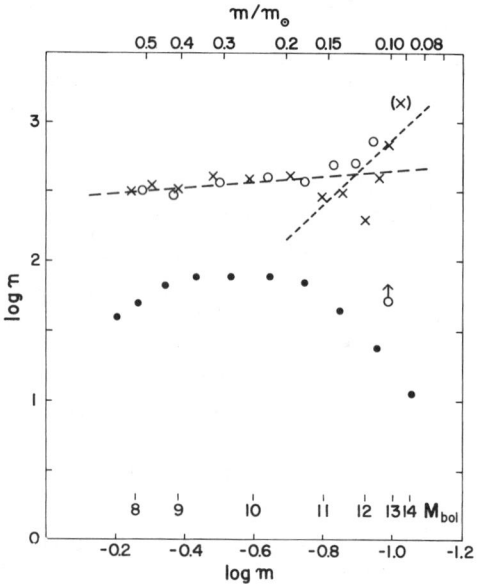

Figure 4 The initial/present-day mass function, shown as the log of the number of stars in a complete volume of radius 20 pc per 0.1 log M vs. log M. The top abscissa axis shows linear mass units, and the approximate location of unit $M_{\rm bol}$ values are indicated above the bottom axis. ○, IMF derived by applying the DM relation to the LF of Reid (1987); X, IMF derived by applying the DM relation to the LF of WJK, as presented in Figure 3; ●, IMF taken from Scalo (1986), in surface density units. Two power-law fits are indicated: ———, $M^{-1.02}$ empirical fit to the entire VLM range; ---, $M^{-2.35}$ (Salpeter law) fit to the rising portion only.

VLM stars. It is convenient to address the issues separately for the middle and end of the VLM regime.

IS THERE A PEAK NEAR 0.3 $M_\odot$? Clearly, there is a strong peak if the flatter empirical M-L relation (Scalo 1986) is adopted. If the theoretical M-L relation of DM is employed with the LF of Reid (1987), the result is a weak, very broad maximum, of questionable significance. Indeed, both the Reid and Scalo IMFs have remarkably flat slopes below 0.4 $M_\odot$ until the sharp divergence occurs below 0.15 $M_\odot$. On the other hand, if the flatter LF of WJK is combined with the theoretical M-L relation of DM, the result is a continuous, and even steepening, rise in the IMF through the regime of VLM stars. Given the reasons for favoring a steepening M-L relation close to the main-sequence mass limit (Section 4.3), our conclusion is that no peak is clearly established in the IMF near 0.3 $M_\odot$ despite the well-defined maximum at the corresponding point of the LF.

DOES THE IMF RISE STEEPLY AT $\lesssim 0.1$ $M_\odot$? Clearly, the answer is yes if the LF shows a sharp increase at $M_{bol} \gtrsim +13$ (Hawkins 1986b, Reid 1987), unless the true M-L relation were to bend oppositely from the predictions of DM. If the DM function is adopted, there is an indicated increase in the IMF slope even if the true LF is flat at $M_{bol} \gtrsim 13$. On balance, the IMF appears to be flat or increasing near the hydrogen-burning mass limit; however, the uncertainties in both the theory and observations are formidable, and this conclusion is fragile.

6.3 *The LF and IMF of the Local Subdwarf M Population*

It is now clear that the halo main sequence in the solar neighborhood (Section 4.4) reaches quite low in luminosity. Hartwick et al. (1984) and C. C. Dahn et al. (in preparation) have shown that a sequence of stars with extreme subdwarf characteristics in colors, spectra, and kinematics extends to $M_V \sim +15$. This sequence includes several stars that are at least 2 magnitudes subluminous according to preliminary Naval Observatory CCD parallaxes.

The preliminary luminosity function derived from applying the $1/V_{max}$ method of Schmidt (1975) suggests a broad maximum in the LF similar to that for disk dwarfs near $M_V \sim +12$, and a rapid decline at $M_V > +14$. Likewise, the failure of P. C. Boeshaar & J. A. Tyson (preprint) to find significant numbers of stars in deep CCD frames with colors appropriate to the least luminous local subdwarfs lends at least qualitative substantiation. Thus, the mass fraction in the solar neighborhood due to visible halo stars is not enhanced appreciably at the lowest masses. The ratio of M dwarfs to extreme subdwarfs of about 250 found by Hartwick et al. (1984) is in good agreement with that derived from a brighter sample by Eggen (1983), with the predictions from modeling the Luyten LHS Catalogue by Dawson (1986), and with the space density of the halo main sequence (3×10^{-4} pc^{-3}) found by Chiu (1980).

Despite uncertainties in parallaxes, effective temperatures, and luminosities, it is clear that the extreme subdwarfs near $M_V = +15$ must have masses close to the hydrogen-burning mass limit. First, the theoretical limit for stable nucleosynthesis increases to 0.095 $M_\odot$ for models with $Z = 0.001$ (D'Antona & Mazzitelli 1982). Second, the luminosities at a given mass are modestly higher than at solar Z. Yet the luminosities assigned to subdwarfs at $M_V \sim 15$ are significantly lower than those of normal Z stars of the same M_V, since the bolometric corrections for the subdwarfs (at higher T_e) are reduced! At $M_{bol} \gtrsim +13$, the lowest luminosity subdwarfs fit theoretical masses below 0.1 $M_\odot$. Since we observe an apparent decline in the observed luminosity function only at this point, and since the shape of the LF is otherwise poorly constrained, there is no useful empirical constraint on the Population II IMF at the "brown dwarf" boundary that can be obtained from study of the interlopers in the solar neighborhood.

6.4 Brown Dwarfs and the Dynamical Disk Mass

Any possibility that the dynamical "missing mass" (Section 2.4) might be accounted for in brown dwarfs hinges on both the assumption of a steepening M-L law and the possible upturn in the LF near the current observational limits (Section 6.2). An empirical mass function fit to the entire VLM range is at best almost flat (Figure 4). Even if extended to ~ 0.01 $M_\odot$, a limit suggested by fragmentation models (Section 2.5), the integrated brown dwarf mass is $\lesssim 0.005$ $M_\odot$ pc^{-3}. Alternatively, an upturn in the LF and corresponding mass function at the lowest observed luminosities can plausibly be fit with a much steeper power law (Figure 4). Note that this implies, in effect, a separate, low-mass mode of star formation, for which there is no prior indication (Section 2.5). If extended to the same limit, the integrated mass becomes ~ 0.1 $M_\odot$ pc^{-3}, with the result dominated by the least massive objects, the most difficult brown dwarfs to observe. Analysis of data from the IRAS survey is beginning to constrain the numbers of such low-temperature brown dwarfs.

An even more ad hoc scenario would place the bulk of the "missing mass" in massive ($\gtrsim 0.05$ $M_\odot$) brown dwarfs. But this requires either a mass spectrum that is discontinuous across the hydrogen-burning boundary or the existence of a significant number of undiscovered VLM stars in the region in which the LF seems reasonably well determined. Observationally, the number of massive and/or young brown dwarfs is constrained (a) by the apparent paucity of luminous, young brown dwarfs in observed samples (Reid 1987); (b) by the failure, so far, to discover such objects as spatially resolved binary companions, especially with the use of new infrared arrays (Jameson et al. 1983, Skrutskie et al. 1986, G. H. Rieke & M. J. Lebofsky, private communication, 1987); and (c) by the absence of much closer companions in the radial velocity searches (Marcy et al. 1986, Campbell et al. 1986).

One may, however, still hypothesize a plausible mass and age distribution of brown dwarfs that provides a significant mass density while satisfying existing observational constraints, although the exercise is becoming more difficult. Recent results from both brown dwarf searches and parameterization of the VLM star population suggest that the crucial test will be thorough surveys for ~ 0.01-$M_\odot$ objects, not for those near the stellar mass limit. At the time of this writing, follow-up observations of IRAS sources have revealed no good candidates for low-luminosity brown dwarfs. However, it remains to be determined whether a useful constraint will be provided by comprehensive analysis of this data base, or whether the definitive answer on the space density of brown dwarfs must await the next generation of space infrared instrumentation.

ACKNOWLEDGMENTS

Space limitations preclude us from thanking all of those who sent us reprints, preprints, and new results. We are especially grateful to Don McCarthy, Conard Dahn, Neill Reid, and Frank Shu for unpublished results and illuminating discussions. We thank Mark Giampapa, John Stauffer, and Hans Zinnecker for critical readings. RP thanks Phil Ianna and Bob O'Connell for first stimulating his interest in low-mass stars; JL thanks the National Science Foundation for continuing support, and Kitt Peak National Observatory for hospitality during the final stages of this collaboration.

Literature Cited

Abt, H. A., Levy, S. G. 1976. *Ap. J. Suppl.* 30: 273
Adams, M. T., Strom, K. M., Strom, S. E. 1983. *Ap. J. Suppl.* 53: 893
Ake, T. B., Greenstein, J. L. 1980. *Ap. J.* 240: 859
Alcaino, G., Thé, P. S. 1982. *Publ. Astron. Soc. Pac.* 94: 335
Alexander, D. R. 1975. *Ap. J. Suppl.* 29: 363
Alexander, D. R., Johnson, H. R., Rypma, R. L. 1983. *Ap. J.* 272: 773
Alexander, J. B., Jones, D. H. P., Sinclair, J. E. 1983. *Bull. R. Greenwich Obs. No. 191*
Allen, D. A., Cragg, T. A. 1983. *MNRAS* 203: 777
Atwood, B. 1975. PhD thesis. Wesleyan Univ., Middletown, Conn.
Atwood, B., Curott, D. R. 1975. *Bull. Am. Astron. Soc.* 7: 337
Auman, J. R., Jr. 1968. In *Low Luminosity Stars*, ed. S. S. Kumar, p. 483. New York: Gordon & Breach
Backman, D. E., Gillett, F. C., Low, F. J. 1986. *Proc. COSPAR Meet., 26th.* In press

Bahcall, J. N. 1984. *Ap. J.* 276: 169
Bahcall, J. N. 1986. *Ann. Rev. Astron. Astrophys.* 24: 577
Bahcall, J. N., Schneider, D. P. 1987. In *Dynamics of Globular Clusters, IAU Symp. No. 126*, ed. J. E. Grindlay, A. G. D. Philip. In press
Bahcall, J. N., Hut, P., Tremaine, S. 1985. *Ap. J.* 290: 15
Baize, M. P. 1966. *J. Obs.* 49: 1
Balega, Y., Bonneau, D., Foy, R. 1984. *Astron. Astrophys. Suppl.* 57: 31
Baliunas, S. L., Vaughan, A. H. 1985. *Ann. Rev. Astron. Astrophys.* 23: 379
Berriman, G., Reid, N. 1987. *MNRAS.* In press
Bessell, M. S. 1979. *Publ. Astron. Soc. Pac.* 91: 589
Bessell, M. S. 1982. *Proc. Astron. Soc. Aust.* 4: 417
Bessell, M. S. 1983. *Publ. Astron. Soc. Pac.* 95: 480
Bessell, M. S. 1986. Preprint
Bidelman, W. P. 1985. *Ap. J. Suppl.* 59: 197
Bidelman, W. P., Lee, S.-G. 1975. *Astron. J.* 80: 239
Biermann, P. 1974. *Astron. Astrophys.* 30: 31
Black, D. C., Scargle, J. D. 1982. *Ap. J.* 263: 854
Blair, M., Gilmore, G. 1982. *Publ. Astron. Soc. Pac.* 94: 742
Bodenheimer, P. 1978. *Ap. J.* 224: 488
Boeshaar, P. C. 1976. PhD thesis. Ohio State Univ., Columbus
Boeshaar, P. C., Tyson, J. A., Seitzer, P. 1986. See Kafatos et al. 1986, p. 76
Bookbinder, J. 1985. PhD thesis. Harvard Univ., Cambridge, Mass.
Bopp, B. W., Noah, P. V., Klimke, A., Africano, J. 1981. *Ap. J.* 249: 210
Boss, A. P. 1986. *Ap. J. Suppl.* 62: 519
Buser, R., Kurucz, R. L. 1978. *Astron. Astrophys.* 70: 555
Byrne, P. B., Rodonò, M., eds. 1983. *Activity in Red-Dwarf Stars, Proc. IAU Colloq. No. 71*. Dordrecht: Reidel
Campbell, B., Walker, G. A. H., Pritchet, C., Long, B. 1986. See Kafatos et al. 1986, p. 37
Campins, H., Rieke, G. H., Lebofsky, M. J. 1985. *Astron. J.* 90: 896
Carbon, D., Gingerich, O., Latham, D. 1968. In *Low Luminosity Stars*, ed. S. S. Kumar, p. 435. New York: Gordon & Breach
Carter, B. S. 1984. MSc thesis. Univ. Cape Town, S. Afr.
Chiu, L. G. 1980. *Astron. J.* 85: 812
Christy, J. W. 1978. *Astron. J.* 83: 1225
Cousins, A. W. J. 1976. *Mem. R. Astron. Soc.* 81: 25
Cousins, A. W. J. 1980a. *Circ. S. Afr. Astron. Obs.* 1: 166
Cousins, A. W. J. 1980b. *Circ. S. Afr. Astron. Obs.* 1: 234
Cousins, A. W. J. 1980c. *Mon. Not. Astron. Soc. S. Afr.* 39: 80
Cousins, A. W. J. 1980c. *Mon. Not. Astron. Soc. S. Afr.* 39: 93
Cousins, A. W. J. 1984. *Circ. S. Afr. Astron. Obs.* 8: 59
Cowley, A. P., Hartwick, F. D. A. 1982. *Ap. J.* 253: 237
Cowley, A. P., Hiltner, W. A., Witt, A. N. 1967. *Astron. J.* 72: 1334
Cox, A. N., Stewart, J. N. 1970. *Ap. J. Suppl.* 19: 243
Cox, A. N., Shaviv, G., Hodson, S. W. 1981. *Ap. J. Lett.* 245: L37
Cram, L. E., Mullan, D. J. 1979. *Ap. J.* 234: 579
Cram, L. E., Woods, D. T. 1982. *Ap. J.* 257: 269
Dahn, C. C. 1985. *Bull. Am. Astron. Soc.* 17: 558
Dahn, C. C., Liebert, J., Kron, R. G., Spinrad, H., Hintzen, P. M. 1977. *Ap. J.* 216: 757
Dahn, C. C., Liebert, J., Boeshaar, P. C. 1983. In *The Nearby Stars and the Stellar Luminosity Function, IAU Colloq. No. 76*, ed. A. G. D. Philip, A. R. Upgren, p. 97. Schenectady, NY: L. Davis
Dahn, C. C., Liebert, J., Harrington, R. S. 1986. *Astron. J.* 91: 621
D'Antona, F., Mazzitelli, I. 1982. *Astron. Astrophys.* 113: 303
D'Antona, F., Mazzitelli, I. 1985. *Ap. J.* 296: 502 (DM)
D'Antona, F., Mazzitelli, I. 1986a. See Kafatos et al. 1986, p. 148
D'Antona, F., Mazzitelli, I. 1986b. *Astron. Astrophys.* 162: 80
Davis, M., Hut, P., Muller, R. A. 1984. *Nature* 308: 715
Dawson, P. C. 1986. *Ap. J.* 311: 984
Dearborn, D. S. P., Liebert, J., Aaronson, M., Dahn, C. C., Harrington, R. S., et al. 1986. *Ap. J.* 300: 314
Dravins, D. 1985. In *Stellar Radial Velocities, IAU Colloq. No. 88*, ed. A. G. D. Philip, D. W. Latham, p. 311. Schenectady, NY: L. Davis
Duerr, R., Imhoff, C. L., Lada, C. J. 1982. *Ap. J.* 261: 135
Dyer, E. R. Jr. 1954a. *Astron. J.* 59: 218
Dyer, E. R. Jr. 1954b. *Astron. J.* 59: 221
Eggen, O. J. 1956. *Astron. J.* 61: 361
Eggen, O. J. 1969. *Ap. J.* 158: 225
Eggen, O. J. 1971a. *Ap. J. Suppl.* 22: 389
Eggen, O. J. 1971b. *Ap. J.* 163: 313
Eggen, O. J. 1974. *Publ. Astron. Soc. Pac.* 86: 697
Eggen, O. J. 1975. *Publ. Astron. Soc. Pac.* 87: 107
Eggen, O. J. 1978. *Ap. J. Suppl.* 37: 251
Eggen, O. J. 1982. *Ap. J. Suppl.* 50: 221

Eggen, O. J. 1983. *Ap. J. Suppl.* 51: 183
Eggen, O. J. 1984. *Astron. J.* 89: 839
Eggen, O. J. 1985. *Astron. J.* 90: 74
Elias, J. H., Frogel, J. A., Matthews, K., Neugebauer, G. 1982. *Astron. J.* 87: 1029
Elias, J. H., Frogel, J. A., Hyland, A. R., Jones, T. J. 1983. *Astron. J.* 88: 1027
Elmegreen, B. G., Mathieu, R. D. 1983. *MNRAS* 203: 305
Engels, D., Sherwood, W. A., Wamsteker, W., Schultz, G. V. 1981. *Astron. Astrophys. Suppl.* 45: 5
Faber, S. M., Burstein, D., Tinsley, B. M., King, I. 1976. *Astron. J.* 81: 45
Fernie, J. D. 1983. *Publ. Astron. Soc. Pac.* 95: 782
Galloway, D. J., Weiss, N. O. 1981. *Ap. J.* 243: 945
Gass, H. 1985. PhD thesis. Univ. Heidelberg, W. Germ.
Gatewood, G. D. 1976. *Icarus* 27: 1
Gatewood, G. D. 1983. *US Patent No. 4,397,559*
Gatewood, G. D., Breakiron, L. A., Goebel, R., Kipp, S., Russell, J. L., et al. 1980. *Icarus* 39: 205
Gatewood, G. D., de Jonge, J. K., Stein, J., DiFatta, C. 1985. In *The Search for Extraterrestrial Life: Recent Developments, IAU Symp. No. 112*, ed. M. D. Papagiannis, p. 65. Dordrecht: Reidel
Gatewood, G. D., de Jonge, J. K., Stein, J., Han, I., Breakiron, L. 1986. See Kafatos et al. 1986, p. 104
Giampapa, M. S. 1985. *Ap. J.* 299: 781
Giampapa, M. S., Liebert, J. 1986. *Ap. J.* 305: 784 (GL)
Giclas, H. L. 1958. *Lowell Obs. Bull.* 4: 1
Giclas, H. L., Burnham, R. Jr., Thomas, N. G. 1971. *The G Numbered Stars.* Flagstaff, Ariz: Lowell Obs.
Giclas, H. L., Burnham, R. Jr., Thomas, N. G. 1972. *Lowell Obs. Bull.* 7: 217
Giclas, H. L., Burnham, R. Jr., Thomas, N. G. 1978. *Lowell Obs. Bull.* 8: 89
Giclas, H. L., Burnham, R. Jr., Thomas, N. G. 1980. *Lowell Obs. Bull.* 8: 157
Gilmore, G., Hewett, P. 1983. *Nature* 306: 669
Gilmore, G., Reid, N. 1983. *MNRAS* 202: 1025
Gilmore, G., Reid, N., Hewett, P. 1985. *MNRAS* 213: 257
Gingerich, O., Latham, D. W., Linsky, J., Kumar, S. S. 1966. In *Colloquium on Late Type Stars*, ed. M. Hack, p. 291. Trieste: Obs. Astron.
Glass, I. S. 1974. *Mon. Not. Astron. Soc. S. Afr.* 33: 53
Glass, I. S. 1983. *Mon. Not. Astron. Soc. S. Afr.* 42: 43
Glass, I. S. 1985. *Ir. Astron. J.* 17: 1
Gliese, W. 1969. *Catalogue of Nearby Stars*, *Veroff. Astron. Rechen-Inst. Heidelberg*, No. 22
Gliese, W. 1974. *Astron. Astrophys.* 34: 147
Gliese, W., Jahreiss, H. 1979. *Astron. Astrophys. Suppl.* 38: 423
Golub, L. 1983. See Byrne & Rodonò 1983, p. 83
Greenstein, J. L., Neugebauer, G., Becklin, E. E. 1970. *Ap. J.* 161: 519
Grossman, A. S., Hays, D., Graboske, H. C. Jr. 1974. *Astron. Astrophys.* 30: 95 (GHG)
Haffner, H., Heckmann, O. 1936. *Naturwissenschaften* 24: 635
Harrington, R. S. 1971. *Astron. J.* 76: 930
Harrington, R. S. 1977. *Publ. Astron. Soc. Pac.* 89: 214
Harrington, R. S. 1986. See Kafatos et al. 1986, p. 3
Harrington, R. S., Behall, A. L. 1973. *Astron. J.* 78: 1096
Harrington, R. S., Kallarakal, V. V., Dahn, C. C. 1983. *Astron. J.* 88: 1038
Hartmann, L., Anderson, C. M. 1977. *Ap. J.* 215: 188
Hartmann, L. W., Noyes, R. W. 1987. *Ann. Rev. Astron. Astrophys.* 25: 271
Hartwick, F. D. A. 1977. *Ap. J.* 214: 778
Hartwick, F. D. A., Cowley, A. P., Mould, J. R. 1984. *Ap. J.* 286: 269
Hawkins, M. R. S. 1986a. See Kafatos et al. 1986, p. 93
Hawkins, M. R. S. 1986b. *MNRAS* 223: 845
Hayes, D. S. 1985. In *Calibration of Fundamental Stellar Quantities, IAU Symp. No. 111*, ed. D. S. Hayes, L. E. Pasinetti, A. G. D. Philip, p. 225. Dordrecht: Reidel
Heintz, W. D. 1969. *J. R. Astron. Soc. Can.* 63: 275
Heintz, W. D. 1976. *MNRAS* 175: 533
Heintz, W. D. 1978. *Ap. J.* 220: 931
Heintz, W. D. 1983. *Lowell Obs. Bull.* 9: 10
Heintz, W. D. 1984a. *Publ. Astron. Soc. Pac.* 96: 439
Heintz, W. D. 1984b. *Astron. J.* 89: 1063
Heintz, W. D. 1985. In *Calibration of Fundamental Stellar Quantities, IAU Symp. No. 111*, ed. D. S. Hayes, L. E. Pasinetti, A. G. D. Philip, p. 71. Dordrecht: Reidel
Heintz, W. D., Borgman, E. R. 1984. *Astron. J.* 89: 1068
Herbig, G. 1985. *Ap. J.* 289: 269
Hershey, J. L. 1982. *Astron. J.* 87: 145
Hershey, J. L., Lippincott, S. L. 1982. *Astron. J.* 87: 840
Hills, J. G. 1985. *Astron. J.* 90: 1876
Hoxie, D. T. 1970. *Ap. J.* 161: 1083
Hoyle, F. 1953. *Ap. J.* 118: 513
Hubbard, W. B., Stevenson, D. J. 1984. In *Saturn*, ed. T. Gehrels, M. S. Matthews, p. 47. Tucson: Univ. Ariz. Press
Ianna, P. A., Bessell, M. S. 1986. *Publ. Astron. Soc. Pac.* 98: 658

Jahreiss, H., Gliese, W. 1985. *Bull. Inform. Cent. Données Stellaires No. 28*, p. 19
Jameson, R. F., Sherrington, M. R., Giles, A. B. 1983. *MNRAS* 205: 39P
Jensen, O. G., Ulrych, T. 1973. *Astron. J.* 78: 1104
Johnson, H. L. 1964. *Bol. Obs. Tonantzintla Tacubaya* 3: 305
Johnson, H. L. 1965a. *Ap. J.* 141: 170
Johnson, H. L. 1965b. *Commun. Lunar Planet. Lab.* 3: 73
Johnson, H. L., Morgan, W. W. 1953. *Ap. J.* 117: 313
Johnson, H. L., Mitchell, R. I., Iriarte, B., Wisniewski, W. Z. 1966. *Commun. Lunar Planet. Lab.* 4: 99
Johnson, H. M. 1981. *Ap. J.* 243: 234
Johnson, H. M. 1983. *Ap. J.* 273: 702
Johnson, H. M. 1986. *Ap. J.* 303: 470
Jones, B. F. 1972. *MNRAS* 159: 3P
Jones, D. H. P. 1973. *MNRAS* 161: 19P
Jones, D. H. P., Sinclair, J. E., Alexander, J. B. 1981. *MNRAS* 194: 403
Jones, T. J., Hyland, A. R. 1982. *MNRAS* 200: 509
Joy, A. H. 1947. *Ap. J.* 105: 96
Joy, A. H., Abt, H. A. 1974. *Ap. J. Suppl.* 23: 1
Joyce, R. R., Probst, R. G., Guetter, H. H. 1984. *Bull. Am. Astron. Soc.* 16: 497
Kafatos, M. C., Harrington, R. S., Maran, S. P., eds. 1986. *Astrophysics of Brown Dwarfs*. London: Cambridge Univ. Press
Keenan, P. C., McNeil, R. C. 1976. *An Atlas of Spectra of the Cooler Stars*. Columbus, Ohio: Ohio State Univ. Press
Kelly, B. D., Cooke, J. A., Emerson, D. 1982. *MNRAS* 199: 239
Koo, D. C., Kron, R. G. 1975. *Publ. Astron. Soc. Pac.* 87: 885
Koornneef, J. 1983a. *Astron. Astrophys. Suppl.* 51: 489
Koornneef, J. 1983b. *Astron. Astrophys.* 128: 84
Kraft, R. H., Preston, G. W., Wolff, S. C. 1964. *Ap. J.* 140: 237
Kron, G. E., Smith, J. L. 1951. *Ap. J.* 113: 324
Kron, G. E., Gascoigne, S. C. B., White, H. S. 1953. *Ap. J.* 118: 502
Kron, G. E., Gascoigne, S. C. B., White, H. S. 1957. *Astron. J.* 62: 205
Kron, G. E., Mayall, N. U. 1960. *Astron. J.* 65: 581
Kuiper, G. E. 1934. *Publ. Astron. Soc. Pac.* 46: 235
Kuiper, G. E. 1936. *Ap. J.* 84: 359
Kuiper, G. E. 1942. *Ap. J.* 95: 201
Kuiper, G. E. 1943. *Ap. J.* 97: 275
Kumar, C. K. 1985. *Publ. Astron. Soc. Pac.* 97: 294
Kumar, C. K. 1986. Preprint
Kumar, S. S. 1963a. *Ap. J.* 137: 1121
Kumar, S. S. 1963b. *Ap. J.* 137: 1126
Lacy, C. H. 1977. *Ap. J.* 218: 444
Landolt, A. U. 1983. *Astron. J.* 88: 439
Larson, K. B. 1973. *MNRAS* 161: 133
Larson, R. B. 1985. *MNRAS* 214: 379
Larson, R. B. 1986. *MNRAS* 218: 409
Lee, S.-G. 1984. *Astron. J.* 89: 702
Leinert, C., Jahreiss, H., Haas, M. 1986. *Astron. Astrophys.* 164: L29
Liebert, J., Dahn, C. C. 1986. See Kafatos et al. 1986, p. 30
Liebert, J., Dahn, C. C., Gresham, M., Strittmatter, P. A. 1979. *Ap. J.* 233: 226
Liebert, J., Boroson, T. A., Giampapa, M. S. 1984. *Ap. J.* 282: 758
Limber, D. N. 1958. *Ap. J.* 127: 387
Linsky, J. 1983. See Byrne & Rodonò 1983, p. 39
Lippincott, S. L. 1955. *Astron. J.* 60: 379
Lippincott, S. L. 1973. *Astron. J.* 78: 303
Lippincott, S. L. 1977. *Astron. J.* 82: 925
Lippincott, S. L. 1978. *Space Sci. Rev.* 22: 153
Lippincott, S. L. 1982. *Astron. J.* 87: 1237
Lippincott, S. L., Hershey, J. L. 1972. *Astron. J.* 77: 679
Lippincott, S. L., Hershey, J. L. 1983. *Lowell Obs. Bull.* 9: 45
Lippincott, S. L., Braun, D., McCarthy, D. W. Jr. 1983. *Publ. Astron. Soc. Pac.* 95: 271
Low, C., Lynden-Bell, D. 1976. *MNRAS* 176: 367
Low, F. 1986. See Kafatos et al. 1986, p. 66
Lunine, J. L., Hubbard, W. B., Marley, M. S. 1986. *Ap. J.* 310: 238
Luyten, W. J. 1938. *Publ. Obs. Univ. Minn.* 2: 123
Luyten, W. J. 1963a. In *Basic Astronomical Data*, ed. K. Aa. Strand, p. 46. Chicago: Univ. Chicago Press
Luyten, W. J. 1963b. *Proper Motion Survey With the 48-inch Schmidt Telescope No. 1*. Minneapolis: Univ. Minn.
Luyten, W. J. 1968. *MNRAS* 139: 221
Luyten, W. J. 1976. *Proper Motion Survey With the 48-inch Schmidt Telescope No. 46*. Minneapolis: Univ. Minn.
Luyten, W. J. 1979a. *LHS Catalogue*. Minneapolis: Univ. Minn. 2nd ed.
Luyten, W. J. 1979b. *NLTT Catalogue*, Vols. 1–4. Minneapolis: Univ. Minn.
Luyten, W. J., Albers, H. 1979. *LHS Atlas*. Minneapolis: Univ. Minn.
Magni, G., Mazzitelli, I. 1979. *Astron. Astrophys.* 72: 134
Manduca, A., Bell, R. A. 1979. *Publ. Astron. Soc. Pac.* 91: 848
Manfroid, J., Heck, A. 1985. In *Calibration of Fundamental Stellar Quantities, IAU Symp. No. 111*, ed. D. S. Hayes, L. E. Pasinetti, A. G. D. Philip, p. 565. Dordrecht: Reidel

Marcy, G. W., Lindsay, V., Bergengren, J., Moore, D. 1986. See Kafatos et al. 1986. p. 50
McAlister, H. A. 1985. In *Calibration of Fundamental Stellar Quantities, IAU Symp. No. 111*, ed. D. S. Hayes, L. E. Pasinetti, A. G. D. Philip, p. 97. Dordrecht: Reidel
McCarthy, D. W. Jr. 1983. In *The Nearby Stars and the Stellar Luminosity Function, IAU Colloq. No. 76*, ed. A. G. D. Philip, A. R. Upgren, p. 107. Schenectady, NY: L. Davis
McCarthy, D. W. Jr. 1986. See Kafatos et al. 1986, p. 9
McCarthy, D. W. Jr., Henry, T. J. 1987. Preprint
McCarthy, D. W. Jr., Probst, R. G., Low, F. J. 1985. *Ap. J. Lett.* 290: L9
McCarthy, D. W. Jr., Cobb, M. L., Probst, R. G. 1986. *Astron. J.* In press
McNamara, B. R., Ianna, P. A., Frederick, L. W. 1987. *Astron. J.* In press
Menzies, J. W., Banfield, R. M., Laing, J. D. 1980. *Circ. S. Afr. Astron. Obs.* 1: 149
Mestel, L., Spitzer, L. 1956. *MNRAS* 116: 503
Mirzoyan, L. V. 1984. *Vistas Astron.* 27: 77
Monet, D. G. 1979. *Ap. J.* 234: 275
Monet, D. G., Dahn, C. C. 1983. *Astron. J.* 88: 1489
Mould, J. 1975. *Astron. Astrophys.* 38: 283
Mould, J. 1976a. *Astron. Astrophys.* 48: 443
Mould, J. 1976b. *Ap. J.* 207: 535
Mould, J. 1976c. *Ap. J.* 210: 402
Mould, J., Hyland, A. R. 1976. *Ap. J.* 208: 399
Mould, J., McElroy, D. B. 1978. *Ap. J.* 220: 935
Mountain, C. M., Leggett, S. K., Selby, M. J., Blackwell, D. E., Petford, A. D. 1985. *Astron. Astrophys.* 151: 399
Mouschovias, T. Ch., Paleologou, E. V. 1981. *Ap. J.* 246: 48
Murray, C. A., Sanduleak, N. 1972. *MNRAS* 157: 273
Nelson, L. A., Rappaport, S. A., Joss, P. C. 1985. *Nature* 316: 42
Nelson, L. A., Rappaport, S. A., Joss, P. C. 1986. *Ap. J.* 311: 226
Oort, J. H. 1960. *Bull. Astron. Inst. Neth.* 15: 45
Pallavicini, R. P., Golub, L., Rosner, R., Vaiana, G. S., Ayres, T. R., et al. 1981. *Ap. J.* 248: 279
Perrier, C., Mariotti, J.-M. 1987. *Ap. J. Lett.* 312: L27
Persson, S. E., Aaronson, M., Frogel, J. A. 1977. *Astron. J.* 82: 729
Pesch, P. 1976. *Astron. J.* 81: 1117
Pesch, P., Dahn, C. C. 1982. *Astron. J.* 87: 122
Pesch, P., Sanduleak, N. 1978. *Astron. J.* 83: 1090

Pettersen, B. R. 1976. *Catalogue of Flare Star Data, Blindern-Oslo Report No. 46*, Inst. Theor. Astrophys., Oslo
Pettersen, B. R. 1980. *Astron. Astrophys.* 82: 53
Pettersen, B. R. 1983. See Byrne & Rodonò, p. 239
Pettersen, B. R., Cochran, A. L., Barker, E. S. 1985. *Astron. J.* 90: 2296
Popper, D. M. 1980. *Ann. Rev. Astron. Astrophys.* 18: 115
Probst, R. G. 1981a. PhD thesis. Univ. Va., Charlottesville
Probst, R. G. 1981b. *Bull. Am. Astron. Soc.* 13: 803
Probst, R. G. 1983a. In *The Nearby Stars and the Stellar Luminosity Function, IAU Colloq. No. 76*, ed. A. G. D. Philip, A. R. Upgren, p. 369. Schenectady, NY: L. Davis
Probst, R. G. 1983b. *Ap. J.* 274: 237
Probst, R. G. 1983c. *Ap. J. Suppl.* 53: 335
Probst, R. G. 1985. *Bull. Am. Astron. Soc.* 17: 553
Probst, R. G. 1986. See Kafatos et al. 1986, p. 22
Probst, R. G., Liebert, J. 1983. *Ap. J.* 274: 245
Rakos, K. D. 1965. *Appl. Opt.* 4: 1453
Rakos, K. D. 1974. *Publ. Astron. Soc. Pac.* 86: 1007
Rakos, K. D. 1983a. *Lowell Obs. Bull.* 9: 220
Rakos, K. D. 1983b. *Lowell Obs. Bull.* 9: 254
Rakos, K. D., Albrecht, R., Jenkner, H., Kreidl, T., Michalke, R., et al. 1982. *Astron. Astrophys. Suppl.* 47: 221
Rappaport, S., Verbunt, F., Joss, P. C. 1983. *Ap. J.* 275: 713
Reid, N. 1982. *MNRAS* 201: 51
Reid, N. 1984. *MNRAS* 206: 1
Reid, N. 1987. *MNRAS* 225: 873
Reid, N., Gilmore, G. 1982. *MNRAS* 201: 73
Reid, N., Gilmore, G. 1984. *MNRAS* 206: 19
Reuyl, D. 1936. *Astron. J.* 45: 133
Reuyl, D. 1943. *Ap. J.* 97: 143
Robertson, T. H. 1983. In *The Nearby Stars and the Stellar Luminosity Function, IAU Colloq. No. 76*, ed. A. G. D. Philip, A. R. Upgren, p. 405. Schenectady, NY: L. Davis
Robertson, T. H. 1984. *Astron. J.* 89: 1229
Rosner, R., Avni, Y., Bookbinder, J., Giacconi, R., Golub, L., et al. 1981. *Ap. J. Lett.* 249: L5
Rosner, R., Golub, L., Vaiana, G. S. 1985. *Ann. Rev. Astron. Astrophys.* 23: 413
Rucinski, S. M. 1984. *Astron. Astrophys.* 132: L9
Ruiz, M. T., Maza, J., Wischnjewsky, M., Gonzalez, L. E. 1986. *Ap. J. Lett.* 304: L25

Russell, J. L., Gatewood, G. D. 1984. *Publ. Astron. Soc. Pac.* 96: 429
Saar, S. H., Linsky, J. L. 1985. *Ap. J. Lett.* 299: L47
Salpeter, E. E. 1955. *Ap. J.* 121: 161
Sanduleak, N. 1965. PhD thesis. Case Inst. Technol., Cleveland, Ohio
Sanduleak, N. 1976. *Astron. J.* 81: 350
Saumon, D., Van Horn, H. M. 1987. *Proc. Workshop on Strongly Coupled Plasmas, Santa Cruz*, ed. F. J. Rogers, H. E. DeWitt. In press
Scalo, J. 1986. *Fundam. Cosmic Phys.* 11: 1
Schmidt, M. 1974. In *Dynamics of Stellar Systems, IAU Symp. No. 69*, ed. A. Hayli, p. 325. Dordrecht: Reidel
Schmidt, M. 1975. *Ap. J.* 202: 22
Schmidt, M. 1983. In *The Nearby Stars and the Stellar Luminosity Function, IAU Colloq. No. 76*, ed. A. G. D. Philip, A. R. Upgren, p. 155. Schenectady, NY: L. Davis
Schmidt, J. H. M. M., Rosner, R., Bohn, H. U. 1984. *Ap. J.* 282: 316
Shipman, H. L. 1986. See Kafatos et al. 1986, p. 71
Shu, F. H. 1983. *Ap. J.* 273: 202
Shu, F. H. 1985. In *The Milky Way, IAU Symp. No. 106*, ed H. van Woerden, W. B. Burton, R. J. Allen, p. 561. Dordrecht: Reidel
Shu, F. H. 1987. In *Interstellar Dust and Related Processes*, ed. S. Aiello. New York: Academic. In press
Shu, F. H., Adams, F. C., Lizano, S. 1987. *Ann. Rev. Astron. Astrophys.* 25: 23
Sibille, F., Chelli, A., Lena, P. 1979. *Astron. Astrophys.* 79: 315
Silk, J. 1977. *Ap. J.* 214: 152
Simon, T., Herbig, G., Boesgaard, A. M. 1985. *Ap. J.* 293: 551
Sinton, W. M., Tittemore, W. C. 1984. *Astron. J.* 89: 1366
Skrutskie, M. F., Forrest, W. J., Shure, M. A. 1986. See Kafatos et al. 1986, p. 82
Skrutskie, M. F., Forrest, W. J., Shure, M. A. 1987. *Ap. J. Lett.* 312: L55
Smethells, W. G. 1974. PhD thesis. Case Western Reserve Univ., Cleveland, Ohio
Smethells, W. G. 1983. In *The Nearby Stars and the Stellar Luminosity Function, IAU Colloq. No. 76*, ed. A. G. D. Philip, A. R. Upgren, p. 421. Schenectady, NY: L. Davis
Smith, R. C. 1983. *Observatory* 103: 29
Sorokin, L. Yu., Tokovinin, A. A. 1985. *Sov. Astron. Lett.* 11: 226
Spruit, H. C., Ritter, H. 1983. *Astron. Astrophys.* 124: 267
Stahler, S. W., Shu, F. H., Taam, R. E. 1980. *Ap. J.* 241: 637
Staller, R. F. A. 1975. *Astron. Astrophys.* 42: 155
Staller, R. F. A. 1979. PhD thesis. Univ. Amsterdam, Neth.
Staller, R. F. A., Thé, P. S., Bochem-Becks, A. Ch. Th. 1981. *Publ. Astron. Soc. Pac.* 93: 728
Stauffer, J. R., Hartmann, L. W. 1986a. *Ap. J. Suppl.* 61: 531 (SH)
Stauffer, J. R., Hartmann, L. W. 1986b. *Publ. Astron. Soc. Pac.* In press
Stevenson, D. J. 1978. *Proc. Astron. Soc. Aust.* 3: 227
Stevenson, D. J. 1982a. *Ann. Rev. Earth Planet. Sci.* 10: 257
Stevenson, D. J. 1982b. *Planet. Space Sci.* 30: 755
Stevenson, D. J. 1986. See Kafatos et al. 1986, p. 218
Straka, W. C. 1971. *Ap. J.* 165: 109
Strecker, D. W., Erickson, E. F., Witteborn, F. C. 1979. *Ap. J. Suppl.* 41: 501
Stringfellow, G. S. 1986. See Kafatos et al. 1986, p. 190
Tarter, J. 1975. PhD thesis. Univ. Calif., Berkeley
Tarter, J. 1986. See Kafatos et al. 1986, p. 121
Thé, P. S., Staller, R. F. A. 1974. *Astron. Astrophys.* 36: 155
Thé, P. S., Karman, C., Alcaino, G. 1981. *Astron. Astrophys. Suppl.* 46: 105
Thé, P. S., Steenman, H. C., Alcaino, G. 1984. *Astron. Astrophys.* 132: 385
Tohline, J. E. 1980. *Ap. J.* 239: 417
Trimble, V. 1987. *Ann. Rev. Astron. Astrophys.* 25: 425
Tsuji, T. 1966. *Publ. Astron. Soc. Jpn.* 18: 127
Tsuji, T. 1968. In *Low Luminosity Stars*, ed. S. S. Kumar, p. 457. New York: Gordon & Breach
Upgren, A. R. 1972. *Ap. J.* 172: 149
Upgren, A. R. 1975. In *Multicolor Photometry and the Theoretical HR Diagram*, ed. A. G. D. Philip, D. S. Hayes, p. 437. Albany, NY: Dudley Obs.
Upgren, A. R. 1978. *Astron. J.* 83: 626
Upgren, A. R. 1985. In *Calibration of Fundamental Stellar Quantities, IAU Symp. No. 111*, ed. D. S. Hayes, L. E. Pasinetti, A. G. D. Philip, p. 31. Dordrecht: Reidel
Upgren, A. R., Grossenbacher, R., Penhallow, W. S., MacConnell, D. J., Frye, R. L. 1972. *Astron. J.* 77: 486
Upgren, A. R., Gliese, W., Jahreiss, H. 1986. In *The Galaxy and the Solar System*, ed. M. Matthews, p. 13. Tucson: Univ. Ariz. Press
Vaiana, G. S., Cassinelli, J. P., Fabbiano, G., Giacconi, R., Golub, L., et al. 1981. *Ap. J.* 245: 163
van Altena, W. F. 1983. *Ann. Rev. Astron. Astrophys.* 21: 131
Van Biesbroeck, G. 1944. *Astron. J.* 51: 61
Van Biesbroeck, G. 1961. *Astron. J.* 66: 528

Van Horn, H. M. 1986. *Mitt. Astron. Ges.* 67: 63
van de Kamp, P. 1939. *Publ. Am. Astron. Soc.* 9: 198
van de Kamp, P. 1967. *Principles of Astrometry*. San Francisco: Freeman
van de Kamp, P. 1981. *Stellar Paths*. Dordrecht: Reidel
van de Kamp, P. 1982. In *Binary and Multiple Stars as Tracers of Stellar Evolution*, ed. Z. Kopal, J. Rahe, p. 81. Dordrecht: Reidel
van de Kamp, P. 1986. *Space Sci. Rev.* 43: 211
VandenBerg, D. A., Hartwick, F. D. A., Dawson, P., Alexander, D. R. 1983. *Ap. J.* 266: 747
Vaughan, A. H., Preston, G. W. 1980. *Publ. Astron. Soc. Pac.* 92: 385
Veeder, G. J. 1974. *Astron. J.* 79: 1056
Vyssotsky, A. N. 1963. In *Basic Astronomical Data*, ed. K. Aa. Strand, p. 192. Chicago: Univ. Chicago Press
Walker, A. R. 1983. *Circ. S. Afr. Astron. Obs.* 7: 106
Wallenquist, A. 1954. *Ann. Uppsala Astron. Obs.* 4: 2
Warren, P. R. 1976. *MNRAS* 176: 667
Weis, E. W. 1982. *Astron. J.* 87: 152
Weis, E. W. 1984. *Ap. J. Suppl.* 55: 289
Weis, E. W. 1986. *Astron. J.* 91: 626
Weistrop, D. 1972. *Astron. J.* 77: 849
Weistrop, D. 1975. *Publ. Astron. Soc. Pac.* 87: 367
Weistrop, D. 1976a. *Astron. J.* 81: 427
Weistrop, D. 1976b. *Astron. J.* 81: 759
Weistrop, D. 1976c. *Ap. J.* 204: 113
Weistrop, D. 1977. *Ap. J.* 215: 845
Weistrop, D. 1980. *Astron. J.* 85: 738
Wielen, R. 1977. *Astron. Astrophys.* 60: 263
Wielen, R., Jahreiss, H., Kruger, R. 1983. In *The Nearby Stars and the Stellar Luminosity Function, IAU Colloq. No. 76*, ed. A. G. D. Philip, A. R. Upgren, p. 163. Schenectady, NY: L. Davis
Wilson, O. C. 1967. *Astron. J.* 72: 905
Wing, R. F. 1973. In *Spectral Classification and Multicolor Photometry, IAU Symp. No. 50*, ed. C. Fehrenbach, B. E. Westerlund, p. 209. Dordrecht: Reidel
Wing, R. F., Dean, C. A. 1983. In *The Nearby Stars and the Stellar Luminosity Function, IAU Colloq. No. 76*, ed. A. G. D. Philip, A. R. Upgren, p. 385. Schenectady, NY: L. Davis
Worden, S. P., Schneeberger, T. J., Giampapa, M. S. 1981. *Ap. J. Suppl.* 46: 159
Worley, C. E. 1962. *Astron. J.* 67: 396
Worley, C. E. 1983. *Lowell Obs. Bull.* 9: 1
Worley, C. E., Heintz, W. D. 1983. *Publ. US Naval Obs. (2nd ser.)*, Vol. 24, pt. 7
Young, A., Skumanich, A., Harlan, E. 1984. *Ap. J.* 282: 683
Zapolsky, H. S., Salpeter, E. E. 1969. *Ap. J.* 158: 809
Zarro, D. M. 1983. *Ap. J. Lett.* 267: L61
Zarro, D. M., Rodgers, A. W. 1983. *Ap. J. Suppl.* 53: 815
Zinnecker, H. 1984. *MNRAS* 210: 43
Zinnecker, H. 1986. See Kafatos et al. 1986, p. 212

THE IRAS VIEW OF THE GALAXY AND THE SOLAR SYSTEM

C. A. Beichman

Infrared Processing and Analysis Center, California Institute of Technology and Jet Propulsion Laboratory, Pasadena, California 91125

> Learned Faustus,
> To find the secrets of astronomy,
> Graven in the books of Jove's high firmament,
> Did mount him up to scale Olympus' top:
> Where sitting in a chariot burning bright
> Drawn by the strength of yoked dragon's necks,
> He views the clouds, the planets and the stars,
> The tropics, zones and quarters of the sky,
> From the bright circle of the horned moon,
> Even to the height of Primum Mobile:
> And the whirling round with the circumference
> Within the concave compass of the poles;
> From East to West his dragons swiftly glide,
> And in eight days did bring him home again.
>
> C. Marlow, *Faustus*, III, iii

1. THE INFRARED SKY

Introduction

The Infrared Astronomical Satellite (IRAS) mapped the sky at 12, 25, 60 and 100 μm for 300 days starting 25 January 1983 and forever changed our view of the sky. The unbiased survey of 96% of the celestial sphere resulted in the recognition of entirely new astrophysical phenomena and increased the number of cataloged astronomical sources by almost 50%. This article concentrates on IRAS results important to Galactic and solar system astronomy. A companion review in this volume by Soifer et al.

(1987) discusses extragalactic sources. Readers interested in the details of the instrument, the survey, or the available IRAS data are referred to Neugebauer et al. (1984) or to *The Explanatory Supplement to the IRAS Catalogs and Atlases* (1984, 1987; hereafter referred to as the *Supplement*).

The infrared emission from the celestial sphere is dominated by the zodiac and the Milky Way. As shown in Figure 1, the relative importance of these two planes varies with the wavelength: At 12 and 25 μm, interplanetary dust dominates the emission, while at 60 and 100 μm, emission from interstellar dust becomes increasingly important. This review begins with the zodiacal cloud and ends with the study of the Galaxy as a galaxy. The key results are introduced in a brief overview: More detailed discussions of these and related topics are given in subsequent sections.

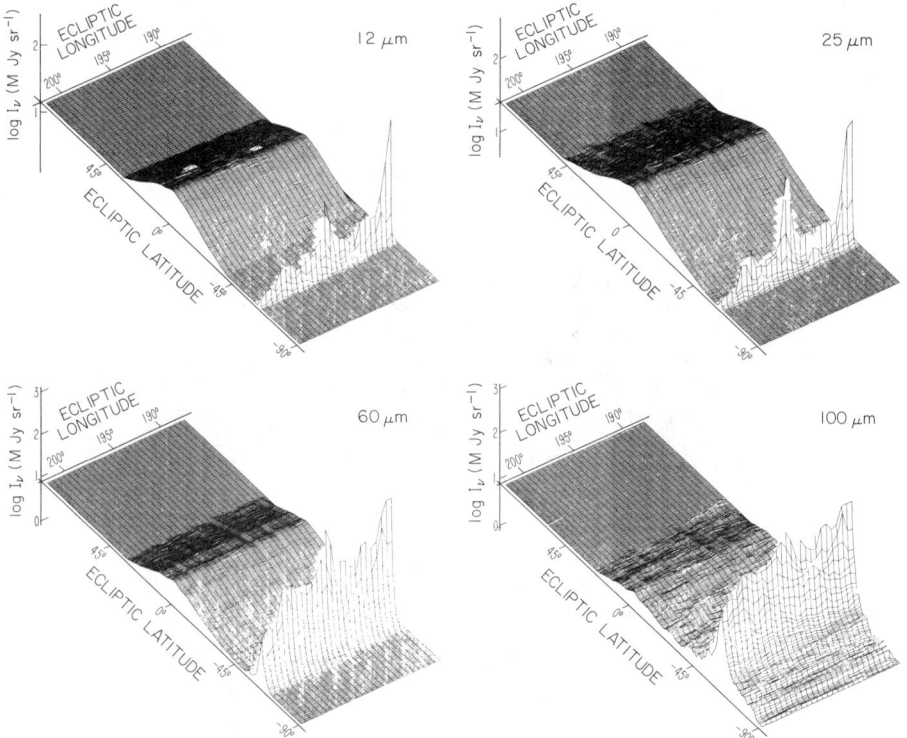

Figure 1 More than 40 1/2° wide scans are shown in this perspective view of the total intensity from the sky in the four IRAS bands. The scans stretch from one ecliptic pole to the other and traverse the Galactic plane near Carina. The relative importance of the zodiacal and Galactic planes changes drastically from 12 to 100 μm.

The Solar System

In addition to measuring the smooth background of thermal emission from interplanetary dust (Figure 1), IRAS found discrete bands of emission that girdle the solar system. The probable origin of these bands in debris created by the collisions of asteroids suggests that such detritus can account for the entire planetary dust cloud. Other solar system topics include the discovery that comets are far dustier than previously thought, and an update on the search for Planet X.

Star Formation

Stars are conceived, gestate, and are born within a cocoon of gas and dust that hides the process from optical telescopes, making the infrared the premier wavelength range for studying the formation of stars. Section 3 discusses how IRAS has helped to bring about a revolution in our understanding of how low-mass stars are formed by finding very young stars, possibly true protostars still accreting nebular material, embedded in small, nearby gas clouds. Circumstellar disks appear to be an important part of the formation of at least low-mass stars.

While IRAS has not yet added directly to our knowledge of how high-mass stars form, IRAS has cataloged thousands of dense, hot cores within molecular clouds and mapped complexes of high-mass star formation throughout the Galaxy. An investigation of the relation between star-forming regions and their environment may ultimately offer clues to the triggers of star formation. Section 7 addresses the link between Galactic star formation and luminous infrared galaxies.

Stars: *Photospheres, Mass Loss, Protoplanetary Disks, and Brown Dwarfs*

Stars of all types account for more than half of the sources in the Point Source Catalog. Section 4 discusses both normal stars as well as those with infrared excesses due to mass loss. An important discovery is that both the rate of mass loss and the elemental abundances of the ejecta can change dramatically in only a few thousand years. Section 5 reviews the startling discovery that the Sun is not the only star with a zodiacal dust cloud. For example, α Lyrae and as many as a third of all dwarf A, F, and G stars have excesses attributable to disks of orbiting solid material; this material may be intimately related to the formation of planetary systems. Section 6 discusses limits to the presence of brown dwarf stars in the IRAS catalog.

Diffuse Galactic Emisson

The total diffuse emission from the Galaxy (Section 7) is far greater than the sum of all the sources in the Point Source Catalog. Dust in the interstellar medium heated either by newly formed high-mass stars or by the ambient interstellar radiation field accounts for approximately one third of the bolometric luminosity of the Galaxy. Much of this emission comes, as expected, at 60 and 100 μm from the previously known population of ~0.1-μm-sized grains, but a surprising aspect of the diffuse emission is its strength at 12 and 25 μm. This short-wavelength radiation accounts for ~25% of the infrared output of the Galaxy and demonstrates that transient heating of very small grains plays a crucial part in the energy balance of the Galaxy.

2. SOLAR SYSTEM OBSERVATIONS
The Zodiacal Dust Cloud

LARGE-SCALE EMISSION   Extended thermal emission from the zodiacal dust cloud is the most prominent component of the sky at 12 and 25 μm; and although the Galactic plane becomes brighter at longer wavelengths, the zodiacal emission remains strong at 60 μm and is visible even at 100 μm (Figure 1). The brightness drops by a factor of three between the ecliptic plane and the poles: Away from the ecliptic plane, the variation is well fitted by a cosecant ($|\beta|$) law (Hauser et al. 1984a).[1]

Good et al. (1986) have unfolded the observed distribution of the interplanetary emission by fitting a general model for the variation with heliocentric distance of density and temperature within the zodiacal cloud. Other authors have derived values for the inclination and longitude of the ascending node of the symmetry plane of the cloud and have demonstrated that the cloud is not a simple axisymmetric volume (Hauser & Gautier 1984, Rickard et al. 1985, Low et al 1984b, Dermott et al. 1985b). The variation with heliocentric distance of these orbital parameters is consistent with Misconi's (1980) suggestion that the inner planets perturb the structure of the zodiacal cloud (Hauser & Houck 1986, Hauser et al. 1985).

The spectral energy distribution (Figure 2) of the material in the ecliptic plane at 90° elongation can be fitted by a 244 ± 44 K blackbody (Hauser et al. 1984a, Dumont & Levasseur-Regourd 1986). The corresponding 12-μm optical depth is 3×10^{-7}, and the implied visual albedo of the dust at 1 AU is 0.07–0.09. This albedo, while quite low, is consistent with some suggested interplanetary materials, such as blackened silicate grains larger than 20 μm, and may vary with increasing heliocentric distance (Hauser

---

[1] In the ecliptic plane at 90° elongation, the zodiacal emission has a surface brightness of 46, 78, 24, 14 MJy sr^{-1} at 12, 25, 60 and 100 μm, respectively (Hauser et al. 1984a, Hauser & Houck 1986). At 12 and 25 μm these values are, for unknown reasons, about 1.4–2.0 times brighter than those determined by Murdock & Price (1985). The discrepancy may be resolved only by the flight of the Cosmic Background Explorer.

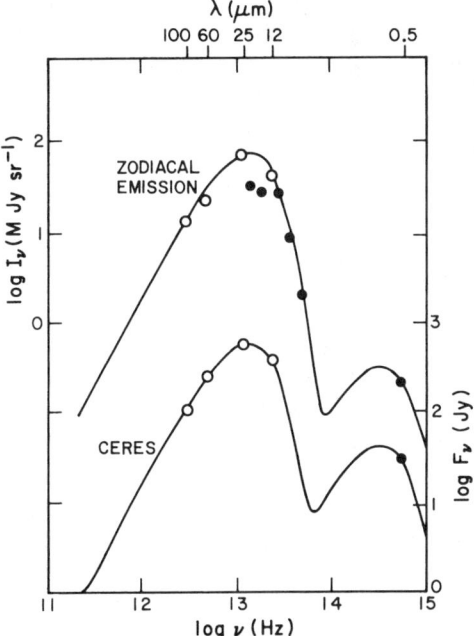

Figure 2 The spectral energy distribution of the zodiacal emission (left-hand scale) is shown for both IRAS data (open circles; Hauser et al. 1984a) and rocket data (filled circles; Murdock & Price 1985). The 0.55-μm point is from Allen (1973). The solid line shows the sum of 5700-K and 244-K blackbodies normalized to the visual and 12-μm data. The flux density of the asteroid Ceres is also shown (right-hand scale). The visual point was calculated for the epoch of the IRAS observations based on data given in the *IRAS Asteroid and Comet Survey*. The solid line shows the sum of a 5700-K and 230-K blackbodies normalized to the visual and 12-μm data.

& Houck 1986, Dumont & Levasseur-Regourd 1986, Levasseur-Regourd & Dumont 1985). If the lower zodiacal intensity measured by Murdock & Price (1985) is correct, then the particle albedo would have a higher and, perhaps, more reasonable value (S. Price, private communication).

ZODIACAL DUST BANDS IRAS found that the zodiacal cloud is not completely smooth (Low et al. 1984a). Three bands of emission seem to circle the ecliptic, two 9° above and below the ecliptic plane and another in the ecliptic. Upon closer examination, the central band breaks up into two bands as well, so that one has to explain two pairs of bands (Dermott et al. 1985b). Each band is only about 5% of the strength of the zodiacal emission near the ecliptic and emits most strongly at 12 and 25 μm. The color temperature of the emission is about 165–200 K, implying a distance from the Sun of 2.2–3.5 AU (Low et al. 1984a), and is consistent with a parallactic measurement of the bands based on IRAS observations at different solar elongation angles (Gautier et al. 1984b).

Orbital mechanics forbids the bands from defining separate planes parallel to the ecliptic. However, a particle in an orbit inclined to the ecliptic spends most of its time at the extrema above or below the ecliptic, so that emission from a cloud of particles sharing common orbital elements except for a random distribution of the longitudes of ascending nodes would produce a pair of bands. The separation of the outer bands from the ecliptic then gives the inclination of the orbits as $\sim 9°$.

A disruptive collision of a single 10–15 km asteroid could produce enough particulate material to account for a set of bands; the Eos, Koronis, and Themis families (Figure 3) have been identified as possible progenitors of the unfortunate asteroids responsible for the observed bands (Burns et al 1984, Dermott et al. 1984, 1985a, Sykes et al. 1984, Sykes & Greenberg 1985, 1986, Gustafson 1986). Estimates of the time scale for formation and dissipation of the bands are a few million years. By integrating the effects of collisions over the entire mass spectrum of asteroids, Sykes & Greenberg (1986) show that "asteroid collisions may provide the bulk, if not all, of the material giving rise to the observed zodiacal thermal emission."

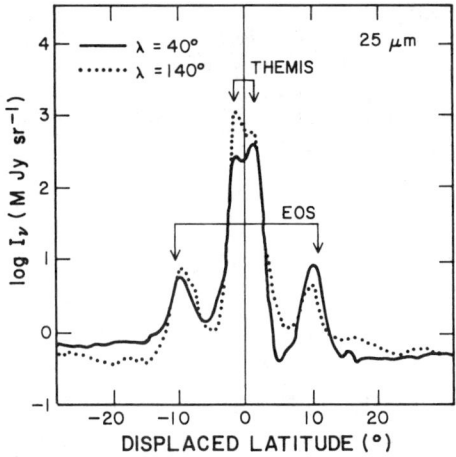

Figure 3 The bands of 25-μm emission that girdle the solar system have been enhanced by high pass filtering of the IRAS total intensity data. Scans at two different ecliptic longitudes are shown. The positions of the bands at ±9° compare favorably with those predicted for debris associated with the Eos family; the center bands agree in position with the Themis and Koronis families. The plotted latitudes have been displaced slightly to account for the differing symmetry planes of the asteroid families (Dermott et al. 1985b).

Comets and Asteroids

The Fast Moving Object program was established at the Chilton (Oxfordshire) ground station during the IRAS mission to provide optical observers with timely notification that IRAS had detected a possible solar system object. As a result of this effort, IRAS was one of the most prolific comet finders in history (Davies et al. 1984, Stewart et al. 1984, Green et al. 1985a, Marsden 1986). With a sensitivity equivalent to a visual magnitude of ~ 17 mag (Davies et al. 1984), IRAS discovered six new comets (Table 1), one minor planet, and over 400 new asteroids. IRAS also detected 1800 previously known asteroids (see Figure 2 for Ceres) and 25 previously known comets (Davies et al. 1984, Walker & Aumann 1984, Marsden 1986, *IRAS Asteroid and Comet Survey* 1986). The three Earth-crossing asteroids found during the mission are consistent with the expected numbers of these objects (Green et al. 1985a).

IRAS first detected Comet IRAS-Araki-Alcock (1983d), on 25 April 1983. Independently, G. Araki and G. Alcock discovered the comet on May 3. Eight days later, the comet passed within 4.5×10^6 km of the Earth, the closest passage by a comet since Comet Lexell in 1770. IRAS-Araki-Alcock produced a dust tail not apparent at visual wavelengths, with a measured length of 200,000 km and an extrapolated length of over 400,000 km (Walker et al. 1984). Spectral energy distributions show that the emitting grains were silicates heated to temperatures of 200–250 K, their size apparently decreasing along the comet's tail from 30 to 5 μm. The particles were ejected from the nucleus at ~ 200 kg s^{-1}. The dust ejection rate measured by IRAS in 11 other comets ranges from 2 to 250 kg s^{-1} and is a sensitive function of heliocentric distance (Walker & Aumann 1984).

The discovery of long *trails* of solid material associated with some

Table 1 IRAS comets[a]

Comet name	Perihelion distance (AU)	Period (yr)	Inclination (°)
1983d P/IRAS-Araki-Alcock	0.99	961	73
1983f IRAS	1.42	—	152
1983j P/IRAS	1.70	13.2	46
1983k IRAS	2.42	—	139
1983o IRAS	2.25	—	121
1983v P/Hartley-IRAS	1.28	21.3	96

[a] Data from Davies et al. (1984) and Marsden (1986).

comets reinforces the idea that comets are far dustier than previously thought. A trail consists of material strung out behind and ahead of the comet in its orbital plane and is distinguished from a *tail* by showing no evidence for diffusion in the plane of the orbit due to radiation pressure. Further, a trail appears to be a permanent structure, unlike a tail, which exists only during perihelion passage.

Although Eaton et al. (1984) first called the 4′ wide and 10° long structure associated with Comet Tempel-2 an "anomalous tail," subsequent examination of the IRAS images by Sykes et al. (1986) showed the structure to be a 0.5-AU-long *trail*. Comet Encke, located at 3.5–4 AU from the Sun, has a particularly remarkable trail of cold debris visible only at 60 μm and extending over 2 AU in length. Other comets with trails include Gunn, Tempel-1, Kopff, and Shoemaker-2; as many as a hundred other trails may yet be identified on the IRAS images (Sykes et al. 1986). The origin of the trail material may be rocky debris ejected by the nucleus at velocities of a few meters per second over dozens of perihelion passages (Sykes et al. 1986).

A final discovery pertaining to comets was of the peculiar object 1983 TB, which may be the defunct nucleus of the cometary progenitor of the Geminid meteor stream (Davies et al. 1984). However, since its visual albedo of 0.11 is lower than expected for a cometary nucleus and its near-infrared colors show no evidence for surface ice, the nature of 1983 TB, whether comet or asteroid, remains uncertain (Green et al. 1985b, Davies 1986, Veeder et al. 1984).

The Search for Planet X

There was much speculation before and during the IRAS mission about the possibility of finding a tenth planet in the Solar System which might have manifested itself as a slowly moving 40–100 K blackbody (Houck et al. 1984). There is, as yet, no compelling candidate for Planet X in the IRAS data; the vast majority of unidentified 60-μm sources turned out to be galaxies (Houck et al. 1985), and the unidentified 12-μm sources turned out to be mostly red giant stars (Chester et al. 1987, Beichman et al. 1987). Nor is there any evidence that any of the 100-μm diffuse emission described below is associated with solar system material (Low et al. 1984b, T. N. Gautier, private communication).

3. PHYSICS OF STAR FORMATION

Low-Mass Star Formation

PROTOSTARS   A protostar is defined here as an object deriving most of its energy from the accretion of infalling material, rather than from either hydrogen (or deuterium) burning or the release of gravitational energy by the quasi-hydrostatic contraction of a pre-main-sequence star (Beichman et al. 1986b). Wynn-Williams (1982) rightly called the search for protostars "the Holy Grail of infrared astronomy" and, indeed, before IRAS the search was somewhat quixotic, since the only detectable protostellar candidates had to be relatively bright and, therefore, luminous. Since massive stars are so few in number and their gestation period so short, the probability of catching an OB star in the accretion phase was very small. IRAS was able to make a complete census of nearby molecular clouds like Taurus and Ophiuchus to the level of 0.1 $L_\odot$ in search of the embedded progenitors of solar-type stars with $T \sim 40$–300 K (Myers et al. 1987, Young et al. 1986). Since solar-type stars are so numerous and have such a leisurely pace of evolution, the likelihood of finding a star in its accretion phase is quite high (Emerson 1985).

Beichman et al. (1986b; see also Clark 1987) examined IRAS data for 95 small (0.1 pc), dense (3×10^4 cm^{-3}), cold (12 K) condensations of molecular gas in the Taurus and Ophiuchus clouds (Myers et al. 1983, Myers & Benson 1983, Benson 1983) and found two characteristic types of objects (Figure 4) associated with more than half of the cores. First are those objects with relatively flat IRAS energy distributions; these objects almost invariably have optical counterparts. Second are those sources whose energy distributions rise steeply to longer wavelengths; these sources mostly do not have optical counterparts. In both cases the source luminosities are in the range 1–50 $L_\odot$; the dust temperatures inferred from the IRAS data range from 30 to a few hundred K, with the optically invisible sources being considerably cooler than the visible ones. Many of the invisible objects are associated with outflowing molecular gas, thought to be an indicator of extreme youth (Myers et al. 1987).

The sources with optical counterparts are either identified with known T Tauri stars or have IRAS colors similar to T Tauri stars (see below). The optically invisible sources could be (a) heavily extinguished T Tauri stars accidentally seen in or behind a core; (b) young, but fully formed, T Tauri stars that have not yet left their birthplaces within the dense core; or (c) considerably younger versions of T Tauri stars still accreting material from the surrounding dense core. Although the fact that the optically invisible sources are found at or near the densest portions of the cores rules out the first possibility (Beichman et al. 1986b, Myers 1985, 1986, Myers et al. 1987), it is considerably harder to decide between the second two possibilities: *One of the major unanswered questions raised by IRAS is whether any of these embedded sources are still accreting material.*

Ground-based infrared (1–20 μm) observations have shown that the deeply embedded IRAS sources are associated with stars having photo-

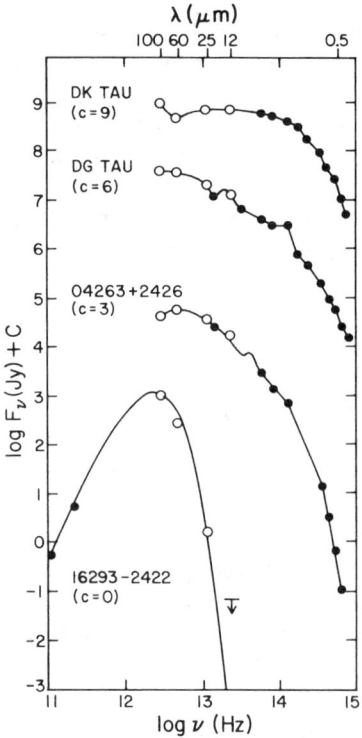

Figure 4 A sample of IRAS sources associated with star formation showing a progression of ever redder, more obscured objects. The visible T Tauri stars DK and DG Tau show a relatively shallow increase of flux density with wavelength. The IRAS data are plotted as open circles; other data are from references in Rucinski (1985). The detailed spectral energy distribution of IRAS 04263+2426 (Haro 6–10) can be fitted by models of a rotating, accreting protostar (Myers et al. 1987, Adams et al. 1986). IRAS 16293–2422 has a very cold spectrum fitted by emission from 42-K dust that is optically thick at 250 μm (Walker et al. 1986, Mundy et al. 1986). The constant c gives the logarithmic scale factor used in plotting the data.

spheric temperatures of 3000–6000 K obscured by $A_V \sim 30$–90 mag (Benson et al. 1984, Myers et al. 1987). The observations further imply that some of the material surrounding these stars must be in the form of a disk, since if one puts enough material within ~ 1 AU of the star in a spherically symmetrical fashion to account for the observed 10–25 μm emission, then that same material would totally obscure the 1–5 μm emission from the star. Myers et al. (1987) assert that an inclined 1–10 AU disk sitting within a cavity 5–50 AU in radius would produce the observed mid-infrared emission, yet would let the near-infrared radiation from the star escape. The existence of disks around T Tauri stars (Grasdalen et al. 1984) and main-sequence stars (Aumann et al. 1984) lends credence to the hypothesis that disks are a natural part of the formation of a star. The disk could be either passive (i.e. heated by either the central star) or active (i.e. heated by infalling material) or some combination of these two cases.

Adams, Shu, and collaborators (see the review by Shu et al. in this volume) have argued on theoretical grounds that a thin accretion disk forms around a rotating protostar. In addition to intercepting and reradiating ~25% of the central star's total energy, the disk can also generate its own luminosity from accretion or viscous effects. The spectral energy distributions predicted by these authors for a number of embedded sources and T Tauri stars agree very well with the 1–100 μm observations (Adams et al. 1986, Shu et al. 1986, Adams & Shu 1986); models without disks do not agree well with the observations (Stahler et al. 1980). However, the theory is quiet about whether the disk is active or passive; one cannot infer from the photometric data alone whether accretion through the disk is responsible for a significant portion of the luminosity in these objects.

One indirect argument suggests that some of the invisible sources must be still accreting. The time for a 0.5–1 $M_\odot$ object to build up at the expected accretion rates is about a free-fall time, or $\sim 2 \times 10^5$ yr (Shu 1977). On the other hand, the ages of the youngest visible pre-main-sequence stars, 10^5 yr (Cohen & Kuhi 1979), give a lower bound to the time it takes a newly formed star to emerge from its placental core. The rough equality of these time scales suggests that a significant fraction of the invisible objects may still be accreting (Beichman et al 1986b, Myers et al. 1987).

An IRAS source found by Walker et al. (1986) in the ρ Oph cloud *may* show the first direct evidence for infall. The object 16293−2422 has a total luminosity of 27 $L_\odot$ and an energy distribution (Figure 4) that is well fitted by emission from ~40-K dust (Walker et al. 1986, Mundy et al. 1986); 16293−2422 is colder and more luminous than most of the core sources found by Beichman et al. (1986b). Emission from 16293−2422 may originate in a 0.25–1.0 $M_\odot$ star, perhaps seen through an edge-on disk (F. Shu, private communication), plus some amount of infall luminosity.

What is most intriguing about 16293−2422 is that the shapes of the line profiles of various transitions of the CS molecule seen toward the source are consistent with material falling inward at $6 \times 10^{-6}\ M_\odot\ \text{yr}^{-1}$. Unfortunately, the millimeter spectra of 16293−2422 and its immediate surroundings are particularly complicated, with both infalling and outflowing material apparently being observed simultaneously (E. Young, private communication). If the interpretation of infall stands the tests of future observations, then as Beichman et al. (1986b; cf. Shu & Tereby 1984, Cassen et al. 1985) suggest, "the traditional notion of steady accretion with all energy obtained from gravitational infall, followed by a period when a pre-main sequence star is obscured by a molecular cloud, may have to be modified. The new view might involve a combination of steady accretion, perhaps through a disk, onto a slowly contracting star with mass outflow proceeding out the poles of the system."

YOUNG STELLAR WINDS   One of the most important discoveries about star formation is that bipolar mass outflow plays an important part in the evolution of young stars (Snell et al. 1980). Manifestations of these collimated stellar winds include high-velocity flows seen in CO and the presence of Herbig-Haro (HH) objects. Margulis & Lada (1986) compared the results of their CO survey of the Monoceros OB1 molecular cloud with IRAS maps of the same region. They found that six of the nine outflow regions they discovered are associated with IRAS sources brighter than their infrared detection limit of $\sim 5\ L_\odot$; fainter infrared sources could account for the other three outflows. On the other hand, 24 of the 30 IRAS sources found in the same region *did not show evidence for high-velocity gas.* The authors suggested that if all the young stars embedded within the Mon OB1 cloud undergo an outflow phase, then that phase can last only about one-fifth of the lifetime of the object as a bright infrared source. Alternatively, the incidence of the outflow phenomenon may have been greatly overestimated, and outflow might occur in only one fifth of young, embedded sources. Additional work is needed to clarify the sense of this crucial observational result.

IRAS images have made the dust associated with at least one outflow directly observable. In the case of the archetypal outflow source L1551, Clark & Laureijs (1985), Edwards et al. (1986), and Clark et al. (1987a) found emission from 20–25 K dust at 60 and 100 μm that extends in two $\sim 20'$-long lobes with roughly the same morphology as the CO emission. Clark et al. (1987b) identified a second possible infrared outflow around the central star of reflection nebula CED 110, although CO observations are needed to demonstrate conclusively the outflow nature of this source.

The radiant luminosity in the two infrared lobes of L1551 is $\sim 18\%$ of the total luminosity of the central source IRS 5 and an order of magnitude larger than the mechanical energy in the gas (Edwards et al. 1986). Radiation pressure from IRS 5 may be insufficient to drive the flow; rather, it may be necessary to invoke a model that channels the gravitational energy of the infalling material directly into the outflow, almost independent of the presence of the central star (Clark et al. 1987).

The heating of the dust in the two lobes poses a separate problem. Since the temperature and surface brightness of the emission do not vary strongly with distance from IRS 5, the dust cannot be centrally heated by direct radiation from IRS 5. Nor can the dust be heated collisionally by the impinging gas (Clark & Laureijs 1985), since the infrared emission extends well outside the boundaries of the molecular flow (Edwards et al. 1986). The dust may be heated by optical photons created at shocks between the flow and the ambient material (Edwards et al. 1986, Clark et al. 1987).

Goldsmith et al. (1986) investigated the role of outflows in clearing a volume around the IRAS sources in the Barnard 5 cloud (Beichman et al. 1984b). Many authors have found the IRAS data useful for locating and estimating the luminosity of the obscured sources responsible for either outflows or Herbig-Haro objects (Emerson et al. 1984, Cohen et al. 1986a, Graham 1986, Pravdo & Chester 1986, Wolstencroft et al. 1986, White & Gee 1986, Hilton et al. 1986, Gear et al. 1986, Takaba et al. 1986).

PRE-MAIN-SEQUENCE STARS Pre-main-sequence stars were known to be farinfrared sources before IRAS (Harvey et al. 1979), but the large sample of young stars with infrared excesses is resulting in a better understanding of the physics and evolutionary status of these sources. Rucinski (1985) and Cohen et al. (1987) found that about two thirds of the optically identified T Tauri stars are IRAS sources, typically emitting 10–20% of their luminosity longward of 12 μm. A variety of energy distributions are observed: flat, slowly rising or falling, and double peaked (Rucinski 1985, Evans et al. 1986, Harris 1985). The infrared excess could be produced by emission from a disk 4–12 AU in size accreting material at a rate of $6 \times 10^{-6} M_\odot$ yr^{-1} (Rucinski 1985, Shu et al. 1986), although more symmetric distributions of material are also possible (Harvey et al. 1979).

High-Mass Star Formation

COMPACT H II REGIONS The energy distributions of compact H II regions show the broad range of temperatures expected from a centrally heated cloud. IRAS colors in conjunction with theoretical models (Crawford & Rowan-Robinson 1986) can be used to distinguish evolutionary effects in H II regions: For example, young compact H II regions have colder 12/25-μm colors than older, more extended ones (Fich & Terebey 1987), while millimeter and submillimeter colors are less sensitive to evolutionary effects (Chini et al. 1984, 1985, 1986a).

Wouterloot & Walmsley (1986) pursued the association between H$_2$O masers and infrared sources and found that masers are preferentially associated with objects having the same range of IRAS colors as the very compact H II regions and are unlike older, more diffuse H II regions. Further, the masers were associated with the most luminous sources: In Orion and Cepheus, 35% of the compact H II regions brighter than 10^4 $L_\odot$ had masers, compared with fewer than 10% of the objects fainter than $10^3 L_\odot$.

TRIGGERS OF STAR FORMATION The IRAS images "[afford] us a uniquely unobscured view of the relationship of recently formed stars to the interstellar medium" (Schwartz 1986). Various authors have tried to demonstrate connections between various phases in the interstellar medium and the formation of stars. IRAS images show limb-brightened rims associated with the boundaries of supernovae remnants, H I shells, and H II regions (Braun 1985, Schwartz 1986, Sandell et al. 1986). IRAS point sources in these rims may be new stars induced to form by the interaction of stellar winds with the ambient interstellar medium (Schwartz 1986, Huang et al. 1986).

A comparison of IRAS and CO surveys of the Galactic plane shows that the efficiency of high-mass star formation decreases or remains constant as the mass of the molecular cloud increases from 2×10^5 to 3×10^6 $M_\odot$ (Scoville & Good 1987, Solomon et al. 1987). This trend argues against internal triggers of star formation, such as sequential star formation (Elmegreen 1986), since the presence of more cloud material should lead to more star formation. Detailed comparisons of the IRAS data with CO surveys for the Galactic plane are underway (N. Scoville, private communication) and should lead to a better understanding of where, within giant molecular clouds, high-mass stars form.

4. STARS IN THE INFRARED

Approximately 65% of the 246,000 sources detected by IRAS are stars, including bare photospheres as well as stars with infrared excesses due to circumstellar shells of gas or dust (Chester 1986). Stars show a marked concentration toward the Galactic plane: half of all of the sources in the catalog are within Galactic latitude $|b| = 7°$ and 85% are within $|b| = 30°$; two thirds of the sources are within $\pm 90°$ of the Galactic center (Figure 5). Above $|b| \sim 30°$, the surface density of stars with $f_v \geq 0.4$ Jy follows a $N_0 \csc(|b|)$ law, with $N_0 \sim 0.6$ stars (sq. deg.)$^{-1}$, roughly independent of Galactic longitude (Rowan-Robinson et al. 1984). Although only about 44,000 of the stars in the IRAS catalog have been identified with previously known objects, it is clear from the associations in the Point Source Catalog that K and M giants, with or without an excess due to mass loss, are the most common IRAS stars. This section addresses the properties of normal stars, stars associated with mass loss, novae, and supernovae. A subsequent section discusses stars surrounded by orbiting solid material (the Vega phenomenon).

Normal Stars

The main sequence provides a standard against which the properties of a given star can be judged. The lower envelope of the points in Figure 6 defines "normal" stars with no infrared excess (Coté & Waters 1986, Waters et al. 1986). The observations of normal stars are in good agreement with photospheric models (Kurucz 1979), except for a $\sim 10\%$ excess of the observed over the predicted emission longward of 12 μm that had been noted by ground-based observers (Rieke et al. 1985). Only a small fraction of stars with spectral type A (13%) to K (1%) have 12-μm excesses greater

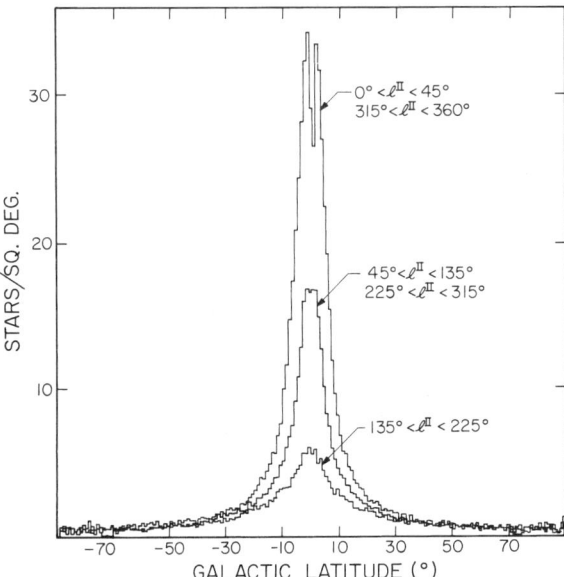

Figure 5 The variation of the number of objects per square degree with Galactic latitude for IRAS sources associated with stars having little or no infrared excess as defined by f_v (12 μm) $\geq f_v$ (25 μm) $\geq f_v$ (12 μm)/3. Three different ranges of Galactic longitude are shown. The dip near the Galactic center is due to the effects of confusion at high source density (see *Supplement*, Chaps. 5, 7).

than 0.5 mag, but a large fraction of B (40%) and M (29%) dwarf and giant stars show more than 0.5 mag of 12-μm excess compared with their expected photospheric temperatures (Waters et al. 1986). Most of the stars with infrared excesses fail to show photospheric colors because of effects due to mass loss associated with evolution through the supergiant phase or the asymptotic giant branch (AGB). Hot stars have excesses due to either dust or ionized gas; cool stars have excesses due to dust (Figure 7).

Hot Stars with Mass Loss

SUPERGIANTS AND WOLF-RAYET STARS Before IRAS it was known that ionized stellar winds or, occasionally, hot 10^3-K dust caused infrared excesses around hot stars (Cassinelli & Hartmann 1977, Barlow & Cohen 1977, Lamers 1986). Emission from these mechanisms becomes much stronger than the photosphere at wavelengths longer than ~ 25 μm (Figure 7). Observations of supergiants such as ζ Puppis (Lamers et al. 1984, Wolfire et al. 1985), P Cygni (Waters & Wesselius 1986), and massive Wolf-Rayet stars (van der Hucht & Olnon 1985, van der Hucht et al.

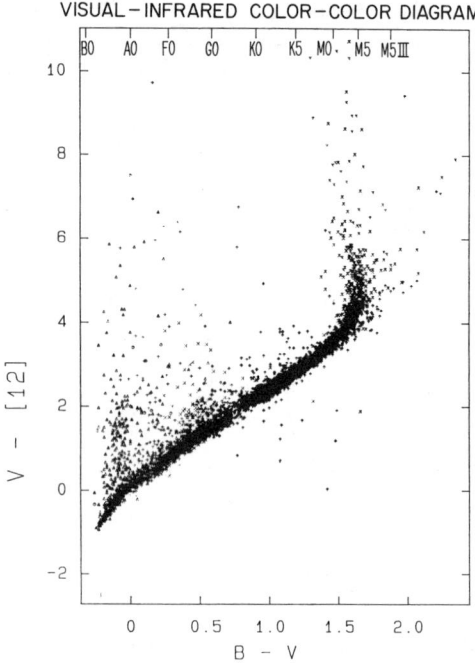

Figure 6 The visual-infrared color-color diagram for stars in the IRAS catalog (Coté & Waters 1986). The lower envelope of points defines the locus of bare photospheres. Points above that line have various amounts of excess. Representative main-sequence spectral types are given across the top.

1985a,b, Stickland et al. 1985) are consistent with mass-outflow rates of 10^{-5}–10^{-4} $M_\odot$ yr^{-1}. In the case of ζ Puppis, there is a controversy over whether or not a corona is required to explain both the infrared and X-ray observations of the star.

The discovery of a hot (1250-K) dust shell toward the B9.5 supergiant HR 4909 suggests that this object is the hottest, earliest example of a protoplanetary nebula (Lamers et al. 1986). The existence of cool dust shells in addition to the expected free-free excess around P Cygni and some Wolf-Rayet stars suggests that some of these objects are also in the process of turning into planetary nebulae (van der Hucht et al. 1985a,b, Waters & Wesselius 1986).

B AND A STARS   Almost all B dwarfs and giants with visual emission lines (Be stars) have infrared excesses due to outflowing shells of ionized gas (Waters 1986a,b,c, Schaefer 1986b), while many Ae stars have shells of 50–400 K dust (Jaschek et al. 1986). For the Be stars, mass loss through

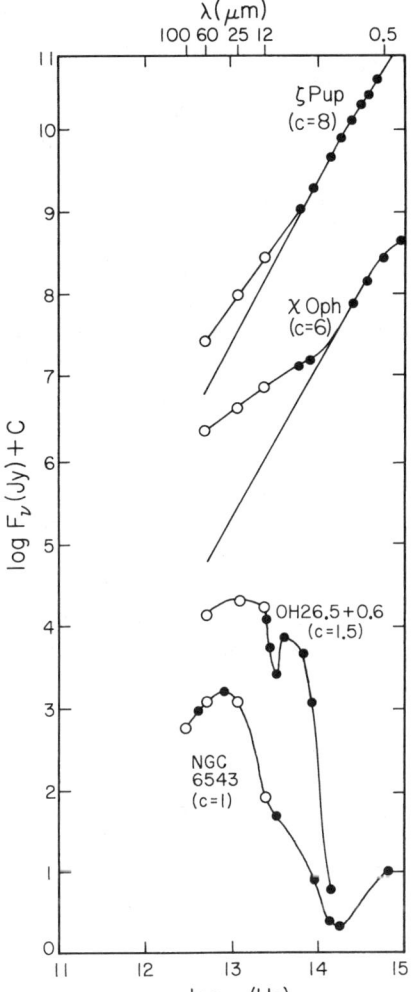

Figure 7 Spectral energy distributions for stars with infrared excesses. The top two stars show long-wavelength excess due to the effects of an ionized stellar wind (Lamers et al. 1984, Wolfire et al. 1985, Waters 1986b). IRAS points are given as open circles. The filled circles give data from the works mentioned in the text. The predicted photospheric emission is represented by a hot blackbody. OH 26.5+0.6 is a classic OH/IR star with a deep silicate feature (data from Herman et al. 1986, Evans & Beckwith 1977). NGC 6543 is a bright planetary nebula (non-IRAS data from Grasdalen 1979, Willner et al. 1972, Mosley 1980, Allen 1973). The constant c gives the logarithmic scale factor used in plotting the data.

an equatorial disk at rates around 10^{-8} $M_\odot$ yr^{-1}, 50 to 100 times greater than the rates implied by UV observations, is required to match the 1–100 μm observations of these stars (Waters 1986b). The existence of such a disk is further suggested by the fact that rapid rotation is a necessary, although insufficient, condition for these stars to have an infrared excess (Waters 1986a). Stars with $v^* \sin(i) \leq 100$–200 km s^{-1} have neither an infrared excess nor optical emission lines; rapid rotators are likely to have both.

PLANETARY NEBULAE Compact planetary nebulae are bright IRAS sources because of emission from dust heated by the central star. Planetaries have distinctive IRAS colors peaking between 25 and 60 μm (Pottasch et al. 1984b), with oxygen-rich planetaries appearing marginally bluer than carbon-rich objects in their 25/60-μm color (Roche & Aitken 1986).

There are marked evolutionary effects in the infrared appearance of planetaries. As the nebulae become older, i.e. larger and more diffuse, the dust emission becomes less prominent because (*a*) the dust temperature falls as a result of a weaker radiation field and (*b*) the dust abundance decreases a hundredfold relative to the gas, presumably as a result of destruction of grains by sputtering (Pottasch et al. 1984b, Pottasch 1986, Kwok et al. 1986). Much of the heating of the dust comes from photons with wavelengths longward of Lα; the typical excess of observed dust luminosity over the available Lα luminosity, denoted IRE, ranges from 1 to 10 and is smallest in large nebulae where the direct heating of the central star is negligible compared with scattered Lα (Pottasch et al. 1984b, Iyengar 1986, Kwok et al. 1986).

Spectra from the Low Resolution Spectrometer (LRS; see *IRAS Catalogues and Atlases: Atlas of Low Resolution Spectra* 1986) and maps in the four bands reveal that as the planetaries become more diffuse, ionic lines become more prominent than dust emission, particularly at 12 and 25 μm. For extremely diffuse planetaries such as NGC 7293, ionic lines can dominate some or all of the four IRAS bands ([S IV] and [Ne II] at 12 μm, [S III] and [O IV] at 25 μm, [O III] and [N III] at 60 μm, and [O III] at 100 μm), resulting in broadband "colors" that are not physical if interpreted in terms of a dust temperature (Leene & Pottasch 1987).

Spectral lines of hitherto unobserved ionization states of neon, e.g. [Ne III], [Ne V], and [Ne VI], have been used to determine abundances, densities, and temperatures in a large number of planetaries (Pottasch et al. 1984a, 1986a,b, Pottasch 1986). The abundance of Ne can vary by a factor of 30 from one planetary to the next and shows a slight increase toward the Galactic center. The surprising presence in NGC 6302 of the 7.65-μm line of [Ne VI] (Figure 8), 126 eV above the ground state, implies the existence of compact clumps of gas with a density of 3×10^5 cm^{-3}, an order of magnitude higher than the gas responsible for producing spectral lines of less highly ionized species (Pottasch et al. 1985).

Mid Spectral Types With Mass Loss

Many spectral type F, G, and K supergiants have infrared excesses due to massive dust shells having temperatures from 80 to 900 K and mass-loss rates of 10^{-8}–10^{-5} $M_\odot$ yr^{-1} (Parthasarathy & Pottasch 1986a,b, Odenwald

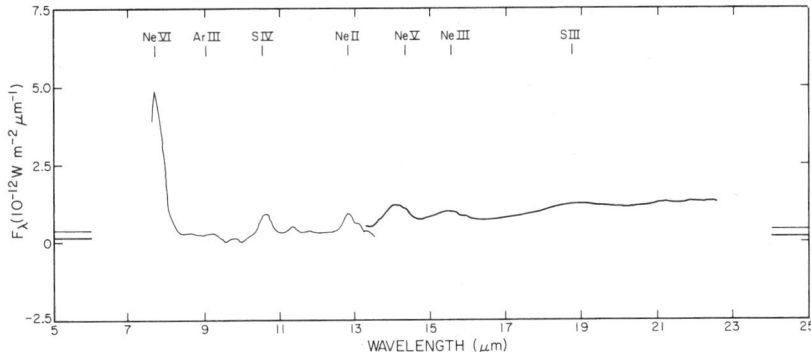

Figure 8 The Low Resolution Spectrometer (LRS) spectrum of the planetary nebula NGC 6302 shows a wide diversity of ionic species. The detection of [Ne VI] at 7.65 μm (Pottasch et al. 1985) implies the presence of very hot, dense gas.

1986a,b, Evans 1985). Many of these objects are RV Tauri variables that have been proposed to be progenitors of planetary nebulae (Jura 1986b). In some stars, the relatively flat shape of the infrared energy distribution has been interpreted to mean that the mass-loss rate may have declined by a factor of 100 during the last 500–1000 yr (Jura 1986a,b, Odenwald 1986a,b). More evidence for sudden changes in the nature of the outflow comes from the discovery of strong silicate emission in the LRS data toward carbon-rich RV Tauri stars. The observed shell material may be left over from an earlier red giant phase (Evans 1985).

Some Cepheids have 2–25 μm excesses due to either free-free emission or dust. The low mass-loss rates of 10^{-7}–10^{-8} $M_\odot$ yr^{-1} are too small to have a significant evolutionary effect on the stars (Deasy & Butler 1986).

Cool Stars With Mass Loss

A crucial part of the evolution of stars of moderate mass (≤ 5 $M_\odot$) is the AGB phase, during which a strong stellar wind ($\sim 10^{-4}$ $M_\odot$ yr^{-1}) returns the majority of a star's mass back to the interstellar medium (Zuckerman 1986). OH masers are often found in association with oxygen-rich stars, and until IRAS, radio surveys were more sensitive to OH/IR stars than were infrared surveys (Olnon et al. 1984, Lewis et al. 1985); however, IRAS could find these stars well beyond the center of the Galaxy, and far-infrared observations of thousands of OH/IR stars have given new insights into the physics and chemistry of mass loss (Olnon et al. 1984) as well as providing a probe of Galactic structure.

PHYSICS OF MASS LOSS The appearance of an AGB star is determined by the opacity of its shell at some reference wavelength, say 9.7 μm, and by the nature of the absorbing dust. Stars with relatively little mass loss ($\sim 10^{-7}\ M_\odot\ \text{yr}^{-1}$) appear as optically bright Mira variables; stars with mass-loss rates increasing up to $\sim 10^{-4}\ M_\odot\ \text{yr}^{-1}$ become progressively less apparent visually and progressively brighter in the infrared. Theoretical calculations of the spectral energy distribution of oxygen-rich stars provide an excellent match to the observations over a large range of shell opacities (Rowan-Robinson & Harris 1983, Bedijn 1986).

An important observational result is that IRAS photometry alone can be used to determine abundances in the mass-loss shell, since oxygen-rich stars are ~ 0.4 mag bluer than carbon-rich stars in the 25/60-μm color (Hacking et al. 1985, Zuckerman & Dyck 1986b, Thronson et al. 1987b). The lack of a comparable effect in the 60/100-μm color led Zuckerman & Dyck (1986b) to attribute the effect to a variation in dust properties between the two types of star, rather than to the presence of extra cold material around the carbon stars as first suggested by Hacking et al. (1985). The IRAS data for carbon stars are suggestive of shells made of amorphous carbon (not graphite) grains (Jura 1986a, Willems 1986, Rowan-Robinson 1986). Correlations are also found between the IRAS colors, the depth of the LRS silicate feature, the luminosity in the radiatively pumped OH lines, and the symmetries of the visual light curves (Herman et al. 1986, van der Veen 1986, Vardya et al. 1986). SiO emission at 8 μm is found toward Miras with SiO masers (Vardya et al. 1986).

Herman et al. (1986) showed that the 12/25-μm color of an OH/IR star can be used to determine both the bolometric correction and the total optical depth through its shell. This result is extremely useful, since the Point Source Catalog contains many such sources for which there are no ground-based observations. OH observations of OH/IR stars yield the expansion velocity of the shell and a kinematic distance to the object. The combined OH and IRAS information yields the luminosity of the star, the mass-loss rate, and the gas-to-dust ratio of the ejecta (e.g. Habing 1987a, Van der Veen 1987) and shows that (*a*) the luminosity function of OH/IR stars peaks between 4000–6000 $L_\odot$ (Habing 1987a), (*b*) mass-loss rates determined in the infrared agree with rates determined by other techniques (CO, near-infrared lines, OH masers) to within a factor of 2 to 3, and (*c*) the gas-to-dust mass ratio deduced from the mass-outflow rates is 160 ± 40 for OH/IR stars and 200 ± 100 for the Miras. Younger, more massive metal-rich stars have lower gas-to-dust ratios than older, smaller metal-poor stars (Herman et al. 1986).

Stellar variability provides the key to understanding the mass-loss process. Habing (1987b) points out that if an M giant is a long-period variable, then it has an infrared excess. Further, the converse is also true; if the star has an excess, then, in the overwhelming number of cases it is variable (for an interesting exception, see Habing et al. 1987). The IRAS data for many OH/IR stars show strong evidence for variability at 12 and 25 μm on a time scale of 6 months (Habing 1987a). This fact lends strength to a theoretical picture in which mass loss begins when the height of atmospheric pulsations becomes large enough to allow dust to condense (Jones et al. 1981, Bedijn 1986); thereafter, radiation pressure on the dust and gas-grain friction drag the gas away from the star. As the mass of the star declines, the scale height of the oscillations increases and the mass-loss rate increases with time. This model makes specific predictions about the luminosity function of mass-loss stars and the numbers of stars with different amounts of mass loss. These predictions have been verified in a number of ways using non-IRAS data (Bedijn 1986), and use of the IRAS data is a logical next step. Some preliminary work shows good agreement is to be expected (Herman et al. 1986).

There is strong evidence that the mass-loss rate and chemistry vary in time as well as from star to star. For example, *silicate* emission seen toward a number of carbon stars implies a change in the abundances of the ejected material within the last 10^2–10^3 yr (Walker 1985, 1986, Little-Marenin 1986, Little-Marenin et al. 1986, Willems & de Jong 1986). The 25/60-μm colors of these objects are also indicative of an oxygen-rich outflow, implying a long (10^3–10^4 yr) duration for the oxygen-rich phase (Zuckerman & Dyck 1986a,b). A few carbon stars show a large 100-μm excess due to cool material in larger amounts than can be accounted for by the present mass-loss rate; episodic mass loss may be responsible (Jura 1986a).

R COR BOR STARS R Coronae Borealis is the prototype of a class of hydrogen-deficient supergiants with irregular drops in brightness due (it is suggested) to patchy clouds of circumstellar matter. Many R Cor Bor and hydrogen-deficient stars show infrared excesses (Walker 1985, 1986, Schaefer 1986a, Rao & Nandy 1986). R Cor Bor itself was already known to have a mid-infrared excess (Stein et al. 1969), but Gillett et al. (1986a) found an 18′ diameter shell (8.5 pc!) of 60- and 100-μm emission centered on the star. This outer shell is completely distinct from the unresolved inner shell responsible for the mid-infrared excess. The mass-loss rate responsible for the outer shell was 10 times larger than the presently inferred rate of mass loss. The larger rate must have ended $\sim 2 \times 10^4$ yr ago. The 60/100-μm color temperature (25–30 K) is constant across the entire shell, ruling out R Cor Bor itself as the heat source. Transient heating of small grains by an intense radiation field or by collisional heating due to halo gas are possible mechanisms.

THE GALACTIC DISTRIBUTION OF MASS-LOSS STARS A number of authors have examined the Galactic distribution of mass-loss stars using somewhat different samples. For oxygen-rich stars, Habing (1987a) derived a horizontal scale length for an assumed exponential distribution of stellar density of 6.5 ± 1.0 kpc. Rowan-Robinson & Chester (1987) obtained a similar value of 6 kpc. Thronson et al. (1987b) obtained 3.7 kpc using their sample; the reason for this discrepancy is unclear. After taking into account their different formalisms for the distribution perpendicular to the Galactic plane, one can show that both Habing and Rowan-Robinson & Chester derived vertical distributions consistent with an exponential falloff having a 200–250 pc scale height.

Thronson et al. (1987b) examined the distribution of carbon stars in the Galaxy using the IRAS data and found that these stars are about nine times less populous than oxygen-rich stars in the solar neighborhood; however, in sharp contrast with the oxygen-rich stars, carbon stars show a relatively flat distribution in galactocentric distance.

Mass loss from both oxygen- and carbon-rich stars is crucial to the chemical enrichment of the Galaxy. Analysis of the IRAS data shows that the oxygen- and carbon-rich giants together return 0.35–0.7 $M_\odot$ yr^{-1} to the interstellar medium (Knapp & Wilcots 1987, Thronson et al. 1987a,b, Jura 1986a). These numbers double previous estimates based on a more limited sample of CO data (Knapp & Morris 1985). The differing radial dependences of the density of carbon- and oxygen-rich stars imply that the ratio of carbon dust to silicate dust will increase with increasing galactocentric distance (Thronson et al. 1987b).

The bulge of the Galaxy is beautifully delineated (Figure 9) by stars detected at 12 and 25 μm (Habing et al. 1985, Habing 1986, 1987a,b, Chester 1986, Rowan-Robinson & Chester 1987). IRAS and ground-based

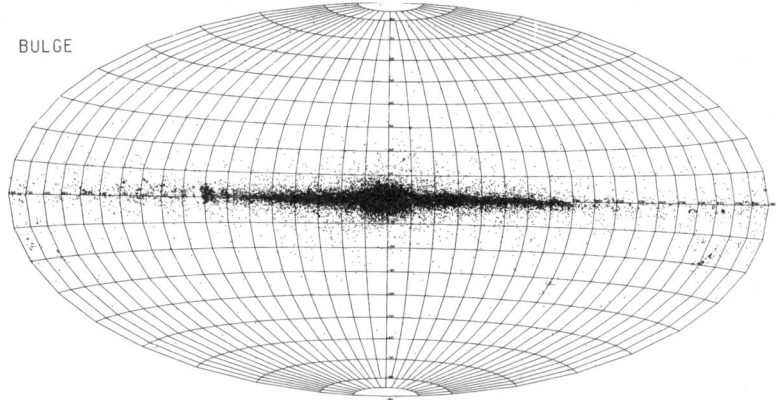

Figure 9 The distribution of stars detected at 12 and 25 μm with color temperatures around 300 K shows clearly the disk and bulge of the Galaxy (Habing et al. 1985, Chester 1986).

observations show that the bulge stars closely resemble local Mira variables and OH/IR stars with bolometric luminosities of ~ 2000–$4000\ L_\odot$ surrounded by dust shells with temperatures of 200–1000 K (Feast 1986, Frogel 1986, Whitelock et al. 1986, Glass 1986). A similar bulge population could account for the 12- and 25-μm emission from the central part of M31 (Rowan-Robinson & Chester 1987, Soifer et al. 1986, Walterbos & Schwering 1986).

The presence of such highly luminous stars presents problems either to our understanding of stellar evolution or to our ideas of the formation of the bulge: If the bulge has the age of the Galaxy (~ 15 Gyr), then why are stars thought to have 1.3–1.7 $M_\odot$ precursors and ages $\ll 10$ Gyr found there (Habing 1987a)? Evolutionary effects due to the metallicity have been suggested as a way out of this difficulty. Enhanced metallicity increases the age at which a star of a given mass reaches a given luminosity on the AGB (Habing 1987a). On the other hand, decreased metallicity results in a greater peak AGB luminosity for a star of a given mass, perhaps resulting in an overestimate of its mass and thus an underestimate of its lifetime (Herman et al. 1986). A completely different solution to the problem is to allow some fraction of the bulge stars to have formed within the last 1 Gyr. The determination of the metallicity of these stars is a crucial follow-up observation.

MISCELLANEOUS SITES OF MASS LOSS   The survey nature of the IRAS data allowed a search for mass loss in a large variety of places. A peculiar object in the globular cluster M22 may be a planetary nebula in an early stage of evolution, still surrounded by a thick dust shell (Gillett et al. 1986b). A number of symbiotic binaries show red IRAS colors suggesting that the red giant member of the pair has a thick, cool shell due to mass loss (Kenyon et al. 1986). On the other hand, Jura (1986c) was unable to find evidence of mass-loss stars in open clusters and inferred that the duration of the mass-loss phase for 5- and 7-$M_\odot$ stars is less than 10^6 and 10^5 yr, respectively.

Two stars were discovered in the Large Magellanic Cloud (LMC) with energy distributions resembling Galactic OH/IR stars (Elias et al. 1986). One of them was detected in OH with an unusually low expansion velocity for its high bolometric luminosity, which suggests that the low metallicity of the LMC has affected the mass-loss process (Wood et al. 1986).

Novae

IRAS detections of classical novae support the idea that fast novae such as GQ Muscae produce free-free emission but no dust emission, while slow novae such as V4077 Sagittarii produce an expanding shell of dust resulting in a bright infrared source (Dinerstein & Robinson 1986, Callus et al 1986, 1987, Evans 1985). V4077 Sgr showed emission from 900-K dust; its decline in brightness by over an order of magnitude when measured by IRAS at 6 and 12 months after its visual maximum was attributed to a decrease in the mass of the dust shell from 4×10^{-6} to 4×10^{-7} $M_\odot$ (Dinerstein 1986).

Three older novae—FH Serpentis, HR Delphini, and DQ Herculis—were all detected by IRAS but with energy distributions that are hard to understand in terms of emission by dust. These objects may be examples where fine-structure lines such as [SIII], [Ne V], [Ne II], and [O III] are responsible for the observed radiation.

IRAS also detected a 12-μm excess around the star RS Ophiuchi *prior* to its 1985 outburst. The near-infrared excess detected *after* the outburst has been attributed to nova ejecta collisionally heating the preexisting mass-loss shell seen by IRAS (Albinson et al. 1986, Schaefer 1986a).

Supernovae

Supernova remnants (SNRs) show a large infrared excess over a synchrotron spectrum that appears to be continuous and smooth between radio and optical wavelengths; the excess is presumably due to emission from hot dust. Observations of SNRs in the Large Magellanic Cloud (Graham et al. 1987) and toward selected Galactic remnants (Dwek et al. 1986) show that cooling of the hot gas in an SNR by gas-grain collisions can dominate the cooling by atomic processes by factors of 10–100; dust must therefore play an important role in the evolution of SNRs (Dwek 1981).

Braun (1985) has made beautiful IRAS maps of the young SNRs Tycho, Kepler, and Cas A, as well as for the evolved SNRs, the Cygnus Loop, and IC 443. The infrared, optical, and X-ray images are well correlated, which implies that the infrared emission originates in interstellar material swept up at the boundary of the SNR and that the emitting dust is heated, perhaps, by collisions in the X-ray-emitting material (Dwek et al. 1987). In the case of IC 443 (Braun 1985, Mufson et al. 1986), the SNR appears to have expanded into a series of preexisting H II regions.

There are two different mechanisms for the heating of the dust in SNRs. Marsden et al. (1984) and Marsden (1985) attributed the excess in the Crab and Tycho SNRs to $5-30 \times 10^{-3}$ $M_\odot$ of 70–90 K dust heated by short-wavelength UV and X-ray radiation. Braun (1985), Dwek et al. (1987), and Dwek (1986) attributed the infrared emission to collisionally heated dust in regions of the interstellar medium shocked by the passage of the supernova blast wave (Draine 1981). In Tycho and Kepler, ~ 13 $M_\odot$ of interstellar matter was swept up by 0.3 $M_\odot$ of ejecta. Including the effects of transiently heated small grains and ionic lines in a model of collisionally heated grains does not change the basic results obtained by Braun (1985) for Cas A (Dwek 1986, Dwek et al. 1986), although ionic lines do appear to be important in IC 443 (Mufson et al. 1986).

5. PROTOPLANETARY DISKS

The Vega Phenomenon

The discovery by Aumann et al. (1984) of an infrared excess around α Lyrae, a main-sequence dwarf with no known mass loss, electrified the scientific and popular imagination because of the proposed origin of the excess in a belt of orbiting solid material—possibly the stuff of which planets are made. It is intriguing to note that this result was presaged by Witteborn et al. (1982), who predicted a detectable 11-μm excess from rocky debris around young stars in the Ursa Major stream.

Gillett (1986) described the four stars that best define the "Vega phenomenon": α Lyr, β Pictoris, α Piscis Austrini, and ε Eridani. Their salient characteristic is an infrared excess starting at a wavelength between 12 and 25 μm with the spectral energy distribution of optically thin *blackbody* grains heated to 60–250 K, which implies emission from grains considerably larger than $\lambda/2\pi \sim 15$ μm (Figure 10). Careful analysis of IRAS

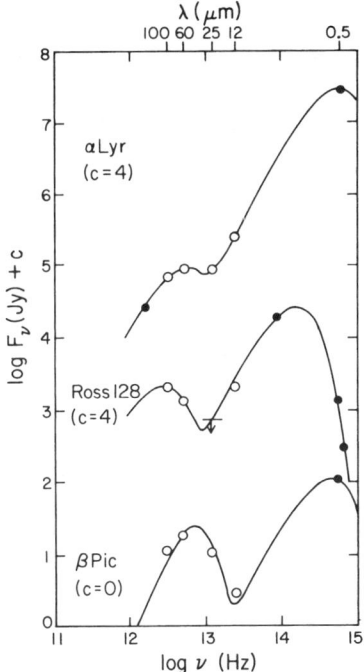

Figure 10 Spectral energy distributions of stars showing the far-infrared excesses due to orbiting solid material. For α Lyr, the solid line gives the sum of 10,000-K and 85-K blackbodies fitted to the visual (filled circles; Allen 1973) and IRAS observations (open circles; Gillett 1986); the 193-μm point is from Harper et al. (1984). For Ross 128 the solid line gives the sum of 2800-K and 46-K blackbodies fitted to the visual (Allen 1973) and IRAS observations (Backman et al. 1986). For β Pic the solid line gives the sum of 8000-K and 100-K blackbodies fitted to the visual (Hoffleit & Jaschek 1982) and IRAS observations (Gillett 1986). The constant c gives the logarithmic scale factor used in plotting the data.

scans has resolved the emitting regions for all of these stars except ε Eri; typical sizes are ~200 AU. For β Pic and α PsA, two orthogonal measurements of the size differ significantly, suggesting the presence of a disklike structure. The measured sizes and deduced temperatures are, to first order, in agreement with the idea of emission from solid material heated by a small fraction of the energy of the central star.

Two important follow-up observations have been made. First, a measurement of (or upper limit to) the 193-μm emission from α Lyr constrains the spectral energy distribution to be narrower than that of an 85-K blackbody (Harper et al. 1984) and implies a maximum grain size of ~30 μm, considerably smaller than the 1-mm size argued on theoretical grounds (Aumann et al. 1984). Second, in a striking confirmation of the IRAS result, Smith & Terrile (1984) made a coronographic observation at 0.89 μm of β Pic that revealed an edge-on disk due to starlight scattered by dust.

The physical structure of these disks is imperfectly defined by the available observations. Diner & Appleby (1986) fitted both the IRAS and coronographic observations of β Pic using realistic scattering properties for the dust and showed that the density of disk material varies with distance from the star, R, in a manner midway between the R^{-1} law appropriate for the Poynting-Robertson effect (Gillet 1986, Buitrago & Mediavilla 1985) and the R^{-3} law deduced from the visual scattering alone (Smith & Terrile 1984). The mass of the disk is perhaps the most poorly determined parameter of these systems, since it depends on the average size of the emitting material. Estimates range from a 10^{-2} $M_\oplus$ for a single grain size to 2×10^2–3×10^3 $M_\oplus$ for an asteroidal distribution of particle sizes (Aumann et al. 1984, Gillett 1986, Weissman 1984). Optical (Hobbs et al. 1985) and ultraviolet (Kondo & Bruhweiler 1985) observations of spectral lines from gaseous species (Ca II H and K, Fe II, Na D and C I) toward β Pic suggest that the mass of gas within ~80 AU of the star may be as high as 1 $M_\oplus$.

Twenty additional stars have been found that meet the various criteria for being considered Vega-like, including [12 μm]–[60 μm] ≥ 1 mag (Aumann 1985, Sadakane & Nishida 1986). The distribution of spectral types showing the effect is sharply peaked around A stars: B type (3 stars), A (16), F (6), G (1), and K (1), with no preference for or against binary systems. This distribution with spectral type may be due to selection effects, since IRAS was able to detect only the hottest, brightest disks. For example, a Vega-like disk around an F or G star instead of an A star would not have been heated enough to produce [12]–[60] ≥ 1 mag. The available data are consistent with as many as 30% of A, F, and G stars having Vega-like disks, but this result is not yet proven. Backman et al.

(1986) found a long-wavelength excess around the nearby stars Ross 128 (M5 V) and τ Ceti (G8 VI).

Surprisingly little has yet been written about the importance of these primordial disks to the question of the formation of planets. Weissman (1984), Harper et al. (1984), and O'Dell (1986) all suggested that because the small 10–30 μm grains observed by IRAS are soon lost to the system because of the effects of radiation pressure, there must be a replenishment mechanism for these grains, such as collisions between (or sublimation from) comets in Oort clouds around these stars. Over 10^8 yr, a reservoir of a few tens to hundreds of Earth masses is required for α Lyr, comparable to the mass of heavy elements in the outer planets of the solar system. But since the gestation period of the outer planets may exceed $\sim 10^9$ yr, longer than the lifetime of an A0 star, α Lyr may be a barren system (Weissman 1984, Harper et al. 1984); the main-sequence lifetime of ε Eri is, on the other hand, over 10 Gyr.

The Sun as Seen From Vega

It is interesting to ask whether the Sun would have an infrared excess as seen from a nearby star, say α Lyr. At a distance of 8.1 pc, the Sun would have an apparent photospheric 12-μm magnitude of ~ 3.0 mag (Allen 1973). If we use the variation with heliocentric distance of the density and temperature of the interplanetary dust inferred from IRAS observations of the zodiacal cloud (Good et al. 1986) and integrate the zodiacal surface brightness from 0.1 to 100 AU, then the 12-μm emission from the zodiacal cloud is only 0.3 mJy, negligible relative to the 1.8 Jy of the photosphere. However, at 100 μm the excess is $\sim 4\%$ of the expected photospheric emission, approximately 1 mJy out of 25 mJy. The existence of a more distant cloud of zodiacal particles would obviously increase this long-wavelength excess. Careful observations of nearby stars by the next generation of orbiting infrared telescopes will be very sensitive to small amounts of interplanetary material.

The Eclipsing System ε Aurigae

The peculiar system ε Aurigae was in one of its rare eclipses during the IRAS mission. The secondary object has a luminosity of 10–50 $L_\odot$ and a temperature of 475–575 K, the exact values depending on assumptions about the viewing geometry and the opacity of the material (Strickland 1985, Backman & Gillett 1985). Geometries can be envisioned by which either a fully opaque solid object or an optically thin screen of ~ 5-μm-sized particles passing in front of the star can account for the observations. Whether this star belongs in the class of mass-loss objects, Vega-like stars, or in a class of its own remains unclear.

6. LIMITS ON BROWN DWARF STARS

Great excitement attended the possibility of IRAS finding the Nemesis star (a putative havoc-wreaking companion star to the Sun) or a brown dwarf (a Jupiter-sized *etoile manquée*, heated only by gravitational contraction to ~ 2500 K). If brown dwarfs supply the missing mass inferred from dynamical studies of the solar neighborhood, then one to five of such objects were predicted to be in the Point Source Catalog (Staller & de Jong 1981, Bahcall 1984).

The search, to date, has been unsuccessful. An examination of some 5000 12-μm sources at $|b| \geq 50°$ yielded no sources that could not be identified with SAO stars, faint M giants, carbon stars, or an infrared-loud quasar (Chester et al. 1987, Beichman et al. 1986a, 1987). MacConnell (1986) classified ~ 6600 12-μm sources in the southern Galactic plane that were not found in the SAO catalog using near-infrared objective-prism plates: about 94% of the objects he examined are stars, mostly of M or carbon type. Shipman (1986) set interesting limits on the existence of brown dwarfs toward a number of nearby white dwarf stars. The lack of brown dwarfs in their sample led Chester et al. (1987) to set an upper limit to the space density of brown dwarfs of 0.4 $M_\odot$ pc^{-3}, 2.5 times the value required to supply the local missing mass (Bahcall 1984).

7. DIFFUSE GALACTIC EMISSION

Emission at High Galactic Latitudes

The extreme sensitivity of IRAS to low-surface-brightness radiation led to the discovery of thermal emission from Galactic dust over the entire sky and, surprisingly, at all wavelengths. Hauser et al. (1984a) and Low et al. (1984b) found a smooth component of the 100-μm emission that followed a csc($|b|$) law. Boulanger & Perault (1987) extended this finding by showing that above $|b| \sim 10°$, the brightness *in all four IRAS wavelengths* follows a csc($|b|$) dependence and correlates well with H I surveys. The scale height of the 100-μm emission is 120 ± 5 pc, comparable to that of H I and greater than the 60 pc characteristic of CO (Burton et al. 1985, 1986). Burton et al. (1985, 1986) and Boulanger & Perault (1987) concluded that the 12–100 μm emission out of the Galactic plane comes from dust associated with the local H I component of the interstellar medium (ISM). We discuss first the association of this emission with various tracers of the ISM and then discuss the implications of the strong 12- and 25-μm components of the background.

INFRARED CIRRUS Low et al. (1984a) coined the name "infrared cirrus" to describe the bright patches of emission seen at high Galactic latitudes.

Yet, in retrospect, cirrus should not have been a surprise. De Vaucouleurs & Freeman (1972) noted faint lanes of Galactic dust toward the LMC, and Sandage (1976) found wisps of dust near M81 and M82 reflecting Galactic starlight. Since such dust must reemit the energy it absorbs, emission from this material was to be expected. What is astonishing is IRAS's sensitivity to such small amounts of material, equivalent to $A_V \leq 0.01$ mag, over the entire sky. The cirrus represents areas of high-surface-brightness emission superposed on the smooth $\csc(|b|)$ component of Galactic emission. Structures range in size from tens of degrees down to the $\sim 2'$ resolution limit of IRAS and are often seen as long filaments. Over this range of sizes, Fourier analysis shows that the spatial amplitudes of the cirrus emission follow a $\sim k^{-1.2}$ dependence, where k is the spatial frequency (Gautier et al. 1987).

TRACERS OF THE INTERSTELLAR MEDIUM Extended emission at high Galactic latitudes is closely linked to other tracers of the ISM. Scattered Galactic light and extinction measured by star counts reveal a direct correlation between cirrus and interstellar dust (de Vries 1986, de Vries & Le Poole 1985, Weiland et al. 1986). The combination of visual observations of scattering and absorption by dust, plus infrared observations of the reradiated thermal emission, gives a powerful tool for investigating the interstellar radiation field (ISRF) and grain properties (de Vries 1986).

There is an excellent correlation between the 100-μm emission and H I emission (Boulanger et al. 1985, Terebey & Fich 1986, Fong et al. 1986, Boulanger & Perault 1987, McGee et al. 1986), although above $A_V \sim 1$ mag, *and thus for brighter cirrus clouds*, the correlation between cirrus and H I breaks down as the hydrogen becomes molecular in form. The relation between cirrus and hydrogen can be carried over to higher column densities, however, if one traces the molecular hydrogen component using the CO molecule. Weiland et al. (1986) found excellent correlations between visual extinction, CO, and 100-μm surface brightness and concluded that in clouds with $A_V \leq 5$ mag, differing gas content (H_2 or H I) makes little difference to either the temperature or emissivity of the cirrus. The association of the infrared emission with CO and interstellar absorption lines yields estimates of 65–150 pc for the distances to some of the closest cirrus clouds (Hobbs et al. 1986, Weiland et al. 1986, de Vries 1986).

Within a given cloud there are tight relations between the 100-μm surface brightness, I_ν, and other tracers of the ISM. Averaging the various determinations of the 100-μm emissivity described above yields 18 ± 3 MJy sr^{-1}/(mag of A_V), corresponding to the $I_\nu(100\ \mu\text{m})/N(\text{H I}) = 1.0 \pm 0.2 \times 10^{-20}$ MJy sr^{-1} (H I cm^{-2})$^{-1}$ (Bohlin et al. 1978). The average value of $I_\nu(100\ \mu\text{m})/I(\text{CO})$ from 10 high-latitude CO clouds is 1.1 MJy sr^{-1} (K·km s^{-1})$^{-1}$. It is important to note, however, that even without considering clouds where optical depth effects or embedded sources are important, there are factor-of-four variations in $I_\nu(100\ \mu\text{m})/N(\text{H I})$ from cloud to cloud around the Galaxy. These differences could be due to variations in the intensity of the ISRF, in the properties of grains, or in the gas-to-dust ratio.

There are a few regions of interstellar gas with enough column density to be detectable according to the above relations, yet which were not detected at 100 μm. For example, no infrared emission was detected toward high-velocity clouds (Wakker & Boulanger 1987) or toward the Magellanic Stream (Fong et al. 1986). Whether this is due to a lack of dust in primeval material or to insufficient heating by a weak radiation field is at present unknown. The existence of diffuse infrared emission without associated H I has been suggested by Heiles et al. (1987); a likely explanation for such a region is a high-latitude CO cloud of the type studied by Weiland et al. (1986).

Finally, it should be mentioned that there is a completely different interpretation for the nature of the high-latitude emission detected by IRAS. Harwit et al. (1986) suggested that cirrus at all wavelengths could be due to fine-structure lines of such species as [O I], [O III], [N III], [Ne II], [S IV], and [S III]. The close associations between cirrus, dust, and neutral H I probably rule out this explanation (Terebey & Fich 1986).

EXTINCTION AT THE POLES The association of a uniform background due to interstellar dust has important cosmological implications because of the debate about visual extinction toward the Galactic poles (Hauser et al. 1984a). The first examination of IRAS data suggests that there is considerable dust at high latitudes. If the cosecant law described above continues all the way to $|b| = 90°$, then a fit to the 100-μm data implies that $A_V \sim 0.09$–0.13 mag at the poles (Boulanger et al. 1985, Boulanger & Perault 1987). This value is equal to the obscuration claimed by de Vaucouleurs & Buta (1983) and is inconsistent with the $A_B = 0$ claimed by Sandage (1973). A separate comparison of visual reddening and the IRAS data near the North Galactic Pole suggests approximately $A_V \sim 0.15$ mag is present (Knude 1986). On the other hand, Fong et al. (1986) used the 100-μm emission around the South Galactic Pole to set a limit of $A_V \sim 0.01$ mag on any patchy visual extinction toward the pole, but their technique subtracted away smooth components larger than $\sim 1°$. Considerably more work remains to be done in this important area, including a careful examination of possible instrumental offsets in the IRAS total intensity data.

SMALL GRAINS IN THE INTERSTELLAR MEDIUM Figure 11 shows the energy distribution of a number of regions of high-Galactic-latitude infrared

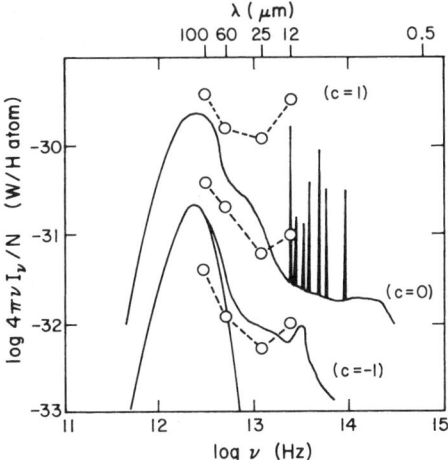

Figure 11 The dust emission per hydrogen atom is shown for three different high-latitude regions: The top set of data (open circles) is from Leene & Beichman (1985), the middle set from Weiland et al. (1986), and the bottom set from Boulanger & Perault (1987). The upper solid line represents the energy distribution predicted by Puget et al. (1985) for a mixture of large grains and PAH (polycyclic aromatic hydrocarbon) molecules. The lower solid lines show the theory of Draine & Anderson (1985) for emission from assemblages of grains with differing amounts of very small grains. The constant c gives the logarithmic scale factor used in plotting the data and theoretical curves.

emission. The ratio of 60/100-μm flux density is generally quite uniform and implies temperatures between 20 and 30 K, depending on the assumed dust emissivity law. However, emission at 12 and 25 μm shows greater variation and is sometimes completely absent. Temperatures derived from the 12/25-μm ratio range from 80 to 500 K (Gautier 1986).

Thermal emission from grains heated by the ISRF can account for the properties of the diffuse high-Galactic-latitude emission, but not without some important modifications to existing models and surprising conclusions (Draine & Anderson 1985, Draine & Lee 1984, Rowan-Robinson 1986). The conundrum posed by the observations is that the observed grain temperatures are significantly hotter than can be accounted for even if one assumes that warm graphite grains rather than cool silicates are observed. At 60 and 100 μm, the problem is not too severe; the typical 60/100-μm temperatures of 20–30 K, derived assuming an emissivity proportional to $v^{1.5}$ (Low et al. 1984a, Gautier 1986), can be achieved by immersing the classical large grains in an ISRF only twice as bright as that found in the solar neighborhood (Rowan-Robinson 1986). But the problem of inadequate heating by the ISRF becomes catastrophic to equilibrium models at 12 and 25 μm, where temperatures ≥ 500 K are observed many parsecs away from obvious sources of adequate heating (Beichman et al. 1984a, Leene & Beichman 1985, Gautier & Beichman 1984, Boulanger et al. 1985, Leene 1985). An excellent example of this anomalous 12-μm emission is seen toward the Pleiades (Figure 12; Castelaz et al. 1986).

The solution to the puzzle may lie in a theory that had been finding currency before IRAS as a way to explain two other puzzling features of the ISM—the anomalous emission from reflection nebulae and the unidentified infrared bands between 3 and 11 μm. Sellgren (1984) first applied the theory that the absorption of a single UV photon by a very small grain (radius $\ll 100$ Å) could transiently heat the grain to hundreds of degrees above its equilibrium temperature (Greenberg 1968, Duley 1973, Purcell 1976) to the problem of anomalously strong 2-μm emission from reflection nebulae (Sellgren et al. 1983). Beichman et al. (1984a) and Leene & Beichman (1985) invoked this theory to account for the diffuse 12- and 25-μm emission seen toward Lynds 255 and R Coronae Australis. Draine & Anderson (1985) were also able to account for the 12–100 μm energy distribution of the high-latitude cirrus using this mechanism (Figure 11).

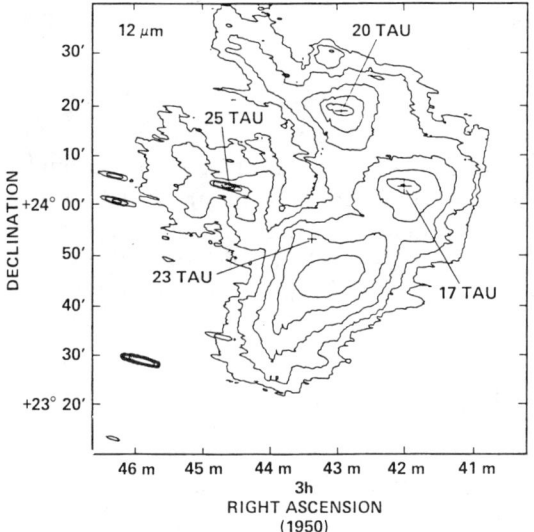

Figure 12 Bright 12-μm emission is seen over an area of more than 1° around the Pleiades reflection nebula (Castelaz et al. 1986). The extent of the emission is far greater than if only classical, large grains were being heated by the radiation field near these stars. Small grains, transiently heated to several hundred kelvin by the absorption of single optical/ultraviolet photons, are responsible for much of the emission.

Puget et al. (1985) and Boulanger et al. (1985) identified the small grains with the macromolecules called polycyclic aromatic hydrocarbons (PAHs), which Léger & Puget (1984) had invoked to account for (a) the unidentified spectral features seen at various wavelengths between 3.3 and 11.3 μm (see also Duley & Williams 1981, Allamandola et al. 1985), (b) the anomalous 2-μm emission from Sellgren's reflection nebulae, and (c) diffuse 12- and 25-μm emission seen around some H II regions by earlier rocket flights (Price 1981). Predictions of the energy distribution of emission from PAHs heated by the ISRF agree qualitatively with observations of the cirrus, although there is a problem in that a true PAH has no continuum emission, only lines (Puget et al. 1985, Désert et al. 1986a,b, Désert 1986). The uncertain optical properties of PAHs result in estimates for the fraction of the cosmic abundance of carbon locked up in PAHs or in small grains that vary from 1 to 15% (Léger & Puget 1984, Puget et al. 1985, Boulanger et al 1985, Draine & Anderson 1985, Boulanger & Perault 1987).

The existence of small and large grains within the ISM results in an anticorrelation between the 12/25-μm and 60/100-μm colors of interstellar emission (Désert 1986). Figure 13 shows that when the ambient radiation

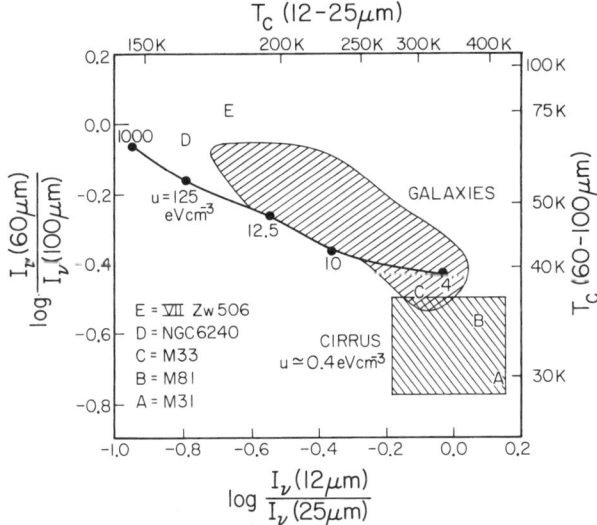

Figure 13 A color-color plot for diffuse emission from dust in the interstellar medium. The solid line gives the colors at different distances from ζ Per in the California Nebula (Boulanger & Perault 1987). The energy density of the radiation field at each position is marked. The area marked "cirrus" represents the locus of colors found for high-latitude emission, as described in the text. The area marked "galaxies" gives colors for non-Seyfert galaxies with differing amounts of star-formation activity corresponding to radiation fields of differing intensity (Helou 1986). A number of individual galaxies are also marked.

field has a low energy density, the 12/25-μm ratio is high while the 60/100-μm color is low. Conversely, at higher energy densities, the 12/25-μm color is cool while the 60/100-μm color is warm. This result can be understood by noting that when the large grains are cool (low 60/100-μm color temperature), only the hot, transiently heated grains emit at 12 and 25 μm, resulting in a hot 12/25-μm color temperature. On the other hand, when the large grains are heated by a strong radiation field (e.g. near an ionizing star), their contribution swamps that of the transiently heated grains, resulting in the cooler 12/25-μm color temperature expected from equilibrium heating. This qualitative model explains IRAS colors observed at various distances from ζ Per in the California Nebula (Boulanger et al. 1985), as well as within galaxies of various luminosities (Helou 1986).

The IRAS data contain many clues to the small-grain physics and chemistry. IRAS maps show that small grains appear in some environments, such as the Pleiades (Castelaz et al. 1986, Cox & Leene 1987; Figure 12), one of the Chamaeleon clouds (Chlewicki et al. 1987), and around embedded sources in Ophiuchus (Young et al. 1986), but not toward other clouds (de Vries 1985, 1986). H II regions such as the one around τ Sco apparently destroy small grains (Puget et al. 1987), yet (as described in Section 4) collisionally excited small grains apparently survive in supernova remnants (Dwek et al. 1986). Whether these effects are due to differences in the abundances of small grains or to excitation effects, or both, is as yet unknown.

LRS spectra have traced the 7.7-, 8.6-, and 11.7-μm features associated with PAHs in many objects and show an emission plateau longward of 11.3 μm that is characteristic of CH bending in PAHs (de Muizon et al. 1986, Cohen et al. 1986b). The lack of association between 12-μm cirrus and stars with the 2175-Å graphite feature is puzzling, but it may be accounted for by PAHs being excited through a strong absorption band at 3000 Å rather than in the ultraviolet (Cox & Leene 1987, Boulanger & Perault 1987).

Emission in the Disk of the Galaxy

Within a few degrees of the Galactic plane, the column density of dust and the energy available for its heating increase drastically compared with higher latitudes. A number of groups have modeled the energy distribution of the Galaxy as the sum of different components. While the quantitative contributions of the various components as derived by different authors do not agree in detail, the order-of-magnitude estimates for the importance of different phases of the interstellar medium to the overall energy budget of the Galaxy are instructive (Figure 14; Cox et al. 1986a,b).

HOT COMPONENT   As shown in Figure 14 there is a broad shoulder of

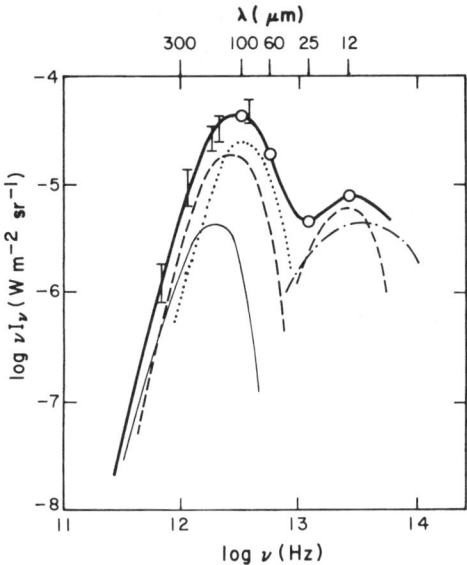

Figure 14 The infrared surface brightness of the Galaxy ($8 \geq R \geq 2$ kpc) can be decomposed into a number of components (Cox et al. 1986a,b, Cox & Mezger 1987): cold material in molecular clouds away from regions of star formation (———), cool material associated with atomic gas (— —), warm material associated with forming OB stars (……..), emission from transiently heated small grains (– – –), emission from OH/IR stars (–·–·–). IRAS data are shown as open circles; other data are from Hauser et al. (1984b), Caux & Serra (1987), and Pajot et al. (1986) as described in Cox & Mezger (1987).

emission around 12 μm in the energy distribution of the Galaxy. After extrapolating the properties of the high-latitude emission to lower latitudes and examining the diffuse mid-infrared emission seen around H II regions (Price 1981, Gautier et al. 1984a), Puget et al. (1985) and Boulanger et al. (1985) inferred that small grains play an important role in the energetics of the entire disk. More than two thirds of the hot dust component in the Galaxy probably comes from PAHs, with the remainder coming from OH/IR stars (Cox & Mezger 1987, Rowan-Robinson & Chester 1987). Thus, emission from small grains in the mid-infrared accounts for 10–25% of the total power radiated by the interstellar medium (Boulanger et al. 1985, 1986a, Boulanger & Perault 1987, Cox & Mezger 1987).

WARM COMPONENT   OB stars in H II regions or embedded in dense molecular clouds heat large dust grains to 30–40 K and account for ~40% of the entire infrared output of the Galaxy (Cox & Mezger 1987). The luminosity-to-mass ratio of this component depends strongly on the proximity of OB stars and ranges from 7 to 30 $L_\odot/M_\odot$ toward hot spots centered on individual H II regions like W51 and M17 (Scoville & Good 1987). The IRE for this component of the interstellar medium varies irregularly with Galactic longitude from 4 to 30 and has an average value of 6 (Sodroski et al. 1987), in close agreement with previous estimates by Myers et al. (1986).

COLD COMPONENT Colder material in H I regions (15–25 K) and in molecular clouds (~ 15 K) away from embedded OB stars is heated predominantly by stars of spectral type B, A, and F and by old disk stars. Dust associated with cold atomic material accounts for about half of the overall infrared luminosity of the Galaxy. The presence of this material is only hinted at by the IRAS 100-μm data, but it dominates the submillimeter band (Drapatz 1979, Cox et al. 1986a,b, Boulanger & Perault 1987, Caux & Serra 1987). Dust in H I regions has an emissivity of 1.4–7 $L_\odot/M_\odot$ (Cox & Mezger 1987, Boulanger & Perault 1987), while dense regions of molecular clouds far from embedded stars have emissivities of 0.5–2.3 $L_\odot/M_\odot$ (Sodroski et al. 1987, Cox & Mezger 1987).

The Galaxy as a Galaxy

ENERGETICS OF THE GALAXY The 60- and 100-μm emission from the Galactic disk is concentrated within 1 kpc of the Galactic center and within the molecular ring ($R = 3$–7 kpc) (Burton et al. 1986, Scoville & Good 1987, Solomon et al. 1987). Cox & Mezger (1987) estimate that the bolometric luminosity of stars in the Galaxy is $\sim 3.6 \times 10^{10} L_\odot$, with one third of that amount, $1.2 \times 10^{10} L_\odot$, emitted between 4 and 100 μm (the distance to the Galactic center has been taken as 8.5 kpc). Scoville & Good (1987) estimate that emission between 1 and 500 μm in the molecular ring accounts for $6 \times 10^9 L_\odot$.

For a total mass of $4.6 \times 10^9 M_\odot$ for the interstellar medium (Scoville & Sanders 1986), the luminosity-to-mass ratio of the Galaxy is $2.6 L_\odot/M_\odot$. This ratio can be as high as 10–20 toward hotspots in selected molecular clouds (Solomon et al. 1987, Scoville & Good 1987). Estimates of the amount of mass going into stars in the Galaxy fall in the range 2.5–7.5 $M_\odot$ yr^{-1}, of which 0.3–5 $M_\odot$ yr^{-1} is irretrievably locked up in low-mass stars. The exact value depends on how many high-mass stars are forming according to the general initial-mass function and how many are forming due to induced processes that preferentially form high-mass stars (Cox & Mezger 1987).

STAR FORMATION AND THE GALAXY Using the infrared as a direct gauge for high-mass star-formation activity is made difficult in our own and many other galaxies because the ISM is relatively transparent on the scale of 100–1000 pc. This means that not all interstellar dust is heated by OB stars, and furthermore that not all of the luminosity emitted by OB stars is converted into infrared luminosity. On the scale of ~ 10 pc within an OB cluster, the correlation between infrared emission and star formation is excellent. For example, 90% of the $2.5 \times 10^5 L_\odot$ emitted from the 9×9 pc ($1° \times 1°$) region centered on OMC 1 is powered by the Trapezium stars (Thronson et al. 1986), but on the larger, 100–200 pc scale the correlation breaks down. For example, only 60% of the total $6.5 \times 10^5 L_\odot$ emitted from a $20° \times 18°$ region centered on OMC 1 is powered by OB stars in Orion A and B; much of the remaining emission in Orion and most of the emission from fainter clouds such as Mon R2 originate in dust heated by the ISRF, not from embedded stars (Boulanger et al. 1986b).

Helou (1986) warns that cirrus (i.e. infrared emission from dust heated by the ambient radiation field within a galaxy) can contaminate estimates of star-formation rates from IRAS observations. But this caveat probably does not apply to a very luminous galaxy with a dense ISM that captures the entire energy output of its massive star clusters. In these galaxies, the formation of OB stars is probably a fundamental energy source. Since the IRAS colors and luminosity-to-mass ratios of giant molecular clouds are similar to those of luminous IRAS galaxies (Sanders et al. 1986), high-mass star formation may be, as Scoville & Good (1987) suggest, a "viable (but certainly not unique) explanation for the high infrared luminosity of these galaxies."

8. CONCLUSIONS

Already the scope of IRAS results, limited to the solar system and the Galaxy, has strained the length of a single article. In years to come, specific astrophysical phenomena will be reviewed in these pages, and IRAS data will be an integral part of their study. In the broadest sense, the IRAS legacy is like that of any of the great sky surveys: Herschel's, Palomar's, the Cambridge Radio surveys, or Uhuru's. The IRAS catalogs and images have revealed the sky in a new light and will be used by future astronomers to learn about objects whose nature has only been hinted at in this review.

ACKNOWLEDGMENTS

I am very grateful to Rosanne Hernandez, Helen Knudsen, and the staff of the IPAC library, whose diligent work was indispensable to the completeness of this review. Thanks are also due to the many authors who sent preprints for inclusion in this review. Drs. G. Neugebauer, T. Soifer, F. Boulanger, and N. Scoville provided critical readings of the manuscript. Drs. F. Boulanger and M. Perault graciously allowed me to quote from their paper in advance of its publication. L. Fullmer, R. Narron, and M. Mehrdad helped prepare a number of figures. Mary Ellen Barba assisted with the preparation of the manuscript. I would also like to thank Drs. P. Encrenaz, J.-L. Puget, and E. Falgarone for the hospitality of the École Normale Supérieure, where much of the reading for this article was carried out; Drs. H. Habing and F. Israel provided both useful discussions and *Genever* during a most pleasant stay in Leiden.

Finally, I can only hope that the breadth of exciting results described in this volume is a source of pride and satisfaction to the hundreds of engineers, technicians, administrators, and secretaries—on both sides of the Atlantic—whose years of hard work and sacrifice made IRAS possible.

IRAS was developed and operated by the US National Aeronautics and Space Administration (NASA), the Netherlands Agency for Aerospace Programs (NIVR), and the UK Science and Engineering Research Council (SERC). The writing of this review was funded in part by NASA's IRAS Extended Mission Program through the Jet Propulsion Laboratory, California Institute of Technology.

Literature Cited

Adams, F. C., Shu, F. H. 1986. *Ap. J.* 308: 836–53

Adams, F. C., Lada, C. J., Shu, F. H. 1986. Preprint

Albinson, J. S., Callus, C. M., Evans, A. 1986. See Israel 1986, pp. 151–52

Allamandola, L. J., Tielens, A. G. G. M., Barker, J. R. 1985. *Ap. J. Lett.* 290: L25–28

Allen, C. W. 1973. *Astrophysical Quantities.* London: Athlone

Aumann, H. H. 1985. *Publ. Astron. Soc. Pac.* 97: 885–91

Aumann, H. H., Gillett, F. C., Beichman, C. A., de Jong, T., Houck, J. R., et al. 1984. *Ap. J. Lett.* 278: L23–27

Backman, D. E., Gillett, F. C. 1985. *Ap. J. Lett.* 299: L99–102

Backman, D. E., Gillett, F. C., Low, F. J. 1986. *Proc. COSPAR Meet., 26th.* In press

Bahcall, J. N. 1984. *Ap. J.* 276: 169–81

Barlow, M. J., Cohen, M. 1977. *Ap. J.* 213: 737–55

Bedijn, P. J. 1986. See Israel 1986, pp. 119–26

Beichman, C., et al. 1987. In preparation

Beichman, C. A., Soifer, B. T., Helou, G., Chester, T. J., Neugebauer, G., et al. 1986a. *Ap. J. Lett.* 308: L1–5

Beichman, C. A., et al. 1984a. *Bull. Am. Astron. Soc.* 16: 522

Beichman, C. A., Myers, P. C., Emerson, J. P., Harris, S., Mathieu, R., et al. 1986b. *Ap. J.* 307: 337–49

Beichman, C. A., et al. 1984b. *Ap. J. Lett.* 278: L45–48

Benson, P. J. 1983. PhD thesis. Mass. Inst. Technol., Cambridge

Benson, P. J., Myers, P. L., Wright, E. L. 1984. *Ap. J. Lett.* 279: L27–30

Bohlin, R. C., Savage, B. D., Drake, J. F. 1978. *Ap. J.* 224: 132–42

Boulanger, F., Baud, B., van Albada, G. D. 1985. *Astron. Astrophys.* 144: L9–12

Boulanger, F., Perault, M. 1987. In preparation

Boulanger, F., Perault, M., Puget, J.-L. 1986a. See Israel 1986, pp. 203–7

Boulanger, F., Maddalena, R. J., Thaddeus, P. 1986b. See Israel 1986, pp. 293–94

Braun, R. 1985. PhD thesis. Leiden Univ., Neth.

Buitrago, J., Mediavilla, E. 1985. *Astron. Astrophys.* 148: L8–10

Burns, J. A., Dermott, S. F., Nicholson, P. O., Houck, J. R. 1984. In *Properties and Interactions of Interplanetary Dust, IAU Colloq. No. 85*, pp. 395–409

Burton, W. B., Deul, E. R., Walker, H. J. 1985. *Proc. ESO-IRAM-Onsala Workshop on Submillimeter Astron.*, ed. P. A. Shaver, K. Kjär, pp. 229–44. Garching: ESO

Burton, W. B., Deul, E. R., Walker, H. J., Jongeneelen, A. A. W. 1986. See Israel 1986, pp. 357–72
Callus, C. M., Albinson, J. S., Evans, A., Bode, M. F. 1986. See Israel 1986, pp. 149–50
Callus, C. M., et al. 1987. Preprint
Cassen, P., Shu, F. H., Terebey, S. 1985. In *Protostars and Planets II*, ed. D. Black, M. S. Matthews, pp. 448–83. Tucson: Univ. Ariz. Press
Cassinelli, J. P., Hartmann, L. 1977. *Ap. J.* 212: 488–93
Castelaz, M. W., Sellgren, K., Werner, M. W. 1986. *Ap. J.* In press
Caux, E., Serra, G. 1987. See Lonsdale 1987, pp. 93–96
Chester, T. J. 1986. See Israel 1986, pp. 3–22
Chester, T. J., et al. 1987. In preparation
Chini, R., Mezger, P. G., Kreysa, E., Gemund, H.-P. 1984. *Astron. Astrophys.* 135: L14–17
Chini, R., Mezger, P. G., Kreysa, E., Gemund, H.-P. 1985. *Astron. Astrophys.* 143: 244
Chini, R., Kreysa, E., Mezger, P. G., Gemund, H.-P. 1986a. *Astron. Astrophys.* 154: L8–11
Chini, R., Kreysa, E., Mezger, P. G., Gemund, H.-P. 1986b. *Astron. Astrophys.* 157: L1–2
Chini, R., Kreysa, E. Krügel, E., Mezger, P. G., Gemund, H.-P. 1986c. See Israel 1986, pp. 29–30
Chlewicki, G., Laureijs, R. J., Clark, F. O., Wesselius, P. R. 1987. See Lonsdale 1987, pp. 113–16
Clark, F. O. 1987. Submitted for publication
Clark, F. O., Laureijs, R. J. 1985. *Astron. Astrophys.* 154: L26–29
Clark, F. O., Laureijs, R. J., Chlewicki, G., Zhang, C. Y., van Oosterom, W., Koster, D. 1987a. See Lonsdale 1987, pp. 83–85
Clark, F. O., Laureijs, R. J., Zhang, C. Y. 1987b. Preprint
Cohen, M., Dopita, M. A., Schwartz, R. 1986a. *Ap. J. Lett.* 302: L55–58
Cohen, M., Emerson, J. P., Beichman, C. A. 1987. In preparation
Cohen, M., Kuhi, L. V. 1979. *Ap. J. Suppl.* 41: 743–843
Cohen, M., Tielens, A. G. G. M., Allamandola, L. J. 1986b. Preprint
Coté, J., Waters, L. B. F. M. 1986. See Israel 1986, pp. 77–78
Cox, P., Leene, A. 1987. Preprint
Cox, P., Krügel, E., Mezger, P. G. 1986a. See Israel 1986, pp. 201–2
Cox, P., Krügel, E., Mezger, P. G. 1986b. *Astron. Astrophys.* 155: 380–96
Cox, P., Mezger, P. G. 1987. See Israel 1986, pp. 23–25
Crawford, J., Rowan-Robinson, M. 1986.

MNRAS 221: 923–29
Davies, J. K., Green, S. F., Stewart, B. C., Meadows, A. J., Aumann, H. H. 1984. *Nature* 309: 315–19
Davies, J. K. 1986. *MNRAS* 221: 19P–23P
Deasy, H., Butler, C. J. 1986. *Nature* 320: 726–28
Dermott, S. F., Nicholson, P. D., Burns, J. A., Houck, J. R. 1984. *Nature* 312: 505–9
Dermott, S. F., Nicholson, P. D., Burns, J. A., Houck, J. R. 1985a. In *Properties and Interactions of Interplanetary Dust*, ed. R. Giese, P. L. Lamy, pp. 395–409. Dordrecht: Reidel
Dermott, S. F., Nicholson, P. D., Wolven, B. 1985b. *Asteroids, Comets, Meteors II*, ed. C-I. Lagerkvist, H. Rickman. Uppsala: Uppsala Astron. Obs.
Désert, F. X. 1986. See Israel 1986, pp. 213–16
Désert, F. X., Boulanger, F., Léger, A., Puget, J. L., Sellgren, K. 1986a. *Astron. Astrophys.* 159: 328–30
Désert, F. X., Boulanger, F., Shore, S. N. 1986b. *Astron. Astrophys.* 160: 295–300
de Muizon, M., Geballe, T. R., d'Hendecourt, L. B., Baas, F. 1986. *Ap. J. Lett.* 306: L105–8
de Vaucouleurs, G., Buta, R. 1983. *Astron. J.* 88: 939–61
de Vaucouleurs, G., Freeman, K. C. 1972. *Vistas Astron.* 14: 163–294
de Vries, C. P. 1986. PhD dissertation. Leiden Univ., Neth.
de Vries, C. P. 1985. *Astron. Astrophys.* 150: L15–17
de Vries, C. P., Le Poole, R. S. 1985. *Astron. Astrophys.* 145: L7–9
Diner, D. J., Appleby, J. F. 1986. *Nature* 322: 436–38
Dinerstein, H. L. 1986. *Astron. J.* 92: 1381–86
Dinerstein, H. L., Robinson, E. L. 1986. See Israel 1986, pp. 145–48
Draine, B. T. 1981. *Ap. J.* 245: 880–90
Draine, B. T., Anderson, N. 1985. *Ap. J.* 292: 494–99
Draine, B. T., Lee, H. M. 1984. *Ap. J.* 285: 89–108
Drappatz, S. 1979. *Astron. Astrophys.* 75: 26–33
Duley, W. W. 1973. *Astrophys. Space Sci.* 23: 43
Duley, W. W., Williams, D. A. 1981. *MNRAS* 196: 269–74
Dumont, R., Levasseur-Regourd, A. C. 1986. See Israel 1986, pp. 45–46
Dwek, E. 1981. *Ap. J.* 247: 614–27
Dwek, E. 1986. *Ap. J.* 302: 363–70
Dwek, E., Petre, R., Szymkowiak, A., Rice, W. 1986. *Ap. J. Lett.* In press
Dwek, E., Dinerstein, H. L., Gillett, F. C.,

Hauser, M. G., Rice, W. L. 1987. *Ap. J.* In press
Eaton, N., Davies, J. K., Green, S. F. 1984. *MNRAS* 211: 15P—19P
Edwards, S., Strom, S. E., Snell, R. L., Jarrett, T. H., Beichman, C. A., Strom, K. M. 1986. *Ap. J. Lett.* 307: L65–68
Elias, J., Frogel, J., Schwering, P. 1986. *Ap. J.* 302: 675–79
Elmegreen, B. 1986. See Israel 1986, 265–76
Emerson, J. P. 1985. In *Star-Forming Regions, IAU Symp. No. 115*, ed. M. Peimbert, J. Jugaku. Dordrecht: Reidel. In press
Emerson, J. P., Harris, S., Jennings, R. E., Beichman, C. A., Baud, B., et al. 1984. *Ap. J. Lett.* 278: L49–52
Evans, A. 1985. *The Observatory* 105: 6–7
Evans, N. J. II, Beckwith, S. 1977. *Ap. J.* 217: 729–40
Evans, N. J. II, Levreault, R. M., Harvey, P. M. 1986. *Ap. J.* 301: 894–900
Explanatory Supplement to the IRAS Catalogs and Atlases. 1984, 1987. Ed. C. A. Beichman, G. Neugebauer, H. J. Habing, P. E. Clegg, T. J. Chester. Washington, DC: US Govt. Print. Off.[2]
Feast, M. W. 1986. See Israel 1986, pp. 339–48
Fich, M., Terebey, S. 1987. See Lonsdale 1987, pp. 63–66
Fong, R., Jones, L. R., Shanks, T., Stevenson, P. R. F., Strong, A. W. et al. 1986. *MNRAS*. In press
Frogel, J. A. 1986. See Israel 1986, pp. 349–56
Gautier, T. N. 1986. See Israel 1986, pp. 49–54
Gautier, T. N., Beichman, C. A. 1984. *Bull. Am. Astron. Soc.* 16: 968
Gautier, T. N., Boulanger, F., Beichman, C. A. 1987. In preparation
Gautier, T. N., Hauser, M. G., Beichman, C. A., Low, F. J., Neugebauer, G., et al. 1984a. *Ap. J. Lett.* 278: L57–58
Gautier, T. N., Hauser, M. G., Low, F. J. 1984b. *Bull. Am. Astron. Soc.* 16: 442
Gear, W. K., Gee, G., Robson, E. I., Ade, P. A. R., Duncan, W. D. 1986. *MNRAS* 219: 835–40
Gillett, F. C. 1986. See Israel 1986, pp. 61–69
Gillett, F. C., Backman, D. E., Beichman, C., Neugebauer, G. 1986a. *Ap. J.* 310: 842–52
Gillett, F. C., Neugebauer, G., Emerson, J. P., Rice, W. 1986b. *Ap. J.* 300: 722–28
Glass, I. S. 1986. *MNRAS* 221: 879–85

[2] The *Supplement* was first published as a Jet Propulsion Laboratory report (D-1855) in 1984. It will be published by the US Government Printing Office in 1987.

Goldsmith, P. F., Langer, W. D., Wilson, R. W. 1986. *Ap. J.* 303: L11–15
Good, J. C., Hauser, M. G., Gautier, T. N. 1986. *Proc. COSPAR Meet., 26th*. In press
Graham, J. A. 1986. *Ap. J.* 302: 352–62
Graham, J. R., Evans, A., Albinson, J. S., Bode, M. F., Meikle, W. P. S. 1987. Preprint
Grasdalen, G. L. 1979. *Ap. J.* 229: 587–92
Grasdalen, G. L., Strom, S. E., Strom, K. M., Capps, R. W., Thompson, D., Castelatz, M. 1984. *Ap. J.* 283: L57–61
Green, S. F., Davies, J. K., Eaton, N., Stewart, B. C., Meadows, A. J. 1985a. *Icarus* 64: 517–27
Green, S. F., Meadows, A. J., Davies, J. K. 1985b. *MNRAS* 214: 29P–36P
Greenberg, J. M. 1968. In *Stars and Stellar Systems*, Vol. 7, ed. B. M. Middlehurst, L. H. Aller, p. 221. Chicago: Univ. Chicago Press
Gustafson, B. A. S. 1986. *Icarus*. In press
Habing, H. J. 1986. See Israel 1986, pp. 329–38
Habing, H. J. 1987a. *Proc. NATO Adv. Study Inst. on the Galaxy, Cambridge, Engl.* In press
Habing, H. J. 1987b. In *The Late Stages of Stellar Evolution*, ed. S. Kwok, S. R. Pottasch. Dordrecht: Reidel. In press
Habing, H. J., Olnon, F. M., Chester, T. J., Gillett, F. C., Rowan-Robinson, M., Neugebauer, G. 1985. *Astron. Astrophys.* 152: L1–4
Habing, H. J., van der Veen, W., Geballe, T. 1987. In *The Late Stages of Stellar Evolution*, ed. S. Kwok, S. R. Pottasch. Dordrecht: Reidel. In press
Hacking, P. L., et al. 1985. *Publ. Astron. Soc. Pac.* 97: 616–33
Harper, D. A., Loewenstein, R. F., Davidson, J. A. 1984. *Ap. J.* 285: 808–12
Harris, S. 1985. *Proc. ESO-IRAM-Onsala Workshop on Submillimeter Astron.*, ed. P. A. Shaver, K. Kjär. Garching: ESO
Harvey, P. M., Thronson, H. A. Jr., Gatley, I. 1979. *Ap. J.* 231: 115–23
Harwit, M., Houck, J. R., Stacey, G. J. 1986. *Nature* 319: 646–47
Hauser, M. G., Gillett, F. C., Low, F. J., Gautier, T. H., Beichman, C. A., et al. 1984a. *Ap. J. Lett.* 278: L15–18
Hauser, M. G., Silverberg, R. F., Stier, M. T., Kelsall, T., Gezari, D. Y., et al. 1984b. *Ap. J.* 285: 74–88
Hauser, M. G., Gautier, T. N. 1984. *Bull. Am. Astron. Soc.* 16: 495
Hauser, M. G., Houck, J. R. 1986. See Israel 1986, pp. 39–44
Hauser, M. G., Gautier, T. N., Good, J. C., Low, F. J. 1985. In *Properties and Interactions of Interplanetary Dust*, ed. R.

Giese, P. L. Lamy, pp. 43–48. Dordrecht: Reidel
Heiles, C., McCarthy, P. J., Reach, W., Strauss, M. A. 1987. See Lonsdale 1987, pp. 553–58
Helou, G. 1986. *Ap. J. Lett.* 311: L33–36
Herman, J., Burger, J. H., Penninx, W. H. 1986. *Astron, Astrophys.* 167: 247–59
Hilton, J., White, G. J., Cronin, N. J., Rainey, R. 1986. *Astron. Astrophys.* 154: 274–78
Hobbs, L. M., Vidal-Madjar, A., Ferlet, R., Albert, C. E., Gry, C. 1985. *Ap. J. Lett.* 293: L29–33
Hobbs, L. M., Blitz, L., Magnani, L. 1986. *Ap. J. Lett.* 306: L109–13
Hoffleit, D., Jaschek, C. 1982. *The Bright Star Catalog.* New Haven, Conn: Yale Univ. Press
Houck, J. R., Soifer, B. T., Neugebauer, G., Beichman, C. A., Aumann, H. H., et al. 1984. *Ap. J. Lett.* 278: L63–66
Houck, J. R., Schneider, D. P., Danielson, G. E., Beichman, C. A., Lonsdale, C. J., et al. 1985. *Ap. J. Lett.* 290: L5–8
Huang, Y.-L., Dickman, R. L., Snell, R. L. 1986. *Ap. J. Lett.* 302: L63–66
IRAS Asteroid and Comet Survey. 1986. Ed. D. L. Matson. Pasadena, Calif: Jet Propuls. Lab.
IRAS Catalogues and Atlases: Atlas of Low Resolution Spectra 1986. Ed. F. M. Olnon, E. Raimond. *Astron. Astrophys. Suppl.* 65: 607–1065
Israel, F. P., ed. 1986. *Light on Dark Matter.* Dordrecht: Reidel
Iyengar, K. V. K. 1986. *Astron. Astrophys.* 158: 89–96
Jaschek, M., Jaschek, C., Egret, D. 1986. *Astron. Astrophys.* 158: 325–28
Jones, T. W., Ney, E. P., Stein, W. A. 1981. *Ap. J.* 250: 324–26
Jura, M. A. 1986a. *Ap. J.* 303: 327–32
Jura, M. A. 1986b. *Ap. J.* 306: 483–89
Jura, M. A. 1986c. *Ap. J.* In press
Kenyon, S. J., Fernandez-Castro, T., Stencel, R. E. 1986. *Astron. J.* 92: 1118–24
Knapp, G. R., Wilcots, E. M. 1987. *Proc. Workshop on the Late Stages of Stellar Evolution*, ed. S. Kwok, S. R. Pottasch. Dordrecht: Reidel
Knapp, G. R., Morris, M. 1985. *Ap. J.* 292: 640–69
Knude, J. 1986. See Israel 1986, pp. 55–58
Kondo, Y., Bruhweiler, F. C. 1985. *Ap. J. Lett.* 291: L1–5
Kurucz, R. L. 1979. *Ap. J. Suppl.* 40: 1–340
Kwok, S., Hrivnak, B. J., Milone, E. F. 1986. *Ap. J.* 303: 451–64
Lamers, H. J. G. L. M. 1986. See Israel 1986, pp. 79–82
Lamers, H. J. G. L. M., Waters, L. B. F. M., Garmany, C. D., Perez, M. R., Waelkens, C. 1986. *Astron. Astrophys.* 154: L20–22
Lamers, H. J. G. L. M., Waters, L. B. F. M., Wesselius, P. R. 1984. *Astron. Astrophys.* 134: L17–20
Leene, A. 1985. *Astron. Astrophys.* 154: 295–302
Leene, A., Beichman, C. A. 1985. In *IAU 87 Nearby Molecular Clouds, IAU Symp. No. 87*, ed. G. Serra, pp. 118–21. Berlin: Springer-Verlag
Leene, A., Pottasch, S. R. 1987. *Astron. Astrophys.* In press
Léger, A., Puget, J.-L. 1984. *Astron. Astrophys.* 137: L5–8
Levasseur-Regourd, A. C., Dumont, R. 1985. *C. R. Acad. Sci. Paris* 300: 109–12
Lewis, B. M., Eder, J., Terzian, Y. 1985. *Nature* 313: 200–2
Little-Marenin, I. R., Stephenson, C. B., Little, S. J. 1986. *Bull. Am. Astron. Soc.* 18: 669
Little-Marenin, I. R. 1986. *Ap. J. Lett.* 307: L15–19
Lonsdale, C. J., ed. 1987. *Star Formation in Galaxies.* Washington, DC: NASA Print. Off.
Low, F. J., Beintema, D. A., Gautier, T. N., Gillett, F. C., Beichman, C. A., et al. 1984a. *Ap. J. Lett.* 278: L19–22
Low, F. J., Neugebauer, G., Gautier, T. N., Gillett, F. C. 1984b. *Bull. Am. Astron. Soc.* 16: 968
MacConnell, D. J. 1986. *A program for spectral classification of the hotter IRAS point sources near the southern Galactic plane.* Jet Propuls. Lab Contract Rep.
Margulis, M., Lada, C. J. 1986. *Ap. J. Lett.* 307: L87–90
Marsden, P. L., Gillett, F. C., Jennings, R. E., Emerson, J. P., de Jong, T., Olnon, F. M. 1984. *Ap. J. Lett.* 278: L29–32
Marsden, P. L. 1985. *The Observatory* 105: 7
Marsden, B. G. 1986. *Q. J. R. Astron. Soc.* 27: 102–18
McGee, R. X., Haynes, R. F., Grognard, R. J.-M., Malin, D. 1986. *MNRAS* 221: 543–52
Misconi, N. Y. 1980. In *Solid Particles in the Solar System*, ed. I. Halliday, B. A. McIntosh, pp. 49–53. Dordrecht: Reidel
Mosley, H. 1980. *Ap. J.* 238: 892–904
Mufson, S. L., McCollough, M. L., Dickel, J. R., Petre, R., White, R., Chevalier, R. 1986. *Astron. J.* 92: 1349
Mundy, L. G., Wilking, B. A., Myers, S. T. 1986. *Ap. J. Lett.* 311: L75–79
Murdock, T. L., Price, S. D. 1985. *Astron. J.* 90: 375–86
Myers, P. C. 1985. In *Protostars and Planets II*, ed. D. Black, M. S. Matthews, pp. 81–103. Tucson: Univ. Ariz. Press
Myers, P. C. 1986. In *Star Forming Regions*,

IAU Symp. No. 115, ed. M. Peimbert, J. Ingaku. Dordrecht: Reidel. In press

Myers, P. C., Benson, P. J. 1983. *Ap. J.* 266: 309-20

Myers, P. C., Dame, T. M., Thaddeus, P., Cohen, R. S., Silverberg, R. F., Dwek, E. Hauser, M. G. 1986. *Ap. J.* 301: 398-422

Myers, P. C., Fuller, G. A., Mathieu, R. D., Beichman, C. A., Benson, P. J., Schild, R. E. 1987. *Ap. J.* In press

Myers, P. C., Linke, R. A., Benson, P. J. 1983. *Ap. J.* 264: 517-37

Neugebauer, G., Habing, H. J., van Duinen, R. Aumann, H. H., Baud, B., et al. 1984. *Ap. J. Lett.* 278: L1-6

O'Dell, C. R. 1986. *Icarus* 67: 71-79

Odenwald, S. F. 1986a. See Israel 1986, pp. 75-76

Odenwald, S. F. 1986b. *Ap. J.* 307: 711-22

Olnon, F. M., Baud, B., Habing, H. J., de Jong, T., Harris, S., Pottasch, S. R. 1984. *Ap. J. Lett.* 278: L41-43

Pajot, F., Boissé, P., Gispert, R., Lamarre, J. M., Puget, J. L., Serra, G. 1986. *Astron. Astrophys.* 157: 393-99

Parthasarathy, M., Pottasch, S. R. 1986a. *Astron. Astrophys.* 154: L16-19

Parthasarathy, M., Pottasch, S. R. 1986b. *Astron. Astrophys.* 161: 417

Pottasch, S. R. 1986. See Israel 1986, pp. 131-42

Pottasch, S. R., Beintema, D. A., Raimond, E., Baud, B., van Duinen, R., et al. 1984a. *Ap. J. Lett.* 278: L33-35

Pottasch, S. R., et al. 1984b. *Astron. Astrophys.* 138: 10-18

Pottasch, S. R., Dennefeld, M., Mo, J-E. 1986a. *Astron. Astrophys.* 155: 397-401

Pottasch, S. R., Preite-Martinez, A., Olnon, F. M., Mo, J.-E., Kingma, S. 1986b. *Astron. Astrophys.* 161: 363-75

Pottasch, S. R., Preite-Martinez, A., Olnon, F. M., Raimond, E., Beintema, D. A., Habing, H. J. 1985. *Astron. Astrophys.* 143: L11-13

Pravdo, S. H., Chester, T. J. 1986. *Ap. J.* In press

Price, S. 1981. *Astron. J.* 86: 193-205

Puget, J.-L., Léger, A., Boulanger, F. 1985. *Astron. Astrophys.* 142: L19-22

Puget, J.-L., Perault, M., Boulanger, F., Falgarone, E. 1987. See Lonsdale 1987, p. 21

Purcell, E. M. 1976. *Ap. J.* 206: 685-90

Rao, N. K., Nandy, K. 1986. *MNRAS*. In press

Rickard, L. J, Dwek, E., White, R. A., Hauser, M. G. 1985. *Bull. Am. Astron. Soc.* 17: 591

Rieke, G. H., Lebofsky, M. J., Low, F. J. 1985. *Astron. J.* 90: 900-6

Roche, P. F., Aitken, D. K. 1986. *MNRAS* 221: 63-76

Rowan-Robinson, M. 1986. *MNRAS* 219: 737-49

Rowan-Robinson, M., Chester, T. J. 1987. Preprint

Rowan-Robinson, M., Clegg, P. E., Beichman, C. A., Neugebauer, G., Soifer, B. T., et al. 1984. *Ap. J. Lett.* 278: L7-10

Rowan-Robinson, M., Harris, S. L. 1983. *MNRAS* 202: 767-96

Rucinski, S. M. 1985. *Astron. J.* 90: 2321-30

Sadakane, K., Nishida, M. 1986. *Publ. Astron. Soc. Pac.* 98: 685-89

Sandage, A. 1973. *Ap. J.* 183: 711-42

Sandage, A. 1976. *Astron. J.* 81: 954-57

Sandell, G., Reipurth, B., Menten, C., Walmsley, M., Ungerechts, H. 1986. See Israel 1986, pp. 295-96

Sanders, D. B., Scoville, N. Z., Young, J. S., Soifer, B. T., Schloerb, F. P., et al. 1986. *Ap. J. Lett.* 305: L45-49

Schaefer, B. E. 1986a. *Publ. Astron. Soc. Pac.* 98: 556-60

Schaefer, B. E. 1986b. Preprint

Schwartz, P. R. 1986. See Israel 1986, pp. 301-2

Scoville, N. Z., Good, J. C. 1987. See Lonsdale 1987, pp. 3-20

Scoville, N. Z., Sanders, D. 1986. In *Interstellar Processes*, ed. H. Thronson, D. Hollenbach. Dordrecht: Reidel. In press

Sellgren, K. 1984. *Ap. J.* 227: 623-33

Sellgren, K., Werner, M. W., Dinerstein, H. L. 1983. *Ap. J. Lett.* 271: L13-17

Shipman, H. L. 1986. In *The Astrophysics of Brown Dwarfs*, ed. M. Kafatos, R. Harrington, S. P. Maran, pp. 71-75. Cambridge: Cambridge Univ. Press

Shu, F. H. 1977. *Ap. J.* 214: 488-97

Shu, F. H., Lizano, S., Adams, F. C. 1986. In *Star Forming Regions, IUA Symp. No. 115*, ed. M. Peimbert, J. Ingaku. Dordrecht: Reidel. In press

Shu, F. H., Terebey, S. 1984. In *Cool Stars, Stellar Systems and the Sun*, ed. S. Baliunas, L. Hartmann, p. 78. Berlin: Springer-Verlag

Smith, B. A., Terrile, R. J. 1984. *Science* 226: 1421-24

Snell, R. L., Loren, R. B., Plambeck, R. L. 1980. *Ap. J. Lett.* 239: L17-22

Sodorski, T. J., Dwek, E., Hauser, M. G., Kerr, F. J. 1987. See Lonsdale 1987, pp. 99-102

Soifer, B. T., Houck, J. R., Neugebauer, G. 1987. *Ann. Rev. Astron. Astrophys.* 25: 187-230

Soifer, B. T., Rice, W. L., Mould, J. R., Gillett, F. C., Rowan-Robinson, M., Habing, H. J. 1986. *Ap. J.* 304: 651-56

Stahler, S. W., Shu, F. H., Taam, R. E. 1980. *Ap. J.* 241: 637-54

Staller, R. F. A., de Jong, T. 1981. *Astron. Astrophys.* 98: 140-48

Stein, W. A., Gaustad, J. E., Gillett, F. C. 1969. *Ap. J. Lett.* 155: L3–L7
Stewart, B. C., Davies, J. K., Green, S. F. 1984. *J. Br. Interplanet. Soc.* 37: 348–52
Stickland, D. J., Lloyd, C., Willis, A. J. 1985. *Astron. Astrophys.* 150: L9–11
Stickland, D. J. 1985. *Observatory* 105: 90
Sykes, M. V., Greenberg, R. 1985. *Publ. Astron. Soc. Pac.* 97: 904
Sykes, M. V., Greenberg, R. 1986. *Icarus* 65: 51–69
Sykes, M. V., Greenberg, R., Hunten, D. M. 1984. *Bull. Am. Astron. Soc.* 16: 690
Sykes, M. V., Lebofsky, L. A., Hunten, D. M., Low, F. J. 1986. *Science* 232: 1115–17
Takaba, H., Fukui, Y., Fujimoto, Y., Sugitani, K., Ogawa, H., Kawabata, K. 1986. Preprint
Terebey, S., Fich, M. 1986. *Ap. J. Lett.* 309: L73–78
Thronson, H. A. Jr., Harper, D. A., Bally, J., Dragovan, M., Mozurkewich, D., et al. 1986. *Astron. J.* 91: 1350–56
Thronson, H. A. Jr., Latter, W. B., Black, J. H., Bally, J. Hacking, P. 1987a. In *Late Stages of Stellar Evolution*, ed. S. Kwok, S. R. Pottasch. Dordrecht: Reidel. In press
Thronson, H. A. Jr., Latter, W. B., Black, J. H., Bally, J., Hacking, P. 1987b. Submitted for publication
van der Hucht, K. A., Olnon, F. M. 1985. *Astron. Astrophys.* 149: L17–19
van der Hucht, K. A., Jurriens, T. A., Olnon, F. M., Thé, P. S., Wesselius, P. R., Williams, P. M. 1985a. *Astron. Astrophys.* 145: L13–16
van der Hucht, K. A., Jurriens, T. A., Olnon, F. M., Thé, P. S., Wesselius, P. R., Williams, P. M. 1985b. In *Birth and Evolution of Massive Stars and Stellar Groups*, ed. W. H. W. M. Boland, H. van Woerden, pp. 167–74. Dordrecht: Reidel
van der Veen, W. E. C. J. 1986. See Israel 1986, pp. 107–8
van der Veen, W. E. C. J. 1987. In *The Late Stages of Stellar Evolution*, ed. S. Kwok, S. R. Pottasch. Dordrecht: Reidel. In press
Vardya, M. S., de Jong, T., Willems, F. J. 1986. *Ap. J.* 304: L29–32
Veeder, G. J., Kowal, C., Matson, D. L. 1984. *Lunar Planet. Sci. Conf. XV*, p. 878 (Abstr.)
Wakker, B. P., Boulanger, F. 1987. Submitted for publication
Walker, C. K., Lada, C. J., Young, E. T., Maloney, P. R., Wilking, B. A. 1986. *Ap. J.* 309: L47–52
Walker, H. J. 1985. *Astron. Astrophys.* 152: 58–62
Walker, H. J. 1986. In *Hydrogen Deficient Stars, IAU Colloq. No. 87*, pp. 407–22. Dordrecht: Reidel
Walker, R. G., Aumann, H. H. 1984. *Adv. Space Res.* 4: 197–201
Walker, R. G., Aumann, H. H., Davies, J., Green, S. F., de Jong, T., Houck, J. R., Soifer, B. T. 1984. *Ap. J. Lett.* 278: L11–14
Walterbos, R. A. M., Schwering, P. B. W. 1986. *Astron. Astrophys.* In press
Waters, L. B. F. M. 1986a. *Astron. Astrophys.* 159: L1–4
Waters, L. B. F. M. 1986b. *Astron. Astrophys.* 162: 121–39
Waters, L. B. F. M. 1986c. See Israel 1986, pp. 83–85
Waters, L. B. F. M., Coté, J., Aumann, H. H. 1986. *Astron. Astrophys.* In press
Waters, L. B. F. M., Wesselius, P. R. 1986. *Astron. Astrophys.* 155: 104–12
Weiland, J. L., Blitz, L., Dwek, E., Hauser, M. G., Magnani, L., Rickard, L. J 1986. Preprint
Weissman, P. R. 1984. *Science* 224: 987–88
White, G. J., Gee, G. 1986. *Astron. Astrophys.* 156: 301–9
Whitelock, P., Feast, M., Catchpole, R. 1986. *MNRAS* 222: 1–9
Willems, F. J. 1986. See Israel 1986, pp. 113–18
Willems, F. J., de Jong, T. 1986. *Ap. J. Lett.* 309: L39–42
Willner, S. P., Becklin, E. E., Visvanathan, N. 1972. *Ap. J.* 175: 699–706
Witteborn, F. C., Bregman, J. D., Lester, D. F., Rank, D. M. 1982. *Icarus* 50: 63–71
Wolfire, M. G., Waldron, W. L., Cassinelli, J. P. 1985. *Astron. Astrophys.* 142: L25–28
Wolstencroft, R. D., Scarrott, S. M., Warren-Smith, R. F., Walker, H. J., Reipurth, B., Savage, B. 1986. *MNRAS* 218: 1P–6P
Wood, P. R., Bessel, M. S., Whiteoak, J. B. 1986. *Ap. J. Lett.* 306: L81–84
Wouterloot, J. G. A., Walmsley, C. M. 1986. *Astron. Astrophys.* In press
Wynn-Williams, C. G. 1982. *Ann. Rev. Astron. Astrophys.* 20: 587–618
Young, E. T., Lada, C. J., Wilking, B. A. 1986. *Ap. J. Lett.* 304: L45–49
Zuckerman, B. 1986. See Israel 1986, pp. 93–100
Zuckerman, B., Dyck, H. M. 1986a. *Ap. J.* 304: 394–400
Zuckerman, B., Dyck, H. M. 1986b. *Ap. J.* 311: 345–59

DYNAMICAL EVOLUTION OF GLOBULAR CLUSTERS

Rebecca Elson and Piet Hut[1]

Institute for Advanced Study, Princeton, New Jersey 08540

Shogo Inagaki

Department of Astronomy, University of Kyoto, Kyoto 606, Japan

1. INTRODUCTION

Globular clusters provide an ideal laboratory for studying long-term effects in stellar dynamics. Interactions between individual stars establish a form of pseudo-equilibrium on a time scale that is typically $\lesssim 10^8$ yr in the core and $\lesssim 10^{10}$ yr at the half-mass radius. In contrast, two-body relaxation time scales in elliptical galaxies generally exceed 10^{10} yr, except perhaps in the inner few parsecs of the nucleus. The instrinsic instability of self-gravitating systems precludes, however, any final equilibrium state in a globular cluster. Two-body encounters drive the outer regions of the cluster to expand, while the inner regions accelerate toward an infinite central density, during what is termed the "collapse phase."

Since singularities indicate the breakdown of model assumptions, we can expect a new physical process to become important, one that somehow releases energy to counteract the gravitational contraction. In stellar evolution this process is nuclear burning. In cluster evolution it is the generation of energy through the formation of binaries and their subsequent "burning" during encounters with single stars, in which the binary tends to become more strongly bound. Another way of heating a cluster is by stellar mass loss, which decreases the total binding energy per unit mass and thereby increases the velocity dispersion relative to the virial equilibrium value. Mass loss may result from stellar evolution or from a runaway merging of stars, which can lead to supernova explosions.

---

[1] Alfred P. Sloan Foundation Fellow.

The onset of energy generation in the core of a cluster marks the end of the collapse phase. Thereafter, the cluster is expected to evaporate on a time scale $\gg 10^{10}$ yr by a combination of three effects: the escape of weakly bound stars, which gain energy through internal two-body encounters and/or external perturbations; the escape of stars catapulted out of the cluster by three- and four-body scattering involving binaries; and mass loss from stellar evolution. This postcollapse phase is not necessarily uneventful: It may involve a series of large-amplitude core oscillations that are driven by the inherently unstable character of self-gravitating systems. At present, an interesting controversy exists concerning the occurrence and interpretation of gravothermal oscillations in real clusters.

Many dynamicists have contributed to our understanding of globular clusters, particularly in the last three decades. An excellent work on the subject is a monograph by Spitzer (1987) dedicated to the dynamical evolution of globular clusters. A general textbook on stellar dynamics, which gives a good feel for the subject as a whole, is that of Binney & Tremaine (1987). Another recent work on stellar dynamics, by Saslaw (1985), approaches the subject from a somewhat more abstract point of view. Other presently available sources are an earlier review of globular cluster evolution by Lightman & Shapiro (1978) and the Proceedings of IAU Symposium No. 113 on the Dynamics of Star Clusters (Goodman & Hut 1985). Recent theoretical reviews have been given by Spitzer (1985) and Inagaki (1987) for precollapse evolution, and by Heggie (1985) and Cohn (1987) for postcollapse evolution.

In this review, we concentrate on the theoretical picture outlined above of core collapse and subsequent evaporation; this study is currently gaining momentum both from recent theoretical developments and from observations indicating that a significant fraction of globular clusters may already have completed their collapse phase and are now entering the equivalent of a main-sequence stage in stellar evolution. Section 2 contains a theoretical discussion of the physical processes that play an important role in the pre- and postcollapse evolution of a globular cluster. In Section 3 we review various computational approaches to cluster evolution, and in Section 4 we summarize relevant observational methods and results. Finally, in Section 5 we outline directions for future theoretical and observational research.

2. THEORY: PHYSICAL PROCESSES

In a stellar system, there are two important time scales on which evolution can occur: the crossing time and the two-body relaxation time. The crossing time is the size of the system divided by a typical stellar velocity and is the shortest time scale on which the system as a whole can react to global changes in its potential. The two-body relaxation time is the time scale in which the cumulative effect of two-body encounters can alter the individual stellar orbits significantly [see Hénon (1973) for an intuitive derivation and a precise definition]. It is the shortest time scale in which energy can be transported through the system, and it is also the time scale that is relevant for most of the evolutionary processes discussed in this review.

2.1 Precollapse Evolution

The overall picture of core collapse is fairly well understood. During the contraction of the inner regions of a cluster, the half-mass radius remains roughly constant (at least for an isolated cluster) while the outer regions expand. This is a direct consequence of energy conservation, which holds approximately, since evaporating stars typically carry away a relatively small amount of energy per unit mass. After core collapse, however, a central energy source is turned on, leading to an expansion of the half-mass radius as well.

The behavior of the inner regions during core collapse is also well understood. In particular, the later stages of core collapse are self-similar. This implies that during collapse $d\rho_c/dt \to c \times \rho_c/t_{cr}$, where ρ_c is the central density and t_{cr} is the central relaxation time. The value of the constant $c \ll 1$ depends somewhat on the assumptions made in the different calculations (cf. Spitzer 1985); for example, Cohn (1979) finds a value $c = 6 \times 10^{-3}$ in a Fokker-Planck simulation in which velocity anisotropy was taken into account, whereas isotropic simulations yield $c = 3.6 \times 10^{-3}$ (Cohn 1980).

2.1.1 FORMATION Globular clusters serve as a fossil record of the earliest stages of the collapse of our Galaxy, and although scenarios for their origin have been proposed (cf. Peebles 1984, Fall & Rees 1985), relevant observations are generally impossible because two-body relaxation has erased most traces of their initial state. The Large Magellanic Cloud (LMC), however, contains many young "globular" clusters, and observations of these may contribute to our understanding of the early stages of evolution of the globular clusters in our own Galaxy (Elson et al. 1987).

A reasonable picture of the formation of a cluster is as follows. A gravitationally bound cloud shatters into stars on a time scale less than a few crossing times. Depending on the initial clumpiness and temperature of the cloud, and on the efficiency of star formation, the protocluster collapses with some degree of dissipation and is mixed through either clump-clump encounters or violent relaxation. This occurs on a few dynamical time scales ($\lesssim 10^8$ yr), and the processes discussed in the remainder of this review only become important long after this stage of evolution is complete.

2.1.2 TIDAL TRUNCATION Globular clusters have a finite radius, beyond which the tidal field of the Galaxy precludes bound orbits. The simplest theoretical model of a near-equilibrium self-gravitating star system with a tidal cutoff follows from the simplifying assumptions of (*a*) equal-mass stars, (*b*) an isotropic velocity distribution, and (*c*) a distribution function that is a solution of the Fokker-Planck equation for a square-well approximation to the potential [cf. Binney & Tremaine (1987) for a detailed derivation]. The resulting velocity distribution is a "lowered Maxwellian." These King (1966a) models are characterized by a single parameter $c = \log(r_t/r_c)$, where r_t is the truncation radius and r_c is the core radius. They provide reasonably good fits to the observed density profiles of most globular clusters (but see Section 4.1 for notable exceptions). A common interpretation of r_t is that it corresponds to the radius beyond which stars are stripped from the cluster by the Galactic tidal field, although there is some controversy concerning the quantitative relation between r_t, the mass of the cluster, and the Galactic tidal field (King 1962, Innanen 1979). Furthermore, the rate and way in which the tidal limit is established are not completely understood (Seitzer 1985, Lee & Ostriker 1987).

2.1.3 STELLAR EVOLUTION Stars can lose a large fraction of their mass during the late stages of their evolution, either by stellar winds or by supernova explosions. The material is blown out of the cluster, which causes the radius to decrease: This can be seen most simply in the case of a cluster in a circular orbit, since the average density of the cluster must remain the same as that of the ambient Galactic material [cf. Mathieu (1983) for a discussion in the context of open clusters]. Mass loss tends to *decrease* the concentration parameter c, defined above (Chernoff 1987). These effects are more important in the first few billion years, when heavier stars for which the fractional mass loss is greater, are present (Angelleti & Giannone 1980, Applegate 1986).

2.1.4 EVAPORATION The tragedy of a globular cluster is that its thermal goal, a state of maximum entropy, is unattainable. Two-body relaxation drives the velocity distribution toward a Maxwellian; however, stars in the high-velocity tail are continuously torn out of the cluster by the tidal field of the Galaxy. The result is a decrease in the radius of the cluster and an *increase* in the concentration parameter c (cf. Chernoff et al. 1986).

Although a cluster never attains a final equilibrium state, its evolution can be represented as a series of near-equilibrium configurations described by King models. This is in reasonable agreement with observations of most globular clusters, as well as with numerical calculations of globular cluster evolution in the Fokker-Planck approximation (Wiyanto et al. 1985), which show that an evolving cluster follows the King sequence closely as long as $c \lesssim 2.1$ (Figure 1; see also Katz 1980). For larger values of c, King models become unstable, as discussed in Section 2.1.6. A limitation of both the King models and the Fokker-Planck calculations is that they assume an isotropic velocity distribution, which is probably unrealistic, especially in the outer parts of a cluster.

2.1.5 TIDAL SHOCKING The rate of evaporation of stars from a cluster is enhanced by tidal shocking of the cluster as it passes through the Galactic plane (Ostriker et al. 1972). Evaporation due to tidal shocking can be as important as evaporation due to two-body relaxation in determining the evolution of globular clusters (Aguilar et al. 1985), and it can cause the concentration parameter c to either increase or decrease (King 1966a, Ostriker et al. 1972, Spitzer & Chevalier 1973, Chernoff et al. 1986).

2.1.6 GRAVOTHERMAL INSTABILITY Two-body relaxation redistributes energy between stars, and since globular clusters are open systems, this redistribution causes some stars to gain positive energy and escape from the cluster. As a result, a globular cluster in an external tidal field will

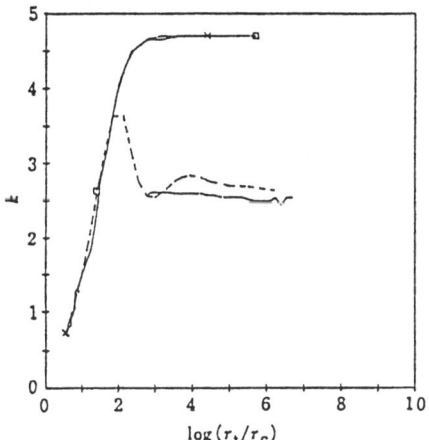

Figure 1 A dimensionless measure of binding energy, plotted as a function of central concentration. Here r_t and r_c are the tidal radius and core radius, respectively, and k is defined through the total energy E of the cluster by $E = -(k+1/2)GM^2/r_t$. The dashed curve represents the one-parameter sequence of King (1966a) models. The solid curves denote the evolutionary sequences obtained from Fokker-Planck calculations. The left curve starts from a relatively unevolved King model, while the flat curve starts from an unstable King model (Wiyanto et al. 1985).

decrease in radius (see Section 2.1.2). The redistribution of energy can also, however, lead to a more catastrophic, internally driven form of evolution.

To illustrate the difference between externally and internally driven evolution, let us consider a cluster that is initially in thermal equilibrium, enclosed within a rigid, reflecting sphere. When stars evaporate from the core into the halo, the energy and temperature (i.e. the velocity dispersion) of the halo increase. The core loses energy, contracts a little, and thereby increases its velocity dispersion. This *increase* in temperature accompanying a *loss* of energy is a manifestation of what may be called the negative heat capacity of a self-gravitating system (although strictly speaking, concepts of statistical mechanics break down in the presence of long-range attractive forces; cf. Lynden-Bell & Wood 1968). Other examples are the case of a satellite orbiting the Earth, where dissipation due to friction with the atmosphere causes the satellite to lose height and simultaneously *increase* its orbital speed; and the case of a protostar, where energy *loss* at the surface results in a contraction, and the virial theorem guarantees an *increase* in temperature to balance the increase in binding energy.

If the core of the cluster is large enough, the increase in temperature will be smaller than the increase in halo temperature; energy will flow back from the halo into the core, and a stable equilibrium will result. However, if the core is too small, the energy absorbed by the halo will not increase the halo temperature sufficiently to keep up with the increase in core temperature. In this case, the temperature gradient between the core and halo will steepen indefinitely while the contraction of the core accelerates.

This runaway process was first discovered, in the somewhat analogous case of a self-gravitating gas confined within a rigid sphere, by Antonov (1962). It was later analyzed in more detail by Lynden-Bell & Wood (1968), who termed it a "gravothermal catastrophe." Antonov found that isothermal equilibrium configurations can be stable only when the density contrast (the ratio between the density near the confining sphere and the central density) is less than a critical value, which he determined to be ≈ 709. He showed that for larger values, a small increase in central temperature leads to an accelerated contraction of the core. Conversely, a small decrease in the central temperature leads to a runaway expansion of the core.

Early numerical calculations of globular cluster evolution indeed pointed toward the occurrence of an accelerated collapse (cf. the review by Spitzer 1975), but they were not sufficiently accurate to follow the evolution far enough to identify the collapse uniquely as the gravothermal catastrophe. Additional insight into the nature of gravothermal collapse was provided by a linear stability analysis of the confined-gas sphere model by Hachisu & Sugimoto (1978), which resulted in eigenfunctions and eigenvalues for entropy perturbations (Figure 2). Hachisu et al. (1978) followed these linear instabilities into the nonlinear regime and found that the collapse proceeded in a nearly self-similar way. The simplifications arising from this self-similarity were exploited by Lynden-Bell & Eggleton (1980), who derived a unique dimensionless model for the late stages of core collapse.

These last three papers greatly improved our understanding of the collapse of a self-gravitating gas sphere, but it was not clear how applicable the main results were to the stellar dynamical case where, in marked contrast with the conducting-gas sphere approximation, the mean-free path of the particles is much larger than the system. Lynden-Bell & Eggleton (1980) gave a detailed discussion of the correspondence between conductivity in a gas and similar effects of two-body relaxation in stellar dynamics, which made a connection between the two types of models plausible.

More direct confirmation of the validity of gas models for stellar dynamical applications was provided by Inagaki (1980) and Cohn (1980). Inagaki performed a linear stability analysis of a stellar dynamical system enclosed in a sphere, using the Fokker-Planck approximation to describe the interactions between the particles, and recovered Antonov's instability criterion for stellar dynamics. Cohn solved the Fokker-Planck equations for diffusion in energy space numerically and closely reproduced Lynden-Bell & Eggleton's self-similar collapse (Figure 3): For example, the latter authors found a logarithmic density gradient $d \ln \rho / d \ln r = -2.21$, in remarkable agreement with Cohn's value of -2.23.

2.1.7 MASS SEGREGATION Simple models of globular clusters assume that all the stars in the cluster have the same mass; however, in a real cluster, a spectrum of stellar masses is present. Equipartition of energy through two-body relaxation causes the more massive stars to sink toward the cluster center, accelerating the collapse of the core by a factor of 1 to 10 (Spitzer 1969, Inagaki 1985, and references therein). An extension of the linear stability analysis of Hachisu et al. (1978) to include mass segregation

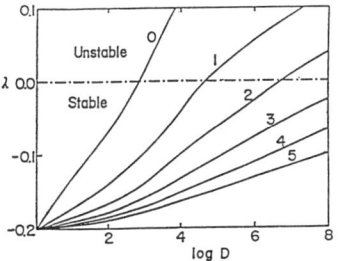

Figure 2 Results from a linear stability analysis of a self-gravitating gas sphere confined by a spherical wall (Hachisu et al. 1978). The eigenvalues of the fundamental (0) and higher modes (1, 2, ...) of the entropy perturbations are plotted as a function of the density contrast D between center and edge. Gravothermal instability ($\lambda > 0$) is found to set in at $D = D_{cr} \simeq 709$.

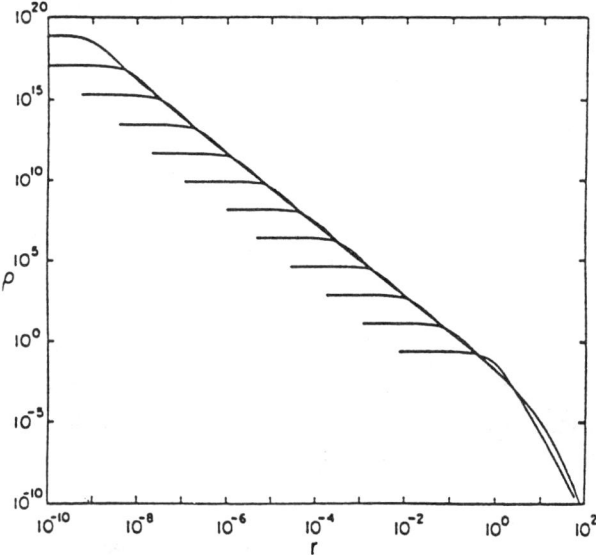

Figure 3 Density profiles during core collapse, from an isotropized, orbit-averaged Fokker-Planck equation (Cohn 1980). The starting point for the computation is a Plummer model. Note the homologous character of the increasingly concentrated profiles.

shows that the critical density contrast needed for the onset of gravothermal collapse is lowered far below the equal-mass value of 709 (Yoshizawa et al. 1978).

While core collapse in a cluster of equal-mass stars is nearly isothermal, that in a cluster with even a modest amount of mass segregation is not. Spitzer (1969) showed that if

$$\chi = (M_2/M_1)(m_2/m_1)^{-1.5}$$

is larger than 0.16, then equipartition between the more massive and the less massive component is impossible for $m_2 \gg m_1$ and $M_2 < M_1$, where m_1 and m_2 are the masses of a less massive star and a more massive star, respectively, and M_1 and M_2 are the total masses of the less massive and more massive stars, respectively. Inagaki & Wiyanto (1984) and Inagaki (1985) confirmed these predictions by numerical Fokker-Planck integration. [Multimass Fokker-Planck calculations had been performed earlier in the context of clusters of galaxies by Merritt (1983, 1984, 1985).] Inagaki & Saslaw (1985) extended the calculations to include 15 mass groups in order to simulate a continuous mass spectrum $dM \propto m^{-\alpha}dm$, where dM is the total mass of all stars in the mass range $(m, m+dm)$. They found that equipartition is impossible for $\alpha \lesssim 6$. This is more stringent than the result $\alpha \lesssim 2.5$ found by Vishniac (1978); however, the latter is based on the somewhat unrealistic assumption of homology of the density profiles for the different mass groups. An extension of Spitzer's analysis to include two-component systems in which the heavier stars are distributed mainly outside the core shows that equilibrium solutions can exist for mass ratios larger than those allowed by Spitzer; however, these solutions are probably not realistic (Merritt 1981).

A remaining uncertainty is the role that the assumption of isotropy plays in these models. Cohn's (1979) anisotropic analysis is more relevant to real clusters, although numerical instabilities prevented the calculations from extending very far. It is generally believed that apart from a quantitative difference in the rate of collapse, neglecting anisotropy does not cause significant qualitative changes in the behavior of a collapsing core [see, however, Heggie (1979) for a different view].

2.1.8 CLOSE ENCOUNTERS: FORMATION OF BINARIES Binary stars can form in a cluster in three distinct ways. First, they may be primordial—that is, a system formed at the time of birth of the two stars. Second, they may form during a three-body encounter if the third star carries away sufficient kinetic energy to leave the other two bound [see Hut (1985) for a review and a quantitative discussion]. Third, they may form in two-body encounters if the stars pass each other within a few stellar radii and tidal dissipation results in capture (Fabian et al. 1975, Press & Teukolsky 1977, Lee & Ostriker 1986, McMillan et al. 1987). The last two types of binary systems are expected to be relatively more numerous in the cores of clusters, both because conditions are more favorable for forming them in higher density regions and because they are on average more massive than single stars and therefore tend to sink toward the cluster center.

Even a small number of binaries can affect the later phases of core collapse in a globular cluster and may lead to its eventual reversal into core expansion [see Section 2.2 (Milgrom & Shapiro 1978, Dokuchaev & Ozernoy 1982a,b, Inagaki 1984, Hut & Inagaki 1985)]. Spitzer & Mathieu (1980) found, however, that even an appreciable fraction of binaries were unable to prevent core collapse; however, their Monte Carlo approach did not allow them to follow the collapse very far.

2.1.9 COLLISIONS: FORMATION OF MASSIVE STARS Encounters that are close enough to form tidal-capture binaries can also lead to the merging of the two stars if the total angular momentum is sufficiently small. Encounters characterized by larger impact parameters can also lead to merging if the excess angular momentum is carried away by a small amount of mass loss. Very little work has been done in this area, and it is not clear what fraction of close encounters will result in mergers as opposed to tidal capture. Furthermore, the distinction between merging and capture becomes blurred in the case where tidal capture gives rise to a common-envelope system.

An interesting consequence of merging is the possibility of a runaway effect in which the merger products, which are heavier than average stars, sink to the core (where the chance for further merging is highest). Repeated merging can lead to the formation of stars with a lifetime short enough to burn up before core collapse is completed, thereby influencing the rate of core collapse (Lee 1987).

2.2 Postcollapse Evolution

The evolution of a globular cluster after core collapse has only recently been studied intensively, and many aspects of our understanding of it remain uncertain and may change in the years following this review. The *mean* behavior of the cluster after core collapse *is*, however, firmly established: The half-mass radius of an isolated cluster expands according to $r_h(t) \propto t^{2/3}$, where t is the time since core bounce, while the velocity dispersion drops according to $v \propto t^{-1/3}$. This relation may be derived from general principles, without any knowledge of the mechanism of energy generation in the core (cf. Hénon 1965, 1975), in a manner analogous to Eddington's (1926) prediction of the mass-luminosity relation for stars, which requires no precise knowledge of the nature of their internal energy generation. The derivation goes as follows: (*a*) The half-mass relaxation time t_{hr} in a self-similar solution scales as $t_{hr} \propto t$, the time since core bounce; (*b*) $t_{hr} \propto N t_{hc}$, where N is the number of stars in the cluster, t_{hc} is the crossing time at the half-mass radius, and we have neglected a factor log N; (*c*) if we neglect the slow change in mass and particle number due to escape, the virial theorem gives $t_{hc} \propto r_h^{3/2}$; and (*d*) by combining these, we find that $t \propto r_h^{3/2}$, which leads to the results quoted above. In contrast, the rate of expansion of the core *does* depend on the details of the central engine (Cohn 1985, Ostriker 1985, Statler et al. 1987).

2.2.1 A CENTRAL ENGINE The contraction of a protostar is halted when the core of the star starts burning nuclear fuel at a rate that balances the energy loss at the surface. Similarly, gravothermal collapse of the core of a globular cluster will be halted when a central source generates more energy than is effectively conducted out of the core by two-body relaxation. This notion of a central energy source was first discussed by Hénon (1961), who suggested that a small subsystem of the core could shrink enough to generate the energy required. For example, even the formation and subsequent hardening of a single close binary with an orbital velocity about an order of magnitude larger than the velocity dispersion in the cluster would release an amount of energy comparable to that of ~ 100 single stars.

Hénon's ideas were confirmed several years later by N-body calculations. For all values of N explored ($10-10^3$), a contraction of the central region produced at least one tight binary with a binding energy comparable to the total binding energy of the system. Formation and hardening of such a binary then fueled the expansion and evaporation of the N-body system (for a review of earlier work, see Aarseth & Lecar 1975).

However, a significant difference between the N-body calculations performed to date and globular clusters is that $N \sim 10^5 - 10^6$ for clusters, whereas $N \leq 10^{3.5}$ for the calculations. The large number of stars in a globular cluster implies that a single hard binary can absorb at most $\sim 1\%$ of the binding energy of the cluster: Attempts to form harder binaries lead either to a merging of the stars or to an escape of the binary from the cluster by the recoil momentum gained in an encounter with a third star. Thus, while in simulations each binary plays a dominant role, in real globular clusters only the cumulative effect of many binaries can change the local energy budget of the cluster significantly.

The first attempt to study the effects of binaries in systems with large N was made by Hénon (1975) using a Monte Carlo Fokker-Planck code. He introduced an artificial energy source in the innermost part of his cluster model, which was tuned to give off just the amount of energy necessary to avoid collapse locally at the inner shell. He found that the cluster reached a maximum central density, after which the collapse was reversed into an overall expansion.

Around the same time X-ray sources were discovered to be remarkably abundant in globular clusters. These sources were thought to be either X-ray binaries comparable to Galactic bulge sources, or heavier black holes in the cluster centers. The latter possibility sparked great interest among theoreticians concerning the evolution of a globular cluster with a central black hole of 10^2-10^3 $M_\odot$; however, an overwhelming amount of subsequent observational evidence has emerged for the identification of globular cluster X-ray sources with X-ray binaries (cf. Lewin 1980, and references therein).

The possibility of X-ray-quiet central black holes, with masses less than a few $\times 10^3$ $M_\odot$, is not excluded by available observations. The central brightness excesses observed in the cores of several globular clusters (Djorgovski & King 1986) are compatible with the presence of a central black hole; however, a more straightforward explanation is that these clusters have undergone core collapse and are currently reexpanding, as discussed above. Since binary formation cannot be avoided during core collapse, whereas the presence of central black holes remains purely speculative, we concentrate on the consequences of the former. For a discussion of central black hole models, we refer the reader to the reviews by Lightman & Shapiro (1978) and Shapiro (1985).

Another development in the mid-1970s was the start of a systematic study of how a central engine can be fueled by purely stellar dynamic processes. In analogy with the nuclear fusion rates needed in stellar evolution calculations, "gravitational binary burning rates" were obtained analytically by Heggie (1975) and numerically by Hills (1975). More recent work in this direction is discussed in Section 3.2.3. Finally, an instructive characterization of binary dynamics is given by Lynden-Bell (1975), who compares binary interactions in star clusters with interactions in human society.

2.2.2 REVERSAL OF GRAVOTHERMAL COLLAPSE Postcollapse evolution of globular clusters was explored little prior to 1982, although some of the essential ingredients for the occurrence of a core bounce during the late stages of core collapse had been suggested much earlier. In contrast, the five years since then have witnessed an explosive growth in our understanding of the nature and character of core reexpansion. Simultaneously, increasingly accurate observations have begun to suggest that a significant fraction of the Galactic globular clusters may have already undergone core collapse (see Section 4), in agreement with more indirect inferences from the statistical distribution of central relaxation times (Cohn & Hut 1984).

Stodółkiewicz (1982) extended Hénon's (1975) calculations of the effect of a central energy source in a large-N system, with several different models for the central energy source. He showed that the rate of expansion after core collapse is indeed independent of the nature of the central energy source.

Several other detailed models of postcollapse evolution have been published by Inagaki & Lynden-Bell (1983), Sugimoto & Bettwieser (1983), Bettwieser & Sugimoto (1984), Goodman (1984), and Heggie (1984). Inagaki & Lynden-Bell constructed a self-similar postcollapse model that can be viewed as a continuation of the self-similar precollapse model of Lynden-Bell & Eggleton (1980). At the moment of core collapse, these two models are identical, with a singular spatial density distribution $\rho \propto r^{-2.21}$. Afterwards, a singular isothermal region expands into the distribution left behind in core collapse, at a rate determined by the local relaxation time at the transition region.

Goodman (1984) constructed a self-similar solution for the asymptotic evolution of a globular cluster long after core collapse; this solution is complementary to that of Inagaki & Lynden-Bell (1983), which applies to the immediate postcollapse stage. Goodman's model, which is essentially a gaseous analogue of the model of Hénon (1965), is singular and isothermal near the center but has a steeper density drop-off in the halo than that of Inagaki & Lynden-Bell, since it has a finite total mass.

Heggie (1984) followed the evolution of a gaseous model numerically, without any assumptions concerning self-similarity, using a technique comparable to that used in stellar evolution. His results agree with the self-similar solution by Inagaki & Lynden-Bell (1983) except for the very inner region, which remains regular in Heggie's model, since he included an extended central energy source that is dependent on the local density and velocity dispersion.

Bettwieser & Sugimoto (1984) performed similar stellar evolution–type calculations, but they found a completely different behavior for the expansion after core collapse. Instead of the monotonic expansion found by previous investigators, they found large oscillations in the size of the core radius, which they interpreted as a new physical phenomenon: gravothermal oscillations.

2.2.3 GRAVOTHERMAL OSCILLATIONS The conflict in results between Bettwieser & Sugimoto (1984), who found oscillations, and Heggie (1985) and Cohn (1985), who did not, triggered a renewed and more careful investigation. Before long, both Heggie (using a gas model) and Cohn & Hut (using a Fokker-Planck model) discovered core oscillations when they lowered their time step to values comparable to the central relaxation time (see Figure 4; cf. Cohn et al. 1986, Heggie & Ramamani 1986). The gravothermal character of the core oscillations reported by Sugimoto & Bettwieser (1983) was confirmed explicitly by Goodman (1987) and is illustrated in the evolution of the core, plotted in a plane showing the central density and velocity dispersion (Figure 5; note the difference in "temperature," i.e. velocity dispersion, between an expanding and contracting phase). Goodman's results include a linear stability analysis of a new regular self-similar model for postcollapse evolution, which he constructed in the same paper.

Gravothermal oscillations are a consequence of the negative heat capacity of gravity. In the previous section we saw how the gravothermal instability can lead to core collapse: If the central temperature is slightly too high, the core will lose more heat than it gains, and this will lead to a contraction and therefore a density increase, which in turn will produce a higher central temperature. After core collapse is reversed into core reexpansion, the opposite may occur, as stressed by Bettwieser & Sugimoto (1984): The expansion may lower the central temperature, leading to an

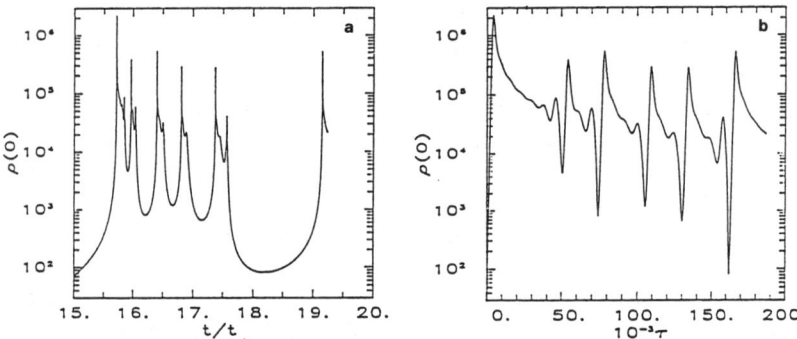

Figure 4 Gravothermal oscillations in the central density from a numerical Fokker-Planck calculation by Cohn et al. (1986). The energy source is modeled by averaging the effects of binaries formed in three-body encounters and interacting with single stars. The density is plotted as a function of (*a*) a linear time scale and (*b*) an elapsed central relaxation time. Here t_{rh} is the initial half-mass relaxation time of the Plummer model from which the computation was started, and τ is defined in the text.

energy flow into the core and in turn to a lowering of both core density and temperature. The result is a runaway expansion that proceeds much faster than dictated by the boundary conditions at the surface of the cluster. The expanding region in the cluster center will grow radially until it reaches a region of radially decreasing temperature. At this point the expansion halts, and the central region starts to collapse again.

These cycles of gravothermal contraction and expansion are clearly visible in Figure 4. Furthermore, on the shoulders of the dominant, strongly nonlinear oscillations, there are smaller oscillations. A similarly chaotic behavior is visible directly in Figure 5. Goodman (1987) classified the different types of behavior according to the type of linear instability: He found that for a total number of stars $N < N_1$ his self-similarly expanding solution is linearly stable, for $N_1 < N < N_2$ the solution is overstable, and for $N > N_2$ the solution is unstable. He estimates that $N_1 \approx 7000$ and $N_2 \approx 40,000$. Although the instability for very large N has indeed the expected character of a gravothermal instability, we do not yet understand the nature of the overstability.

Although our understanding of the fundamental character and the gravothermal nature of gravitational instabilities has improved significantly, it is not yet clear to what extent they apply to real globular clusters, as opposed to gas spheres and equal-mass point-particle models. Bettwieser & Sugimoto (1984) suggest that many globular clusters may already have undergone an initial core collapse and are presently oscillating gravo-

DYNAMICS OF GLOBULAR CLUSTERS 579

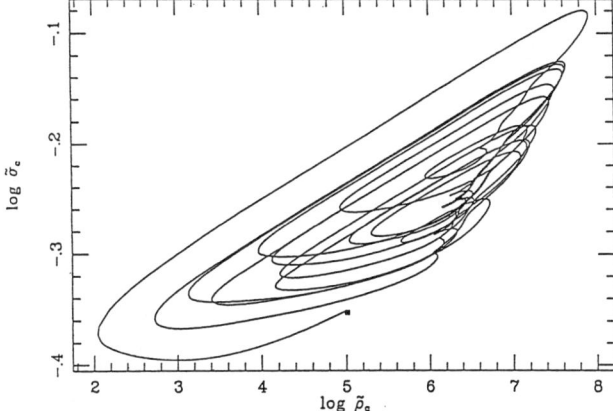

Figure 5 The evolution of core oscillations in a conducting gas sphere, starting from Goodman's self-similar solution (indicated by the small square below the center). Evolution is shown in the plane of central velocity dispersion vs. central density (Goodman 1987).

thermally. The nonlinear nature of these oscillations would imply that the cores spend most of their time near the maximally expanded stages, during which the clusters may have a normal appearance with finite-size cores (cf. Bettwieser 1985, Sugimoto 1985, Bettwieser & Fritze 1984).

Inagaki (1986) has argued that gravothermal oscillations may not apply to real clusters because the finite number of particles implies that the number of particles in the core, and therefore the temperature of the core, will fluctuate with time. (All studies that show oscillations implicitly apply continuous energy diffusion.) He performed N-body simulations with $N = 1000$ and 3000 and found that the effects of fluctuations are indeed important, and that it is impossible to produce the coherent inverse temperature gradient assumed in previous studies of gravothermal instability. In addition, McMillan & Lightman (1984b) and McMillan (1986) have not found gravothermal oscillations in their hybrid code calculations, which simulate the evolution of a globular cluster with a total particle number well above the Goodman criterion for instability.

Both Inagaki (1986) and McMillan (1986) do find modest oscillations with central density amplitudes $\lesssim 10$, which they ascribe to the individual formation and escape of binaries and to the corresponding jitter in the energy input of the central engine. Makino et al. (1986), however, have found oscillations that they interpret as being gravothermal in their N-body calculations; these calculations employ a softened potential and are for a star cluster with $N = 100$, confined by a fixed sphere.

3. A SURVEY OF COMPUTATIONAL MODELS AND RESULTS

Numerous numerical models have been designed to study the evolution of globular clusters, and each includes some kind of statistical approximation to keep the computer time required within reasonable bounds. Three classes of simulations can be distinguished, as well as a hybrid form that combines two of these.

3.1 *N-Body Integration*

Direct N-body calculations, in which the equations of motion describing all $N(N-1)/2$ mutual interactions are solved by what seems to be "brute force" methods, are conceptually the simplest. In practice, however, a great deal of ingenuity is required to attain sufficient accuracy and speed for systems with $N > 100$. A state-of-the-art review of this subject, which includes discussions of polynomial orbit extrapolation, two-body regularization, and two-time-scale methods such as neighbor schemes, has been published by Aarseth (1985). In addition, vectorization can result in a gain of nearly an order of magnitude in computer time on supercomputers (Makino 1986, Aarseth & Inagaki 1986).

3.1.1 MONTE CARLO METHODS: SCALING TO LARGER N The maximum number of particles that has been used in direct N-body calculations of star cluster evolution is $N = 3000$ (Inagaki 1986). While appropriate for some open clusters, this number is about two orders of magnitude too small for modeling typical globular clusters. However, we can apply some of the results of direct N-body calculations to globular clusters if we take each particle to represent 100–1000 stars. This Monte Carlo approach, in which only a small fraction of all physical objects is considered in a calculation, is useful in the precollapse stage of evolution because we can scale the results back to the original physical system using the known dependency of two-body relaxation on particle number (Chandrasekhar 1942). One of the important applications of such scaling has been a direct check of weak-scattering Fokker-Planck models (Aarseth et al. 1974).

Several fundamental characteristics of the evolution of N-body systems have been discovered using direct integration for small N. Examples are the tendency for the core to collapse in a few half-mass relaxation times, and the formation of binaries around the time of collapse, which can provide the energy needed for reexpansion (Aarseth & Lecar 1975). However, important limitations remain, especially in the postcollapse stages of globular cluster evolution. Even with $N = 3000$ and a realistic mass spectrum, a binary composed of two of the heavier stars in a cluster quickly reaches a binding energy that is a considerable fraction of the total binding energy of the cluster—a very unrealistic situation. This suggests that the role played by binaries in present N-body calculations cannot simply be scaled to apply to globular clusters. Clearly we need to use statistical approximations to learn more about the evolution of globular clusters, at least as long as a direct approach is too expensive.

3.1.2 DIRECT MODELING FOR REALISTIC VALUES OF N A calculation with 10^5 equal-mass particles that are followed to core collapse with Aarseth's (1985) most sophisticated N-body code would require a few dozen years, even on the fastest computers currently being built (cf. Makino & Hut 1987). However, a significant gain in speed may result from the use of "tree codes," in which the interparticle forces are calculated explicitly only for nearby particle pairs, while more distant particles are lumped together in increasingly large groups that are approximated as point masses. These hierarchical force calculations are accomplished most efficiently by grouping all particles in a tree (Appel 1985, Jernigan 1985, Porter 1985, Barnes & Hut 1986). Even with improved algorithms, direct modeling of globular cluster evolution for $N = 10^5$–10^6 seems impossible until computers become orders of magnitude faster and until an efficient algorithm is designed for a highly parallel execution of collisional stellar dynamics calculations. With the important but time-consuming role that individual binaries play, this will not be easy to accomplish.

3.2 *Weak-Scattering Approximations: Fokker-Planck Methods*

A statistical treatment of interparticle interactions is required to arrive at a thermodynamic description of the evolution of a globular cluster in terms of its density and temperature (i.e. velocity dispersion) as a function of radius. Such a treatment is provided by the Fokker-Planck approximation, where all two-body encounters are approximated as weak, small-angle scattering events. Excellent introductions to the use of Fokker-Planck methods can be found in the reviews by Spitzer (1987) and Binney & Tremaine (1987).

The first self-consistent solution of the Fokker-Planck equations for globular cluster evolution driven by two-body relaxation was given by Hénon (1961). He found a self-similar solution of the isotropized orbit-averaged Fokker-Planck equation by converting the integro-differential equation to an ordinary differential equation. At that time it was impossible to integrate the Fokker-Planck equation directly because of limitations in computer speed.

3.2.1 MONTE CARLO METHODS FOR APPROXIMATING FOKKER-PLANCK INTEGRATION Two different approaches to the use of Monte Carlo methods for solving the Fokker-Planck equations were introduced by Spitzer & Hart (1971) and Hénon (1971). The former followed the orbits of individual stars, adding perturbations scaled appropriately with total particle number. The latter followed the diffusion, governed by orbit-averaged perturbations, of representative "super stars" in energy–angular momentum space. Each method has its merits, and both have been shown to lead to equivalent results that are in reasonable agreement with those of N-body integration (Aarseth et al. 1974, Hénon 1975). For a recent review of Monte Carlo methods, see Shapiro (1985).

3.2.2 DIRECT INTEGRATION OF THE FOKKER-PLANCK EQUATIONS The first direct numerical solution of the Fokker-Planck equations, with no Monte Carlo approximation, was presented in two papers by Cohn (1979, 1980). The first paper solves two-dimensional orbit-averaged diffusion in energy–angular momentum space, while the second one solves the simpler one-dimensional diffusion in energy alone. The latter solution was significantly more accurate than the former and could be followed arbitrarily deep into core collapse, confirming the self-similar nature of the collapse (see Section 2). The techniques used by Cohn were developed by Cohn & Kulsrud (1978) for solving the Fokker-Planck equation in a fixed gravitational potential in order to study the effects of a central black hole (see Section 3).

Further improvements were made by Inagaki & Wiyanto (1984), who included two-mass components in a one-dimensional Fokker-Planck computation; by Cohn (1985), who reported results from multimass component one-dimensional calculations as well as from equal-mass two-dimensional calculations; and by Inagaki & Saslaw (1985), who also developed multiple-mass methods. For a recent review of multimass Fokker-Planck solutions, see Inagaki (1985).

Another way of making globular cluster evolutionary calculations more realistic is to include finite star-size effects, such as two-body dissipational encounters. Statler et al. (1987) included these effects in their Fokker-Planck calculations. They found that binaries formed by two-body tidal capture could reverse core collapse, although the subsequent reexpansion showed a density profile developing in a different way than observed in the case where all binaries are formed by three-body capture (cf. Ostriker 1985). In this case too, they found gravothermal oscillations when the time steps in the calculations were sufficiently small (Cohn et al. 1986, Cohn 1987).

Even more realistic calculations are reported by Lee & Ostriker (1986), who include tidal truncation in their cluster model, in which tidally captured binaries reverse core collapse. Complementary calculations have been performed by Lee (1987) in which every tidal encounter results in a merger. The heavier stars form after repeated merging, evolve relatively fast, and can give rise to supernovae and therefore to mass loss and expansion of the cluster. In Lee's calculations, three-body heating eventually dominates the evolution and reverses core collapse.

3.2.3 INCLUDING THE EFFECTS OF BINARIES In a weak-scattering approximation, strong interactions between binaries and single stars may be included statistically, in a manner resembling the use of reaction rates in stellar evolution models that include continuous nuclear energy sources rather than individual nuclear reactions. (In the latter case, individual scattering events are indeed computed or observed in the laboratory, but they are used only to derive statistical reaction rates.) Determination of gravitational binary reaction rates dates back to Heggie (1975), who used analytical methods, and to Hills (1975), who modeled individual scattering events through numerical orbit calculations. Earlier work concerning three-body interactions, but not formulated in terms of statistical cross sections and reaction rates, has been reviewed by Heggie (1975), and additional references may be found in the recent review by Anosova (1986).

Recent calculations of binary–single star scattering as well as binary-binary scattering have been reviewed by Hut (1985). Hut & Bahcall (1983) have reported the results of a few million three-body scattering experiments, and Hut (1984) has published a description of later, more detailed experiments in the form of an atlas of cross sections for different types of hard-binary–single-star scattering processes. Extensive numerical calculations of binary-binary scattering have been performed by Mikkola (1983a,b, 1984a,b).

3.3 *Approximating Stellar Dynamics: Conducting Gas Spheres*

The near-equilibrium structure of a cluster may be described surprisingly well by the equations of gas-dynamics, which include a short mean-free path, as compared with those of stellar dynamics, where the mean-free path of the stars is much larger than the radius of the system (see Section 3 and the discussion by Inagaki 1980). Gas-dynamical approximations to stellar-dynamical systems were pioneered by Hachisu et al. (1978), who followed the gravothermal catastrophe of a gaseous sphere using an expression for the conductivity normally used in plasmas. Their solutions are very similar to those of Larson (1970), which resulted from the more complicated moment equations of the Fokker-Planck equation (cf. Bettwieser 1983).

Although plasma physics is similar to stellar dynamics in many respects, Lynden-Bell & Eggleton (1980) showed that the expression for plasma conductivity is not directly applicable to stellar systems. They modified the effective conductivity, to be used in gas-dynamical approximations, by requiring the rate of energy transport through a stellar system to be inversely proportional to the two-body relaxation time. They showed that the implied opacity κ has a dependence on density ρ and temperature T (which is proportional to the square of the velocity dispersion) of the form $\kappa \propto T^{3.5}\rho^{-2}$.

Comparing the effective globular cluster opacity with Kramers' opacity in stellar interiors ($\kappa \propto T^{-3.5}\rho$), we immediately note two opposite tendencies: Increasing the temperature decreases the opacity of a gas but increases that of a stellar system, whereas increasing the density increases the opacity of a gas but decreases that of a stellar system. The underlying reason for this is the dependence of two-body relaxation on density and velocity dispersion: Higher densities increase the normally rather weak coupling by two-body encounters, whereas higher velocity dispersions decrease the coupling by lowering the angles of deflection in individual encounters.

At present it is not clear how far the analogy between gaseous models and stellar systems can be extended: At least in the case of multimass models, serious discrepancies have been found (Bettwieser & Inagaki 1985).

3.4 Hybrid Methods

The statistical methods in the previous two subsections rely on a two-body relaxation description of globular cluster evolution, which is appropriate during the core collapse phase of a cluster. To halt the collapse and to drive the reexpansion of the core, binaries play an essential role. The most direct way to incorporate the energy generation of the binaries and still retain some of the advantages of a predominantly statistical description of the cluster is to adopt a hybrid approach, as pioneered by McMillan & Lightman (1984a). The core region, containing about 100 stars, is modeled by direct N-body integration, using a version of Aarseth's code. It is enveloped by a large region in which the distribution of both single and binary stars is treated in a Monte Carlo Fokker-Planck fashion, and this is in turn enveloped by a region of direct Fokker-Planck integration. One of the severe complications of this type of code arises at the boundaries between the different domains, where stars lose and gain their individuality as they are immersed into, and emerge from, statistical anonymity. Extensive tests have shown, however, that these difficulties can indeed be conquered (McMillan & Lightman 1984a).

4. OBSERVATIONS

The theoretical models of the structure and evolution of globular clusters discussed in the previous sections either predict or assume different distributions of the positions and motions of the cluster stars. Isolating the most realistic models therefore requires detailed observations of the density profile and of the dependence of stellar velocities on radius within a cluster.

Three important mechanisms affect the structure of a globular cluster: (a) The external gravitational field strips away stars in the outer parts of the cluster, producing a rapid decrease in density at some limiting radius; (b) star-star encounters drive the system toward equipartition of energy and therefore mass segregation, and for each mass group, the stellar velocities approach a distribution that resembles a lowered Maxwellian; and (c) core collapse may result in a central surface brightness exceeding that predicted by King (1966a) models that are based on lowered Maxwellians. An additional factor influencing a cluster's structure is the modification of its stellar content through stellar evolution, preferential escape of low-mass stars in the presence of mass segregation, and the formation of binaries. While all these processes tend to obliterate the initial structure of a cluster, vestiges of the early stages of its evolution may be preserved in the outer parts if the time scale for two-body relaxation is long compared with the age of the cluster, or if tidal stripping is not complete.

Observations of the surface brightness profiles of globular clusters are discussed in Section 4.1 below, and observations of their internal velocity distributions are discussed in Section 4.2. Section 4.3 summarizes observations of the stellar content of globular clusters.

4.1 Surface Brightness Profiles

4.1.1 OBSERVATIONAL TECHNIQUES Typically the surface density of a globular cluster varies by a factor of at least 10^3 between the inner parts where the structure of the core is revealed, and the outer parts where the effects of tidal truncation become apparent. The optimal method for measuring the surface brightness therefore depends on the radial range of interest. Standard techniques include integrated photometry with concentric apertures (multiaperture photometry), small-aperture scans of the cluster's integrated light (driftscans), and star counts and surface photometry from CCD images and photographic plates. Qualitative discussions of these types of observations and their relative merits and disadvantages are given by King (1975, 1985) and Djorgovski (1987). More comprehensive discussions, including quantitative analyses of the errors inherent in each approach, may be found in papers by King (1966b), King et al. (1968), Illingworth & Illingworth (1976), and Hanes & Brodie (1985).

In general, a combination of techniques is necessary to obtain a full profile for a cluster. One fundamental problem, however, in constructing a composite profile is that photometric measurements are generally dominated by the cluster giants and subgiants, while star counts are dominated by the subgiants and stars on the upper main sequence. Thus, if mass segregation is present, the different methods may be sensitive to structurally different subsystems. In practice, this is not generally the case because of the small range of observable masses present, and a mixture of methods may be used.

4.1.2 OUTER STRUCTURE The shape of the outer part of a globular cluster profile is determined by the degree of tidal stripping, velocity anisotropy, and mass segregation. Single-component, isotropic, tidally truncated models (King 1966a; Section 2.1.2) provide adequate representations of cluster profiles in many cases; however, in many other cases, departures from these models indicate that one or more of the underlying assumptions are inappropriate (cf. Illingworth & Illingworth 1976; Figure 6). Multi-component models (Da Costa & Freeman 1976) allow the inclusion of mass classes with different degrees of central concentration and produce profiles that are more power law–like at intermediate radii than single-component models (Figure 7). Figure 8 shows the effect of including radial anisotropy in the outer parts of a King model: The result is essentially indistinguishable from that obtained by including mass segregation. Surface brightness profiles are therefore of limited use for determining the degree of mass segregation or anisotropy in the outer parts of a cluster, unless one or the other effect can be constrained by independent observations.

Da Costa (1982) uses star counts from photographic plates of various limiting magnitudes to look for mass segregation in 47 Tuc, NGC 6397, and NGC 6752. He finds that in 47 Tuc, the most massive stars (~ 0.8 $M_\odot$) are more centrally concentrated than the least massive observable stars (~ 0.5 $M_\odot$) at radii $\lesssim 50 r_c$. The other two clusters show no mass segregation, but this is probably because of the small range of masses and/or the limited radial range over which the observations extend and should not be interpreted as an indication of an absence of mass segregation.

Finally, very extensive profiles and accurate determinations of the background are needed to verify whether a cluster is indeed truncated at the radius predicted by theory. For example, most of the profiles by King et al. (1968) stop far short of the expected tidal limit, and even the deep profiles by Illingworth & Illingworth (1976) do not necessarily preclude halos of unbound stars. Elson et al. (1987) present profiles for 10 young clusters in the LMC, some of which are manifestly not tidally truncated and appear to have unbound halos.

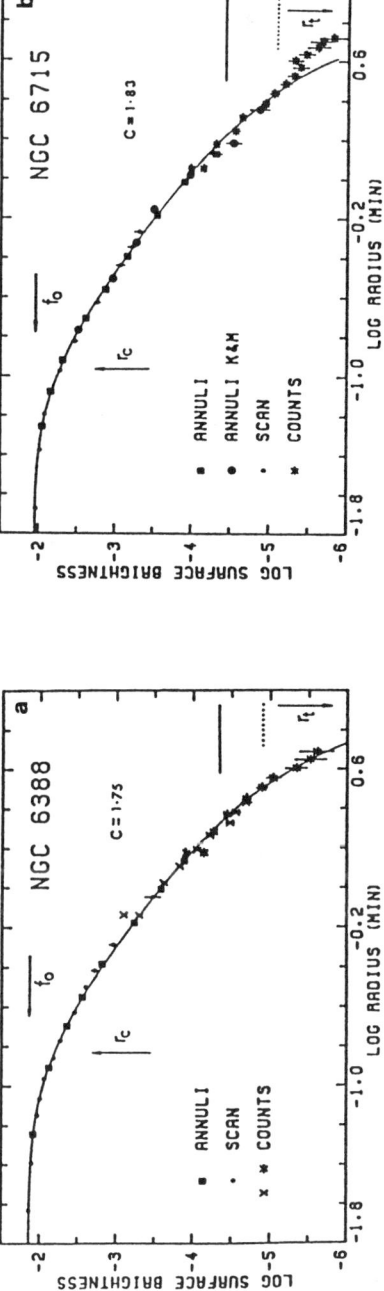

Figure 6 Surface brightness profiles for two typical globular clusters, (*a*) NGC 6388 and (*b*) NGC 6715 (Illingworth & Illingworth 1976). The best-fitting single-component King model is plotted for each cluster. Annuli, Scan, and Counts refer to data derived from centered-aperture measurements, photoelectric scans, and star counts, respectively. The central surface brightness f_0, core radius r_c, and tidal cutoff r_t are indicated. The surface brightness units are $V = 10.00$ mag arcsec^{-2}. Note that in (*a*) the King model fits well at all radii, whereas in (*b*) the model falls off too rapidly at large radii.

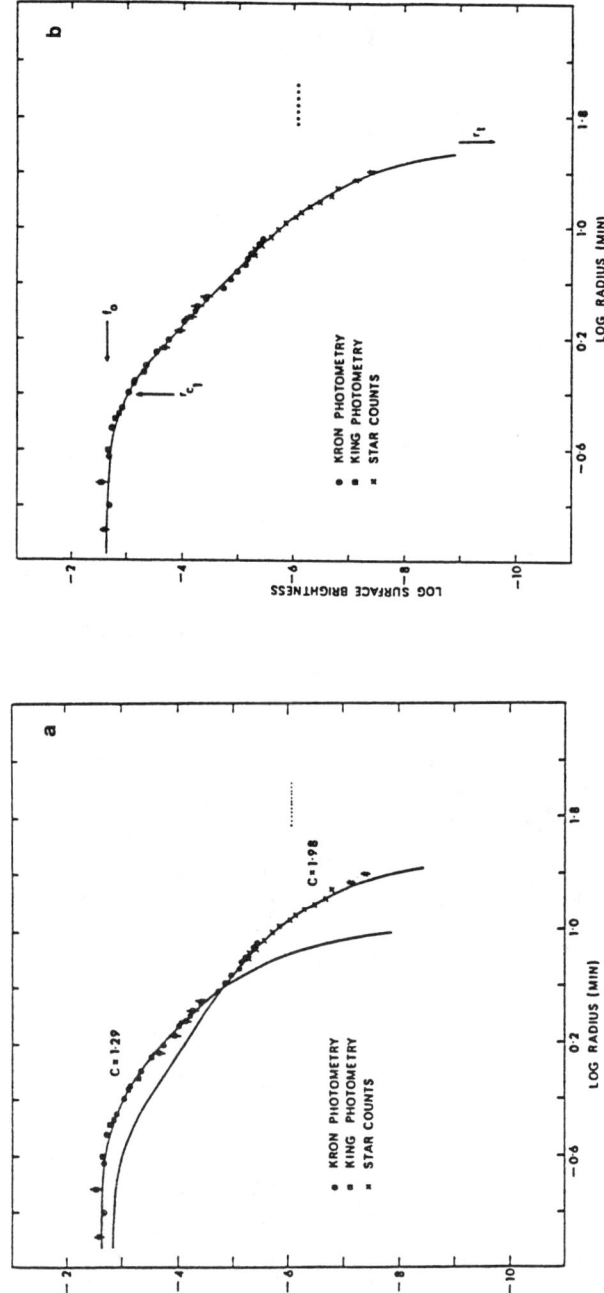

Figure 7 Comparison of the observed surface brightness profile for M3 with the surface density distribution of (*a*) single-mass models and (*b*) a multimass model (Da Costa & Freeman 1976). The curves in (*a*) are labeled with the concentration parameter $c = \log r_t/r_c$ and the surface brightness units are $V = 10.00$ mag arcsec^{-2}.

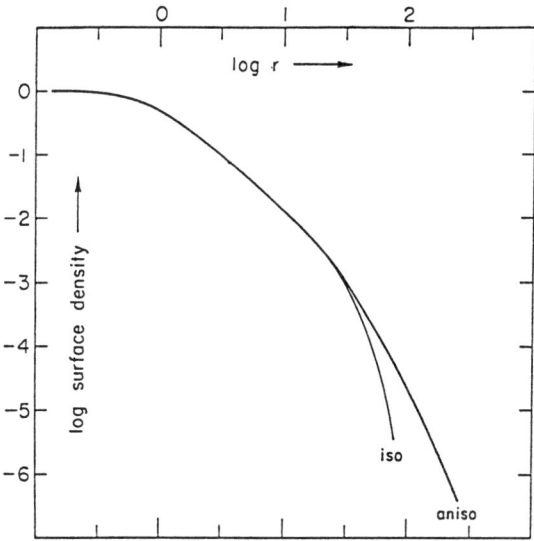

Figure 8 Projected density vs. radius, in two models with the same core, to illustrate the effect of including an anisotropic velocity distribution (King 1975).

4.1.3 CORE STRUCTURE King models have long been believed to provide accurate fits to the cores of globular clusters (Figures 6, 7), with one notable exception: The core of M15 is significantly brighter than predicted by the King models, which best represent its profile at larger radii (Newell & O'Neil 1978; Figure 9). At the time this central peak was discovered, several explanations were advanced, including a central 800-$M_\odot$ black hole (Newell et al. 1976) and a centrally concentrated population of neutron stars (Illingworth & King 1977). Recent developments in the theory of cluster evolution have, however, prompted the alternative interpretation that M15 is an example of a cluster undergoing core collapse.

If this is so, then there should be many other clusters with core profiles similar to M15. Techniques specifically suited for observing the structure of globular cluster cores include star counts and density scans from short-exposure, high-resolution photographic or electrographic plates or CCD images. The observations are inherently difficult because excessive crowding prohibits star counts and because small-aperture photometric measurements are limited by statistical sampling effects; however, the latter effects are minimized in U-band measurements, where the light is less dominated by individual bright giants.

An extensive survey of the core structure of 113 globular clusters by Djorgovski & King (1986) is currently underway: Their preliminary results

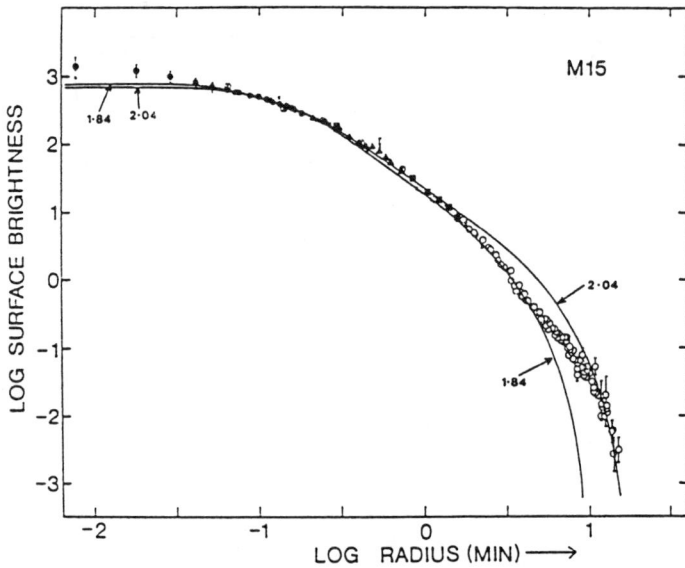

Figure 9 Composite B surface brightness profile for M15 (Newell & O'Neil 1978). The abscissa is the logarithm of the radial distance from the center of the cluster (in arcminutes), and the ordinate is the logarithm of surface brightness on an arbitrary scale. Filled symbols are from electrographic or photoelectric observations, and open symbols are from star counts. Representative error bars are shown. The solid lines represent the computed surface brightness distribution for two single-mass modified isothermal models; each curve is labeled with the concentration parameter $c = \log r_t/r_c$. In this diagram, $\log(\text{S.B.}) = 0.0$ is equivalent to $B = 22.63$ mag arcsec^{-2}. Note the central brightness excess.

indicate that about one fifth of the globular clusters show signs of a collapsed core. Table 1 contains 33 clusters with central surface brightness in excess of that of the best-fitting King model, both from the survey by Djorgovski & King and from earlier work. Djorgovski (1987) discusses trends in the properties of clusters with collapsed cores, including location, concentration, and metallicity.

4.2 Velocity Measurements

4.2.1 CENTRAL VELOCITY DISPERSIONS The central line-of-sight velocity dispersion for a globular cluster may be obtained from a high-dispersion spectrum of its integrated light either by direct matching of the cluster spectrum and that of a suitable template star, broadened by an appropriate amount, or by matching the slopes of the Fourier transforms of the same spectra (Illingworth 1976). These methods rely on finding a template star with approximately the same distribution of line strengths as the cluster

Table 1 Clusters with singular cores[a]

NGC	Name	Possible[b]	Definite[b]
362		1	
1851		3	
1904	M79	1	
4147		1	
5272	M3	8	
5946			1
6256	Ter 12		1
6266	M62	1	9
6284			1
6293			1
6325			1
6342			1
6355			1
6380	Ton 1	1	
6397			1
6453			1
6522			1
6544		1	
6558			1
6624			1, 5, 6
6626		1	
6642		1	
6652		1	
6681	M70		1
6717	Pal 9	1	
6752			1
7078	M15		1, 2, 5, 6, 7, 9
7099	M30		1, 4
—	Ter 1		1
—	Ter 2		1
—	Ter 6		1
—	Ter 9		1
—	HP		1

[a] For a tabulation of preliminary results of high-resolution observations of the cores of 113 globular clusters, see (1); only the clusters from their table marked c or $c?$ have been included here.

[b] References: (1) Djorgovski & King 1986, (2) Newell & O'Neil 1978, (3) Bahcall et al. 1977, (4) Auriere et al. 1985, (5) Lugger et al. 1985, 1986, (6) Hertz & Grindlay 1985, (7) Auriere & Cordoni 1981, (8) Auriere & Cordoni 1983, (9) Kron et al. 1984.

and on the assumption that the distribution of stellar velocities in the cluster is Gaussian. This latter assumption is likely to be satisfied, since the integrated light comes mainly from the cluster core, which is dynamically relaxed. The Fourier method is more sensitive to the small velocities found in the centers of globular clusters (typically 5–10 km s^{-1}), and relies less on finding a template star with a good match of relative line strengths of the most prominent spectral features. Central velocity dispersions have been obtained this way for 10 Galactic globular clusters (Illingworth 1976) and one old cluster in the LMC (Elson & Freeman 1985).

Alternatively, central velocity dispersions may be derived from radial velocity measurements of a sample of stars near the cluster center. Radial velocities of the stars may be determined by cross correlating their spectra with those of field stars with known radial velocities (Da Costa et al. 1977, Gunn & Griffin 1979). Peterson & Latham (1986) have derived central line-of-sight velocity dispersions for four low-concentration Galactic globular clusters from measurements of 12–19 stars per cluster.

In conjunction with a cluster's structural parameters, the central velocity dispersion may be used to estimate the central mass-to-light ratio, which in turn provides an indication of the cluster's stellar content. Illingworth (1976) found $M/L_V \approx 1-3$ for the globular clusters in his sample, which have $1.6 < c < 2$. For the low-concentration clusters ($1.1 < c < 1.5$), Peterson & Latham found $M/L_V \approx 1 \pm 1$. The old LMC cluster ($c = 1.5$) has $M/L_V \approx 0.4$. These values are consistent with the global values found from the velocity profiles discussed in the next section.

4.2.2 VELOCITY DISPERSION PROFILES   The initial stages of the collapse of a protocluster are expected to produce predominantly radial stellar motions. Two-body encounters will eventually randomize these velocities in the central regions; however, the time scale for this is sufficiently long in the outer parts of some clusters that a considerable degree of anisotropy may persist. Furthermore, two-body relaxation may enhance radial anisotropy in the halo of a cluster (cf. Lightman & Shapiro 1978). The velocity dispersion as a function of radius within a cluster, the degree of anisotropy in the outer parts, and the overall rotation of the cluster may all in principle be estimated by observing the radial velocities of a sufficiently large sample of stars over a range of radii.

Figures 10a and b show the dispersion as a function of radius for an isotropic and an anisotropic model with identical light profiles, and the observed velocity profile of M3 (Gunn & Griffin 1979). Unfortunately, the difference between the isotropic and anisotropic models is generally sufficiently small that very large samples of stars with accurate velocities are required to distinguish between them. Velocity profiles can, however,

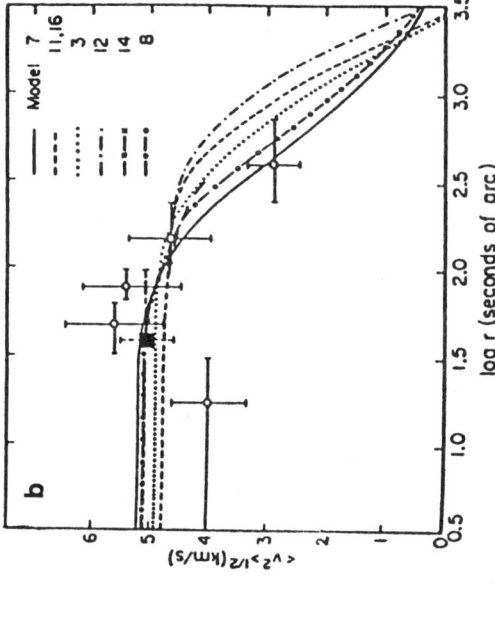

Figure 10 Radial velocity dispersion profiles (from Gunn & Griffin 1979). (*a*) An isotropic model (dashed curve) and an anisotropic model (solid curve), both of which have indistinguishable luminosity profiles. (*b*) Observed velocity dispersion for M3: Models 3, 11, and 12 are isotropic, and the rest are anisotropic.

provide indirect evidence of mass segregation. This is illustrated in Figure 11, where a multimass King model is required to fit the velocity dispersion of 47 Tuc (Da Costa & Freeman 1985).

Measurements of the velocity dispersion profiles of six Galactic globular clusters are summarized in Table 2. There is some evidence for anisotropic velocity distributions in the outer parts of most of them, and in general, rotation appears to be dynamically important. In some clusters it appears that a significant amount of dark matter, perhaps in the form of white dwarfs, is necessary in order to fit the density and velocity data with multicomponent thermal equilibrium models; however, in other cases only "missing mass" in the form of low-mass stars ($m < 0.5 \ M_\odot$) need be invoked.

4.2.3 PROPER MOTIONS Ideally one would like to observe not just the line-of-sight velocities of a large sample of stars, but also their transverse velocities, so that the three-dimensional velocity distribution of the cluster

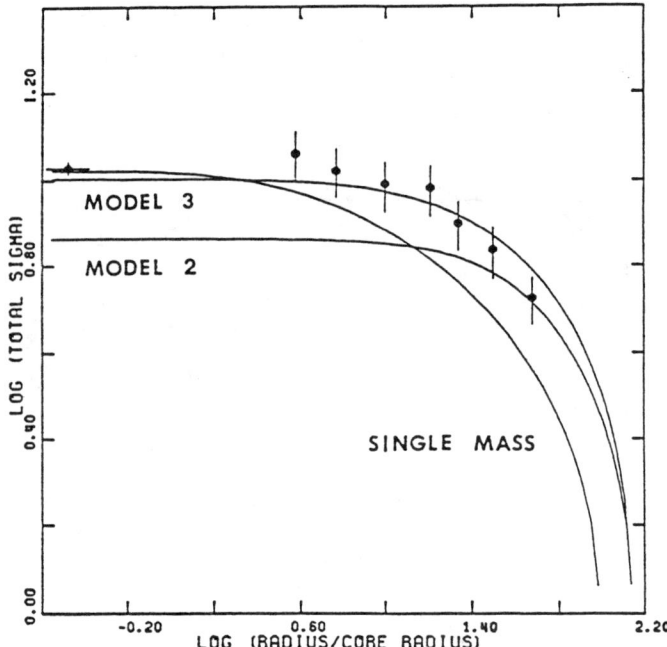

Figure 11 Line-of-sight velocity dispersion as a function of radius in 47 Tuc. Error bars are $\pm 1\sigma$. Curves are isotropic single-mass or, in the case of models 2 and 3, multimass King models (Da Costa & Freeman 1985). Note that the single-mass model, scaled to fit the observed central velocity dispersion, does not represent the data well.

Table 2 Velocity dispersion profiles for globular clusters

NGC	Name	N^a	Error[b] (km s^{-1})	Anisotropy[c] (r_c)	Rotational V_{max}^d (km s^{-1})	Dark mass[e] (%)	Source[f]
104	47 Tuc	135	~3	—	6	30–40	4
104	47 Tuc	272	0.6	slight	6.5	10–40	1
5139	ω Cen	318	0.9	slight	8	~40	1
5272	M3	111	1.0	15	0	No	2
6205	M13	147	~1	5	5	No	3
6341	M92	49	~1	20	~3	No	3
7089	M2	69	~1	5–10	~3	3–14	5

[a] Number of stars for which radial velocities have been determined.
[b] Accuracy of velocity measurements.
[c] Radius at which anisotropy becomes important.
[d] Maximum rotation velocity.
[e] Is dark matter, including heavy remnants and white dwarfs but not low-mass stars, required to make multicomponent thermal equilibrium models fit?
[f] References: (1) Meylan & Mayor 1986, (2) Gunn & Griffin 1979, (3) Lupton et al. 1987, (4) Da Costa & Freeman 1985, (5) Pryor et al. 1986.

may be constructed. This requires proper motions, which for stars in globular clusters are typically $\lesssim 10^{-4}$ arcsec yr^{-1}. Proper motions have been measured for seven globular clusters: M3 (Cudworth 1979a), M5 (Cudworth 1979b), M13 (Cudworth & Monet 1979), M15 (Cudworth 1976a), M22 (Cudworth 1985), M71 (Cudworth 1986), and M92 (Cudworth 1976b). The measurements are not of high enough accuracy to use directly, but they can be analyzed statistically, provided the errors are understood. Lupton et al. (1987) discuss the velocity distribution of M13 based on Cudworth & Monet's measurements of proper motions for 330 cluster members. They find isotropic velocities in the central regions and anisotropic velocities beyond about $7r_c$, in good agreement with the estimate of the anisotropy radius based on the radial velocity dispersion and surface brightness profile.

4.3 Stellar Content

4.3.1 INITIAL-MASS FUNCTIONS A knowledge of the stellar content of a globular cluster is important in understanding its dynamical evolution because the relative proportion of stars of different mass influences the rate at which various evolutionary processes occur. In addition, the presence of a dark component can lead to a misinterpretation of the luminous data. Recently, main-sequence luminosity functions based on CCD photometry have been published for nine globular clusters (Harris & Hesser 1985, Lupton & Gunn 1986, McClure et al. 1986, and references therein; Figure

12). These functions sample the magnitude range $+4.0 < M_V < +9.5$, which for a cluster of intermediate metallicity ([Fe/H] ~ 1.5) corresponds to a range in mass from $\approx 0.45\ M_\odot$ to the turnoff at $\approx 0.8\ M_\odot$. In most cases the number of stars at faint magnitudes increases monotonically to the limit of the observations.

There appears to be a significant variation in the slope of the luminosity function among clusters. This could in part be due to the effect of mass segregation, which is expected to be present in globular clusters (see Section 4.1.2), since the luminosity functions are in general drawn from different radii in each cluster. However, McClure et al. (1986) find a correlation between luminosity function slope and cluster metallicity in the sense that high-metallicity clusters have relatively fewer low-mass stars.

For a general review of mass functions both in globular clusters and elsewhere, see Scalo (1986).

4.3.2 BINARY STARS AND UNSEEN COMPONENTS The formation of binary stars in the core of a globular cluster can significantly affect the dynamical evolution of the cluster, as discussed in Sections 2 and 3. High- and low-

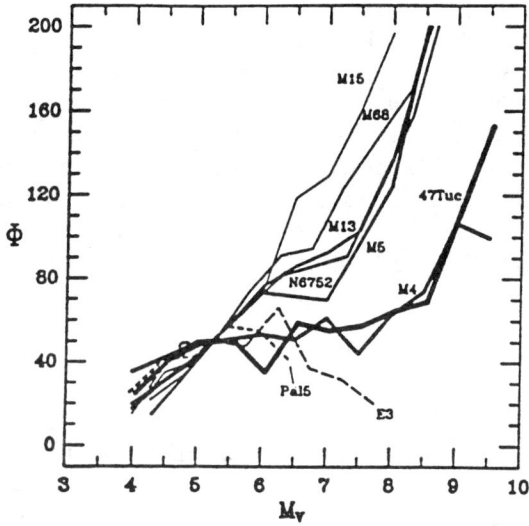

Figure 12 Luminosity functions for globular clusters, obtained from CCD observations. Here Φ is the number of stars in a magnitude interval $\Delta M_V = 0.5$ mag. The thickness of the curves increases with the metal abundance of the cluster. The observations for the two sparse clusters Pal 5 and E3 were obtained in the cluster centers and are almost certainly severely affected by dynamics (McClure et al. 1986).

luminosity X-ray sources in globular clusters indicate the presence of binaries containing a neutron star in the former case (cf. Lewin 1980) and a white dwarf in the latter (Hertz & Grindlay 1983, Grindlay et al. 1984, Hertz & Wood 1985). Aside from the X-ray candidates, two likely close binaries have been identified spectroscopically in M5 and M30 by Margon et al. (1981) and Margon & Downes (1983). Other spectroscopic observations have, however, yielded little evidence for binary stars in globular clusters. Gunn & Griffin (1979) found no spectroscopic binaries in M3 among the 111 giants they observed. Harris & McClure (1983) demonstrated, however, that this should not be interpreted as an indication of a deficiency of binaries, since a significant number could have gone undetected. Nevertheless, another more detailed search for spectroscopic binaries in M3 yielded only one candidate out of 112 giants (Pryor et al. 1985). Shara et al. (1985) searched for cataclysmic variables in M3 between 4 and $30 r_c$ and found none. In other clusters, Mayor et al. (1984) found no spectroscopic binaries in 47 Tuc at $r > 5 r_c$, and Lupton & Gunn (1986) found no evidence for a secondary main sequence corresponding to equal-mass binaries in M13; however, these observations should not be considered as evidence for the absence of binaries in globular clusters.

The presence of currently unobservable components in a globular cluster can also have a significant effect on the cluster's structure and evolution. Larson (1984) considers the observational consequences of black hole remnants with $M \sim 2.5\ M_\odot$, and Peebles (1984) discusses observational tests for globular cluster halos composed of primordial dark matter.

5. FUTURE DIRECTIONS

From a purely theoretical point of view, the occurrence of gravothermal oscillations in the idealized models analyzed so far has been convincingly established. This is an important result for the area of mathematical physics, where a statistical treatment of long-range forces has been notoriously difficult. The solutions of the equations of motion for the most symmetric, isolated, self-gravitating N-body system are surprisingly rich in revealing unexpected oscillations (Bettwieser & Sugimoto 1984), which may be understood in terms of a linear stability analysis (Goodman 1987). However, many problems remain. For example, the interpretation of the overstable solutions found by Goodman must be clarified, the relationship between gaseous and stellar-dynamic models must be established, and the effects of granularity in N-body systems (which have so far been neglected in statistical studies) must be investigated.

From a phenomenological point of view, it is not yet clear whether core oscillations actually occur in globular clusters. The simplifications inherent in the mathematical models with oscillatory solutions may not be essential, but this has not yet been proven, and no consensus has been reached. This question provides an important stimulus for devising more realistic models of globular cluster evolution. Progress along these lines has been impressive in the last five years and is expected to continue in the coming few years. A more remote goal is a realistic N-body calculation in which every star is modeled by a point mass with additional internal degrees of freedom; this, however, will require $\gtrsim 10^{17}$ floating-point operations, i.e. at least several years on present-day supercomputers.

From an observational point of view, detailed measurements of the structure of the cores of a large sample of globular clusters both in our own and nearby galaxies are needed to clarify the role of gravothermal instabilities in the evolution of clusters (cf. Djorgovski & King 1986). Accurate radial velocity measurements and proper motions for large samples of stars in globular clusters will increase our understanding of their internal dynamics. Present observations of the mass functions of globular clusters are limited to stars above $\sim 0.5\ M_\odot$; observations of low-mass stars and evolved stars would be of great use in constraining models of globular clusters in which the total amount of mass in low-mass stars and remnants is presently unknown. While progress in some of these areas is possible from ground-based telescopes, other observations must await the Hubble Space Telescope (cf. Bahcall 1985, 1987).

ACKNOWLEDGMENTS

We thank John Bahcall, Josh Barnes, Ken Freeman, Jeremy Goodman, Robert Lupton, Jeremiah Ostriker, and Lyman Spitzer for helpful comments on the manuscript. SI thanks the Institute for Advanced Study, Princeton, where this work was completed, for its hospitality. Part of this work was supported by the Alfred P. Sloan Foundation, by the National Science Foundation through grant PHY-8240263, and by the Ministry of Education, Science, and Culture of Japan through grant 60540159. SI also thanks the Yamada Science Foundation for supporting his travel to the IAS. RAWE acknowledges support from NSF grant PHY-8217352 and the Corning Glass Works Foundation.

Literature Cited

Aarseth, S. J. 1985. In *Multiple Time Scales*, ed. J. U. Brackhill, B. I. Cohen, p. 377. New York: Academic

Aarseth, S. J., Hénon, M., Wielen, R. 1974. *Astron. Astrophys.* 37: 183

Aarseth, S. J., Inagaki, S. 1986. In *The Use*

of Supercomputers in Stellar Dynamics, ed. P. Hut, S. McMillan, p. 203. Berlin: Springer-Verlag
Aarseth, S. J., Lecar, M. 1975. Ann. Rev. Astron. Astrophys. 13: 1
Aguilar, L. A., Hut, P., Ostriker, J. P. 1985. Bull. Am. Astron. Soc. 16: 947
Angeletti, L., Giannone, P. 1980. Astron. Astrophys. 85: 113
Anosova, J. P. 1986. Astrophys. Space Sci. 124: 217
Antonov, V. A. 1962. Vestn. Leningr. Univ. 7: 135
Appel, A. 1985. SIAM J. Sci. Stat. Comput. 6: 85
Applegate, J. 1986. Ap. J. 301: 132
Auriere, M., Cordoni, J. P. 1981. Astron. Astrophys. 100: 307
Auriere, M., Cordoni, J. P. 1983. Astron. Astrophys. Suppl. 52: 383
Auriere, M., Cordoni, J. P., LeFevre, O., Terzan, A. 1985. See Goodman & Hut 1985, p. 63
Bahcall, J. 1985. See Goodman & Hut 1985, p. 481
Bahcall, J. 1987. In Globular Cluster Systems in Galaxies, IAU Symp. No. 126, ed. J. Grindlay, A. G. D. Philip, p. 455. Dordrecht: Reidel
Bahcall, N. A., Lasker, B. M., Wamsteker, W. 1977. Ap. J. Lett. 213: L105
Barnes, J., Hut, P. 1986. Nature 324: 446
Bettwieser, E. 1983. MNRAS 203: 185
Bettwieser, E. 1985. See Goodman & Hut 1985, p. 219
Bettwieser, E., Fritze, U. 1984. Publ. Astron. Soc. Jpn. 36: 403
Bettwieser, E., Inagaki, S. 1985. MNRAS 213: 473
Bettwieser, E., Sugimoto, D. 1984. MNRAS 208: 439
Binney, J. J., Tremaine, S. D. 1987. Galactic Astronomy, Vol. 2: Galactic Dynamics. Princeton, NJ: Princeton Univ. Press
Chandrasekhar, S. 1942. Principles of Stellar Dynamics. Chicago: Univ. Chicago Press
Chernoff, D. F. 1987. In Globular Cluster Systems in Galaxies, IAU Symp. No. 126, ed. J. Grindlay, A. G. D. Philip, p. 283. Dordrecht: Reidel
Chernoff, D. F., Kochanek, C. S., Shapiro, S. L. 1986. Ap. J. 309: 183
Cohn, H. 1979. Ap. J. 234: 1036
Cohn, H. 1980. Ap. J. 242: 765
Cohn, H. 1985. See Goodman & Hut 1985, p. 161
Cohn, H. 1987. In Globular Cluster Systems in Galaxies, IAU Symp. No. 126, ed. J. Grindlay, A. G. D. Philip, p. 379. Dordrecht: Reidel
Cohn, H., Hut, P. 1984. Ap. J. Lett. 277: L45
Cohn, H., Kulsrud, R. M. 1978. Ap. J. 226: 1087
Cohn, H., Wise, M. W., Yoon, T. S., Statler, S. T., Ostriker, J. P., Hut, P. 1986. In The Use of Supercomputers in Stellar Dynamics, ed. P. Hut, S. McMillan, p. 206. Berlin: Springer-Verlag
Cudworth, K. M. 1976a. Astron. J. 81: 519
Cudworth, K. M. 1976b. Astron. J. 81: 975
Cudworth, K. M. 1979a. Astron. J. 84: 1312
Cudworth, K. M. 1979b. Astron. J. 84: 1866
Cudworth, K. M. 1985. Astron. J. 90: 65
Cudworth, K. M. 1986. Astron. J. 92: 348
Cudworth, K. M., Monet, D. G. 1979. Astron. J. 84: 774
Da Costa, G. S. 1982. Astron. J. 87: 990
Da Costa, G. S., Freeman, K. C. 1976. Ap. J. 206: 128
Da Costa, G. S., Freeman, K. C. 1985. See Goodman & Hut 1985, p. 69
Da Costa, G. S., Freeman, K. C., Kalnajs, A. J., Rodgers, A. W., Stapinski, T. E. 1977. Astron. J. 82: 810
Djorgovski, S. G. 1987. In Globular Cluster Systems in Galaxies, IAU Symp. No. 126, ed. J. Grindlay, A. G. D. Philip, p. 333. Dordrecht: Reidel
Djorgovski, S. G., King, I. R. 1986. Ap. J. Lett. 305: L61
Dokuchaev, V. I., Ozernoy, L. M. 1982a. Astron. Astrophys. 111: 1
Dokuchaev, V. I., Ozernoy, L. M. 1982b. Astron. Astrophys. 111: 16
Eddington, A. S. 1926. The Internal Constitution of the Stars. Cambridge: Cambridge Univ. Press
Elson, R. A. W., Fall, S. M., Freeman, K. C. 1987. Ap. J. In press
Elson, R. A. W., Freeman, K. C. 1985. Ap. J. 288: 521
Fabian, A. C., Pringle, J. E., Rees, M. J. 1975. MNRAS 172: 15P
Fall, S. M., Rees, M. 1985. Ap. J. 298: 18
Goodman, J. 1984. Ap. J. 280: 298
Goodman, J. 1987. Ap. J. 313: 576
Goodman, J., Hut, P., eds. 1985. Dynamics of Star Clusters, IAU Symp. No. 113. Dordrecht: Reidel
Grindlay, J., Hertz, P., Steiner, J. E., Murray, S. S., Lightman, A. P. 1984. Ap. J. Lett. 282: L13
Gunn, J. E., Griffin, R. F. 1979. Astron. J. 84: 752
Hachisu, I., Nakada, Y., Nomoto, K., Sugimoto, D. 1978. Prog. Theor. Phys., Kyoto 60: 393
Hachisu, I., Sugimoto, D. 1978. Prog. Theor. Phys., Kyoto 60: 123
Hanes, D. A., Brodie, J. P. 1985. MNRAS 214: 491
Harris, W. E., Hesser, J. E. 1985. See Goodman & Hut 1985, p. 81
Harris, W. E., McClure, R. D. 1983. Ap. J. Lett. 265: L77

Heggie, D. C. 1975. *MNRAS* 173: 729
Heggie, D. C. 1979. *MNRAS* 188: 525
Heggie, D. C. 1984. *MNRAS* 206: 179
Heggie, D. C. 1985. See Goodman & Hut 1985, p. 139
Heggie, D. C., Ramamani, N. 1986. In preparation
Hénon, M. 1961. *Ann. Astophys.* 24: 369
Hénon, M. 1965. *Ann. Astophys.* 28: 62
Hénon, M. 1971. *Astrophys. Space Sci.* 14: 151
Hénon, M. 1973. In *Dynamical Structure and Evolution of Stellar Systems, 3rd, Adv. Course Swiss Soc. Astron. Astrophys.*, ed. L. Martinet, L. Mayor, p. 224. Sauverny: Geneva Obs.
Hénon, M. 1975. In *Dynamics of Stellar Systems, IAU Symp. No. 69*, ed. A. Hayli, p. 133. Dordrecht: Reidel
Hertz, P., Grindlay, J. E. 1983. *Ap. J.* 275: 105
Hertz, P., Grindlay, J. E. 1985. *Ap. J.* 298: 95
Hertz, P., Wood, K. S. 1985. *Ap. J.* 290: 171
Hills, J. G. 1975. *Astron. J.* 80: 809
Hut, P. 1984. *Ap. J. Suppl.* 55: 111
Hut, P. 1985. See Goodman & Hut 1985, p. 231
Hut, P., Bahcall, J. N. 1983. *Ap. J.* 268: 319
Hut, P., Inagaki, S. 1985. *Ap. J.* 298: 502
Illingworth, G. 1976. *Ap. J.* 204: 73
Illingworth, G., Illingworth, W. 1976. *Ap. J. Suppl.* 30: 227
Illingworth, G., King, I. R. 1977. *Ap. J. Lett.* 218: L109
Inagaki, S. 1980. *Publ. Astron. Soc. Jpn.* 32: 213
Inagaki, S. 1984. *MNRAS* 206: 149
Inagaki, S. 1985. See Goodman & Hut 1985, p. 189
Inagaki, S. 1986. *Publ. Astron. Soc. Jpn.* 38: 853
Inagaki, S. 1987. In *Globular Cluster Systems in Galaxies, IAU Symp. No. 126*, ed. J. Grindlay, A. G. D. Philip, p. 367. Dordrecht: Reidel
Inagaki, S., Lynden-Bell, D. 1983. *MNRAS* 205: 913
Inagaki, S., Saslaw, W. C. 1985. *Ap. J.* 292: 339
Inagaki, S., Wiyanto, P. 1984. *Publ. Astron. Soc. Jpn.* 36: 391
Innanen, K. A. 1979. *Astron. J.* 84: 960
Jernigan, G. 1985. See Goodman & Hut 1985, p. 275
Katz, J. 1980. *MNRAS* 190: 497
King, I. R. 1962. *Astron. J.* 67: 471
King, I. R. 1966a. *Astron. J.* 71: 64
King, I. R. 1966b. *Astron. J.* 71: 276
King, I. R. 1975. In *Dynamics of Stellar Systems, IAU Symp. No. 69*, ed. A. Hayli, p. 99. Dordrecht: Reidel
King, I. R. 1985. See Goodman & Hut 1985, p. 1
King, I. R., Hedemann, E., Hodge, S. M., White, R. E. 1968. *Astron. J.* 73: 456
Kron, G. E., Hewitt, A. V., Wasserman, L. H. 1984. *Publ. Astron. Soc. Pac.* 96: 198
Larson, R. B. 1970. *MNRAS* 147: 323
Larson, R. B. 1984. *MNRAS* 210: 763
Lee, H. M. 1987. Submitted for publication
Lee, H. M., Ostriker, J. P. 1986. *Ap. J.* 310: 176
Lee, H. M., Ostriker, J. P. 1987. *Ap. J.* In press
Lewin, W. H. G. 1980. In *Globular Clusters*, ed. D. Hanes, B. Madore, p. 315. Cambridge: Cambridge Univ. Press
Lightman, A. P., Shapiro, S. L. 1978. *Rev. Mod. Phys.* 50: 437
Lugger, P. M., Cohn, J. E., Grindlay, J. E. 1985. See Goodman & Hut 1985, p. 89
Lugger, P. M., Cohn, J. E., Grindlay, J. E., Bailyn, C., Hertz, P. 1986. Preprint
Lupton, R., Gunn, J. E. 1986. *Astron. J.* 91: 317
Lupton, R., Gunn, J. E., Griffin, R. F. 1987. *Astron. J.* 93: 1114
Lynden-Bell, D. 1975. In *Dynamics of Stellar Systems, IAU Symp. No. 69*, ed. A. Hayli, p. 443. Dordrecht: Reidel
Lynden-Bell, D., Eggleton, P. P. 1980. *MNRAS* 191: 483
Lynden-Bell, D., Wood, R. 1968. *MNRAS* 138: 495
Makino, J. 1986. In *The Use of Supercomputers in Stellar Dynamics*, ed. P. Hut, S. McMillan, p. 151. Berlin: Springer-Verlag
Makino, J., Tanekusa, J., Sugimoto, D. 1986. *Publ. Astron. Soc. Jpn.* 38: 865
Makino, J., Hut, P. 1987. Preprint
Margon, B., Downes, R. A. 1983. *Ap. J. Lett.* 274: L31
Margon, B., Downes, R. A., Gunn, J. E. 1981. *Ap. J. Lett.* 247: L89
Mathieu, R. D. 1983. *Ap. J. Lett.* 267: L97
Mayor, M., Imbert, M., Andersen, J., Ardberg, A., Benz, W., et al. 1984. *Astron. Astrophys.* 134: 118
McClure, R. D., Vandenberg, D. A., Smith, G. H., Fahlman, G. G., Richer, H. B., et al. 1986. *Ap. J. Lett.* 307: L49
McMillan, S. L. W. 1986. *Ap. J.* 307: 126
McMillan, S. L. W., Lightman, A. P. 1984a. *Ap. J.* 283: 801
McMillan, S. L. W., Lightman, A. P. 1984b. *Ap. J.* 283: 813
McMillan, S. L. W., McDermott, P. N., Taam, R. E. 1987. *Ap. J.* In press
Merritt, D. 1981. *Astron. J.* 86: 318
Merritt, D. 1983. *Ap. J.* 264: 24
Merritt, D. 1984. *Ap. J.* 276: 26
Merritt, D. 1985. *Ap. J.* 289: 18
Meylan, G., Mayor, M. 1986. *Astron. Astrophys.* 166: 122

Mikkola, S. 1983a. *MNRAS* 203: 1107
Mikkola, S. 1983b. *MNRAS* 205: 733
Mikkola, S. 1984a. *MNRAS* 207: 115
Mikkola, S. 1984b. *MNRAS* 208: 75
Milgrom, M., Shapiro, S. L. 1978. *Ap. J.* 223: 991
Newell, B., Da Costa, G. S., Norris, J. 1976. *Ap. J. Lett.* 208: L55
Newell, B., O'Neil, E. J. 1978. *Ap. J. Suppl.* 37: 27
Ostriker, J. P. 1985. See Goodman & Hut 1985, p. 347
Ostriker, J. P., Spitzer, L., Chevalier, R. A. 1972. *Ap. J. Lett.* 176: L51
Peebles, P. J. E. 1984. *Ap. J.* 277: 470
Peterson, R. C., Latham, D. W. 1986. *Ap. J.* 305: 645
Porter, D. 1985. PhD thesis. Physics Dept., Univ. Calif., Berkeley
Press, W. H., Teukolsky, S. A. 1977. *Ap. J.* 213: 183
Pryor, C. P., Latham, D. W., Hazen-Liller, M. L. 1985. See Goodman & Hut 1985, p. 99
Pryor, C., McClure, R. D., Fletcher, J. M., Hartwick, F. D. A., Kormendy, J. 1986. *Astron. J.* 91: 546
Saslaw, W. C. 1985. *Gravitational Physics of Stellar and Galactic Systems*. Cambridge: Cambridge Univ. Press
Scalo, J. 1986. *Fundam. Cosmic Physics* 11: 1
Seitzer, P. 1985. See Goodman & Hut 1985, p. 343
Seitzer, P., Freeman, K. C. 1986. Preprint
Shapiro, S. L. 1985. See Goodman & Hut 1985, p. 373
Shara, M. M., Moffat, A. F. J., Hanes, D. A. 1985. See Goodman & Hut 1985, p. 103
Spitzer, L. Jr. 1969. *Ap. J. Lett.* 158: L139
Spitzer, L. Jr. 1975. In *Dynamics of Stellar Systems, IAU Symp. No. 69*, ed. A. Hayli, p. 3. Dordrecht: Reidel
Spitzer, L. Jr. 1985. See Goodman & Hut 1985, p. 109
Spitzer, L. 1987. *Dynamical Evolution of Globular Clusters*. Princeton: Princeton Univ. Press
Spitzer, L., Chevalier, R. A. 1973. *Ap. J.* 183: 563
Spitzer, L., Hart, M. H. 1971. *Ap. J.* 166: 483
Spitzer, L., Mathieu, R. D. 1980. *Ap. J.* 241: 618
Statler, T. S., Ostriker, J. P., Cohn, H. 1987. *Ap. J.* 316: 626
Stodółkiewicz, J. S. 1982. *Acta Astron.* 32: 63
Sugimoto, D. 1985. See Goodman & Hut 1985, p. 207
Sugimoto, D., Bettwieser, E. 1983. *MNRAS* 204: 19P
Vishniac, E. T. 1978. *Ap. J.* 223: 986
Wiyanto, P., Kato, S., Inagaki, S. 1985. *Publ. Astron. Soc. Jpn.* 37: 715
Yoshizawa, M., Inagaki, S., Nishida, M. T., Kato, S., Tanaka, Y., Watanabe Y. 1978. *Publ. Astron. Soc. Jpn.* 30: 279

THE GALACTIC SPHEROID AND OLD DISK

K. C. Freeman

Mount Stromlo and Siding Spring Observatories, The Australian National University, Canberra, A.C.T. 2606, Australia

1. INTRODUCTION

In the last few years, there has been some progress in understanding the structure of the Galaxy and the motions of stars in its various components. Some of this progress comes from the recent activity in automated stellar photometry and star counts, which has been recently reviewed by Bahcall (3). Some of it comes from the relative ease with which accurate radial velocities and metallicities of faint stars can now be measured. This is now a very active field of research. The goal is to know the present dynamical and chemical state of the Galaxy, which is an essential step in understanding how it formed.

In this review, I discuss some of the major Galactic components, such as the old disk, the thick disk, the metal-weak halo, the globular cluster system, and the Galactic bulge. The emphasis is on the structure, kinematics, and general chemical properties of these components. I do not attempt to review the properties of the young disk or the dark corona, nor do I discuss the chemical evolution of the Galaxy in any detail, although these topics occasionally come up.

Although this review is broken down into sections that deal with individual components, the distinction between the various components is not always clean and well understood. The thin disk *is* well defined: It is flat and close to rotational equilibrium, and its formation history was surely dissipative. The metal-weak halo is also fairly well defined: It is spheroidal and rotates slowly, and it is supported by the random motions of its stars. The central bulge and the metal-weak halo are together known as the spheroidal component of the Galaxy. However, the dynamical and cosmogonical relationship between the bulge and the halo is not yet clear. The thick disk, which was identified only recently, was originally believed to be an intermediate component of the spheroid. Now it seems more natural to see it as an immediate precursor of the thin disk itself. The globular cluster system probably belongs partly to the halo and partly to this thick disk. Despite this uncertainty about the nature of some of the components, I have broken the review up in this way because these structural components are now well recognized. Without detailed information about the structure and stellar motions of each component, it is difficult to understand how the components fit in with the dynamics and formation history of the Galaxy.

For some basic concepts in Galactic dynamics, see Binney & Mihalas (11) and Freeman (39). Oort's review (80) is essential reading. The subject of Galactic populations has been reviewed by Delhaye (23), Blaauw (14), and Sandage (99). In the last few years, there have been several conferences (35, 50, 106, 125) that have discussed various aspects of Milky Way structure and dynamics. Two conferences in 1986 (46, 73) are particularly relevant to the topics covered in this review.

Some notation: (R, φ, z) are cylindrical coordinates with the origin at the Galactic center; $z = 0$ is the Galactic plane. Alternatively, (r, φ, θ) are spherical coordinates, with $\theta = 90°$ in the Galactic plane. Stellar velocities (U, V, W) are relative to the local standard of rest; the components are positive toward $(l = 180°, b = 0)$, $(l = 90°, b = 0)$, and $(b = 90°)$, respectively. The rotational velocity for a population of stars, $v_{\rm rot}$, is the mean φ component of their velocities relative to a nonrotating frame. The velocity dispersion components in the (R, φ, z) directions are denoted $(\sigma_R, \sigma_\varphi, \sigma_z)$.

2. THE OLD DISK

2.1 *The Age-Metallicity and the Age-Velocity Relations*

The Galactic disk is a flat, rapidly rotating component that contains most of the Galactic luminous mass (about 6×10^{10} $M_\odot$). It includes stars covering a wide range of ages, abundances, and kinematical properties. The oldest stars may be almost as old as the globular clusters (29, 100). The [Fe/H] values for disk stars appear to lie between about -0.5 and 0.1 (e.g. 52). The chemical and kinematical properties of nearby disk stars are correlated with their age, in the sense that the oldest stars have the lowest [Fe/H] values, the largest velocity dispersions, and therefore the largest scale heights. The increase in velocity dispersion with age is believed to come from scattering by massive gas clouds (111, 112), although it is not clear that such scattering can produce the entire velocity dispersion of the oldest disk stars (66), or by interaction with transient spiral waves and smaller scale instabilities (8, 105, 128).

The precise forms of the age-metallicity relation (AMR) and the age-velocity relation (AVR) are not yet well established. For example, Twarog's derivation of the AMR (115) shows a fairly steady rise of [Fe/H] from about −0.6 at ages of 12 Gyr to about the solar value at 5 Gyr; from this point, the AMR is then approximately constant to the present time. The AMR derived more recently by Nissen et al. (72) has similar form. These relations contrast with that of another study (20), using a subset of Twarog's stars, showing a more gentle rise from about −0.2 at 15 Gyr to about 0.1 for the youngest stars. The differences between the two investigations of Twarog's stars come from systematic differences in the metallicity calibrations and in the isochrones used to derive ages.

There are also systematic differences between various derivations of the AVR. Wielen's (127) AVR shows a steady monotonic rise of the velocity dispersion with age: For example, the velocity dispersion components (σ_R, σ_z) rise from (15, 5) km s^{-1} at age 0.2 Gyr to (45, 22) km s^{-1} at 10 Gyr. Mayor's (68) AVR is fairly similar. This form of the AVR fits well with the predictions of a diffusive heating model (128, 129). On the other hand, Carlberg et al. (20) find that the tangential velocity dispersion for a sample of F stars at the south Galactic pole (SGP) rises steadily to a value of about 40 km s^{-1} at age 5 Gyr and is then constant for older stars. This form of the AVR is well fit by the spiral-wave heating model (105). These two forms of the AVR come from separate samples of stars; the differences between them are probably at least in part due to the different procedures for estimating the ages of the stars.

The history of the star formation rate in the disk has been discussed in several studies (e.g. 20, 69, 116). These studies conclude that the star formation rate has been approximately constant, within about a factor of two. Most of the stars in the disk are thus older than a few billion years.

In the solar neighborhood, the younger stars (ages $\lesssim 1$ Gyr) are not dynamically well mixed. In the (U, V) plane, the principal axes of the velocity distribution for the younger stars are rotated away from the (U, V) axes. Figure 1, from (28), illustrates this vertex deviation for nearby F stars. The velocity distribution in Figure 1a is dominated by a few clumps or moving groups that produce the vertex deviation. The nature of these moving groups is not yet fully understood; they are probably the unmixed debris of dissolving stellar aggregates. The older stars in Figure 1b do not show this effect. [There are also several known moving groups among the old-disk stars in the solar neighborhood (33). Although they are dynamically very interesting, their membership is fairly sparse, and thus they do not contribute significant structure to the distribution of old-disk stars in the (U, V) plane.]

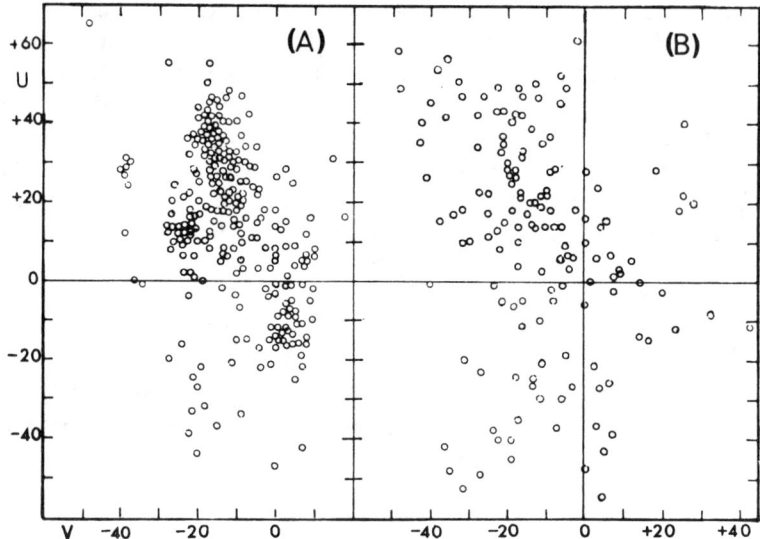

Figure 1 Distribution of the (U, V) velocities of 430 F stars (a) younger and (b) older than NGC 752 [from (28)].

The two major clumps in Figure 1a are the Hyades and Sirius moving groups. Using a simple theory of dissolving aggregates that predicts the radial velocities of more distant group members, Wilson (130) attempted to trace the path of these two groups away from the immediate solar neighborhood. The theory predicts that group members at a distance of about 500 pc will be found in the longitude range 240° to 305°, so he measured radial velocities for several hundred K giants in this longitude range, near the Galactic plane and out to the 500-pc distance. Figure 2 shows the radial velocity vs. longitude distribution for stars with $-0.2 < [Fe/H] < 0.3$ (the younger stars) and with $-0.7 < [Fe/H] < -0.2$. Again, the distribution of the younger stars is dominated by the two moving groups or streams, whose predicted velocity-longitude loci are shown by the continuous lines. The older stars are smoothly distributed. Figure 2 shows that disk stars older than about 1 Gyr ($[Fe/H] < -0.2$) are dynamically well mixed. Wilson's stars give another measure of the age-velocity dispersion relation in the form of σ_φ vs. [Fe/H]: σ_φ increases smoothly with decreasing [Fe/H] to about 28 km s^{-1} at $[Fe/H] = -0.6$. [We note that σ_R and σ_φ are not independent; epicyclic theory gives the relation $(\sigma_\varphi/\sigma_R)^2 = -B/(A-B)$, where A and B are the local Oort (80) constants.] An earlier demonstration of the velocity dispersion–metallicity relation for nearby disk stars was given by Janes (59) from radial velocities and proper motions of about 800 G and K giants.

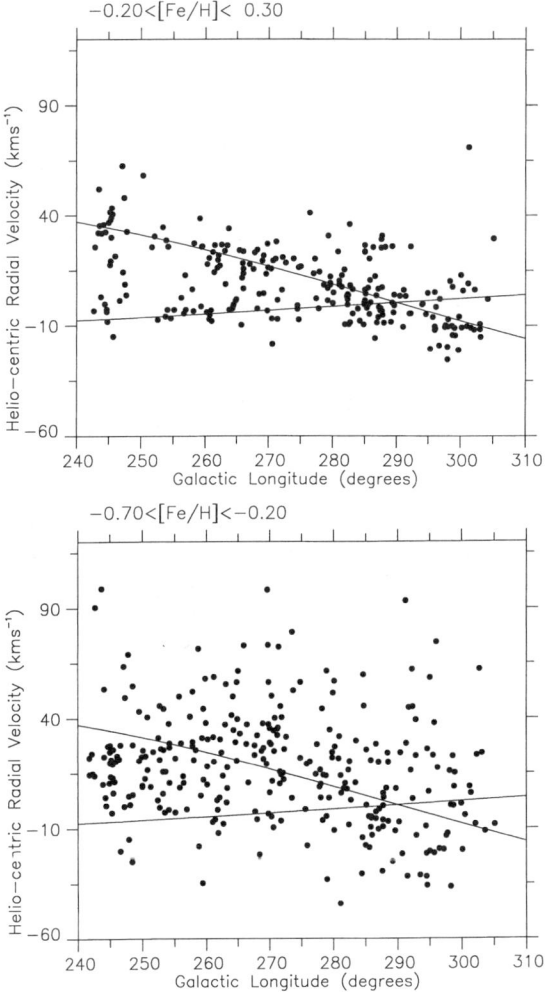

Figure 2 Distribution in the radial velocity–Galactic longitude plane of K giants near the Galactic plane. (*top*) Stars with −0.2 < [Fe/H] < 0.3. The continuous curves show the predicted loci for the Hyades (curve rising to the left) and Sirius (curve slowly rising to the right) groups. (*bottom*) Stars with −0.7 < [Fe/H] < −0.2. The curves are as above [from (130)].

2.2 *The Structure and Stability of the Old Disk*

For a recent review of the structural properties of the Galaxy, see (119). From the previous section, we know that the old disk is a flat, rapidly rotating system that contains most of the Galactic luminous mass. The density distribution of the Galactic disk cannot be measured directly.

However, surface photometry of other spiral systems shows that almost all of them have exponential surface density distributions (24, 38). In the vertical direction, star counts toward the Galactic poles show that the density distribution is again exponential (e.g. 86). Therefore, the density distribution of the Galactic disk is usually represented by $\rho(R,z) \propto \exp(-R/h_R - z/h_z)$, as in (7). [The distribution $L(z) \propto \text{sech}^2(z/z_0)$, which is asymptotically exponential, is often used to model the vertical distribution of light in edge-on spirals (124).] The vertical scale height h_z comes from star counts and is in the range 300 to 350 pc (47, 86). The radial scale length is probably between 3.5 and 5.5 kpc. The shorter estimate comes both from comparison of the calculated surface brightness of the nearby disk with the standard central surface brightness of the disks of other spirals and from the radial distribution of H I and CO (26). The longer value is given by van der Kruit's (120) analysis of the Pioneer 10 photometry.

The old disk is supported by its velocity dispersion in the z direction and mainly by its rotation in the radial direction. Near the Sun, its velocity dispersion components $(\sigma_R, \sigma_\varphi, \sigma_z)$ are about (40, 25, 20) km s^{-1} (52, 127). The circular velocity is about 220 km s^{-1} [see (24a) for references]. The asymmetric drift (the lag between the circular velocity and the mean rotational velocity of the old disk due to the contribution of the velocity dispersion to its radial support) is about 15 km s^{-1} (23).

The vertical and radial equilibrium of the disk can be used to estimate its local density and surface density. The local density can be derived in principle from the vertical equilibrium of a tracer population, such as F stars and K giants, whose vertical distribution $\rho(z)$ and vertical variation of velocity dispersion $\sigma_z(z)$ are known [see (2, 80) for reviews and recent work]. The local mass density ρ_0 derived in this way is typically between 0.15 and 0.20 $M_\odot$ pc^{-3}. This value is significantly larger than the observed mass in the form of stars, gas, and dust, which is about 0.10 $M_\odot$ pc^{-3}. The nature of the "missing mass" is not yet understood: Stellar remnants and faint dwarfs are possible contenders. Its vertical distribution is not known; the estimates of ρ_0 are therefore model dependent (2). The total surface density Σ_0 derived from these models is about 70 $M_\odot$ pc^{-2}, but this value is again model dependent (2). An upper limit to the total surface density of the disk near the Sun comes from the circular velocity. The disk, the spheroid, and the dark corona all contribute to the radial component of the gravitational field near the Sun, which determines the circular velocity. With a minimum contribution from the dark corona (consistent with a flat rotation curve out to about 50 kpc) and from the spheroid, the local surface density of the disk could be as high as about 110 $M_\odot$ pc^{-2} (L. Ferrario, unpublished). However, the disk may then be unstable to axisymmetric modes (see below). The problem of the local mass density and the missing matter is obviously very important. Several groups are now working on the kinematics and distributions of new samples of stars at the Galactic poles, and some further estimates of ρ_0 will soon be available.

The old disk near the Sun is relatively cold (mean rotational velocity $\gg$ velocity dispersion). Is the disk cold everywhere? The answer to this question is worth knowing because it bears on the stability of the Galactic disk. Cold disks are locally unstable to axisymmetric modes if Toomre's (113) parameter $Q = \sigma_R \kappa/(3.36 G \Sigma) < 1$, where κ is the local epicyclic frequency and Σ is the surface density. Disks may still be unstable to bar modes if $Q > 1$. Near the Sun, the value of Q is about 1.3. The bar modes can be suppressed in the following ways:

1. The Galaxy may contain a significant fraction of spherically distributed matter in its inner regions (27, 83). The inner bulge component and the dark corona may contribute to the stability of the disk. In the Galaxy, the dynamical importance of the dark corona in the inner parts is not yet clear because the local surface density of the disk is still uncertain. In other galaxies, the dark coronas that give the flat rotation curves are probably dynamically important in the outer parts only (19, 117).
2. The disks may be hotter in the inner parts (61). For example, Athanassoula & Sellwood (1) showed that a Kuzmin disk with $Q \gtrsim 2.5$ in the inner parts is stable. This option is not obviously attractive, because the Galactic disk is fairly cold near the Sun.

However, the surface photometry of edge-on galaxies suggests that the inner parts of their disks may not be so cold. Photometry by van der Kruit & Searle (124) shows that the vertical scale height of edge-on disks is approximately constant with radius. From the vertical equilibrium, we would then expect the vertical velocity dispersion to rise toward the Galactic center, following $\sigma_z \propto \exp(-R/2h_R)$. This trend of σ_z has been observed for the old disks of a few face-on spirals (121). Then, *if* the ratio σ_R/σ_z is approximately constant with radius, the radial component of the velocity dispersion would also follow $\sigma_R \propto \exp(-R/2h_R)$. In the inner parts of the Galaxy, σ_R would then rise to about 110 km s^{-1}, a value that is similar to the velocity dispersion of the bulge and the halo (see below).

The radial variation of the σ_R component is not easy to measure in other disk galaxies. For σ_R to be measurable, the galaxy must not be seen face-on, and then the observed line-of-sight velocity dispersion is the projection of at least two of the three dispersion components and the variation of the mean rotational velocity, integrated along the line of sight. However, in the Galaxy, it is possible to measure σ_R directly from the radial velocities of individual disk stars that lie in windows in the Galactic absorption.

Lewis (67) has measured radial velocities for about 600 K giants in several windows where the reddening is known and is relatively low. The distances to individual stars in these windows, which have Galactic latitudes $|b| < 4°$, were estimated from photometric parallaxes. The K giants in Baade's window and an anticenter window show directly that σ_R for the old disk varies smoothly over the interval $2 < R < 18$ kpc and follows closely the form $\sigma_R(R) = 106 \exp[-R(\text{kpc}/2h_R]$ km s^{-1}, where $h_R = 4.4$ kpc (Figure 3). This is what we would expect for a disk of uniform vertical scale height and uniform ratio of σ_R/σ_z, as discussed above. The good agreement of this kinematically estimated h_R value with the structural determinations mentioned above suggests that the anisotropy σ_R/σ_z for the old disk does not change significantly with radius.

From this observed run of $\sigma_R(R)$, the radial behavior of Toomre's Q parameter can be calculated if we assume that the surface density is exponential and calculate $\kappa(R)$ from the Galactic circular velocity curve (18). Figure 4 shows the $Q(R)$ curve for the old disk. In the inner parts of the disk, Q rises above 2.5. From comparison with the theoretical results of Athanassoula & Sellwood (1), it seems likely that the disk is hot enough in its inner regions to be stable against barlike modes.

The asymmetric drift of an exponential disk with an exponential $\sigma(R)$ is interesting. Lewis (67) also measured the azimuthal component of the velocity dispersion, $\sigma_\varphi(R)$, which he then used to calculate the expected run of the asymmetric drift with radius. For such a disk, the asymmetric drift is everywhere less than 20 km s^{-1}, even near the center where the random motions are large. Lewis's observations of the mean rotational motion of the old disk show directly that the asymmetric drift is indeed small, at least in the region $3 < R < 12$ kpc where the mean rotation is well determined. In contrast, an isothermal isotropic population with density distribution $\rho(R) \propto R^{-3.5}$ (the density law of the Galactic halo; see Section 4) in a potential with a circular velocity of 220 km s^{-1} is in

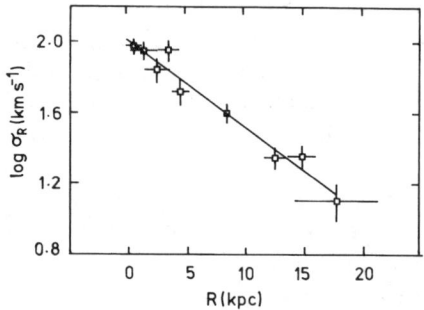

Figure 3 The radial variation of σ_R for the old disk [from (67)].

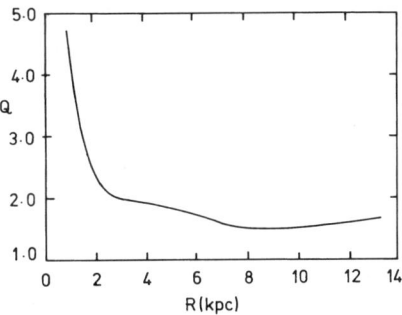

Figure 4 The radial variation of Q for the old disk [from (67)].

equilibrium with zero mean rotation (asymmetric drift of 220 km s^{-1}) if its σ_R value is about 120 km s^{-1}.

We see later that the velocity dispersion of stars in the inner part of the Galactic bulge is also about 110 to 120 km s^{-1}. If the velocity dispersions of the disk and bulge are similar near the Galactic center, then what distinguishes them dynamically in this region? The two important properties are the density gradient (much steeper for the bulge, which therefore has a much larger asymmetric drift) and the large anisotropy ($\sigma_R/\sigma_z \approx 2$) of the disk's velocity dispersion, which contributes to its flatness.

With these results for the Galaxy and guided by data for other galaxies, we can put together a likely overall picture of the density distribution and internal motions in the old disk of the Galaxy. The surface density distribution is approximately exponential, with a scale length between 4 and 5 kpc. The vertical density distribution is again approximately exponential, with a radially uniform vertical scale height of about 300 to 350 pc. The radial component of the velocity dispersion σ_R increases exponentially toward the Galactic center, from about 40 km s^{-1} near the Sun to about 110 km s^{-1} at the center, with the anisotropy ratio σ_R/σ_z remaining approximately constant with radius. The radial equilibrium of the disk is dominated by its rotation: The asymmetric drift is everywhere less than about 20 km s^{-1}.

A key problem now is to understand the heating processes that led to the exponential rise of σ_R and σ_z toward the Galactic center, with approximately uniform anisotropy σ_R/σ_z. At this time, an explanation combining the process of heating in the plane by transient instabilities and vertical scattering by giant molecular clouds looks promising (105, 129).

3. THE THICK DISK

Van der Kruit & Searle (122-124) measured the vertical (z) surface brightness profiles for several edge-on spirals. In those spirals with negligible bulges, the profiles showed only the thin-disk component. However, for spirals like the Galaxy, with small bulges, the profiles also showed a second, more extended flat component. This *thick disk* is structurally like the thick disks of S0 galaxies discovered earlier by Burstein (16) and Tsikoudi (114). The Galaxy appears to have a similar thick-disk component. Gilmore & Reid (47) derived the vertical density profile for the Galaxy from stellar photometry at the SGP. They interpreted their density profile as the sum of two components: the thin disk with scale height $h_z \approx 300$ pc and a thick disk with scale height ≈ 1500 pc. The local density of this thick-disk component is about 2% of the density of the thin disk.

Galactic models, which predict the distribution of apparent magnitude and color for stars in different directions, have been used extensively over the last few years to constrain the stellar content and structural parameters of the Galaxy. There has been some controversy about the need to include this thick-disk component in such Galactic models. Does a two-component model (thin disk plus spheroid) already give an adequate representation of the observed stellar content of the Galaxy? This question is not yet entirely settled: For example, see (3, and references therein) and (45, 48, 90, 132) for different viewpoints.

If the local density of the thick disk is so low, then it is difficult to study its properties from nearby stars unless there is some fairly unambiguous way of recognizing the nearby thick-disk members among the far more numerous thin-disk stars. As the thick disk has a larger scale height than the thin disk, it seems safer to study its properties in regions that are high enough above the Galactic plane ($z > 2$ kpc) for the thick disk to be the dominant population. In an important paper, Hartkopf & Yoss (52; hereafter HY) observed the kinematics and abundances of G and K giants at the galactic poles, at distances of up to about 5 kpc. Within a few hundred parsecs of the Galactic plane, almost all their stars have abundances in the range $0 > [Fe/H] > -0.5$, and their velocity dispersion σ_z is 22 km s^{-1}. These stars belong to the old thin disk, whose scale height is about 300 pc. At greater heights there are many metal-weak stars ([Fe/H] < -1), which belong to the Galactic halo (see below). However, there are also about 30 relatively metal-rich giants ([Fe/H] > -1) far from the Galactic plane ($2 < z < 5$ kpc), and these stars could not belong to a thin exponential disk population with a scale height of about 300 pc. They are instead tentatively identified with the thick disk. The vertical velocity dispersion of these 30 stars is 39 ± 6 km s^{-1}. [See (49) for further discussion of the change in abundance distribution with height above the Galactic plane.]

This σ_z value of about 40 km s^{-1} is characteristic of the thick-disk population at the solar distance from the Galactic center. It is found for several samples of stars whose abundances lie between the characteristic range for the thin disk ($0 > $ [Fe/H] $ > -0.5$) and the low values associated with the halo stars ([Fe/H] < -1). For example, the nearby stars in the HY sample ($z < 1$ kpc) with abundances between -0.5 and -1 have $\sigma_z = 43 \pm 7$ km s^{-1}. As another example, Rose (93) identified a sample of nearby luminous G stars as red horizontal branch (RHB) stars with typical abundances of about -0.7. Their σ_z value is about 40 km s^{-1}, and Rose associated them with the thick disk. However, Norris (76) has argued that these RHB candidates may in fact be old, relatively metal-weak ([Fe/H] < -0.4) clump giants of the *thin* disk. This possibility raises the difficult question of the discreteness of the thin and thick disks, which we discuss later.

Is a σ_z value of about 40 km s^{-1} for the thick disk consistent with estimates of its scale height from star counts? Yoshii et al. (132) have recently fitted a refined three-component Galactic model (thin disk, thick disk, and spheroid) to their new *UBV* photometry of 18,000 stars with $12 < V < 18$ at the north Galactic pole (NGP). In their best model, the exponential scale height of the thick disk is 950 pc, and its local density is 2% of the thin disk's density. We can show that this scale height is consistent with a velocity dispersion of 40 km s^{-1}. For example, consider the vertical equilibrium of two coexisting disks, both with exponential density distributions in the z-direction and with scale heights of 300 and 900 pc. If the 300-pc thin disk dominates the potential, then the expected ratio of the σ_z values of the disks is about 1:2 in the symmetry plane, as observed.

In its vertical structure and kinematics, the thick disk is intermediate between the rapidly rotating thin disk ($\sigma_z \approx 20$ km s^{-1}) and the slowly rotating metal-weak halo ($\sigma_z \approx 75$ km s^{-1}; see below). Is the thick disk also rotationally an intermediate population? K. U. Ratnatunga & K. C. Freeman (unpublished data; hereafter RF) measured abundances and kinematics of K giants in a field at $l = 270°$, $b = 39°$, which is well situated for estimating the mean rotation of the different populations. The sample includes stars out to distances of about 25 kpc and with abundances in the range $0 > $ [Fe/H] $ > -2$. In particular, it includes metal-rich stars ([Fe/H] > -1), between 2 and 5 kpc from the Galactic plane, that must belong to the same population as the HY thick-disk stars at the Galactic poles. The kinematical properties of the RF stars show a fairly sharp break at [Fe/H] ≈ -0.8. The more metal-weak stars are members of a nonrotating halo population with a line-of-sight velocity dispersion of about 125 km s^{-1}. The more metal-rich stars (which lie in the region $2 < z < 5$ kpc) have a mean velocity (relative to the local standard of rest) of only 36 ± 14 km s^{-1} and belong to a rapidly rotating population. Their line-of-sight velocity dispersion is 50 ± 10 km s^{-1}. Again, these metal-rich stars are identified with the thick disk because they are so far from the Galactic plane.

From these observations, RF could estimate the σ_R component of the velocity dispersion and the asymmetric drift for this thick-disk population. It was necessary to assume that (a) these metal-rich stars have the same σ_z as the corresponding HY stars with $2 < z < 5$ kpc at the Galactic poles, (b) the Galactic rotation curve is flat between the solar radius and the mean galactocentric distance for the sample (9.3 kpc), and (c) the velocity dispersion σ_φ followed the relation $\sigma_\varphi = \sigma_R/\sqrt{2}$, which is roughly consistent with (b). It turned out that the thick-disk stars have $\sigma_R = 69\pm 18$ km s^{-1} and that their asymmetric drift is only 29 ± 19 km s^{-1}. These numbers are compatible dynamically if the radial scale length of the thick disk is between 4 and 6 kpc. (There is no direct information yet on the radial scale length of the thick disk.) Similar values for the velocity dispersion and asymmetric drift of the thick disk have been estimated by Sandage & Fouts (102) from their study of the kinematics of nearby high-proper-motion stars, and by G. Gilmore & R. Wyse from their unpublished kinematical survey of G dwarfs out to distances of a few kiloparsecs in several fields. Also, Norris (75) has pointed out that the asymmetric drift for the most metal-weak ([Fe/H] ≈ -0.6), highest velocity dispersion ($\sigma_z \approx 35$ km s^{-1}) group of bright G and K giants is similarly low. Somewhat higher values for the velocity dispersion and asymmetric drift were derived earlier by Wyse & Gilmore (131) from proper-motion data [see Norris (75) for discussion of the effects of proper-motion zero-point errors on their conclusions]. We see that the thick disk is kinematically hotter than the thin disk, but that both are rapidly rotating populations. Table 1 compares present estimates of some parameters for the thin and thick disks.

The metallicity distribution of thick-disk stars, as given by the HY and RF in situ samples of metal-rich giants with $2 < z < 5$ kpc, includes stars with [Fe/H] values at least as high as the solar value. However, the mean metallicity of the thick disk is probably somewhat lower than that of the thin disk. Gilmore & Wise (49) showed that the mean abundance of HY

Table 1 Comparison of parameters for the thin and thick disks

Disk	h_z (pc)	σ_z (km s^{-1})	σ_R (km s^{-1})	Asymmetric drift (km s^{-1})
Thin	350	20	40	15
Thick	1000	40	70	30–40

stars with $1 < z < 2$ kpc is about -0.6. Sandage & Fouts (102) extracted a subsample of nearby thick-disk stars from their high-proper-motion sample, using the kinematic criteria $|V| < 100$ km s^{-1}, $40 \lesssim |W| \lesssim 79$ km s^{-1}. These stars have similar velocity dispersions and asymmetric drifts to those derived for the high-z stars of HY and RF and cover a similar abundance interval: Their mean [Fe/H] is again about -0.6.

RF showed directly that the population of metal-rich stars with $z > 2$ kpc is rotating almost as rapidly as the thin disk. Here we have associated this population with the Gilmore-Reid (GR) thick disk because (a) these stars are too far from the Galactic plane to be part of a strictly exponential thin disk, if the GR thick/thin disk normalization of 2% is correct; and (b) their σ_z value is consistent with the estimated scale height of about 1 kpc (132). However, the relationship between the thin and thick disks is not yet clear. Are they discrete components, or is the thick disk just the higher energy, slightly metal-weaker tail of the thin-disk population, as argued recently by Norris (75, 76). He showed (75) that GR's vertical density distribution can be reproduced by a simple four-component model for the Galactic disk, in which the four components have different vertical velocity dispersions (from 15 to 33 km s^{-1}) and [Fe/H] values (from -0.05 to -0.57, respectively).

The question of whether the thin and thick disks are discrete components is not just semantic. Its answer is important, because different pictures for the formation of the thick disk have different implications for the discreteness of the two components. Did it form during the Galactic collapse, when the Galaxy was already close to centrifugal equilibrium but had not completely settled in the vertical direction? Or was it produced *after* the thin disk had begun to form? Here we present two possibilities for the second picture:

1. Is the thick disk just the old, high-velocity, metal-weak tail of the thin disk's AVR and AMR (see Section 2.1) (75)? In this picture, the stars of the thick disk were formed in the early, relatively metal-weak *thin* disk and were then heated by the same secular processes that produced the observed AVR for the thin disk.
2. Was the thick disk formed from the early thin disk by a relatively short-lived heating process, such as heating by a transient bar or by the accretion of satellites (87), that occurred early in the life of the thin disk, perhaps in its first 1 or 2 Gyr? At this time, the disk was still mainly gaseous and the mean metal abundance of the disk stars that had already formed was still low (say about -0.6; see Section 2.1). In this picture, the thick disk is a *discrete* relic of the early stellar thin disk. After the heating event that produced the thick disk, the gaseous thin disk then continued its dissipative evolution undisturbed.

Further data on the vertical structure and kinematics of the thin and thick disks, as determined from a homogeneous sample of stars, would be helpful in deciding whether or not the two components are discrete. More precise information on the shape of the velocity ellipsoid for the thick-disk stars would be useful in choosing between the possible formation mechanisms. For example, the apparent similarity of the anisotropy ratio σ_R/σ_z for the thin disk and the thick disk (see Table 1) may suggest that similar heating mechanisms have acted on both components. It would be particularly useful to know whether *the thick-disk stars lie on the same AVR and AMR as the thin disk stars*. Such a relationship would be difficult to establish, but it would constrain the choice of formation mechanisms.

Thick disks and bulges appear to go together: Disk galaxies with negligible bulges do not appear to have the second, longer scale height component that is recognized as representing the thick disk (122–124). The absence of thick disks in such galaxies may argue against some of the obvious mechanisms for thick-disk formation, such as the heating of the thick disk by the normal secular heating processes that are still going on now in the thin disk. We need to understand why thick disks and bulges are associated. It is possible that the bulge and the thick disk are part of the same dynamical component, at least in systems like the Galaxy with small bulges (60). Some observational support for this view comes from Bahcall & Kylafis (5), who showed that a thin-disk plus thick-disk model gives an excellent representation of the surface photometry of the edge-on galaxy NGC 891, which has a small bulge and is probably similar to the Galaxy (118).

Consider a thick disk in which the radial component of the velocity dispersion rises toward the Galactic center, reaching a central value similar to that observed for the inner parts of the thin disk (about 110 km s^{-1}; see Section 2.2). If its velocity dispersion near the center becomes more nearly isotropic, then the central regions of this thick disk would no longer be disklike, but rather they would probably have the appearance of a small bulge. G. Rowley (unpublished) has made some preliminary models of a thick-disk population in the potential of a dominant thin disk. These models are able to reproduce many of the known structural and kinematical properties of the thick disk and small bulge of systems like the Galaxy.

When the thick disk was first identified in other disk galaxies, it was believed to be part of the outer spheroidal component, flattened by the flat potential of the thin disk (16, 40, 123). This explanation is less plausible now. In the Galaxy, the thick disk is rotating rapidly and is relatively metal rich. In its dynamical and chemical properties, it looks *qualitatively* like an extension of the thin disk's AVR and AMR to larger σ and lower [Fe/H]. The stars of the thick disk probably formed either just before or just after the beginning of thin-disk formation as part of a dissipated structure that was already close to rotational equilibrium. The thick disk should be recognized as part of that disklike structure.

4. THE METAL-WEAK HALO

Most of the Galactic metal-weak stars ([Fe/H] < −1) are in a spheroidal, slowly rotating system. Its volume density distribution, as defined by the globular clusters and RR Lyrae stars, is fairly well represented by an $r^{-3.5}$ law (56, 94, 135). Near the Sun, the density of this component relative to the thin disk is about 0.1 to 0.2% (4, 6, 132). Its surface density distribution, as derived mainly from the globular clusters, is close to an $r^{1/4}$ law (26) with an effective radius of about 2.7 kpc. The globular cluster distribution in M31 also follows approximately an $r^{1/4}$ law (26). The metal-weak halo of the Galaxy shows no clear evidence for an abundance gradient: At any radius, there are stars and clusters with a wide range of [Fe/H] values (94, 97, 103, 135, and references therein).

The velocity dispersion and mean rotation of nearby halo stars can be measured for samples of halo stars found kinematically (from high-proper-motion catalogs) or nonkinematically (from spectroscopic and photometric surveys and from variability of RR Lyrae stars). The apparent velocity distribution of kinematically selected stars may be significantly biased, depending on the selection criteria, and some correction for this bias may be necessary. For example, Bahcall & Casertano (4) applied Monte Carlo simulations to Eggen's catalog (30, 32) and derived $v_{\rm rot} = 66 \pm 23$ km s^{-1} and $(\sigma_R, \sigma_\varphi, \sigma_z) = (140 \pm 12, 100 \pm 12, 76 \pm 6)$ km s^{-1} as their best estimate of the true kinematical parameters for the halo. The simulations show how a kinematically selected sample of stars can have a significantly lower value of $v_{\rm rot}$ and a significantly higher value of σ_R than the population from which it is drawn; there are also smaller effects on σ_φ and σ_z. Some typical values of the errors induced by kinematical selection in these simulations are about 70 km s^{-1} in $v_{\rm rot}$ and 40 km s^{-1} in σ_R. The actual values for a particular kinematically selected sample will clearly depend on the precise parameters of the selection procedure (see also 74, 77).

Estimates of the velocity dispersion and mean rotation for nearby non-kinematically selected samples are more straightforward. Norris et al. (79) found $v_{\rm rot} = 27 \pm 22$ and $(\sigma_R, \sigma_\varphi, \sigma_z) = (125 \pm 11, 96 \pm 9, 88 \pm 7)$ km s^{-1} for a sample of about 70 spectroscopically selected stars with [Fe/H] < −1.2 and with good UVW velocities. From a more extended compilation of nonkinematically selected stars with [Fe/H] < −1.2, which includes RR Lyrae stars, blue horizontal branch (BHB) stars, and metal-weak giants and dwarfs, Norris (77) finds $v_{rot} = 37 \pm 10$ and $(\sigma_R, \sigma_\varphi, \sigma_z) = (131 \pm 6, 106 \pm 6, 85 \pm 4)$ km s^{-1}. These values agree well with those derived from the kinematically selected sample (4). The metal-weak halo near the Sun is slowly rotating and has a significantly anisotropic velocity dispersion.

The metal-weak halo shows no obvious abundance gradient. Are the kinematical properties of this population near the Sun then also independent of abundance, as one might expect? This is an important question, because different models of galaxy formation make different predictions about how the local kinematics and abundance are related (34, 103). Norris's (77) nonkinematically selected sample shows no significant change in v_{rot} and σ with [Fe/H] for stars with [Fe/H] < -1.2. On the other hand, the kinematically selected sample of Sandage & Fouts (102) does show a steady increase of σ and a decrease of v_{rot} with decreasing [Fe/H] for these metal-weak stars. The reason for the difference between these two large samples, selected in different ways, is not yet properly understood [see (78, 101) for a discussion of this problem].

When Eggen et al. (34) wrote their famous paper on the relation between the orbital properties of stars and their abundances, all the metal-weak stars ([Fe/H] < -1.2) in their kinematically selected sample had orbital eccentricities $e > 0.4$. (The orbital eccentricity used here is calculated from the stellar U and V velocities only and is therefore the eccentricity of the orbit projected onto the Galactic plane, assuming that the (R, φ) and z components of the motion are independent.) This observation led to their picture of a rapid early collapse of the Galaxy. Larger kinematically selected samples that came later included a few metal-weak stars with lower eccentricities (102, 133; B. Carney & D. Latham, unpublished). However, in the nonkinematically selected sample of Norris et al. (79), about 20% of the metal-weak stars have $e < 0.4$. The detailed kinematics of these stars are interesting. Their mean rotation is high, as one would expect from their low eccentricities: Their rotational velocity is $v_{rot} = 183 \pm 8$ km s^{-1} compared with $v_{rot} = 27 \pm 22$ km s^{-1} for the whole metal-weak sample. However, their vertical velocity dispersion σ_z is only 44 ± 10 km s^{-1} (excluding one high-velocity star) compared with 88 ± 7 km s^{-1} for the whole sample. To which population do these low-eccentricity, low-[Fe/H] stars belong? Their σ_z value of 44 km s^{-1} is characteristic of the thick disk and suggests that these stars could be the metal-weak tail of the thick disk. The presence of such stars in the thick disk could be a useful pointer to its formation history. On the other hand, these stars could just be the kinematical tail of the slowly rotating halo population with $\sigma_\varphi \approx 100$ km s^{-1}. The φ velocities of the low-e stars of this halo population would be about two standard deviations from the mean of the population; if the velocity distribution of halo stars is more or less ellipsoidal, then these

THE GALACTIC SPHEROID AND OLD DISK   619

low-e stars would have predominantly low z velocities [particularly if the energy distribution of the halo stars is truncated, as in a King (62) model]. However, the high fraction of these low-e stars may argue against this interpretation.

Now we turn to the kinematics of more distant halo objects. Data are available now for metal-weak giants, RR Lyrae stars, BHB stars, and globular clusters out to large galactocentric distances. The observed velocity dispersion of these objects is remarkably isothermal (at about 120 km s^{-1}) from near the Galactic center out to at least 35 kpc (41, 77), except for fields near the Galactic poles, where the dispersion is lower. What are the shape and orientation of the velocity ellipsoid for stars far from the Galactic plane? Near the Sun, we know that the mean velocity dispersion components for metal-weak stars are $(\sigma_R, \sigma_\varphi, \sigma_z) \approx (140, 100, 75)$ km s^{-1} (i.e. the velocity ellipsoid is anisotropic and elongated in the radial direction). Does the velocity ellipsoid remain anisotropic far from the plane, and what is its orientation? Does it point toward the Galactic center or toward the axis of rotation? There are enough data available now on the kinematics of halo stars in different directions and at different distances from the Sun to test some simple kinematical models of the halo.

K. U. Ratnatunga & K. C. Freeman (unpublished) have tested two such kinematical models, using their data (88) and those of Beers et al. (10) for metal-weak giants and the data of Pier (84) and Sommer-Larsen & Christensen (110) for BHB stars. The models that they tested held the velocity ellipsoid constant, first in spherical coordinates [i.e. $(\sigma_r, \sigma_\varphi, \sigma_z) = (140, 100, 75)$ km s^{-1} everywhere] and then in cylindrical coordinates [i.e. $(\sigma_R, \sigma_\varphi, \sigma_z) = (140, 100, 75)$ km s^{-1} everywhere]. [See (88) for a previous version of this test.] Near the SGP, the observed velocity dispersion for these stars appears to be approximately constant at about 75 ± 10 km s^{-1} up to about 25 kpc from the plane (Figure 5). This is what one would expect for a system that is isothermal in the cylindrical sense as defined above. The velocity dispersion data in directions away from the Galactic poles are also well fit by the isothermal cylindrical model; the model that is isothermal in spherical coordinates is not a good fit to the data.

The constant value of $\sigma_z = 75$ km s^{-1} given above is somewhat less than the global value of 85 ± 4 km s^{-1} derived for nearby, nonkinematically selected metal-weak stars (77) [see also (21)]. However, Norris (77) points out that for 48 objects within 10° of the Galactic poles and farther than 4 kpc from the Galactic plane, the σ_z value is only 64 ± 7 km s^{-1}. (The mean z-distance of these stars is 15.7 kpc.) The reason for the difference between the global and polar values of σ_z is not clear. Norris suggests that it may be associated with small-scale clumping of halo stars in velocity space.

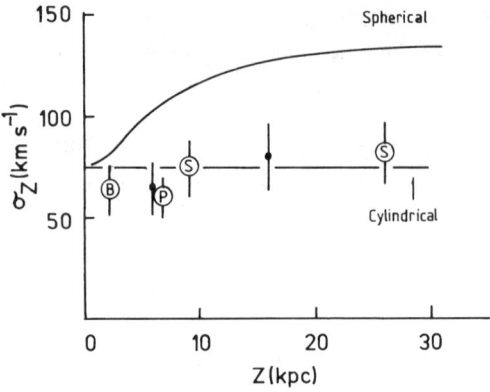

Figure 5 The variation of velocity dispersion with height above the Galactic plane for metal-weak stars near the SGP. Filled symbols and B denote metal-weak giants [from K. Ratnatunga & K. Freeman (unpublished) and (10)]. P and S denote BHB stars [from (84) and (110), respectively]. The curves show the predictions of the spherical and cylindrical kinematical models described in the text.

Although the isothermal cylindrical model gives an adequate description of the kinematics of metal-weak giants and BHB stars out to about 25 kpc, it does no more than that. It is certainly not a unique way to represent the data (126). The two mean points for the more distant BHB stars in Figure 5 may oversimplify the real situation. The sample (110) contains only 27 BHB stars, so the σ_z values depend on how the sample is partitioned: Inspection of the data suggests that σ_z may decrease beyond about 25 kpc from the plane. Also, some other samples of halo stars do not conform to the model. Pier's (85) unpublished data for RR Lyrae stars at the NGP show a constant velocity dispersion, at about 110 km s^{-1}, from a few kiloparsecs to over 20 kpc from the plane. The σ_z values for distant halo carbon stars are also significantly higher than the values for the giants and BHB stars (55, 71).

K. Ratnatunga & K. Freeman (unpublished) also used their velocity data for distant metal-weak giants in a field at $l = 270°$, $b = 39°$ to see if the kinematical properties of the outer halo change with [Fe/H]. The stars between 10 and 25 kpc from the Sun were split into two groups: $-1 > $ [Fe/H] > -1.4, and [Fe/H] < -1.4. The line-of-sight kinematics are very similar: $\langle v_{los}\rangle \approx 170$ km s^{-1} and $\sigma_{los} \approx 125$ km s^{-1}.

In summary, the metal-weak halo rotates slowly and is roughly isothermal. It has an anisotropic velocity dispersion, and the long axis of the velocity ellipsoid probably points toward the axis of Galactic rotation. Some of these conclusions, however, depend on fairly small samples of distant halo stars; larger samples of such stars are needed.

The flattening of the metal-weak halo, which we have not yet discussed, is an interesting problem at this time because the kinematical evidence and structural evidence on the shape of the halo appear to be in conflict. We return to the metal-weak halo in Section 6 to review this problem.

5. THE GLOBULAR CLUSTER SYSTEM

Zinn (135) showed that the Galactic globular clusters appear to fall into two families with different spatial, chemical, and kinematical properties. The abundance distribution of the clusters is bimodal, with the two modes separating at [Fe/H] ≈ -1. The clusters in the more metal-rich mode lie in a rotating disk, with $v_{rot} = 152 \pm 29$ km s^{-1} and a characteristic velocity dispersion $\sigma = 71 \pm 12$ km s^{-1}. The scale height of this disk is not well determined; however, the kinematical parameters are similar to those of the thick disk, as discussed above, and it is tempting at this stage to identify the metal-rich clusters with the thick disk. The metal-weak clusters belong to a slowly rotating spheroidal population, with $v_{rot} = 50 \pm 23$ km s^{-1} and $\sigma = 114 \pm 9$ km s^{-1}. These values are similar to those for the metal-weak stellar halo.

The system of metal-weak clusters shows no clear evidence for an abundance gradient, at least between about 7 and 35 kpc from the Galactic center, although there is a wide range of metallicity (135). However, the second parameter phenomenon (the presence of red horizontal branch stars in clusters of low [Fe/H]) is most marked for clusters at large distances from the Galactic center (134). Although the nature of the second parameter effect is not yet fully understood, it is clearly related to the large-scale structure of the halo. One possibility for the second parameter is age: If this is correct, then the outer halo clusters formed later and over a longer period of time than the inner metal-weak clusters.

Frenk & White (43) made a kinematic solution for the motions of the clusters and found that the velocity dispersion of the cluster system is close to isotropic. This is an interesting result, because we know that the velocity ellipsoid for the field halo stars is highly anisotropic. If the clusters and the halo stars have different kinematics, then the clusters and the stellar halo may have had different dynamical histories, or perhaps the clusters in the most elongated orbits have by now been tidally destroyed. However, Norris (77) has recently solved for a larger sample of metal-weak clusters with [Fe/H] < -1.2. Although the three components of the velocity ellipsoid are approximately equal, as found in (43), the formal errors of the solution are large, and it is not yet possible to say with confidence whether the clusters are indeed kinematically different from the halo stars.

Rodgers & Paltoglou (92; hereafter RP) investigated the rotation of the globular cluster system in more detail. They noted that globular cluster formation has continued to the present time in the disks of smaller galaxies like the Large Magellanic Cloud (LMC) and M33 (22, 58), and that even the oldest globular clusters in the LMC lie in its disk (42). RP considered the possibility, suggested by Searle & Zinn (103), that the halo globular clusters formed in small disklike satellite galaxies that were then accreted by the Galaxy. Although the satellite itself would be tidally disrupted [its stars would presumably become part of the metal-weak halo star population; see also (107)], the globular clusters would probably survive because they are more tightly bound. Clusters that came from a common parent may then be in a restricted abundance range, and they may still have common kinematics. RP therefore made rotation solutions for clusters in several abundance intervals. The v_{rot} values for most of these abundance intervals were between 40 and 100 km s^{-1} (except for the more rapidly rotating metal-rich clusters). However, the 30 clusters with $-1.3 > [\text{Fe/H}] > -1.7$ showed retrograde rotation, with $v_{\text{rot}} \approx -70$ km s^{-1}. This result is marginally significant.

Quinn & Goodman (87) have studied the dynamics of the accretion of satellites by a disk galaxy. They showed that satellites, which are initially within several disk scale lengths of the parent galaxy and in direct orbits inclined at less than about 60° to the plane of its disk, are dragged down toward the plane by dynamical friction in a few orbital times before being tidally disrupted. Satellites in more highly inclined orbits just lose orbital energy until they are disrupted, without being pulled down into the plane. The work of Quinn & Goodman also suggests that satellites in retrograde orbits will suffer a similar fate but on a somewhat longer time scale. This process could lead naturally to the observed structural and kinematic discontinuity between the slowly rotating metal-weak halo (formed in this way from the debris of accreted satellites) and the rapidly rotating thin and thick disks. It would also have implications for the flattening of the metal-weak halo, which we discuss in Section 6. As mentioned in Section 3, it is possible that the thick disk itself was formed from the early thin disk by the heating that took place during this satellite accretion phase.

On the other hand, the formation of halo globular clusters may have occurred naturally from the collapsing gas of a protogalaxy. Fall & Rees (36) argued that this gas is liable to fragment when the cooling time is comparable with the free-fall time. If the abundance is low ([Fe/H] < -2), then the smallest gravitationally unstable clouds will have masses and mean densities similar to those of globular clusters. We need to find observational ways to test these processes for cluster formation. More information on the dynamics of accretion, the nature of the second parameter effect, and the orbital properties of globular clusters would be valuable. Another important question concerns the abundance distribution for the clusters and for the field halo stars—namely, are they similar? This question is not settled yet: For different viewpoints, see (53) and B. Carney et al. (unpublished).

A few moving groups of metal-weak stars are known. They are probably the debris of broken-up globular clusters or accreted satellite galaxies. In the solar neighborhood, Kapteyn's star group (31) is the best known, although its reality is still contentious. This group has [Fe/H] ≈ -1.8, and its orbit carries it close to the Galactic center. This is at least consistent with the view that the group comes from some small, tidally disrupted system. Recently, a group of six BHB stars was discovered by Sommer-Larsen & Christensen (109) in a small region of space, about 4 kpc from the Sun. The radial velocities of the members correlate tightly with their distance. The dynamical nature of moving groups, particularly old ones, is not yet well understood. However, with some simple assumptions (similar values of the integrals of motion for the member stars), it can be shown that the orbit of this group also passes close to the Galactic center. Another prominent old group is the Arcturus group (31). With a V velocity of -118 km s^{-1} and [Fe/H] ≈ -0.6, it probably belongs to the thick disk rather than to the halo, and it may be the debris of a disrupted disk globular cluster. G. Gilmore & R. Wyse (unpublished), who are making a large survey of G dwarfs within a few kiloparsecs of the Sun, have found a concentration of stars with a V velocity of about -118 km s^{-1}. These stars are presumably more distant members of the Arcturus group.

The Galactic globular cluster system provides important information about the early history of the Galaxy. However, the properties of this globular cluster system may not be typical of those for other large spirals. L. Searle (oral communication) showed that the cluster system of M31 is different: Its abundance distribution is not bimodal, and the kinematical and chemical properties of the system are not correlated as they are in the Galaxy. Burstein et al. (17) compared integrated light spectra of the M31 and Galactic globular clusters. They found that the metal-rich M31 clusters have significantly enhanced CN and Hβ relative to their Galactic counterparts. These differences are not yet understood. However, they suggest that there may not be a unique process for the formation of globular cluster systems in spiral galaxies.

6. THE SHAPE OF THE HALO

Evidence on the flattening of the metal-weak halo is apparently contradictory. Until recently, most of the work on star counts and on the spatial distribution of halo objects suggested that the halo is almost spherical. On the other hand, the shape of the velocity ellipsoid for halo stars indicates that the halo is significantly flattened. Observations of edge-on galaxies similar to the Galaxy would give us no guidance here because the surface brightness of the metal-weak component would be very low. For example, say the metal-weak halo follows an $r^{-3.5}$ density law and has a local density of 0.002 of the disk's density. Then, if the Galaxy were viewed edge on, the surface brightness of the metal-weak component at the solar radius would be only about 26 B mag arcsec^{-2}. In this section, the halo is assumed to be oblate, and the estimates of its axial ratio are given as c/a (<1).

The work on star counts and spatial distributions has been reviewed by Bahcall (3) and is summarized briefly here. Faint J and F photometry (63, 65) in SA 57 (NGP) and SA 68 ($l = 110°$, $b = -46°$) shows the expected bimodal distribution in color. The blue mode comes almost entirely from distant halo stars (10 to 20 kpc), so the star counts in these two fields give an estimate of the axial ratio of the outer halo: The c/a value is 0.80 ($+0.20$, -0.05). In another star-count analysis, Pritchet (86) fitted a halo plus disk model to star counts in 12 regions with $111° < l < 173°$ and also at the NGP; he found that $c/a > 0.9$ for the halo component. Oort & Plaut (82) studied the distribution of RR Lyrae stars in the inner parts of the Galaxy (1 to 5 kpc from the center). They used the period-amplitude relation to identify the metal-weak stars ($\Delta S > 5$, [Fe/H] < -1.3). For stars with latitudes $|b| > 8°$, the axial ratio of this population is between 0.8 and 1.0. For the outer RR Lyrae stars ($R < 60$ kpc), Hawkins (56) finds that the axial ratio is about 0.9. The distribution of blue horizontal branch stars in three fields—the SGP, ($l = 37°$, $b = -51°$), and ($l = 352°$, $b = 52°$)—was used by Sommer-Larsen (108) to measure the axial ratio of this population in the inner halo ($R < 10$–15 kpc): His value of c/a is 0.80 ± 0.15. For the globular cluster system, Frenk & White (44) estimated that $c/a = 0.85 \pm 0.13$. These estimates all agree in giving an axial ratio of about 0.8 for the metal-weak halo.

Zinn's (135) kinematical data for the metal-weak clusters are also consistent with this value of c/a. The velocity dispersion for the metal-weak cluster system is close to isotropic (43), and its systemic V/σ value is 0.4 ± 0.2. (Here V and σ are the rotation and velocity dispersion of the cluster system, respectively.) The inferred c/a value from the standard (V/σ-ε) diagram for oblate, isotropic, rotationally flattened systems (13) is then 0.87 ($+0.09$, -0.13). (We note that the V/σ-ε diagram is not strictly applicable to the cluster system, which is not self-gravitating; it lies in a somewhat flattened imposed potential.)

On the other hand, we know from Section 4 that the velocity dispersion for halo stars is significantly anisotropic. In the solar neighborhood, the velocity dispersion depends somewhat on the particular sample of metal-weak stars, but it is typically $(\sigma_R, \sigma_\varphi, \sigma_z) = (140, 100, 75)$ km s^{-1}. Dynamical models show that such a population must be relatively flat, with c/a values of about 0.3 to 0.5, depending on the details of the model; see (12) and also (108, 126) if $\sigma_z(z)$ remains small at all z (Section 4).

How can these two conflicting estimates of the flattening of the metal-weak halo be reconciled? There is already a suggestion from star counts that the flattening of the metal-weak component increases with decreasing galactocentric distance (48). Star counts in three Galactic bulge fields by Rodgers et al. (91) also indicate that the flattening between 3 to 5 kpc from the Galactic center is greater than near the Sun. Hartwick (54) has recently analyzed the space distribution of RR Lyrae stars with [Fe/H] < -1 and $|b| > 30°$ and has found that their distribution can be described in terms of two components: a moderately flattened inner component with $c/a \approx 0.6$, which is dominant near the Sun, and a spherical outer component. The space distribution of the metal-poor globular clusters is also consistent with this model: The metal-poor clusters with $R < 8$ kpc appear to form a flatter distribution than those with $R > 8$ kpc. However, rotation solutions for these two sets of clusters give similar low rotational velocities (30 to 40 km s^{-1}) and similar velocity dispersions (110 to 120 km s^{-1}), so there is no suggestion that this inner metal-*weak* population is rotationally flattened. It must be flattened by the anisotropy of its velocity dispersion. In this sense, it is distinct from Zinn's inner metal-*rich* cluster system, which has a V/σ value of about 2 and does appear to be part of a fairly rapidly rotating disk population. Hartwick's (54) moderately flat inner component also appears to be distinct from the population of metal-weak stars of low orbital eccentricity found by Norris et al. (79; see Section 4); the latter population is again relatively rapidly rotating and may be the metal-weak tail of a disk or thick-disk system.

The two-component picture for the metal-weak halo may explain why the star count and spatial distribution data give an approximately spherical halo, whereas the local velocity dispersion for metal-weak stars suggests a flattened system. The local kinematics are dominated by Hartwick's inner flattened component, while the star count and spatial distribution data come mainly from objects at galactocentric distances of 10 to 20 kpc and therefore pertain mainly to the spherical outer component.

Hartwick showed that the inner and outer subsamples of globular clusters have fairly similar abundance distributions. From this, he argued that the dominant flattening mechanism is due to the dynamical evolution of an extant population of stars and clusters, rather than to some process involving dissipation and star formation with the accompanying metal enrichment. For example, Binney & May (12) showed that the slow formation of a disk within an initially spherical stellar spheroid can significantly flatten the distribution of spheroid stars, particularly those stars on elongated orbits. Alternatively, the two-component nature of the metal-weak halo is qualitatively consistent with the formation of the halo by accretion of metal-weak satellites *after* the disk had already settled. Quinn & Goodman (87) studied the evolution of satellite orbits under the effects of the dynamical friction of the disk. They found that satellite orbits that are initially inclined at less than about 60° to the disk plane are dragged down toward the plane by dynamical friction before the satellites are finally disrupted. Their debris would then form a flattened system with an anisotropic velocity dispersion, whereas the debris of satellites in orbits that were initially more highly inclined would produce a more nearly spherical halo population.

Some support for Hartwick's picture comes from Sommer-Larsen's (108) study of the orbits of 143 nearby nonkinematically selected stars with [Fe/H] < -1.2. He compared the motions of the 33 stars with $-1.2 >$ [Fe/H] > -1.5 and the 110 stars with [Fe/H] < -1.5. The mean rotational velocities of the two samples are both low (67 ± 19 and 29 ± 11 km s^{-1}, respectively). Their velocity dispersion components ($\sigma_R, \sigma_\varphi, \sigma_z$) are ($136\pm 17, 106\pm 14, 59\pm 8$) and ($147\pm 10, 109\pm 7, 102\pm 7$) km s^{-1}, respectively. The velocity dispersion for the metal-weaker stars is in good agreement with the values derived by Carney & Latham (21) for a partly independent set of nonkinematically selected stars with [Fe/H] < -1.5. Although it has often been pointed out that σ_z for halo stars increases with decreasing [Fe/H] (e.g. 34, 96), it is interesting that *none* of the other kinematical parameters for these two subsamples is (dynamically) significantly different. Sommer-Larsen's results again point to a two-component picture for the metal-weak halo: He suggests that the two components are differentiated by abundance rather than by galactocentric distance (as proposed in Hartwick's analysis). Both components are again slowly rotating. The metal-stronger component is flattened by its anisotropic velocity dispersion (as for giant ellipticals), whereas the metal-weaker component has a more nearly isotropic velocity dispersion and is therefore more nearly spherical.

To understand the origin of the two components, Sommer-Larsen (108) examined the distribution of orbital inclinations for his sample of stars. He adopted the Binney & May (12) picture and studied the evolution of the stellar orbits backward in time by slowly removing the disk. He concluded that if the population of metal-weak stars formed before the Galactic disk, then most of the stars of the metal-weaker component formed in an approximately spherical halo. Some of them, however, formed in an initially flattened, slowly rotating population. From the distribution of initial orbital inclinations for these stars, Sommer-Larsen argued that the two components are discrete rather than being part of a continuous distribution. Most of the metal-stronger halo stars also formed in this initially flattened population. However, among the stars with $-1.2 > \text{[Fe/H]} > -1.5$, those that have the lowest inclination orbits show a relatively rapid systemic rotation of 147 ± 19 km s^{-1}; this subsample includes the metal-weak, low-eccentricity stars found by Norris et al. (79). Again, the presence of two components, now differentiated by [Fe/H], can also be qualitatively understood within the picture of halo formation by satellite accretion. The more massive satellites will be less metal weak, their orbits will be more affected by dynamical friction, and as their orbits shrink, they will survive farther in toward the Galactic center before they are tidally disrupted. Their debris would therefore form an inner, flattened, less metal-weak component.

The two-component picture for the metal-weak halo supplies an attractive working hypothesis for interpreting some discrepant data on the structure and kinematics of the halo. However, there may still be some problems. For example, for nonkinematically selected stars with [Fe/H] < -1.2, Norris (77) showed that σ_z is smaller (at the 2σ level) for stars with galactocentric distances $R > 8$ kpc than for stars with $R < 8$ kpc. This would not be expected if the spherical component becomes more dominant with increasing R. The limited data on the kinematics of halo stars far from the Sun suggest that the anisotropy observed in the solar neighborhood persists to large R (see Section 4); if this is correct, the anisotropy may not fit readily into the two-component picture. The apparent absence of an abundance gradient in the metal-weak halo in the R and z directions may be a problem: How can it be reconciled with a two-component picture in which the flatter component is more metal rich?

To put the two-component picture on a firm dynamical basis, we need to know the run of velocity dispersion with distance for metal weak stars out to large distances in several directions, particularly toward the Galactic poles. However, different kinds of stars give somewhat different results (see Section 4), and more data would be very useful. Also, following Hartwick's (54) comments on the spatial and chemical distributions of globular clusters, we need to find out whether the inner, moderately flattened, metal-weak component (as defined by the field stars) has a metallicity distribution similar to that of the outer spherical component. Is the inner flattened component seen directly in the spatial distribution of other halo tracers, such as BHB stars and metal-weak giants? To answer these questions, further systematic survey programs for such objects are needed in order to establish the spatial and abundance distributions of these objects out to distances of about 20 kpc.

7. THE BULGE

If the Galaxy were seen edge on, from outside, its appearance would probably be fairly similar to that of the spiral galaxy NGC 891 (95, 118). The spheroidal component would be seen on photographs only as a small inner bulge; the metal-weak halo would only negligibly contribute to the surface brightness. The bulge component in the Galaxy is the subject of this section; it has been left to the end so that it can be discussed in the context of the thick disk and the metal-weak halo. The visible bulge is probably the inner part of the thick disk or halo, but which one is not yet clear.

The inner bulge is metal rich. Studies of K giants in Baade's window (89) show abundances over the range $-1 < [Fe/H] < 1$, with a peak at about 0.3. Stellar photometry in several bulge fields (91) indicates that the mean abundance is significantly lower at distances of 2 to 5 kpc from the center.

Surface photometry (B and V) of the diffuse bulge light (26) in the latitude range $10° < |b| < 45°$ is well represented by an $r^{1/4}$ law with an effective radius of 2.7 kpc. This effective radius is very similar to that for the globular cluster system (25), which suggests that the diffuse bulge should be associated dynamically with the halo. However, this suggestion may not be correct because the surface photometry for the comparison galaxy NGC 891 can be equally well fit by a thin-disk plus thick-disk distribution as by a thin-disk plus $r^{1/4}$-law distribution (5). We recall from Section 3 that thick disks and bulges go together in other galaxies, and it may be that they are both part of the same dynamical component.

The IRAS 12-μm Galactic source counts show a box-shaped structure in the inner bulge, which is particularly clear in the distribution of the weaker and redder sources (51). However, the effective radius of this IRAS bulge is only 0.7 kpc, which is much shorter than the effective radius for the optical light. The surface density of late-M stars also falls off much more rapidly than does the optical light (15). These late-type objects, some of which are probably significantly younger than the Galactic halo, will be found preferentially in regions of high metallicity, and thus the steep gradients of their distributions may simply reflect an abundance gradient in the inner bulge.

The bulges of other spirals rotate rapidly enough to be flattened by their rotation (64). Galaxies with box-shaped bulges show cylindrical rotation, i.e. the rotational velocity at a given radius remains constant with height above the Galactic plane. The IRAS data and earlier 2.4-μm surface photometry (57) suggest that the Galactic bulge is itself box-shaped. This may favor the dynamical association of the bulge with the thick disk because the thick disk is in approximately cylindrical rotation, whereas the metal-weak halo is known to rotate slowly. However, the rotation of the Galactic bulge has not yet been studied in detail (although several groups are working on this problem). In particular, little is known yet about the rotation of the metal-weak component in the inner parts of the Galaxy, at radii comparable to those at which bulge rotation is observed in other galaxies.

Velocity dispersions have been measured for different kinds of bulge objects, but they give little guidance about the nature of the bulge because the velocity dispersions of the different old Galactic components are all fairly similar in the inner parts. In Baade's window, the velocity dispersion for M stars is 113 ± 11 km s^{-1} (70), and for K stars it is about 104 km s^{-1} (89). Mira-type variable stars near the Galactic center have a dispersion of about 112 km s^{-1} (37). For OH/IR sources (9) and planetary nebulae (81; A. Kalnajs, private communication) near the center, the velocity dispersion is about 130 km s^{-1}. For stars in directions close to the Galactic center, there is always some uncertainty about whether they belong to the old disk or to the bulge itself. A direct measurement of the velocity dispersion from the integrated bulge light in two bulge fields, whose lines of sight pass within about 0.7 kpc of the Galactic center, gave $\sigma = 125 \pm 10$ km s^{-1} (K. Freeman et al., unpublished). These values are all fairly close to the approximately isothermal velocity dispersion of the halo (about 125 km s^{-1}; see Section 4) and to the central dispersion of the old disk (about 110 km s^{-1}; see Section 2.2).

Data on the rotation of the bulge will be very helpful in understanding how the bulge fits in with the rest of the Galaxy. The objects used to measure bulge rotation should be far enough above the Galactic plane so that pollution by the inner disk is negligible. It would be particularly useful to know whether the metal-weak stars in the inner bulge region are rotating slowly, as they do in the solar neighborhood.

8. CONCLUSION

There are still many observational things that we do not know about our Galaxy. These include the following:

1. What are the disk heating processes? A definitive age-velocity relation for the Galactic disk is needed to answer this question.
2. Is the thick disk a discrete component? This question will be difficult to answer, even with many more data on the structure, kinematics, and chemical properties of the thin and thick disks.
3. What parameters define the two components of the metal-weak halo,

and are these two components discrete? How does the velocity distribution of halo stars change with position in the halo?
4. Why are thick disks and bulges associated in other galaxies? Is the bulge of the Galaxy dynamically related to the thick disk or to the metal-weak outer halo?
5. What are the orbital properties of the metal-weak globular clusters? This question will soon be within reach, and its answer may be very helpful in understanding the origin of the halo cluster system.

In this review, I have emphasized studies attempting to elucidate the present dynamical state of the Galaxy. This knowledge is needed before we can hope to proceed with any confidence to the next step: understanding the chain of events that occurred during the formation of the Galaxy.

ACKNOWLEDGMENTS

It is a pleasure to thank J. Bahcall, B. Carney, G. Gilmore, F. D. A. Hartwick, J. Lewis, J. Norris, P. Quinn, K. Ratnatunga, A. Rodgers, G. Rowley, A. Sandage, J. Sommer-Larsen, P. van der Kruit, G. Wilson, and R. Wyse for discussions and unpublished material.

Literature Cited

1. Athanassoula, E., Sellwood, J. A. 1986. *MNRAS* 221: 213–32
2. Bahcall, J. N. 1984. *Ap. J.* 287: 926–44
3. Bahcall, J. N. 1986. *Ann. Rev. Astron. Astrophys.* 24: 577–611
4. Bahcall, J. N., Casertano, S. 1986. *Ap. J.* 308: 347–56
5. Bahcall, J. N., Kylafis, N. 1985. *Ap. J.* 288: 252–58
6. Bahcall, J. N., Schmidt, M., Soneira, R. 1983. *Ap. J.* 265: 730–47
7. Bahcall, J. N., Soniera, R. 1980. *Ap. J. Suppl.* 44: 73–110
8. Barbanis, B., Woltjer, L. 1967. *Ap. J.* 150: 461–68
9. Baud, B., Habing, H., Mathews, H., Winnberg, A. 1981. *Astron. Astrophys.* 95: 171–76
10. Beers, T. C., Preston, G. W., Shectman, S. A. 1985. *Astron. J.* 90: 2089–2102
11. Binney, J., Mihalas, D. 1981. *Galactic Astronomy*. San Francisco: Freeman. 597 pp.
12. Binney, J., May, A. 1986. *MNRAS* 218: 743–60
13. Binney, J. 1978. *MNRAS* 183: 501–14
14. Blaauw, A. 1965. In *Galactic Structure (Stars and Stellar Systems, Vol. 5)*, ed. A. Blaauw, M. Schmidt, p. 435. Chicago: Univ. Chicago Press
15. Blanco, V., Blanco, B. 1986. *Astrophys. Space. Sci.* 118: 365–66
16. Burstein, D. 1979. *Ap. J.* 234: 829–36
17. Burstein, D., Faber, S. M., Gaskell, C. M., Krumm, N. 1984. *Ap. J.* 287: 586–609
18. Burton, W. B., Gordon, M. A. 1978. *Astron. Astrophys.* 63: 7–27
19. Carignan, C., Freeman, K. C. 1985. *Ap. J.* 294: 494–501
20. Carlberg, R. G., Dawson, P. C., Hsu, T., VandenBerg, D. 1985. *Ap. J.* 294: 674–81
21. Carney, B., Latham, D. 1986. *Astron. J.* 92: 60–71
22. Christian, C. A., Schommer, R. A. 1982. *Ap. J. Suppl.* 49: 405–24
23. Delhaye, J. 1965. In *Galactic Structure (Stars and Stellar Systems, Vol. 5)*, ed. A. Blaauw, M. Schmidt, p. 61. Chicago: Univ. Chicago Press
24. de Vaucouleurs, G. 1959. In *Handbuch der Physik*, ed. S. Flugge, 53: 311–72. Berlin: Springer-Verlag
24a. de Vaucouleurs, G. 1983. *Ap. J.* 268: 451–67
25. de Vaucouleurs, G., Buta, R. 1978. *Astron. J.* 83: 1383–89
26. de Vaucouleurs, G., Pence, W. 1978. *Astron. J.* 83: 1163–73

27. Efstathiou, G., Lake, G., Negroponte, J. 1982. *MNRAS* 199: 1069–88
28. Eggen, O. J. 1969. *Ap. J.* 155: 701–7
29. Eggen, O. J. 1970. In *Vistas in Astronomy*, ed. A. Beer, 12: 367–414. Oxford: Pergamon
30. Eggen, O. J. 1979. *Ap. J. Suppl.* 39: 89–101
31. Eggen, O. J. 1979. *Ap. J.* 229: 158–74
32. Eggen, O. J. 1980. *Ap. J. Suppl.* 43: 457–68
33. Eggen, O. J. 1971. *Proc. Astron. Soc. Pac.* 83: 251–85
34. Eggen, O. J., Lynden-Bell, D., Sandage, A. 1962. *Ap. J.* 136: 748–66
35. Faber, S. M., ed. 1987. *Nearly Normal Galaxies*. New York: Springer-Verlag. In press
36. Fall, S. M., Rees, M. J. 1985. *Ap. J.* 298: 18–26
37. Feast, M. W., Robertson, B., Black, C. 1980. *MNRAS* 190: 227–35
38. Freeman, K. C. 1970. *Ap. J.* 160: 811–30
39. Freeman, K. C. 1975. In *Galaxies and the Universe (Stars and Stellar Systems*, Vol. 9), ed. A. Sandage, M. Sandage, J. Kristian, p. 409. Chicago: Univ. Chicago Press
40. Freeman, K. C. 1980. In *Photometry, Kinematics and Dynamics of Galaxies*, ed. D. S. Evans, p. 85. Austin: Univ. Tex. Press
41. Freeman, K. C. 1985. See Ref. 125, pp. 113–22
42. Freeman, K. C., Illingworth, G. D., Oemler, A. 1983. *Ap. J.* 272: 488–508
43. Frenk, C. S., White, S. D. M. 1980. *MNRAS* 193: 295–311
44. Frenk, C. S., White, S. D. M. 1982. *MNRAS* 198: 173–92
45. Gilmore, G. 1984. *MNRAS* 207: 223–40
46. Gilmore, G., Carswell, R., eds. 1987. *The Galaxy*. Dordrecht: Reidel. In press
47. Gilmore, G., Reid, N. 1983. *MNRAS* 202: 1025–47
48. Gilmore, G., Reid, N., Hewett, P. 1985. *MNRAS* 213: 257–78
49. Gilmore, G., Wyse, R. 1985. *Astron. J.* 90: 2015–26
50. Grindlay, J. E., Philip, A. G. D., eds. 1987. *Globular Cluster Systems in Galaxies*. Dordrecht: Reidel
51. Habing, H. 1986. See Ref. 46. In press
52. Hartkopf, W. I., Yoss, K. M. 1982. *Astron. J.* 87: 1679–1709
53. Hartwick, F. D. A. 1983. *Mem. Soc. Astron. Ital.* 54: 51–64
54. Hartwick, F. D. A. 1986. See Ref. 46. In press
55. Hartwick, F. D. A., Cowley, A. 1985. *Astron. J.* 90: 2244–48
56. Hawkins, M. R. S. 1984. *MNRAS* 206: 433–48
57. Hayakawa, S., Matsumoto, M., Murakami, H., Uyama, K., Yamagami, T., Thomas, J. A. 1979. *Nature* 279: 510–12
58. Hodge, P. 1961. *Ap. J.* 133: 413–19
59. Janes, K. A. 1975. *Ap. J. Suppl.* 29: 161–83
60. Jones, B. J. T., Wyse, R. 1983. *Astron. Astrophys.* 120: 165–80
61. Kalnajs, A. 1972. *Ap. J.* 175: 63–76
62. King, I. R. 1966. *Astron. J.* 71: 64–75
63. Koo, D. C., Kron, R. G. 1982. *Astron. Astrophys.* 105: 107–19
64. Kormendy, J., Illingworth, G. D. 1982. *Ap. J.* 256: 460–80
65. Kron, R. G. 1980. *Ap. J. Suppl.* 43: 305–25
66. Lacey, C. G. 1984. *MNRAS* 208: 687–707
67. Lewis, J. R. 1986. PhD thesis. Aust. Natl. Univ., Canberra
68. Mayor, M. 1974. *Astron. Astrophys.* 32: 321–27
69. Miller, G. E., Scalo, J. M. 1979. *Ap. J. Suppl.* 41: 513–47
70. Mould, J. 1983. *Ap. J.* 266: 255–62
71. Mould, J. R., Schneider, D. P., Gordon, G. A., Aaronson, M., Liebert, J. W. *Publ. Astron. Soc. Pac.* 97: 130–37
72. Nissen, P. E., Edvardsson, B., Gustafsson, B. 1985. In *ESO Workshop on Production and Distribution of C, N, O Elements*, ed. I. Danziger, F. Matteucci, K. Kjär, pp. 131–49. Garching: ESO
73. Norman, C., Renzini, A., Tosi, M. eds. 1987. *Stellar Populations*. Cambridge: Cambridge Univ. Press. 245 pp.
74. Norris, J. E. 1987. *Ap. J.* In press
75. Norris, J. E. 1987. *Ap. J. Lett.* In press
76. Norris, J. E. 1987. *Astron. J.* In press
77. Norris, J. E. 1986. *Ap. J. Suppl.* 61: 667–98
78. Norris, J. E. 1986. See Ref. 46. In press
79. Norris, J. E., Bessell, M. S., Pickles, A. J. 1985. *Ap. J. Suppl.* 58: 463–92
80. Oort, J. H. 1965. In *Galactic Structure (Stars and Stellar Systems*, Vol. 5), ed. A. Blaauw, M. Schmidt, p. 455. Chicago: Univ. Chicago Press
81. Oort, J. H. 1976. *Publ. Astron. Soc. Pac.* 88: 596–97
82. Oort, J. H., Plaut, L. 1975. *Astron. Astrophys.* 41: 71–86
83. Ostriker, J. P., Peebles, P. J. E. 1973. *Ap. J.* 186: 467–80
84. Pier, J. R. 1983. *Ap. J. Suppl.* 53: 791–813

85. Pier, J. R. 1986. See Ref. 46. In press
86. Pritchet, C. 1983. *Astron. J.* 88: 1476–88
87. Quinn, P. J., Goodman, J. 1986. *Ap. J.* 309: 472–95
88. Ratnatunga, K. U., Freeman, K. C. 1985. *Ap. J.* 291: 260–69
89. Rich, M. 1986. PhD thesis. Calif. Inst. Technol., Pasadena
90. Robin, A., Creze, M. 1986. *Astron. Astrophys.* 157: 71–90
91. Rodgers, W. A., Harding, P. H., Ryan, S. 1986. *Astron. J.* 92: 600–9
92. Rodgers, A. W., Paltoglou, G. 1984. *Ap. J. Lett.* 283: L5–7
93. Rose, J. 1985. *Astron. J.* 90: 787–802
94. Saha, A. 1985. *Ap. J.* 289: 310–19
95. Sandage, A. 1961. *The Hubble Atlas of Galaxies.* Washington DC: Carnegie Inst. Washington
96. Sandage, A. 1969. *Ap. J.* 158: 1115–36
97. Sandage, A. 1981. *Astron. J.* 86: 1643–57
98. Deleted in proof
99. Sandage, A. 1986. *Ann. Rev. Astron. Astrophys.* 24: 421–58
100. Sandage, A. 1982. *Ap. J.* 252: 574–81
101. Sandage, A. 1986. See Ref. 46. In press
102. Sandage, A., Fouts, G. 1987. *Astron. J.* In press
103. Searle, L., Zinn, R. 1978. *Ap. J.* 225: 357–79
104. Deleted in proof
105. Sellwood, J. A., Carlberg, R. G. 1984. *Ap. J.* 282: 61–74
106. Shuter, W., ed. 1983. *Kinematics, Dynamics and Structure of the Milky Way.* Dordrecht: Reidel. 392 pp.
107. Silk, J., Norman, C. 1979. *Ap. J.* 234: 86–99
108. Sommer-Larsen, J. 1986. Licentiate thesis. Copenhagen Univ., Den.
109. Sommer-Larsen, J. Christensen, P. 1986. See Ref. 108
110. Sommer-Larsen, J., Christensen, P. 1986. *MNRAS* 219: 537–46
111. Spitzer, L., Schwarzschild, M. 1951. *Ap. J.* 114: 385–97
112. Spitzer, L., Schwarzschild, M. 1953. *Ap. J.* 118: 106–12
113. Toomre, A. 1964. *Ap. J.* 139: 1217–38
114. Tsikoudi, V. 1980. *Ap. J. Suppl.* 43: 365–77
115. Twarog, B. A. 1980. *Ap. J.* 242: 242–59
116. Twarog, B. A. 1980. *Ap. J. Suppl.* 44: 1–29
117. van Albada, T. S., Bahcall, J. N., Begeman, K., Sancisi, R. 1985. *Ap. J.* 295: 305–13
118. van der Kruit, P. 1984. *Astron. Astrophys.* 140: 470–75
119. van der Kruit, P. 1986. See Ref. 46. In press
120. van der Kruit, P. 1986. *Astron. Astrophys.* 157: 230–44
121. van der Kruit, P., Freeman, K. C. 1986. *Ap. J.* 303: 556–72
122. van der Kruit, P. C., Searle, L. 1981. *Astron. Astrophys.* 95: 105–15
123. van der Kruit, P. C., Searle, L. 1981. *Astron. Astrophys.* 95: 116–26
124. van der Kruit, P. C., Searle, L. 1982. *Astron. Astrophys.* 110: 61–78
125. van Woerden, H., Allen, R., Burton, W. B., eds. 1985. *The Milky Way Galaxy.* Dordrecht: Reidel. 660 pp.
126. White, S. D. M. 1985. *Ap. J. Lett.* 294: L99–102
127. Wielen, R. 1974. In *Highlights of Astronomy*, ed. G. Contopoulos, 3: 395–407. Dordrecht: Reidel
128. Wielen, R. 1977. *Astron. Astrophys.* 60: 263–75
129. Wielen, R. 1986. See Ref. 46. In press
130. Wilson, G. A. 1986. PhD thesis. Aust. Natl. Univ., Canberra
131. Wyse, R., Gilmore, G. 1986. *Astron. J.* 91: 855–69
132. Yoshii, Y., Ishida, K., Stobie, R. 1987. *Astron. J.* In press
133. Yoshii, Y., Saio, H. 1979. *Publ. Astron. Soc. Jpn.* 31: 339–68
134. Zinn, R. 1980. *Ap. J.* 241: 602–17
135. Zinn, R. 1985. *Ap. J.* 293: 424–44

VARIATIONS OF SOLAR IRRADIANCE DUE TO MAGNETIC ACTIVITY

G. A. Chapman

San Fernando Observatory and Department of Physics and Astronomy, California State University, Northridge, California 91330

1. INTRODUCTION

Variations in the Sun's luminosity have become, almost overnight, an active and productive field of astrophysics. There is now a nearly unbroken time series of precise data, from two separate satellites, beginning in late 1978 and running to the time of this writing. The study of these solar variations has jumped to a level of activity that might be expected in any field suddenly presented with a three-decade improvement in data quality. This improvement has been created largely by Willson's (90–93) Active Cavity Irradiance Monitor (ACRIM) on the Solar Maximum Mission (SMM). This was the first satellite to have been repaired in space by a manned mission; the repair was tremendously important for the study of solar variability. The variations so far seen have been as large as approximately 0.4% over a time scale of a week. The irradiance has also exhibited a steady decline since 1980 of the order of 0.02% per year. A review of temporal variations is given by Fröhlich (31a).

In this review, we concentrate on solar luminosity variability due to magnetic activity. The precision of the ACRIM while operating in its normal mode approached 10 ppm of the mean solar irradiance, when averaged over a day. At this level of precision, the Sun's output is almost always changing. Another experiment, the Earth Radiation Budget (ERB) on the Nimbus-7 satellite, has produced an even longer record of solar variability measurements, though with less precision (35). The data from the two spacecraft have revolutionized the field within the past six or seven years.

There are several important questions that have been raised by these observations. One question, which is addressed in this review, is whether or not the trend in, and the fluctuations of, the solar irradiance are caused by solar magnetic fields—in particular, by sunspots and faculae. Most studies have been limited to the effect of active regions on the solar irradiance (47); the contribution of the photospheric network to solar variations is uncertain. If the decrease in the solar irradiance, detected by ACRIM and Nimbus-7, is related to the solar cycle, then we will be able to estimate the irradiance from Earth-based observations. If, on the other hand, the irradiance variations are only partly caused by magnetic activity, then our ability to predict or model the solar irradiance, based on cyclic solar magnetic phenomena, will be severely limited, with obvious geophysical implications.

An important astrophysical problem has to do with the formation and stability of sunspots and their role in the flow of energy in and around an active region (cf. 69). There is presently no completely acceptable explanation of the solar cycle in general and the sunspot phenomenon in particular. If our understanding of the solar cycle is incomplete, then we are less sure of other aspects of the solar interior and of astrophysical plasmas in general. There are other stars that have magnetic cycles that are much more vigorous than the Sun's. Our understanding of these objects must be considered quite provisional.

From the point of view of solar physics, we would like to have a better idea of the structure of the flux tubes that form sunspots and faculae. We do not yet understand why the magnetic flux appears in two different structures, one dark and one bright, with magnetic flux density differing by roughly a factor of two. A detailed understanding of the magnetic connectivity in active regions would greatly help in the study of solar flares, for example, which can cause nuclear reactions at the surface of the Sun and can interfere with man's activities on Earth and in space. Although the variations in solar luminosity seem quite small (approximately 0.1% from maximum to minimum), this is of the same order of magnitude as the longer period insolation changes that give rise to the major ice ages (65). The 11-yr solar cycle has been seen in measurements of the Earth's Length Of Day (LOD) (17), which provides evidence that the Sun affects the Earth's moment of inertia. Ancient evidence shows that the solar cycle may have had an even greater effect than now on weather patterns in particular terrestrial locales (89).

This review concentrates on what has been and can be learned from observations of the effect of solar magnetic fields on the solar irradiance. Newkirk (62) has written a general review of possible causes of solar variability. A recent review of solar variability and oscillations has been written by Hudson (42). Earlier NASA workshops, reviewing solar variability, are available as Conference Proceedings (47, 80).

The review is organized in the following way. Section 2 discusses the use of proxy data—that is, the position and area or some other property that is not in itself a photometric measure of intensity. Section 3 reviews direct photometric measurements, and Section 4 describes instrumentation and ongoing observing programs. Finally, Section 5 is a brief discussion of a few of the key theoretical problems that are posed by the observed short- and long-term solar irradiance variations.

2. PROXY DATA

Comparisons of magnetic features with satellite irradiance data have begun with nonphotometric data obtained from synoptic programs. The two features of concern are sunspot and facular area. The data are used in an algorithm as a proxy for photometry. Modeling the irradiance from sunspot data is easier because of a sunspot's greater contrast compared with that of faculae. Further, a sunspot's area should be easier to measure because of its simpler shape. We discuss modeling the irradiance using sunspots first.

2.1 *Area Proxy*

2.1.1 SUNSPOTS Since solar irradiance variations appear to be caused mostly by sunspots and faculae, models have been developed to give photometric fluctuations based on their areas. These data become a "proxy" for actual photometry. One reason for using proxy data is obvious: Areas and positions of active regions are published for most periods of interest when spacecraft data are available, permitting a thorough comparison to be made. Furthermore, it might be possible to reconstruct the solar irradiance for periods before the availability of space observations. An estimate of past values of the solar irradiance based on proxy data was presented for several intervals by Eddy (21). However, these modeled irradiances are based on an inadequate description of the contrast of faculae and so are probably in error. This and related points have been discussed in the literature [see (20, 47, 63, 73, 74, 81)].

Since models of irradiance fluctuations from proxy data involve only the area and position of sunspots and plage or faculae, their use is quite simple. Unfortunately, there may be a great deal of error, both random and systematic, in the data. In an attempt to eliminate systematic errors, Oster et al. (67) used as sunspot areas the largest values seen during a spot's disk transit. This choice resulted in an overestimate of the area of spots, but since the contrast was a free parameter, the overall error on solar irradiance fluctuations may be small (40). In spite of these uncertainties, these authors found that faculae were an important source of irradiance variations, roughly balancing sunspot effects.

In fact, measurements of sunspot areas published by different observatories sometimes disagree seriously, although the random errors for the whole disk appeared to be less than 15% during part of 1980 (20). Hoyt et al. (40) studied the relative errors between sunspot areas published by several observatories. They found systematic errors of about 9% between different observatories, but on 5% of the days random errors in areas caused errors in the proxy irradiance of 500 ppm. These estimates of systematic errors were based on internal agreement; the actual systematic errors need to be confirmed by a digital analysis of high-resolution images. Furthermore, the use of only the area of a sunspot or facular region assumes that the photometric properties are always the same from one region to another. There is evidence that not all sunspots are alike. It is well known that the ratio of umbral to penumbral area for sunspots is variable (56). Furthermore, the umbral/penumbral area ratio may change with the solar cycle (39), as may the darkness of the umbra (1).

A good presentation of the derivation of a proxy irradiance signal based on the area and solar disk position can be found in Appendix A of Foukal (22), although different mathematical forms suggested by others (13, 76) may be more appropriate.

We discuss a proxy irradiance function for sunspots first, since it is simpler than that for faculae. We adopt the use of ψ to represent what has been called PSI (the Photometric Sunspot Index). This quantity can be calculated for each sunspot on the disk of the Sun for a given day, and the values for each sunspot can then be summed, giving a value for the entire visible hemisphere. Following the scheme outlined by Hudson in Willson et al. (90), we write

$$\psi = \text{PSI} = -C'_s \Sigma A_s \mu (\Delta I/I) I(\mu)/I(0), \qquad 1.$$

where A_s is the sunspot area (in millionths of the hemisphere), C'_s is an empirically determined coefficient, $\Delta I/I$ is the local contrast of the sunspot (assumed constant), $\mu = \cos\theta$ is the foreshortening, and $I(\mu)/I(0)$ is the quiet Sun limb darkening, with the sum taken over all regions. It is common to adopt a gray approximation for the relative intensity of the quiet Sun, where $I(\mu)/I(0) = (3\mu+2)/5$. Then Equation 1 becomes

$$\text{PSI} = -C_s A_s \mu (3\mu+2). \qquad 2.$$

More accurate functions could be derived, based on observed limb darkening and the center-to-limb contrast of sunspots. However, the gray description of limb darkening is probably a reasonable approximation to the bolometric limb darkening. The contrast of sunspots is only weakly dependent on disk position (1). The value of C_s determined in (13) is -0.164, which is consistent with the average value of -0.33 in (90), which should be equal to $2C_s$. Although the expressions for the quiet Sun limb darkening and the contrast function of sunspots could be improved over those that have been used, the uncertainties in the proxy areas and their conversion to irradiance fluctuations probably dominate the errors.

Figure 1 shows the actual irradiance data from ACRIM until the end of 1984. The figure is from Willson et al. (93). The proxy parameter, ψ, is able to account for about 50% of the rapid variance in the daily ACRIM signal in 1980. The fraction of the variance explained seems to be similar from one analysis to another. A similar fraction is mentioned in other studies (26, 43, 76). How much of the remaining variance is caused by noise in the sunspot data and how much is due to the facular component is an important, unresolved question. To determine the random errors in published sunspot areas, we are comparing them with digitized spot areas from full-disk photographs (8). Several months of full-disk images from 1980 have been digitized and will be converted to daily values of PSI. Figure 2 gives an example of one of these images. These values of area and PSI will be compared to digital results from Catania and from others who may have digital areas of sunspots.

2.1.2 FACULAE Photospheric faculae are bright features associated with the magnetic fields of an active region. They are known to be the white-light manifestation of the photospheric network. Developing a proxy irradiance function for faculae is considerably more difficult than for sunspots, partly because these features are not visible near the center of the solar disk. Even the detailed correspondence between faculae and the network at high spatial resolution is not so easily understood (46, 64, 82, 98). The contrast of flux tubes, which are the basic constituents of a facular region, is a subject of current research.

The area of photospheric faculae, determined from photographs made at the Royal Greenwich Observatory, has been used in earlier studies of solar variability (20, 21, 23, 24). These studies reached rather diverse conclusions about the importance of faculae to solar irradiance. Part of the reason for these differences is the difficulty of determining the area of low-contrast features that can only be seen when they are foreshortened and near the limb.

Use of facular area for a long-term proxy irradiance index has become more difficult because the Greenwich program no longer reports daily facular areas. Furthermore, it is possible that these areas may be systematically in error owing to the occultation of the solar limb (12). Thus

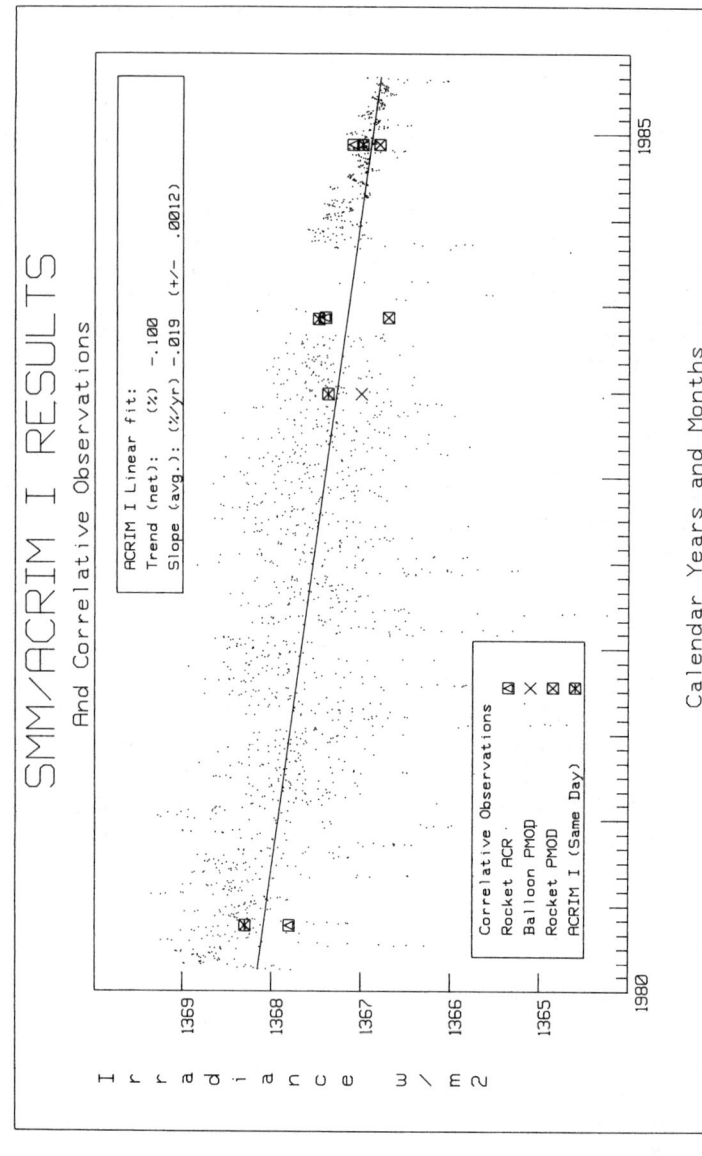

Figure 1 Solar irradiance seen by the ACRIM on the Solar Max satellite (93). The downward trend in irradiance is confirmed by the ERB on the Nimbus-7 satellite (35). Most of the fluctuations are real and presumably due to solar magnetic regions, faculae, and sunspots.

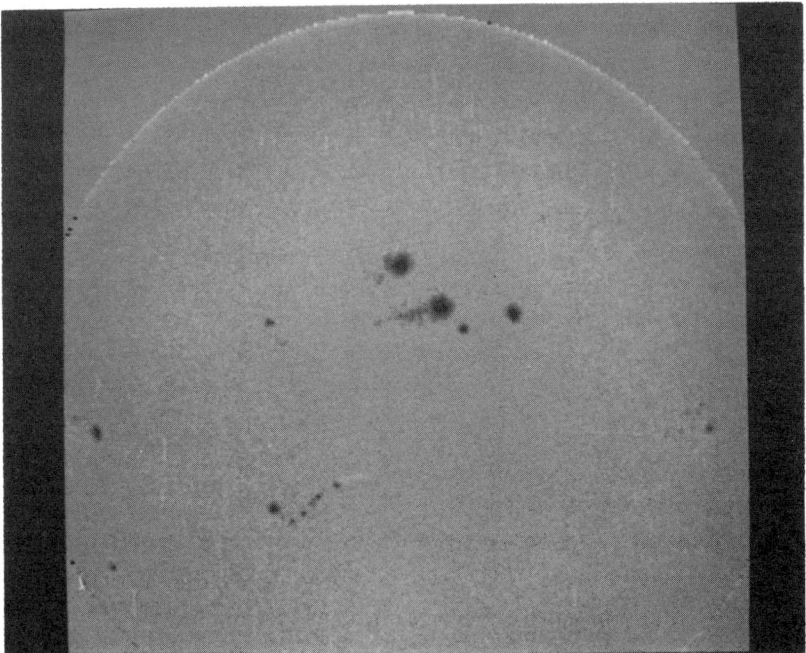

Figure 2 Computer display of a digitized photograph from Sacramento Peak Observatory for 26 May 1980. The photograph has been converted to an intensity scale using solar limb darkening; then the quiet Sun limb darkening has been removed (8) and the resulting contrast map searched for sunspots.

the facular to sunspot area ratio reported by Eddy et al. (20) may be anomalously low, especially at solar maximum, when the extent of facular regions is the greatest. Furthermore, facular area in white light is probably very sensitive to seeing, since the contrast of faculae is much lower than that of sunspots.

It seems more reliable to use the calcium plage area as a proxy for facular area and to find a conversion factor that converts plage area into facular area. Plage and plage fragments have much higher contrast than do white-light faculae. Thus, small elements are less likely to be missed. However, plage contrast, and hence the area, may be a sensitive function of the wavelength and bandpass of the filter or spectroheliograph. This sensitivity to instrumental parameters will mean a different conversion factor for different observatories or for a different filter at a particular observatory (76). In addition, the center-to-limb behavior of the contrast of white-light faculae is quite different than that of calcium plage. Calcium plages are visible across the solar disk, whereas white-light faculae are only seen beyond $\sin\theta \gtrsim 0.6$. Thus, a proxy irradiance index based on calcium plage area and position will be very different than one based on facular area and position. However, in either case, if only the area is to be used, some model of the center-to-limb contrast function of faculae will have to be used [see the discussion on different facular contrast functions (37)].

Two problems with using a Photometric Facular Index (PFI) to represent the facular component of irradiance fluctuations are the greater uncertainty in facular area and the uncertainty in the facular limb darkening. Chapman & Meyer (13) have developed a PFI that is functionally quite simple. This index can be derived by analogy with Equation 1 for the PSI:

$$\text{PFI} = C_p A_p \mu (\Delta I/I) I(\mu)/I(0),$$

where C_p is an empirically determined coefficient, A_p is the published area of a calcium plage, and the remainder of the expression is the facular contrast function and the gray approximation for the limb darkening of the quiet Sun. For the facular contrast function, we have used (6)

$$\Delta I/I = b(\mu^{-1} - a), \qquad 3.$$

where b and a are empirically determined constants. With $a = 1$, we have (7)

$$\text{PFI} = C_p A_p (\mu - 3\mu^2 + 2). \qquad 4.$$

The value of a is determined from observations at disk center. By adopting $a = 1$, we assume that the disk center contrast of faculae is zero. Most values reported for this quantity are near zero (28, 29, 38). If we wish to parameterize the contrast at the disk center, we can write

$$\text{PFI} = C_p A_p [(3 - 2a)\mu - 3a\mu^2 + 2]. \qquad 5.$$

Figure 3 shows the facular indices that result from several authors. Using a value for the contrast of 0.1% (28, 29), we find that $a = 0.99$. This value is almost certainly wavelength dependent (28, 31) because of the temperature structure of faculae and the wavelength dependence of the hydrogen opacity. J. K. Lawrence et al. (in preparation) have found that the disk center contrast is about 0.4% at 8643 Å. This contrast, when corrected for the low filling factor, might lead to a value of a in Equation 5 of 0.7. This would make active regions able to account for the change in irradiance measured by the ACRIM since launch. The value of the white-light contrast is not accurately known. This uncertainty is discussed in Sections 3 and 5. The value of b is determined by fitting PFI to photometry near the limb. From (13) we obtain $b = 0.09$, corresponding to a wavelength of approximately 5300 Å. Equation 3 with this value of b

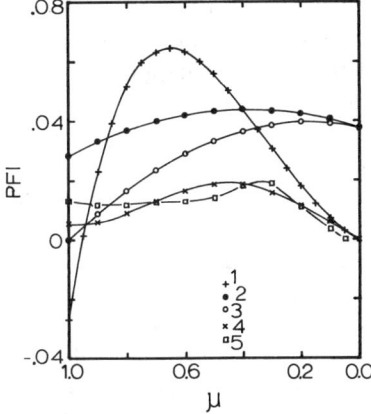

Figure 3 Plot of the facular proxy irradiance fluctuation, called the Photometric Facular Index (PFI), vs. $\mu = \cos\theta$ for several authors. The curves are normalized to unit facular or plage area. If the author did not specify the quiet Sun limb darkening, the gray approximation (Equation 2) was used. The numbered symbols 1 through 5 correspond to the following references: 1 (22), 2 (this review), 3 (13), 4 (81), and 5 (37).

predicts a contrast at $\mu = 0.14$ of 0.6, a value that is not inconsistent with that from Stratoscope (72).

Although the contrast function in Equation 3 is not well behaved at the limb, photometric measurements are foreshortened by μ, and the product is well behaved across the disk. The rise in the facular contrast toward the limb is a matter of some controversy. Observations showing an increasing contrast as the limb is approached have been reported by Chapman (6), Chapman & Klabunde (10), and Hirayama et al. (38). On the other hand, Libbrecht & Kuhn (53) have argued that the modified Princeton oblateness telescope shows that the facular contrast declines near the limb. Figure 4 shows the mean contrast of faculae in the green ($\lambda = 0.53$ μm) obtained with the Extreme Limb Photometer (ELP) at the San Fernando Observatory. The data from 1982 were taken from Oscas (66) and have been scaled to fit the data from 1975 and 1979 (10). The data shown are from all days on which data were obtained; there has been no data selection.

The determination of C_p in Equation 4 has been carried out using

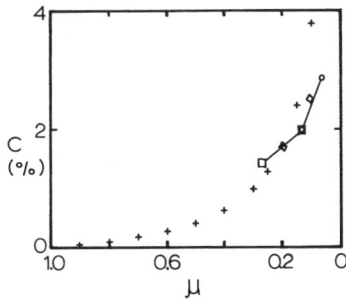

Figure 4 Mean facular contrast versus $\mu = \cos\theta$ for data from the Extreme Limb Photometer (ELP). The circles (1979 data) and squares (1975 data) are from (10). The diamonds are 1982 results from (66) that have been normalized at $\mu = 0.2$ to the earlier data. The crosses represent the contrast obtained from Equation 3 with $a = 1.0$ and $b = 0.00425$.

photometry obtained with the ELP (8). This photometry permits us to check the form of Equation 4 and to determine the average of the product $C_p A_p$. From observations obtained in 1980 and 1982, Chapman & Meyer (13) obtained a value of $C_p = 0.019$. This value has a formal standard error of ±15% and pertains to the average plage area for the observing interval. This coefficient has a t-statistic greater than 6 for 45 observed regions and a correlation coefficient of 0.72 for a linear fit between the PFI and the region's contribution to the fluctuation of the solar irradiance, $\Delta B/B$, measured by the ELP. This agreement suggests that the form for the facular contrast in Equation 3 is not seriously in error.

A very important reason for trying to calibrate proxy data in terms of irradiance variations is that the mathematical model incorporated in the proxy function can be integrated over angle to obtain a model luminosity fluctuation for the active regions on the disk for a particular day. This is an important astrophysical quantity that cannot be obtained by direct observation. To do so would require that the Sun be ringed by irradiance monitors. Since flux is the integral of specific intensity over solid angle, the integral of the irradiance fluctuation from a specific active region is related to the variation in flux for that region summed over the area of the region, assuming that the extent of the region represents a small heliocentric angle. The flux is given by

$$F = 2\pi \int I\mu \, d\mu. \qquad 6.$$

The fluctuation in this quantity due to an active region is given by

$$\Delta F = 2\pi \int \Delta I\mu \, d\mu. \qquad 7.$$

We can evaluate the case for a gray atmosphere using Equation 4 and normalize the flux variation for different assumed facular or sunspot contrast functions. (The contrast is defined in terms of the specific intensity I.) Hirayama et al. (37) have calculated the flux variation (i.e. flux per unit area) for several different assumed facular contrast functions. For Equation 4 we obtain a flux excess of $\Delta F/F = 0.07$, which is about two times higher than that obtained by Hirayama et al. (37). This value of $\Delta F/F$ corresponds to a flux excess of 4.4×10^9 erg s^{-1} cm^{-2} and refers to the facular flux tube, not the area of the calcium plage. If the contrast of the faculae is lower at the limb than specified by Equation 3, then the flux excess will be less. For example, if the contrast is assumed to go to zero at $\mu = 0.25$, then the flux excess decreases by only 3%.

The flux deficit of sunspots is -0.79 of the quiet Sun, which gives

-5.0×10^{10} erg s^{-1} cm^{-2}, according to Hirayama et al. (37). (All areas are in millionths of the solar hemisphere, unless stated otherwise.) The value for the spot flux deficit from the PSI of Equation 2 is -0.83 in close agreement with the Hirayama et al. result. This value is determined by $-2C_s$ divided by the normalized flux from the quiet Sun, giving a value of 0.4.

We can integrate these flux variations over the sunspot or calcium plage to obtain the total flux deficit or excess. We can then measure or estimate the lifetime of the respective feature to obtain the energy deficit or excess. This is the procedure described by Hirayama et al. (37), who obtained an energy deficit of -0.9×10^{36} erg and a facular energy excess of 1.1×10^{36} erg for a group of active regions. From a combination of photometry and proxy data, Chapman et al. (12) found that the sunspots of a sequence of active regions, making up an activity complex, were associated with an energy deficit of approximately -1×10^{37} erg. They further found that the facular energy excess was approximately the same as the spot deficit (but see the more detailed discussion in Section 3).

Using proxy data calibrated with ELP photometry, Chapman (7) and Chapman & Meyer (13) calculated an approximate energy balance for 45 active regions at various positions on the disk. This conclusion was based on the luminosity fluctuation determined by integrating equations 2 and 4 over solid angle, as indicated by Equations 6 and 7, an empirical determination of the coefficients in Equations 2 and 4, and the average ratio of the area of calcium plages and sunspots. As pointed out by Eddy (21) and Lawrence (50, 51), this ratio is probably a function of the solar cycle. Lawrence found that the ratio of published calcium plage to sunspot area varies from about 25 near activity minimum to a value of about 10 near the maximum of the solar cycle. Figure 5 shows the variation of spot and plage areas over a solar cycle. We need to know whether or not the coefficients in Equations 2 and 4 are also cycle dependent.

Schatten et al. (76) evaluated various forms of both PFI and PSI in matching the ACRIM and Nimbus-7 data. They found that these proxy signals were significantly correlated with the spacecraft data, and the rms differences in the fits were all in the range of $r = 0.6$ W m^{-2}. Hudson et al. (44) found correlation coefficients of 0.73 ($r^2 \simeq 0.5$) for 276 days of ACRIM data fit to a PSI signal only, where r^2 is the coefficient of variation (i.e. the fraction of the variance that is explained by the statistical model).

To investigate the errors in published sunspot areas, one could compare values of PSI from several measurements of sunspot areas, some of which are digital. By assuming that the ACRIM fluctuations are due only to solar activity, and that the PSI is an accurate proxy for the sunspot irradiance deficit, we can subtract the PSI from the ACRIM to derive an

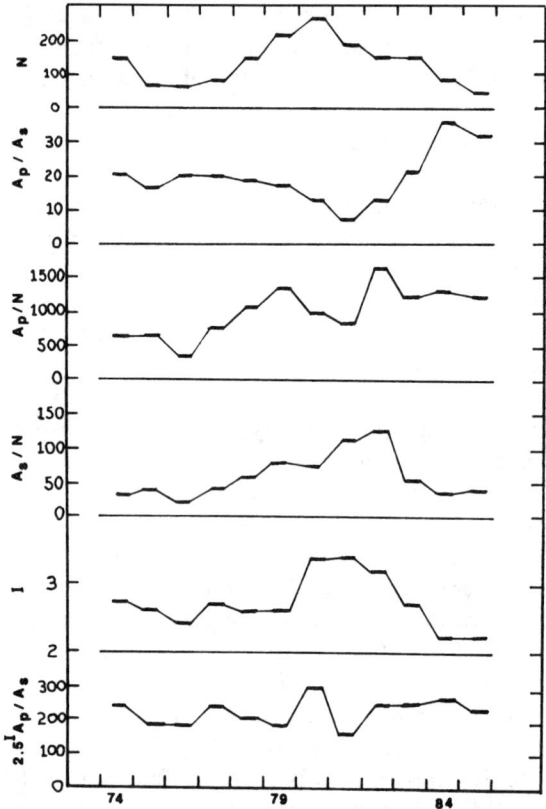

Figure 5 Variation of published sunspot and Ca plage areas sampled each summer season as a function of the solar cycle. The largest plage to spot area ratios occur at solar minimum (50, 51).

estimate of the facular irradiance excess. For the one-week period centered on the "big dipper" of April 1980, caused by two large sunspot groups (Hale Nos. 16747 and 16752), our group found a difference (ACRIM − PSI) of approximately 800 ppm. From the calcium plage data published by NOAA and from Equation 4, we found an average PFI of about 400 ppm. Furthermore, the rms disagreement between the PSI from sunspot areas, from SFO (digitized Sacramento Peak film), from Catania, and from *The Solar-Geophysical Bulletin* was approximately 200 ppm (G. A. Chapman et al., in preparation). If the value of a in Equation 3 is set to 0.7, then the proxy irradiance fluctuations from active regions may come close to explaining the drop in irradiance from 1980 to 1985 seen by ACRIM. This value of a corresponds to a facular contrast at the center of the disk of 2.7%. J. K. Lawrence et al. (in preparation) find a contrast at disk center of about 0.4% at 8643 Å, with a pixel size of 1.9″ × 1.9″. A filling factor of 0.15 would correspond to a contrast of 2.7%. On the other hand, Muller & Keil (61) found a contrast of 30–50% for resolved faculae at a wavelength of 5750 Å at disk center. Other results on flux tube models and contrast can be found in Muller (59) and Spruit & Zwaan (83). The resolution was 0.25″. This contrast is much higher than that quoted by most authors, although it is difficult to compare facular contrast measurements that correspond to such differing spatial resolutions. The large uncertainty in the contrast of faculae leads to a large uncertainty in the modeled irradiance fluctuation of active regions. Thus, there are serious problems with the use of proxy data at this time.

Nonetheless, proxy data allow us to overcome the special place of the Earth with respect to the solar equator and central meridian and to obtain estimates of luminosity and energy fluctuations from ground-based observations.

2.2 Magnetic Proxy

The modeling of irradiance variations, caused by sunspots and faculae, might be based on the magnetic field associated with these phenomena rather than their area. The advantage in using the magnetic field is that where there is no field, there is no model irradiance variation. In other words, one looks for the polarized light that is associated with the magnetic field. The quiet Sun has no polarized signal, so there is no contribution to the activity-caused irradiance fluctuation. There is no need to make accurate limb-darkening fits to the quiet Sun (5). The problem is, of course, that one is measuring magnetic flux rather than irradiance. Thus, an algorithm must be developed that will convert magnetogram signals into a proxy for irradiance fluctuations.

Early attempts to construct an irradiance fluctuation from the magnetogram signals were made in 1983 with magnetograms from the Kitt Peak vacuum telescope (G. A. Chapman & H. P. Jones, unpublished). The high spatial resolution (1″ pixel size) made the data-management problem difficult. A similar analysis was carried out with data from the Mount Wilson daily full-disk magnetograms from the 150-ft tower telescope. The Mount Wilson data set is fairly homogeneous. However, with a pixel size of approximately 12″, there is some concern that strong field concentrations of faculae may be mistaken for small sunspots. This problem may be avoided by using the intensity in the wing of the 5250-Å line to separate dark structures from bright ones. The low spatial resolution makes this procedure less certain than would be the case for high-spatial-resolution data. A three-year portion of these data was used by Bruning

& LaBonte (5) to form an irradiance fluctuation from the intensity signal in the wing of the 5250-Å line that is used in forming the magnetogram signal. The intensity signal is also proportional to the area and filling factor of a facular region. Unfortunately, the signal is the intensity in the wing of the 5250-Å line and does not have the same limb-darkening characteristics of photospheric faculae, i.e. the faculae at 5250 Å can be seen all across the solar disk (5), whereas white-light faculae are seen only beyond $\sin \theta \gtrsim 0.6$. This difference will tend to exaggerate the effect of faculae, especially near disk center. Perhaps this explains why Bruning & LaBonte concluded that faculae produced an irradiance excess that on average was balanced by an irradiance deficit due to sunspots. This, in turn, led them to conclude that there was a poleward excess in the Sun's net irradiance. Again, these conclusions are based on observations of faculae in the wing of the 5250-Å line. In the continuum, faculae are not easily seen at disk center (28, 29), and one would thus not expect faculae to contribute so strongly to irradiance fluctuations near disk center. However, faculae can still be quite important to luminosity fluctuations, because the luminosity weights solid angle away from the solar vertical, whereas the irradiance is a uniform weighting with zenith angle (see the factor of μ in Equations 6 and 7).

The conversion of the Mount Wilson daily magnetograms into a "pseudo-irradiance" fluctuation was carried out by Chapman & Boyden (9). The Mount Wilson data are uniform and exist for more than a solar cycle, but they have low spatial resolution, which complicates their conversion to irradiance fluctuations. The magnetic flux levels corresponding to sunspots and faculae were determined by comparing the shapes to white-light photographs at the same scale. It was determined that the best matches were contours of 10 and 100 G for faculae and sunspots, respectively. In the author's work with H. P. Jones, using Kitt Peak data, it was found that improvements in the magnetic proxy can be made by using the line intensity signal recorded along with the magnetic signal to discriminate between small sunspots and strong facular areas. This procedure would probably help in the analysis of the Mount Wilson data, which have lower resolution than those of Kitt Peak.

Figure 6 shows the Mount Wilson pseudo-irradiance for the better part of solar cycle 21 and the ACRIM signal that was used to calibrate the coefficients used to convert the Mount Wilson magnetograph signal to the pseudo-irradiance. An interesting feature of the Mount Wilson signal is that it is highest at the time of the maximum in the solar activity cycle. The calibration of the pseudo-irradiance was carried out with small parts of the data set showing large fluctuations, rather than with the entire ACRIM signal. The average peak Mount Wilson facular irradiance signal

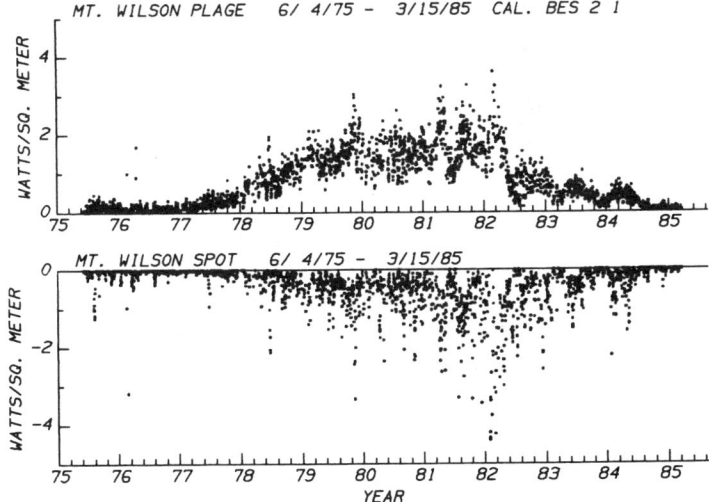

Figure 6 Magnetic proxy irradiance, called pseudo-irradiance, from the Mount Wilson 150-ft tower telescope. The upper plot shows the facular component of the pseudo-irradiance, and the lower plot shows the sunspot component of the pseudo-irradiance after calibration against the larger fluctuations in the ACRIM signal (9).

($1.7\ \mathrm{W\ m^{-2}}$) is larger than the value of $0.5\ \mathrm{W\ m^{-2}}$ determined by Schatten et al. (76). This may indicate that the photospheric network is important to fluctuations in the solar irradiance, as was suggested by LaBonte (48). However, because there may have been some misidentification of spots or faculae, this analysis should be repeated with higher resolution data or reanalyzed using the wing intensity as a discriminator between faculae and sunspots. Only then can we be confident in the magnetic proxy method.

3. DIRECT MEASUREMENTS: PHOTOMETRY OF ACTIVE REGIONS

A major goal in the study of the flux deficit of sunspots is to determine whether or not some of this missing flux might reappear as a faint bright ring around the spot. Since this ring would be superimposed on the granulation pattern, its detection would require careful photometry. In the past, photometry of magnetic active regions has been attempted using photographic techniques. However, there are problems in using photographic film, especially in its limited dynamic range and its difficult calibration. In addition, there are adjacency effects, where underexposed regions such as exist in sunspots can cause an overdevelopment to occur in adjacent, normally exposed regions. Such effects could be especially important around sunspots.

Hirayama & Okamoto (36) found, using photographic photometry, a weak bright ring around sunspots that approximately balanced the flux deficit from the spots themselves. They checked their photographic results with photoelectric scans and found the weak emission to be a continuum, not weakened absorption lines. Further analysis and new data (37) showed that this bright ring was apparently a photographic artifact, indicating the difficulty of using photographic photometry in the neighborhood of large intensity gradients. A photographic analysis of sunspot deficits by Bray (4) also shows the difficulty of using photographic photometry.

Foukal and several collaborators (28–31) have carried out a photoelectric program of photometry using the vacuum telescope and magnetograph at Kitt Peak. Fowler et al. (31) found no evidence for a bright ring such as was reported by Hirayama & Okamoto (36). Instead they found an extended bright area having a contrast of 0.2–0.3%, which they believed was a region of faculae rather than the hypothetical smooth bright ring. Foukal (26) speculated that this emission may have been due to continuum bright points associated with faculae around the sunspots that they studied. On the other hand, they found that this ring did not increase in contrast as the limb was approached, unlike the expected behavior of faculae.

The search for a diffuse bright ring is motivated by the view that a sunspot is a barrier to the flow of heat from beneath, and that this redirected heat should show up at the surface as a diffuse region of slightly higher temperature. The failure to convincingly find any bright ring around sunspots suggests that a spot may not be simply a barrier at the top of the convection zone. Furthermore, if the magnetic flux tubes of an active region blocked the flow of heat while rising to the surface, then one might find "thermal shadows" at the photosphere. These shadows would be expected to occur at the location of a new active region prior to the appearance of spots. Observations by Foukal et al. (30) failed to show any evidence of a thermal shadow, with an rms upper limit of $\Delta T \lesssim 1.5$ K. This sets a lower limit to the thermal diffusion coefficient in the convection zone, in the view of these thermal blocking models. The distinction of any possible diffuse bright ring from faculae is an important but difficult task. The upper limit to the brightness of any diffuse, nonfacular bright ring appears to be less than 1.5 K and may be zero (i.e. no bright ring).

Quite aside from the question of diffuse bright rings, the disk center contrast of faculae is important to proxy models of facular effects on the solar irradiance. Recent work by Foukal (25) has shown that the contrast of faculae near disk center is about 0.1%. This value supersedes the results of Foukal et al. (29) in which faculae were found to be slightly darker than the photosphere at disk center. The contrast of faculae is certainly a function of wavelength (as well as position on the disk), and further work on the facular contrast spectrum is needed. Results from a new photometer (38) operated at a wavelength of approximately 5450 Å shows a facular contrast near the center of the disk of less than 0.1%, perhaps as low as 0.05%, although the one-sigma error is about 0.2%. Although the pixel size is 2.75" in the radial direction, the observations were averaged over nearly two hours, making the effective resolution about 20". This resolution is substantially poorer than that corresponding to the Kitt Peak observations of Foukal et al. Analysis of dual-wavelength observations (J. K. Lawrence, unpublished, 1986) have shown a facular contrast near disk center of about 0.4% at a wavelength of 8643 Å. The facular pixels were identified by the selection of the brighter pixels (contrast greater than three standard deviations) in an image made in the wing of the Ca II 8662 Å line, simultaneously with the continuum image at 8643 Å. The spatial resolution of these dual-channel images is typically 4 or 5" from data originally obtained with 1" pixels.

LaBonte (48) has reported on low-resolution spectroscopy of calcium plage regions, which are the chromospheric manifestations of facular magnetic fields. He found, from spectra covering 56 Å around the H and K lines, a 4% plage contrast due to the weakening of photospheric lines. Based on line blanketing, this contrast is reduced to 0.4% relative to the quiet Sun. This contrast is due only to lines and assumes that the plage continuum has the same intensity as the quiet Sun. This broadband contrast may be more representative than the contrast in the continuum of a facular region's contribution to the total solar irradiance. The total facular contribution includes the lines and continuum. The mid-visible facular contrasts for the continuum, mentioned above, range from 0 to 1%. Thus the broadband facular contrast may range from a few tenths of a percent to 1.5%, including lines and continuum. The disk center contrast is important because features there are not foreshortened and their irradiance contribution occurs where the quiet Sun has its highest intensity.

An important point discussed by LaBonte is the relative contribution to irradiance fluctuations of active regions (defined by the calcium plage) as compared with that of the photospheric network. He suggests that the contribution of the network may be about the same as that from active regions. This is an important question with regard to energy balance; to answer it will require high-resolution global observations of irradiance and magnetic flux.

Photometry of solar active regions was begun at the San Fernando Observatory in 1980 as part of the Solar Maximum Mission. This work was carried out with an Extreme Limb Photometer (ELP). The ELP rotates a slit over the photosphere, sweeping out an annular swath 39″ wide (10). The ELP was designed primarily for Sun-centered operation, with the aperture filled partly by the Sun and partly by the sky. By pointing the telescope to a position away from the center of the solar disk, however, one can scan any active region on the disk. For active regions away from the limb and not foreshortened, it is often necessary to repoint the telescope several times in order to completely scan the entire region. The usual practice is to move the telescope in a north-south direction for each subsequent scan in order that each swath cover as much of the active region as possible and thus minimize the number of separate pointings of the telescope.

The results of some of this work have been reported by Chapman & Meyer (13). The ELP has only a single photodetector and so is unable to make an image. However, near the solar limb, the contrast structure of an active region in azimuthal angle is easy to see because the intensity of the quiet Sun, averaged over the 39″ slit, is relatively constant as the slit rotates around the image at constant radial distance. Figure 3 in (8) has three adjacent scans that cover an active region near the solar limb, showing that the positive facular contrast exceeds the negative contrast of the sunspot near the limb. This is a common situation. The low contrast of the sunspot is due to the small filling factor of the spot in the ELP slit, which results from foreshortening near the limb and the facular emission that surrounds the sunspot. To show that the ELP measures large negative contrasts when the filling factor is large, we show a scan that crossed a very large sunspot on 25 July 1981. This sunspot was Mount Wilson No. 22411 in Hale No. 17751 and had an area of about 1500 millionths of the solar hemisphere. This day was associated with the second largest dip ever seen in the ACRIM signal. Figure 7 shows one ELP scan on 25 July 1981. The ACRIM fluctuation was -2200×10^{-6}. The ELP indicated a fluctuation of -899 millionths, which for this one sunspot is almost half of the ACRIM fluctuation.

In 1982 systematic photometry of active regions was begun with a linear photodiode array having 512 elements. Using a 28-cm vacuum telescope, each diode corresponds to a spatial size of 0.94″. Figure 8 shows a computer display of such an image. Results from this program have been reported by Chapman et al. (11, 12) and Lawrence et al. (49). Data-reduction efforts have been directed toward the study of the energy balance of solar activity complexes (occurrences at the same longitude of a succession of active regions). The lifetime of such an activity complex is estimated to be about six months. It seems useful to begin our analysis of the question of the energy budget of magnetic regions by studying an

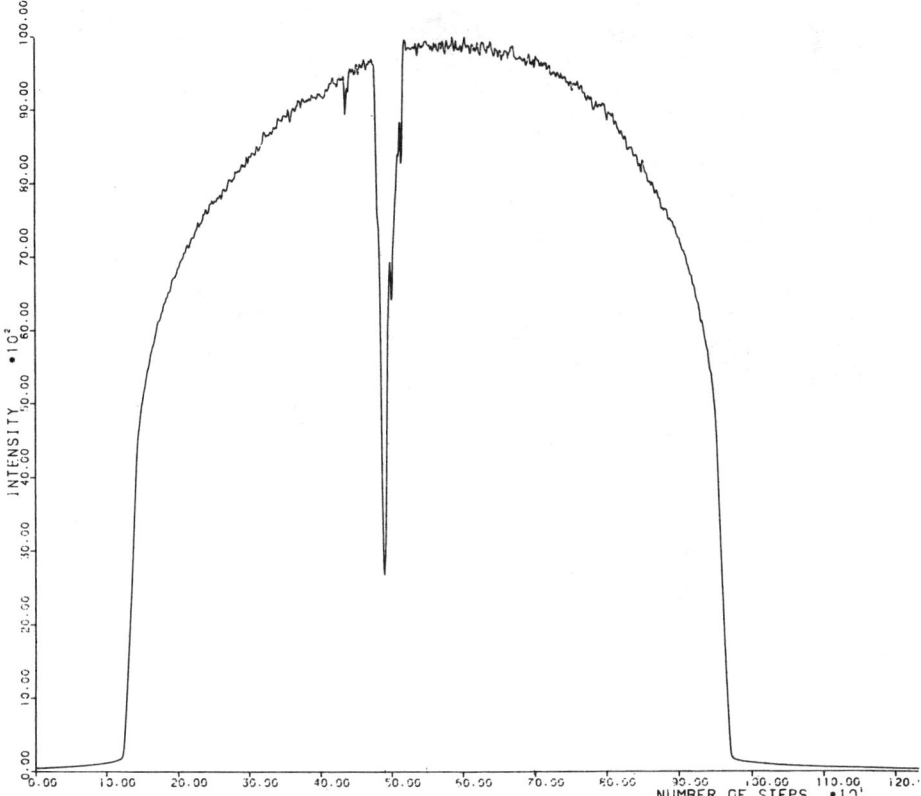

Figure 7 Extreme Limb Photometer (ELP) scan of a large spot on 25 July 1981, showing the large negative contrast when a sunspot extends nearly across the entire 39″ length of the ELP aperture. The wavelength was 0.53 μm. The ordinate is the mean intensity in arbitrary units, and the abscissa is in steps of 3″.

activity complex, because the proportion of observations lost as a result of the rotation of the Sun should be less than for a short-lived active region. By calibrating proxy data in terms of actual photometry, it should now be possible to determine the energy budget of individual active regions, as long as they are on the visible hemisphere for their entire lifetimes.

One well-studied activity complex first appeared in June 1982 and disappeared in December 1982. The sequence of active regions that appeared at the same longitude is considered to be part of this complex. [The concept of an activity complex is elaborated by Gaizauskas et al. (32) and Castenmiller et al. (5a)]. This complex began with a very prominent sunspot

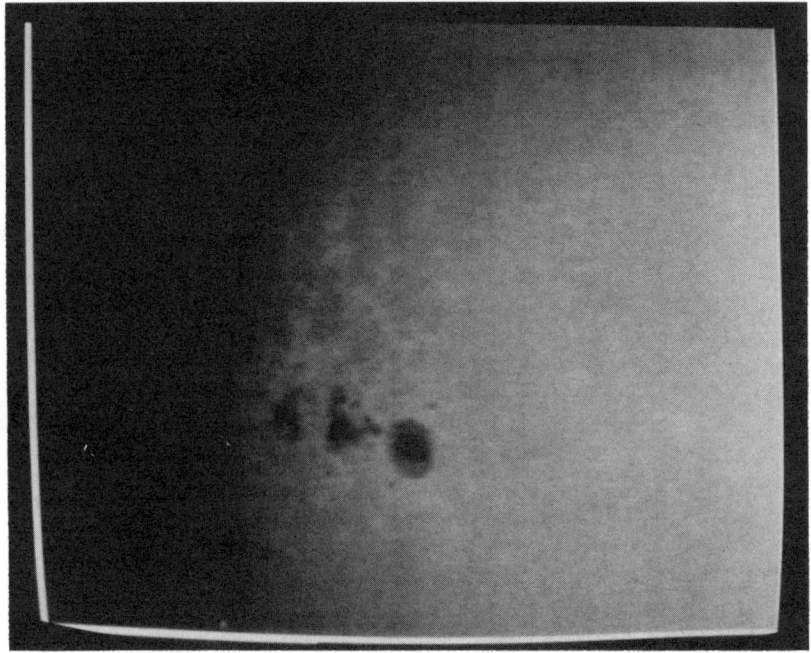

Figure 8 Diode array image of BBSO No. 18511 near the east limb. The width of the image is 480″ and was obtained with the 28-cm vacuum telescope at the San Fernando Observatory on 4 August 1982 at $\lambda = 6264$ Å.

in Big Bear Region No. 18422. Although there was some magnetic activity at this longitude before this date, it was relatively minor in area compared with the June outburst. The last sunspots in this region were visible in September 1982. After this date, only calcium plage was visible and no NOAA/AFGL number was assigned, although a Big Bear number was assigned to the calcium plage. The latest analysis of these data was published by Chapman et al. (11). The photometry from San Fernando Observatory was combined with irradiance fluctuations from proxy data where photometric data were unavailable. The irradiance fluctuations were transformed to flux changes which were then numerically integrated to obtain a luminosity change for the east and west hemispheres. These were then averaged to obtain a mean luminosity fluctuation for each two-week interval that each region was seen. It was found, after correcting the data for the occultation by the solar limb, that the facular energy excess was comparable to the sunspot energy deficit. Figure 9 shows the measured luminosity fluctuations for this activity complex (12). The facular energy excess was 95 ± 25 per cent of the sunspot's energy deficit. This result

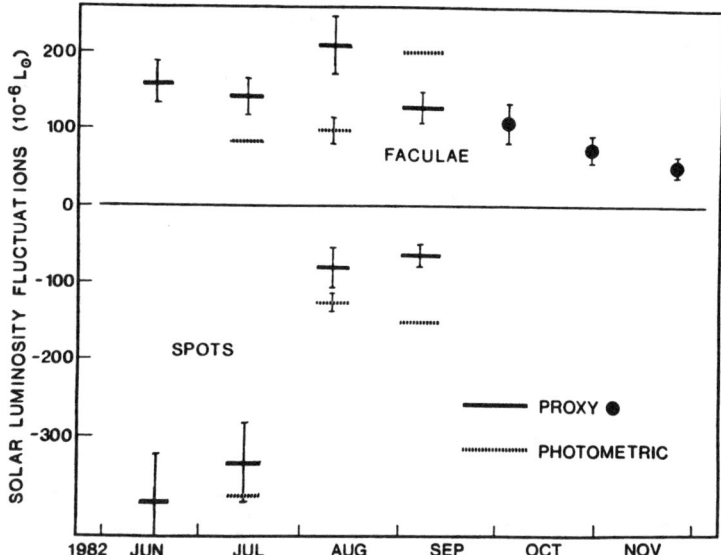

Figure 9 The luminosity fluctuations of the activity complex that began in June 1982 and ended in December 1982. Each line segment corresponds to the disk transit of an active region. The lower and upper symbols correspond to sunspots and faculae, respectively. The integration over time of the two curves give the energy deficit or excess. The excess energy of the faculae is 95% of the energy deficit of the spots (12).

includes an estimated 5% contribution from the UV below 300 nm. If the UV is much more important than this, then the facular energy excess will be larger than the sunspot energy deficit.

One should remember that the energy fluctuation is the luminosity fluctuation integrated over time. The luminosity fluctuation is the flux excess or deficit integrated over the area of the active region. This is, in turn, an integral over angle of the intensity excess or deficit. In other words, from Earth, one does not observe the flux excess or deficit directly but only the irradiance fluctuation. Even a network of spacecraft stationed around the Sun at different locations would not easily resolve this problem observationally, unless there were coverage in latitude as well as longitude. It is not a simple matter to go from observations to the energy budget of an active region.

The measurements of solar irradiance carried out by Bruning & LaBonte (5) using the Mount Wilson 150-ft tower telescope have already been discussed in Section 2. Although they analyzed full-disk images of the Sun for irradiance variations, their data set was not representative of the whole solar spectrum, since it was obtained in the wing of the 5250-Å absorption line. This line is known to weaken greatly in facular regions (14). It would be very interesting to reanalyze these data so as to more closely correspond to the white-light appearance of faculae and sunspots by using models for the center-to-limb behavior of the contrast. Bruning & LaBonte (5) reached the conclusion that there was an approximate irradiance balance between faculae and sunspots. This conclusion appears to be directly contradicted by the record from both ACRIM and Nimbus-7, which shows a drop in irradiance during the waning sunspot activity from 1980 to 1985. The contradiction is probably due to the use of a narrow-wavelength band in the wing of a temperature-sensitive absorption line.

Since the presence of nonspot magnetic fields alters the upper photosphere and chromosphere much more than the base of the photosphere, monitoring the strength of absorption lines should track the area of magnetic active regions. Measurements of the strength of the Ca II K line have been reported by Keil & Worden (45) and Livingston & Wallace (54). Both of these groups report a clear dependence of the K line on the solar cycle, with maximum emission in the core of the K line coinciding with the maximum sunspot number. Livingston & Wallace also report on the change in strength of several other weak absorption lines. The change in strength of the Mn I line at 5394.7 Å also seems to show fluctuations similar to those of the K line, although Livingston & Wallace claim that the variability of this line is not confined to plage regions alone. Furthermore, they show that the variation in the strength of the C I line at 5380.3 Å, which is indicative of T_{eff} and its changes, is similar to the ACRIM signal. If this line is not affected by magnetic active regions, its correspondence to the change in the ACRIM signal will be difficult to reconcile with moderately good agreement between the ACRIM signal and various measures of the irradiance fluctuations due to sunspots and faculae.

4. OBSERVING PROGRAMS AND INSTRUMENTATION

A thorough understanding of solar irradiance variations will require an improved data base over that available now. New observing programs at several institutions should help in acquiring improved observations that may produce a nearly continuous photometric record of the effects of magnetic active regions. To obtain relatively high photometric accuracy requires the use of photoelectric detectors, which have a linear transfer from instrument response to intensity and a higher dynamic range than film.

From the ground, it seems exceedingly difficult to obtain absolute accuracy comparable to that from spacecraft, such as from SMM or Nimbus-7. Instead, one should accurately measure differences in irradiance from the quiet Sun and compare this signal with that from spacecraft. The quiet Sun affords a convenient photometric standard. However, since such relative measurements are referenced to the quiet Sun, we must have a precision radiometer or photometer in space that measures the absolute solar irradiance. A space-borne imaging photometer would be desirable, but a distributed system of ground-based photometric telescopes would be a cost-effective alternative. For relative photometry, the Earth's atmosphere is less of a problem than is the weather. A network of telescopes around the world, as planned for the Global Oscillation Network Group (GONG) by the National Solar Observatory, could secure a nearly unbroken record of irradiance fluctuations due to active regions.

The program of daily filtergrams of the Ca II K line is continuing in the US at Big Bear Solar Observatory, at the National Solar Observatory at the Sacramento Peak station, and at the Mees Observatory on Haleakala, Maui. These filtergrams will be very helpful in estimating the facular irradiance signal, and hopefully they will be supplemented by data from other longitudes. Existing programs at Meudon and Kodaikainal (and any others) are needed to fill in data gaps. Methods for processing the great amount of information contained in a filtergram will be essential parts of any synoptic program that hopes to provide proxy irradiance signals. Great care must be given to the processing and handling of those data that are on film in order to obtain the most stable calibration.

A new program, which hopefully is continuing, is the daily magnetograms obtained with the 150-ft tower on Mount Wilson. These magnetic signals are converted into daily irradiance fluctuations (9). The magnetic signals in the daily magnetograms obtained by the National Solar Observatory at Kitt Peak, Arizona, should also be processed to obtain a proxy irradiance fluctuation. The higher resolution Kitt Peak data should provide a more accurate proxy irradiance, and with two programs operating there would be fewer gaps due to weather or equipment problems.

Luttermoser & Jones (55) have presented results from continuum photometry using the magnetograph at Kitt Peak. Photoelectric scans were obtained in the continuum at 5256.3 Å. This instrument will soon have two-dimensional detectors that will permit the simultaneous photometry of the solar disk both in a spectrum line and in the continuum.

Hirayama et al. (38) have developed a photometer much like the instrument at San Fernando Observatory described below. This photometer scans an image of the Sun in a rotary fashion, using a linear array of photodiodes aligned along a radius of the image. Each diode scans an annulus that is 2.75″ wide and has 256 samples per rotation, giving a pixel size at the limb of 2.75″ by 24″. This type of photometer is designed to avoid the need to interpolate between pixels on the usual Cartesian grid when the quiet Sun intensity is rapidly changing near the limb. The data near disk center are greatly oversampled, but they can be averaged in accordance with their distance from the rotation axis. The results in (38) are from measurements with this photometer on 266 active regions, and they show a definite increase in facular contrast toward the limb.

Two new imaging photometers have been developed at the San Fernando Observatory in order to obtain images of the Sun that can give irradiance fluctuations from active regions. One of these is the Cartesian Full-Disk Telescope (CFDT). This instrument is an integral telescope and photometer having a Reticon 512-G diode array and a 1-m focal length telescope. The linear diode array is aligned in a north-south direction and is scanned in the east-west direction by the rotation of the Earth. This idea is from a suggestion by Hudson (41) based on a similar idea from G. Hurford. The time to scan the Sun and the surrounding sky is about three minutes. An image of the Sun from the CFDT has a diameter of about 377 pixels, giving an image with a pixel size of 5″, centered in a 512 × 512 pixel frame. Synoptic observations began in the spring of 1985. Initially, the images were at $\lambda = 6678$ Å. Figure 10 shows a cathode-ray tube representation of a CFDT image from 12 June 1985 that has had the quiet Sun limb darkening removed. Figure 10 is thus a contrast image. The current observational program is to obtain a minimum of two images per day, one each at 6723 Å and 3920 Å.

Changes in sky transparency are measured by a detector that monitors the intensity of the disk of the Sun and the nearby sky. These data are digitized after each line of the image and permit removal of sky changes during later processing. Each image contains 0.6 megabytes of data and represents a compromise between resolution and data volume. It is hoped that these data will permit an accurate determination of sunspot and facular irradiance variations that, when compared with spacecraft observations, will allow one to remove that component of solar variability caused by magnetic solar active regions. There will still be the question of scattered facular elements, i.e. the photospheric network. The measurement of this component may well require higher spatial resolution than can be obtained with the CFDT. Furthermore, near the limb, the low spatial resolution of the CFDT will be inadequate to obtain an accurate facular signal.

To overcome the problem of foreshortening at the limb, another instrument, the Rotating Full-Disk Photometer (RFDP), has been constructed.

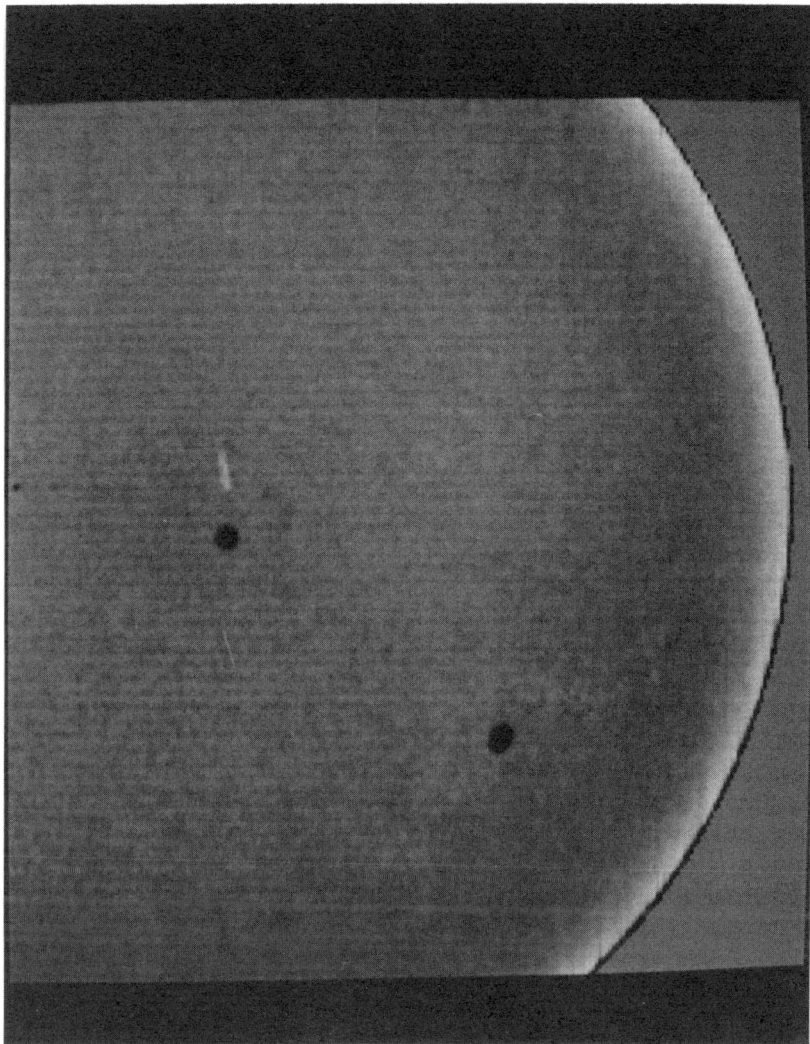

Figure 10 An enhanced computer display of part of a full-disk image obtained with the Cartesian Full-Disk Telescope at a wavelength of 6678 Å on 12 June 1985. The pixel size is 5″. North is up, west to the right.

The RFDP rotates a 512-G Reticon like the "second hand" of a clock, with one end at the disk center and the other end in the sky, beyond the limb. This instrument is an improvement over the ELP and is similar in some respects to the rotating photometer described above (38). The RFDP has a resolution of 2.5″ at the limb. At present, the RFDP is operating at reduced resolution in order to lower the data rate. One scan of the solar disk requires 2.5 minutes and produces 2.5 megabytes, which are currently reduced to 625 kilobytes and written directly onto magnetic tape. Data have been recorded at 6723 Å and 5320 Å.

These two instruments are complementary; the CFDT produces undistorted images of the Sun center, and the RFDP supplies higher resolution, undistorted images near the solar limb. Together, they form a system for daily monitoring of the position and irradiance fluctuation of sunspots and faculae. Figure 11 shows a contour plot of the corrected intensity of

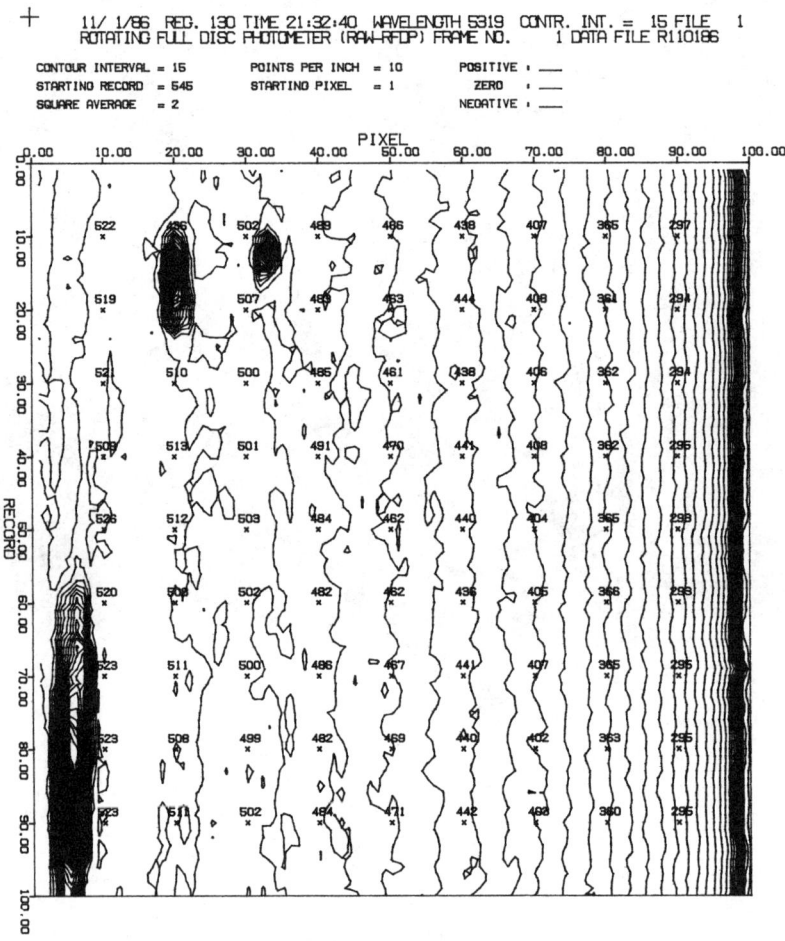

Figure 11 A contour plot of a portion of an image from the Rotating Full-Disk Photometer (RFDP) obtained on 1 November 1986 at a wavelength of 0.53 μm. The plot is at a reduced resolution of 9″ × 9″ per pixel. The left edge is near the disk center, and the right edge shows the solar limb.

part of the solar disk from the RFDP. Each record (horizontal line) represents a scan of limb darkening. Each vertical line represents the intensity of a particular diode at a fixed radial distance from the Sun center. It should be a relatively simple task to remove the quiet Sun limb darkening to obtain the irradiance fluctuation of an active region. In turn, it should also be straightforward to remove the effects of active regions to obtain more than 1000 limb-darkening measurements from each image and thus to produce an accurate, globally averaged determination of quiet Sun limb darkening.

These instruments will be checked by photometry at San Fernando Observatory with the vacuum telescope and spectroheliograph. Simultaneous observations will be obtained at several different wavelength pairs to calibrate the lower resolution synoptic observations from the CFDT and RFDP.

Although not a new program, the integrated-light photometry in the Ca II K line (45, 54, 88a) that has been carried out at the National Solar Observatory will continue to provide important data on the level of magnetic activity to be compared with both proxy and photometric measurements of the Sun's magnetic activity. For example, measurements of the calcium plage should be able to match the integrated-light K line photometry.

5. THEORETICAL CONSIDERATIONS

The precision record of the total solar irradiance provided by the ACRIM challenges us to attempt to understand the luminosity fluctuations in terms of magnetic activity recorded as sunspots, plage, or magnetic field strength.

Proxy estimates of irradiance fluctuations based on area and position of sunspots and faculae, do not seem capable of duplicating the ACRIM signal. It is too soon to say whether this discrepancy is caused by errors in the mathematical models, by errors in measurements of spot and plage areas, or by omission of the network component. With regard to the latter possibility, magnetic proxy signals may help to include the contribution of the network. More work is needed with magnetic proxy signals using high-resolution magnetic and intensity data to separate spot from facular signals. More accurate sunspot and plage measurements are needed in time for the coming solar cycle. In addition to plage area, the relation between plage and facular area must be determined, since it is the facular excess that must be modeled. Variations in the ultraviolet, which are not measured at the ground, can cause some of the disagreement between satellite irradiance observations and ground-based measurements, whether proxy or direct photometric measurements of irradiance fluctuations. Estimates of the contribution of the UV seem quite uncertain (19, 27, 52).

A second major issue has to do with energy balance. How much can irradiance variations tell us about the formation and structure of sunspots? There is now some observational evidence suggesting a reasonably close balance in energy between sunspots and faculae, at least for the groups of active regions termed activity complexes. How close is this balance, and does it hold also for large and small active regions? For small plage regions, sometimes no sunspots are reported. Is this a detection problem, or were there never any sunspots or pores associated with these plage regions? If there is a balance of energy over the life of an active region that is related to its magnetic evolution, this relationship could be complicated by magnetic connection to distant sunspots, even across the solar equator. The problem here is to trace magnetic flux from its appearance to its disappearance—that is, to determine where the scattered network flux comes from.

If energy balance is generally true, it would seem difficult to explain using standard models of flux tubes. The need to explain energy balance as a coincidence, as required by purely thermal models (29, 82), seems unconvincing and may hide some interesting insights into the formation of magnetic fields. One observational difficulty with the standard view of a rising flux tube forming a sunspot is the lack of any detectable thermal shadow. Furthermore, although large concentrations of magnetic flux lying horizontally are known to cause a darkening around a sunspot, which we call the penumbra, why do we not see some similar dark structure between every growing sunspot group? Although dark penumbral-type structures are occasionally seen, they do not clearly connect the major concentrations of flux. This is not to say that there is no evidence for arched flux tubes. The signature of an emerging flux region in the chromosphere is a set of small arches with bright Hα emission at their footpoints. The appearance of such a region in the photosphere, especially as the region grows in flux, does not fit the picture of a large tube of flux in a collection of tightly packed fibrils. See the recent review by Zwaan (100) on photospheric magnetic fields. High-resolution observations of an active region by Brants and collaborators (2, 3) show the complexity of magnetic flux emergence within an active region. See also (32–34). In general, although Hα images show arches in erupting flux regions, it is not clear that sufficient flux crosses the photosphere to account for the maximum flux of the region. This is a problem that requires vector magnetic field observations. Another curiosity is the lack of any detectable temperature perturbation in the photosphere around a sunspot. This problem has been the inspiration for several new theories in recent years to explain sunspots (68, 75, 94–97).

See the review by Moore & Rabin (58) of the dynamics of sunspots. The absence of a bright ring may set limits on the thermal diffusivity or may indicate the depth of the sunspot, but it may be that thermal blocking models are simply not appropriate to this problem. More careful searches for bright "rings" are needed that completely avoid any facular emission. Any redirected thermal emission detected should have no magnetic flux threading it.

An issue related to energy balance is that of storage. If there is a delay between the emission of energy, which is emitted from faculae rather than from sunspots, the energy must have been stored somewhere. The storage time for an activity complex has been found to be of the order of two months (12). This is probably an upper limit, since an activity complex has a large scale and much magnetic flux. What is the lower limit for the storage time? When we consider that bright facular features are often seen at approximately the same time as emerging flux, it is possible that there is no storage or shorter term storage on the small scale. If this were true, one would still expect to see irradiance dips from sunspots because of the differing radiation pattern of faculae and sunspots. Investigating the expected appearance using Equations 2 and 4, we find that behavior similar to what is seen would be expected. Perhaps the maximum storage period is the time from the maximum area of the sunspots to the maximum area of the faculae, which is loosely a function of the size of the active region. Hudson (42) has suggested that the storage time could reveal itself as a hysteresis as a region forms and dissipates. The lack of significant hysteresis indicates that storage times may be short (perhaps on the order of hours or a few days), depending on the depth of the flux tube and the average flux density, given (possibly) by the depth at which the magnetic fibrils of the sunspot merge [adopting, for the moment, the picture of a sunspot given by Parker (70)].

The submergence of magnetic flux has been mentioned as a way of making flux disappear at the photosphere (58). This process, if it commonly occurs, should give clear signs (elongated granulation, horizontal magnetic fields, etc.) at the photosphere in the reverse order from flux eruption. If these effects are not seen in a majority of cases, and if the magnetic flux of an active region disappears within the region, then perhaps the magnetic energy is also being lost.

The rate of loss of magnetic flux determined from observations is in the range of 1.2×10^{11} Wb s^{-1} (1.2×10^{17} Mx s^{-1}) for low-resolution observations, characteristic of active regions, to 3×10^8 Wb s^{-1} (3×10^{14} Mx s^{-1}) for higher resolution observations (53a). It is interesting to note that the magnetic energy contained in a magnetic flux tube (an element of faculae or filigree) is about 5×10^{19} J (for a tube of 4.4 GWb, and a volume of 6×10^{15} m^3). For the radiative flux for the facula given in Section 2.1, we find that this amount of energy could be radiated away in about 300 s. This is only a factor of two or three less than the lifetime of facular points (60). Some of the radiative losses from faculae are thought to be supplied by thermal diffusion to the wall of the flux tube. However, the amount of radiative loss predicted by theoretical models is uncertain (15, 18, 46, 64, 82). It will be difficult to separate, on the basis of observations, the source of thermal energy, i.e. whether it is from a local generation or from the photosphere by diffusion. The propagation of waves in flux tubes is discussed by Thomas (85).

The lifetime of small flux tubes can be estimated from the following relation given by Priest (71):

$$\tau = l^2/\eta,$$

where l is the size of the tube, and η is the magnetic diffusivity. Considering a tube of 4.4 GWb, containing a magnetic flux density of 150 mT (1500 G), and using the diffusion coefficient found by Sheeley et al. (78) of 7×10^8 m^2 s^{-1}, we obtain a lifetime $\tau = 15$ s. Although this diffusivity may be too large, even smaller tubes may exist (88). If these tubes, which have a flux of about 10^{16}–10^{17} Mx contain kilogauss fields, then even with a diffusion coefficient as small as 2.5×10^7 m^2 s^{-1}, their lifetime against diffusion is about 300 s. The problem with this apparent coincidence is that observers report that most of the magnetic flux of an active region disappears "in place"—that is, without moving a significant distance away from its origin at the photosphere. For example, Simon & Wilson (79, 99) report instances of magnetic flux concentrations, probably small collections of magnetic flux tubes, that appear to decay in place. Similar events are reported by Topka et al. (86) and Wallenhorst & Topka (87). Perhaps the magnetic flux density decreases rapidly below the detectability of present-day magnetographs. Other schemes for dissipation are discussed by Meyer (57).

Another problem is determining the source of the flux seen away from quiet regions. Is the network and internetwork flux the residue from nearby active regions, or is it "created" in place and not in any way connected with active regions? Is it possible, as suggested by J. K. Lawrence (private communication) that there is an energy imbalance, in the sense that faculae radiate more energy than that blocked or stored by sunspots? This situation might correspond to there being more magnetic flux in faculae than in sunspots, in which case the faculae would draw on the thermal energy in the convection zone. This would seem to be the opposite situation from the usually imagined one in which magnetic fields block the flow of convective energy that is stored in the convection zone. Alternate views of the situation are provided by the models of Chiang & Foukal (15), Schatten & Mayr (75), and Schatten et al. (76a).

An important datum relating to energy balance is the distribution of magnetic flux within active regions between sunspots and faculae, and between active regions and the network. There appears to be no recent information on the mean magnetic flux density in active regions. Does the flux that erupts in active regions account for most of the magnetic flux seen on the Sun, or instead does a large amount exist in the network? A study that should be repeated with higher sensitivity is that of Harvey (33) and Harvey et al. (34) in which they find that about one half of the flux appearing on the Sun is in ephemeral active regions (small active regions with lifetimes less than about one day). Is there a continuum in flux between ephemeral regions and the network?

The most outstanding question from the standpoint of solar physics is whether or not there exists an energy connection between the dark features (sunspots and pores) and the bright features (plage, faculae, and the network). To answer this question will require improved observations of the highest resolution that cover the entire solar disk. Do all faculae arise from pores and sunspots, or can magnetic flux erupt without first passing through the sunspot phase? The answer to this question will require high spatial resolution observations that are continuous in time so as to provide a complete temporal history of the photometric properties of an active region. Such observations are not possible at the present time and will require a global network of telescopes similar to the GONG project for the study of global oscillations. An approximate energy balance has been found for an activity complex in 1982 (12). Can the uncertainty in the energy balance be reduced below the 25% reported by these authors? Furthermore, can the energy balance be established for other activity complexes and active regions? If energy balance exists for active regions and activity complexes, then it becomes difficult to understand how this could be without some energy storage within the magnetic fields of the active region or complex. Arguments (26) that energy balance between sunspots and faculae is a coincidence seem unconvincing. One must ask why the lifetimes of faculae and their excess flux should equal the energy deficit of sunspots, if this does prove to be the case for most active regions.

The simplest and most stringent test for any model that claims to explain the solar irradiance is to match the trend in irradiance seen by the ACRIM. The solar irradiance has decreased by approximately 1.33 $W m^{-1}$, or 0.1%, between 1980 and the end of 1984. Only the Mount Wilson pseudo-irradiance, based on full-disk magnetograms, shows a change of this magnitude. Unfortunately, this signal does not have the same slope as that of the ACRIM. Furthermore the modeled minimum solar irradiance is not as low as the data from the ACRIM.

Alternatively, using the proxy irradiance functions (Equations 2 and 4), we can account for only about one half of this 5-yr ACRIM decrease. The discrepancy could be explained by (a) increasing the ratio of plage to spot area, (b) adding a contribution from the photospheric network approximately equal to that of active regions, or (c) using a larger facular contrast at disk center than usual. A mid-visible contrast of about 3.5% would suffice. It seems that the plage to spot area ratio is lower than average at solar maximum, so the first possible explanation seems unlikely. Of the last two possibilities, it seems that the latter is perhaps more likely, although both effects may be responsible for increasing the proxy signal. To settle this question, photometric measurements of the solar disk must be carried out with high enough spatial resolution to measure the irradiance from the network. Such observations should also be able to determine the facular contrast at disk center in active regions.

6. SUMMARY

The ACRIM has shown a decline in the solar irradiance of about 0.1% in five years, roughly in phase with the decline in sunspot area. This decline is almost surely due to the changing magnetic flux seen primarily as sunspots and faculae. However, attempts to model the short- and long-range fluctuations in irradiance using proxy irradiance models based on area and position of these features has met with only partial success. Improvements to the models and observational data are needed. Planning for the next solar cycle should be well underway by now.

An improved understanding of solar variations in terms of magnetic phenomena observable from the ground may permit the prediction of the solar output, assuming that the quiet Sun does not change. Alternatively, accurate modeling could help to reveal any luminosity variations not associated with magnetic activity.

A detailed match of the ACRIM signal may permit a better understanding of the formation and destruction of sunspots and faculae. The solar dynamo is not yet well understood, and irradiance variations give us information on the flow of energy in the convection zone that must accompany a dynamical model of solar magnetism.

The degree of our observational knowledge of energy balance between sunspots and faculae will depend on our ability to measure the irradiance fluctuations due to sunspots, faculae in active regions, and the photospheric network. The difficulty in making these measurements will be affected by the temporal behavior of magnetic fields as they appear, diffuse, and disappear from the solar surface.

In closing, it seems fairly clear that if the variations in solar irradiance seen so far are due to magnetic activity, then the lower terrestrial temperatures associated with the latter part of the Little Ice Age might indeed have been caused by the low solar activity of the Maunder Minimum.

ACKNOWLEDGMENTS

I thank H. S. Hudson, J. K. Lawrence, and K. H. Schatten for a critical reading of early drafts of this review. Thanks also to J. W. Harvey, K. L. Harvey, H. P. Jones, and P. R. Wilson for discussions of various topics included in this review. The author benefitted from his participation in the first Solar Cycle Workshop held at the Big Bear Solar Observatory. This work was partially supported by grants from NASA and the National Science Foundation.

Literature Cited

1. Albregtsen, F., Jôras, P. B., Maltby, P. 1984. *Sol. Phys.* 90: 17–30
2. Brants, J. J. 1985. *Sol. Phys.* 98: 197–218
3. Brants, J. J., Steenbeck, J. C. M. 1985. *Sol. Phys.* 96: 229–52
4. Bray, R. J. 1981. *Sol. Phys.* 69: 3–8
5. Bruning, D. H., LaBonte, B. J. 1983. *Ap. J.* 271: 853–58
5a. Castenmiller, M. J. M., Zwaan, C., van der Zahm, E. B. J. 1986. *Sol. Phys.* 105: 237–55
6. Chapman, G. A. 1980. *Ap. J. Lett.* 242: L45–48
7. Chapman, G. A. 1984. *Nature* 308: 252–54
8. Chapman, G. A. 1984. See Ref. 47, pp. 73–87
9. Chapman, G. A., Boyden, J. E. 1986. *Ap. J.* 302: L71–73
10. Chapman, G. A., Klabunde, D. P. 1982. *Ap. J.* 261: 387–95
11. Chapman, G. A., Lawrence, J. K., Herzog, A. D., Shelton, J. C. 1984. *Ap. J. Lett.* 282: L99–101
12. Chapman, G. A., Lawrence, J. K., Herzog, A. D. *Nature* 319: 654–55
13. Chapman, G. A., Meyer, A. D. 1986. *Sol. Phys.* 103: 21–31
14. Chapman, G. A., Sheeley, N. R. Jr. 1968. *Sol. Phys.* 5: 442–61
15. Chiang, W.-H., Foukal, P. 1985. *Sol. Phys.* 97: 9–20
16. Cram, L. E., Thomas, J. H. eds. 1981. *The Physics of Sunspots.* Sunspot, NM: Sacramento Peak Obs.
17. Currie, R. G. 1981. *Science* 211: 386–89
18. Deinzer, W., Hensler, G., Schüssler, M., Weisshaar, E. 1984. *Astron. Astrophys.* 139: 435–49
19. Donnelly, R. F., Heath, D. F., Lean, J. L., Rottman, G. J. 1984. See Ref. 47, pp. 233–49
20. Eddy, J. A., Hoyt, D. V., Gilliland, R. L. 1982. *Nature* 300: 689–93
21. Eddy, J. A. 1984. See Ref. 47, pp. 213–29
22. Foukal, P. 1981. See Ref. 16, pp. 391–423
23. Foukal, P. V., Mack, P. E., Vernazza, J. E. 1977. *Ap. J.* 215: 952–59
24. Foukal, P., Vernazza, J. 1979. *Ap. J.* 234: 707–15
25. Foukal, P. 1984. See Ref. 47, pp. 97–117
26. Foukal, P. V. 1987. *J. Geophys. Res.* 92: 801–7
27. Foukal, P., Lean, J. 1986. *Ap. J.* 302: 826–35
28. Foukal, P., Duvall, T. Jr. 1985. *Ap. J.* 296: 739–45
29. Foukal, P., Duvall, T. Jr., Gillespie, B. 1981. *Ap. J.* 249: 394–99

30. Foukal, P. V., Fowler, L. A., Livshits, M. 1983. *Ap. J.* 267: 863–71
31. Fowler, L. A., Foukal, P. V., Duvall, T. 1983. *Sol. Phys.* 84: 33–48
31a. Fröhlich, C. 1987. *J. Geophys. Res.* 92: 796–800
32. Gaizauskas, V., Harvey, K. L., Harvey, J. W., Zwaan, C. 1983. *Ap. J.* 265: 1056–65
33. Harvey, K. L. 1985. *Aust. J. Phys.* 38: 875–83
34. Harvey, K. L., Harvey, J. W., Martin, S. F. 1975. *Sol. Phys.* 40: 87–102
35. Hickey, J. R., Alton, B. M. 1984. See Ref. 47, pp. 43–57
36. Hirayama, T., Okamoto, T. 1981. *Sol. Phys.* 73: 37–43
37. Hirayama, T., Okamoto, T., Hudson, H. S. 1984. See Ref. 47, pp. 59–71
38. Hirayama, T., Hamana, S., Mizugaki, K. 1985. *Sol. Phys.* 99: 43–54
39. Hoyt, D. V. 1979. *Clim. Change* 2: 79–92
40. Hoyt, D. V., Eddy, J. A., Hudson, H. S. 1983. *Ap. J.* 275: 878–88
41. Hudson, H. S. 1984. See Ref. 47, p. 297
42. Hudson, H. S. 1987. Submitted for publication
43. Hudson, H. S. 1987. Submitted for publication
44. Hudson, H. S., Silva, S., Woodard, M., Willson, R. C. 1982. *Sol. Phys.* 76: 211–19
45. Keil, S. L., Worden, S. P. 1984. *Ap. J.*, 276: 766–81
46. Knolker, M., Schüssler, M., Weisshaar, E. 1985. See Ref. 77, pp. 195–97
47. LaBonte, B. J., Chapman, G. A., Hudson, H. S., Willson, R. C., eds. 1984. *Solar Irradiance Variations on Active Region Time Scales. NASA Conf. Publ. CP-2310*. Washington, DC: NASA Off. Space Sci.
48. LaBonte, B. J. 1986. *Ap. J. Suppl.* 62: In press
49. Lawrence, J. K., Chapman, G. A., Herzog, A. D., Shelton, J. C. 1985. *Ap. J.* 292: 297–308
50. Lawrence, J. K. 1987. *J. Geophys. Res.* 92: 813–17
51. Lawrence, J. K. 1987. Submitted for publication
52. Lean, J. L., White, O. R., Livingston, W. C., Heath, D. F., Donnelly, R. F., Skumanich, A. 1982. *J. Geophys. Res.* 87: 10,307–17
53. Libbrecht, K. G., Kuhn, J. R. 1984. *Ap. J.* 277: 889–96
53a. Livi, S. H. B., Wang, J., Martin, S. F. 1985. *Aust. J. Phys.* 38: 855–73
54. Livingston, W., Wallace, L. 1987. *Ap. J.* In press
55. Luttermoser, D. G., Jones, H. P. 1985. *Bull. Am. Astron. Soc.* 17: 639–40
56. McIntosh, P. S. 1981. See Ref. 16, pp. 7–54
57. Meyer, F. 1985. See Ref. 77, pp. 251–60
58. Moore, R., Rabin, D. 1985. *Ann. Rev. Astron. Astrophys.* 23: 239–66
59. Muller, R. 1983. *Sol. Phys.* 85: 113–21
60. Muller, R., 1985. *Sol. Phys.* 100: 237–55
61. Muller, R., Keil, S. 1983. *Sol. Phys.* 87: 243–50
62. Newkirk, G. Jr. 1983. *Ann. Rev. Astron. Astrophys.* 21: 429–67
63. Newkirk, G. Jr. 1984. See Ref. 47, pp. 133–35
64. Nordlund, A. 1985. See Ref. 77, pp. 101–19
65. North, G. R. 1981. See Ref. 80, pp. 1–9
66. Oseas, J. 1985. MS thesis. Calif. State Univ., Northridge
67. Oster, L., Sofia, S., Schatten, K. 1982. *Ap. J.* 256: 768–73
68. Parker, E. N. 1974. *Sol. Phys.* 36: 249–78
69. Parker, E. N. 1979. *Ap. J.* 230: 905–13
70. Parker, E. N. 1984. *Ap. J.* 280: 423–27
71. Priest, E. R. 1984. *Solar Magnetohydrodynamics*. Dordrecht: Reidel
72. Rogerson, J. B. Jr. 1961. *Ap. J.* 134: 331–36
73. Schatten, K. H. 1984. *Geophys. Res. Lett.* 11: 156–57
74. Schatten, K. H. 1984. See Ref. 47, pp. 125–131, 145
75. Schatten, K. H., Mayr, H. G. 1985. *Ap. J.* 299: 1051–62
76. Schatten, K. H., Miller, N., Sofia, S., Endal, A. S., Chapman, G., Hickey, J. 1985. *Ap. J.* 294: 689–96
76a. Schatten, K. H., Mayr, H. G., Omidvar, K., Maier, E. 1986. *Ap. J.* 311: 460–73
77. Schmidt, H. U., ed., 1985. *Theoretical Problems in High Resolution Solar Physics, Proc. MPA/LPARL Workshop, München, MPA 212*. München: Max-Planck-Inst. Phys. Astrophys.
78. Sheeley, N. R. Jr., Boris, J. P., Young, T. R. Jr., DeVore, C. R., Harvey, K. L. 1983. See Ref. 84, pp. 273–76
79. Simon, G. W., Wilson, P. R. 1985. *Ap. J.* 295: 241–57
80. Sofia, S., ed., 1981. *Variations of the Solar Constant, NASA Conf. Publ. CP-2191*. Washington, DC: NASA Off. Space Sci.
81. Sofia, S., Oster, L., Schatten, K. 1982. *Sol. Phys.* 80: 87–98

82. Spruit, H. C. 1976. *Sol. Phys.* 50: 269–95
83. Spruit, H. C., Zwaan, C. 1981. *Sol. Phys.* 70: 207–28
84. Stenflo, J. O., ed. 1983. *Solar and Stellar Magnetic Fields: Origins and Coronal Effects, IAU Symp. No. 102.* Dordrecht: Reidel
85. Thomas, J. H. 1985. See Ref. 77, pp. 126–47
86. Topka, K. P., Tarbell, T. D., Title, A. M. 1986. *Ap. J.* 306: 304–16
87. Wallenhorst, S. G., Topka, K. P. 1985. *Sol. Phys.* 81: 33–46
88. Wang, J., Zirin, H., Shi, Z. 1985. *Sol. Phys.* 98: 241–54
88a. White, O. R., Livingston, W. C., Wallace, L. 1987. *J. Geophys. Res.* 92: 823–27
89. Williams, G. E., Sonett, C. P. 1985. *Nature* 318: 523–27
90. Willson, R. C., Gulkis, S., Janssen, M., Hudson, H. S., Chapman, G. A. 1981. *Science* 211: 700–2
91. Willson, R. C. 1982. *J. Geophys. Res.* 87: 4319–26
92. Willson, R. C. 1984. See Ref. 47, pp. 1–40
93. Willson, R. C., Hudson, H. S., Fröhlich, C., Brusa, R. W. 1986. *Science* 234: 1114–17
94. Wilson, P. R. 1971. *Sol. Phys.* 21: 101–15
95. Wilson, P. R. 1972. *Sol. Phys.* 27: 354–62
96. Wilson, P. R. 1972. *Sol. Phys.* 27: 363–72
97. Wilson, P. R. 1981. See Ref. 16, pp. 83–97
98. Wilson, P. R. 1981. *Sol. Phys.* 69: 9–14
99. Wilson, P. R., Simon, G. W. 1983. *Ap. J.* 273: 805–21
100. Zwaan, C. 1985. *Sol. Phys.* 100: 397–414

SUBJECT INDEX

A

A stars, disks around, 237–38, 523, 536, 545–47
Abell 2029, 436
ACRIM (Active Cavity Irradiance Monitor), 633–34, 637, 640, 643–46, 650, 654, 659, 663–64
Active galactic nuclei, 436
  infrared emission from, 198, 207–8, 220–21
AD Leonis, 285
Alpha Lyrae, 335, 523, 545–47
Alpha Ophiuchi, 334
Alpha Persei Cluster, 275–76, 278
Alpha Piscis Austrini, 545–46
Andromeda Galaxy, 442–43, see also M31
Arcturus star group, 623
Arp 220, 218–22
Asteroids, 523, 526–28
Asymptotic Giant Branch phase, 535, 539–43

B

B stars, infrared emission in, 535–37
Barnard's star, 492
Baryons, 450, 452–55, 460–62
DD I 26°730, 298
Becklin-Neugebauer object, 63, 73
Beta Canis Majoris, 325
Beta Pictoris, 237–38, 545–46
Binary stars
  formation of, 26–27, 53, 71
  formation of, in globular clusters, 565, 573, 575, 579–84, 596–97
  very low mass, 488–94, 498–99, 503–5, 508–9
  Wolf-Rayet, 122–23, 125–28
BL Lacertae objects, 195, 198, 209–12
Black hole at Galactic center, possibility of, 403, 406–20
Black holes in active galactic nuclei, 198, 221–22
Black holes possibility
  in cores of globular clusters, 575–76
  as dark matter, 455

Brown dwarfs, 499, 510
  atmospheres of, 478
  evolution of, 476–77
  as missing-mass possibility, 454–55, 478–79, 512–13, 523, 548
  searches for, 484–85, 489–91
  as solar companion possibility, 478–79, 548

C

Calcium II emission measurements, 281, 285–91
Calcium II emission relationship with magnetic activity cycle, 295–98
Carbon stars, 540–42
Cassiopeia A, 144, 544–45
Cepheid binaries, 362–63
Cepheids, 345–75
  distances of, 345–75
  extragalactic, 353–54, 369–72
  and extragalactic distance scale, 369–72
  in Galactic clusters, 354–58, 364–65
  infrared observations of, 351–53, 368, 370–72, 539
  interstellar reddening in, 345–47, 351–54
  long-period, 368
  in Magellanic clouds, 346–53, 363–69
  metallicity in, 349–52
  in M31, 363–64, 371
  period-luminosity-color relationship in, 345–72
  period-luminosity relationship in, 345–72
  pulsation parallaxes of, 359–62
  radius determinations of, 359–62
  statistical parallaxes of, 359
  ultraviolet observations of 362–63
CFDT (Cartesian Full-Disk Telescope), 656–59
Chi Bootes A, 282
Chiron, 264–65
Collisionless Boltzmann codes, 167, 183
Coma cluster of galaxies, 445
Cometesimals, 234–37

Comets, 231–69
  albedos of, 241–43
  CO in, 235–36, 249–50, 253–54
  coma production rates for, 256–63
  comae of, 249–63
  composition of, 234–36, 239, 249–55
  infrared detection of, 523, 527–28
  infrared observations of, 241, 247, 250–53
  interstellar medium and formation of, 233–39
  masses of, 244–45
  mass loss in, 256–62
  nuclei of, 239–49
  origin of, 232–39
    cataclysmic theories of the, 238–39
    star-formation theory of the, 233–38
  oxygen production in, 258–63
  rotation of, 240–42
  ultraviolet measurements of, 253–54, 258
  water molecules in, 234–35, 249–52, 258, 261–65
Comets, individual
  Arend-Rigaux, 240–41
  D'Arrest, 240
  Encke, 244, 528
  Giacobini-Zinner, 231, 246, 263–64
  Gum, 528
  Halley, 231,
    composition of, 250–55
    jets in, 247–49
    nucleus of, 236, 240–49
    production rates in, 258–62
    radio spectra of, 254
  Humason, 236
  IRAS-Araki-Alcock, 236–37, 250, 254, 260, 527
  Kohoutek, 239
  Kopff, 528
  Morehouse, 236
  Neujmin 1, 240–41
  Shoemaker-2, 528
  Tago-Sato-Kosaka, 239
  Tempel-1, 528
  Tempel-2, 528
  West, 236, 253

669

670 SUBJECT INDEX

Copernicus satellite observations, 333, 335
Cosmic-ray ionization, 40–41
Cosmic rays, 142, 316, 324, 331–32
Cosmological constant, 450, 456–57
Cosmology, IRAS survey importance for, 222–25
Crab Nebula supernova, 544
CRAF (Comet Rendezvous and Asteroid Flyby), 265–66
Cygnus Loop, 314–15, 544

D

Dark matter, 425–72. See also Missing mass
 in clusters of galaxies, 444–49
 cold, 459–61
 in elliptical galaxies, 426, 436–39, 451
 hot, 458–60
 in Milky Way, 451
 in planetary nebulae, 437
 in spiral galaxies, 427–36
Distance scale, Cepheids as indicators of, 345–75
Dynamos, stellar, 271–301

E

Einstein satellite observations, 122, 384–85, 503
ELP (Extreme Limb Photometer), 641–43, 650
Epsilon Aurigae, 547
Epsilon Eridani, 545–47
EXOSAT observations, 336
Extragalactic distance scale, 369–72
Extragalactic radio sources, 190–95, 205, 214–16, 427
Extragalactic sky
 infrared observations of, 187–230
 luminosity functions of objects in, 189–92, 224–25

F

F stars
 dwarf
  disks around, 523
  rotation in, 278–79, 294–98
 in Galactic disk, 428, 605–6
 subgiant, 295
Fornax cluster, 441

G

G stars, dwarf,
 Ca II emission in, 287–88

disks around, 523
emission-rotation relation in, 291–97
rotation in, 276–78
Galactic bulge, 603, 616, 625
 IRAS observations of, 542–43, 628
 metal-rich stars in, 628
 relationship with thick disk, 628–30
 rotation of, 628–29
 velocity dispersion of stars in, 611, 629
Galactic center, 377–423
 central source in, 387–90
 circumnuclear material in, 390, 394–98
 distance to, 378, 429
 gamma-ray emission from 384–87
 gas dynamics in, 394–98, 407–15, 419
 giant molecular clouds in, 380–81
 infrared emission from, 380–84, 387–407, 418–19
 infrared sources in, 378, 382–83, 389, 402–4, 409, 412, 415–19
 interstellar matter in, 390–406, 419
 ionized gas clouds in, 399–404, 412, 414–15
 ionized gas streamers in, 398–404, 409–12, 414–15, 419
 mass distribution in, 406–15, 419–20
 possibility of black hole at, 403, 406–20
 possibility of starburst in, 387–90
 radio continuum emission in, 379–81, 387, 399–400, 419–20
 radio source in, 381–83, 389, 400, 415–20
 stellar cluster in, 378–79, 407, 412–15
 supernova remnant in, 381
 ultraviolet emission from, 388, 400–1, 414, 418–20
 X-ray emission from, 384–87, 404–6
Galactic disk, 603–11, 615
 age-metallicity relation of stars in, 604–6, 615–16
 age-velocity relation of stars in, 604–6, 615–16, 629
 density distribution in, 608–11
 missing mass in, 608–9
 mass of, 428–29, 454
 infrared emission in, 554–56

star formation in, 480–81
 star-formation rate in, 605
 structure of, 607–11
Galactic disk, thick, 604, 611–17, 628–30
 density of, 612
 formation of, 615
 metal-rich giant stars in, 612–17
Galactic emission, 549, 556
Galactic emission, diffuse, 222, 523–24
 infrared cirrus, 548–49, 552, 557
 and interstellar radiation field, 549–54
Galactic globular cluster system, 603–4, 617, 619, 621–25, 627, 630
Galactic halo, 603, 613, 617–21, 628–30
 mass of, 430–32
 metal-weak stars in, 613, 617–30
 shape of, 623–27
 star counts in, 624–25
 velocity disperion of stars in, 617–20
Galactic spheroid, 603–4, 617–30
Galaxies
 clusters of
  dark matter in, 438, 442–51
  N-body simulations for, 151–52
 with companions
  N-body simulations for, 152, 160–61, 164–65
 disks of, 616, 622, 626
 infrared observations of, 187–230
 interacting, 217–18
 masses of, 425–28
 origin of luminosity in, 195–98
 N-body simulations for, 151–84
 number counts for infrared sources in, 223–24
 small groups of,
  dark matter in, 444, 451
  masses of, 443–44
 superclusters of, 446–48
 ultraluminous infrared, 216–22
Galaxies, barred spiral, 202, 435
Galaxies, binary
 dark matter in, 442, 444, 451
 masses of, 441–43
Galaxies, blue compact, 203–4, 440

SUBJECT INDEX 671

Galaxies, dwarf
 infrared observations of, 203–4
 masses of, 430, 439–40
Galaxies, dwarf spheroidal, 440–41, 451
Galaxies, elliptical
 dark matter in, 451
 infrared observations of, 188–89, 205–7
 masses of, 426, 436–39
 N-body simulations for, 151–54, 166
 with shells, 152, 161–62
Galaxies, irregular
 dark matter in, 440
 infrared observations of, 203–4
Galaxies, ring spiral, 202
Galaxies, spheroidal, 161, 164
Galaxies, spiral
 dark matter in, 426, 432–36, 451
 infrared observations of, 188, 197–202
 N-body simulations for, 152–54, 160–61, 164, 181–82
 star formation in, 23–81
Galaxies, S0, 188, 205, 209
Galaxy, see also Milky Way
 dark corona of, 608–9
 infrared observations of 521–63
 star formation in, 23–81, 528–34
 Wolf-Rayet stars in, 116–17, 123–25, 141–44
Galaxy formation, 458–59
Gamma^{2}Velorum, 118, 123, 128–29, 134, 142, 145
Gamma-ray emission, from Galactic center, 384–87
Giotto spacecraft, 243, 246–48, 254
Gl 65, 498–99
Gl 234, 498–99
Gl 473, 498–99
Gl 623, 498–99
Gl 748, 498–99
Globular cluster system, Galactic, 603–4, 617, 619, 621–25, 627, 630
Globular clusters
 binary formation and evolution in, 565, 573–75, 579–84, 596–97
 black holes in cores of, 575–76
 central engines in, 574–76
 core structure of, 589–90, 598
 evaporation of stars from, 568–70
 formation of, 567–68, 622

initial-mass functions in, 595–96
 mass loss of stars in, 565, 568
 mass segregation in, 571–73, 585–86, 594–96
 masses of, 427, 430–31
 massive-star formation in, 573–74
 metal-rich, 621
 metal-weak, 621, 623–25, 630
 N-body calculations for, 575–81, 584, 597–98
 proper motions of stars in, 594–95, 598
 stellar evolution in, 565, 568, 585
 stellar content of, 595–97
 surface brightness of, 585–90
 velocity measurements of, 435–41, 451, 590–95, 598
 X-ray sources in, 575, 596–97
Globular clusters, dynamical evolution of, 595–601
 core collapse phase of, 565–74, 589–90
 energy generation in, 566, 574–76
 Fokker-Planck approximations in, 567–75, 581–84
 gravothermal collapse in, 574–76
 gravothermal instability in, 569–71
 gravothermal oscillations in, 566, 577–79, 597–98
 postcollapse phase of, 566, 574–79
 central energy source in, 574–76
 precollapse phase of, 567–74
GQ Muscae, 543
G 77−61, 501

H

H II regions
 near Cepheids, 349
 near Galactic center, 380–81, 399–401
 infrared observations of, 200, 203–4
 Wolf-Rayet stars and, 125, 144
H II regions, compact, 380, 533
H and K emission measurements, 286–91, 295–98
HD 50896, 122, 137, 141, 145
HD 93162, 123
HD 165763, 129
HD 192163, 115, 118

HD 192641, 120
HD 193793, 119–20, 122
HD 197406, 127
HEAO-1, HEAO-3, HEAO-C, 386
Herbig-Haro objects, 60, 532–33
Hierarchical fragmentation, 24, 27–28, 479–80
HL Tauri, 66, 70–71
HR 4909, 536
Hubble constant, 426
Hyades cluster, 278–79, 606
HZ 43, 331, 336–37

I

IC 443, 544–45
IC 1613, 125
ICE (International Cometary Explorer), 231, 246, 263–64
Infrared cirrus, 548–50, 552, 557
Infrared emission
 from Cepheids, 539
 from Galactic center, 380–84, 387–90, 393–94, 399–401
 in Galactic disk, 554–56
Infrared emission, extragalactic
 dust responsible for, 195–96, 203, 207–14
 mechanisms for, 195–99
 photospheric, 195–207
 synchrotron, 195, 198, 209–16
Infrared observations
 of Beta Pictoris, 237–38, 545–46
 of comets, 241, 247, 250–53, 523, 527–28
 of Galaxy, see IRAS observations
 of stellar magnetic fields, 285
Infrared observations, extragalactic, see IRAS observations
Infrared photometry
 of Cepheids, 351–53, 368, 370–72
 of very low mass stars, 486–87, 490, 494
Infrared sources, see also IRAS observations
 in molecular clouds, 30, 59–60, 63, 72
 spiral galaxies as, 198–202
 Wolf-Rayet stars as, 118–22, 129–30
Initial-mass function, 25, 74, 454–55
 in globular clusters, 595–96
 for very low mass stars, 478–81, 507–12

672 SUBJECT INDEX

Interstellar matter,
  in Galactic center, 390–406, 419
  infrared detection of extragalactic, 196–97, 200, 207
Interstellar medium, Galactic
  formation of comets in, 233–39
  IRAS observations of, 548–57
  small grains in, 550–54
  star formation and, 523, 533–34
  Wolf-Rayet stars and, 141–42
Interstellar medium, local (LISM), 303–44, see also Local Bubble, Local Cloud, Local Fluff, Loop I Bubble
Interstellar medium, very local (VLISM), 308, 328–41
Interstellar radiation field, 549–54
Interstellar reddening, Cepheids and, 345–47, 351–54
Iota Virginis, 295
IRAS observations, 187–230, 521–63
  of active galactic nuclei, 207–16
  of asteroids, 526–28
  of BL Lacertae objects, 195, 209–11
  of comets, 523, 527–28
  extragalactic, 187–230
  of Galactic bulge, 628
  of Galactic emission, 523–24, 548–57
  of galaxies
    barred spiral, 202
    blue compact, 203
    dwarf, 203
    elliptical, 188–89, 205–7
    irregular, 203–4
    Markarian, 190, 213, 216
    radio, 208–9
    ring spiral, 202
    Seyfert, 188–89, 195, 212–15
    spiral, 188, 197–202
    S0, 188, 205–7
    ultraluminous, 216–22
  of the Galaxy and solar system, 521–63
  of the Magellanic Clouds, 203–4, 543
  of molecular clouds, 30, 48, 59, 523, 528–34, 555–57
  of novae, 543–44
  of OVV quasars, 195, 207, 209–12
  of planetary nebulae, 538
  of quasars, 195, 207, 215–16
  of stars, 523, 534–45

asymptotic giant branch, 539–41
B and A, 536–37
brown dwarf, 548
carbon, 540–42
main-sequence, 534
with mass loss, 539–43
with protoplanetary disks, 237, 545–47
supergiant, 535–36, 538, 541
T Tauri, 65, 533
Wolf-Rayet, 118, 120, 535–36
  of supernova remnants, 544–45
  of zodiacal dust cloud, 524–26
IRAS survey, cosmological importance of, 222–25
IRAS 13349+2438, 216
IRS 1, 402, 409
IRS 5, 62, 532
IRS 7, 378, 389, 412
IRS 16 complex, 382–89, 400–4, 409, 415–19
IUE observations
  of Cepheid binaries, 362–63
  of comets, 236, 254, 258
  of nearby stars, 335

J

Jets, cometary, 247–49
Jets, stellar, 60, 62–63, 73–74

K

K stars
  dwarf, rotational velocities of, 276–77, 294–98
  giant, 606, 610, 612
    distribution of, in Galactic bulge, 628–29
    infrared observations of, 534
    in thick disk of Galaxy, 612–15
    velocity dispersions of, 428, 441
  supergiant, infrared observations of, 538–39
Kapteyn's star, 501
Kapteyn's star group, 623
Kepler supernova remnant, 544
Kuiper Airborne Observatory, 220–21

L

Large Magellanic Cloud
  Cepheids in, 346–53, 367–69

globular clusters in, 567, 586, 622
IRAS observations of, 203–4, 543
stars with mass loss in, 543
Wolf-Rayet stars in, 116, 124, 145
LHS 2397a, 499
LHS 2924, 499
Local Bubble, 304–28, 339–42
  boundary of, 310–13, 316–20
  clouds in, 306–7, 320–23, 326–28, 339–42
  density of, 304–5, 310, 317–23, 328
  interstellar pressure in, 305–6, 310, 315–23
  models for, 317–23
  soft X-ray background in, 304–7, 310–20, 324–28, 331–32, 336–38, 341
  as supernova remnant possibility, 304–6, 309, 313–16, 339
  thermal evaporation in, 306, 320–23, 339–42
Local Cloud, 328–41
  density of, 329–33
  ionization sources in, 337–39
  solar backscatter observations in, 328–33
Local Fluff, 307–8, 328–39, 341. See also Local Cloud
Local Group of galaxies, 441–43, 451
Loop I Bubble, 325–26, 341
L204 molecular cloud, 36
L1551, 60–62, 532

M

M stars
  dwarf, see also Stars, very low mass
    ages of, 505–7
    binaries among, 488–93
    infrared excesses in, 535
    kinematics of, 505–6
    magnetic activity in, 295–97, 502–7
    photometry of, 486–87, 493–94
    rotational velocities of, 276, 278–79, 295–97
    search for, 483
    spectral classification of, 487–88
  dwarf emission, 285, 504
  giant, 534–35
  subdwarf
    initial-mass function of, 512

SUBJECT INDEX 673

luminosity function of, 511–12
  metal abundances of, 500–1
Magellanic Clouds, see also Large Magellanic Cloud, Small Magellanic Cloud
  Cepheids in, 346–53, 363–69
  IRAS observations of, 203–4
Magnetic activity, solar
  chromospheric indicators of, 87–96
  cycle of, 94, 98–103, 296–97, 634, 643
  and variations in solar irradiance, 633–67
Magnetic activity, stellar, 271–301, 502–7
  determination of periods of, 295–98
  direct measurements of, 282–85
  emission-line measurements of, 285–95
  tracers of, 281–98
Magnetic field(s), solar, 83–111, 633–67. See also Magnetic flux, solar; Magnetic structure, solar
  measurements of, 84–87, 646–59
Magnetic fields, molecular-cloud, 33–49, 62–63, 71–75
Magnetic fields, stellar, see also Magnetic activity, stellar
  direct measurements of, 282–85
  rotation and, 271–301
  in very low mass stars, 502
Magnetic flux, solar
  cancellation of, 105–6
  density of, 85–87, 91, 100–2, 634, 642–43, 663–64
  disappearance of, 83, 103–7, 109, 661–62
  emergence of, 83, 96–98, 107–9, 660
  transport of, 83, 103–7
Magnetic structure, solar
  faculae in, see Sun, faculae in
  filigree in, 89–90
  flux tubes in, collapse of, 96–98
  granulation in, 91–92, 96–97
  intranetwork field in, 84, 94–96, 98, 105–9
  knots in, 89
  network in, 86, 89, 91–92, 94–96, 100–7
  network clusters in, 94–95, 100, 105, 108
  plages in, see Sun, plages in

sunspots in, see Sunspots
Main-sequence stars, 271–301
  Ca II emission measurements for, 286–91
  emission-rotation relation in, 291–95, 298
  as infrared sources, 285, 534–35
  magnetic activity in, 271–301
  magnetic cycle periods in, 295–98
  rotation in, 271–301
  rotational evolution in 273–81
Markarian galaxies, 190, 209, 213, 215–16, 220–22
Mass loss, 523, 538–43
  cool stars with, 539–43
  Galactic distribution of stars with, 541–43
  globular cluster stars with, 565, 568
  hot stars with, 535–38
  planetary nebulae with, 538
Mass-luminosity relation, 479, 488, 492, 495–98, 507–13
Microwave background radiation, 222–23, 448–49
Milky Way
  dark matter in, 428–32, 451
  rotation curves for, 428–29, 451
  satellite dwarf galaxies in, 430–31
Missing mass, see also Dark matter
  in Galactic disk, 428–29, 608–9
  in globular clusters, 594, 597
  possibility of brown dwarfs as, 478, 512–13, 548
Molecular clouds
  Alfvénic turbulence in, 34–39
  ambipolar diffusion in, 41–43, 45, 47–51, 65, 71, 74
  cloud contraction in, 44–45
  fluid turbulence in, 33–34, 62
  gravitational collapse of cores of, 49–53, 71–72
  IRAS observations of cores of, 30, 48, 59, 200, 528–34, 555–57
  lifetimes of, 32
  magnetic fields in, 33–49, 62–63, 71–75
  masses of, 427
  protostar formation and evolution in, 53–60
  rotation in, 39–40, 52–53
  star formation in, 23–81
  temperatures of, 44
Molecular clouds, dwarf, 32–33

Molecular clouds, giant, 29, 32, 46–47
  formation of very low mass stars and, 480–81
  at Galactic center, 380
  and origin of comets, 234, 238–39
Molecular clumps, 29–49, 71
Monoceros OB1 molecular cloud, 532
MR 80, 120
M3, 592, 595, 597
M5, 595, 597
M13, 454, 595
M15, 427, 589, 595
M22, 543, 595
M30, 597
M31
  dark matter in, 432–35, 442–43
  distance to, 363–64, 367, 371
  globular clusters in, 617, 623
  IRAS observations of, 198–202, 206–7, 543
  Wolf-Rayet stars in, 125
M33
  globular clusters in, 622
  IRAS observations in, 198–202
  Wolf-Rayet stars in, 125, 143, 145
M71, 595
M81, 371–72
M82, 216
M87, 436–38
M92, 595
M101, 198, 354, 372
Mrk 171, 220
Mrk 231, 216, 221–22
Mrk 501, 209
Mrk 1014, 215

N

N-body calculations, for globular clusters, 575–81, 584, 597–98
N-body simulation codes, checks for, 182–83
N-body simulations, 151–86
  Cartesian grids in, 165–70, 175, 181
  cloud-in-cell (CIC) method, 171–73
  grid methods for, 156, 165–67
  particle-mesh techniques in, 167–76
    grid noise in, 172–73
  particle-particle-mesh method in, 167, 173–75
  polar grids in, 171, 175–76, 181

SUBJECT INDEX

quiet-particle mesh method in, 173
quiet starts in, 181–82
relaxation in, 152–59, 167
softening in, 156–60, 164
spatial resolution in, 159–60
spherical harmonic codes in, 176–78
stellar wakes in, 154–55
techniques for, 160–80
three-dimensional systems, 152–53, 156–57, 165–66, 169–70
time integration scheme in, 179–80
tree schemes in, 162–63, 178–79
triangular-shaped-cloud method, 172–73
two-dimensional systems, 153, 157, 165–69, 175–76
NGC 253, 216
NGC 720, 438
NGC 891, 616, 628
NGC 1068, 221
NGC 1275, 214
NGC 1705, 440
NGC 1866, 369
NGC 2403, 353, 370–72
NGC 3198, 433
NGC 3992, 433, 435
NGC 4278, 209, 437
NGC 5128, 436
NGC 5666, 437
NGC 6240, 218–20
NGC 6302, 538
NGC 6397, 586
NGC 6543, 223–24
NGC 6752, 586
NGC 6822, 125
NGC 7097, 437
NGC 7293, 538
Nimbus 7 satellite, 633–34, 643, 654
Novae, 543–44
Nucleosynthesis, 453–55

O

O stars
  atmospheres of, 135
  companions to Wolf-Rayet stars, 116, 120, 125–28
  Galactic distribution of, 123
  as progenitors of Wolf-Rayet stars, 142, 144–45
  X-ray observations of, 123
OB stars
  formation of, 24, 26, 32, 45–46, 54
  at Galactic center, 388, 418
  in Galactic disk, 555–57

mass loss in, 140, 143
photometry of, 355
X-ray emission in, 132
OH-IR stars
  in Galactic center, 412–14, 629
  in Magellanic clouds, 204
  mass loss in, 539–43
OJ 287, 211
Oort cometary cloud, 233–39
Orion molecular cloud, 24, 39, 557
OVV (optically violent variable) quasars, 195, 207–12

P

P Cygni, 535–36
Period-luminosity-color relationship, 345–72
Period-luminosity relationship, 345–72
PG 1351+640, 215, 221
Photometric Facular Index (PFI), 640–45
Photometric Sunspot Index (PSI), 636–37, 643–44
Photometric surveys for very low mass stars, 483–84
PKS 1345+125, 214
Planet X, search for, 523, 528
Planetary nebulae, 437
  in Galactic bulge, 629
  IRAS observations of, 538
Planetary systems, formation of, 71, 73, 545–47
Pleiades cluster, 24
  distance modulus of, 355–58, 364
  rotational velocities of stars in, 276–78
Polycyclic aromatic hydrocarbons (PAHs), 553–55
Proper-motion surveys, for very low mass stars, 481–83
Protostars
  accretion disks around, 52, 530–31
  IRAS search for, 59, 528–31
Protostellar evolution, 52–55, 58–63
PSR 0950+08, 337–38
PSR 1929+10, 338
Pulsars, 337–38

Q

Quarks, 456–57
Quasars, 195, 207–12, 215–16, 221
QSOs, 435, 446

R

R Coronae Australis, molecular cloud in, 31, 36
R Coronae Borealis stars, 541
Radio galaxies, 208–9, 212–16
Radio continuum emission, from Galactic center, 379–84, 387–88, 399–400, 404, 419
Radio emission, extragalactic, 189–92, 197, 204–5
Radio sources, see also Sgr A
Radio sources, extragalactic, 193–94, 204–5, 208–9, 214–16
Radio sources, Wolf-Rayet stars as, 118–22, 131–32, 138
Rho Ophiuchi molecular cloud, 31, 36, 39, 46, 48, 531
Rotating Full-Disk Photometer (RFDP), 656–59
RR Lyrae stars
  distance moduli of, 363–65
  in Galactic halo, 617, 619–20, 624–25
  masses of, 430–31
RS Ophiuchi, 544
RV Tauri variables, 539

S

Sagittarius A, 379–89, 400–1, 404, 409, 415–20
Sagittarius B2, 378
Scorpio-Centaurus association, 304, 341
Seyfert galaxies, 188, 195, 212–15, 220–21
Skylab, 101–2
Small Magellanic Cloud
  Cepheids in, 346–53, 369
  IRAS observations of, 203
  Wolf-Rayet stars in, 125, 145
SMM (Solar Maximum Mission), 633, 649
Solar activity cycle, 98–103, 196–97, 634, 643
Solar backscatter, 308, 328–33
Solar flares, 634
Solar irradiance, variations in, 633–67
  magnetic proxy for measuring, 645–47, 659
Solar neighborhood, dark matter in, 428, 451, 478, 511–13
Solar system, IRAS observations of, 521–28
Solar wind, 101, 109, 332
Sombrero galaxy, 435–36, 439
SPARTAN mission, 385
Star formation, 23–81
  bimodal, 24–25, 44–45

SUBJECT INDEX   675

deuterium burning in, 56–58, 72
at Galactic center, 388–89, 418
by hierarchical fragmentation, 24, 27–28, 479–80
high-mass, 24–27, 40, 44–46, 53–55, 73–74, 533–34, 556–57
IRAS observations of evidence of, 194, 196–204, 213, 216–25, 523, 528–34, 556–57
low-mass, 24–30, 44, 46–48, 54–58, 63–65, 74, 528–33
very-low-mass, 479–81, 512
Star-formation rate, in Galactic disk, 605
Star plages, 284
Starbursts, 187, 195–98, 213, 216–25, 224–25
Stars
A-type, see A stars
B-type, see B stars
binary, see Binary stars
blue horizontal branch, 617–20, 623–24
carbon, see Carbon stars
chromospheres of, 282–95, 475, 502–7
coronae of, 285–86, 294, 475, 502–7
disks around, 65–73, 530, 533, 545–47
IRAS observations of, 534–45
K-type, see K stars
magnetic activity in, see Magnetic activity, stellar
main-sequence, see Main-sequence stars
masses of
in globular clusters, 571–73
origin of, 23–24
metal-rich
in Galactic bulge, 628
in Galactic disk, 612–17
metal-weak, in Galactic halo, 613, 617–21, 623–27, 629–30
N-body simulations for large aggregates of, 151–86
pre-main-sequence, 55–57
IRAS observations of, 531, 533
rotational evolution of, 271–76, 290
supergiant, 535–36, 538–43
Wolf-Rayet, see Wolf-Rayet stars
Starspots, 284

Stellar evolution
in globular clusters, 565, 568, 585
Wolf-Rayet stars as phase of, 113, 143–46
Stellar flares, 505–6
Stellar rotation, 271–301, 502–5
evolution of, 271–81
relation to magnetic activity, 281–82, 291–99
Stellar winds, 24, 54–63, 72–73, 272, 276, 531–33, 535, 539
in Wolf-Rayet stars, 113–45, 535
Sun
active nests in, 99–103, 107, 109
active region of, see also Magnetic field(s), solar
photometry of, 647–59, 664
brown dwarf companion, possibility of, 478–79, 548
coronal holes in, 101–3, 108
energy balance between faculae and sunspots in, 660–63
energy storage in activity complex in, 661
ephemeral (active) regions in, 92–95, 98–103, 106–8
faculae in, 89–90, 106, 634–65
photometric proxy for irradiance function of, 637–45, 659, 664
photometry of, 647–54
magnetic field(s) of, see Magnetic field(s), solar; Magnetic structure, solar
plage area of, 96, 106, 108
photometry of, 639–45, 649, 652
relation to faculae, 659–60, 664
structure of, 86, 90–93
rotational velocity-activity cycle relation in, 279, 296–97
variations in luminosity of, 633–67
Sunspots
dissolution of, 103–7
formation of, 90–92, 96–99, 634, 660
influence on solar irradiance, 634–37, 642–43, 660–63
nests of, 99–103
photometric proxy for, 635–37, 664
photometry of, 643, 647–48

structure of, 83, 87–89, 634, 660
Sunspot pores, 96–97, 103, 105, 108
Supernova 1983k, 144
Supernova 1983n, 144
Supernova 1985f, 144
Supernova remnants, 381, 544–45
and local interstellar medium, 304–6, 309, 313–17, 323, 339
Supernovae, Wolf-Rayet star evolution and, 144
S106, 61, 73

T

T Tauri stars
disks around, 66–73, 530–31
emission activity in, 290
formation and evolution of, 24, 27, 30, 47, 56, 64–65
infrared excesses of, 66–68
IRAS observations of, 529–33
rotational velocities of, 63, 273–78
Taurus molecular cloud, 24, 29, 38–39, 45–47, 52
47 Tucanae, 586, 597
Tycho supernova, 544

U

Ultraviolet emission
from Galactic center, 388, 400–1, 405, 414, 418–20
from Local Bubble, 308, 315, 325, 335–38
Ultraviolet observations
of Cepheids, 362–63
of comets, 253–54, 258
of Wolf-Rayet stars, 115, 128–29, 136–37
Universe, dark matter in the, 425–72

V

VB 8, 490
VB 10, 489
Vega, 335, 523, 545–47
Vega-like stars, 545–47
Vega spacecraft, 246–48
Very low mass stars, 473–519
atmospheres of, 477–78
binaries among, 488–94, 498–99, 503–5, 508–9
chromospheres of, 475, 502–7
coronae of, 475, 502–7
effective temperatures of, 493–94

equation of state for, 475–77
formation of, 479–81
identification of, 481–85
infrared photometry of, 486–87, 490, 494
initial-mass function of, 478–80, 507
luminosity function of, 507–12
magnetic activity in, 502–7
masses of, 494–95
opacity tables for, 475–77
photometry of, 483–87
radial velocities of, 488
rotation-chromospheric activity relation in, 504–5
rotation-coronal activity relation in, 502–3
trigonometric parallaxes for, 485
X-ray emitters among, 498, 503–6
Virgo cluster of galaxies, 438, 442, 445–47
VLA observations
of the Galactic center, 379, 398, 404
of galaxies, 192, 437, 440
of Wolf-Rayet stars, 118–19
VLBI, 378
Voyager observations, 336, 338
V444 Cygni, 122, 134–35, 138–40
V4077 Sagittarii, 543–44

W

White dwarfs, 455, 478, 482
WIMPs, 459–61
Wolf-Rayet stars, 113–50

abundances in, 114–15, 128
as binary stars, 122–28
collapsed companions of, 126–27
colors of, 116–17
dust in, 118, 120, 142
effective temperatures of, 136–38
evolution of, 124, 143–46
expanding atmospheres of, 133–34
Galactic distribution of, 123–24, 145
infrared observations of, 115, 118, 120–22, 129–30, 535–36
and the interstellar medium, 141–42
luminosities of, 138–40
magnitudes of, 116–17
masses of, 127–28, 145
mass loss in, 118–19, 128, 131–32, 140–45
model atmospheres for, 132–41
nonthermal emission in, 118–20, 123
in other galaxies, 124–25
as radio sources, 118–22, 131–32, 138
spectral types of, 114–16, 145–46
ultraviolet observations of, 115, 128–29, 136–37
variability of, 116–18, 127
winds in, 113, 118–22, 127–45, 535
X-ray observations of, 122–23, 126–27, 132
W3, 46
W49, 46

XYZ

X-ray background, soft, 304–7, 310–20, 324–28, 331–32, 336–38, 341
X-ray emission, from Galactic center, 384–87, 404–6
X-ray observations
of galaxies, 197, 207, 213–14, 438–39, 446, 451
of very low mass stars, 498, 503–6
X-ray sources
in globular clusters, 575, 597
Wolf-Rayet stars as, 122–23, 126–27, 132
Young stellar objects (YSOs), 24, 60–71
bipolar outflows from, 60–63, 73–74
Zeeman splitting, 84–85
Zero-age main sequence, 275, 355, 364, 495–98
Zeta Puppis, 135, 535–36
Zodiacal infrared emission, 522–26, 547

MISCELLANEOUS
3C 48, 215, 221
3C 273, 207, 212
3C 345, 209–11
3C 390.3, 208–9, 212
3C 446, 211–12
3C 459, 215
0421+040P06, 214–15
0857+561, 446
1983 TB, 528
16293-2422, 531

CUMULATIVE INDEXES

CONTRIBUTING AUTHORS, VOLUMES 15–25

A

Abbott, D. C., 25:113–50
Abt, H. A., 21:343–72
Adams, F. C., 25:23–81
Aizenman, M. L., 16:215–40
Akasofu, S.-I., 20:117–38
Allen, R. J., 16:103–39
Ambartsumian, V. A., 18:1–13
Angel, J. R. P., 16:487–519; 18:321–61
Athanassoula, E., 23:147–68

B

Backer, D. C., 24:537–75
Bahcall, J. N., 16:241–64; 24:577–611
Bahcall, N. A., 15:505–40
Balick, B., 20:431–68
Baliunas, S. L., 23:379–412
Baym, G., 17:415–43
Beckwith, S., 20:163–90
Beichman, C. A., 25:521–63
Bignami, G. F., 21:67–108
Binney, J., 20:399–429
Boesgaard, A. M., 23:319–78
Böhm-Vitense, E., 19:295–318
Borra, E. F., 20:191–220
Bosma, A., 23:147–68
Bracewell, R. N., 17:113–34
Bradt, H. V. D., 21:13–66
Brault, J. W., 22:291–317
Bridle, A. H., 22:319–58
Brown, R. L., 16:445–85; 22:223–65
Burns, J. A., 15:97–126

C

Carbon, D. F., 17:515–49
Carswell, R. F., 19:41–76
Cassen, P., 15:97–126
Cassinelli, J. P., 17:275–308
Caughlan, G. R., 21:165–76
Cesarsky, C. J., 18:289–319
Chapman, C. R., 16:33–75
Chapman, G. A., 25:633–67
Chevalier, R. A., 15:175–96
Chincarini, G. L., 22:445–70
Chiosi, C., 24:329–75
Chupp, E. L., 22:359–87
Conti, P. S., 16:371–92; 25:113–50

Coroniti, F. V., 15:389–436
Coulman, C. E., 23:19–57
Cowie, L. L., 24:499–535
Cowling, T. G., 19:115–35; 23:1–18
Cox, A. N., 18:15–41
Cox, D. P., 25:303–44

D

Davidson, K., 23:119–46
Davis, M., 21:109–30
Deubner, F.-L., 22:593–619
Dravins, D., 20:61–89
Dressler, A., 22:185–222
Dulk, G. A., 23:169–224
Dupree, A. K., 24:377–420

E

Edmunds, M. G., 19:77–113
Elliot, J. L., 17:445–75
Ellis, G. F. R., 22:157–84
Elson, R., 25:565–601

F

Faber, S. M., 17:135–87
Feast, M. W., 25:345–75
Fesen, R. A., 23:119–46
Ford, W. K. Jr., 17:189–212
Forman, W., 20:547–85
Fowler, W. A., 21:165–76
Freeman, K. C., 19:319–56; 25:603–32
Fujimoto, M., 24:459–97

G

Gallagher, J. S., 16:171–214; 17:135–87; 22:37–74
Gault, D. E., 15:97–126
Genzel, R., 25:377–423
Giffard, R. P., 16:521–54
Gillett, F. C., 19:411–56
Giovanelli, R., 22:445–70
Goldreich, P., 20:249–83
Golub, L., 23:413–52
Gott, J. R. III, 15:235–66
Gough, D., 22:593–619
Greenstein, J. L., 22:1–35
Gursky, H., 15:541–68

H

Habing, H. J., 17:345–85
Hansen, C. J., 16:15–32
Harris, M. J., 21:165–76
Harris, W. E., 17:241–74
Hartmann, L. W., 25:271–301
Hartmann, W. K., 16:33–75
Haynes, M. P., 22:445–70
Heckman, T. M., 20:431–68
Hellings, R. W., 24:537–75
Hermsen, W., 21:67–108
Hillas, A. M., 22:425–44
Ho, P. T. P., 21:239–70
Hoag, A. A., 17:43–71
Hodge, P. W., 19:357–72
Hollenbach, D. J., 18:219–62
Holt, S. S., 20:323–65
Houck, J. R., 25:187–230
Howard, R., 15:153–73; 22:131–55
Hoyle, F., 20:1–35
Hunter, D. A., 22:37–74
Hurford, G. J., 20:497–516
Hut, P., 25:565–601

I

Iben, I. Jr., 21:271–342
Inagaki, S., 25:565–601
Ionson, J. A., 19:7–40
Israel, F. P., 17:345–85

J

Jones, C., 20:547–85
Joss, P. C., 22:537–92
Joyce, R. R., 19:411–56

K

Kaler, J. B., 23:89–117
Kellermann, K. I., 19:373–410
Kennel, C. F., 15:389–436
Kleinmann, S. G., 19:411–56
Knapp, R. R., 16:445–85
Kraft, R. P., 15:309–43
Kuperus, M., 19:7–40

L

Labeyrie, A., 16:77–102
Lada, C. J., 23:267–317
Landstreet, J. D., 20:191–220

677

Larson, H. P., 18:43–75
Lebofsky, M. J., 17:477–511
Leovy, C. B., 17:387–413
Lesh, J. R., 16:215–40
Liebert, J., 18:363–98; 25:473–519
Linsky, J. L., 18:439–88
Liszt, H. S., 22:223–65
Lizano, S., 25:23–81
Lockman, F. J., 16:445–85
Lubow, S. H., 19:227–93

M

Mackay, C. D., 24:255–83
Maeder, A., 24:329–75
Manchester, R. N., 15:19–44
Margon, B., 22:507–36
Mariska, J. T., 24:23–48
Marov, M. Ya., 16:141–69
Marsh, K. A., 20:497–516
Mathews, W. G., 24:171–203
Mathis, J. S., 17:73–111
McAlister, H. A., 23:59–87
McClintock, J. E., 21:13–66
McCray, R., 17:213–40; 20:323–65
McCrea, W. H., 25:1–22
McKee, C. F., 18:219–62
Merrill, K. M., 17:9–41
Mestel, L., 20:191–220
Miley, G., 18:165–218
Moore, R., 23:239–66
Moran, J. M., 19:231–76
Morris, M., 20:517–45
Morrison, D., 20:469–95
Mould, J. R., 20:91–115

N

Narayan, R., 24:127–70
Ness, N. F., 20:139–61
Neugebauer, G., 25:187–230
Newkirk, G. Jr., 21:429–67
Nicholls, R. W., 15:197–234
Nityananda, R., 24:127–70
Norris, J., 19:319–56
Noyes, R. W., 15:363–87; 25:271–301

O

Oort, J. H., 15:295–362; 19:1–5; 21:373–428
Öpik, E. J., 15:1–17
Osterbrock, D. E., 24:171–203

P

Pagel, B. E. J., 19:77–113
Parker, E. N., 15:45–68
Pauliny-Toth, I. I. K., 19:373–410
Payne-Gaposchkin, C., 16:1–13
Pearson, T. J., 22:97–130
Peebles, P. J. E., 21:109–30
Perley, R. A., 22:319–58
Pethick, C., 17:415–43
Pettengill, G. H., 16:265–92
Phillips, T. G., 20:285–321
Pipher, J. L., 16:335–69
Podosek, F. A., 16:293–334
Pollack, J. B., 22:389–424
Popper, D. M., 18:115–64
Pringle, J. E., 19:137–62
Probst, R. G., 25:473–519

R

Rabin, D., 23:239–66
Racine, R., 17:241–74
Rappaport, S. A., 22:537–92
Raymond, J. C., 22:75–95
Readhead, A. C. S., 22:97–130
Rees, M. J., 22:471–506
Reid, M. J., 19:231–76
Renzini, A., 21:271–342
Reynolds, R. J., 25:303–44
Rickard, L. J, 20:517–45
Rickett, B. J., 15:479–504
Ridgway, S. T., 17:9–41; 22:291–317
Rieke, G. H., 17:477–511
Rosner, R., 16:393–428; 23:413–52

S

Salpeter, E. E., 15:267–93
Sandage, A., 24:421–58
Savage, B. D., 17:73–111
Schwartz, D. A., 15:541–68
Schwartz, R. D., 21:209–37
Sellwood, J. A., 25:151–86
Shu, F. H., 19:277–93; 25:23–81
Shull, J. M., 20:163–90
Smith, A. G., 17:43–71
Smith, M. G., 19:41–76
Snow, T. P. Jr., 17:213–40
Sofue, Y., 24:459–97
Soifer, B. T., 16:335–69; 21:177–207; 25:187–230
Songaila, A., 24:499–535
Spicer, D. S., 19:7–40
Spinrad, H., 25:231–69
Spite, F., 23:225–38
Spite, M., 23:225–38
Starrfield, S., 16:171–214
Steigman, G., 23:319–78
Stein, W. A., 21:177–207
Stern, D. P., 20:139–61
Stinebring, D. R., 24:285–327
Stockman, H. S., 18:321–61
Strom, R. G., 15:97–126
Strömgren, B., 21:1–11

Sunyaev, R. A., 18:537–60
Svalgaard, L., 16:429–43
Swings, P., 17:1–7
Syrovatskii, S. I., 19:163–229

T

Taylor, J. H., 15:19–44; 24:285–327
Thomas, H.-C., 15:127–51
Toomre, A., 15:437–78
Townes, C. H., 21:239–70; 25:377–423
Tremaine, S., 20:249–83
Trimble, V., 25:425–72
Tsuji, T., 24:89–125
Tyson, J. A., 16:521–54

V

Vaiana, G. S., 16:393–428; 23:413–52
van Altena, W. F., 21:131–64
van der Kruit, P. C., 16:103–39
Vauclair, G., 20:37–60
Vauclair, S., 20:37–60
Vaughan, A. H., 23:379–412

W

Wagner, W. J., 22:267–89
Walker, A. R., 25:345–75
Wannier, P. G., 18:399–437
Watson, W. D., 16:585–615
Weaver, T. A., 24:205–53
Weedman, D. W., 15:69–95
Weiss, R., 18:489–537
Wetherill, G. W., 18:77–113
Weymann, R. J., 19:41–76
Whitford, A. E., 24:1–22
Wielebinski, R., 24:459–97
Wilcox, J. M., 16:429–43
Williams, J. G., 16:33–75
Withbroe, G. L., 15:363–87
Woodward, P. R., 16:555–84
Woody, D. P., 20:285–321
Woolf, N. J., 20:367–98
Woosley, S. E., 24:205–53
Wynn-Williams, C. G., 20:587–618

Y

York, D. G., 20:221–48
Yorke, H. W., 24:49–87

Z

Zel'dovich, Ya. B., 18:537–60
Zimmerman, B. A., 16:165–76
Zuckerman, B., 18:263–88
Zwaan, C., 25:83–111

CHAPTER TITLES, VOLUMES 15–25

PREFATORY CHAPTER

About Dogma in Science and Other Recollections of an Astronomer	E. J. Öpik	15:1–17
The Development of Our Knowledge of Variable Stars	C. Payne-Gaposchkin	16:1–13
A Few Notes on My Career as an Astrophysicist	P. Swings	17:1–7
On Some Trends in the Development of Astrophysics	V. A. Ambartsumian	18:1–13
Some Notes on My Life as an Astronomer	J. H. Oort	19:1–5
The Universe: Past and Present Reflections	F. Hoyle	20:1–35
Scientists I Have Known and Some Astronomical Problems I Have Met	B. Strömgren	21:1–11
An Astronomical Life	J. L. Greenstein	22:1–35
Astronomer by Accident	T. G. Cowling	23:1–18
A Half-Century of Astronomy	A. E. Whitford	24:1–22
Clustering of Astronomers	W. H. McCrea	25:1–22

SOLAR SYSTEM ASTROPHYSICS

Mercury	D. E. Gault, J. A. Burns, P. Cassen, R. G. Strom	15:97–126
Jupiter's Magnetosphere	C. F. Kennel, F. V. Coroniti	15:389–436
The Asteroids	C. R. Chapman, J. G. Williams, W. K. Hartmann	16:33–75
Results of Venus Missions	M. Ya. Marov	16:141–69
Physical Properties of the Planets and Satellites from Radar Observations	G. H. Pettengill	16:265–92
Isotopic Structures in Solar System Materials	F. A. Podosek	16:293–334
Martian Meterology	C. B. Leovy	17:387–413
Stellar Occultation Studies of the Solar System	J. L. Elliot	17:445–75
Infrared Spectroscopic Observations of the Outer Planets, Their Satellites, and the Asteroids	H. P. Larson	18:43–75
Formation of Terrestrial Planets	G. W. Wetherill	18:77–113
Planetary Magnetospheres	D. P. Stern, N. F. Ness	20:139–61
The Dynamics of Planetary Rings	P. Goldreich, S. Tremaine	20:249–83
The Satellites of Jupiter and Saturn	D. Morrison	20:469–95
Origin and History of the Outer Planets: Theoretical Models and Observational Constraints	J. B. Pollack	22:389–424
Comets and Their Composition	H. Spinrad	25:231–69

SOLAR PHYSICS

The Origin of Solar Activity	E. N. Parker	15:45–68
Large-Scale Solar Magnetic Fields	R. Howard	15:153–73
Mass and Energy Flow in the Solar Chromosphere and Corona	G. L. Withbroe, R. W. Noyes	15:363–87
Recent Advances in Coronal Physics	G. S. Vaiana, R. Rosner	16:393–428
A View of Solar Magnetic Fields, the Solar Corona, and the Solar Wind in Three Dimensions	L. Svalgaard, J. M. Wilcox	16:429–43
On the Theory of Coronal Heating	M. Kuperus, J. A. Ionson, D. S. Spicer	19:7–40
High Spatial Resolution Solar Microwave Observations	K. A. Marsh, G. J. Hurford	20:497–516
Variations in Solar Luminosity	G. Newkirk, Jr.	21:429–67
Solar Rotation	R. Howard	22:131–55
Coronal Mass Ejections	W. J. Wagner	22:267–89

679

CHAPTER TITLES

High-Energy Neutral Radiations From the Sun	E. L. Chupp	22:359–87
Helioseismology: Oscillations as a Diagnostic of the Solar Interior	F.-L. Deubner, D. Gough	22:593–619
Radio Emission From the Sun and Stars	G. A. Dulk	23:169–224
Sunspots	R. Moore, D. Rabin	23:239–66
The Quiet Solar Transition Region	J. T. Mariska	24:23–48
Elements and Patterns in the Solar Magnetic Field	C. Zwaan	25:83–111
Variations of Solar Irradiance due to Magnetic Activity	G. A. Chapman	25:633–67

STELLAR PHYSICS

Recent Observations of Pulsars	J. H. Taylor, R. N. Manchester	15:19–44
Secular Stability: Applications to Stellar Structure and Evolution	C. J. Hansen	16:15–32
Theory and Observations of Classical Novae	J. S. Gallagher, S. Starrfield	16:171–214
The Observational Status of β Cephei Stars	J. R. Lesh, M. L. Aizenman	16:215–40
Mass Loss in Early-Type Stars	P. S. Conti	16:371–92
Magnetic White Dwarfs	J. R. P. Angel	16:487–519
Infrared Spectroscopy of Stars	K. M. Merrill, S. T. Ridgway	17:9–41
Stellar Winds	J. P. Cassinelli	17:275–308
On the Nonhomogeneity of Metal Abundances in Stars of Globular Clusters and Satellite Subsystems of the Galaxy	R. P. Kraft	17:309–43
Physics of Neutron Stars	G. Baym, C. Pethick	17:415–43
Model Atmospheres for Intermediate- and Late-Type Stars	D. F. Carbon	17:513–49
The Masses of Cepheids	A. N. Cox	18:15–41
Stellar Masses	D. M. Popper	18:115–64
Envelopes Around Late-Type Giant Stars	B. Zuckerman	18:263–88
White Dwarf Stars	J. Liebert	18:363–98
Stellar Chromospheres	J. L. Linsky	18:439–88
Mass, Angular Momentum, and Energy Transfer in Close Binary Stars	F. H. Shu, S. H. Lubow	19:277–93
The Effective Temperature Scale	E. Böhm-Vitense	19:295–318
Element Segregation in Stellar Outer Layers	S. Vauclair, G. Vauclair	20:37–60
Photospheric Spectrum Line Asymmetries and Wavelength Shifts	D. Dravins	20:61–89
Magnetic Stars	E. F. Borra, J. D. Landstreet, L. Mestel	20:191–220
The Search for Infrared Protostars	C. G. Wynn-Williams	20:587–618
The Optical Counterparts of Compact Galactic X-Ray Sources	H. V. D. Bradt, J. E. McClintock	21:13–66
Galactic Gamma-Ray Sources	G. F. Bignami, W. Hermsen	21:67–108
Herbig-Haro Objects	R. D. Schwartz	21:209–37
Asymptotic Giant Branch Evolution and Beyond	I. Iben, Jr., A. Renzini	21:271–342
Normal and Abnormal Binary Frequencies	H. A. Abt	21:343–72
Observations of Supernova Remnants	J. C. Raymond	22:75–95
High Angular Resolution Measurements of Stellar Properties	H. A. McAlister	23:59–87
Planetary Nebulae and Their Central Stars	J. B. Kaler	23:89–117
Radio Emission From the Sun and Stars	G. A. Dulk	23:169–224
The Composition of Field Halo Stars and the Chemical Evolution of the Halo	M. Spite, F. Spite	23:225–38
Stellar Activity Cycles	S. L. Baliunas, A. H. Vaughan	23:379–412
On Stellar X-Ray Emission	R. Rosner, L. Golub, G. S. Vaiana	23:413–52
Molecules in Stars	T. Tsuji	24:89–125
The Physics of Supernova Explosions	S. E. Woosley, T. A. Weaver	24:205–53
Recent Progress in the Understanding of Pulsars	J. H. Taylor, D. R. Stinebring	24:285–327
The Evolution of Massive Stars With Mass Loss	C. Chiosi, A. Maeder	24:329–75

Mass Loss From Cool Stars	A. K. Dupree	24:377–420
The Population Concept, Globular Clusters, Subdwarfs, Ages, and the Collapse of the Galaxy	A. Sandage	24:421–58
Pulsar Timing and General Relativity	D. C. Backer, R. W. Hellings	24:537–75
Star Formation in Molecular Clouds: Observation and Theory	F. H. Shu, F. C. Adams, S. Lizano	25:23–81
Wolf-Rayet Stars	D. C. Abbott, P. S. Conti	25:113–50
Rotation and Magnetic Activity in Main-Sequence Stars	L. W. Hartmann, R. W. Noyes	25:271–301
Very Low Mass Stars	J. Liebert, R. G. Probst	25:473–519

DYNAMICAL ASTRONOMY

Theories of Spiral Structure	A. Toomre	15:437–78
Astrometry	W. F. van Altena	21:131–64
Dynamical Evolution of Globular Clusters	R. Elson, P. Hut, S. Inagaki	25:565–601
The Galactic Spheroid and Old Disk	K. C. Freeman	25:603–32

INTERSTELLAR MEDIUM

The Interaction of Supernovae with the Interstallar Medium	R. A. Chevalier	15:175–96
Formation and Destruction of Dust Grains	E. E. Salpeter	15:267–93
Interstellar Scattering and Scintillation of Radio Waves	B. J. Rickett	15:479–504
Radio Recombination Lines	R. L. Brown, F. J. Lockman, G. R. Knapp	16:445–85
Gas Phase Reactions in Astrophysics	W. D. Watson	16:585–615
Observed Properties of Interstellar Dust	B. D. Savage, J. S. Mathis	17:73–111
The Violent Interstellar Medium	R. McCray, T. P. Snow, Jr.	17:213–40
Compact H II Regions and OB Star Formation	H. J. Habing, F. P. Israel	17:345–85
Interstellar Shock Waves	C. F. McKee, D. J. Hollenbach	18:219–62
Cosmic-Ray Confinement in the Galaxy	C. J. Cesarsky	18:289–319
Nuclear Abundances and Evolution of the Interstellar Medium	P. G. Wannier	18:399–437
Interstellar Molecular Hydrogen	J. M. Shull, S. Beckwith	20:163–90
Herbig-Haro Objects	R. D. Schwartz	21:209–37
Interstellar Ammonia	P. T. P. Ho, C. H. Townes	21:239–70
Observations of Supernova Remnants	J. C. Raymond	22:75–95
The Influence of Environment on the H I Content of Galaxies	M. P. Haynes, R. Giovanelli, G. L. Chincarini	22:445–70
Planetary Nebulae and Their Central Stars	J. B. Kaler	23:89–117
Cold Outflows, Energetic Winds, and Enigmatic Jets Around Young Stellar Objects	C. J. Lada	23:267–317
The Dynamical Evolution of H II Regions—Recent Theoretical Developments	H. W. Yorke	24:49–87
High-Resolution Optical and Ultraviolet Absorption-Line Studies of Interstellar Gas	L. L. Cowie, A. Songaila	24:499–535
Star Formation in Molecular Clouds: Observation and Theory	F. H. Shu, F. C. Adams, S. Lizano	25:23–81
The Local Interstellar Medium	D. P. Cox, R. J. Reynolds	25:303–44

SMALL STELLAR SYSTEMS

Consequences of Mass Transfer in Close Binary Systems	H.-C. Thomas	15:127–51
Theoretical Models of Star Formation	P. R. Woodward	16:555–84
Mass, Angular Momentum, and Energy Transfer in Close Binary Stars	F. H. Shu, S. H. Lubow	19:277–93
The Chemical Composition, Structure, and Dynamics of Globular Clusters	K. C. Freeman, J. Norris	19:319–56
Normal and Abnormal Binary Frequencies	H. A. Abt	21:343–72

Dynamical Evolution of Globular Clusters	R. Elson, P. Hut, S. Inagaki	25:565–601

THE GALAXY

The Galactic Center	J. H. Oort	15:295–362
Cosmic-Ray Confinement in the Galaxy	C. J. Cesarsky	18:289–319
The Chemical Composition, Structure, and Dynamics of Globular Clusters	K. C. Freeman, J. Norris	19:319–56
Stellar Populations in the Galaxy	J. R. Mould	20:91–115
Gas in the Galactic Halo	D. G. York	20:221–48
The Optical Counterparts of Compact Galactic X-Ray Sources	H. V. D. Bradt, J. E. McClintock	21:13–66
Galactic Gamma-Ray Sources	G. F. Bignami, W. Hermsen	21:67–108
Sagittarius A and Its Environment	R. L. Brown, H. S. Liszt	22:223–65
Neutron Stars in Interacting Binary Systems	P. C. Joss, S. A. Rappaport	22:537–92
Star Counts and Galactic Structure	J. N. Bahcall	24:577–611
Physical Conditions, Dynamics, and Mass Distribution in the Center of the Galaxy	R. Genzel, C. H. Townes	25:377–423
The IRAS View of the Galaxy and the Solar System	C. A. Beichman	25:521–63
The Galactic Spheroid and Old Disk	K. C. Freeman	25:603–32

EXTRAGALACTIC ASTRONOMY

Seyfert Galaxies	D. W. Weedman	15:69–95
Clusters of Galaxies	N. A. Bahcall	15:505–40
The Kinematics of Spiral and Irregular Galaxies	P. C. van der Kruit, R. J. Allen	16:103–39
Masses and Mass-to-Light Ratios of Galaxies	S. M. Faber, J. S. Gallagher	17:135–87
Globular Clusters in Galaxies	W. E. Harris, R. Racine	17:241–74
Infrared Emission of Extragalactic Sources	G. H. Rieke, M. J. Lebofsky	17:477–511
The Structure of Extended Extragalactic Radio Sources	G. Miley	18:165–218
Optical and Infrared Polarization of Active Extragalactic Objects	J. R. P. Angel, H. S. Stockman	18:321–61
Absorption Lines in the Spectra of Quasistellar Objects	R. J. Weymann, R. F. Carswell, M. G. Smith	19:41–76
Abundances in Stellar Populations and the Interstellar Medium in Galaxies	B. E. J. Pagel, M. G. Edmunds	19:77–113
Compact Radio Sources	K. I. Kellermann, I. I. K. Pauliny-Toth	19:373–410
Dynamics of Elliptical Galaxies and Other Spheroidal Components	J. Binney	20:399–429
Extranuclear Clues to the Origin and Evolution of Activity in Galaxies	B. Balick, T. M. Heckman	20:431–68
Molecular Clouds in Galaxies	M. Morris, L. J Rickard	20:517–45
X-Ray-Imaging Observations of Clusters of Galaxies	W. Forman, C. Jones	20:547–85
Dust in Galaxies	W. A. Stein, B. T. Soifer	21:177–207
Superclusters	J. H. Oort	21:373–428
Structure and Evolution of Irregular Galaxies	J. S. Gallagher, III, D. A. Hunter	22:37–74
The Evolution of Galaxies in Clusters	A. Dressler	22:185–222
Extragalactic Radio Jets	A. H. Bridle, R. A. Perley	22:319–58
Black Hole Models for Active Galactic Nuclei	M. J. Rees	22:471–506
Shells and Rings Around Galaxies	E. Athanassoula, A. Bosma	23:147–68
Emission-Line Regions of Active Galaxies and QSOs	D. E. Osterbrock, W. G. Mathews	24:171–203
Global Structure of Magnetic Fields in Spiral Galaxies	Y. Sofue, M. Fujimoto, R. Wielebinski	24:459–97
The IRAS View of the Extragalactic Sky	B. T. Soifer, J. R. Houck, G. Neugebauer	25:187–230
Cepheids as Distance Indicators	M. W. Feast, A. R. Walker	25:345–75

Existence and Nature of Dark Matter in the Universe	V. Trimble	25:425–72
OBSERVATIONAL PHENOMENA		
Extragalactic X-Ray Sources	H. Gursky, D. A. Schwartz	15:541–68
Masses of Neutron Stars and Black Holes in X-Ray Binaries	J. N. Bahcall	16:241–64
Infrared Spectroscopic Observations of the Outer Planets, Their Satellites, and the Asteroids	H. P. Larson	18:43–75
Optical and Infrared Polarization of Active Extragalactic Objects	J. R. P. Angel, H. S. Stockman	18:321–61
Measurements of the Cosmic Background Radiation	R. Weiss	18:489–537
Preliminary Results of the Air Force Infrared Sky Survey	S. G. Kleinmann, F. C. Gillett, R. R. Joyce	19:411–56
Spectra of Cosmic X-Ray Sources	S. S. Holt, R. McCray	20:323–65
Galactic Gamma-Ray Sources	G. F. Bignami, W. Hermsen	21:67–108
The Evolution of Galaxies in Clusters	A. Dressler	22:185–222
Observations of SS 433	B. Margon	22:507–36
Recent Developments Concerning the Crab Nebula	K. Davidson, R. A. Fesen	23:119–46
The IRAS View of the Extragalactic Sky	B. T. Soifer, J. R. Houck, G. Neugebauer	25:187–230
Existence and Nature of Dark Matter in the Universe	V. Trimble	25:425–72
GENERAL RELATIVITY AND COSMOLOGY		
Recent Theories of Galaxy Formation	J. R. Gott III	15:235–66
Gravitational-Wave Astronomy	J. A. Tyson, R. P. Gifford	16:521–54
Measurements of the Cosmic Background Radiation	R. Weiss	18:489–537
Microwave Background Radiation as a Probe of the Contemporary Structure and History of the Universe	R. A. Sunyaev, Ya. B. Zel'dovich	18:537–60
The Extragalactic Distance Scale	P. W. Hodge	19:357–72
Evidence for Local Anisotropy of the Hubble Flow	M. Davis, P. J. E. Peebles	21:109–30
Alternatives to the Big Bang	G. F. R. Ellis	22:157–84
Big Bang Nucleosynthesis: Theories and Observations	A. M. Boesgaard, G. Steigman	23:319–78
Pulsar Timing and General Relativity	D. C. Backer, R. W. Hellings	24:537–75
Existence and Nature of Dark Matter in the Universe	V. Trimble	25:425–72
INSTRUMENTATION AND TECHNIQUES		
Stellar Interferometry Methods	A. Labeyrie	16:77–102
Instrumentation for Infrared Astronomy	B. T. Soifer, J. L. Pipher	16:335–69
Advances in Astronomical Photography at Low Light Levels	A. G. Smith, A. A. Hoag	17:43–71
Computer Image Processing	R. N. Bracewell	17:113–34
Digital Imaging Techniques	W. K. Ford, Jr.	17:189–212
Millimeter- and Submillimeter-Wave Receivers	T. G. Phillips, D. P. Woody	20:285–321
High Resolution Imaging from the Ground	N. J. Woolf	20:367–98
Astrometry	W. F. van Altena	21:131–64
Image Formation by Self-Calibration in Radio Astronomy	T. J. Pearson, A. C. S. Readhead	22:97–130
Astronomical Fourier Transform Spectroscopy Revisited	S. T. Ridgway, J. W. Brault	22:291–317
Fundamental and Applied Aspects of Astronomical "Seeing"	C. E. Coulman	23:19–57

High Angular Resolution Measurements of Stellar Properties	H. A. McAlister	23:59–87
Maximum Entropy Image Restoration in Astronomy	R. Narayan, R. Nityananda	24:127–70
Charge-Coupled Devices in Astronomy	C. D. Mackay	24:255–83
The Art of N-Body Building	J. A. Sellwood	25:151–86

PHYSICAL PROCESSES

Transition Probability Data for Molecules of Astrophysical Interest	R. W. Nicholls	15:197–234
Interstellar Shock Waves	C. F. McKee, D. J. Hollenbach	18:219–62
Nuclear Abundances and Evolution of the Interstellar Medium	P. G. Wannier	18:399–437
The Present Status of Dynamo Theory	T. G. Cowling	19:115–35
Accretion Discs in Astrophysics	J. E. Pringle	19:137–62
Pinch Sheets and Reconnection in Astrophysics	S. I. Syrovatskii	19:163–229
Masers	M. J. Reid, J. M. Moran	19:231–76
Interaction Between a Magnetized Plasma Flow and a Strongly Magnetized Celestial Body with an Ionized Atmosphere: Energetics of the Magnetosphere	S.-I. Akasofu	20:117–38
Interstellar Molecular Hydrogen	J. M. Shull, S. Beckwith	20:163–90
The Optical Counterparts of Compact Galactic X-Ray Sources	H. V. D. Bradt, J. E. McClintock	21:13–66
Thermonuclear Reaction Rates, III	M. J. Harris, W. A. Fowler, G. R. Caughlan, B. A. Zimmerman	21:165–76
The Origin of Ultra-High-Energy Cosmic Rays	A. M. Hillas	22:425–44
Observations of SS 433	B. Margon	22:507–36
The Physics of Supernova Explosions	S. E. Woosley, T. A. Weaver	24:205–53

Annual Reviews Inc.
A NONPROFIT SCIENTIFIC PUBLISHER

4139 El Camino Way
P.O. Box 10139
Palo Alto, CA 94303-0897 • USA

ORDER FORM
Now you can order
TOLL FREE
1-800-523-8635
(except California)

Annual Reviews Inc. publications may be ordered directly from our office by mail or use our Toll Free Telephone line (for orders paid by credit card or purchase order, and customer service calls only); through booksellers and subscription agents, worldwide; and through participating professional societies. Prices subject to change without notice.   ARI Federal I.D. #94-1156476

- **Individuals:** Prepayment required on new accounts by check or money order (in U.S. dollars, check drawn on U.S. bank) or charge to credit card — American Express, VISA, MasterCard.
- **Institutional buyers:** Please include purchase order number.
- **Students:** $10.00 discount from retail price, per volume. Prepayment required. Proof of student status must be provided (photocopy of student I.D. or signature of department secretary is acceptable). Students must send orders direct to Annual Reviews. Orders received through bookstores and institutions requesting student rates will be returned. You may order at the Student Rate for a maximum of 3 years.
- **Professional Society Members:** Members of professional societies that have a contractual arrangement with Annual Reviews may order books through their society at a reduced rate. Check with your society for information.
- **Toll Free Telephone orders:** Call 1-800-523-8635 (except from California) for orders paid by credit card or purchase order and customer service calls only. California customers and all other business calls use 415-493-4400 (not toll free). Hours: 8:00 AM to 4:00 PM, Monday-Friday, Pacific Time.

Regular orders: Please list the volumes you wish to order by volume number.
Standing orders: New volume in the series will be sent to you automatically each year upon publication. Cancellation may be made at any time. Please indicate volume number to begin standing order.
Prepublication orders: Volumes not yet published will be shipped in month and year indicated.
California orders: Add applicable sales tax.
Postage paid (4th class bookrate/surface mail) **by Annual Reviews Inc.** Airmail postage or UPS, extra.

ANNUAL REVIEWS SERIES		Prices Postpaid per volume USA & Canada/elsewhere	Regular Order Please send:	Standing Order Begin with:
			Vol. number	Vol. number
Annual Review of ANTHROPOLOGY				
Vols. 1-14	(1972-1985)	$27.00/$30.00		
Vols. 15-16	(1986-1987)	$31.00/$34.00		
Vol. 17	(avail. Oct. 1988)	$35.00/$39.00	Vol(s). _____	Vol. _____
Annual Review of ASTRONOMY AND ASTROPHYSICS				
Vols. 1-2, 4-20	(1963-1964; 1966-1982)	$27.00/$30.00		
Vols. 21-25	(1983-1987)	$44.00/$47.00		
Vol. 26	(avail. Sept. 1988)	$47.00/$51.00	Vol(s). _____	Vol. _____
Annual Review of BIOCHEMISTRY				
Vols. 30-34, 36-54	(1961-1965; 1967-1985)	$29.00/$32.00		
Vols. 55-56	(1986-1987)	$33.00/$36.00		
Vol. 57	(avail. July 1988)	$35.00/$39.00	Vol(s). _____	Vol. _____
Annual Review of BIOPHYSICS AND BIOPHYSICAL CHEMISTRY				
Vols. 1-11	(1972-1982)	$27.00/$30.00		
Vols. 12-16	(1983-1987)	$47.00/$50.00		
Vol. 17	(avail. June 1988)	$49.00/$53.00	Vol(s). _____	Vol. _____
Annual Review of CELL BIOLOGY				
Vol. 1	(1985)	$27.00/$30.00		
Vols. 2-3	(1986-1987)	$31.00/$34.00		
Vol. 4	(avail. Nov. 1988)	$35.00/$39.00	Vol(s). _____	Vol. _____

ANNUAL REVIEWS SERIES	Prices Postpaid per volume USA & Canada/elsewhere	Regular Order Please send:	Standing Order Begin with:
		Vol. number	Vol. number

Annual Review of **COMPUTER SCIENCE**
| Vols. 1-2 | (1986-1987).................. | $39.00/$42.00 | | |
| Vol. 3 | (avail. Nov. 1988)............. | $45.00/$49.00 | Vol(s). _____ | Vol. _____ |

Annual Review of **EARTH AND PLANETARY SCIENCES**
Vols. 1-10	(1973-1982)..................	$27.00/$30.00		
Vols. 11-15	(1983-1987)..................	$44.00/$47.00		
Vol. 16	(avail. May 1988).............	$49.00/$53.00	Vol(s). _____	Vol. _____

Annual Review of **ECOLOGY AND SYSTEMATICS**
Vols. 2-16	(1971-1985)..................	$27.00/$30.00		
Vols. 17-18	(1986-1987)..................	$31.00/$34.00		
Vol. 19	(avail. Nov. 1988).............	$34.00/$38.00	Vol(s). _____	Vol. _____

Annual Review of **ENERGY**
Vols. 1-7	(1976-1982)..................	$27.00/$30.00		
Vols. 8-12	(1983-1987)..................	$56.00/$59.00		
Vol. 13	(avail. Oct. 1988).............	$58.00/$62.00	Vol(s). _____	Vol. _____

Annual Review of **ENTOMOLOGY**
Vols. 10-16, 18-30	(1965-1971; 1973-1985)........	$27.00/$30.00		
Vols. 31-32	(1986-1987)..................	$31.00/$34.00		
Vol. 33	(avail. Jan. 1988).............	$34.00/$38.00	Vol(s). _____	Vol. _____

Annual Review of **FLUID MECHANICS**
Vols. 1-4, 7-17	(1969-1972, 1975-1985)........	$28.00/$31.00		
Vols. 18-19	(1986-1987)..................	$32.00/$35.00		
Vol. 20	(avail. Jan. 1988).............	$34.00/$38.00	Vol(s). _____	Vol. _____

Annual Review of **GENETICS**
Vols. 1-19	(1967-1985)..................	$27.00/$30.00		
Vols. 20-21	(1986-1987)..................	$31.00/$34.00		
Vol. 22	(avail. Dec. 1988).............	$34.00/$38.00	Vol(s). _____	Vol. _____

Annual Review of **IMMUNOLOGY**
Vols. 1-3	(1983-1985)..................	$27.00/$30.00		
Vols. 4-5	(1986-1987)..................	$31.00/$34.00		
Vol. 6	(avail. April 1988).............	$34.00/$38.00	Vol(s). _____	Vol. _____

Annual Review of **MATERIALS SCIENCE**
Vols. 1, 3-12	(1971, 1973-1982).............	$27.00/$30.00		
Vols. 13-17	(1983-1987)..................	$64.00/$67.00		
Vol. 18	(avail. August 1988)...........	$66.00/$70.00	Vol(s). _____	Vol. _____

Annual Review of **MEDICINE**
Vols. 1-3, 6, 8-9 11-15, 17-36	(1950-1952, 1955, 1957-1958) (1960-1964, 1966-1985)........	$27.00/$30.00		
Vols. 37-38	(1986-1987)..................	$31.00/$34.00		
Vol. 39	(avail. April 1988).............	$34.00/$38.00	Vol(s). _____	Vol. _____

Annual Review of **MICROBIOLOGY**
Vols. 18-39	(1964-1985)..................	$27.00/$30.00		
Vols. 40-41	(1986-1987)..................	$31.00/$34.00		
Vol. 42	(avail. Oct. 1988).............	$34.00/$38.00	Vol(s). _____	Vol. _____

FROM

NAME _____

ADDRESS _____

_____ ZIP CODE _____

Annual Reviews Inc. A NONPROFIT SCIENTIFIC PUBLISHER
4139 El Camino Way
P.O. Box 10139
Palo Alto, CALIFORNIA 94303-0897

PLACE
STAMP
HERE

ANNUAL REVIEWS SERIES	Prices Postpaid per volume USA & Canada/elsewhere	Regular Order Please send: Vol. number	Standing Order Begin with: Vol. number
Annual Review of PUBLIC HEALTH			
Vols. 1-6 (1980-1985)	$27.00/$30.00		
Vols. 7-8 (1985-1987)	$31.00/$34.00		
Vol. 9 (avail. May 1988)	$39.00/$43.00	Vol(s). _____	Vol. _____
Annual Review of SOCIOLOGY			
Vols. 1-11 (1975-1985)	$27.00/$30.00		
Vols. 12-13 (1986-1987)	$31.00/$34.00		
Vol. 14 (avail. Aug. 1988)	$39.00/$43.00	Vol(s). _____	Vol. _____

Note: Volumes not listed are out of print.

SPECIAL PUBLICATIONS	Prices Postpaid per volume USA & Canada/elsewhere	Regular Order Please Send:

The Excitement and Fascination of Science

Volume 1	(published 1965)	Clothbound . . . available on request	_____ Copy(ies).
Volume 2	(published 1978)	Hardcover . . . . available on request	_____ Copy(ies).
		Softcover . . . . . available on request	_____ Copy(ies).
Volume 3	(published 1988)	Hardcover . . . . available on request	_____ Copy(ies).

Intelligence and Affectivity:
Their Relationship During Child Development, by Jean Piaget

(published 1981)   Hardcover . . . . $8.00/$9.00   _____ Copy(ies).

Telescopes for the 1980s

(published 1981)   Hardcover . . . . $27.00/$28.00   _____ Copy(ies).

TO: **ANNUAL REVIEWS INC.,** a nonprofit scientific publisher
4139 El Camino Way
P.O. Box 10139
Palo Alto, CA 94303-0897 USA

Please enter my order for the publications checked above. **California orders, add sales tax.**
Prices subject to change without notice.

Institutional purchase order No. _____

Amount of remittance enclosed $ _____

Charge my account ☐ VISA
☐ MasterCard   ☐ American Express

INDIVIDUALS: Prepayment required in U.S. funds or charge to bank card below. Include card number, expiration date, and signature.

Acct. No. _____

Exp. Date _____   _____
                                                    Signature

Name _____
        Please print

Address _____
        Please print

_____ Zip Code _____ Date _____

_____ Send free copy of current ***Prospectus*** ☐

Area(s) of Interest   ARI Federal I.D. #94-1156476